Computer Aided Optimal Design: Structural and Mechanical Systems

NATO ASI Series

Advanced Science Institutes Series

A series presenting the results of activities sponsored by the NATO Science Committee, which aims at the dissemination of advanced scientific and technological knowledge, with a view to strengthening links between scientific communities.

The Series is published by an international board of publishers in conjunction with the NATO Scientific Affairs Division

A Life Sciences **B Physics**	Plenum Publishing Corporation London and New York
C Mathematical and Physical Sciences	D. Reidel Publishing Company Dordrecht, Boston, Lancaster and Tokyo
D Behavioural and Social Sciences **E Applied Sciences**	Martinus Nijhoff Publishers Boston, The Hague, Dordrecht and Lancaster
F Computer and Systems Sciences **G Ecological Sciences** **H Cell Biology**	Springer-Verlag Berlin Heidelberg New York London Paris Tokyo

Series F: Computer and Systems Sciences Vol. 27

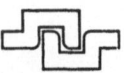

Computer Aided Optimal Design: Structural and Mechanical Systems

Edited by

Carlos A. Mota Soares

Centro de Mecânica e Materiais da Universidade
Técnica de Lisboa – C E M U L
Instituto Superior Técnico
Avenida Rovisco Pais
1096 Lisboa Codex/Portugal

Springer-Verlag
Berlin Heidelberg New York London Paris Tokyo
Published in cooperation with NATO Scientific Affairs Divison

Proceedings of the NATO Advanced Study Institute on Computer Aided Optimal Design: Structural and Mechanical Systems held in Tróia, Portugal, June 29–July 11, 1986

ISBN-13:978-3-642-83053-2 e-ISBN-13:978-3-642-83051-8
DOI: 10.1007/978-3-642-83051-8

Library of Congress Cataloging-in-Publication Data. NATO Advanced Study Institute on Computer Aided Optimal Design: Structural and Mechanical Systems (1986 : Tróia, Portugal) Computer aided optimal design. (NATO ASI series. Series F: Computer and systems sciences ; vol. 27) "Proceedings of the NATO Advanced Study Institute on Computer Aided Optimal Design: Structural and Mechanical Systems held in Tróia, Portugal, June 29 – July 11, 1986"—T.p. verso. "Published in cooperation with NATO Scientific Affairs Division." 1. Computer-aided design—Congresses. 2. Engineering design—Data processing—Congresses. I. Soares, Carlos A. Mota, 1945-. II. North Atlantic Treaty Organization. Scientific Affairs Division. III. Title. IV. Series: NATO ASI series. Series F, Computer and system sciences ; no. 27.
TA174.N365 1986 620'.00425'0285 87-382
ISBN-13:978-3-642-83053-2 (U.S.)

Softcover reprint of the hardcover 1st edition 1987

2145/3140-543210

NATO/NASA/NSF/USAF ADVANCED STUDY INSTITUTE

COMPUTER AIDED OPTIMAL DESIGN

Structural and Mechanical Systems

TROIA, PORTUGAL, 29th June - 11th July, 1986

MAIN SPONSORS

NATO: North Atlantic Treaty Organization
NASA: The U. S. National Aeronautics and Space Administration
NSF: The U. S. National Science Foundation
USAF: The U. S. Air Force

OTHER SPONSORS

CEMUL: Centro de Mecânica e Materiais da Univ. Técnica de Lisboa
INIC: Instituto Nacional da Investigação Científica
Secretaria de Estado do Ensino Superior
Comissão Permanente **INVOTAN**
JNICT: Junta Nacional de Investigação Científica e Tecnológica
IST: Instituto Superior Técnico
UTL: Universidade Técnica de Lisboa
MAGUE: Construções Metalomecânicas, S.A.R.L.
PECHINEY
European Research Office, United States Army
SOREFAME: Sociedades Reunidas de Fabricações Metálicas,S.A.R.L.
Escola Náutica Infante D. Henrique
BPA: Banco Português do Atlântico
TAP Air Portugal
CARRIS: Companhia dos Caminhos de Ferro de Lisboa
RN: Rodoviária Nacional
MAN Technologie Gmbh

DIRECTOR : C. A. Mota Soares

ORGANIZING COMMITTEE

L. BERKE, Chief Scientist for Structures, NASA-Lewis, USA
C. FLEURY, Associate Professor, UCLA, USA
E. HAUG,Professor, Univ. of Iowa, USA
H. HORNLEIN, Research Scientist, MBB, FRG
C. A. MOTA SOARES,Professor, Tech. Univ. of Lisbon, PORTUGAL
J. SOBIESKI,Deputy Head Interdiscipl. Res. Office, NASA-Langley,USA
J. TAYLOR, Professor, Univ. of Michigan, USA

ORGANIZED BY

CEMUL: Centro de Mecânica e Materiais da Univ. Técnica de Lisboa
IST: Instituto Superior Técnico
UTL: Universidade Técnica de Lisboa

NATO/NASA/NSF/USAF ADVANCED STUDY INSTITUTE

COMPUTER AIDED OPTIMAL DESIGN

Structural and Mechanical Systems

TROIA, PORTUGAL, 29 June - 11 July, 1986

DIRECTOR : C. A. Mota Soares

ORGANIZING COMMITTEE

L. BERKE , Chief Scientist for Structures, NASA-Lewis,USA
C. FLEURY , Associate Professor, UCLA,USA
E. HAUG , Professor, Univ. of Iowa,USA
H. HORNLEIN , Research Scientist, MBB,FRG
C. A. MOTA SOARES , Professor, Tech. Univ. of Lisbon,Portugal
J. SOBIESKI, Deputy Head Interdiscipl. Res. Office, NASA-Langley,USA
J. TAYLOR, Professor, Univ. of Michigan,USA

LECTURERS

P. BECKERS, Associate Professor, Laboratoire Tech. Aero. et Spatiales, Belgium
L. BERKE, Chief Scientist for Structures, NASA-Lewis,USA
K. CHOI, Associate Professor, Univ. of Iowa,USA
C. CINQUINI, Professor, Univ. of Pavia,Italy
C. FLEURY, Associate Professor, UCLA,USA
D. GRIERSON, Professor, Univ. of Waterloo,Canada
R. HAFTKA, Professor, Virginia Poly. Institute & St. Univ.,USA
E. HAUG, Professor, Univ. of Iowa,USA
H. HORNLEIN, Research Scientist, MBB,FRG
N. KIKUCHI, Professor, Univ. of Michigan,USA
A. MORRIS, Professor, Cranfield Inst. of Technology,UK
C. A. MOTA SOARES, Professor, Tech. Univ. of Lisbon,Portugal
Z. MROZ,Professor, Academy of Science of Poland,Poland
N. OLHOFF,Professor, Univ. of Aalborg,Denmark
P. PAPALAMBROS, Associate Professor, Univ. of Michigan,USA
J. SOBIESKI, Deputy Head Interdiscipl. Res. Office, NASA-Langley,USA
J. TAYLOR, Professor, Univ. of Michigan,USA
G. VANDERPLAATS, Professor, Univ. of California, Santa Barbara,USA

PARTICIPANTS

A. ADÃO DA FONSECA, Associate Professor, Univ. of Oporto,Portugal
M. AHMAD, Lecturer, Limerick Tech. Univ.,Ireland
P. AMABILE, Structural Engineer, Alitalia,Italy
J. AMBROSIO, Ph. D. Student, Univ. of Arizona,USA
B. ANDERSSON, Professor, Aeronautical Res. Inst. of Sweden,Sweden
J. ARORA, Professor, Univ. of Iowa,USA
E. ATREK, Research Engineer, EMRC,USA
S. AZARM, Assist. Professor, Univ. of Maryland,USA
J. BARBOSA, Assist. Professor, Nautical School,Portugal
A. BARROS, Ph. D. Student, Tech. Univ of Lisbon,Portugal
B. BARTHELEMY, Ph. D. Student, Virginia Poly. Inst.,& St. Univ.,USA
J. BARTHELEMY, Research Scientist, NASA-Langley,USA
P. BARTHOLOMEW, Research Engineer, Royal Aircraft Est.,UK
T. BELEK, Associate Professor, Istanbul Tech. Univ.,Turkey

M. BENDSOE, Associate Professor, Tech. Univ. Of Denmark,Denmark
R. BENEDIT, Research Engineer, GOODYEAR,USA
K. BLETZINGER, Research Engineer, Univ. of Stuttgart,FRG
T. BRAMA, Research Engineer, SAAB-SCANIA AB,Sweden
R. CARNEIRO BARROS, Assist. Professor, Univ. of Oporto,Portugal
D. CHENAIS, Assist. Professor, Univ. of Nice,France
R. CONTRO, Professor, Polytechnic of Milan,Italy
R. COSTA, Engineer, PROFABRIL,Portugal
R. CRUZ, Engineer, MAGUE,Portugal
H. CUDNEY, Ph. D. Student , State Univ. of New York,USA
R. CUNTZE, Head of Dept. Structural Analysis, MAN,FRG
M. DARKINS, Research Engineer, British Aerospace,UK
K. DEMS, Assist. Professor, Lodz Tech. Univ.,Poland
J. DOMINGUEZ, Professor, Univ. of Sevilla,Spain
P. DRAKATOS, Professor, Univ. of Patras,Greece
F. EL-YAFI, Research Engineer, CETIM,France
Y. EROTOKRITOS, Senior Research Engineer, ISVR,UK
B. ESPING, Research Scientist, Royal Inst. of Technology,Sweden
F. FAGUNDO, Associate Professor, Univ. of Florida,USA
L. FERNANDES, Ph. D. Student, Tech. Univ. of Lisbon,Portugal
F. FORTSCH, Research Associate, Univ. of Aachen,FRG
G. FROSI, Manager, IBM - ITALIA,Italy
A. GAGO, Ph. D. Student, Tech. Univ. of Lisbon,Portugal
J. GARCIA-LOMAS, Professor, Univ. of Sevilla,Spain
E. GOMES, Ph. D. Student, Tech. Univ. of Lisbon,Portugal
C. GUEDES SOARES, Assist. Professor, Port. Naval Acad.,Portugal
M. GURGOZE, Associate Professor, Istanbul Tech. Univ.,Turkey
R. HABER, Associate Professor, Univ. of Illinois,USA
P. HAJELA, Assist. Professor, Univ. of Florida,USA
D. HARTMANN, Professor, Univ. of Dortmund,FRG
S. HERNANDEZ, Reader, Univ. of Zaragoza,Spain
J. HERSKOVITS, Assoc. Professor, Univ. of Rio de Janeiro,Brazil
D. HOELTZEL, Assist. Professor, Columbia Univ.,USA
J. HOU, Assist. Professor,Old Dominion Univ.,USA
M. JAKIELA, Ph. D. Student, Univ. of Michigan,USA
B. KAFTANOGLU, Professor, Middle East Tech. Univ.,Turkey
M. KAMAT, Professor, Georgia Inst. of Technology,USA
M. KAPOOR, Professor, Indian Inst. of Tech.,India
B. KARIHALOO, Professor, Univ. of Sidney,Australia
S. KHOT, Aero. Engineer, WPAFB,USA
U. KIRSCH, Dean, Israel Inst. of Tech.,Israel
C. KNOPF-LENOIR, Research Engineer, GRADIENT,France
J. KRAMMER, Engineer, MBB,FRG
S. KURITZ, Ph. D. Student, UCLA,USA
B. KWAK, Professor, K.A.I.S.T.,Republic of Korea
R. LATIF, Structural Analysis Eng., British Aerospace,UK
R. LEAL, Ph. D. Student, Univ. of Coimbra,Portugal
G. LECINA, Senior Engineer, Avions Marcel Dassault,France
J. MARCELIN, Research Associate, INSA,France
P. MARTIN, Ph. D. Student, State Univ. of Ghent,Belgium
H. MELZER, Engineer, MBB,FRG
R. MERIC, Assoc. Professor, Res. Inst. Basic Sciences,Turkey
P. MIKOLAJ, Engineer, ERNO,FRG
H. MLEJNEK, Research Engineer, Inst. Comp. Applications,FRG
P. MONTRULL, Assist. Professor, Univ. Poly. Valencia,Spain
S. MORIANO, Ph. D. Student, Univ. of Nice,France
C. M. MOTA SOARES, Assist. Prof.,Tech. Univ. of Lisbon,Portugal

V. NGUYEN, Professor, Univ. of Namur,Belgium
M. NO, Professor, Univ. of Navarra,Spain
R. NOTENBOOM, Staff Scientist , Tech. Univ. of Delft,Netherland
G. PALASSOPOULOS, Professor, Military Academy of Greece,Greece
PAULI PEDERSEN, Professor, Tech. Univ. of Denmark,Denmark
PREBEN PEDERSEN, Professor, Tech. Univ. of Denmark,Denmark
J. PIMENTEL, Engineer, MAGUE,Portugal
M. REITMAN, Professor, Univ. of Illinois,USA
S. RIZZO, Professor, Univ. of Palermo,Italy
H. RODRIGUES, Ph. D. Student, Univ. of Michigan,USA
B. ROUSSELET, Professor, INRIA,France
G. ROZVANI, Professor, Univ. of Essen,FRG
J. SANTOS, Ph. D. Student, Univ. of Iowa,USA
A. SCHWAB, Assist. Professor , Tech. Univ. of Delft,Netherland
M. SERNA, Professor, Univ. of Navarra,Spain
L. SIMÕES, Assist. Professor, Univ. of Coimbra,Portugal
P. STRONA, Research Engineer, FIAT,Italy
D. TEMPLEMAN, Reader, Univ. of Liverpool,UK
S. THOMSEN, Ph. D. Student, Tech. Univ. of Denmark,Denmark
B. TOPPING, Lecturer, Univ. of Edinburgh,UK
L. TRABUCHO CAMPOS, Assist. Prof.,Tech. Univ. of Lisbon,Portugal
J. TRIGO CORTÊS, Engineer, SOREFAME,Portugal
P. TROMPETTE, Professor, Univ. of Grenoble,France
V. VENKAYYA, Aero. Engineer, WPAFB,USA
M. VINCENT, Engineer, Aerospatiale,France
P. WARD, Director of Research and Development, SDRC,UK
R. WATKINS, Ph. D. Student, Cranfield Inst. of Technology,UK
K. WELLEN, Engineer, MBB,FRG
Y. YOO, Assist. Professor, K.A.I.S.T.,Republic of Korea
R. ZIMMERMANN, Research Scientist, DFVLR,FRG
M. ZYCZKOWSKI, Professor, Tech. Univ. of Krakow,Poland

PREFACE

This book contains the edited version of lectures and selected papers presented at the NATO ADVANCED STUDY INSTITUTE ON COMPUTER AIDED OPTIMAL DESIGN: Structural and Mechanical Systems, held in Tróia, Portugal, 29th June to 11th July 1986, and organized by CEMUL - Center of Mechanics and Materials of the Technical University of Lisbon. The Institute was attended by 120 participants from 21 countries, including leading scientists and engineers from universities, research institutions and industry, and Ph.D. students. Some participants presented invited and contributed papers during the Institute and almost all participated actively in discussions on scientific aspects during the Institute. The Advanced Study Institute provided a forum for interaction among eminent scientists and engineers from different schools of thought and young reseachers.

The Institute addressed the foundations and current state of the art of essential techniques related to computer aided optimal design of structural and mechanical systems, namely: Variational and Finite Element Methods in Optimal Design, Numerical Optimization Techniques, Design Sensitivity Analysis, Shape Optimal Design, Adaptive Finite Element Methods in Shape Optimization, CAD Technology, Software Development Techniques, Integrated Computer Aided Design and Knowledge Based Systems. Special topics of growing importance were also presented.

This book is organized in eight parts, each one addressing a technical aspect of the field of Computer Aided Optimal Design:

Part I : Variational Methods in Optimal Design
Part II : Numerical Methods in Optimal Design
Part III : Shape Optimal Design
Part IV : Multilevel and Interdisciplinary Optimal Design
Part V : Optimal Design of Mechanical Systems
Part VI : Knowledge Based Systems in Optimal Design
Part VII : Integrated CAD/FEM/OPTIMIZATION Techniques and Applications
Panel Discussion : Trends in Computer Aided Optimal Design

The foundations and recent developments of variational and finite element methods and mathematical programming techniques, applied to the optimal design and control of elastic and nonlinear structures, are presented by leading scientists.

Several contributors address different methods for shape optimal design of structures, including recent research on boundary element methods in shape optimal design, design sensitivity analysis and optimal design of nonlinear structures, adaptive finite element methods for shape optimization and the practical implementation of shape optimal design in commercially available software.

In this book special emphasis is placed on the integration of CAD techniques for geometric modelling, finite element analysis and optimization methods. Several academic and industrial specialists reviewed the current state of development. A critical review of the available commercial codes is presented. Some researchers have integrated all these techniques in codes and applied them to the design of structures in aerospace, aircraft and car industries. These applications show that the integration of these techniques into standard tools for practical design is a major factor for industrial usage of computers in design of structural systems.

Other papers presented or submitted to the Advanced Study Institute, but not included in this book, are published in a Special Issue of the Journal of Engineering Optimization in 1987.

Without the sponsorship and financial support of the Scientific Affairs Division of NATO, National Aeronautics and Space Administration, National Science Foundation and United States Air Force the Institute and this book would not have been possible. The financial support of all other sponsors also contributed decisively to the success of the Institute.

The Editor deeply appreciated all the advice and help in organizing the Institute given by Dr. Craig Sinclair and the late Dr. M. di Lullo, both of the Scientific Affairs Division of NATO. I am indebted to all members of the Organizing Committee (Prof. J.E. Taylor, Prof. E.J. Haug, Dr. J. Sobieski, Dr. L. Berke, Prof. C. Fleury and Dr. H. Hornlein) for the outstanding work that led to a very successful Institute. I am also grateful to all authors for their effort in writing the lectures and papers in time, allowing this book to be published as planned.

Special thanks to CEMUL staff, Ms Glória Ramos, Ms. Alexandra Confeiteiro and Mr. Amândio Rebelo, for their effort in administrative planning and support of the Institute.

I am very grateful to my family, Maria do Rosário and our daughter Joana Sofia, for all the support during the organization of the Institute and of this book. Special thanks for my uncle Hermano Cabral for the video and magnetic tape of the Panel Session of the Institute

Lisbon, December 1986

Carlos A. Mota Soares

CONTENTS

PART I

VARIATIONAL METHODS IN OPTIMAL DESIGN

DISTRIBUTED PARAMETER OPTIMAL STRUCTURAL DESIGN: SOME BASIC PROBLEM FORMULATIONS AND THEIR APPLICATION

J. E. Taylor
The University of Michigan
Ann Arbor, Michigan - 48109

True Economy . . .
transcends material and
economic costs !

to **WILLIAM PRAGER**
- creative, thoughtful, gentleman -
in fond memory

BASIC FORMULATIONS FOR DISTRIBUTED PARAMETER STRUCTURAL OPTIMIZATION

Introduction

As a part of the broad development that has taken place in the field of structural optimization in recent years, analytical modelling for the design of continuum structures has been extended to cover a variety of new applications. Thus there are formulations available now for the optimization with respect to various modes of response or measures of performance, for most types of structural form, to be optimal relative to material distribution, shape, choice of materials, prestress, and so on. Only a modest part out of the comprehensive list of topics is to be covered in these lectures. The reader will find a good many of the major areas of application e.g., 'design for dynamic response,' 'shape design', 'grid optimization,' and 'sensitivity analysis' – to name a few , treated in separate lectures given elsewhere within the institute. (Citations to other lectures in this collection are identified by the authors name with an asterisk attached to it.) Our effort is directed more toward an *exposition of methods* for the interpretation of design problems into a form convenient for analysis. This is to be done mainly within the perspective of well known results from the mathematics of optimization. The material presented here is comprised for the most part of formal problem statements, listings and interpretation of necessary conditions, and the presentation of example applications.

In another way, the purpose of these notes/lectures is to provide a presentation of *formal procedures* for a variational formulation of structural optimization problems, and to furnish exemplification of their application. The objective is to make the coverage as general as possible, given the present restrictions on time and space. With this in mind, the word 'Basic' in the title given above is intended to refer mainly to simple fundamental forms of design problem. Also the methods are demonstrated

NATO ASI Series, Vol. F27
Computer Aided Optimal Design: Structural and Mechanical Systems. Edited by C. A. Mota Soares

generally through the vehicle of simplistic example problems. The hope is that by limiting the scope to be simple in these ways, our effort to provide a clear statement to the *fundamental ideas* that underly the analysis will benefit. At the same time, it should be practical as a result to cover a larger range of types of problems than would otherwise be possible. The discussion to follow covers certain classical results among variational formulations for structural optimization, and the extensions of them that are needed more generally for the treatment of problems with global and/or local constraints. Brief descriptions are provided for a number of special topics, among them the design of optimal elastic foundations, the formulation for optimization in problems with simple contact between elastic bodies, and a method for the relaxation of constraints. A useful scalar formulation for *multicriterion problems* is given in a separate section; its application is exemplified there through a variety of sample problems. Fail-safe optimal design and the design of optimal structural remodeling also are treated separately as special topics.

It might contribute to the understanding of the material in these lecture notes if the following ideas are kept in mind.

1. The prediction of optimal structural design and the (traditional) task of structural analysis – are aspects of *one problem*. Stated differently, the vector of unknowns in the structural optimization problem is comprised of design and state variables (or their equivalent) as components. Thus once any form of decoupling has been introduced into the modeling of design problems, it must be recognized as a specific, i.e., less than general, interpretation of the problem. This applies to the approaches that one may identify with iterative or sequential-step solution methods, for example. For present purposes, the developments are expressed for the most part in a form consistent with the fully coupled problem.

Optimization:
Analysis and Design
Modeled together ...

2. The important matter of how to judge whether or not a problem is properly posed – is not addressed directly in the treatments that follow. This issue may be resolved without much trouble for special categories of design problem, but general methods for dealing with the question are lacking. On the side of being practical, we note that an ill-posed problem might sometimes be reinterpreted into tractable form either by enlarging the space of feasible designs (as a means to achieve G-closure), or otherwise in certain cases through the imposition of additional constraints.

Perfect Solution
exist? If not, widen the
choices - or narrow them!

Finally, it is hoped that the material of these lectures should prove to be meaningful in one way or another beyond the scope of the lectures themselves. Regrettably, it has not been possible in this writing to develop all ties with other works in the field wherever such connections might sensibly have been made. Indeed, given the high level of research, development, and applications activity in structural optimization, it has become a serious challenge just to maintain broad contact with overall progress in the subject. On the other hand, contemporary work generally is well documented and fortunately the literature is quite well indexed, at least into the year 1982. Exceptionally extensive listings are given in the book by Carmichael (on the order of 700 items), and in the Komkov and Haug translation of Banichuk's book (Original listing with 246 entries, extended to 531 in the translation) – these books in themselves, and also the surveys by

Lev and by Kruzelecki and Zyczkowski and the proceedings edited by Morris, provide unusually broad coverage of the field, each in its own way. The interested reader is referred to the list provided at the end of this set of notes for these and other books, and for the many other useful resources in the form of proceedings, reviews, and surveys as well (This part of the reference list is taken mostly from Olhoff and Taylor). It should be noted that, with the exception of the resources just named, for the most part the citation of references in these notes is narrow, i.e., the references listed are limited to only a small part of the literature that relates to the immediate topics covered in the text. Of course the material covered in these lectures depends more broadly on the past efforts of others, and the author wishes to acknowledge his debt of gratitude to the colleagues and co-workers whose contributions may not have been cited specifically.

A Classical Variational Form

In cases where the measure of the criterion for an optimization problem relates directly to the quantity that appears as an argument in the associated *minimum principle of mechanics*, the optimal design problem can be stated in a particularly simple way. Examples are design for *maximum Euler load* or for *maximum normal mode frequency* (so long as the solution is unimodal!), design for *minimum compliance*, design for *maximum collapse load* (assuming 'perfectly plastic behavior'), and design for *maximum creep strength* (for a linear creep law). Numerous papers were written on variational formulations for these problems, mainly over the years from 1955 through the 1970's. Except for a few applications, the results of such interpretations for optimal design problems are of little practical value today; however the material has historical significance as a part of the overall development in structural optimization during the recent decades.

The design of axially loaded elastic bars for minimum compliance is described here, as an example of the more or less classical style for variational formulation. For specified load p(x) and admissible deformation $\eta(x)$, compliance is measured by $\int_\Omega p\eta dx$. The equilibrium state for the elastic bar is identified with a minimum with respect to $\eta(x)$ of the potential energy π given by

$$\pi[B(x); \eta(x)] = \int_\Omega [1/2\, EB(\eta')^2 - p\eta]dx \qquad (1)$$

B(x) represents cross-sectional area, and the structure is initially free of stress. Note that for the actual deformation , say w(x), the value of compliance equals the negative of twice the value of potential energy:

$$\int_\Omega pw\, dx = -2\pi[B;w] \qquad (2)$$

In this case , the necesary conditions for the optimum design problem

$$\begin{aligned} &\min (\int_\Omega p\eta \; dx) \\ &\text{subject to:} \qquad (3) \\ &\text{Conditions of Equilibrium} \\ &\underline{A} - B \leq 0 \\ &\int_\Omega B\, dx - R \leq 0 \end{aligned}$$

(value R, representing available resource, and the lower bound $\underline{A}$ to the cross section are specified) are equivalent to the conditions for stationarity with respect to B(x) and $\eta(x)$ of the functional

$$L(B,\eta,\lambda,\Lambda) = \pi(B,\eta) - \int_\Omega \lambda(\underline{A} - B)dx - \Lambda[\int Bdx - R] \qquad (4)$$

These necessary conditions are (A(x) and u(x) represent the optimal design and its associated equilibrium state):

$$
\begin{array}{lll}
(EAu')' + p = 0 & x \text{ in } \Omega & \\
EAu' = 0 \quad \text{or} \quad u = 0 & x = 0,L & \\
1/2\, E(u')^2 - \Lambda + \lambda = 0; & x \text{ in } \Omega & (5) \\
\Lambda[\int A dx - R] = 0; \quad \Lambda \geq 0; & \int A dx - R \leq 0 & \\
\lambda(\underline{A} - B) = 0; \quad \lambda \geq 0; & \underline{A} - B \leq 0 &
\end{array}
$$

Domain Ω of the optimum structure is comprised of intervals where $A > \underline{A}$ $\Rightarrow \lambda = 0$, and sections in $\Omega_o = \Omega - \Omega_A$ where $A = \underline{A}$. An example solution is shown in Figure 1 for the bar supported at $x = 0$ and subject to a uniform load.

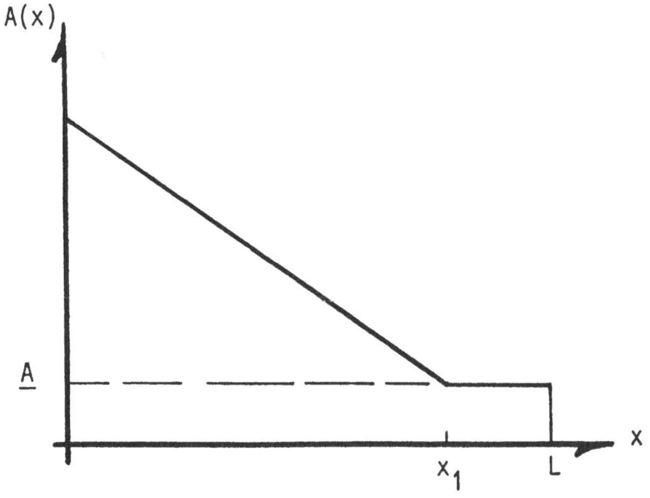

Figure 1. Design of an Elastic Bar for Minimum Compliance

As mentioned earlier, historically problems of the kind designated as 'classical' have been treated extensively, even though the type of criterion associated with such problems is sometimes of limited practical interest. The reader may wish to consult the literature (mainly from the period 1955-80) for other example applications within the classical form, e.g., optimal Euler columns and plates – design to maximize natural frequency for bars, beams and plates – and minimum compliance design of beams, plates, frames and disks.

With the addition of the *'corner conditions'*, the system of equations (5) are sufficient as well as necessary. The corner conditions require that the solution design A(x) must be continuous over the domain of the structure.

Making use of the 'minimum principle for the mechanics', sufficiency may be demonstrated quite simply. Let w(x) represent the equilibrium deformation for arbitrary design B(x) , and recall that u(x) symbolizes the equilibrium state for the putative optimal design A(x) . Then since solution u(x) is admissible for the design B(x), we have from the minimum potential energy statement and equation (2):

$$- 1/2 \int_{\Omega} pw\, dx = \int_{\Omega} [1/2EB(w')^2 - pw]dx \leq \int_{\Omega} [1/2EB(u')^2 - pu]dx \qquad (6)$$

The right side of the inequality is interpreted as

$$\int_{\Omega} 1/2[EB(u')^2 - pu]dx = \int_{\Omega}[1/2\, EA(u')^2 - pu]dx + 1/2 \int_{\Omega} E(B-A)(u')^2 dx \qquad (7)$$

Substitute $(u')^2 = \Lambda - \lambda$ from equation (5) into the last term of equation (7):

$$\int_{\Omega} 1/2E(B-A)(u')^2 dx = \Lambda E \int_{\Omega} (B-A)dx - E \int \lambda\, (B-A)dx$$

But since $\int_{\Omega} A\, dx = R$, $\int_{\Omega} B\, dx \leq R$, and $\Lambda > 0$, $\Lambda \int_{\Omega} (B - A)\, dx \leq 0$. Also $\lambda = 0$ in the design domain, and $\lambda \geq 0$; $(B - A) \geq 0$ in Ω_0, so that

$$\int_{\Omega} \lambda(B - A)dx = \int_{\Omega_0} \lambda(B - A)dx \geq 0$$

whereby

$$\int_{\Omega} E(B - A)(u')^2\, dx \leq 0$$

Thus from equations (6) and (7) we have

$$- 1/2 \int_{\Omega} pw\, dx \leq \int_{\Omega} [1/2\, EA(u')^2 - pu]dx \overset{d}{=} \pi(A;u)$$

But since u(x) is the actual state for design A(x) ,

$$2\pi(A;u) = -\int_\Omega pudx$$

and so from equation (7)

$$\int_\Omega pwdx \geq \int_\Omega pudx$$

to complete the proof.

Sufficiency may be verified through similar argument for various examples of classical structural optimization problem which are convex. At the same time, given the general result from analysis that for convex problems the (generalized) KKT conditions are sufficient as well as necessary for the optimal solution, such exercises may be redundant.

More General Variational Forms and Their Applications

We set out to establish in this section a variational formulation where the criterion and/or constraint functions are represented in general form. It is the intention to accommodate criteria or constraints having either global or local measure. (As an example of the latter, the criterion may have the form *max of a specified function* , where the maximum is taken with respect to the domain variable. In such cases it is not uncommon that differentiability of the criterion becomes an issue; a detailed discussion of this point is provided below.) Thus within the limitations of linearly elastic small deformation mechanics stated earlier, the results presented here facilitate the handling of a broad variety of design problems. The development is expressed first for various problems of *single purpose* structural optimization. Essentially parallel treatments for multicriterion problems (e.g., multipurpose or multimodal design) are discussed in a later lecture of this series.

It is particularly convenient for present purposes to work with the 'weak form' statement of equilibrium conditions, i.e., with the virtual displacement equation. Considering first problems with global criterion, the criterion is expressed in general form as

$$J = \int_{\Omega} F(B;\eta)dx$$

for argument $F(B;\eta)$ specified over fixed domain Ω of the structure. The argument F is supposed to be differentiable w.r.t. design $B(x)$ and state field $\eta(x)$ and derivitives of η to the required order. The optimal design problem may be stated symbolically as:

$$\min_{B(x)} \quad [J = \int_{\Omega} F(B,\eta)\, dx] \tag{8}$$

subject to:
* The virtual displacement equation
* Performance and/or design constraints
* $\int_{\Omega} B\, dx - R < 0$

The value R in the isoperimetric (resource) constraint is specified, as are the load, structural form, constraint bounds, and so on.

The formal treatment of design problems within the context of (8) is demonstrated via a specific example. For simplicity, we choose a generalization of the minimum compliance design problem for the one dimensional structure used earlier, namely the axially loaded bar. Generalized compliance is expressed for specified weight function $\phi(x)$ as:

$$J = \int_{\Omega} \phi(x)\, u(x)\, dx \tag{9}$$

The equilibrium requirement, for load $p(x)$, Young's modulus E, and design $B(x)$, is represented by:

$$\int_{\Omega} [EB\,\eta'\,\zeta' - p\zeta\,]dx = 0 \quad \text{for all admissible } \zeta \tag{10}$$

i.e., the function among kinematically admissible functions η that satisfies this equation for all admissible ζ is the equilibrium solution.

Supposing that there are no performance constraints on the problem, for this example the specific statement corresponding to equation (8) is

$$\min_{B(x)} \; [J = \int_{\Omega} \phi\eta \, dx] \tag{11}$$

subject to:

$$\int_{\Omega} [EB\, \eta' \zeta' - p\, \zeta\,]dx = 0 \quad \text{for all admissible } \zeta$$

$$\underline{A} - B \leq 0 \qquad \text{all x in } \Omega$$

$$\int_{\Omega} B\, dx - R \leq 0.$$

The necessary conditions for this problem may be identified with the Lagrangian

$$L = \int_{\Omega} [\phi\eta - \Lambda_e(EB\, \eta' \zeta' - p\zeta) + \lambda(\underline{A}\text{-}B) + \Lambda[\int_{\Omega} B\, dx - R] \tag{12}$$

Multiplier Λ_e is taken to have unit value, without restriction on the generality of what follows. The necessary conditions themselves (the generalized Karash-Kuhn-Tucker (KKT) conditions) are:

$$(EAu')' + p = 0$$

$$\text{all x in } \Omega$$

$$(EAv')' + \phi = 0$$

and their respective boundary conditions (13)

$$Eu'v' = \Lambda - \lambda$$

$$\lambda(\underline{A} - A) = 0; \qquad \lambda \geq 0; \qquad \underline{A} - A \leq 0$$

$$\Lambda[\int_{\Omega} A\, dx - R] = 0 \;\; ; \quad \Lambda \geq 0 \;\; ; \quad \int_{\Omega} A\, dx - R \leq 0$$

Here A(x), u(x), and v(x) represent the solution functions for design, state, and adjoint state. Let Ω_A represent the design domain, i.e., the set of intervals within the domain $\Omega : x \varepsilon (0,L)$ of the entire structure for which

$A(x) > \underline{A}$. Then

$$Eu'v' = \begin{cases} \Lambda & \text{all } x \text{ in } \Omega_A \\ \Lambda - \lambda & \text{all } x \text{ in } \Omega_0 \overset{d}{=} \Omega - \Omega_A \end{cases} \tag{14}$$

In other words, the positive constant Λ bounds unit mutual strain energy $Eu'v'$ of the optimal structure, and the value of unit mutual energy equals the bound in the design intervals. Outside of Ω_A the optimal design is given by $A(x) = \underline{A}$. This much of the interpretation of the necessary conditions is summarized symbolically in Figure 2.

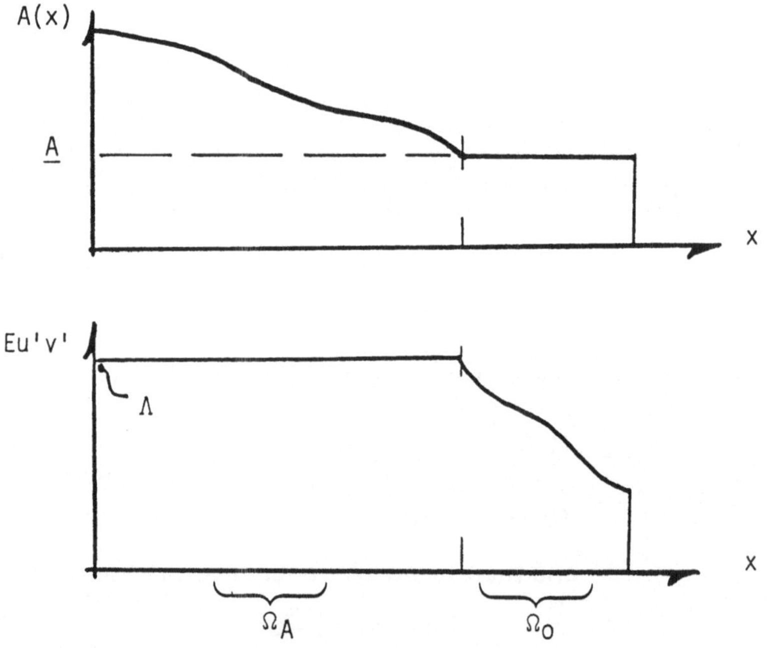

Figure 2.

The solution for the optimal design may be completed using the first two of equations (13), and the equations $Eu'v' = \Lambda$ and (since $\Lambda \neq 0$) $\int_\Omega A dx - R = 0$, together with corner conditions. The example is not particularly interesting so the details are not pursued here. Note, however, that for $\phi(x) = p(x)$ this problem is equivalent to the conventional 'minimum compliance design' problem that is described in the prior section; this equivalence can be verified through a comparison of the sets of necessary conditions for the two versions of the problem.

While it is tempting to consider detailed treatments of some of the many interesting types of design problems that are covered within the generalized form for global constraints (several examples are covered elsewhere within presentations of the institute), for the sake of conserving time and space we will make do with a few remarks instead. For one, note that it might be instructive to examine the necessary conditions as they would appear for the general (i.e., unspecified) form for the argument $F(B;\eta)$ in the criterion functional. The formal statement of the Euler-Lagrange equations sufffices for this purpose, so long as the usual restrictions on argument F are kept in mind. Having these equations available makes it possible to observe directly the effect on the analysis for design problems of making certain changes in the criterion functional. These effects are identified with the term that corresponds to 'load' in the adjoint equilibrium equation, and with terms that show up in the equation(optimality condition) reflecting variation with respect to design, as a result of explicit dependence of argument function F on the design. As an example of the latter, one might note the effect of including an account of *self weight* in the criterion for the compliance design problem.

Design Problems of
minmax type - make the best of
worst criteria !

We consider next the formulation of structural optimization problems where the criterion is expressed in terms of some *local measure*, for example the design to minimize the maximum value of stress or displacement over the structure. In such situations the usual differentiability properties, i.e., the properties required to support the common form of statement for necessary conditions, may not be available. In fact, in the field of structural optimization this result is not exceptional. Consider the following example: for an axially loaded bar comprised of two uniform segments (see Figure 3), determine the design that minimizes the maximum absolute value of stress. The lengths l_i of the segments are specified.

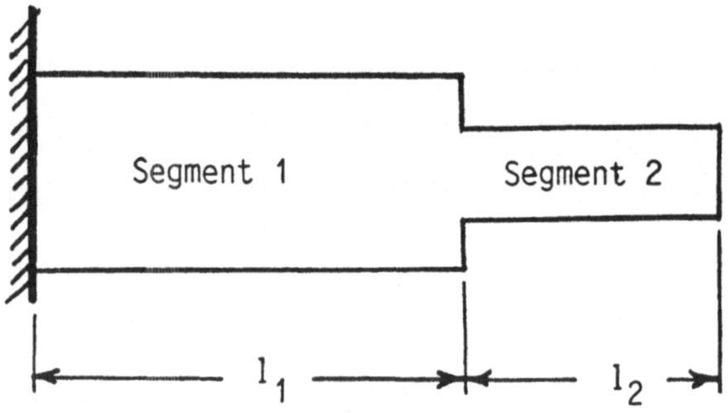

Figure 3.

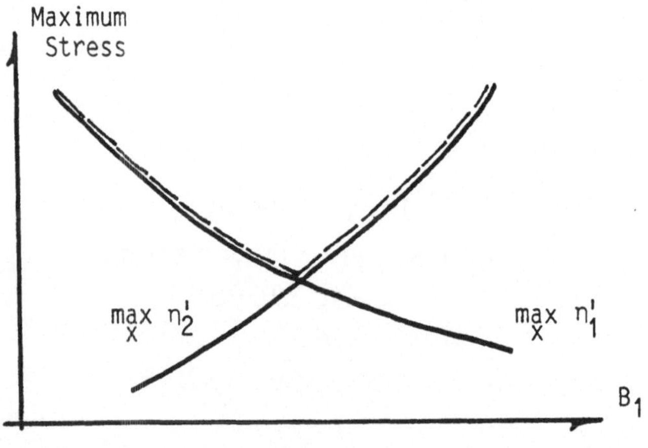

Figure 4.

Note that for this structure $\sigma_i \sim \eta'_i$; then the design problem can be stated as:

$$\min_{B_1,B_2} \; [\max_{i=1,2} \; (\max_{x\varepsilon\Omega_i} |\eta'_i|)]$$

subject to:

$$\int_{\Omega_i} [EB_i\eta'_i\zeta'_i - p\zeta_i]dx = 0 \qquad \text{all admissible } \zeta_i \quad i=1,2 \qquad (15)$$

$$\sum_{i=1}^{2} l_i B_i - R \leq 0$$

For designs that make full use of the resource R (the optimal solution lies within this set), the maximum strains in the two segments vary with opposite sense (Figure 4). Thus the argument max | u'(x) | (indicated by the dashed curve in the Figure) has a corner at the optimal solution.

Technically this difficulty may be circumvented by use of the 'p-norm' to simulate the argument 'max(local measure).' For the problem of Equation (15) for example, one would seek the design to minimize

$$[\sum_i \int_{\Omega_i} | \eta_i | ^p dx]^{1/p}$$

for sufficiently large value of p. However, computation on the p-norm quickly becomes poorly conditioned as the value of p is increased, so that in fact the method is not very practical.

As a more viable alternative, we next consider an interpretation of the minmax problem (see, e.g., Taylor-Bendsφe) into a form that reflects minimization of *a bound on the local measure*. (This form is identified in these notes as the *bound formulation..*) Suppose that the criterion is expressed as max |f(B(x); η(x)|, so that the problem statement along the lines of equation (8) appears as:

$$\min_{B(x)} \left[\max |f(B;\eta)|\right]$$

subject to:

*the virtual displacement equation (16)

* performance and/or design constraints

* $\int_\Omega B dx - R \leq 0.$

As before, B(x) and η(x) represent admissible design and displacements, respectively. We intoduce an additional scalar unknown, say β, and interpret the problem of equation (16) in the form:

$$\min_{B(x)} (\beta)$$

subject to:

* $|f(B;\eta)| - \beta \leq 0$ (17)

* (constraints as in (16))

Within moderate restrictions on the form of the argument f(B;η), necessary conditions for the minmax problem (16) now may be obtained via formal procedures, i.e., as the generalized KKT conditions for the problem (17). Equivalence of the two problems is easily verified from the set of necessary conditions; in general the supremum β of argument |f| equals the maximum value of |f| .

The 'bound formulation' comprises a convenient device for the interpretation of a variety of important basic problems in structural optimization. Several examples of its application in single purpose design problems are given in the material immediately following. Use of the formulation as the basis for a broadly applicable statement of multicriteria and fail-safe design problems is described in later sections of these lectures. We observe that problems where the criterion is expressed in terms of a combination of global and local measures may be interpreted as well using effectively the same means as those already described.

Design of A Segmented Structure

As a first example , the bound formulation is applied to the above problem of design of a bar with two uniform segments, i.e., we seek to determine the design A_i from among admissible B_i that will minimize the max (max |u'|) . // Is it 'sensible' to treat this problem within a discussion under the name "distributed parameter" design? // Noting that

$$\max_i \left(\max_{x \varepsilon \Omega_i} | \eta_i' | \right) = \max_{\substack{x \varepsilon \Omega_i \\ i=1,2}} | \eta_i' |$$

we write in terms of the bound β on $|\eta_i'|$:

$$\min_{\substack{B_i \\ i=1,2}} (\beta)$$

subject to: (18)

$$| \eta_i' | - \beta \leq 0 \qquad \text{all } x \text{ in } \Omega_i \qquad i=1,2$$

$$\int_{\Omega_i} (EB_i \, \eta' \zeta'_i - p\zeta_i) dx = 0 \qquad \text{all admissible } \zeta_i$$

$$\sum_i B_i l_i - R \leq 0$$

Necessary conditions for the problem (18) are identified with stationarity of the Lagrangian

$$L = \beta + \sum_{i=1}^{2} \int_{\Omega_i} [- EB_i \eta_i \zeta'_i + p\zeta_i + \lambda^+_i(\eta'_i - \beta) + \lambda^-_i(-\eta' - \beta)] dx + \Lambda [\sum_{i=1}^{2} B_i l_i - R] \tag{19}$$

The optimal design A , and associated response u'_i and adjoint state v'_i satisfy the system

$$1 + \sum_i \int_{\Omega_i} (-\lambda^+_i - \lambda^-_i)dx = 0$$

$$(EA\,u')' + p = 0$$

$$(EA\,v')' - (\lambda^+_i - \lambda^-_i)' = 0 \qquad \text{all x in } \Omega_i$$

$$-(1/l_i)\int_{\Omega_i} Eu'v'_i dx + \Lambda = 0 \qquad i=1,2 \qquad (20)$$

$$\lambda^+_i(u'_i - \beta) = 0 \qquad \lambda^+_i \geq 0 \qquad u'_i - \beta \leq 0$$

$$\lambda^-_i(-u'_i - \beta) = 0 \qquad \lambda^-_i \geq 0 \qquad -u'_i - \beta \leq 0 \qquad \text{all x in } \Omega_i$$

$$\Lambda(\sum_i B_i l_i - R) = 0 \qquad \Lambda \geq 0 \qquad \sum_i B_i l_i - R \leq 0$$

Since $u_i' - \beta \neq 0$ almost everywhere (AE) in Ω_i, $\lambda_i^{+,-} = 0$ AE ; according to this property together with the first of equations (20) , the $\lambda_i^{+,-}$ are Dirac functions, say $\lambda_i^{+,-} = \lambda_{ik}\delta(x-x_k)$. The x_k identify points where a stress constraint is tight . For the present example problem, if $p(x) > 0$ throughout Ω then $x_k = \{0, l_1\}$. The optimal solution is obtained directly from the system (20). For instance, if p(x) has constant value, say p_o, the solution is given by

$$A_1 = p_o(l_1 + l_2)/E \qquad\qquad A_2 = p_o l_2/E$$

$$\beta = p_o(l_1 + l_1 l_2 + l_2)/ER$$

Maximum strains in the two segments have equal value β. Note also that for the optimal solution the average value of *unit mutual strain energy* has constant value from segment to segment over the domain, as indicated by the fourth of equations (20).

Beam Design for MinMax (Stress)

Taking the height h(x) of a beam with solid rectangular cross section as the design variable, the objective in this example problem is to minimize the maximum stress.* In this case, the stress is proportional to hw" so that the design problem is stated:

$$\min_{h(x)} \; [\max_{x} |f = hw''|]$$

The corresponding min problem is

$$\min_{h} \; (\beta)$$

subject to:

$$\begin{aligned} & hw'' - \beta \leq 0 \\ & -\mu hw'' - \beta \leq 0 \\ & \int_0^1 [\alpha h^3 w'' \bar{w}'' - p\bar{w}]dx = 0 \quad \text{for all admissible } \bar{w} \\ & h_{min} - h \leq 0 \\ & b\int_0^1 h\,dx - R \leq 0 \end{aligned} \tag{21}$$

The function αh^3 (α = constant) equals beam rigidity and w symbolizes the beam deformation. Admissible designs h(x) are defined in terms of the last two constraints; R, the bound on total resource, and h_{min}, the bound on local measure of the design, and the factor μ represent specified non-negative values. Width b of the beam is taken to be constant.

For this problem the Lagrangian has the specific form:

$$L = \beta + \int_0^1 \{\eta_1(hw''-\beta) + \eta_2(-\mu hw''-\beta) - \alpha h^3 w''\bar{w}'' + p\bar{w}$$

$$+ \eta_3(h_{min}-h)\}dx + \eta_4[b\int_0^1 h\,dx - W]. \tag{22}$$

Stationarity of L requires (solution functions for design and for state are not distinguished from the admissible functions here):

---

*The following material is taken from Taylor-Bendsøe.

$$1 - \int_0^1 (\eta_1 + \eta_2)dx = 0 \tag{23}$$

$$- 3\alpha h^2 w''\bar{w}'' - \eta_3 + \eta_4 b + (\eta_1 - \mu\eta_2)w'' = 0 \tag{24}$$

$$- (\alpha h^3 \bar{w}'') + [(\eta_1 - \mu\eta_2)h]'' = 0 \tag{25}$$

$$\eta_1(hw'' - \beta) = 0 \tag{26}$$

$$\eta_2(-\mu hw'' - \beta) = 0 \tag{27}$$

$$\eta_3(h_{min} - h) = 0 \tag{28}$$

$$\eta_4(b \int_0^1 h\, dx - W) = 0 \tag{29}$$

The solution is governed by these equations together with the original constraints and the (Kuhn-Tucker) conditions $\eta_i \geq 0$. The first equation represents a normalization of the multipliers (adjoint load) η_1 and η_2. The second and third equations are simply the specific forms of optimality condition and adjoint state equation for this example. Note that from equations (23, 26, 27), constraints $hw'' - \beta \leq 0$ and/or $-\mu hw'' - \beta \leq 0$ are tight at least somewhere in the domain (0,1). In other words, in the solution for stationarity of L, the value of criterion measure hw'' (or $-\mu hw''$) equals β at its maximum. This substantiates the identification of problem (21) with the associated min-max problem, or, in the more general form it serves to identify the problem (17) with its min-max problem.

From the fourth and fifth equations we have that η_1 and η_2 are orthogonal, $\eta_1 \cdot \eta_2 = 0$. Additional interpretation of the system provides that $\eta_4 \neq 0$, whereby the resource constraint is tight. Also, by the switching equation $\eta_3(h_{min}-h) = 0$, the domain, (0,1) is covered in intervals with either $h > h_{min}$; $\eta_3 = 0$ (design intervals), or $h = h_{min}$; $\eta_3 \geq 0$. In the design intervals $\eta_1 \neq 0, \eta_2 = 0$ and $hw'' - \beta = 0$, or $\eta_2 \neq 0$, $\eta_1 = 0$ and $-\mu hw'' - \beta = 0$. Making use of these results the entire system can be reduced by algebraic manipulation to a substantially simpler form.

In fact the solution itself is obtained directly for simple loads; for $p(x) = p_0 x/l$ as an example, the shape of the optimal beam in design intervals is given by $h(x) = [p_0(x^3 + c_1 x + c_2)/6l\mu\beta]^{1/2}$, where c_1 and c_2 are constants. An iterative method is used to determine the boundaries (x_1 and x_2 in Figure 1) of design intervals, and the values of β, integration constants, etc. Details of the solution for a simply supported beam under the cited load and for μ = 1 are shown in Figure 5.

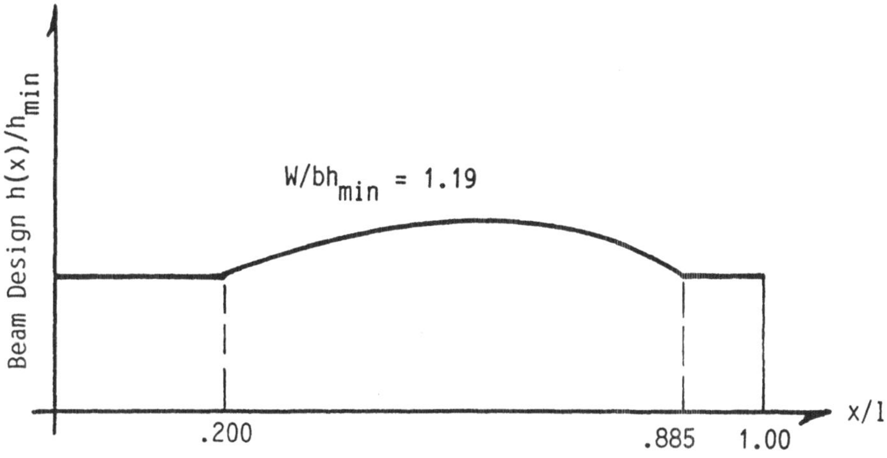

Figure 5. Beam Design to Minimize the Maximum Stress

Design of an Elastic Foundation

As a second example, we sketch a treatment for the optimal design of an elastic foundation supporting an elastic beam. The goal is to predict the distribution of foundation stiffness, represented through the foundation modulus function k(x), that minimizes the foundation pressure. The magnitude of pressure is given by |kw| (w(x) symbolizes the beam deformation), so the design problem is stated as:

$$\min\,[\max |kw|\,]$$

Note that in this example the criterion function again depends explicitly on the state and on the design function. The 'minimize on a bound' form for this problem is:

$$\min_{k(x)} (\beta)$$

Subject to:

$$kw - \beta \leq 0$$

$$-\mu\, kw - \beta \leq 0$$

$$\int_0^1 [R w'' \bar{w}'' + (kw - p)\bar{w}]dx = 0$$

$$k - k_{max} \leq 0$$

$$K - \int_0 k\,dx \leq 0$$

where beam stiffness is symbolized by R.

In contrast to the prior example, here the measure of global resource is bounded from below while the design k(x) is limited locally from above (the system does not admit solutions with $k \leq 0$).

The Lagrangian is formed for this problem, and the solution may be established from the associated necessary conditions, in much the same way as was indicated for the first example (for brevity the details are omitted). A solution is sketched in Figure 6 for the case $p \equiv 0$, hinged supports, and with prescribed, equal-valued displacement of the beam ends into the foundation. As indicated in the figure, for the optimal solution the magnitude of foundation pressure is constant in the design intervals ($|kw| = \beta$).

The combined design of beams and their foundations is considered by Plaut.*

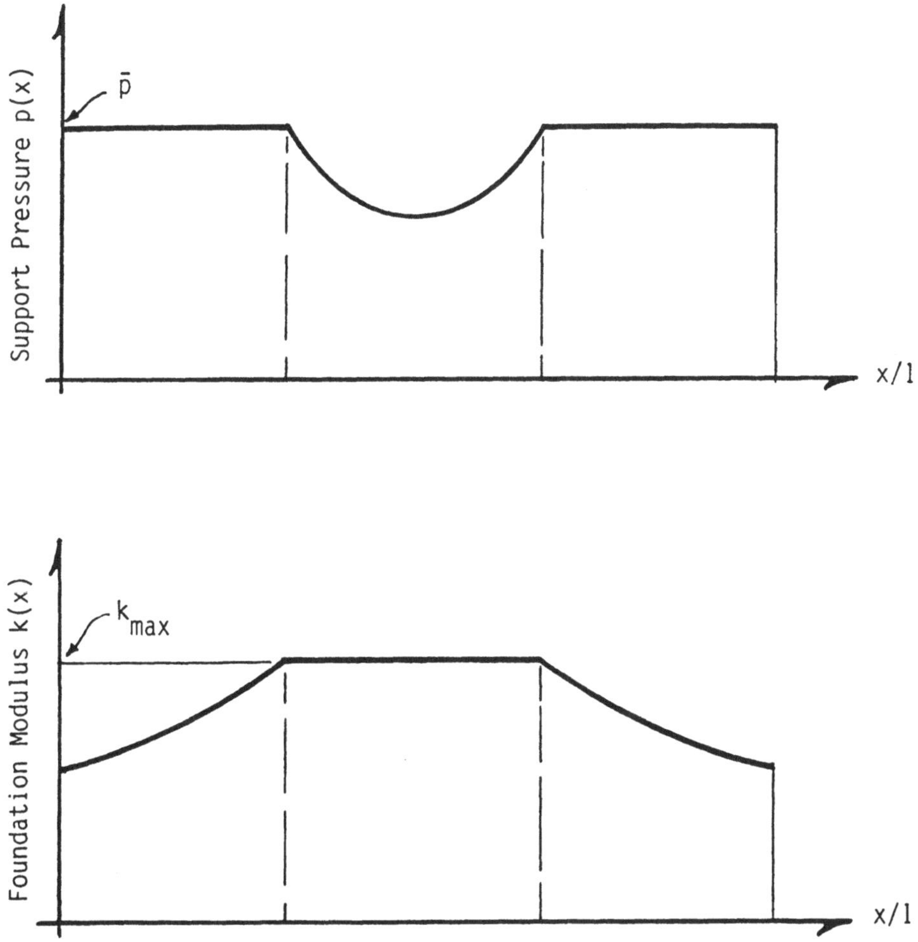

Figure 6. Design and Foundation Pressure for the Optimal Elastic Foundation

Minimize the Maximum Displacement and a Relaxed Form for Min-Max Problems

In problems where the criterion function is directly a measure of state or in bending problems of its first derivative, the adjoint loads (multipliers on the 'criterion constraints') are in general singular. This property is demonstrated for the former case, i.e., $\min_D [\max |w(x)|]$, where w(x) represents beam or plate deflection. The Lagrangian for the problem is stated as:

$$L = \beta + \int_\Omega [\eta_1(w-\beta) + \eta_2(-\mu w-\beta)]d\Omega - a_D(w,\bar{w}) + b(\bar{w}) + \psi, \quad (30)$$

Here the state equation is represented in terms of the energy bilinear form a_D and the load linear functional b. The associated necessary conditions related to multipliers η_1 and η_2 are:

$$1 - \int_\Omega(\eta_1 + \eta_2) = 0 \quad (31)$$

$$\begin{aligned} &\eta_1(x) = 0 \text{ if } w(x) < \beta; && \eta_2(x) = 0 \text{ if } -\mu w(x) < \beta \\ &\eta_1(x) \geq 0 \text{ if } w(x) = \beta; && \eta_2(x) \geq 0 \text{ if } -\mu w(x) = \beta. \end{aligned} \quad (32)$$

Typically the deflection function cannot have constant value over any interval of positive measure, whereby the stated conditions (31,32) dictate that η_1 and η_2 must be certain linear combinations of Dirac δ-functions. (Haug substantiates this result; also in his variational formulation Cinquini identifies the adjoint load as a Dirac-function, but the above normalization (31) is not present in his treatment.)

According to this result, the determination of (adjoint state) $\bar{w}$ requires the solution of a beam or plate boundary value problem with singular loads. As an alternative, by treating this type of problem in a slightly modified form such singularities may be avoided. The modification amounts to a globally-bounded relaxation relative to the original constraint on local measure f(D,w). Thus the constraint for criterion f(D,w), i.e., $|f(D,w)| - \beta$

≤ 0, is rewritten in terms of relaxation $\varepsilon(x)$ as $|f(D,w)| - (\beta+\varepsilon) \leq 0$. In place of the original minmax problem, we now consider the relaxed problem stated as:

$$\min_{D} \beta \tag{33}$$

subject to:

$$\left.\begin{array}{r} f - (\beta+\varepsilon) \leq 0 \\ -\mu f - (\beta+\varepsilon) \leq 0 \end{array}\right\} \tag{34}$$

$$a_D(w,\bar{w}) - b(\bar{w}) = 0 \tag{35}$$

$$D_{min} - D \leq 0 \tag{36}$$

$$\int_{\Omega} D \, d\Omega - W \leq 0 \tag{37}$$

$$-\varepsilon \leq 0 \tag{38}$$

$$\int \varepsilon dx - E \leq 0 \tag{39}$$

Again, state equation (35) is expressed in terms of (energy) bilinear form a_D and the (load) linear functional $b(\bar{w})$. Design is symbolized by D, and c, E, D_{min}, and W represent specified nonnegative numbers. Thus the relaxed problem corresponds still to minimization with respect to design of the bound value β, but now with an admissible violation $\varepsilon(x)$ (the admissible set is defined by (38,39)) of the bound on f, where the total measure of violation is not to exceed the value of E.

It may be verified that the original min-max problem is recovered for $E \Rightarrow 0$. The relationships among w, ε, and β are indicated in the sketch of Figure 7 for the case $f = w$, as an example.

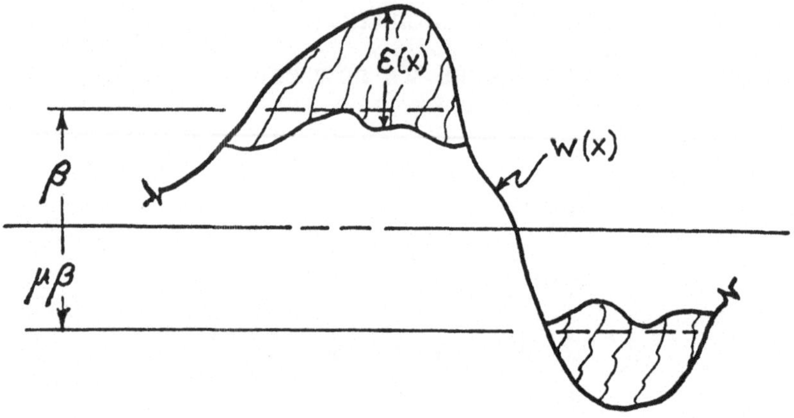

Figure 7. Constraint Relaxation Function ε(x)

The optimal relaxation from among admissible functions ε(x) is associated with equality in one of the constraints (34). In other words, in the solution of the relaxed problem, the difference f-ε (or -μf-ε) has constant value equal to the bound β.

For problems with E > 0 the load in the adjoint problem is no longer singular; it is verified below that multipliers η_1 and η_2 in fact have constant value wherever they differ from zero. Note also that the introduction of a relaxation regularizes the problem in the sense that constraints (34) are regular even through their counterparts in the original min-max problem may lack regularity.

An augmented functional for the relaxation problem (33-39) has the form

$$L^* = \beta - a_D(w,\bar{w}) + b(\bar{w}) + \int_\Omega [\eta_1(f-\beta-\varepsilon) + \eta_2(-\mu f-\beta-\varepsilon) + \eta_3(D_{min}-D)$$

$$+ \eta_5(-\varepsilon)]d\Omega + \eta_4[\int_\Omega D d\Omega - W] + \eta_6[\int_\Omega \varepsilon d\Omega - E] \qquad (40)$$

where the terminology is the same as in the prior examples except for the additional multipliers η_5 and η_6 associated with the constraints that define admissible $\varepsilon(x)$. Stationarity of L^* with respect to β,w,ε require satisfaction of equations similar to (23) and (24) of the 'Design for Stress' min-max problem, and in addition the equation reflecting variation w.r.t. $\varepsilon(x)$:

$$-(\eta_1+\eta_2)-\eta_5+\eta_6=0 \qquad (41)$$

The solution must also satisfy the conditions:

$$\left.\begin{array}{l}\eta_1=0 \text{ if } f<\beta+\varepsilon \\ \text{or} \\ \eta_1\geq 0 \text{ if } f=\beta+\varepsilon\end{array}\right\} \qquad (42)$$

$$\left.\begin{array}{l}\eta_2=0 \text{ if } -f<\mu\beta+\varepsilon \\ \text{or} \\ \eta_2\geq 0 \text{ if } -f=\mu\beta+\varepsilon\end{array}\right\} \qquad (43)$$

$$\left.\begin{array}{l}\eta_5=0 \text{ if } -\varepsilon<0 \\ \text{or} \\ \eta_5\geq 0 \text{ if } -\varepsilon=0\end{array}\right\} \qquad (44)$$

$$\left.\begin{array}{l}\eta_6=0 \text{ if } \int\varepsilon\, d\Omega=E<0 \\ \text{or} \\ \eta_6\geq 0 \text{ if } \int\varepsilon d\Omega-E=0\end{array}\right\} \qquad (45)$$

The condition for stationarity with respect to design of L of the original problem remains unchanged in the relaxed problem identified with the functional L^*.

The following properties are apparent:

i) The multipliers η_1 and η_2 are elements of $L^2(\Omega)$, since constraint equations (34) and (35) must be in $L^2(\Omega)$ when ε is in $L^2(\Omega)$. In other words, η_1 and η_2 are elements of the dual of $L^2(\Omega)$, i.e., $L^2(\Omega)$: Therefore Dirac functions do not appear in the 'load' for the adjoint problem in the relaxed or ε-min-max problem.

ii) The constraint (39) $\int \varepsilon d\Omega \leq E$ is active, since otherwise equation (45) requires that $\eta_6 = 0$, which leads in turn to the requirement from equation (41) that the non-negative functions η_1, η_2, η_5 have zero value almost everywhere. The latter condition would violate the necessary condition that (η_1,η_2) have nonzero measure.

iii) From equation (44), $\varepsilon > 0$ implies $\eta_3 = 0$, whereby from (41) $\eta_1 + \eta_2 = \eta_6$ has constant value for this case. Furthermore, since (from equations (42) and (43)) $\eta_1 \cdot \eta_2 = 0$ it follows that when $\varepsilon > 0$ either ($\eta_1 = \eta_6$ and $\eta_2 = 0$) or ($\eta_1 = 0$ and $\eta_2 = \eta_6$). Thus either the upper or the lower constraint on f is active when ε is non-zero, and therefore the load in the adjoint equation has constant value. Combining this with the fact that from (ii) $\int \varepsilon d\Omega = E$ leads to the conclusion that the optimal colution takes 'maximum advantage' of the possibility for f to exceed the value β, as afforded by the presence of the relaxation function $\varepsilon(x)$.

The main features of the relaxed formulation are illustrated via the treatment of design of a beam to minimize maximum displacement, i.e., the case f = w cited at the beginning of this section. Suppose beam cross-section A(x) is the design variable, and for this example the beam rigidity is taken to be proportional to A^k. Then the state equation is given by

$$\int_0^1 (r\, A^k w'' \bar{w}'' - p\bar{w})dx = 0, \qquad \text{for all admissible } \bar{w}(x),$$

and L^* is obtained from (40) with f, D, and D_{min} replaced by w, A, and A_{min}. To proceed toward a specific solution, for a cantilevered beam supported at x = 0 under uniform load, w(x) increases monotonically with x. Thus there is a value, say x_0, such that

$$\varepsilon(x) = 0 \quad \text{for } 0 \le x \le x_0; \qquad \varepsilon(x) > 0 \quad \text{for } x_0 < x \le 1$$

The corresponding load η_1 in the adjoint problem is (the notation for multipliers η_1 through η_6 is the same as in equation (40); also since $w(x) \ge 0$, $\eta_2 = 0$):

$$\eta_1(x) = \begin{cases} 0 & \text{for } 0 \le x \le x_0 \\ \dfrac{1}{1 - x_o} & \text{for } x_0 < x \le 1 \end{cases}$$

Note that η_1 has constant value wherever it differs from zero, whereas without the constraint relaxation $\varepsilon(x)$, the adjoint load is given by the Dirac δ-function at x = L.

Integrating the state and adjoint equations leads to

$$A^k w'' \stackrel{d}{=} F(x) = \tfrac{1}{2} P_o (1-x)^2$$

$$A^k \bar{w}'' = G(x) = \begin{cases} \tfrac{1}{2}(1 + x_0) - x & \text{for } 0 \le x \le x_0 \\ (1-x)^2/2(1-x_0) & \text{for } x_0 < x \le 1 \end{cases}$$

The design itself may be expressed as

$$A(x) = \begin{cases} A_{min} & \text{for } x_1 \le x \le 1 \\ {}^{k+1}/\sqrt{k\, F(x) G(x)/\eta_6} & \text{for } 0 \le x \le x \end{cases}$$

where the value x_1 is obtained from

$$k\,F(x_1)G(x_1) = \eta_6\,A_{min}^{k+1}$$

which is the optimality condition (27) evaluated for the point x_1. Taking specific values $l = 1$, $k = 1$, $A_{min} = 1$, $p_o = 2$ and $V = 5.15$, the complete design is given by $x_o = 1/2$, $x_1 = 3/4$ and:

$$A(x) = \begin{cases} 1 & \text{for } 3/4 \le x \le 1 \\ 16(1-x)^2 & \text{for } 1/2 \le x \le 3/4 \\ 16(1-x)\sqrt{34-x} & \text{for } 0 \le x \le 1/2. \end{cases}$$

This design is pictured in Figure 8.

Other Applications

The convenience afforded by the transformation of min-max design problems to simple min problems and by the relaxation of constraints, as demonstrated on the beam examples, might be realized in the context of various other problems in mechanics and optimal design. Other types of design problems discussed briefly in this section include a problem of design for elastic bodies in contact, a formulation for optimal remodel design, and a form of mesh optimization for finite element grids.

Minimize Pressure
between bodies in Contact.
The optimum fit!

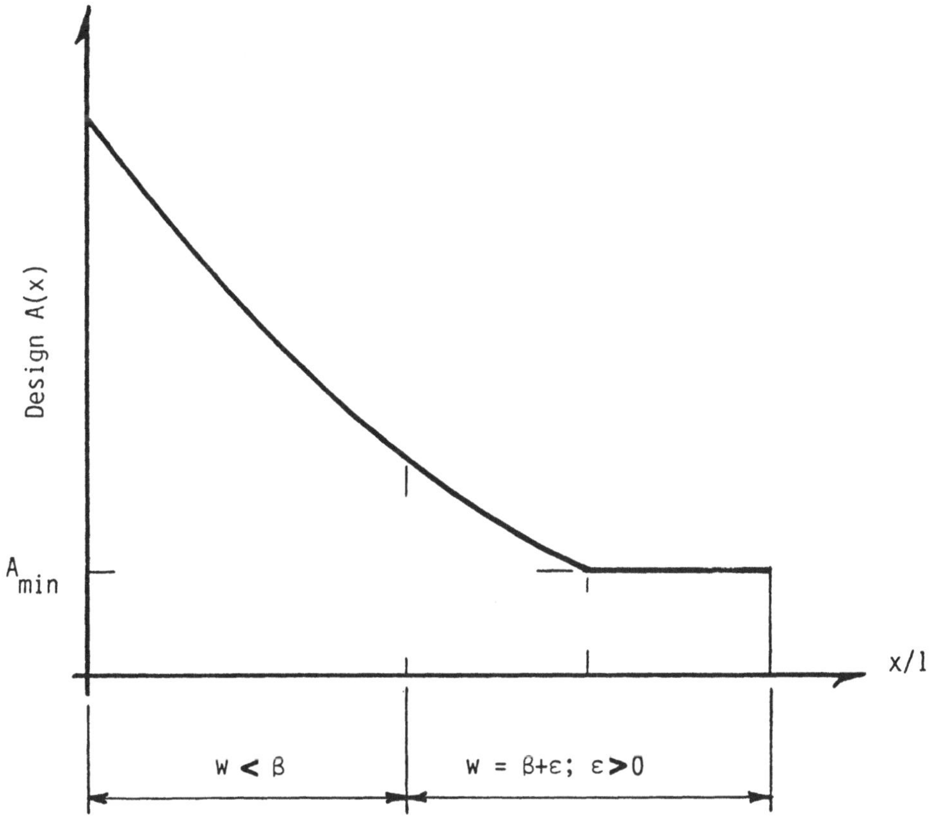

Figure 8. Solution for the Relaxed Problem to Minimize the Maximum Displacement

The objective to *minimize* relative to design the *maximum value of contact pressure* may be taken as the basis for optimization in contact problems. The case where the purpose is to design the initial gap, say $g_o(x)$, between the bodies is used as a particular example. Thus the problem is stated as

$$\min_{g_o \,\varepsilon\, \Gamma} \; (\max_{x \,\varepsilon\, X} p(x))$$

subject to:

'state equations'

$$- g_o(x) \leq 0$$

$$\int g_o dx - G \leq 0$$

The corresponding min problem, stated in terms of bound $\bar{p}$ on pressure, has the form , i.e.:

$$\min_{g_o \,\varepsilon\, \Gamma} \bar{p}$$

subject to:

'state equations'

$$p - \bar{p} \leq 0$$

$$- g_o \leq 0$$

$$\int g_o dx - G \leq 0.$$

Pressure p(x) is of course a function of state. (In its present usage, the term 'state equations' is intended to reflect the usual constraints for contact problems, as well as the governing field equations and boundary conditions. In the procedure of Benedict-Taylor, for example, the gap constraint

$$g(x) = g_o(x) - \overset{I}{u}(x) - \overset{II}{u}(x) \geq 0$$

was appended to the potential energy for the two bodies; then the associated multiplier turns out to be contact pressure. Here u^I and u^{II} represent outer normal displacement of the two bodies evaluated along the contact boundary interval).

The necessary conditions for the min problem reproduce results obtained earlier, but with the clear advantage that the Lagrangian for the problem in this form directly reflects the original design objective. Also, the introduction of a constraint relaxation may be useful in the analysis for contact problems, e.g., the presence of a relaxation $\varepsilon(x) \neq 0$ in the constraint gap g(x) assures that ordinary functions will suffice to express the contact pressure.

The formulation of optimal remodel design, which is treated independently in a separate section of these lectures, represents a quite different example application. The name refers to the type of problem where the purpose is to predict the design for optimum modification of a given structure, rather than the overall optimal design. For the problem in its general form, the solution may represent a combination of reinforcement (added material) over some parts, and lightening (removed material) over other parts of the domain of the structure.

The purpose of discussing the problem here is to point out that the general remodel problem can be represented quite simply with the use of a relaxation in the local constraint on design variable d(x), where d represents reinforcement to a given structure. The problem is stated:

$$\min_{d(x)} \beta$$

subject to:

'state equations'

$$f - \beta \leq 0$$
$$-\mu f - \beta \leq 0$$
$$-(d + \varepsilon) \leq 0$$
$$-\varepsilon \leq 0$$
$$E - \int_{\Omega} \varepsilon dx \leq 0$$
$$\int_{\Omega} (d + \varepsilon) dx - D \leq 0$$

Here $\varepsilon(x)$ appearing as a relaxation of $-d(x) \leq 0$ serves to represent removal of material. For the case $D > 0$ and $E = 0$, the problem statement corresponds to 'reinforcement only' modifications. On the other hand, a value $E > 0$ prescribes the global measure of material removed. It may be verified from the necessary conditions for this problem statement that if both $D > 0$ and $E > 0$, the design modification $d(x)$ must be negative over intervals of the design domain and positive over others, i.e., $d(x)$ represents a general remodel.

Often in the analysis and computational work done using finite element methods, there is a stringent need for a high level of precision in the determination of a local measure such as stress. This need becomes apparent in the treatment of contact problems for elastic bodies , as an example, where the location of 'contact boundaries' may be rather sensitive to changes in contact pressure. Similarly, for certain situations in the design of optimum shape the determination of shape is exceptionally sensitive to the level of precision achieved in the evaluation of stresses. The determination of stresses in the area of crack tips, or near the corners of a sharp punch also suffer from the same sort of sensitivity. The implementation of devices for mesh adaptation into the finite element scheme make it possible to obtain improvements in the authenticity of computational results. One should note, however, that in all cases such as those mentioned here, the adaptation must be responsive to some index of *local error* in order to be effective.

Computational
Adaptation - minimize the local error.

We consider an adaptation scheme based on the objective to minimize the maximum value of local error , where the choice of the measure of error is up to the user. This minmax formulation provides the framework in analysis for a grid optimization method that may be tailored to the particular application. Kikuchi* provides an extensive treatment of this approach, including many applications; thus we limit the present discussion to a brief statement of the problem and a few remarks.

For the local measure of error designated E_e , the problem min $[\max(E_e)]$ w.r.t. node locations x_k is expressed as

$$\min_{x_k} (\beta)$$

subject to:

* $|E_e| - \beta < 0$
* Constraints to define admissible grid modification

In a grossly oversimplified view, the necessary conditions for this problem furnish guidance for the construction of a grid modification scheme. We note that where it might be desirable to use grid adaptation in conjunction with structural optimization, an adaptation step might be called for periodically along the process controlled by the design algorithm. Again, the reader is referred to Kikuchi* for the demonstration of such methods. Viewed differently, there should be some advantage to an approach that is consistent with simultaneous structural and grid optimization. The justification for this claim is that the two aspects of the problem may be strongly coupled. We return to this issue in the discussion that appears at the end of the section on multicriterion optimal design.

We end this section with the comment that the minmax approach is possibly more familiar in its many applications outside structural optimization, e.g., for the interpretation of data, in curve fitting (as generalized Chebyshev approximations),...

Minimize local
error - defines a 'best fit' .
Yes , minmax again !

And while it is possible to imagine applying the idea more widely, one is advised to consider carefully the possible implications...

Minimize the worst
Discomfort : Utopian
in concept , but Boring !

Closure

It is most fitting to commemorate William Prager on the occasion of this Advanced Study Institute - to pay tribute to him, and to acknowledge here again his exceptional contributions to the field of structural optimization. Prager's pioneering studies on analytical modeling set the stage for contemporary developments in several major aspects of interest in our field. Certainly the influence of his work is strongly evident still in much of the material scheduled for presentation in this meeting. I look forward to George Rozvany's special lecture in honor of William Prager, with the expectation of being treated to a most informative and interesting exposition.

One might be informed about some of Prager's earlier and basic work through the recently published book by Save and Prager. The book is a useful resource for additional information on the subject of these lectures as well. Also, the reference list given in this book is helpful as a guide (with notable exceptions) to literature on certain of the topics that are not covered here or in the book. The list identifies published material on Rozvany's singular developments for grillage design and on other topics, for example. Leads may be found there to various works by Cinquini, Masur, Mroz,and Lamblin (to name a few) as well. Papers by Masur on his modeling of problems for segmented structures, on design for strength and stiffness, and on singular problems (with Mroz) are listed with these notes, as are samples of the varied contributions by Mroz, and by Dems and Mroz. Historically significant schools of earlier work ,e.g., those in Poland (Wasiutynski), in Denmark (Niordson), and in Italy, the USSR and the US are described in the various survey articles cited here too.

Directly within the ASI we benefit to hear from Olhoff* (eigenvalue problems, solid plates, sensitivities), Mroz* (shape design, nonlinear problems), Plaut* (shallow shells), Bendsøe* (appropriate design spaces, structure plus control, and Cinquini* (elasto-plastic structures) for additional coverage on topics related more or less to the business of problem formulation and the associated analysis. Haug* presents important results from his extensive work on sensitivity analysis, and Choi* covers sensitivity in relation to shape design. Braibant* provides a clear and useful exposition on the variational modeling for sensitivitiy with respect to shape change. Haftka* and Fleury* each discuss aspects of the implementation of finite element methods for the computational solution of structural optimization problems. Interpretations of finite element grid optimization, treatments of shape design w.r.t. local criteria, and design for bodies in contact (all interpreted for design w.r.t local criteria using the minmax modeling) are reported in the lectures by my colleague Noboru Kikuchi*.

Lastly, the important works on analysis by Lurie, Fedorov and Cherkaev, and of Kohn and Strang are cited here; it is regrettable that their work is not better represented at this meeting.

OPTIMAL REMODEL DESIGN

The optimal remodel design problem is formulated using the same means as those employed in the the treatments just described. However, the design problem itself is substantially more general. As it is presented here, the objective is to predict the optimal reinforcement to and/or removal from an arbitrary specified starting design, within a prescribed limit on the total change in the structure. A variety of practical applications lie within this general problem statement. One may wish simply to determine the best way to improve upon an existing structure that is to be used for some new purpose, for example. To identify another use of the formulation, the benefit of prior experience may be introduced into the design process by specifying the starting design in the form of the best structure from among known examples for problems of the given kind. Also, a sequence of optimal remodel designs may be interpreted to represent a form of evolution in design. As an example, the evolution might be identified with a set of incremental steps in the procedure used for the computational solution of optimal design problems. In this case, each step would itself be associated with a well posed optimization problem.

The better 'next step'
depends on what follows, and
also on what's past ...

The optimal remodel formulation is applicable to almost all forms of structuctural design problems. In this formulation, the optimally modified design is represented schematically in terms of a specified initial design D_o with boundary Γ_o as

$D = D_o + d^+ - d^-$

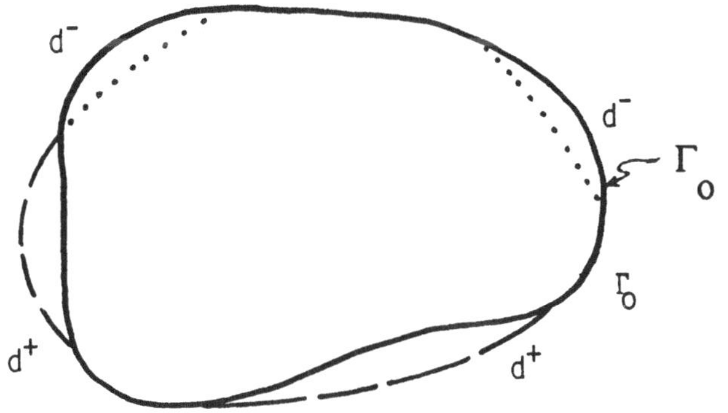

where $d^+ > 0$ symbolizes addition or reinforcement to D_o and $d^- > 0$ stands for a removal or lightening of the initial structure. Limits on the extent of modification may be expressed through the isoperimetric (resource) constraints as

$$\int_\Omega d^+ dx - \mathbf{R}^+ \leq 0 \quad ; \quad \int_\Omega d^- dx - \mathbf{R}^- \geq 0$$

Ω represents the space of admissible designs. In this case the 'addition' and 'removal' are controlled separately through specification of the bounds R^+ and R^-. For example, either simple reinforcement or simple lightening is predicted by setting R^- or R^+ equal to zero. As another possibility, remodel design within the constraints

$$\int_\Omega d^+ dx - \mathbf{R}^+ \leq 0 \quad ; \quad \int_\Omega (d^+ - d^-) dx \leq 0$$

serves to predict the best 'relocation of resource' while preserving the total volume or weight of the original structure. For the problem so stated, in general there exists a value for the bound R^+ such that the solution to the optimal remodel problem for that value is identical to the solution for the conventional optimal design problem. In other words , the conventional problem is imbedded in the set of optimal remodel problems.

The similarity between the variational treatments for optimal remodel problems and for the conventional design problems is made apparent here by presenting the remodel version for the earlier example of design to minimize the maximum stress for the axially loaded elastic bar. Taking into account that $\sigma \sim u'$ and making use of the 'bound interpretation' for the minmax problem, the remodel design problem has the form

$$\min_{b^+,b^-} (\beta)$$

subject to: (46)

$$\int_0^L [E(A_0 + b^+ - b^-)\eta'\zeta' - p\zeta]dx = 0 \qquad \text{all } \zeta \text{ in } Z$$

$$\eta' - \beta \leq 0$$

$$-\eta' - \beta \leq 0$$

$$-b^+ \leq 0$$

$$-b^- \leq 0$$

$$\int_0^L b^+ \, dx - \mathbf{R}^+ \leq 0$$

$$\int_0^L (b^+ - b^-)dx \leq 0$$

The Lagrangian associated with this problem is given by

$$L = \beta + \int_0^L \{- [E(A_0+b^+-b^-)\eta'\zeta' - p\zeta] + \lambda^+(\eta'-\beta) + \lambda^-(-\eta'-\beta) - \mu^+b^+ - \mu^-b^-\}dx \qquad (47)$$

$$+ \Lambda [\int_0^L b^+ dx - R^+] + \Gamma [\int_0 (b^+ - b^-) dx]$$

The optimal remodel functions a^+ and a^- and the associated multipliers must satisfy the ('optimality') equations

$$Eu'v' = \Lambda + \Gamma - \mu^+ \qquad (48)$$

$$Eu'v' = \Gamma + \mu^- \qquad (49)$$

as well as the KKT conditions

$$\mu^+ a^+ = 0 \qquad a^+ \geq 0 \qquad \mu^+ \geq 0$$

$$\mu^- a^- = 0 \qquad a^- \geq 0 \qquad \mu^- \geq 0 \qquad (50)$$

(Only these necessary conditions are needed for the immediate purposes). If the quantity $(Eu'v' - \Gamma)$ is eliminated between equations (48) and (49) we have

$$\mu^+ + \mu^- - \Lambda = 0$$

For segments of the bar in the set, say Ω^+, where $a^+ > 0 \Rightarrow \mu^+ = 0$, from equation (48)

$$Eu'v' = \Lambda + \Gamma$$

Similarly, for segments of the bar within Ω^- where $a^- > \Rightarrow \mu^- = 0$, from equation (49)

$$Eu'v' = \Gamma$$

In other words, the unit mutual strain energy has constant value $\Lambda + \Gamma$ or Γ over the design intervals in Ω^+ and Ω^- respectively. In general the value in segments where material is added differs from the value where material is removed, whereby $\Lambda \neq 0$. This result combined with equation (50) provides the orthogonality relation

$$a^+ a^- = 0$$

Furthermore, since $\mu^+ \geq 0$; $\mu^- \geq 0$, the constants Γ and $\Gamma + \Lambda$ bound the value of the unit mutual strain energy over the entire domain of the structure, as depicted in the Figure 9.

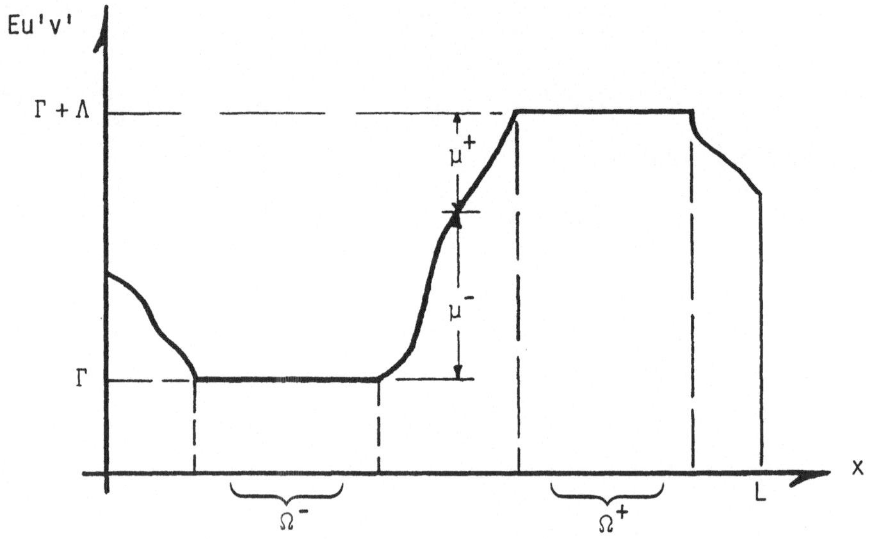

Figure 9. Unit Mutual Strain Energy (Typical)

It would be a relatively simple matter to run through the details for the actual solution of an example remodel design problem, say for particular loads and boundary conditions. However, the procedure is quite similar to what has already been described in the earlier example problems, and at this point it would most likely be boring to read through such material . At the same time it may be useful to note the following features of such problems. The extended set of necessary conditions provides that the *design domain* is identified with the union of segments for which $\lambda^+ > 0$ and $\lambda^- > 0$ (implies $|u'| = \beta$,ie., where the stress constraint is tight). The intersections between design intervals and segments where the original design A_0 remains unchanged may be determined with the help of the *corner conditions*. The same comment applies for the intersection of intervals in Ω^+ and Ω^-. Finally, the entire optimally modified design is made up of segments that alternate among the domains Ω_0, Ω^+ and Ω^-.

The model is easily extended to accommodate a lower bound, say $\underline{A}$, on the optimal design. In the results for this case, there may occur intervals within segments where $A_0 \leq \underline{A}$ with $b^- > 0$; $b^+ = 0$ and $A_0 - b^- = \underline{A}$, and in such intervals the value of the unit mutual energy may be less than Γ, i.e., $Eu'v' < \Gamma$ (Compare to Figure 9.). In the contrasting case, wherever $A_0 < \underline{A}$ within the domain Ω, then the constraint requires that $b^+ > 0$, and here too one may find that $Eu'v' < \Gamma$. Such results for the constrained remodel problem follow directly from the appropriately extended set of KKT conditions.

We mention here another important application of the ideas of optimal remodel, namely its use in the context of redesign for structures that are already under load (or preload). The treatment of such cases was reported in the recent paper by Garstecki. It is important to realize that the remodel of a stress-free structure in general differs from the redesign of a structure with preload or with prestress.

The Optimum Path
through designs or states - either
by best remodel ...

Applications

As suggested in the introduction, the optimal remodel formulation is basic, and therefore one might consider its use in the context of any structural design problem. The examples described here are limited to two applications in optimal shape modification (this material appeared in Na, et al.).

In the first example problem, the objective is to design the optimal hole in the cross-section of a torsion bar with specified external shape. 247 elements were used in the FEM model for the cross-section, and the design boundary is represented by twenty nodes. One-quarter of the cross-section is shown (see Figure), and results are given for optimally shaped holes corresponding to five, ten, and fifteen percent of the original area.

In the second example, the optimal shapes for external boundaries along two free edges of a sheet are predicted for 10, 20, and 30 percent required reduction from the original area (see Figure). The left and lower edges of the sheet are fixed, and load is applied at the truncated corner identified with nodes (7,8,9).

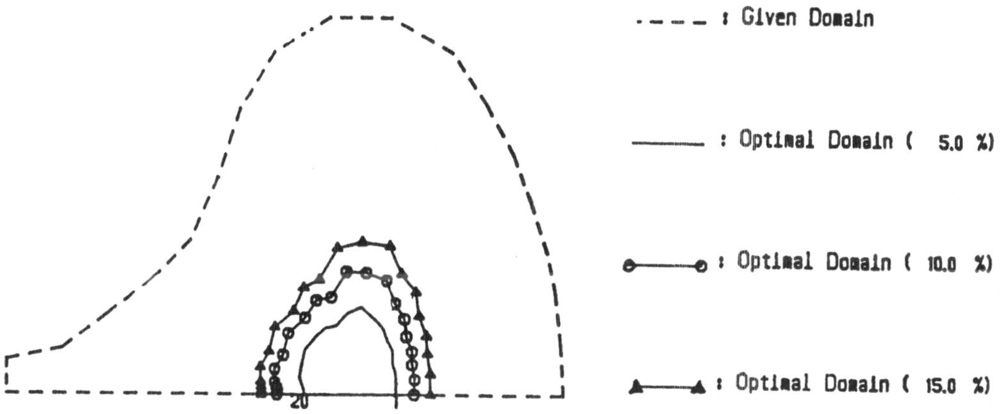

Figure 10. Hole Generation Inside an Originally Simply Connected Domain ... for Maximum Torsional Rigidity

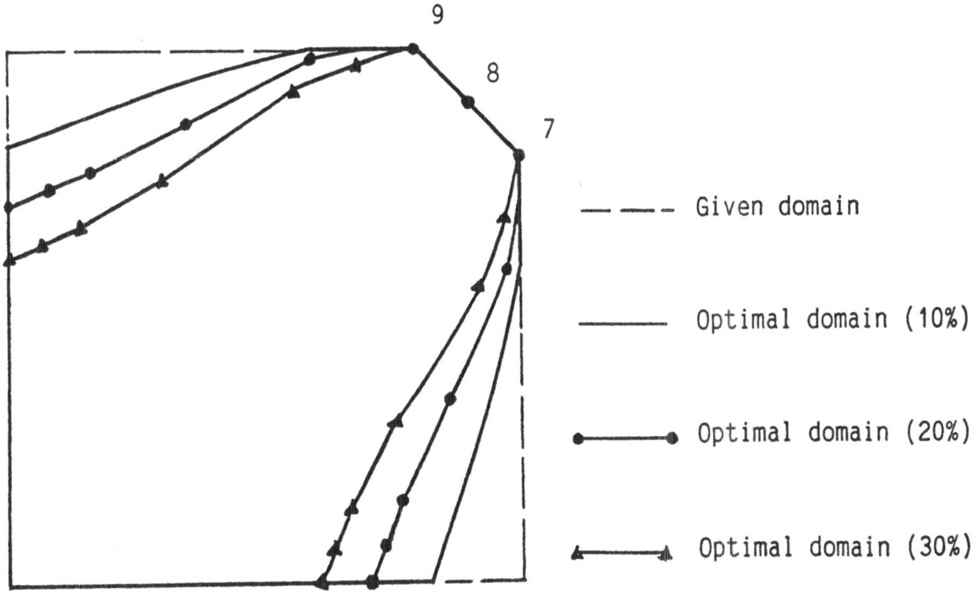

Figure 11. Optimal Shapes for Free Edges of a Sheet-Design for Maximum Stiffness.

MULTICRITERION PROBLEMS IN STRUCTURAL OPTIMIZATION

Most of our goals and
Aspirations: their merit
measured in trade-off.

Introduction

The goal with the presentation in this chapter is to provide an interpretation for *multicriterion* or *vector* structural optimization. Most practical problems in structural design could be characterized properly as multicriterion problems - for example, multipurpose design: the design of a structure to be optimal relative to a set of loads, or multimodal design: where the structure is to perform optimally in different response modes (the two generally are not independent) - so the importance of being equipped to handle problems of this kind is clear. Fortunately the multicriterion problems are treated quite reasonably through an interpretation into scalar form. We follow closely the style and content of Bendsφe-Olhoff-Taylor for the material that follows; thus the scalar counterpart used here for the multicriterion problem is a *minmax* problem, and the minmax problem is resolved through use of the bound formulation as described in earlier applications to minmax optimization. This approach provides the basis for a quite general model of multicriterion structural optimization, and so we can expect to have available the means to handle multicriterion problems where the separate criteria may cover up to the full range of problem types discussed earlier in these notes.

The distinct criteria in a multicriterion problem might reflect design requirements expressed in terms of stress or displacement (or some function thereof), compliance, eigenvalue, or any other measure of structural performance. Given appropriate rules for ordering within and operating on the set of criteria, the problem may be represented properly in a form that corresponds to minimization on a vector argument. Viewed in the context of

Pareto optimization, the optimal solution to such a vector problem is an element of the Pareto set. Also, in one approach the solution itself may be identified via an interpretation on the Pareto set. However, this generally requires the introduction of additional information such as weighting, or the provision of an appropriate basis for making judgment. The procedure tends to be cumbersome and, particularly in structural optimization where one typically has the benefit of substantial physical insight, it is more common to make an a priori interpretation of the multicriterion problem into a scalar form (in this treatment as a minmax problem). The relationship between the Pareto model and multicriterion optimization is discussed in more detail at the end of this chapter.

An extensive description of the mathematical background for vector optimization is given in the survey paper by Stadler.[a] Stadler[b] also provides a comprehensive compilation from the literature on applications, in a second part of this survey. As pointed out there, only rather few cases are known from published works where analysis has been applied to vector structural optimization problems. The report by Koski furnishes a concise description of variational problem statements and their application to truss design problems. Brief summaries on methods of solution are given in the survey paper by Osyczka and Koski; useful listings of the recent literature are included in this paper.

The variational formulation proposed in the present paper corresponds to the min-max problem for a minimum with respect to design of the maximum value among a set of (weighted) criteria. This min-max problem is interpreted as a simple min problem for the minimum of a bound value, where (the measures of) the criteria are related to the bound through constraints. The min problem is also constrained by a bound on the total design resource, e.g. weight of structural material. It appears that the analysis and treatment for solution may be accomplished with less difficulty for this scalar problem, compared to existing forms such as other min-max statements, or constraint, or global criteria methods. Also, the present bound formulation may be compared to other approaches where weighting factors are introduced, such as utility function methods. Relationships among various methods for vector optimization problems are discussed in some detail below.

The developments described herein apply equally well for discrete or continuous structures, and to situations where the design itself has discrete or distributed parameter form. Multipurpose design is covered as well, where the name refers to problems for which different structural purposes (criteria) are identified with distinct load configurations and possibly with different associated response modes. Various example problems are given to demonstrate how these considerations are handled.

Formulation

The analytical formulation is demonstrated for distributed parameter design, expressed in a somewhat general form with regard to objective. As indicated in the introduction, the development applies equally well for discrete systems; a discrete design problem is treated in one of the examples.

It is assumed that the separate criteria g_i of the set $\{g_i\}$ are conditioned to be consistent with direct minimization on each of them. Then the multicriterion (vector) problem is interpreted in the form:

$$\min_{D} \left[\max_{i} (a_i\, g_i) \right]$$

subject to:

$$\begin{aligned} &(\text{state equations})_{\alpha} \\ &D_{min} - D \le 0 \qquad\qquad (1) \\ &\int_{\Omega} D\, d\Omega - W \le 0 \end{aligned}$$

where D represents the design variable constrained from below by (specified) value D_{min}, and W is the specified value for the bound on resource. The $a_i \geq 0$ denote imposed weighting or utility factors. Index α on the representation of 'state equations' is included to accommodate cases where there are constraints reflecting independent response states, as would occur for multiple loads (multipurpose), or where various response modes are to be taken into account.

The development is to be performed in a way to include elements among the g_i that represent a max over the domain of the structure for some local measure of performance. (Min with respect to design of the max stress or displacement are examples of problems with this type of max argument.) For the sake of clarity, distinct symbols are introduced for criteria with global measure on the one hand, and those associated with a 'max argument' on the other. Thus criteria that reflect global objectives such as compliance or eigenvalue (i.e. negative of eigenvalue) are represented by f_i ; $i = 1, 2, \ldots, M$, while the symbol $\tilde{f}_j$ is used to denote the argument in a criterion

$$\max_{x \varepsilon \Omega} \tilde{f}_j(x) \quad ; \; j = 1, 2, \ldots, N$$

Here the $\tilde{f}_j$ may measure stress or displacement, for example.

Using the new notation in place of g_i for the criteria, equation (1) has the form

$$\min_{D} \left[\max_{i,j} \left\{ a_i f_i \, ; a_j \max_{x \varepsilon \Omega} \tilde{f}_j(x) \right\} \right] . \tag{2}$$

The max-max part of this statement condenses to simply a max, whereby (2) may be written as

$$\min_{D} \left[\max_{i,(x)_j} \{ a_i f_i \, ; \, a_j \tilde{f}_j (x) \} \right] . \tag{3}$$

The notation $(x)_j$ is to indicate that for the collection of criteria $a_j \tilde{f}_j$ the max is to cover $x \varepsilon \Omega$.

The problem of minimizing the argument in equation (3) is recast as follows in the form of a simple min problem (Taylor and Bendsoe) that reflects minimization of an upper bound β on all of the separate criteria:

$$\min_{D} \beta$$

subject to:

$$\begin{aligned} & a_i f_i - \beta \leq 0 && i = 1, \ldots, M \\ & a_j \tilde{f}_j (x) - \beta \leq 0 && x \varepsilon \Omega \, ; \, j = 1, \ldots, N \\ & (\text{state equations})_{\alpha} && \\ & -D + D_{min} \leq 0 && \\ & \int_{\Omega} D \, d\Omega - W \leq 0 \, . && \end{aligned} \tag{4}$$

A problem statement similar to the first part of Eqs. (4) was considered in a different context by Gembicki, under the name "goal attainment method."

It is informative to form and to operate on the Lagrangian associated with problem (4). Note that if a criterion f_i (or $-f_i$) can be expressed as a maximum with respect to state, the state equation in (4) corresponding to this f_i can be dropped when f_i is replaced by the functional for which f_i is an extremum (e.g. eigenvalue replaced by Rayleigh expression) : the state equation will automatically follow as a condition of stationarity of the Lagrangian with respect to state.

The necessary conditions obtained from (4) , together with Kuhn-Tucker conditions, comprise a substantial basis for an approach toward solving the multicriteria design problem. The procedure is demonstrated in detail for the several examples described in the next section.

The form (4) of the minimization problem is convenient when considering the issue of existence of solutions. It follows, as demonstrated in Taylor & Bendsøe that for a G-closed design space and functionals f_i , $\tilde{f}_j$ that are explicitly continuous and convex in design and state, the existence of solutions to problem (4) is guaranteed. It also follows from this formulation of the min-max problem that the problem of minimizing the maximum of the components $a_i f_i$; $a_j \tilde{f}_j (x)$ with a constraint on resource is equivalent to minimizing resource subject to constraints on the separate criteria. The equivalent formulation has the form:

$$\min_{D} \int D \, d\Omega$$

$$\text{subject to:} \quad a_i f_i - \beta^* \leq 0 \qquad i = 1, \ldots, M$$

$$a_j f_j (x) - \beta^* \leq 0 \qquad x \, e \, W \; , \; j = 1, \ldots, N \qquad (5)$$

$$(\text{state equations})_{\alpha}$$

$$- D + D_{min} \leq 0$$

where (specified) value β^* is the solution for problem (4). The equivalence may be verified by comparing necessary conditions for the two problems.

Design of a Simply Supported Beam for a Global and a Local Objective

As an illustration of the analysis in Section 2 a vector optimization problem for a simply supported beam will be considered. To illustrate the usefulness of the formulation in treating global as well as local performance indices, we will use compliance and maximum value of deflection as two criteria.

The beam is subjected to a load p increasing linearly along its length l , the deflection of the beam is w and the cross-sectional area D of the beam is the design variable. For simplification the rigidity of the beam is set to be equal to D and the beam is supported at $x = 0$ and $x = l$.

The problem is then

$$\min_{D} \max \{ a \max_{x} | w(x) | \; ; \; b \int_0 p\, w dx \}$$

subject to:

$$(D w'')'' = p$$

$$\int_0^1 D\, dx \leq W \tag{6}$$

$$0 \leq D_{min} \leq D,$$

where a and b are weighting factors on the criteria and W is an imposed upper bound on volume. It is readily seen that the scalar criterion to be optimized can be rewritten as

$$\max_{x} \{ b \int_0^1 p\, w\, d\Omega \, ; a\, w(x) \, ; - a\, w(x) \} \, . \tag{7}$$

A reformulation of the min-max problem as a min problem is stated as:

$$\operatorname*{Min}_{D(x)} \beta$$

subject to:

$$
\begin{aligned}
& b \int_0^1 p\,w\,dx - \beta \le 0 \\
& a\,w(x) - \beta \le 0 \\
& -a\,w(x) - \beta \le 0 \\
& (D\,w'')'' = p \\
& D_{min} - D \le 0 \\
& \int_0^1 D\,dx - W \le 0 \;.
\end{aligned}
\tag{8}
$$

With Lagrangian multipliers η_1, η_2, η_3, $\bar{w}$, γ and Λ, augmented functional (Lagrangian) for this problem has the form

$$
\begin{aligned}
L = \beta + \int_0^1 & [\eta_2 (a\,w - \beta) + \eta_3(-a\,w - \beta) - \bar{w}\,((D\,w'')'' - p) \\
& + \gamma(D_{min} - D)]\,dx \\
& + \eta_1 [b \int_0^1 p\,w\,dx - \beta] + \Lambda[\int_0^1 D\,dx - W] \;.
\end{aligned}
\tag{9}
$$

Stationarity of L requires:

$$1 - \eta_1 - \int_0^1 (\eta_2 + \eta_3)\,dx = 0 \tag{10}$$

$$a\eta_2 - a\eta_3 + \eta_1 b p = (D\,\bar{w}'')'' \tag{11}$$

$$-w''\,\bar{w}'' - \gamma + \Lambda = 0 \tag{12}$$

with associated Kuhn-Tucker conditions.

For the problem at hand it is known that $w \ge 0$, so $\eta_3 = 0$, and as is usual for problems involving the maximum deflection as a performance index, the multiplier η_2 is a Dirac δ - function or it is zero everywhere. As w has one local maximum along the beam, the adjoint $\bar{w}$ is a deflection corresponding to the beam being subjected to a distributed load η_1 bp and a point load of size $a(1-\eta_1)$ at a point x_0 somewhere along the beam; this follows from equations (10) and (11).

With $p = p_0 x/l$ the state equation and the adjoint equation (11) are integrated to obtain:

$$Dw'' = F(x) = C(z^3 - z) \tag{13}$$

where

$$C = p_0 l^2/6$$

$$z = x/l$$

and

$$D\bar{w}'' \overset{d}{=} G(x) = \begin{cases} \eta_1 bCz^3 + z(a(1-\eta_1)(x_0-1)-\eta_1 bC) \\ \qquad\qquad \text{for } 0 \le x \le x_0 \\ \eta_1 bCz^3 + (z-1)a(1-\eta_1)x_0 - \eta_1 bCz \\ \qquad\qquad \text{for } x_0 \le z \le 1 \end{cases} \tag{14}$$

The optimal design can then be verified via equation (12) to be

$$D(x) = \begin{cases} D_{min} & 0 \le x \le x_1 \;,\; x_2 \le x \le l \\ \sqrt{F \cdot G/\Lambda} & x_1 \le x \le x_2 \;, \end{cases} \tag{15}$$

where x_1, x_2 are the solutions to $F(x_i) \cdot G(x_i) = \Lambda \cdot D^2{}_{min}$, $i = 1$, 2. The multiplier Λ is given by the volume constraint being active, and η_1,

η_2, x_0 are governed by

$$(\eta_1 = 0 \ \& \ \eta_2 = 1) \quad \text{or} \quad (\eta_1 = 1 \ \& \ \eta_2 = 0) \quad \text{or}$$

$$\int_0^1 b\,p\,w\,dx = a\,w(x_0) \ \& \ \eta_1 + \eta_2 = 1 \tag{16}$$

$$w(x_0) = \max_x w(x) . \tag{17}$$

Figure 1 shows the optimal area function for weights $a = 3$, $b = 1.5$, length $l = 1$, load $p = 6x$ and constraints $D_{min} = 0.27$ and $W = 0.356$. The optimal value of the criteria is 0.29 and this value is obtained by both the weighted maximum deflection and the weighted compliance. For a uniform beam of the same volume the value of the performance index is 0.33 , so in this example optimization results in about a thirteen-percent improvement.

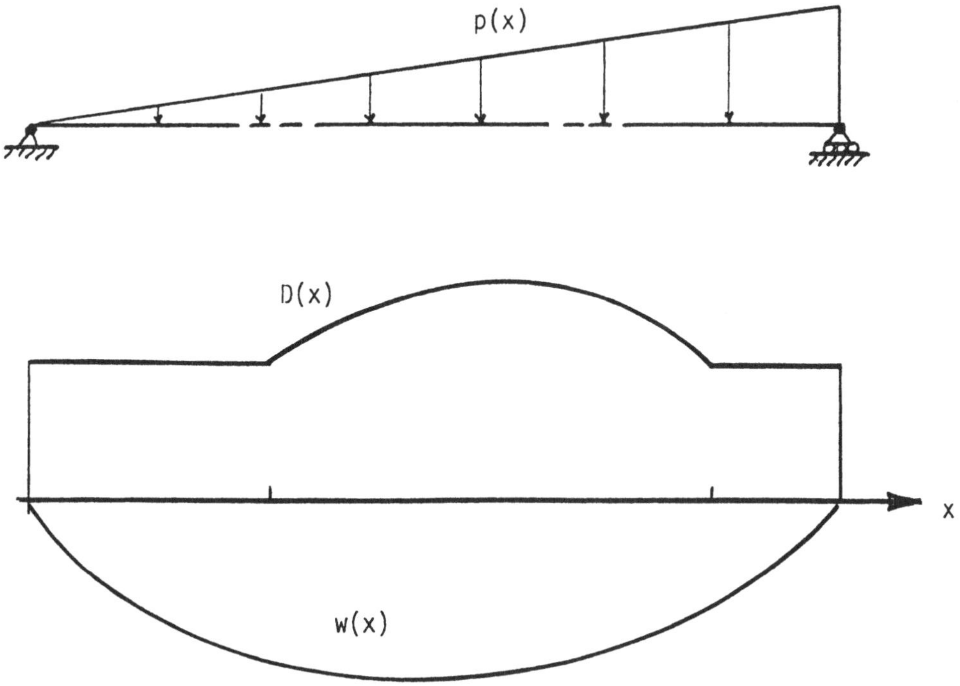

Figure 1 . Design of a Simple Beam Under Two Criteria

Design of a Simple Truss Under Two Criteria

The design of an elastic two-bar truss subject to two criteria suffices to demonstrate some interesting features of multi-criteria design of a discrete structure. The criteria are taken to be weighted measures of displacement in two directions. Cross-sectional areas A_i are the design variables. For components $\{P_1, P_2\}$ of load and $\{u_1, u_2\}$ of displacement, the problem statement is:

$$\min_{A_1} \left[\max_k |a_k u_k| \right]$$

subject to:

$$\begin{aligned} & P_k - \sum_j K_{kj} u_j = 0 ; \quad k = 1, 2 \\ & A_{min} - A_i \leq 0 \quad i = 1,2 \\ & \sum_i l_i A_i - W \leq 0 \end{aligned} \tag{18}$$

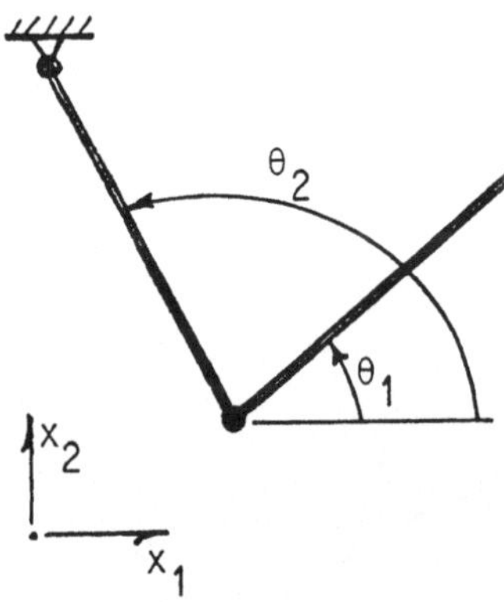

Figure 2 - Truss Layout

Subscript i identifies the members, l_i represent member lengths, and the member directions are given by θ_i as shown in Fig. 2. The stiffness matrix K is

$$[K] = \begin{bmatrix} \sum_i s_i \cos^2 \theta_i & \sum_i s_i \sin \theta_1 \cos \theta_1 \\ \sum_i s_i \sin \theta_i \cos \theta_i & \sum_i s_i \sin^2 \theta_i \end{bmatrix} \tag{19}$$

where $s_i = (A_i E / l_i)$ are the member stiffnesses.

The above min-max problem is solved in the form of the min problem:

$$\min_{A_i} (\beta)$$

subject to:

$$\begin{aligned} & a_k u_k - \beta \le 0 \\ & -(a_k u_k) - \beta \le 0 \\ & P_k - \sum_j K_{kj} u_j = 0 \\ & A_{min} - A_i \le 0 \\ & \Sigma l_i A_i - W \le 0 . \end{aligned} \tag{20}$$

The Lagrangian associated with this problem is

$$\begin{aligned} L = \beta + \sum_K \bar{u}_k \left(P_k - \sum_J K_{kj} u_j \right) + \sum_i \mu_i (A_{min} - A_i) \\ + \sum_K \left\{ \eta_k [a_k u_k - \beta] + \upsilon_k [-(a_k u_k) - \beta] \right\} + \Lambda_i \left(\Sigma A_i l_i - W \right). \end{aligned} \tag{21}$$

Corresponding necessary conditions are (in part) :

$$1 - \Sigma(\eta_k + \upsilon_k) = 0 \tag{22}$$

$$-\Sigma K_{kj} \bar{u}_j + a_k (\eta_k - \upsilon_k) = 0 \tag{23}$$

$$-\sum_k \sum_j (\partial K_{kj}/\partial A_i) u_k \bar{u}_j - \mu_i + \Lambda = 0 \tag{24}$$

and the original equilibrium and design constraints, as well as the 'switching equations.' Equation (22) enforces the result that at least one among the displacement constraints is tight for the solution. Multipliers (η_k, υ_k) on the displacement constraints appear as the loads in (the adjoint) equation (23) as expected. Design sensitivity, given by the first term of equation

(24), is expressed as *unit mutual energy* - mutual between original and adjoint displacements. Given specified values for W, a_k and A_{min}, and using the KKT conditions on the multipliers, the optimal solution may be determined from this system.

Two particular cases are examined, both within the simplifications $a_i = 1$, $l_i = l$, $P_i = P$, and $A_{min} = 0$. For values $\theta_i = \pi/2 ; \theta_2 = \pi$, the solution is easily determined to be $\eta^*_1 = \upsilon^*_2 = 1/2$, $\eta^*_2 = \upsilon^*_1 = 0$, $u^*_1 = u^*_2$, $A^*_1 = A^*_2 = W/2l$, and $u^*_1 = -u^*_2 = 2Pl^2/WE$. Note that for this example *every admissible design is Pareto optimum*, i.e., starting with any design it is not possible to improve on either of the criteria without sacrifice on the remaining one (see the next Section). The comment still applies for the more general problem with $a_1 \pm a_2$; $P_1 \pm P_2$.

In the second example case, θ_i have the values $\theta_1 = \pi/4 ; \theta_2 = 3\pi/4$ The equations predict $K^*_{12} = 0$, $A^*_1 = A^*_2 = W/2l$, $u^*_1 = -u^*_2 = 2Pl^2/EW$. Here, the condition $K^*_{12} = 0$ identifies the optimal truss as the one whose *principal directions* coincide with the directions of the load. For this problem there is *only one Pareto optimal design*. Given the results for these two cases as the extremes, one may appreciate that in general the multicriteria design problem might have anywhere from one to an infinity of Pareto optimal designs.

Eigenvalue Problems

The multicriterion formulation provides an appropriate means for the treatment of eigenvalue design problems as well, particularly those where the structural design depends on separate response modes. Design of an elastic bar to be optimal with respect to both axial and bending vibrations is an example. Also all cases where the response for the optimal design is *multimodal* are managed properly using the minmax (or rather maxmin) statement for multicriterion problems. Consider the design problem with the objective to *maximize the minimum eigenvalue.* The multicriterion problem statement expressed for vector $\{\lambda_i\}$, $i = 1, 2, \ldots$ of eigenvalues associated with design $D(x)$ is simply

$$\max_{D(x)} (\min_{i} \{ \lambda_i \}) ,$$

with a set of constraints much like the ones associated with the example problems already described. A fully detailed treatment of problems of this kind is furnished in the lectures presented by Olhoff* , so the reader is referred there for the rest of the story on multimodal design relative to eigenvalues.

Discussion

Results from the min-max formulation given above may be related to the problem of finding the set of Pareto optimal points for the vector valued objective, and to other scalar forms that make use of a weighted sum of criteria. For convenience the discussion of these issues is stated here in the form associated with discrete problems. However, the arguments apply for distributed parameter problems as well.

An admissible design is represented by element $D: (D_1, D_2, \ldots D_n) \in \mathbf{R}^n$ subject to the constraints (dependence on state is implicit in these expressions) :

$$g_j(D) \leq 0 \qquad j = 1, 2, \ldots, k. \tag{25}$$

The vector criterion to be minimized may be expressed as

$$f(D) = \left[f_1(D), f_2(D), \ldots, f_m(D) \right]. \tag{26}$$

A Pareto optimal design D^* is a design among admissible designs D that satisfies

$$f(D) \leq f(D^*) \Rightarrow D = D^*. \tag{27}$$

Here $\leq$ is the partial ordering of $\mathbf{R}^m$ inherited from the total ordering of $\mathbf{R}$. Values of the criterion in terms of vector f might not be comparable using the ordering $\leq$. The criterion for Pareto optimum implies that $f(D^*)$ is a minimum among those values of the criterion for which a comparison using $\leq$ is possible. A consequence of this is that a design giving rise to a value of the criterion that cannot be compared to any other value is optimal. As observed with the truss example in Section 6, for a problem with two conflicting (i.e., monotone varying with opposite sign) criteria any admissible design is Pareto optimal. This property might be represented qualitatively as shown in Fig. 3. Considering a design in the interval indicated there, any change in design results in an increase in the value of at least one among the Fig. 3 criteria f_i.

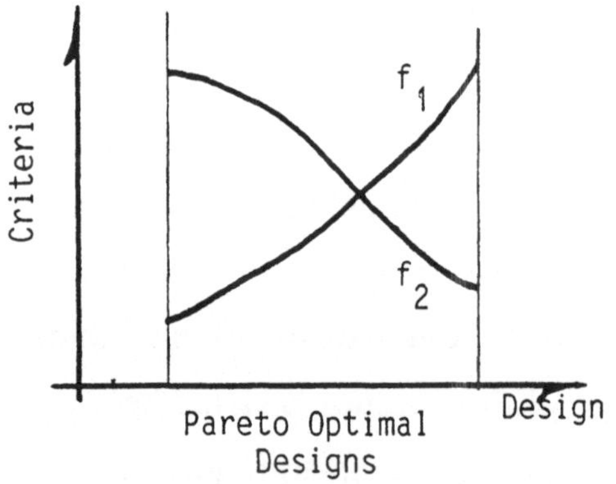

Figure 3

In order to discuss relationships among the several approaches for treating multi-criteria problems, recall that for the present min-max formulation the objective for minimization is the scalar

$$\max_i \{ a_i f_i \}. \tag{28}$$

In another common approach, minimization is performed on a weighted sum of criteria f_i, say

$$\sum_i b_i f_i \qquad (29)$$

with weights b_i (note that for the discussion a_i, b_i and f_i are taken to be non-negative).

Considering the problem of minimization we now state and compare necessary conditions for local minima for the problems involving the criteria (26), (28) or (29) and the constraints (25). For the Pareto problem the condition is (see e.g. Koski):

$$\sum_{i=1}^{m} \eta_i(\partial f_l/\partial D_l) + \sum_{j=1}^{k} \mu_j (\partial g_j/\partial D_l) = 0, \qquad l = 1, \ldots, n$$

$$\eta_i \geq 0, \quad \mu_j \geq 0, \quad \mu_j g_j = 0, \qquad (30)$$

where η_i and μ_j are Lagrangian multipliers. Stating the problem of minimizing the functional (28) as a minimization problem for a bound β on the components $a_i f_i$ leads to a necessary condition for the min-max problem of the form

$$\sum_{i=1}^{m} \eta_i a_i(\partial f_i/\partial D_l) + \sum_{j=1}^{k} \mu_j(\partial g_j/\partial D_l) = 0, \qquad l = 1, \ldots, n$$

$$\eta_i \geq 0, \quad \mu_j \geq 0, \quad \sum_{i=1}^{m} \eta_i = 1, \qquad (31)$$

$$\eta_i (a_i f_i - \beta) = 0, \quad \mu_j g_j = 0,$$

where η_i and μ_j are multipliers. Finally for the weighted sum (29) we obtain

$$\sum_{i=1}^{m} b_i (\partial d_i/\partial D_l) + \sum_{j=1}^{k} \mu_j(\partial g_j/\partial D_l) = 0, \quad l = 1, \ldots, n$$

$$\mu_j \geq 0 , \qquad \mu_j g_j = 0 , \tag{32}$$

where μ_j are Lagrangian multipliers.

From these equations it is seen that a design D* that satisfies the necessary conditions (31) for the min-max problem also satisfies the necessary condition for the weighted sum if the weights b_i are chosen to be $\eta_i a_i$ and the values of the two criteria (28) and (29) are equal at this design. Conversely, if a design D* satisfies (32) it also satisfies (31) when the weights a_i are chosen as $\beta/f_i(D^*)$, $\beta = \sum b_i f_i(D^*)$, and the performance for the two problems is the same (equal to β). Thus the min-max problem and the weighted sum problem, both solved for all combinations of utility factors, generate the same set of designs, and from (30) it is seen that this set is exactly the set of designs that satisfy the necessary condition for Pareto minimality.

The introduction of a scalar criterion of the form of (28) or (29) implies that the designer decides *a priori* on how to weight different components of the vector valued performance criteria. In contrast, finding the full set of Pareto optimal points leaves this decision to be taken at a later stage of the design process. As most methods of finding the full set of Pareto optimal points involves solving a whole class of scalar optimization problems, for many practical cases it is more sensible to consider only a well-chosen scalar criterion, for example of the max-type presented in this paper, or perhaps a limited set of them such as would be identified with a practical range of values for the weight factors.

Closure

We remark that while the example problems presented are (as usual) simple, the method employed is quite general. Thus one can imagine a similar approach as it would apply to design problems with a broad variety of types of criteria or constraints, with local or global measure, covering a variety of response modes, various purposes, and so on. One example of such 'broader' type of problem is described in the section of these notes on Fail-Safe design. The multicriterion formulation provides a useful basis for the treatment of problems in the simultaneous design of structure and control (this subject is discussed in separate presentations given by Bendsφe*, by Khot*, and by Venkaya*), as another example.

Lastly, the model for multicriterion optimization might be applied to obtain an interpretation for the combined 'design of computational means for solving problems' together with 'structural optimization.' This general idea can be illustrated with the example where the finite element method is used for the mechanics analysis in a problem of shape optimal design. Suppose the objective for shape design is to minimize the maximum value of a local measure, say f_D , and let E_h represent a bound on the local finite element error. If both measures of error can be evaluated over the same set of points x_k , and with shape determined by parameters y_i and the finite element grid points symbolized by z_g, the combined (vector) design problem is expressed as:

$$\min_{y_i, z_g} \left[\max_{x_k} \{f_D, E_h\} \right]$$

For an analysis developed according to this statement, the shape design and grid adaptation problems are fully coupled. Coupled forms such as this may be warranted in cases where the original problem is particularly sensitive, i.e., where the prediction of shape is sensitive to imprecision in the finite element model computations.

FAIL- SAFE DESIGN AND OTHER APPLICATIONS

This section is dedicated mainly to the presentation of a model for the optimal design of fail-safe structures. The consideration of fail-safe design requires that two or more distinct structural forms must be taken into account at once in the design process, and this is the feature that makes the category of problems unique. Earlier studies of such problems are reported in the book by Haug and Arora, and in the papers by Haftka, and by Sun, et al., for example. In the treatment given here, means for accommodating multimodal response and multipurpose loading are incorporated into the model for fail-safe design, so the model is in this sense comprehensive. This model was described (Taylor) at the 16th ICTAM Congress, Copenhagen, 1984.

The Fail-Safe Structure -
Optimum to begin with,
after damage too !

According to the concept of *fail-safe design,* a system is required to meet one or more sets of performance requirements beyond those dictated for its primary purpose, under circumstances where the structure itself may have different forms for the alternative requirements. Generally the altered structure is designated in way to reflect damage or some other form of degradation to the primary structure (damaged or diminished structures and their associated loads and/or modes are labeled 'secondary'). Otherwise, the fail-safe criterion may be identified with the possible occurrence of one or more adverse changes in load, for example (such alternate loads also are labeled 'secondary'). For present purposes it is assumed that where the secondary context is related to structural damage or degradation, the form

D_s of the diminished structure is defined somehow in terms of the primary structure. As a simple example, the diminished structure might be given as a (not necessarily spatially uniform) prescribed percentage reduction of the primary structure. Or as another possibility, prescribed portions of the original structure or its supports might be removed to characterize the damaged form.

Our purpose here is, as indicated, to provide a relatively broad statement of the optimization problem for fail-safe design. Fail-safe optimal design is properly characterized as a multicriterion optimization problem. Thus the variational treatment follows directly according to the minmax interpretation already given for multicriterion problems. Once again, the *bound form* for a problem with the set of (properly conditioned) criteria G_j ; $j = 1, 2, \ldots N$ weighted by factors a_i may be stated as:

$$\min_{D} (\beta)$$

subject to:

* $a_j G_j - \beta < 0$ $\qquad j = 1,2, \ldots N$
* performance and design constraints

The main requirement for application to the fail-safe problem is to have sensible identification of criteria G_j and the 'state' and 'design' constraints with the respective primary and secondary contexts already mentioned. This bookeeping is accomplished simply enough with the introduction of appropriate notation. We introduce symbols $D(x)$, $\bar{u}_\alpha(x)$, and $G_j(D;u)$ to represent the primary design, associated state, and criteria respectively, and the parallel notation D_s, $\hat{\bar{u}}_{s\alpha}$, and G_{si} to represent the like quantities for the secondary contexts. For distributed parameter structures the vector valued functions $\bar{u}_\alpha$ (or $\hat{\bar{u}}_{s\alpha}$) represent displacement fields of the appropriate dimension. Subscript α on the state function $\bar{u}_\alpha$, $\alpha = 1,2...,M$ serves to identify states associated for different 'purposes' or 'modes' with the single primary design D; thus the value of M equals the total number of primary 'modes plus purposes.' Subscript s, on the other hand, identifies the separate secondary contexts. Thus D_s, $s = 1,2,...,\Sigma$ represents the various forms of secondary (damaged) structure, Σ, in total number, and $\bar{u}_{s\hat{\alpha}}$ symbolizes the state for the $\hat{\alpha}^{th}$ purpose or mode, for the sth secondary structure. Altogether the problem statement for fail-safe design of a structure with N_c primary criteria and N_{sc} secondary criteria, $s = 1,2, ...,\Sigma$ is stated:

$$\min_D (\beta)$$

subject to:

$$\left.\begin{array}{ll} a_j G_j(D,\bar{u}_d) - \beta \le 0 & \\ a_{s\hat{j}} G_{s\hat{j}}(D_s,u_{s\hat{\alpha}}) - \beta \le 0 & \\ D_s - f_s(D) \le 0 & j = 1,2...,N_c \\ & \hat{j} = 1,2,...,N_{sc} \\ \text{State}_\alpha (D,\bar{u}_\alpha,p_\alpha) & s = 1,2,...,\Sigma \\ \text{State}_{s\hat{\alpha}} (D_s,\bar{u}_{s\hat{\alpha}},p_{\hat{\alpha}}) & \hat{\alpha} = 1,2,...,M \\ & \alpha = 1,2,...,M \\ \text{Design Constraints} & \end{array}\right. \qquad (2)$$

Weight factors a_j and a_{sj} have specified values. In the constraint written as $D_s - f_s(D) \le 0$, which is intended to define the secondary structure in terms if the primary one, f_s may be thought of as a 'damage function.' In fact the matter is not quite so simple as it might appear, and care must be taken with the implication of dependence in the prediction of the optimum solution on the form of this statement. It is simpler to appreciate that in the case of a secondary context which relates just to a different load and its associated set of criteria, $D_s = D$. Note also that a different 'purpose' is represented simply by a distinct element of the vector p_α (or $p_{s\hat{\alpha}}$ for a secondary structure), whereas the accounting of different response modes is reflected in the choice of the appropriate 'state equations.' Functions $\bar{u}_\alpha$ and $\bar{u}_{s\hat{\alpha}}$ represent the corresponding 'state fields' in either case.

Details for the treatment of problems according to the statement of equation (2) are by now mostly routine, i.e., necessary conditions can be listed and then interpreted along much the same lines as was done for the developments described earlier. Indeed there may not be much reason to trouble over example problems at all, except perhaps to demonstrate that it is possible to make sense out of the mess of notation. Examples are described for the fail-safe optimal design of trusses and of beams. (We make use of the truss design example because the 'discrete structure' model is in some ways more convenient as a means to demonstrate the formulation and the use of notation. With respect to these aspects of modelling the fail-safe problems, the approaches for discrete structures and for distributed parameter problems are quite the same.)

Fail-Safe Truss Design

Suppose that the truss layout is given; then the design is expressed in terms of the member areas A_i, i=1,2,...,P for the truss with 'P' members. A damaged structure represented by A_{si} is defined relative to the original A_i by the constraints

$$A_{si} - r_i A_i \leq 0 \qquad i=1,2,\ldots,P \qquad (3)$$

Values for the factors r_i are specified; they should be non-negative, and to reflect damage or degradation $r_i < 1$. We consider a single primary purpose with constraints on stresses and on displacements, and a single-purpose secondary context with just one constraint to reflect a limit strength for the damaged members. The problem is summarized in the statement (patterned after (2)):

$$\begin{array}{lll}
\underset{A_i}{\text{Min}}\ (\beta) & & \\
\text{subject to:} & & \\
|\sigma_i(u^{(\gamma)})| - \beta \leq 0 & & \\
a|u^{(\gamma)}| - \beta \leq 0 & i=1,2,\ldots,P & \\
a_s|\sigma_{si}(u_s^{(\gamma)})| - \beta \leq 0 & s=1,2,\ldots,\Sigma & (4) \\
A_{si} - r_i A_i \leq 0 & \gamma=1,2,\ldots,N_u & \\
\{u\}^T[K(A)]\{v\} - \{P\} = 0 & \text{for all } v \text{ in } V & \\
\{u_s\}^T[K_s(A_s)]\{v_s\} - [P_s\} = 0 & \text{for all } v_s \text{ in } V_s & \\
A_{min} - A_i \leq 0 & & \\
\sum_{i-1}^{P} A_i l_i - R \leq 0 & &
\end{array}$$

Components $u^{(\gamma)}$ and $P^{(\gamma)}$ of the discrete vector fields $\{u\}$ and $\{P\}$ appearing in the virtual displacement (state) equations represent nodal displacements and nodal loads (subscript α does not appear here because both the primary and the secondary contexts are taken to be single purpose/single modal). Matrix [K] for the system stiffness is interpreted in terms of the dependence of its elements, say $K_{\alpha\beta}$, on the design variables A_i. Displacements $\{u\}$ and $\{v\}$ belong to the set, say V, of admissible displacement fields. Value R in the isoperimetric constraint is the measure of available structural resource, e.g., the volume of material available for the truss.

The Lagrangian for problem (4), say L, depends on β, $u^{(\gamma)}$, $u_s^{(\gamma)}$, A_i, $v^{(\gamma)}$, $v_s^{(\gamma)}$ and the multipliers with the list of constraints. The 'optimality conditions' for the problem are obtained simply as the conditions

$$\partial L/\partial A_i = 0 \qquad\qquad i=1,2,\ldots,P$$

for stationarity of the Lagrangian with respect to design. Requirements for stationarity of L w.r.t. $u^{(\gamma)}$ and $u_s^{(\gamma)}$ are the primary and secondary adjoint state equations. The adjoint loads which appear in these equations are expressed in terms of the multipliers associated with the criterion constraints. The full set of KKT conditions are necessary *and* sufficient for this problem, and so they provide the means to solve for the minimal fail-safe design.

For an illustrative example, the primary and secondary strucures are taken to have the forms shown in the following figure.

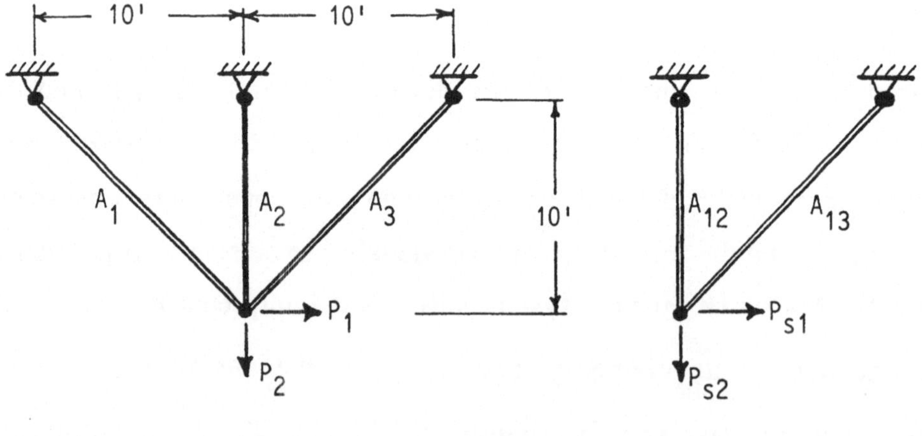

Primary and Secondary Truss Structures

One member of the 'primary' three-bar truss is completely removed in the secondary context, while the other two members remain unchanged (undamaged). Accordingly, the 'damage matrix' r of equation (4) has the form:

$$r = \begin{bmatrix} 0 & 0 & 0 \\ 0 & 1 & 0 \\ 0 & 0 & 1 \end{bmatrix}$$

The detailed solution is presented here for the case with a single primary load $\{P^{(1)}, P^{(2)}\}^T$, one secondary load $\{P_1^{(1)}, P_1^{(2)}\}^T$, and where the displacement constraint of (4) is inactive. Suppose that the multipliers on the stress constraints are symbolized by λ_i and λ_{1i} for the primary and secondary contexts, respectively. Then the results for 'switching equations' associated with these constraints are:

$$\lambda_1[E(u^{(1)} + u^{(2)})/20 - \beta] = 0 \quad \text{(a)}$$

$$\lambda_2[Eu^{(2)}/10 - \beta] = 0 \quad \text{(b)}$$

$$\lambda_3[-E(-u^{(1)} + u^{(2)})/20 - \beta] = 0 \quad \text{(c)}$$

$$\lambda_{12}[a_1Eu_1^{(2)}/10 - \beta] = 0 \quad \text{(c)}$$

$$\lambda_{13}[-a_1E(-u_1^{(1)} + u_1^{(2)})/20 - \beta] = 0 \quad \text{(e)}$$

The optimal design may or may not be influenced by the secondary context, represented here in the latter two of the five equations, depending on the value of weight factor a_1; if

$$a_1 > \min[10\beta/Eu_{1P}^{(2)};\ 20\beta/E(u_1^{(1)}-u_{1P}^{(2)}) \stackrel{d}{=} a_1^*]$$

the secondary context does affect the solution. The range of possible optimal solutions, limited here for simplicity to non-negative values for specified loads $P^{(\gamma)}$ and $P_1^{(\gamma)}$, is summarized in the following two tables, the first one covering results for $a_1 < a_1^*$ and the second table for $a_1 > a_1^*$.

Specific numerical results are given for optimal solutions corresponding to the values $P^{(1)} = 20$ ksi, $P^{(2)} = 30$ ksi, $A_{min} = .1$ in^2, and $E = 10^7$ psi, as an example. For the situation $a_1 < a_1^*$, the solution is

$$A_1 = 1.13 \text{ in}^2, \qquad A_2 = 0.40 \text{ in}^2, \qquad A_3 = 0.1 \text{ in}^2$$

$$\beta = 2.5 \text{ ksi}, \qquad R = 21.4 \text{ in}^3$$

This result is identified with Case 1 of Table 1. On the other hand, with the additional secondary context $P_1^{(1)} = 1$ ksi, $P_1^{(2)} = 11$ ksi, and $a_1 = .893$, the optimal solution belongs in Case 2 of Table 2 and we find:

$$A_1 = 1.126 \text{ in}^2, \qquad A_2 = 0.429 \text{ in}^2, \qquad A_3 = 0.1 \text{ in}^2$$

$$\beta = 2.5 \text{ ksi}, \qquad R = 21.6$$

Table 1 - Summary of Optimal Truss Designs

for $a_1 < a_1^*$ ... Primary Context Dominates

Case	Active Stress Constraints	Optimal Designs	Solution Values for β
1	(a) and (b)	$A_1 = \sqrt{2}\, P^{(1)}/\beta$ $A_2 = (P^{(2)} - P^{(1)})/\beta$ $A_3 = A_{min}$	$10(P^{(1)} + P^{(2)})/$ $\cdot\, (R - 10\sqrt{2}\ A_{min})$
2	(b) and (c)	$A_1 = A_{min}$ $A_2 = (P^{(1)} + P^{(2)})/\beta$ $-2\sqrt{2}\, A_{min}$ $A_3 = \sqrt{2} P^{(1)}/\beta - 2A_{min}$	$(30P^{(1)} + 10P^{(2)})/$ $\cdot (R + (20+10\sqrt{2})A_{min}$
3	(a) and (c)	$A_1 = \sqrt{2}\,(P^{(1)} + P^{(2)})/2\beta$ $A_2 = A_{min}$ $A_3 = \sqrt{2}\,(P^{(1)} - P^{(2)}/2\beta$	$20\, P^{(1)}/(R - 10A_{min})$

Table 2- Summary of Optimal Solutions for

$a_1 > a_1^*$ ... Design Influenced by Secondary Context

Case	Active Stress Constraints	Optimal Solutions
1	(a),(b) and (e)	$A_1 = \sqrt{2}\, P^{(1)}/\beta$ $A_2 = (P^{(2)} - P^{(1)}/\beta$ $A_3 = \sqrt{2}\, a_1\, P_1{}^{(1)}/\beta$ $\beta = 10(P^{(1)} + P^{(2)} + 2a_1P_1{}^{(1)})/\mathbf{R}$
2	(a) and (d)	$A_1 = [(\sqrt{2}A_1 + A_3)P^{(1)} + A_3P^{(2)} - A_2A_3\beta]$ $\cdot (A_2 + \sqrt{2}\, A_3)\beta$ where $A_2 = a_1(P_1{}^{(1)} + P_1{}^{(2)})/\beta$ $A_3 = A_{min}$ $\beta = - G + \sqrt{G^2 - 4FH} /2F$ $F = 200\, A^2{}_{min} - 10\sqrt{2}\, R\, A_{min}$ $G = 100\sqrt{2}\, A_{min}(P^{(1)} + P^{(2)})$ $+ a_1(200\, A_{min} - R)(P_1{}^{(1)} + P_1{}^{(2)})$ $H = 20P^{(1)}\, a_1(P_1{}^{(1)} + P_1{}^{(2)})$ $+ 10\sqrt{2}\;\; a_1{}^2\, (P_1{}^{(1)} + P_1{}^{(2)})^2$
3	(a), (d) and (e)	$A_1 = \kappa/\beta$ $A_2 = a_1(P_1{}^{(1)} + P_1{}^{(2)})/\beta$ $A_3 = \sqrt{2}\, a_1P_1{}^{(1)}/\beta$ where $\kappa = \sqrt{2}\, [2P^{(1)}P_1{}^{(1)} + P^{(1)}P_1{}^{(2)} +$ $P^{(2)}P_1{}^{(1)} - a_1P_1{}^{(1)}(P_1{}^{(1)} + P_1{}^{(2)}]$

The secondary loads applied to the Table 1 design would result in a 7% violation of the stress constraint for member two.

Example of Fail-Safe Beam Design

The primary system for this example is comprised of a propped cantilevered beam under a single uniform lead. In the secondary context the same beam is required to carry a lesser load but with the prop support

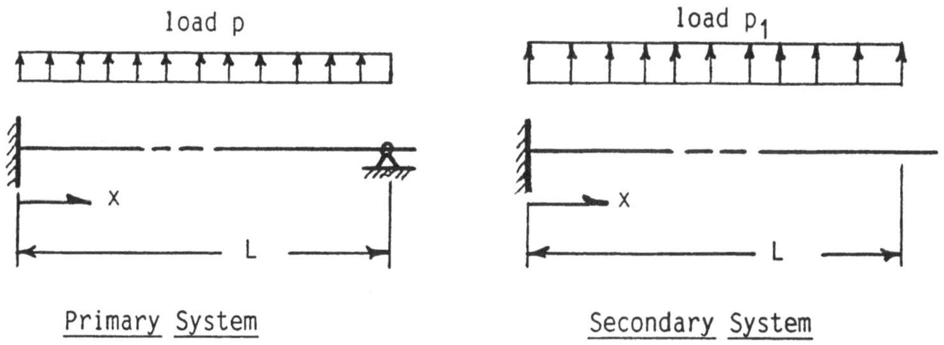

removed. (Note that in this case sets V and V_s of admissible displacement (cf. equation (4)) are distinct.) The design is required to meet a single criterion in the form of a bound on bending stress in each of the primary and secondary contexts. The width b(x) of a beam cross-section with uniform depth is the design variable. Problem statement (2) simplified for this example has the form:

$$\min_{b(x)} (\beta)$$

subject to:

$$|w''| - \beta \leq 0$$

$$\alpha|w''_1| - \beta \leq 0 \qquad \text{all x in (0,L)}$$

$$b_{min} - b \leq 0$$

$$\int_o^L [ebw''v'' - pv]dx = 0 \qquad \text{all v in V}$$

$$\int_o^L [eb\, w''_1 v''_1 - p_1 v_1]dx = 0 \qquad \text{all v in } V_1$$

$$\int_o^L b\, dx - \mathbf{R} \leq 0$$

Here flexural rigidity has been represented by cb(x), c = constant, and the (prescribed) ratio of primary to secondary allowable stress is given by constant a.

As in the prior example, the KKT conditions are sufficient as well as necessary for the optimal fail-safe beam design, so the solution can be determined directly from this system of equations. For the sake of brevity, we bypass the detailed analysis for the problem and merely point out that (a) the optimal solution may be affected by the primary or the secondary contexts alone, or by them in combination, depending on the values prescribed for R, α, p, p_1, and b_{min}, and (b) generally the solution is comprised of segments over the beam length, where the segments have functionally unique form depending on one or another of the local constraints in equation (5).

The specific result for the optimal design and the associated stresses obtained for the value $\alpha p_1/p = 1/6$ are shown in the figures. Of the five intervals covering the length of the beam, in the first and fourth the primary criterion dominates the design, the secondary criterion governs in the second interval, and $b = b_{min}$ in the remaining segments.

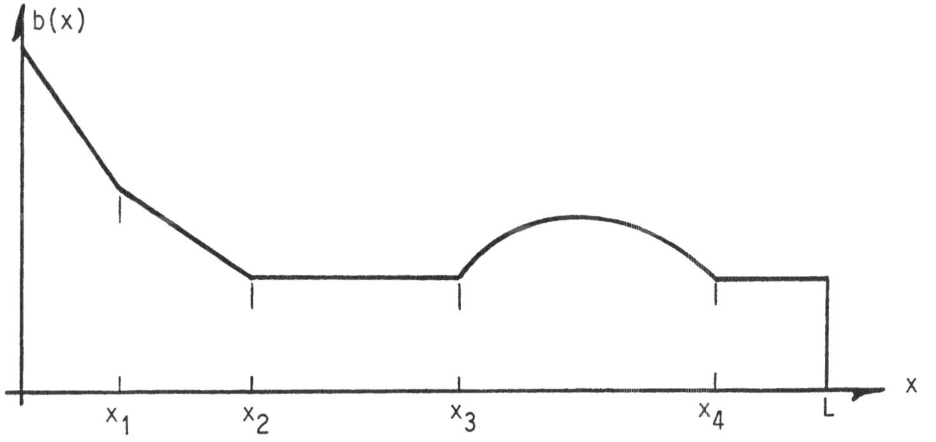

Optimal Fail-Safe Beam Design: $\alpha p_1/p = 1/6$

Closure

Wherever safety or dependability of a system is at issue, the fundamental notions of fail-safe design must be represented in the process that leads to the prediction of practical designs. Thus the subject is quite basic, and yet there is apparently only relatively moderate interest in basic research on the modelling of such problems. At the same time, it seems that the further development and application of methods for fail-safe design should be within reach with reasonable effort, and therefore one might expect an increase of activity in this important subject. Several aspects of the subject clearly could benefit from attention. Surely there is a need to create effective means for the computational solution of fail-safe design problems. Additional work is required in order to have an understanding of how properly to characterize damage or structural degradation in tractable form. Also there is a need for clarification on the relationship between worst-case design and fail-safe design, and on the modelling for sets of continuously varying damage states.

REFERENCES

Reviews and Surveys:

Ashley, H., "On Making Things the Best - Aeronautical Uses of Optimization," *J. Aircraft,* Vol. 19, pp. 5-28, 1982.

Kruzelecki, J. and Zyczkowski, M., "Optimal Structural Design of Shells - A Survey," SM Archives 10, pp. 101-170, 1985.

Lev, O. E., (Ed.), *Structural Optimization, Recent Developments and Applications,* American Society of Civil Engineers, New York, USA, 1981.

Niordson, F. I., and Pedersen, P., "A Review of Optimal Structural Design" in *Proc. 13th Int. Cong. Th. Appl. Mech.* (Eds. E. Becker and G. K. Mikhailov), Moscow 1972, Springer-Verlag, pp. 264-278, 1973.

Olhoff, N., "Optimal Design with Respect to Structural Eigenvalues" in *Proc. 15th Int. Cong. Th. Appl. Mech.* (Eds. F. P. J. Rimrott and B. B. Tabarrok), Toronto 1980, North-Holland, pp. 133-149, 1980.

Olhoff, Niels and Taylor, J. E., "On Structural Optimization," *J. Appl. Mech.* Anniv. Edn., Dec. 1983.

Qian, L., "Structural Optimization Research in China" in *Proc. Int. Conf. Finite Element Methods,* Shanghai, China, pp. 16-24, August 1982.

Rozvany, G. I. N., and Mroz, Z., "Analytical Methods in Structural Optimization," *Appl. Mech. Rev.,* Vol. 30, pp. 1461-1470,1977.

Schmit, L. A., "Structural Synthesis - Its Genesis and Development," *AIAA Journal,* Vol. 19, pp. 1249-1263, 1981.

Sheu, C. Y., and Prager, W., "Recent Developments in Optimal Structural Design," *Appl. Mech. Rev.,* Vol. 21, pp. 985-992, 1968.

Vanderplaats, G. N., "Structural Optimization - Past, Present, and Future," *AIAA Journal,* Vol. 20, pp. 992-1000, 1982.

Venkayya, V. B., "Structural Optimization: A Review and Some Recommendations," *Int. J. Num. Meth. Engrg.,* Vol. 13, pp. 205-228, 1978.

Wasiutynski, Z., and Brandt, A., "The Present State of Knowledge in the Field of Optimum Design of Structures," *Appl. Mech. Rev.,* Vol. 16, pp. 344-350,1963.

Proceedings:

Eschenauer, H., and Olhoff, N. (Eds.), *Optimization Methods in Structural Design,* Proc. Euromech Colloquium 164, Siegen, FRG, 1982, Bibliographisches Institut, Mannheim, 1983.

Gallagher, R. H. (Ed.), *Proc. Int. Symp. Optimum Structural Design,* University of Arizona, Tucson, Arizona 1981.

Haug, E. J., and Cea, J. (Eds.), *Optimization of Distributed Parameter Structures,* Vol. 1 and 2, Proc. Nato ASI Meeting, Iowa City 1980, Noordhoff, The Netherlands, 1981.

Morris, A. J. (Ed.), *Foundations of Structural Optimization: A Unified Approach,* Proc. Nato ASI Meeting, Liege, Belgium 1980, Chichester, 1982.

Sawczuk, A., and Mroz, Z. (Eds.), *Optimization in Structural Design,* Proc. IUTAM Symp., Warsaw 1973, Springer-Verlag, Berlin 1975.

Textbooks:

Banichuk, N. V., *Optimization of the Shapes of Elastic Bodies* (in Russian), Nauka, Moscow, 1980, English Translation by V. Komkov (E. J. Haug, ed.) Plenum Press, New York/London, 1983.

Brandt, A. M. (ed), *Criteria and Methods of Structural Optimization* , PWN, Warszawa/Martinus Nijhoff, The Hague, 1984.

Bensoussan, A., Lions, J. L., and Papanicolaou, G., *Asymptotic Analysis for Periodic Structures,* North-Holland, Amsterdam, 1978.

Carmichael, D. G., *Structural Modelling and Optimization,* Ellis Horwood Ltd., Chichester, 1981.

Grinev, W. B., and Filippov, A. P., *Optimization of Beams Governed by Eigenvalue Problems* (in Russian), Naukova Dumka Kiev, 1979.

Haftka, R. and Kamat M., *Elements of Structural Optimization,* Martinus-Nijhoff, Dordrecht, Netherlands, 1985.

Haug, E. J., and Arora, J. S., *Applied Optimal Design,* Wiley, New York, 1979.

Haug, E. J., Choi, K. K. and Komkov, V., *Structural Design Sensitivity Analysis,* Academic Press, New York, 1986.

Kirsch, U., *Optimum Structural Design,* McGraw-Hill, New York, 1981.

Lepik, U., *Optimal Design of Inelastic Structures Under Dynamic Loading,* (in Russian; extended summary in English), Walgus, Tallinn, 1982.

Lurie, K. A., *Applied Optimal Control of Distributed Systems,* Plenum Press, New York (to appear).

Mangeron, D., Poterasu, V. F. and Vulpe, A., "Teoria Optimizarii Strucţurilor-Cu Aplicatii," (in Romanian) Junimea, IASI, Romania, 1984.

Peterasu, V. F. and Nicolae Florea, "Practica Optimizarii Structurilor," (in Romanian), Junimea, IASI, Romania, 1984.

Prager, W., *Introduction to Structural Optimization;* Int. Centre for Mech. Sci., Udine, No. 212, Springer-Verlag, Vienna, 1974.

Rozvany, G. I. N., *Optimal Design of Flexural Systems,* Pergamon Press, Oxford, 1976.

Save, M., and Prager, W., *Structural Optimization - Vol. 1, Optimality Criteria, (W. H. Warner, ed.),* Plenum Press, New York/London, 1985.

Troitskij, W. A., and Petuchov, L. W., *Optimal Design of Elastic Bodies* (in Russian), Nauka, Moscow, 1982.

<u>**Papers and Reports**</u>:

Bendsoe, M. P., Olhoff, N., and Taylor, J. E., "A Variational Formulation for Multicriteria Structural Optimization," *J. Struct. Mech.,* Vol. 11, No. 4, 1983.

Bendsoe, M. P., *Generalized Plate Models and Optimal Design,* Proc. IMA Workshop on "Homogenization and Effective Moduli," University of Minnesota, October 1984.

Benedict, R. L. and J. E. Taylor, "Optimal Design for Elastic Bodies in Contact," Proc. of NATO-ASI, Iowa City, Iowa, May 20-June 4, 1980. *Optimization of Distributed Parameter Structures - Vol. 2,* (Eds. E. J. Haug and J. Cea), Sijthoff and Noordhoff, Netherlands, 1553-1599, 1981.

Cinquini, C., "Optimal Elastic Design for Prescribed Maximum Deflection," *J. Struct. Mech.,* 7(1), 21-34, 1979.

Dems, K. and Mroz, Z., "Multiparameter Structural Shape Optimization by the Finite Element Method," *Int. J. Numer. Methods Eng.,* 13, 247-263 1978.

Dems, K. and Mroz, Z., "Variational Approach by Means of Adjoint Systems to Structural Optimization and Sensitivity Analysis with Boundary Variation," *Int. J. Solids & Struct.* 20(6), 527-542, 1984.

Garstecki, A., "Optimal Redesign of Elastic Structures in the State of Initial Loading," *J. Struct. Mechs.,* 12(3), pp. 279-302, 1984.

Gembicki, F., "Vector Optimization for Control with Performance and Parameter-Sensitivity Indices," Ph.D. Dissertation, Case Western Reserve University, 1973.

Haftka, Raphael T., "Damage Tolerant Design using Collapse Techniques," *AIAA J.,* 21(10), pp. 1462-1466, 1983.

Haug, E. J., "A Unified Theory of Optimization of Structures with Displacement and Compliance Constraints," *J. Struct. Mech.,* 9(4), 415-437, 1981.

Kohn, R. V. and Strang, G., "Optimal Design for Torsional Rigidity," in: Atlari, S. N., Gallagher, R. H., and Zienkiewicz, O. C. (eds.), *Hybrid and Mixed Finite Element Methods,* Chichester, England, Wiley and Sons, 1983.

Koski, J., "Truss Optimization with Vector Criteria," Tampere Univ. of Technology, Publ. No. 6, Tampere, Finland, 1979.

Lurie, K. A. and Cherkaev, A. V., "G-Closure of Some Particular Sets of Admissible Material Characteristics for the Problem of Bending of Thin Plates," *DCAMM Report No. 214,* The Technical University of Denmark, June 1981.

Lurie, K. A., Fedorov, A. V., and Cherkaev, A. V., "Regularization of Optimal Design Problems for Bars and Plates, I, II," *J. Optimiz. Theory Appl. 37,* 499-521, 523-543, 1982.

Masur,[a] E. F., "Optimum Stiffness and Strength of Elastic Structures," *J. Eng. Mech. Div./ASCE,* 96 (EM5), pp. 621-640, 1970.

Masur,[b] E. F., "Optimal Placement of Available Sections in Structural Eigenvalue Problems," *J. Opt. Th. & Applics.,* 15, pp. 69-84, 1975.

Masur, E. F. and Mroz, Z., "Non-Stationary Optimality Conditions in Structural Design," *Int. J. Solids Structures,* Vol. 15, pp. 503-512, 1979.

Masùr, E. F. and Mroz, Z., "Singular Solutions in Structural Optimization Problems," in *Proc. IUTAM Symp. Variational Meth. in Mech.,* Evanston, Ill., 1978, Nemat-Nasser, S., (ed.), Pergamon Press, Oxtord, pp. 337-343, 1980.

Mroz, Z., "Optimal Design of Structures of Composite Materials," *Int. J. Solids and Structures,* 6, pp. 859-879, 1970.

Mroz, Z., "Multiparameter Optimization of Plates and Shells," *J. Struct. Mech.,* 1(3), pp. 371-392, 1973.

Mroz, Z. and Mironov, A., "Optimal Design for Global Mechanical Constraints," *Arch. Mech.,* Warsaw, 32(4), pp. 505-516, 1980.

Na, Moon-Soo, Kikuchi, N., and Taylor, J. E., "Optimal Modification of Shape for Two-Dimensional Elastic Bodies," *J. Struct. Mech.,* 11(1), pp. 111-135, 1983.

Na, Moon-Soo, Kikuchi, N., and Taylor, J. E., "Optimal Shape Remodelling of Linearly Elastic Plates Using Finite Element Methods," *Int. J. Num. Meth. Eng.,* 19, 1984.

Olhoff, N. and Taylor, J. E., "On Optimal Structural Remodeling," *J. Opt. Th. & Applics.,* 27, pp. 571-582, 1979.

Olhoff, N., Lurie, K. A., Cherkaev, A. V., and Fedorov, A. V., "Sliding Regimes and Anisotrophy in Optimal Design of Vibrating Axisymmetric Plates," *Int. J. Solids Structures,* Vol. 17, pp. 931-948, 1981.

Osyczka, A. and Koski, J., "Selected Works Related to Multicriteria Optimization Methods for Engineering Design," *Proc. Euromech Colloq. 164 - Optimization Methods in Structural Design,* Siegen, F.R. Germany, Oct. 12-14, 1982, B.I.-Wissenschaftsverlag, Mannheim, FRG, pp. 175-181, 1983.

Prager, W. and Taylor, J. E., "Problems of Optimal Structural Design," J. Appl. Mechs., Vol. 35, pp. 102-106, March, 1968.

Rozvany, G. I. N., Olhoff, N., Cheng, K.-T., and Taylor, J. E., "On the Solid Plate Paradox in Structural Optimization," *J. Struct. Mech.,* Vol. 10, pp. 1-32, 1982.

Stadler,[a] W., "A Survey of Multicriteria Optimization or the Vector Maximum Problem, Part I: 1776-1960," *JOTA,* Vol. 29, pp. 1-52, 1979.

Stadler,[b] W., "Applications of Multicriteria Optimization in Engineering and the Sciences (A Survey)," in *MCDM-Post Decade and Future Trends,* Zeleny M., (ed.), JAI Press, Greenwich, Conn., 1983.

Sun, P. F., Arora, J. S., and Haug, E. J., "Fail Safe Optimal Design of Structures," *J. Engin., Opt.,* 2(1), pp. 45-53, 1976.

Taylor, J. E. and Bendsoe, M. P., "An Interpretation for Min-Max Structural Design Problems Including a Method for Relaxing Constraints," *Int. J. Solids Structures,* 20(4), pp. 301-314, 1984.

Taylor, J. E. and Kikuchi, N., "Optimal Fail-Safe Design: A Multicriterion Formulation," *Proc. 16th ICTAM Congress,* Copenhagen, 1984.

STRUCTURAL OPTIMIZATION BY VARIATIONAL METHODS

NIELS OLHOFF

Department of Mechanical Engineering and Energy Technology,
University of Aalborg, DK-9220 Aalborg, Denmark

INTRODUCTION

This chapter contains notes for lectures delivered as part of the Advanced Study Institute "Computer Aided Optimal Design: Structural and Mechanical Systems" organized in Troia, Portugal, 30 June - 11 July 1986, by the Center of Mechanics and Materials of the University of Lisbon. The author gratefully acknowledges Professor Carlos Mota Soares and his collagues for a most successful arrangement of the meeting and the other lecturers for several extremely interesting contributions.

The chapter is subdivided into seven sections with the following titles:

1. Optimal Design of One-Dimensional, Conservative, Elastic Continuum Systems with Respect to a Fundamental Eigenvalue.
2. Optimal Design of Elastic Columns Against Buckling. Bimodal Optimization.
3. Optimization of Transversley Vibrating Beams and Rotating Shafts with Respect to the Fundamental Natural Frequency or Critical Speed.
4. Optimization with Respect to Higher Order Eigenfrequencies.
5. Optimal Design of Viscoelastic Structures under Forced Steady State Vibration.
6. Optimal Design of Solid, Elastic, Axisymmetric Plates.
7. References.

NATO ASI Series, Vol. F27
Computer Aided Optimal Design: Structural and Mechanical Systems. Edited by C. A. Mota Soares

Throughout the chapter the usefulness of adopting variational methods for the mathematical formulation of a broad class of structural optimization problems is emphasized. Thus, problems of geometrically constrained and unconstrained optimal design are formulated by variational analysis and discussed for linearly elastic {Sections 1 - 4,6} and viscoelastic {5} structures whose deflections are governed by ordinary or partial differential eigenvalue or boundary value problems via an extremum principle.

Primarily, problems of optimal material distribution in beam {1,3-5}, column {1,2}, shaft {1,3,4} and plate {6} elements of fixed size are considered, and the optimality criteria include minimum weight, maximum static stiffness, maximum Euler buckling load, maximum fundamental or higher order natural vibration frequencies, maximum gap between adjacent natural vibration frequencies or critical whirling speeds, and maximum dynamic stiffness under external vibrational excitation.

A quite extensive list of references connected with the topics discussed is given by the end {7} of the chapter. Comprehensive general reviews of the field of optimal structural design have been published by Wasiutyński and Brandt [1] in 1963, Sheu and Prager [2] in 1968, Prager [3] in 1971, Niordson and Pedersen [4] in 1973, Rozvány and Mróz [5] in 1977, Venkayya [6] in 1978, Schmit [7] in 1981, Vanderplaats [8] in 1982, and by Olhoff and Taylor [9] in 1983. This rapid sequence of reviews and the publication of a number of textbooks [10-26] witness the recent progress and increasing importance of the field.

SECTION 1

OPTIMAL DESIGN OF ONE-DIMENSIONAL, CONSERVATIVE, ELASTIC CONTINUUM SYSTEMS WITH RESPECT TO A FUNDAMENTAL EIGENVALUE

1.1 INTRODUCTION

For one-dimensional, conservative structural systems made of linearly elastic material, problems of optimal design with respect to a fundamental eigenvalue have by now been studied quite intensively. In this section, we consider eigenvalues of self-adjoint and full-definite structural eigenvalue problems, as for example the first frequency of free axial, torsional or transverse vibrations of rods and beams, the first critical speed of rotating shafts (excluding gyroscopic effects), the critical Euler buckling load of columns, or the critical torsional divergence velocity of a wing. A unified variational formulation for optimal design with respect to a fundamental eigenvalue will be presented in the following, and we outline the basic concepts and characteristics of such problems.

Optimal design with respect to structural eigenvalues was already considered by Lagrange [27] in 1770, and later by Clausen [28] in 1851, but it was Keller's paper [29] from 1960 that provided inspiration for the considerable contemporary research efforts in the area. Refs. [27-29] deal with optimal design of columns against Euler buckling, i.e. optimization under static loads.

Optimal design with respect to eigenvalues of structures under dynamic loading conditions is generally more complex, because the loading changes with changes in the design. This field of optimization was opened by the significant paper [30] published by Niordson in 1965. Surveys have since been prepared by Ashley and McIntosh [31], Pierson [32], Reitman and Shapiro [33], Rao [34], and by the author [35,36].

1.2 PRELIMINARY CONSIDERATIONS

We follow Ref. [36], and consider a straight, one-dimensional, single-purpose structure with a coordinate axis x embedded. The structure has given length L and variable cross-sections with common directions for principal axes of inertia. Let $A(x)$ denote the cross-sectional area of the structural material, that is, linearly elastic material contributing to the specific structural stiffness $s(x)$, where the term specific refers to unit length of the structure. We then consider relationships between $s(x)$ and $A(x)$ in the form

$$s(x) = c\,E\,A^p(x) \qquad 0 \le x \le L \tag{1.1}$$

where Young's modulus E, the factor c and the power p are positive constants, that are assumed to be given.

Although (1.1) restricts the cross-sectional variation, it covers a large class of structural types and behaviour. Thus, axial deformation is covered by $c = p = 1$. In torsional divergence or vibration, thin-walled cross-sections of variable thickness but constant planform are covered by $p = 1$, and geometrically similar solid cross-sections, by $p = 2$. For the bending associated with transverse vibration of beams, whirling instability of rotating shafts, and Euler buckling of columns, Eq. (1.1) models the following types of cross-sections: Fixed width sandwich cross-sections with uniform thickness cores of zero stiffness covered by two identical thin face sheets of variable thickness are modeled for $p = 1$, and so are solid cross-sections of constant thickness and varying width. Solid cross-sections of variable size but fixed shape correspond to $p = 2$, and solid rectangular cross-sections of fixed width and variable thickness are modeled for $p = 3$.

A generalized expression for the Rayleigh quotients associated with conservative problem formulations for free axial, torsional or transverse vibrations, whirling instability, divergence or buckling instability, respectively, of our one-dimensional structure can be written in the form

$$\Lambda = \frac{\int_0^L cEA^p(x)\, e[y(x)]dx}{\int_0^L \left\{b\rho A^r(x) + q(x)\right\} f[y(x)]dx + \sum_i Q_i\, f[y(x_i)]} , \qquad (1.2)$$

where the stiffness representation (1.1) is used. In (1.2), Λ denotes the square of the fundamental angular frequency ω for a particular vibration problem, the square of the first critical angular speed ω for a whirling problem, the Euler load P for a buckling problem, or the square of the critical free-stream velocity v multiplied by a factor for a torsional divergence problem for a wing, and $y(x)$ represents the corresponding deflection mode. According to Rayleigh's principle for self-adjoint and full-definite eigenvalue problems, Λ is stationary and equal to the fundamental eigenvalue at the fundamental mode $y(x)$ among all other kinematically admissible deflection functions.

The deflection $y(x)$ and its derivatives are contained in positive definite quadratic forms in the symbols $e[y]$ and $f[y]$, which must be interpreted according to the particular type of problem, cf. Table 1. The expressions given for $e[y]$ in the cases of transverse vibrations, whirling instability and buckling instability, respectively, are those consistent with Bernoulli-Euler beam theory. It is assumed by the form of (1.2) that linearly independent, homogeneous boundary conditions and conditions at possible interior supports are specified for the structure. Elastic supports are excluded for brevity.

Type of problem	Characteristics	e[y]	f[y]	Differential equation
Torsional divergence of a straight wing	$p=1$, $r=0$, $q(x)\equiv 1$, $b=0$, $Q_i=0$	y'^2	y^2	$-\{EAy'\}'=\Lambda y$
Axial vibration of rods	$p=r=b=c=1$	y'^2	y^2	$-\{EAy'\}'=\Lambda\{\rho A+q\}y$
Torsional vibration of rods	$p=r\geq 1$	y'^2	y^2	$-\{cEA^p y'\}'=\Lambda\{b\rho A^p+q\}y$
Transverse vib. of beams or whirling instability of rotating shafts	$p\geq 1$, $r=b=1$	y''^2	y^2	$\{cEA^p y''\}''=\Lambda\{\rho A+q\}y$
Buckling of columns	$p\geq 1$, $r=0$, $q(x)\equiv 1$, $b=0$, $Q_i=0$	y''^2	y'^2	$-\{cEA^p y''\}''=\Lambda y''$

Table 1.

The problem of torsional divergence instability of a straight airplane wing with elastic axis perpendicular to the airstream and constant cross-sectional profile along the span, is covered by (1.2) with $q(x)\equiv 1$, $b=0$, $Q_i=0$ and $p=1$, if aerodynamic strip theory is used and the dominating contribution to the torsional stiffness comes from the skin, the (small) thickness of which is assumed to vary along the span.

For vibration problems, the term $b\rho A^r(x)$ in (1.2), with $p\geq r\geq 1$, b a given positive constant and ρ the mass density, represents the specific structural mass for rectilinear vibration types, or the specific structural polar mass moment of inertia for torsional vibration. The former types are associated with $r=b=1$, whereas $r=p$ for the latter. Similarly, $q(x)$ represents the specific mass or specific polar mass moment of inertia of distributed dead mass loading and/or non-structural material (e.g. core filler in sandwich structures). Eq. (1.2) also covers vibrating structures carrying lumped dead mass loads at specified points $x=x_i$. The constants Q_i identify their masses or mass moments of inertia. Both Q_i and $q(x)$ are assumed to be given.

In the classical problems of buckling instability under external axial compressive forces, the own weight distribution of the structure can be disregarded, and it is then characteristic that $A(x)$ does not appear in the denominator of the Rayleigh quotient. The problem is included in (1.2) if we take $q(x) \equiv 1$, $b = 0$ and $Q_i = 0$. Allowance can be made for buckling caused by own structural weight [37] and/or lumped dead weights attached along the column span, by changing the denominator of (1.2) to a slightly more general expression, but it is omitted here for reasons of brevity.

1.3 MAXIMUM FUNDAMENTAL EIGENVALUE FOR PRESCRIBED VOLUME

For given structural length and material, cross-sectional style, given type of single-modal behaviour and given boundary conditions, the optimal design problem may be stated in the form:

> With the cross-sectional area $A(x)$ of structural material as the *design variable* and the fundamental eigenvalue Λ as the *objective function*, determine the design that *maximizes* Λ subject to the integral constraint of *given structural volume* $V = \int_0^L A(x)dx$, and subject to the *geometric constraint* that $A(x)$ may nowhere be less than a prescribed *minimum* value $\bar{A}$, i.e. $A(x) \geq \bar{A}$.

Using variational formulation, the optimal design $A(x)$, its associated mode $y(x)$ and optimal eigenvalue Λ are identified with stationarity of the following augmented form of the functional (1.2),

$$\Lambda^* = \frac{\int_0^L cEA^p(x)e[y(x)]dx}{\int_0^L \left\{b\rho A^r(x) + q(x)\right\}f[y(x)]dx + \sum_i Q_i f[y(x_i)]} - \kappa\left\{\int_0^L A(x)dx - V\right\} - \int_0^L \beta(x)\left\{g^2(x) - A(x) + \bar{A}\right\}dx \; . \tag{1.3}$$

Here, the quantities κ and $\beta(x)$ are Lagrangian multipliers, and the geometric minimum constraint $A(x) \geq \bar{A}$ has been converted to an equality constraint by means of the real slack variable $g(x)$ defined by $g^2(x) =$

$A(x) - \bar{A}$. In optimal design problems, minimum cross-sectional area constraints were first considered by Taylor [38] in 1968.

Now, Rayleigh's principle and the introduction of the Lagrangian multipliers permit the variation of Λ^* with respect to variations of $y(x)$, $A(x)$ and $g(x)$ to be taken independently.

Noting that the stationarity of Λ^* (1.3) with respect to all admissible variation of $y(x)$ is equivalent to the stationarity of the Rayleigh quotient with respect to variation of $y(x)$, we first obtain the differential equation (cf. Table 1) and the natural boundary conditions for the problem, along with the jump conditions at inner supports and at the points x_i where the concentrated loads Q_i are attached to the structure.

Next, stationarity of Λ^* for arbitrary admissible variation of $g(x)$ gives

$$\beta(x)g(x) = 0 \tag{1.4}$$

Finally, stationarity of Λ^* with respect to all admissible variation of the design variable $A(x)$ yields the *optimality condition*

$$pcEA^{p-1}(x)e[y(x)] - rb\rho\Lambda A^{r-1}(x)f[y(x)] = \kappa(1 - \beta(x)) \tag{1.5}$$

after applying (1.2) and redefining the Lagrangian multiplier κ by dividing it by the denominator in expression (1.2).

In order to formulate the governing equations without explicit appearance of $\beta(x)$ and $g(x)$, we exploit that either $g(x) = 0$ or $g(x) \neq 0$. Denoting by x_c the (unions of) sub-intervals in which $g(x) = 0$ may take place, and denoting by x_u the remaining sub-intervals (where we have $g(x) \neq 0$) , Eq. (1.4) gives us that $A(x) = \bar{A}$ (constrained) for $x \in x_c$ and that $A(x) > \bar{A}$ (unconstrained) for $x \in x_u$. In the latter sub-interval(s), Eq. (1.4) can only be satisfied if $\beta(x) = 0$, which clearly reduces the optimality condition (1.5) for $x \in x_u$.

A complete set of *governing equations for optimality*, which by their derivation are *necessary* conditions for a *possible* optimal solution, may now for convenience be listed as follows,

Rayleigh quotient expression for Λ , Eq. (1.2) (1.6a)

Differential equation for problem type, cf. Table 1 $\quad 0 \leq x \leq L$ (1.6b)

$$pcEA^{p-1}(x)e[y(x)] - br\rho\Lambda A^{r-1}(x)f[y(x)] = \kappa \qquad x \in x_u \tag{1.6c}$$

$$A(x) > \bar{A} \quad x \in x_u \quad , \quad A(x) = \bar{A} \quad x \in x_c \tag{1.6d}$$

$$\int_0^L A(x)dx = V . \tag{1.6e}$$

Note that the form (1.6c) of the so-called optimality condition is restricted to the unions of a priori unknown sub-intervals x_u where the design variable $A(x)$ is unconstrained. Clearly, the unions of sub-intervals x_u and x_c make up the entire interval $0 \leq x \leq L$. Specific forms of the optimality condition for particular problems are easily identified with the help of Table 1. It is noteworthy that the second term on the left-hand side of the optimality equation is not present in torsional divergence and classical buckling optimization, for which $b = 0$. The first term of the optimality condition is interpreted as the average strain energy density in the design fibres, i.e. the fibres that are affected by a change in the design. Constancy of this energy density is found to be a general principle in geometrically unconstrained optimal design under *static loads*, see Masur [39].

Along with the boundary conditions and the other conditions mentioned above, Eqs. (1.6a-e) constitute a coupled, non-linear, ordinary integro-differential eigenvalue problem, where the unknowns to be determined are the optimal eigenvalue Λ , the optimal distribution of structural material $A(x)$ (which includes determination of the sub-intervals x_c and x_u), the associated fundamental mode $y(x)$, and the Lagrangian multiplier κ .

It is noted that the cross-sectional constants p and r play a fundamental role in the coupling and the non-linearities of the equations. Evidently, cases of $p = r = 1$ are the easiest to deal with, because such problems are linear in the design variable $A(x)$, which even vanishes from the optimality condition (1.6c). Although (1.6c) remains non-linear in the deflection, it is often possible to obtain analytical solutions to problems with $p = r = 1$, see for example Prager and Taylor [40]. For buckling problems, where $b = 0$, so that terms involving r drop out, analytical solutions have even been obtained for $p = 2$, as for example in [29]. For vibration optimization problems associated with values of p

other than unity, however, the coupling and the non-linearities of the governing equations generally only permit numerical solution.

1.4 MINIMUM VOLUME DESIGN FOR PRESCRIBED FUNDAMENTAL EIGENVALUE

An alternative, *dual* formulation for optimal design is the following:

> With the cross-sectional area $A(x)$ of structural material as the *design variable* and the structural volume $V = \int_0^L A(x)dx$ as the *objective function*, determine the design that *minimizes* V subject to the *behavioural constraint of specified fundamental eigenvalue* Λ and the *geometric minimum constraint* $A(x) \geq \bar{A}$ where $\bar{A}$ is given. Again, the structural length and material, the cross-sectional style, the type of single-modal behaviour, and the boundary conditions, are assumed to be given.

The set of necessary governing equations for a possible optimal solution to this formulation are easily derived by variational analysis of the functional

$$V^* = \int_0^L A(x)dx - \gamma\left\{\frac{\int_0^L cEA^p(x)e[y(x)]dx}{\int_0^L \left\{b\rho A^r(x) + q(x)\right\}f[y(x)]dx + \sum_i Q_i f[y(x_i)]} - \Lambda\right\}$$
$$- \int_0^L \mu(x)\left\{g^2(x) - A(x) + \bar{A}\right\}dx \ , \qquad (1.7)$$

where the behavioural constraint and the geometric minimum constraint have been adjoined to the functional V by means of Lagrangian multipliers γ and $\mu(x)$, respectively.

By variation of $A(x)$ we find that the optimality condition takes the form

$$\gamma\left\{pcEA^{p-1}(x)e[y(x)] - rb\rho\Lambda A^{r-1}(x)f[y(x)]\right\} = 1 - \mu(x) \qquad (1.8)$$

which for $x \in x_u$ reduces to

$$\gamma\left\{pcEA^{p-1}(x)e[y(x)] - rb\rho\Lambda A^{r-1}(x)f[y(x)]\right\} = 1 \quad x \in x_u \tag{1.9}$$

since the Lagrangian multiplier $\mu(x)$ vanishes in the sub-interval(s) where the design variable is geometrically unconstrained. The form of Eqs. (1.8) and (1.9) involves a redefinition of μ in the same manner as the Lagrangian multiplier κ in Eqs. (1.5) and (1.6c).

As the result of the complete variational analysis, we find that the present min.V fixed Λ formulation is also governed by Eqs. (1.6a-e), with the only exception that the Lagrangian multiplier κ in the optimality condition (1.6c) is replaced by $1/\gamma$. In the present formulation, the unknowns to be determined are the minimum structural volume V, the optimal distribution $A(x)$ of structural material, the fundamental mode $y(x)$, and the Lagrangian multiplier γ.

1.5 ON EQUIVALENCE OF DUAL OPTIMIZATION PROBLEMS

The system of equations governing the max.Λ fixed V problem in Section 1.3 and the governing equations for the min.V fixed Λ problem considered in the foregoing section only differ by the Lagrangian multipliers κ and γ and corresponding slightly different appearences of the optimality conditions. It is therefore not surprising that these dual optimization problems, generally speaking, are equivalent. However, exceptions exist.

Comparing the optimality conditions, (1.6c) and (1.9), it is obvious that *a min.*V *fixed* Λ *solution will at the same time be a solution to a max.*Λ *fixed* V *problem, with* $\kappa = 1/\gamma$. Note that vanishing of the Lagrangian multiplier γ is excluded by the form of Eq. (1.9).

Furthermore, *a max.*Λ *fixed* V *solution associated with a non-zero Lagrangian multiplier* κ *is at the same time a solution to a min.*V *fixed* Λ *problem, with* $\gamma = 1/\kappa$.

However, a max.Λ fixed V solution with $\kappa = 0$ is not a min.V fixed Λ solution, because it is unable to satisfy the optimality condition (1.9) for a problem of the latter type.

Now, generalizing an approach of Brach [41], let us state precisely for which type of problems the equivalence may be lost. First, we multiply Eq. (1.5) by $A(x)$ and integrate over the interval $0 \le x \le L$. Using Eqs. (1.6d-e) and employing that $\beta(x) \equiv 0$ for $x \in x_u$, we find

$$\int_0^L pcEA^p(x)e[y(x)]dx - \int_0^L rb\rho\Lambda A^r(x)f[y(x)]dx = \kappa\left\{V - \bar{A}\int_{x_c} \beta(x)dx\right\} . \tag{1.10}$$

Then, multiplying Eq. (1.6a), i.e. Eq. (1.2), by the product of r and the denominator on the right-hand side, and subtracting the resulting equation from Eq. (1.10), we obtain

$$\kappa\left\{V - \bar{A}\int_{x_c} \beta(x)dx\right\} = (p-r)\int_0^L cEA^p(x)e[y(x)]dx + r\Lambda\left\{\int_0^L q(x)f[y(x)]dx + \sum_i Q_i\, f[y(x_i)]\right\} . \tag{1.11}$$

Here, the first integral on the right-hand side (representing twice the potential energy of the structure) is positive. Furthermore, we have $p \geq r$ for the types of problems considered, cf. Table 1. The subsequent terms on the right-hand side of (1.11) are non-negative.

Thus, κ can only vanish, i.e. *the equivalence of the dual optimization problems can only be lost if both* $p = r$, $q(x) \equiv 0$ *and* $Q_i = 0$. Note that, among the types of problems considered here, the equivalence can only be lost for vibration optimization problems.

Taylor [42] was the first to establish the equivalence of dual formulations for optimal design. The subject has also been considered in papers by Vavrick & Warner [43] and Seiranyan [44].

1.6 GEOMETRICALLY UNCONSTRAINED OPTIMAL DESIGN

We may drop the minimum constraint in the formulations considered above by setting $\beta(x) \equiv 0$ and $\mu(x) \equiv 0$, respectively, in Eqs. (1.3) and (1.7). The governing equations for the resulting *geometrically unconstrained* optimization problem are then obtained as a special case of Eqs. (1.6a-e) associated with $\bar{A} = 0$ and validity of the optimality condition (1.6c) in the entire interval $0 \leq x \leq L$ (which x_u becomes identical to). Examples of geometrically unconstrained optimal solutions are illustrated in dimensionless form in Fig. 1 for columns and in Fig. 7 for transversely vibrating beams.

A geometrically unconstrained optimal design is in general associated with maximum obtainable merit. However, geometrically unconstrained solutions must often be regarded as limiting solutions from the point of view of practical design, in other words, geometrically constrained optimal designs are preferable in practice. Nevertheless, it is evident that knowledge of maximum obtainable efficiency is of both theoretical and practical importance, and that geometrically unconstrained solutions direct the designer towards maximum economy of material.

It is a general feature that geometrically unconstrained optimization of one-dimensional structures results in *statically determinate* solutions *when a single mode formulation is used.* This fact is connected with the occurrence of points in the optimal solutions with vanishing structural material. At these points, the derivative of the deflection appearing in the optimality condition may exhibit discontinuous behaviour in problems with $p = 1$, and in cases of $p = 2$ and $p = 3$, points of vanishing structural cross-section are associated with significant singular behaviour of the deflection and/or its derivatives. In optimal design of transversely vibrating beams with $p = 2$ or 3, the cross-section may vanish in two essentially different ways at singular points; either in a way that is found at a hinge, a so-called *Type I singularity*, see e.g. Fig. 7a, or in a manner found at a free structural end (Fig. 7b) and at an inner separation (Fig. 7c), which is called a *Type II singularity*. In optimal columns, the governing equations only admit Type I singularities (cf. Figs. 1a,b,c) if the point is under axial compression, which by the way seems obvious from physical grounds.

Now, when *a priori* statically indeterminate structures are optimized without geometric minimum constraint on the basis of a single mode formulation, it is the automatic formation of hinges or separations in the structure that reduces this to a statically determinate one, cf. Figs. 1c and 7c. The singular behaviour at points of zero cross-section is studied quite intensively in [30,45-47] for different one-dimensional problems.

1.7 SUFFICIENT CONDITIONS OF OPTIMALITY

The optimality equations (1.6a-e) are derived as necessary conditions for a possible optimal solution, and they do not, in general, state sufficient conditions for global optimality.

For problems associated with $p = r = 1$, sufficiency can be shown in specific cases, however. This was first demonstrated by Taylor [42], who proved the global optimality of a solution obtained by Turner [48]. Shortly after, sufficiency of possible solutions to geometrically unconstrained vibration problems associated with $p = r = 1$ was shown by Prager and Taylor [40]. A proof for corresponding constrained problems does not seem to be available, but might, for example, be established along lines indicated in [49]. For vibration optimization problems with $p > 1$, sufficiency is generally not ensured.

As to buckling optimization, sufficiency is proved by Taylor and Liu [50] for a geometrically constrained, statically determinate $p = 1$ column, and a proof for statically determinate $p = 2$ columns is available in Tadjbakhsh and Keller [51]. The latter authors claimed validity of their proof independently of the column boundary conditions, but Masur [52] and Popelar [53] have since noted that the proof breaks down for statically indeterminate cases, and [54] provides an illustration of this.

1.8 EXISTENCE OF SOLUTIONS

Existence of optimal solutions cannot generally be assured a priori. Consequently, their possible existence cannot be demonstrated until the actual solution is arrived at.

Non-existence of optimal solutions on the contrary, can in some cases be shown. For example, *no min.V fixed* Λ *solutions exist to the types of vibration optimization problems considered, if* $p = r$ *and both geometrical constraints, non-structural material, and external dead loading are absent**). In this case, the eigenvalue problem defined by the differential equation (1.6b), cf. Table 1, and the boundary conditions considered, is linear and homogeneous in both $A^p(x)$ and $y(x)$. Thus, denoting by $\tilde{A}(x)$ a design associated with the prescribed value of Λ and satisfying (1.1) along with a vibration mode $\tilde{y}(x)$, the eigenvalue Λ is maintained by a design $C\tilde{A}(x)$, where C is an arbitrary constant. The volume of this design, however, can be made arbitrarily small by choosing a sufficiently small value of C .

---

*) Note, in view of the discussion in Section 1.5, that the dual problem would be associated with $\kappa = 0$, Eq. (1.11).

Considering the dual problem of maximum Λ at fixed V and L specially for transversely vibrating $p = r = 1$ beams with $q(x) \equiv 0$, $Q_i = 0$ and no minimum constraint, Brach [55] demonstrated non-existence for the case of a cantilever, but existence of an optimal solution for a simply supported beam - results that have later been confirmed in [46,56]. It is interesting to note, as an illustration of the result cited at the end of Section 1.5, that for these cases of $p = r$, $q(x) \equiv 0$ and $Q_i = 0$, the equivalence is lost for the dual simply supported beam problems, whereas the equivalence holds for the cantilever problems in the sense that no optimal solution exists for either of the dual formulations.

SECTION 2

OPTIMAL DESIGN OF ELASTIC COLUMNS AGAINST BUCKLING. BIMODAL OPTIMIZATION

2.1 INTRODUCTION

We consider the problem of determining the optimal design of a thin, elastic column such that the Euler buckling load attains a maximum possible value for given material volume, length and boundary conditions. We first assume the optimum buckling load to be a simple eigenvalue, and obtain the governing equations for the problem from the general theory of section 1. The type of singular behaviour that may occur in geometrically unconstrained problems is discussed, and conditions for optimal location of inner singular points are stated.

For structures of some complexity or statical indeterminacy, optimization against buckling must be conducted with bimodal or even multimodal optimal buckling loads in perspective. This trend, which requires an extended formulation for optimal design, already manifests itself in the case of a doubly clamped column, and we shall discuss this problem in detail.

2.2 SINGLE MODE FORMULATIONS FOR OPTIMAL DESIGN

Consider a thin, straight, elastic column which has the volume V, length L and Young's modulus E, and is subjected to an axial compressive force, the value of which is P at buckling. The cross-section of the column is permitted to vary along the column axis according to (1.1) with $s(x) = EI(x)$, the bending stiffness of structural material. In (1.1), $A(x)$ is the cross-sectional area of structural material, and c and p are given constants. The cross-sectional styles corresponding to $p = 1$, 2 and 3 are described in Section 1.2. We consider columns of Bernoulli-Euler type with given support conditions and exclude flexible supports for brevity.

In *geometrically constrained* form, our optimization problem consists in determining the cross-sectional area distribution $A(x)$ that maximizes the fundamental buckling load P for given values of V, L, $\bar{A}$, E, p and c. Assuming P to be a *simple eigenvalue*, we easily obtain the following governing equations for this problem from Table 1 and the general optimality equations (1.6a-e),

$$P = \frac{\int_0^L cEA^p y''^2 dx}{\int_0^L y'^2 dx} \tag{2.1a}$$

$$(cEA^p y'')'' = -Py'' \qquad 0 \le x \le L \tag{2.1b}$$

$$pcEA^{p-1} y''^2 = \kappa \qquad x \in x_u \tag{2.1c}$$

$$A(x) > \bar{A} \qquad x \in x_u \quad , \quad A(x) = \bar{A} \qquad x \in x_c \tag{2.1d}$$

$$\int_0^L A(x)dx = V . \tag{2.1e}$$

Specification of a minimum constraint for the cross-sectional area of a column was introduced by Taylor & Liu [50], and has later been done in Refs. [40,54,57-62], for example. If V is minimized at fixed P, such a constraint is equivalent to a constraint on the maximum prebuckling stress.

Let us now, for convenience, nondimensionalize the coordinate x by division by L and introduce a dimensionless cross-sectional area $\alpha(x)$ and buckling load λ by

$$\alpha(x) = A(x)L/V \tag{2.2}$$

$$\lambda = \frac{PL^{p+2}}{EcV^p} \tag{2.3}$$

The *geometrically unconstrained* problem (where $\bar{\alpha} = \bar{A}L/V = 0$) , is thus governed by the following dimensionless equations, obtainable from Eqs. (2.1a-e) where (2.1d) drops out and the optimality condition (2.1c) becomes valid in the entire interval $0 \le x \le 1$ for the dimensionless variable:

$$\lambda = \frac{\int_0^1 \alpha^p y''^2 dx}{\int_0^1 y'^2 dx} \tag{2.4a}$$

$$(\alpha^p y'')'' = -\lambda y'' \qquad 0 \le x \le 1 \tag{2.4b}$$

$$\alpha^{p-1} y''^2 = \kappa \qquad 0 \le x \le 1 \tag{2.4c}$$

$$\int_0^1 \alpha(x) dx = 1 \tag{2.4d}$$

These equations expose the Rayleigh quotient, the differential equation for Euler buckling, the optimality condition (where κ has been redefined) and the volume constraint, respectively, in dimensionless form. Normalizing the deflection $y(x)$ such that the denominator in (2.4a) is set equal to unity,

$$\int_0^1 y'^2 dx = 1 , \tag{2.5}$$

multiplying (2.4c) by $\alpha(x)$ and integrating over the interval $0 \le x \le 1$, taking (2.4d) and (2.5) into account, we find that the Lagrangian multiplier κ is simply given by

$$\kappa = \lambda . \tag{2.6}$$

Hence, for cases of $p = 2$ or $p = 3$, Eq. (2.4c) gives us the optimal cross-sectional area function $\alpha(x)$ in the form

$$p = 2\,,3 : \quad \alpha(x) = \left(\frac{\lambda}{y''^2}\right)^{\frac{1}{p-1}}\,, \tag{2.7a}$$

while, for $p = 1$, Eq. (2.4e) states that the curvature $y''(x)$ of the deflection is constant throughout, except for possible sign shifts,

$$p = 1 : \quad y''(x) = \pm\sqrt{\lambda}\,. \tag{2.7b}$$

The continuity conditions for a column are as follows. At all points free from kinematic constraint (column supports), the bending moment $m(x) = \alpha^p y''$ and the function $(\alpha^p y'')' + \lambda y'$ are continuous:

$$\langle \alpha^p y'' \rangle = 0 \tag{2.8}$$

$$\langle (\alpha^p y'')' + \lambda y' \rangle = 0 \tag{2.9}$$

The function $(\alpha^p y'')' + \lambda y'$ represents, for the buckled column, the force component of the stress resultants in the direction perpendicular to the x-axis, while the shear force $t(x) = (\alpha^p y'')'$ is the force component perpendicular to the deflected column axis.

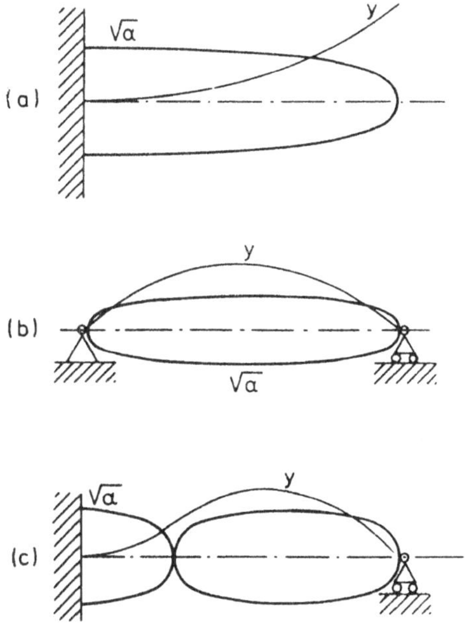

Fig. 1.

The geometrically unconstrained problem is seen to be quite simple for $p = 1$, and solutions can be obtained analytically, see e.g. [5,40,45,63,64]. In cases of $p = 2$ and $p = 3$, where the problem becomes non-linear in $y(x)$, it is still possible to apply analytical methods of solution provided that the boundary conditions are sufficiently simple [2,3,51,65]. Else, numerical methods are available, see for example [16,37,54,58,61,66,67].

Optimal shapes $\pm\sqrt{\alpha}$ and corresponding buckling modes y are indicated in Fig. 1 for (a) cantilevered [51], (b) simply supported [29], and (c) clamped-simply supported [51] $p = 2$ columns. The buckling loads of the optimal solutions are increased by (a) 1/3, (b) 1/3 and (c) 35.1% when compared with the buckling loads of correspondingly supported *uniform* columns of the same volume, length and material [29,51].

2.3 SINGULARITIES IN GEOMETRICALLY UNCONSTRAINED SOLUTIONS

We now consider the behaviour of a solution to the geometrically unconstrained, single mode formulation at a point $x = x_j$ of zero bending moment $m = \alpha^p y''$. For the case of $p = 1$, zero bending moment must imply, in view of (2.7b), that α also vanishes. For $p = 2$ and $p = 3$, Eq. (2.7a) and the relation $m = \alpha^p y''$ show that the cross-section α also vanishes and that y'' tends to infinity at a point $x = x_j$ of vanishing bending moment m.

If the point $x = x_j$ of $m = \alpha = 0$ is an interior point in the interval for x, then a discontinuity of the slope y' of the deflection is possible at $x = x_j$, but the deflection $y(x)$ and the function $(\alpha^p y'')' + \lambda y'$ are continuous. Therefore, the point $x = x_j$ corresponds, physically and kinematically, to an *inner hinge* of the optimal column. Detailed information on the singular behaviour of the functions at an inner hinge is available in Ref. [45] (see also Ref. [18], pp. 156-158).

In problems of optimizing statically determinate columns by means of a single mode, geometrically unconstrained formulation, the locations of singular points of zero bending moment are known beforehand. Thus, simply supported or free end points are predetermined to be singular. However, in *a priori* statically indeterminate problems of the type mentioned, singularities may occur at inner points. The positions of such points can be prescribed for a particular column to be optimized, while in other problems, we may consider the locations $x = x_j$, $j = 1,\dots,S$ of the singularities (hinges) to be additional design variables. Problems of the latter type were for the first time considered by Masur [68].

For columns, the condition for optimal location $x = x_j$ of an inner hinge is [45]

$$\textit{either} \qquad y'(x_j^+) = y'(x_j^-) \tag{2.10a}$$

$$\textit{or} \qquad y'(x_j^+) = -y'(x_j^-) - 2\,\frac{y(x_j) - y(x_{j+1})}{x_{j+1} - x_j}\,, \tag{2.10b}$$

where, in (2.10b), x_{j+1} denotes the position of an adjacent inner hinge or hinged end point of the column, and it is presumed that no beam support or additional hinge are placed between the points x_j and x_{j+1}.

The reader is referred to Ref. [45] (or Ref. [18], pp. 158-163) concerning the derivation of the above condition.

2.4 DISCUSSION

The buckling load of the clamped-simply supported optimal column shown in Fig. 1c and originally determined in [51], is actually bimodal. For this type of column, we obtain the same optimal design independently of whether we use (2.10a) or (2.10b) to govern the location of the inner hinge of zero bending moment. However, as will be illustrated next, it is necessary to pay full attention to both conditions in other geometrically unconstrained problems.

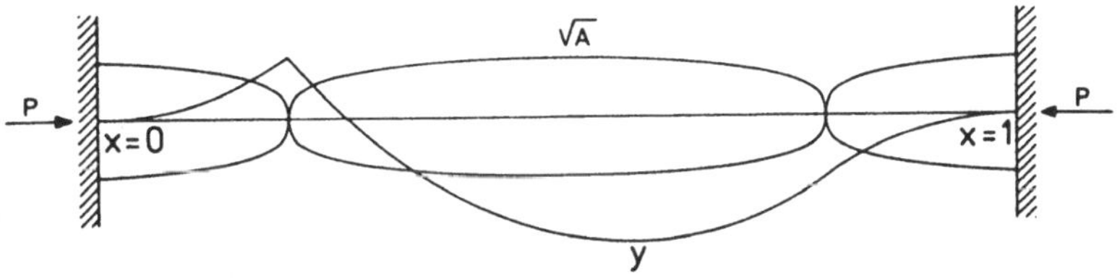

Fig. 2.

Fig. 2 shows the geometrically unconstrained optimal design of a doubly clamped column $(p = 2)$ with two inner hinges, Refs. [54,45]. This design is also bimodal, but in this case, the optimal position of the left hand hinge is governed by condition (2.10b), while the position of the right hand hinge is governed by condition (2.10a). If, for example, (2.10a) were used for both inner hinges, their positions would change, and a slightly different design be obtained. This design would maximize the *second* buckling eigenvalue (with a symmetric mode), see Ref. [54], but it would have a much lower fundamental Euler buckling eigenvalue than the design shown in Fig. 2, and hence not be optimal in the sense of maximizing the buckling load. In fact, the doubly clamped column solution published in [51] is subject to this mistake.

2.5 MULTIMODAL FORMULATION FOR OPTIMAL DESIGN

In this subsection, we consider a problem in which a bimodal rather than a single mode formulation is necessary in order to arrive at the correct optimal design. For reasons of generality, we present a multimodal formulation of the problem.

The example problem consists in maximizing the Euler buckling eigenvalue λ , Eq. (2.3), of a doubly clamped, solid elastic column $(p = 2)$ of given volume and length. This problem was first considered in [51], but as is shown in [54], an erroneous solution was arrived at. The design shown in Fig. 2 constitutes the correct solution within the premises of a single mode formulation of the problem, and it replaces the design from [51], which was also obtained on the basis of a single mode formulation.

However, the design shown in Fig. 2 is only optimal within the class of doubly clamped columns with *two inner hinges*. It is quite obvious that the column would obtain a greater Euler buckling load for the same volume, length and material, if it were made to buckle in a symmetric fundamental buckling mode with a continuous slope throughout. This could easily be achieved by restributing the given material slightly so that the hinges became locked. The problem is, however, that the field equations of the geometrically unconstrained column *do predict* zero cross-section (singular behaviour) at points of vanishing bending moment, and two such points are necessarily present in a clamped-clamped column whenever a single mode formulation is used.

This clearly indicates that the single mode formulation is inadequate for the problem under consideration, and motivates a reformulation of the problem in [54], leading to a new optimality condition, that does not necessarily lead to vanishing cross-section at points of zero bending moment.

Let us now reformulate and expand the initial formulation of our optimization problem in Section 2.2 by following [69]. The eigenvalues λ_i , $i = 1$, ... , ∞ , of our elastic column are expressed in terms of the modes $y_i(x)$ by

$$\lambda_i = \int_0^1 \alpha^p y_i''^2 dx \qquad i = 1 , \ldots , \infty , \tag{2.11}$$

provided that the modes are normalized by

$$\int_0^1 y_i'^2 dx = 1 \qquad i = 1 , \ldots , \infty . \tag{2.12}$$

Also the eigenfunctions for $\lambda_i \neq \lambda_j$ are orthogonal, i.e.,

$$\int_0^1 y_i' y_j' dx = 0 \qquad \text{for } \lambda_i \neq \lambda_j \tag{2.13}$$

but this need not be the case if $\lambda_i = \lambda_j$, $i \neq j$. However, let us take the entire set of modes $\{y_k(x)\}$ to be orthonormalized according to

$$\int_0^1 y_i' y_j' dx = \delta_{ij} \qquad i,j = 1 , \ldots , \infty , \tag{2.14}$$

where δ_{ij} is Kronecker's delta.

The condition of given volume for the column is expressed by

$$\int_0^1 \alpha dx = 1 , \tag{2.15}$$

and to formulate the problem in some generality, we will consider a geometric minimum constraint for the design variable $\alpha(x)$, namely that

$$\alpha(x) \geq \bar{\alpha} \tag{2.16}$$

throughout, assuming the minimum allowable value $\bar{\alpha}$ $(0 \leq \bar{\alpha} \leq 1)$ to be given.

The design problem may now be stated as

$$\max_{\alpha(x)} \left[\min_i (\lambda_i) \right] \tag{2.17}$$

This max-min problem is non-differentiable, however. In order to circumvent this difficulty, we use a bound formulation [69,70], which consists in introducing an extra parameter β which ensures that we have a standard differentiable problem even if multimodal eigenvalues occur. Hence, we transform the problem (2.17) into the problem of maximizing a bound β subject to the constraints $\lambda_i \geq \beta$, $i = 1, \ldots, \infty$. In this way the parameter β replaces a non-differentiable functional and is to be maximized over a constraint set in an enlarged space. The points of non-differentiability correspond to "corners" in the constraint set of the enlarged space and arise from intersections of differentiable constraints.

The bound formulation for our problem has the form:

$$\max_{\alpha(x)} \beta$$

$$\text{subject to} \quad \beta \leq \lambda_i \quad \Longleftrightarrow \quad \beta - \lambda_i + h_i^2 = 0 \qquad i = 1, \ldots, \infty$$

$$\lambda_i = \int_0^1 \alpha^p y_i''^2 dx \qquad i = 1, \ldots, \infty \tag{2.18}$$

$$\int_0^1 y_i' y_j' dx = \delta_{ij} \qquad i = 1, \ldots, \infty, \quad j = 1, \ldots, \infty$$

$$\int_0^1 \alpha dx = 1$$

$$\alpha(x) \geq \bar{\alpha} \quad \Longleftrightarrow \quad \bar{\alpha} - \alpha(x) + g^2(x) = 0.$$

Here the symbols h_i and $g(x)$ designate real slack variables that convert the inequality constraints to equality constraints.

To solve the problem (2.18), we construct an augmented Lagrangian

$$L = \beta - \sum_{i=1}^{\infty} \eta_i \left[\beta - \int_0^1 \alpha^p y_i''^2 dx + h_i^2\right]$$

$$- \sum_{i=1}^{\infty} \sum_{j=1}^{i} \rho_{ij} \left(\int_0^1 y_i' y_j' dx - \delta_{ij}\right) \tag{2.19}$$

$$- \Gamma\left(\int_0^1 \alpha dx - 1\right) - p\int_0^1 \sigma(x)\left(\bar{\alpha} - \alpha(x) + g^2(x)\right)dx$$

where η_i, ρ_{ij}, Γ and $\sigma(x)$ are Lagrangian multipliers. Note that the multipliers ρ_{ij} are only defined for $i \geq j$.

The condition of stationarity of L with respect to variation of β and h_i gives

$$\sum_{i=1}^{\infty} \eta_i = 1 \tag{2.20}$$

where

$$\eta_i = 0 \quad \text{if} \quad \lambda_i > \beta \ , \quad \eta_i \geq 0 \quad \text{if} \quad \lambda_i = \beta \ , \quad i = 1, \ldots, \infty \ . \tag{2.21}$$

Variation of $\alpha(x)$ and $g(x)$ yields the so-called optimality condition

$$\alpha^{p-1} \sum_{i=1}^{\infty} \eta_i y_i''^2 = \kappa - \sigma(x) \tag{2.22}$$

where $\kappa = \Gamma/p$, and

$$\sigma(x) = 0 \quad \text{if} \quad \alpha(x) > \bar{\alpha} \ , \quad \sigma(x) \geq 0 \quad \text{if} \quad \alpha(x) = \bar{\alpha} \ . \tag{2.23}$$

Finally, stationarity of L with respect to variation of the i-th mode y_i is expressed by

$$2\eta_i (\alpha^p y_i'')'' + 2\rho_{ii} y_i'' + \sum_{j=1}^{i-1} \rho_{ij} y_j'' + \sum_{j=i+1}^{\infty} \rho_{ji} y_j'' = 0 \tag{2.24}$$

after integration by parts, using the boundary conditions. In (2.24) and in the following, summation is only to be carried out over repeated indices when explicitly stated.

Let us now assume that a total number of N Lagrangian multipliers η_i are greater than zero, which is the same as assuming that the fundamental eigenvalue is (at least) N-fold, c.f. Eq. (2.21). Moreover, let us renumber our variables, i.e., use the first N values of an index n , i.e., $n = 1, \ldots, N$, to identify the Lagrangian multipliers η_n and modes $y_n(x)$ that are associated with the N-fold eigenvalue $\beta = \lambda_n$, $n = 1, \ldots, N$. Then Eqs. (2.20) - (2.24) become

$$\sum_{n=1}^{N} \eta_n = 1 \tag{2.25}$$

$$\left.\begin{array}{lll} \eta_n > 0 \ , \ \lambda_n = \beta & n = 1,\dots,N \\ \eta_n = 0 \ , \ \lambda_n \geq \beta & n = N+1,\dots,\infty \end{array}\right\} \qquad (2.26)$$

$$\left.\begin{array}{lll} \alpha^{p-1} \sum_{n=1}^{N} \eta_n y_n''^2 = \kappa & (\text{if } \alpha > \bar{\alpha}) & x \in x_u \\ \alpha(x) = \bar{\alpha} & & x \in x_c \end{array}\right\} \qquad (2.27)$$

$$2\eta_n(\alpha^p y_n'')'' + 2\rho_{nn} y_n'' + \sum_{j=1}^{n-1} \rho_{nj} y_j'' + \sum_{j=n+1}^{\infty} \rho_{jn} y_j'' = 0,$$

$$n = 1,\dots,\infty \ . \qquad (2.28)$$

Eqs. (2.27) are readily obtained from Eqs. (2.22) and (2.23). The symbols x_u and x_c denote the unions of sub-intervals in which we have $\alpha(x) > \bar{\alpha}$ (*unconstrained* cross-sectional area) and $\alpha(x) = \bar{\alpha}$ (*constrained* area), respectively.

In order to determine the Lagrangian multipliers ρ_{nm}, we first multiply (2.28) by y_n, integrate by parts over the interval in applying of the boundary conditions, and use (2.11) and (2.14) to obtain $\rho_{nn} = \lambda_n \eta_n$, $n = 1,\dots,\infty$. In view of (2.26) we thus have

$$\rho_{nn} = \begin{cases} \beta\eta_n \ , & n=1,\dots,N \ , \\ 0 \ , & n=N+1,\dots,\infty \ . \end{cases} \qquad (2.29)$$

To determine the remaining components of ρ_{nm} (i.e., those associated with $n > m$), we first write Eq. (2.28) with index n replaced by m. Then, we multiply this equation by y_n and Eq. (2.28) by y_m, integrate both equations by parts using the boundary conditions, assume $n > m$, and apply (2.14). Subtracting and adding the two resulting equations, we finally obtain

$$\rho_{nm} = (\eta_n + \eta_m) \int_0^1 \alpha^p y_n'' y_m'' dx \quad , \quad n > m \quad , \quad m = 1,\dots,\infty \ , \qquad (2.30)$$

$$(\eta_n - \eta_m) \int_0^1 \alpha^p y_n'' y_m'' dx = 0 \ , \quad n > m \quad , \quad m = 1,\ldots,\infty \ , \tag{2.31}$$

respectively.

For values of $n > N$ we have $\eta_n = 0$ by (2.26) and it is then easily seen from (2.30) and (2.31) that

$$\rho_{nm} = 0 \ , \quad n > m \ , \quad n = N+1,\ldots,\infty \ . \tag{2.32}$$

This equation implies together with (2.26) and (2.29) that only the modes associated with the N-fold fundamental eigenvalue enter Eqs.(2.28), and that we only need to consider Eqs.(2.28) for $n = 1,\ldots,N$.

Consider now any of the equations (2.28) associated with a given $n \leq N$, and write its solution y_n as $y_n = w_n + z_n$ where the function w_n designates the solution to the eigenvalue problem consisting of the given set of boundary conditions and the differential equation $(\alpha^p w_n'')'' + \beta w_n'' = 0$, which is constructed from (2.28) by setting the two first terms on the left hand side equal to zero and using (2.29). The function z_n is then due to the terms under the summation signs in (2.28). The Rayleigh quotient associated with the aforementioned eigenvalue problem is defined by $R[u] = \int_0^1 \alpha^p u''^2 dx / \int_0^1 u'^2 dx$, where u is an admissible function, and Rayleigh's minimum principle implies that $\beta = R[w_n] \leq R[w_n + z_n] = R[y_n] = \lambda_n$, where the last relationship follows from Eqs.(2.11), (2.12) and the definition of R . Now, Eqs.(2.26) require strict equality of β and λ_n , i.e., $\beta = \lambda_n$ for $n = 1,\ldots,N$. Hence, we must have $y_n(x) \equiv w_n(x)$, i.e., $z_n(x) \equiv 0$, $n = 1,\ldots,N$, and as is shown in the Appendix of Ref. [69] this requires vanishing of the Lagrangian multipliers,

$$\rho_{nm} = 0 \ , \quad n > m \ , \quad n = 2,\ldots,N \ , \tag{2.33}$$

in Eq.(2.28)

By means of (2.26), (2.29), (2.32) and (2.33), we may now write Eqs. (2.28) as the familiar differential equations for buckling

$$(\alpha^p y_n'')'' + \lambda_n y_n'' = 0 \ , \quad n = 1,\ldots,N \ , \tag{2.34}$$

where $\lambda_n = \beta$, $n = 1,\ldots,N$. It is also worth noting that Eqs.(2.30), (2.26) and (2.33) imply vanishing of the integrals

$$\int_0^1 \alpha^p y_n'' y_m'' dx = 0 \ , \quad n > m \ , \quad n = 2,\ldots,N \ , \qquad (2.35)$$

that represent the mutual bending energy of the modes y_n and y_m.

2.6 THE BIMODAL CASE

For the type of problem treated above, the multiplicity N of the fundamental eigenvalue is not known beforehand, but must be determined as the highest possible number of positive Lagrangian multipliers η_n, which, together with their associated linearly independent modes y_n, is admitted by the optimality condition (2.27). Up to now the highest multiplicity found for an optimal Euler column buckling eigenvalue is $N=2$, cf. [54] for the case of a doubly clamped column. $N=1$ is usual for columns with other boundary conditions.

If $N=1$, Eq. (2.25) gives $\eta_1 = 1$ and it is readily seen that Eq. (2.27) reduces to the traditional single mode optimality condition, Eq. (2.4c).

For the case of $N=2$, Eqs. (2.25) and (2.27) may be combined into the following condition where we write $1-\gamma$ in place of η_1:

$$\alpha^{p-1}\left\{(1-\gamma)y_1''^2 + \gamma y_2''^2\right\} = \kappa \qquad (\text{if } \alpha > \bar{\alpha}) \qquad x \in x_u \qquad (2.36)$$

This condition is identical with the optimality condition derived by Olhoff and Rasmussen [54]. It follows directly from (2.25) and (2.26) that in (2.36) γ must lie in the interval

$$0 \leq \gamma \leq 1 \ . \qquad (2.37)$$

This condition has earlier been established as a sufficient condition for local optimality by Masur and Mróz [71,72].

Together with the condition $\alpha(x) = \bar{\alpha}$ for $x \in x_c$, Eq. (2.36) may be solved for $\alpha(x)$ to give

$$p = 2,3 : \alpha(x) = \begin{cases} \left(\dfrac{\kappa}{(1-\gamma)y_1''^2 + \gamma y_2''^2}\right)^{\frac{1}{p-1}} & (\text{if } > \bar{\alpha}) \quad x \in x_u \\ \bar{\alpha} & x \in x_c \end{cases} \qquad (2.36a)$$

for $p = 2$ or $p = 3$. For the case of $p = 1$, we obtain

$$p = 1 : \quad \begin{aligned} &(1-\gamma)y_1''^2 + \gamma y_2''^2 = \kappa \quad (\text{if } \alpha > \bar{\alpha}) && x \in x_u \\ &\alpha(x) = \bar{\alpha} && x \in x_c \end{aligned} \tag{2.36b}$$

As is the case for Eq. (2.7b), we note that for $p = 1$, the optimality condition for the geometrically unconstrained sub-interval(s) does not contain the design variable α , which must therefore be determined from the buckling differential equation. A method of solution for $p = 1$ is presented in [71] and the case of $p = 1$ will not be considered further here.

For exemplification, let us follow [54] and derive convenient expressions for the Lagrangian multipliers κ and γ and the optimum buckling eigenvalue λ for the more complex cases of $p = 3$ and $p = 2$. First, we substitute (2.36a) into the volume constraint (2.15), thereby obtaining an explicit expression for κ ,

$$p = 2,3 : \kappa = \left[\frac{1 - \bar{\alpha}\displaystyle\int_{x_c} dx}{\displaystyle\int_{x_u} \frac{dx}{\{(1-\gamma)y_1''^2 + \gamma y_2''^2\}^{1/(p-1)}}} \right]^{p-1} \tag{2.38}$$

Then, subtracting the two equations comprised in (2.11) for $i = 1$ and $i = 2$, substituting $\alpha(x)$ from (2.36a) and using (2.38), we find the following implicit equation for γ ,

$$p = 2,3 : \int_{x_u} \frac{y_1''^2 - y_2''^2}{\{(1-\gamma)y_1''^2 + \gamma y_2''^2\}^{p/(p-1)}} dx + \bar{\alpha}^p \left[\frac{\displaystyle\int_{x_u} \frac{dx}{\{(1-\gamma)y_1''^2 + \gamma y_2''^2\}^{1/(p-1)}}}{1 - \bar{\alpha}\displaystyle\int_{x_c} dx} \right]^p \int_{x_c} (y_1''^2 - y_2''^2)\, dx = 0 \tag{2.39}$$

Finally, substitution of (2.36a) and (2.38) into first of Eqs. (2.11) gives an explicit expression for λ ,

$$p = 2,3 : \lambda = \bar{\alpha}^p \int_{x_c} y_1''^2 dx + \left[\frac{1 - \bar{\alpha}\int_{x_c} dx}{\int_{x_u} \frac{dx}{\{(1-\gamma)y_1''^2 + \gamma y_2''^2\}^{1/(p-1)}}} \right]^p \times$$

$$\int_{x_u} \frac{y_1''^2}{\{(1-\gamma)y_1''^2 + \gamma y_2''^2\}^{p/(p-1)}} dx \; . \tag{2.40}$$

Equations $(2.11)_{i=1,2}$, $(2.12)_{i=1,2}$, $(2.34)_{n=1,2}$, (2.36a), (2.38)-(2.40) comprise the complete set of necessary equations governing the bimodal optimal design problem for $p = 3$ and $p = 2$, and they constitute a strongly coupled, non-linear integro-differential eigenvalue problem. The unknowns to be determined are the optimal buckling eigenvalue λ , the optimal column cross-sectional area function $\alpha(x)$ (and thereby the sub-intervals x_u and x_c), eigenfunctions y_1 and y_2 , and the Lagrangian multipliers κ and γ , respectively. The solutons depends in general on the minimum constraint $\bar{\alpha}$, which is the only specified quantity in the non-dimensional formulation.

A method of numerical solution based on successive iterations is presented in [54]. In that paper, the modes y , and y_2 were not taken to be mutually orthogonal. The results are exposed in the next subsection. Other examples where bimodality of optimal eigenvalues occur, may be found in Refs. [73-76].

The bimodal formulation for optimal design described above contains geometrically unconstrained optimization and/or single mode optimization as special cases. The principal advantage of the new formulation is that while the optimality condition (2.4c) of the single mode formulation predicts formation of hinges at points of zero bending moment in a geometrically unconstrained formulation of optimal design, the bimodal optimality condition (2.36) does not necessarily lead to zero cross-section and formation of hinges at points of zero bending moment.

2.7 EXAMPLE: BIMODAL OPTIMIZATION OF A DOUBLY CLAMPED $p = 2$ COLUMN [54]

Fig. 3 illustrates optimal designs $\pm\sqrt{\alpha}$ and assosiated fundamental single or double modes corresponding to selected values of a geometric minimum constraint $\bar{\alpha}$ on the cross-sectional area α of a doubly clamped $p = 2$ column. In Fig. 3a, $\bar{\alpha} = 0.7$ and $\lambda = 48.690$ is a simple. In Fig. 3b,

$\bar{\alpha} = 0.4$ and $\lambda = 51.775$ is simple. In Fig. 3c, $\bar{\alpha} = 0.25$ and the optimal buckling load $\lambda = 52.349$ is bimodal. Fig. 3d shows the optimal solution corresponding to any value of $\bar{\alpha}$ belonging to the interval $0 \le \bar{\alpha} \le 0.226$, where the constraint is no longer active in the design. The corresponding optimal buckling load $\lambda = 52.3563$ is bimodal.

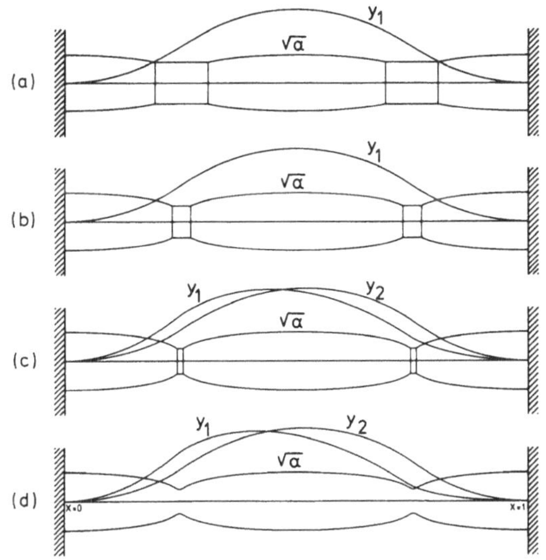

Fig. 3.

In Fig. 4, curve ABCD is based on a number of solutions and shows λ as a function of the geometric minimum constraint $\bar{\alpha}$. For $0.280 < \bar{\alpha} \le 1$, the optimal designs are associated with a simple fundamental eigenvalue λ, given by curve CD. Curve CE shows the second order eigenvalues λ_2 of the simple optimal eigenvalue designs behind curve CD. At point C, the two curves are seen to coalesce at the value 0.280 for $\bar{\alpha}$, and for $0 \le \bar{\alpha} \le 0.280$, the optimal designs are associated with a *bimodal fundamental eigenvalue*, cf. modes y_1 and y_2 in Figs. 3c and d. All the designs obtained are symmetrical (this was not assumed in the solution procedure), and purely symmetrical and antisymmetrical linear combinations of double modes y_1 and y_2 can be constructed.

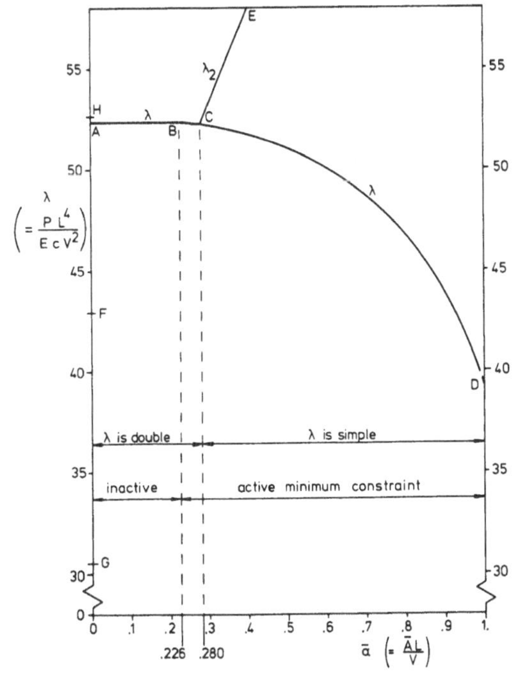

Fig. 4.

As shown by curve DCB of Fig. 4, the optimal buckling eigenvalue λ increases with decreasing constraint for $0.226 \leq \bar{\alpha} \leq 1$, and for these values of $\bar{\alpha}$ the constraint is *active* in the optimal designs, cf. Figs. 3a, b and c. However, for values of $\bar{\alpha}$ belonging to the interval $0 \leq \bar{\alpha} \leq 0.226$, the minimum constraint is *inactive* in the optimal design, and the associated bimodal fundamental buckling eigenvalue is *constant*, cf. AB in Fig. 4. For these values of $\bar{\alpha}$, $0 \leq \bar{\alpha} \leq 0.226$, the optimal design, see Fig. 3d, is the *same*, and it has *finite* variable cross-section throughout, with a minimum magnitude of $\alpha = 0.226$.

This $\bar{\alpha}$ independent bimodal optimal design in Fig. 3d is the *solution to the geometrically unconstrained optimization problem for a doubly clamped column* of $p = 2$. Its fundamental, double buckling eigenvalue λ is 32.62% higher than the fundamental eigenvalue of a corresponding uniform column of the same volume, length and material. The bimodal optimal design replaces not only the solution arrived at in [51], but also the geometrically unconstrained, candidate design in Fig. 2.

The result provides a noteworthy example of a *statically indeterminate* solution to a geometrically unconstrained, one-dimensional, single purpose, structural optimization problem: it constitutes an abvious exception to the "general rule" that solutions to the broad class of all such problems will always be statically determinate.

SECTION 3

OPTIMIZATION OF TRANSVERSELY VIBRATING BEAMS AND ROTATING SHAFTS WITH RESPECT TO THE FUNDAMENTAL NATURAL FREQUENCY OR CRITICAL SPEED

3.1 INTRODUCTION

This section deals with problems of determining the distribution of structural material in transversely vibrating beams or rotating circular shafts, such that maximum values of natural frequencies or critical whirling speeds are obtained for a prescribed amount of material, length, and boundary conditions for the beam or shaft. Equivalently, we minimize the volume of structural material for a given vibration frequency or critical speed.

The practical significance of such problems is that they provide designs of minimum weight (or cost) of material against beam vibrational resonance due to external excitation of given frequency, and against failure due to whirling instability at service speeds, within a large range from zero and up to the particular fundamental frequency or first critical speed.

3.2 GEOMETRICALLY CONSTRAINED OPTIMAL DESIGN

An elastic Bernoulli-Euler beam of length L and structural volume V vibrates at its fundamental angular frequency ω of free transverse vibrations. The beam is made of a material with Young's modulus E and the mass density ρ, and it has variable but similarly oriented cross-sections with the relationship $I = cA^p$ between the area moment of inertia I and the area A, cf. Eq. (1.1). We restrict ourselves to the cases of $p = 2$ (geometrically similar, solid cross-sections) and $p = 3$ (solid cross-sections of fixed width and variable height), because the case of $p = 1$ (sandwich cross-sections) is often degenerate for the types of problems to be considered, cf. the discussion in Sections 1.5, 1.8, and Refs. [35,41,46,47]. The constant c for the cross-sectional shape and the value of p ($p = 2$ or $p = 3$) are assumed to be given.

For the *particular case of* $p = 2$, we may conceive the structure to be a *shaft* of circular cross-sections, that rotates at its fundamental critical angular whirling speed ω, if we neglect gyroscopic effects.

We shall assume that our vibrating beam (or rotating shaft) carries no *distributed* nonstructural mass, but that a number K of given non-structural masses (or circular disks) Q_i, $i = 1,\ldots,K$, are attached to the beam/shaft at specified points $X = X_i$, where X is the beam/shaft coordinate (which is denoted by x in Chapter 1).

Identifying Λ of Section 1 as ω^2, and introducing non-dimensional quantities,

$$x = X/L \qquad 0 \le x \le 1 \tag{3.1a}$$

$$\alpha(x) = A(x)L/V \quad , \quad \bar{\alpha} = \bar{A}L/V \tag{3.1b}$$

$$q_i = \frac{Q_i}{\rho V} \qquad x = x_i \quad , \quad i = 1,\ldots,K \tag{3.1c}$$

$$\lambda = \omega^2 \frac{\rho L^{3+p}}{E c V^{p-1}} \quad , \tag{3.1d}$$

i.e. coordinate x , cross-sectional area function $\alpha(x)$, non-structural masses (or circular disks) q_i and fundamental eigenvalue λ , respectively, the non-dimensional beam or shaft will have unit volume and unit length.

Our *dimensionless, geometrically constrained optimization problem* then consists in determining the function $\alpha(x) \geq \bar{\alpha}$ that maximizes λ for given minimum allowable beam/shaft cross-sectional area $\bar{\alpha}$, given positions x_i and magnitudes q_i of non-structural masses/disks, and given, homogeneous boundary conditions.

Using Eqs. (3.1a-d), the governing equations for the optimization problem are easily obtained as a special case of the general set of optimality equations (1.6a-e) with the help of Table 1. Normalizing the vibration/whirling mode according to

$$\int_0^1 \alpha y^2 dx + \sum_i q_i y^2(x_i) = 1 \ , \tag{3.2}$$

such that the denominator of the dimensionless Rayleigh quotient equals unity, the set of governing equations takes the following form

$$\lambda = \int_0^1 \alpha^p y''^2 dx \tag{3.3a}$$

$$(\alpha^p y'')'' = \lambda \alpha y \qquad 0 \leq x \leq 1 \tag{3.3b}$$

$$p\alpha^{p-1} y''^2 - \lambda y^2 = \kappa \qquad x \in x_u \tag{3.3c}$$

$$\alpha(x) > \bar{\alpha} \quad x \in x_u \quad , \quad \alpha(x) = \bar{\alpha} \quad x \in x_c \tag{3.3d}$$

$$\int_0^1 \alpha(x)\, dx = 1 \ , \tag{3.3e}$$

where we have redefined the Lagrangian multiplier κ in the optimality condition (3.3c).

In addition to Eqs. (3.2) and (3.3a-e), our optimization problem must satisfy (i) the boundary conditions, (ii) the condition of continuity of the bending moment $m = \alpha^p y''$ except at possible points of prescribed y', and (iii) the conditions of continuity of the shear force $t(x) = -(\alpha^p y'')'$ except at possible points of prescribed y and at the points $x = x_i$, $i = 1,\ldots,K$ with attached nonstructural masses q_i. At the latter points, the jumps of t are identified as $\langle t\rangle_{x_i} = -\lambda q_i y(x_i)$.

Due to the nonlinearity and coupling of the governing equations, closed form solutions cannot be expected for $p = 2$ and $p = 3$, and numerical solution procedures must therefore be applied. Such procedures are available in Refs. [78-80]. Refs. [78] and [80] present results for geometrically constrained $p = 2$ cantilevers with and without nonstructural masses, respectively, and Ref. [79] offers results for $p = 2$ and $p = 3$ beams with various other boundary conditions.

Fig. 5 shows results obtained in [78], namely cantilever beams of geometrically similar cross-sections, $p = 2$, (or rotating circular

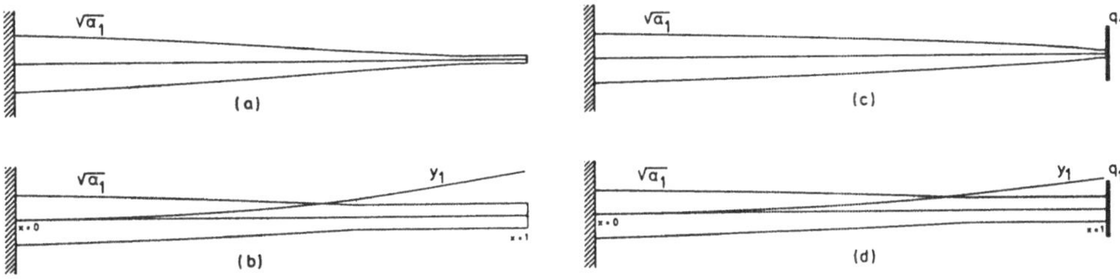

Fig. 5.

cantilever shafts) optimized with respect to the fundamental natural transverse vibration frequency (or first critical speed) $\omega = \omega_1$. The beams are illustrated by optimal *shapes* $\pm\sqrt{\alpha_1}$, where $\alpha_1(x) = \alpha(x) = A(x)L/V$, and the solutions in (a), (c) and (b), (d), respectively, correspond to minimum constraints $\bar{\alpha} = 0.05$ and 0.5. The dimensionless nonstructural tip mass in (c) and (d) is $q_1 = Q_1/\rho V = 0.1$. The fundamental frequencies ω_1 of the optimal designs are increased by (a) 279%, (b) 88%, (c) 81% and (d) 57%, respectively, in comparison with those corresponding to uniform designs of the same volume, length, material, and, for (c) and (d), tip mass.

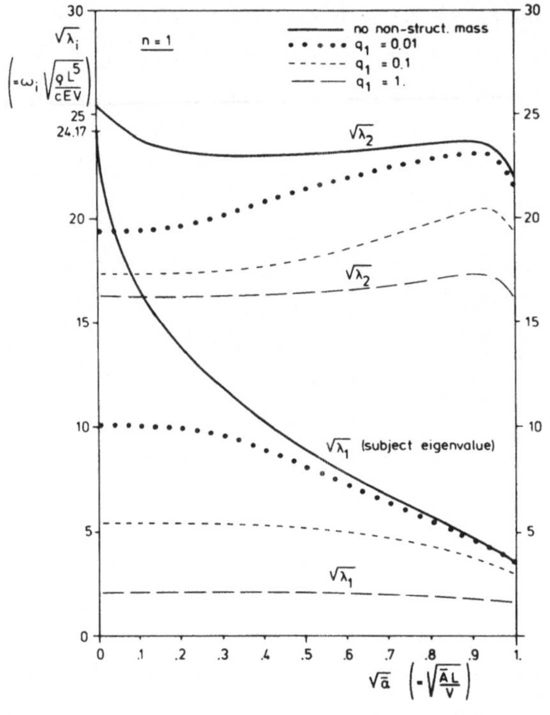

Fig. 6.

Fig. 6 shows the square root $\sqrt{\lambda_1}$ of the fundamental eigenvalue $\lambda = \lambda_1$ and the square root $\sqrt{\lambda_2}$ of next eigenvalue λ_2 as functions of the dimensionless cross-sectional area constraint $\sqrt{\bar{\alpha}} = \sqrt{\bar{A}L/V}$ for optimal cantilevers of the type in Fig. 5. Note that $\bar{\alpha} = 0$ and $\bar{\alpha} = 1$ correspond to geometrically unconstrained and fully constrained (uniform) designs, respectively, and that the square root eigenvalues are proportional to the first and second vibration frequencies (critical speeds) ω_1 and ω_2, respectively, for given beam volume, length and material. The solid curves in Fig. 6 represent optimal $\lambda = \lambda_1$ designs without nonstructural mass, see for example Fig. 5a,b, while other curves are for optimal $\lambda = \lambda_1$ designs with a dimensionless tip mass $q_1 = Q_1/\rho V$, see e.g. Fig. 5c,d.

Fig. 6 clearly illustrates that geometrically unconstrained designs are associated with maximum obtainable merits in comparison with corresponding geometrically constrained designs, cf. the discussion in Section 1.6.

3.3 GEOMETRICALLY UNCONSTRAINED OPTIMIZATION

We now consider the case where no geometric constraint is specified for the cross-sectional area function $\alpha(x)$ in the process of optimization. This constitutes a special case of the formulation considered in Section 3.2, and corresponds to setting $\bar{\alpha} = 0$ and the interval x_u equal to the entire interval $0 \leq x \leq 1$. Doing this, the system of Eqs. (3.2), (3.3a-e) reduces to the following system for geometrically unconstrained optimization,

$$\lambda = \int_0^1 \alpha^p y''^2 dx \tag{3.4a}$$

$$\int_0^1 \alpha y^2 dx + \sum_i q_i y^2(x_i) = 1 \tag{3.4b}$$

$$\int_0^1 \alpha dx = 1 \tag{3.4c}$$

$$(\alpha^p y'')'' = \lambda \alpha y \qquad 0 \leq x \leq 1 \tag{3.4d}$$

$$\alpha(x) = \left[\frac{\kappa + \lambda y^2}{p y''^2}\right]^{\frac{1}{p-1}} \qquad 0 \leq x \leq 1 \tag{3.4e}$$

Here the optimality condition (3.4e) is now valid in the entire interval $0 \leq x \leq 1$. If we write this equation in the form $p\alpha^{p-1} y'' - \lambda y^2 = \kappa$, multiply it by α , integrate over the interval, and use Eqs. (3.4a-c), we find

$$\kappa = \lambda\left[p - 1 + \sum_i q_i y^2(x_i)\right] , \tag{3.5}$$

i.e., Lagrangian multiplier κ is always positive for problems with $p > 1$.

Geometrically unconstrained solutions obtained numerically by successive iterations are available in Refs. [30,46,47,56,81]. Ref. [46] presents optimal $p = 2$ and $p = 3$ cantilevers with and without tip mass, and $p = 2$ solutions without nonstructural mass are available in Refs. [30,47] for other boundary conditions.

Fig. 7 shows examples of geometrically unconstrained optimal design of $p = 2$ beams with respect to the fundamental natural vibration frequency $\omega = \omega_1$, when no nonstructural masses are considered. The design in Fig. 7a is the solution for a simply supported beam [30], Fig. 7b shows the optimal cantilever design [46], and Fig. 7c illustrates the optimal design of a doubly clamped beam [47]. The design in Fig. 7b may be compared with the constrained designs in Figs. 5a-b, and it should be noted that its optimal characteristics (maximum ω_1 for given V and L , or minimum V for given ω_1 and L) are represented by the value indicated for the solid $\sqrt{\lambda_1}$ curve at $\sqrt{\bar{\alpha}} = 0$ in Fig. 6. It is also worth mentioning that the design in Fig. 7b is at the same time the optimal design of a clamped-simply supported beam [47].

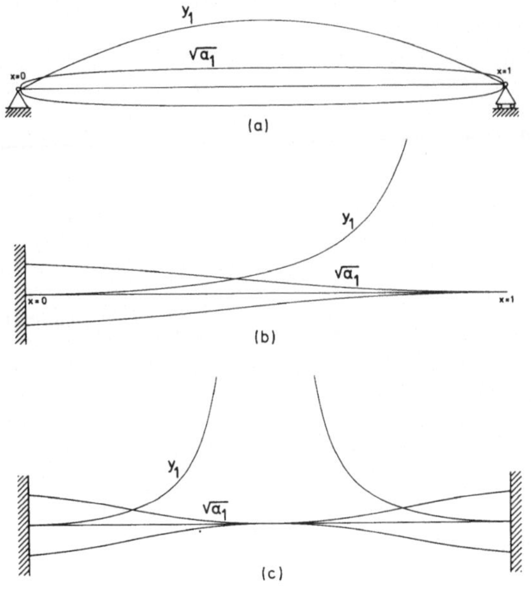

Fig. 7

The fundamental frequencies ω_1 of the optimal solutions in Fig. 7 are increased by (a) 6.6%, (b) 588% and (c) 332%, respectively, when compared with the frequencies of corresponding uniform beams of the same volume, length and material. Comparing (b) with a uniform clamped-simply supported beam, its fundamental frequency ω_1 is increased by 57%.

3.4 TYPES OF SINGULAR BEHAVIOUR

In problems of optimizing transversely vibrating Bernoulli-Euler beams or rotating shafts without geometric constraint, there may occur two different types of singular behaviour, both of which are associated with zero bending moment $m(x) = \alpha^p y''$ and cross-section α, but in one type (I) the shear force is finite, while in the other type (II) the shear force vanishes at the singularity. Physically, an inner Type I singularity corresponds to an *inner hinge*, and an inner Type II singularity to an *inner separation* of the beam. The types of the singularities are independent of whether the beam (or shaft) is optimized with respect to the fundamental frequency (first critical speed) or a higher order eigenfrequency (higher order critical speed).

The singular behaviour can be determined analytically by expanding solutions to Eqs. (3.4d) and (3.4e) in power series near the singular point $x = x_j$, which may either be an inner point or an end point of the beam. Detailed information is available in Refs. [45,47] and Ref. [18] pp. 185-189. The conditions that govern optimal location of Type I and Type II singularities can also be found in the references cited.

SECTION 4

OPTIMIZATION WITH RESPECT TO HIGHER ORDER EIGENFREQUENCIES

4.1 INTRODUCTION

Here, we shall consider an extension of the types of optimization problems studied in Section 3. Thus, instead of maximizing the fundamental natural frequency (or critical speed), we will deal with the problem of maximizing a particular *higher order* natural frequency (or critical speed) of given order n $(n > 1)$ for a transversely vibrating Bernoulli-Euler beam (or rotating shaft) of prescribed structural volume, length, material, and boundary conditions.

This problem is governed by a set of dimensionless equations consisting of Eqs. (3.2) and (3.3a-e) for geometrically constrained optimal design or Eqs. (3.4a-e) for geometrically unconstrained design, with λ, $\alpha(x)$ and $y(x)$ subscribed as λ_n, $\alpha_n(x)$ and $y_n(x)$ (indicating reference to the given order n of the subject eigenfrequency), and the *additional equations*

$$\int_0^1 \alpha_n y_n y_j dx + \sum_i q_i y_n(x_i) y_j(x_i) = 0 \quad , \quad j = 1,\dots,n-1 \tag{4.1a}$$

$$(\alpha_n^p y_j'')'' = \lambda_j \alpha_n y_j \quad , \quad j = 1,\dots,n-1 \tag{4.1b}$$

see Refs. [47,78]. Eqs. (4.1a) are conditions of orthogonality of $y_n(x)$ against the lower modes $y_j(x)$, $j = 1,\dots,n-1$, and Eqs. (4.1b) constitute together with the boundary conditions $n-1$ eigenvalue problems for the lower modes $y_j(x)$ of the optimal design α_n .

As will be discussed in the following, a geometrically unconstrained solution α_n to the problem coincidently constitutes the optimal design to the problem of maximizing the *difference* between two adjacent natural frequencies (or critical speeds) ω_n and ω_{n-1} for given volume and length of the beam (or rotating shaft), see [47]. It is not surprising, therefore, that geometrically *constrained* solutions also exhibit large gaps between two adjacent frequencies [78], and that these gaps are very close to the maximum obtainable gaps. The point is that the geometrically constrained problem of maximum, single, higher order natural frequency is simpler to solve than problem formulated in terms of maximum difference between two adjacent natural frequencies. The latter type of problem has been solved in Refs. [83,84] (see also Ref. [85]).

Thus, the type of problem to be considered is directly related to practical design against resonance of beams due to external excitation and whirling instability of rotating shafts: if the structure is designed such that the external excitation frequency or the service speed is isolated in a broad gap between two consecutive higher order natural frequencies or higher order critical speeds, considerable weight savings are possible compared with designs where the excitation frequency or service speed is placed between zero and a high value of the fundamental natural frequency or first critical speed.

Another direct and very important advantage of considering optimization (geometrically constrained as well as unconstrained) with respect to a single, higher order natural frequency (or critical speed) of given order n, is that the resulting optimal design is, at the same time, the optimal design to the problem of optimizing with respect to the *fundamental* natural frequency (or first critical speed), assuming the positions of $n-1$ available interior supports to be design variables in addition to the cross-sectional area distribution [47,78]. According to Mróz and Rozvany [86] and Rozvany [87], zero support reaction is a necessary condition for optimum location of an interior simple support. This implies [47,78] that the optimal positions for available interior supports in a problem of optimal design with respect to the fundamental frequency, are simply identified with the $n-1$ nodal points of the vibration mode $y_n(x)$ of the higher order frequency optimal design.

Fig. 8 illustrates as an example the geometrically unconstrained design with optimal positions of four available inner supports that maximize the fundamental frequency of a transversely vibrating $p = 2$

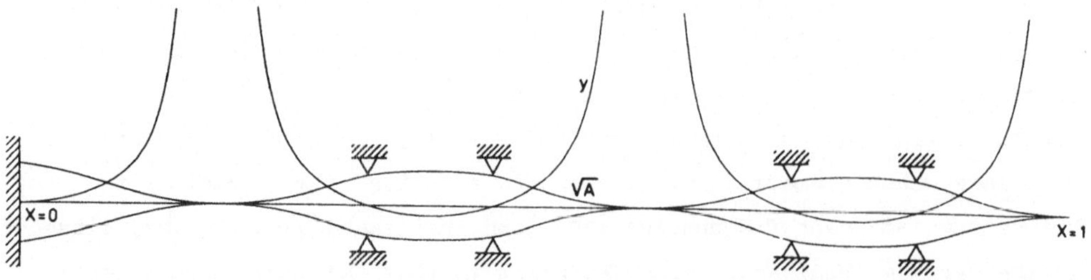

Fig. 8.

beam with clamped and free end points. The design is determined by optimizing the beam without inner supports with respect to its fifth eigenfrequency, and the inner supports are subsequently placed optimally at the four nodal points of the corresponding mode.

4.2 RESULTS OF GEOMETRICALLY CONSTRAINED PROBLEMS

Let us now consider some examples of geometrically constrained solutions for transversely vibrating Bernoulli-Euler cantilever beams $(p = 2)$ and rotating, cantilevered circular shafts from Ref. [78]. Results for Timoshenko beams are available in [88].

Fig. 9 shows cantilever beams optimized with respect to the third natural frequency, ω_3. The optimal designs are associated with $p = 2$ and are geometrically constrained with $\bar{\alpha} = \bar{A}L/V = 0.05$ for (a) and (c), and $\bar{\alpha} = 0.5$ for (b) and (d), respectively. Designs (c) and (d) are equipped with a dimensionless tip mass, $q_1 = Q_1/\rho V = 0.1$. The first four vibration modes of the optimal designs (a) and (c) are also shown in the figure. The natural frequencies ω_3 of the optimal beams are increased by (a) 129%, (b) 39%, (c) 82% and (d) 28%, respectively, when compared with the same frequency of uniform beams of the same volume, length, material, and, for (c) and (d), tip mass. The frequency differences $\omega_3 - \omega_2$ of the optimal beams are (a) 228%, (b) 53%, (c) 156% and (d) 41% higher than the corresponding frequency differences of the uniform beams.

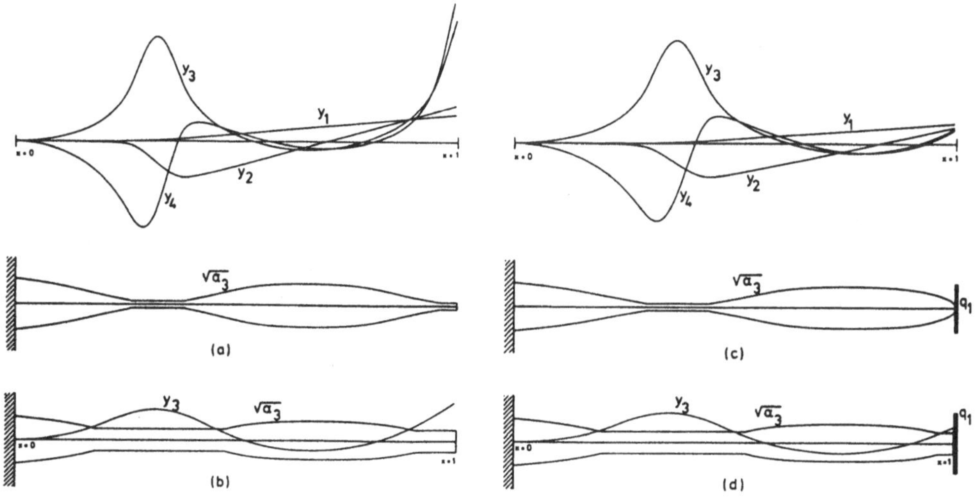

Fig. 9.

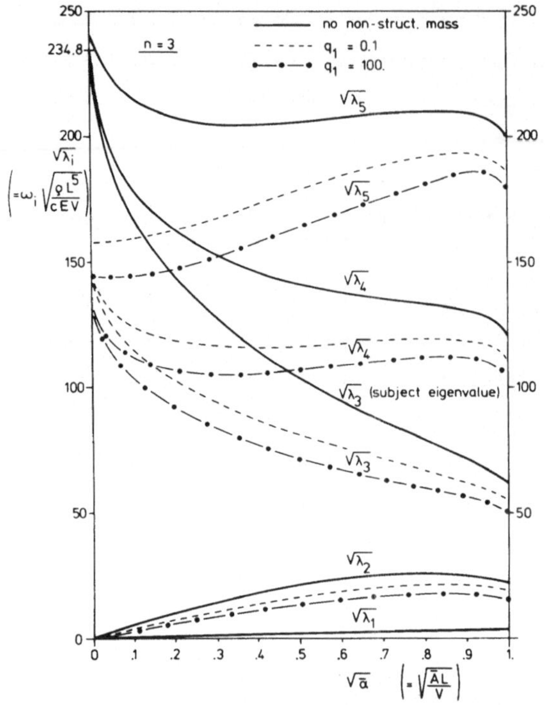

Fig. 10.

Fig. 10 summarizes, for $n = 3$, the results for a number of optimal solutions of the types shown in Fig. 9. Adopting the concept of given volume V and length L, the square roots of the eigenvalues in Fig. 10 directly represent the natural *frequencies* of the optimal designs associated with $n = 3$. The results in Fig. 10 may be compared with corresponding results for $n = 1$ in Fig. 6, noting especially the different ordinate scales and the different type of behaviour of the subject eigenfrequencies of the mass-carrying beams, substantiating advantages of optimizing with respect to higher order eigenfrequencies. Similarly, the optimal $n = 3$ designs in Fig. 9 may be compared with the $n = 1$ designs in Fig. 5.

As is illustrated for $n = 3$ in Fig. 10, we find for any value of $n > 1$ that both the optimal natural frequency ω_n and the distance between the consecutive natural frequencies ω_n and ω_{n-1} increase with decreasing geometric constraint. In fact, the absolute values of the differences between ω_n and ω_{n-1} increase with increasing n, and this, irrespective of whether the beams are equipped with nonstructural mass or not.

Optimizing cantilevers for $n > 3$, not only the subject frequency ω_n but also the closest subsequent natural frequencies ω_{n+1}, ω_{n+2}, etc. are pushed upwards, and the sub-spectrum consisting of these natural frequencies becomes significantly condensed for optimal solutions associated with small geometric constraint. In the limiting case of geometrically unconstrained optimization, the subject frequency may be increased to the extent that it coalesces with one or more of the subsequent natural frequencies [78], see e.g. Fig. 10. For the problems considered in [78], coalescence of the subject eigenvalue with one or more higher order eigenvalues is always found to take place in the limiting case of geometrically unconstrained design $(\bar{\alpha} = 0)$. This implies the advantage that *it is not necessary here, to apply a bi- or multimodal formulation*, in order to obtain the correct optimal design.

Another characteristic feature of optimizing with respect to a higher order natural frequency ω_n , $n \geq 2$ is that all the lower natural frequencies ω_j , $j = 1, \ldots, n-1$, are kept small by this process, and that they tend toward a multiple zero eigenvalue as the geometric minimum constraint tends towards zero, see Fig. 10.

4.3 GEOMETRICALLY UNCONSTRAINED OPTIMIZATION

Geometrically unconstrained solutions are important because they constitute *limiting designs* for corresponding geometrically unconstrained designs, and their associated optimal eigenvalues constitute *upper bound values* for corresponding eigenvalues of all similar beams with and without non-structural mass. In the following we shall consider some geometrically unconstrained optimal designs obtained in [47] for transversely vibrating Bernoulli-Euler beams with $p = 2$.

The solutions are determined numerically in [47] on the basis of a formal integration of the geometrically unconstrained formulation for optimal design, with possible types of singular behaviour appropriately allowed for in the numerical solution procedure. The positions as well as the types (I and II) of the singularities are additional design variables in the optimization process, and the optimal solutions are, in fact, determined via a path through a class of geometrically unconstrained sub-optimal solutions.

Fig. 11 provides an illustration of this in the case of optimizing a cantilever for $n = 3$. The solutions in Fig. 11a are both sub-optimal

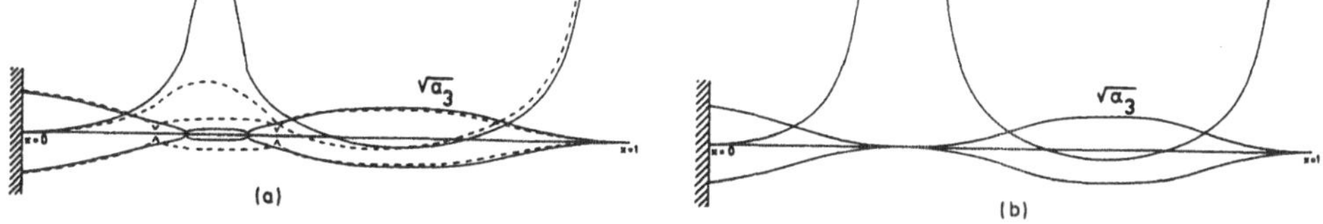

Fig. 11.

solutions. The dashed solution has inner Type I singularities at $x = 0.218$ and $x = 0.418$, and the eigenvalue λ_3 is $\lambda_3 = 1.77 \cdot 10^4$. For the solid solution in Fig. 11a, the type I singularities are placed at $x = 0.268$ and $x = 0.368$, and the eigenvalue is increased, $\lambda_3 = 2.87 \cdot 10^4$. Fig. 11b shows the resulting optimal solution with $\lambda_3 = 5.511 \cdot 10^4$. This solution has an inner Type II singularity, which is optimally placed at $x = 0.321$, and the optimal frequency ω_3 of the design is increased by 280% in comparison with the corresponding frequency of a uniform cantilever of the same volume, length and material. It should be noted that it is the design in Fig. 11b which lies behind the result for $\bar{\alpha} = 0$ in Fig. 10, and that it may be compared to corresponding constrained optimal designs in Figs. 9a,b.

Now, each Type I and Type II singularity introduces, respectively, one and two degrees of kinematic freedom to a geometrically unconstrained beam, and when optimizing with respect to the n'th natural frequency ω_n $(n > 1)$, it turns out [47] that the optimal design exactly possesses $n-1$ degrees of kinematic freedom to perform *rigid body motions*. Thus, all natural frequencies lower than the subject natural frequency of a geometrically unconstrained optimal design correspond to rigid body motions*) and attain zero value, cf. Fig. 10 (with $\bar{\alpha} = 0$).

This clearly implies that a geometrically unconstrained solution to the problem of optimizing the n'th natural frequency ω_n $(n > 1)$ is coincidently the solution to the problem of optimizing the *difference* between the two adjacent frequencies ω_n and ω_{n-1}.

Figs. 12a,b show optimal cantilevers for $n = 2$ and $n = 4$, respectively. Figs. 13a,b show optimal simply supported beams for $n = 2$ and $n = 3$, respectively, and Figs. 14a,b illustrate free beams optimized for $n = 3$ and $n = 4$, respectively. (The first two natural frequencies of free beams correspond to rigid body motions, and are always zero).

---

*) Note, in Fig. 9 above, the tendency of the lower vibration modes y_1 and y_2 to become rigid body motions in the limiting case of zero constraint, and that these modes indicate the two degrees of kinematic freedom associated with an inner Type II singularity, namely jumps in both deflection and slope.

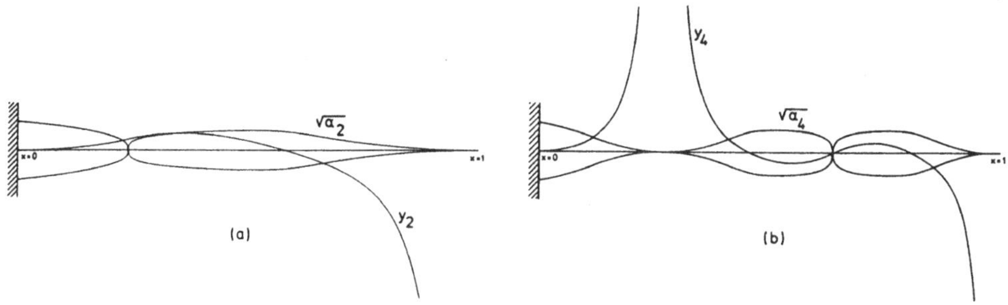

Fig. 12.

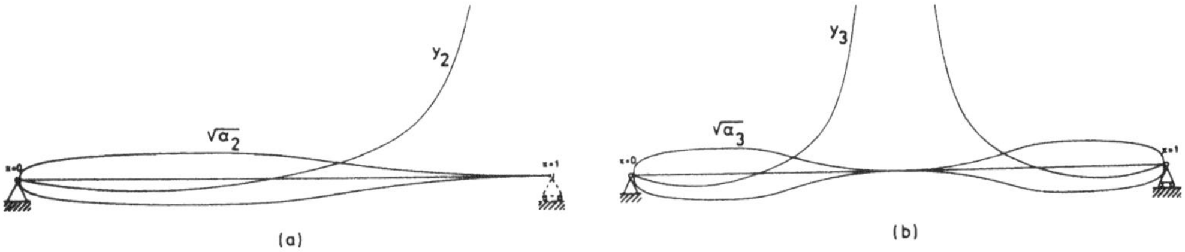

Fig. 13.

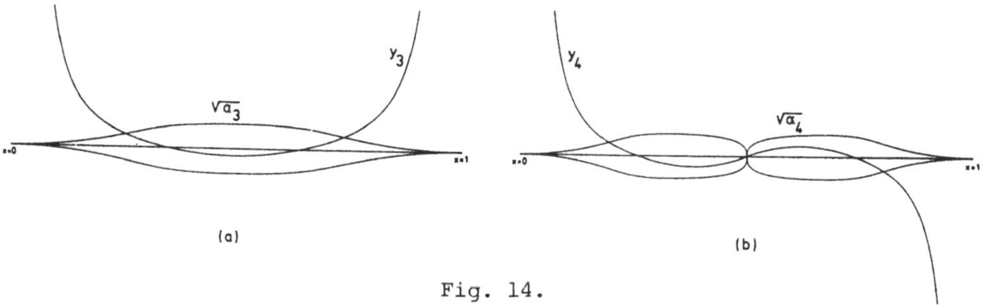

Fig. 14.

The study in Ref. [47] reveals the notable feature that *two* Type I singularities may *coalesce*, thereby forming *one* Type II singularity. Fig. 11 provides an illustration of such coalescence at an inner point. Note also that the inner beam separation formed at the resulting Type II singularity of the optimal $n = 3$ cantilever in Fig. 11b in fact divides this beam into a *scaled* optimal cantilever corresponding to $n = 1$, see Fig. 7b, and a scaled optimal free beam corresponding to $n = 3$, as shown in Fig. 14a. The coalescence may also take place at a beam end point; for example, the Type II singularity of the simply supported, optimal $n = 2$ beam in Fig. 13a results from an original inner Type I singularity coalescing with an a priori singularity of Type I at the beam end.

The formation of Type II singularities is found to contribute considerably to large subject eigenvalues, and singularities of this type consequently play a predominant role in geometrically unconstrained optimal designs. Thus, no more than *one* inner Type I singularity is found in any of the designs [47].

In fact when optimizing beams of a given type of end conditions, it is only necessary to apply the numerical solution procedure for values of n up to a particular value, N, where the first Type II singularity occurs in the corresponding optimal design. As discussed and outlined in [47], the inner Type II singularities open up the possibility of determining the optimal designs associated with higher values of n simply by assembling optimal beam elements obtained numerically for small values of n. In fact, in Ref. [47], a so-called "Method of scaled beam elements" is developed by means of which optimal designs subject to any higher value of n are determined for beams with any combination of clamped, simply supported, and free end conditions.

SECTION 5

OPTIMAL DESIGN OF VISCOELASTIC STRUCTURES UNDER FORCED STEADY STATE VIBRATION

5.1 INTRODUCTION

This section concerns non-selfadjoint problems of optimal design. We consider viscoelastic, one-dimensional structures, such as rods, bars and beams, that are subjected to forced, steady state vibration by excitation by given, harmonically varying, external loads. The structural vibrations may be axial, torsional or transverse vibrations, and the external excitation frequency is assumed to be smaller than the fundamental natural frequency for the particular type of vibration. It is our objective to derive a general set of governing equations for the optimal structural design that is associated with minimum dynamic response for given volume, length and boundary conditions, and given viscoelastic material of the structure.

The structural material is assumed to follow the well-known three-parameter constitutive law for viscoelasticity, and the concept of complex stiffness modulus is adopted for simplifying the analysis. Hence, complex variables are used throughout, but the correspondence principle can be employed.

The contents of this section is based on the paper [89] by Lekszycki and the author. Optimal design of purely *elastic* structures under forced steady state vibration due to harmonically varying external loading has earlier been considered by Mróz [90], Icerman [91] and Johnson *et al.* [92].

5.2 VIBRATION OF ONE-DIMENSIONAL VISCOELASTIC STRUCTURES UNDER HARMONIC EXTERNAL EXCITATION

We consider a straight, one-dimensional structure, such as a rod, bar or beam, which is made of a solid, viscoelastic material with the three-parameter constitutive relationship

$$\sigma + p_1\dot{\sigma} = q_0\varepsilon + q_1\dot{\varepsilon} \tag{5.1}$$

between uniaxial normal (or shear) stresses, strains, and -rates. The constitutive parameters p_1 , q_0 and q_1 are assumed to be given for the particular type of deformation in question.

A coordinate axis x is embedded in the structure, which has given length ℓ and variable cross-sections with common directions y and z for principal axes of inertia. The real function $\alpha(x)$ is assumed to characterize the structural design. The structure is subjected to forced, steady state harmonic vibration, such as axial, torsional or transverse vibration with a mode $U(x, t)$ under given external loading $P(x, t)$. In order to simplify the analysis and computation, the mode and the loading will be taken to be complex functions of the real independent variables x and (time) t , i.e., they will be considered in the form

$$U(x, t) = u(x)e^{i\omega t} = [u_1(x) + iu_2(x)]e^{i\omega t} \tag{5.2}$$

$$P(x, t) = p(x)e^{i\omega t} = [p_1(x) + ip_2(x)]e^{i\omega t} , \tag{5.3}$$

where i denotes the imaginary unit. The relevant uniaxial strain and stress components will then have the form

$$\begin{aligned} \varepsilon(x, y, z, t) &= \varepsilon^*(x, y, z)e^{i\omega t} \\ \sigma(x, y, z, t) &= \sigma^*(x, y, z)e^{i\omega t} , \end{aligned} \tag{5.4}$$

where $\varepsilon^* = \varepsilon_1^* + i\varepsilon_2^*$ and $\sigma^* = \sigma_1^* + i\sigma_2^*$ are time-independent, complex strains and stresses, respectively.

Substituting Eqs. (5.4) into Eq. (5.1), we obtain the following simple and convenient time-independent stress-strain relationship

$$\sigma^*(x, y, z) = k(\omega)\varepsilon^*(x, y, z) , \tag{5.5}$$

where the real and imaginary parts of the complex stiffness modulus

$$k(\omega) = k_1(\omega) + ik_2(\omega) \tag{5.6}$$

are given by the frequency ω of the loading and the material parameters p_1, q_0 and q_1, as follows,

$$k_1 = \frac{q_0 + p_1 q_1 \omega^2}{1 + \omega^2 p_1^2} , \quad k_2 = \frac{\omega(q_1 - q_0 p_1)}{1 + \omega^2 p_1^2} . \tag{5.7}$$

Due to the simple form of the stress-strain relationship (5.5), the equations governing the motion of our one-dimensional viscoelastic structure can now be written in precisely the same form as in a purely elastic case, provided that the structural stiffness, loading and deflection, and the coefficients of relevant spatial and temporal differential operators, are conceived to be complex. Hence, let the equation of motion have the form

$$M(\alpha)U(x, t) = P(x, t) \qquad 0 \le x \le \ell , 0 \le t \le T \tag{5.8}$$

with boundary conditions

$$A(\alpha)U(x, t) = Q(x, t) \quad x = 0 , x = \ell , 0 \le t \le T \tag{5.9}$$

and initial conditions

$$T(\alpha)U(x\,,0) = r(x) \qquad 0 \le x \le \ell\;,\; t = 0 \tag{5.10}$$

Here, $Q(x\,,t)$ on the right-hand side of Eq. (5.9) is assumed to have the harmonic form $q(x)e^{i\omega t}$, where $q(x)$ is complex.

Now, for the case of steady state harmonic vibration under consideration, we may eliminate time-dependence by substituting Eq. (5.2) for U, Eq. (5.3) for P, and the similar equation for Q, into Eqs. (5.8)-(5.10). Hence, we obtain the purely spatial state equation

$$L(\alpha)u(x) = p(x) \qquad 0 \le x \le \ell \tag{5.11}$$

and boundary conditions

$$B(\alpha)u(x) = q(x) \qquad x = 0\;,\; x = \ell\;, \tag{5.12}$$

to govern our analysis problem. In Eqs. (5.11) and (5.12), $L(\alpha)$ and $B(\alpha)$ denote linear, homogeneous, complex, differential operators. In the following, the operator $L(\alpha)$ will be considered in the form

$$L(\alpha)u(x) \equiv \sum_{\nu=0}^{m}\left[g_\nu\left(k(\omega)\,,\alpha\,,x\right)u^{(\nu)}(x)\right]^{(\nu)} \tag{5.13}$$

which covers a number of practical applications. In (5.13), $f^{(\nu)} \equiv d^\nu f/dx^\nu$, and the order of the differential operator L is $2m$. The operator $B(\alpha)$ in (5.12) is assumed to have the form

$$B(\alpha)u = \sum_{\nu=0}^{2m-1}\left[a_\nu\left(k(\omega)\,,\alpha\right)u^{(\nu)}(0) + b_\nu\left(k(\omega)\,,\alpha\right)u^{(\nu)}(\ell)\right]. \tag{5.14}$$

5.3 TWO OBJECTIVE FUNCTIONALS FOR MINIMIZATION OF DYNAMIC STRUCTURAL RESPONSE

An objective functional I_1, which is of relevance for the types of optimization problems under consideration, is defined as follows by Mróz [93]

$$I_1 = \int_0^{\ell} C_1(u)\,dx \quad ; \quad C_1(u) = \begin{cases} 0 & \text{if } |u| < u_o \\ (|u| - u_o)^n & \text{if } |u| \geq u_o \, . \end{cases} \tag{5.15}$$

Here, n is a positive real exponent and u_o is a real, positive constant or function of x. Both n and u_o are assumed to be given. Furthermore,

$$|u(x)| = \sqrt{u_1^2(x) + u_2^2(x)} \, . \tag{5.16}$$

Since $|u(x)|$ represents the maximum value of the total deflection of any point x, $0 \leq x \leq \ell$, the functional I can be interpreted as a global measure of the difference between the actual deflection amplitudes and a given comparison amplitude u_o.

An alternative functional I_2 to be considered, represents a measure of the work done by the external forces: Although the loading $P(x, t)$ and deflection $U(x, t)$ are dealt with in the general form (5.3) and (5.2), that is

$$P(x, t) = p(x)e^{i\omega t} = \left(p_1(x)\cos\omega t - p_2(x)\sin\omega t\right) + i\left(p_1(x)\sin\omega t + p_2(x)\cos\omega t\right) \tag{5.3a}$$

$$U(x, t) = u(x)e^{i\omega t} = \left(u_1(x)\cos\omega t - u_2(x)\sin\omega t\right) + i\left(u_1(x)\sin\omega t + u_2(x)\cos\omega t\right) , \tag{5.2a}$$

we may assume that an actual loading $P^*(x, t)$ only varies with cosine in time such that $p_2(x) = 0$ and

$$P^*(x, t) = p_1(x)\cos\omega t , \tag{5.17}$$

whereby the associated deflection $U^*(x, t)$ becomes

$$U^*(x, t) = u_1(x)\cos\omega t - u_2(x)\sin\omega t \, . \tag{5.18}$$

The work done by the external forces during a cycle $T = 2\pi/\omega$ is then

$$\int_0^T \int_0^\ell U^*(x , t)P^*(x , t)dxdy = \frac{\omega}{\pi}\int_0^\ell p_1(x)u_1(x)dx , \qquad (5.19)$$

which provides a basis for selecting functional I_2 as

$$I_2 = \int_0^\ell c_2 dx = \int_0^\ell p_1(x)u_1(x)dx . \qquad (5.20)$$

5.4 DERIVATION OF A GENERAL SET OF GOVERNING EQUATIONS FOR OPTIMAL DESIGN

We now consider a quite general formulation [89] for optimal design of our viscoelastic structure under given, harmonically varying external loading. Using the real function $\alpha(x)$ as the design variable, it is our objective to minimize the real part of the functional

$$I = \int_0^\ell C(\alpha , u , x)dx \qquad (5.21)$$

(viz. I_1 or I_2 of the preceeding subsection) subject to the differential constraint of Eq. (5.11) and the integral constraint of given total structural volume

$$V = \int_0^\ell f(\alpha , x)dx . \qquad (5.22)$$

We assume that a geometrical minimum constraint value α_{min} is specified for the design variable $\alpha(x)$, such that we must everywhere have $\alpha(x) \geq \alpha_{min}$, which may be expressed as

$$\alpha(x) - \alpha_{min} = \alpha_s^2(x) \qquad (5.23)$$

by means of the real slack variable $\alpha_s(x)$.

To derive the governing equations for the optimization problem stated above, we adjoin conditions (5.22), (5.23) and (5.11) to the functional (5.21) by means of Lagrangian multipliers μ , $\eta(x)$ and $\lambda(x) = \lambda_1(x) + i\lambda_2(x)$, respectively, where $\lambda(x)$ is the so-called adjoint variable, and obtain

$$I^* = \int_0^{\ell} C(\alpha\,,\,u\,,\,x)\,dx + \mu\left(\int_0^{\ell} f(\alpha\,,\,x)\,dx - V\right)$$
$$+ \{\eta\,,\,\alpha - \alpha_{min} - \alpha_s^2\} + \{\lambda\,,\,Lu - p\}\;. \tag{5.24}$$

Here, the scalar product*) $\{f\,,\,g\}$ of the functions f and g, which are generally complex and which belong to the $\overline{L^2}$ space, is defined as

$$\{f(x)\,,\,g(x)\} = \int_0^{\ell} \overline{f(x)}\;g(x)\,dx\;. \tag{5.25}$$

Now, the stationarity condition for the real part of the augmented functional I^* defined in Eq. (5.24) can be written

$$\mathrm{Re}(\delta I^*) = \mathrm{Re}\left(\delta_\lambda I^* + \delta_u I^* + \delta_\alpha I^* + \delta_\mu I^* + \delta_\eta I^* + \delta_{\alpha_s} I^*\right) = 0\;, \tag{5.26}$$

which expresses the variation of I^* due to variations of the complex variables λ and u, and of the real variables α, μ, η and α_s, respectively. These variables may be considered to be mutually independent, and varying them independently, the stationarity requirement (5.26) then leads to the state equation for the actual structure, the state equation for the adjoint structure, the optimality condition, and some additional conditions. These equations will be derived in some detail in the following.

---

*) The following properties of the scalar product are frequently used in the sequel,

$$\{f(x)\,,\,g(x)\} = \{\overline{g(x)}\,,\,\overline{f(x)}\}\;, \tag{5.25a}$$

$$\{af(x) + bh(x)\,,\,g(x)\} = \bar{a}\{f(x)\,,\,g(x)\} + \bar{b}\{h(x)\,,\,g(x)\} \tag{5.25b}$$

where a and b are complex constants, and

$$\{f(x)\,,\,g(x)\} = \Big(\{\mathrm{Re}(f(x))\,,\,\mathrm{Re}(g(x))\} + \{\mathrm{Im}(f(x))\,,\,\mathrm{Im}(g(x))\}\Big)$$
$$+\, i\Big(\{\mathrm{Re}(f(x))\,,\,\mathrm{Im}(g(x))\} - \{\mathrm{Im}(f(x))\,,\,\mathrm{Re}(g(x))\}\Big)\;. \tag{5.25c}$$

5.4.1 State equation for the actual structure. Firstly, let us consider variation of the complex adjoint variable $\lambda(x)$. With $\delta_\lambda I^* = \{\delta\lambda , Lu - p\}$, we obtain by means of (5.25c) the stationarity condition

$$\mathrm{Re}(\delta_\lambda I^*) = \{\mathrm{Re}(\delta\lambda) , \mathrm{Re}(Lu - p)\} + \{\mathrm{Im}(\delta\lambda) , \mathrm{Im}(Lu - p)\} = 0 , \quad (5.27)$$

which, for arbitrary variation $\delta\lambda_1$ and $\delta\lambda_2$, gives

$$\mathrm{Re}(Lu - p) = 0 \quad (5.28a)$$

$$\mathrm{Im}(Lu - p) = 0 . \quad (5.28b)$$

In complex form, these equations can simply be written as

$$Lu - p = 0 , \quad (5.28)$$

which is the *state equation for the actual structure*. It has been assumed here that the complex deflection function (state variable) $u(x)$ satisfies appropriate boundary conditions.

5.4.2 State equation for the adjoint structure. We now take the variation of $\mathrm{Re}(I^*)$ with respect to variation of the complex variable $u(x)$. As a first step, we obtain

$$\delta_u I^* = \{\lambda , L\delta u\} + \int_0^\ell \frac{\partial C}{\partial u} \delta u dx , \quad (5.29)$$

but in order to express our final result in a convenient form, let us define the *adjoint operator* L^a corresponding to the operator L by the equation

$$\{\lambda , L\delta u\} = \{L^a\lambda , \delta u\} . \quad (5.30)$$

With the form of L given by Eq. (5.13) and the deflection (5.25) of the scalar product, the left-hand side of (5.30) can, after integration by parts, be written as

$$\{\lambda , L\delta u\} = \int_0^{\ell} \sum_{\nu=0}^{m} \overline{\left[\bar{g}_\nu(k , \alpha , x)\lambda^{(\nu)}\right]^{(\nu)}} \delta u dx$$

$$+ \sum_{\nu=1}^{m} \sum_{j=0}^{\nu-1} \left\{(-1)^j \bar{\lambda}^{(j)} \left[g_\nu(k , \alpha , x)\delta u^{(\nu)}\right]^{(\nu-j-1)}\right\}\Big|_0^{\ell} \tag{5.31}$$

$$+ \sum_{\nu=1}^{m} \sum_{j=0}^{\nu-1} \left\{(-1)^{2\nu-j-1} \delta u^{(j)} \left[g_\nu(k , \alpha , x)\bar{\lambda}^{(\nu)}\right]^{(\nu-j-1)}\right\}\Big|_0^{\ell} ,$$

where the symbol $z|_0^{\ell}$ identifies the difference $z(\ell) - z(0)$ of an argument z . Appropriate boundary conditions for $\lambda(x)$ are obtained from conditions of vanishing Dirichlet boundary terms in (5.31) at $x = 0$ and $x = \ell$. Satisfaction of these conditions implies that the adjoint differential operator L^a defined in (5.30) can be identified as

$$L^a \equiv \sum_{\nu=0}^{m} \frac{d^\nu}{dx^\nu}\left[\bar{g}_\nu(k , \alpha , x) \frac{d^\nu}{dx^\nu}\right] \tag{5.32}$$

by means of (5.25). Now, rewriting Eq. (5.29) in the form

$$\delta_u I^* = \{L^a\lambda , \delta u\} + \{\overline{\left(\frac{\partial C}{\partial u}\right)} , \delta u\} , \tag{5.33}$$

and making use of (5.25c) and the well known rule

$$\frac{\partial F}{\partial \xi} = \frac{\partial \mathrm{Re}(F)}{\partial \xi_1} + i\frac{\partial \mathrm{Im}(F)}{\partial \xi_1} = \frac{\partial \mathrm{Im}(F)}{\partial \xi_2} - i\frac{\partial \mathrm{Re}(F)}{\partial \xi_2}$$

for partial differentiation of a complex function F with respect to a complex argument ξ , the stationarity condition $\mathrm{Re}(\delta_u I^*) = 0$ becomes

$$\mathrm{Re}(\delta_u I^*) = \{\mathrm{Re}\left[L^a\lambda + \overline{\left(\frac{\partial C}{\partial u}\right)}\right] , \mathrm{Re}(\delta u)\} + \{\mathrm{Im}\left[L^a\lambda + \overline{\left(\frac{\partial C}{\partial u}\right)}\right] , \mathrm{Im}(\delta u)\}$$

$$= \{\mathrm{Re}\left(L^a\lambda + \frac{\partial C}{\partial u_1}\right) , \mathrm{Re}(\delta u)\} + \{\mathrm{Im}(L^a\lambda) + \mathrm{Re}\left(\frac{\partial C}{\partial u_2}\right) , \mathrm{Im}(\delta u)\} = 0 \quad . \tag{5.34}$$

Since δu_1 and δu_2 are arbitrary, Eq. (5.34) is equivalent to the two equations

$$\mathrm{Re}\left(L^a\lambda + \frac{\partial C}{\partial u_1}\right) = 0 \tag{5.35a}$$

$$\mathrm{Im}(L^a\lambda) + \mathrm{Re}\left(\frac{\partial C}{\partial u_2}\right) = 0 \tag{5.35b}$$

which in complex form may be written as

$$L^a\lambda = -\left[\mathrm{Re}\left(\frac{\partial C}{\partial u_1}\right) + i\,\mathrm{Re}\left(\frac{\partial C}{\partial u_2}\right)\right] , \tag{5.35}$$

i.e., the *state equation for the adjoint structure*. Here, complex $\lambda(x)$ can be interpreted as the state variable (deflection function) of the adjoint structure. As was mentioned earlier, $\lambda(x)$ must satisfy appropriate boundary conditions that make the boundary terms vanish in Eq. (5.31).

It is worth noting that when the coefficients $g_\nu(k, \alpha, x)$, $\nu = 0,\ldots,m$, which define the L-operator via Eq. (5.13), are independent or depend linearly on the complex stiffness modulus k, see Eqs. (5.5)-(5.7), then the adjoint operator L^a given by Eq. (5.32) reduces to

$$L^a \equiv \sum_{\nu=0}^{m} \frac{d^\nu}{dx^\nu}\left[g_\nu(\bar{k}, \alpha, x)\frac{d^\nu}{dx^\nu}\right] . \tag{5.36}$$

This implies that *the adjoint structure has precisely the same design as the actual structure*. Moreover, it consists of a material which is also the same, except for that its viscous coefficients p_1 and q_1 are negative, cf. Eqs. (5.36), (5.13), (5.7) and (5.1). Thus, *the material of the adjoint structure has negative damping, but is otherwise the same as that of the actual structure*. It follows from (5.35) that *the loading on the adjoint structure depends on the specific cost function* C *and is generally different from the loading on the actual structure*.

In cases where the optimization is based on the functional I_1 with the associated cost function $C = C_1$ defined by Eq. (5.15), the loading on the adjoint structure $\partial C_1/\partial u_1$ and $\partial C_1/\partial u_2$ to be used in Eq. (5.35), are easily found to be given by

$$j = 1,2 : \frac{\partial C_1}{\partial u_j} = \begin{cases} 0 & \text{if } |u| < u_o \\ n(|u| - u_o)^{n-1}\dfrac{u_j}{|u|} & \text{if } |u| \geq u_o \end{cases} \tag{5.37}$$

by means of Eqs. (5.15) and (5.16).

If the actual loading is taken in the form of (5.17), i.e. with $p_2(x) = 0$, and the optimization is based on the functional I_2 and specific cost function $C = C_2$ defined by Eq. (5.20), then the loading to be used in Eq. (5.35) for the adjoint structure is simply given via

$$\frac{\partial C_2}{\partial u_1} = p_1(x) \quad , \quad \frac{\partial C_2}{\partial u_2} = 0 \ . \tag{5.38}$$

This implies that the analysis problems for the actual and the adjoint structure become identical except for the difference between the L and the L^a operator discussed above.

Let us note that in the special case of a purely elastic structure we have $L \equiv L^a$ (selfadjointness of the L operator).

5.4.3 Optimality condition. The variation of I^* in Eq. (5.24) with respect to variation of the real design variable $\alpha(x)$ is given by

$$\delta_\alpha I^* = \int_0^\ell \frac{\partial C}{\partial \alpha}\delta\alpha dx + \mu\int_0^\ell \frac{df}{d\alpha}\delta\alpha dx + \{\eta \ , \delta\alpha\} + \{\lambda \ , \frac{\partial Lu}{\partial \alpha}\delta\alpha\} \ . \tag{5.39}$$

With differential operator L of the form given by Eq. (5.13), the last term in Eq. (5.39) can be written as

$$\{\lambda \ , \frac{\partial Lu}{\partial \alpha}\delta\alpha\} = \int_0^\ell \bar{\lambda} \sum_{\nu=0}^{m}\left[\frac{\partial g_\nu(k \ , \alpha \ , x)}{\partial \alpha}\delta\alpha u^{(\nu)}\right]^{(\nu)} dx$$

$$= \int_0^\ell \sum_{\nu=0}^{m} (-1)^\nu \bar{\lambda}^{(\nu)} \frac{\partial g_\nu(k \ , \alpha \ , x)}{\partial \alpha} u^{(\nu)}\delta\alpha dx \tag{5.40}$$

$$+ \sum_{\nu=1}^{m}\sum_{j=0}^{\nu-1} (-1)^j \bar{\lambda}^{(j)}\left[\frac{\partial g_\nu}{\partial \alpha} u^{(\nu)}\delta\alpha\right]^{(\nu-j-1)}\Bigg|_0^\ell$$

after integration by parts. Assuming vanishing of the boundary terms,

$$\sum_{\nu=1}^{m}\sum_{j=0}^{\nu-1} (-1)^j \bar{\lambda}^{(j)}\left[\frac{\partial g_\nu(k \ , \alpha \ , x)}{\partial \alpha} u^{(\nu)}\delta\alpha\right]^{(\nu-j-1)}\Bigg|_0^\ell = 0 \ , \tag{5.41}$$

due to the boundary conditions, and using the definition (5.25) for scalar products, we can write Eq. (5.39) in the form

$$\delta_\alpha I^* = \{\frac{\partial \bar{C}}{\partial \alpha}, \delta\alpha\} + \{\mu\frac{df}{d\alpha}, \delta\alpha\} + \{\eta \ , \delta\alpha\} + \{S(\bar{u} \ , \lambda) \ , \delta\alpha\} \ , \tag{5.42}$$

where $S(\bar{u} \ , \lambda)$ identifies

$$S(\bar{u} \ , \lambda) = \sum_{\nu=0}^{m} (-1)^\nu \lambda^{(\nu)} \frac{\partial \bar{g}_\nu(k \ , \alpha \ , x)}{\partial \alpha}\bar{u}^{(\nu)} \ . \tag{5.43}$$

The requirement of stationarity of $Re(\delta_\alpha I^*)$ with respect to arbitrary variation $\delta\alpha$ then yields

$$Re\left[\frac{\partial\bar{C}}{\partial\alpha} + S(\bar{u}\,,\,\lambda) + \mu\frac{df}{d\alpha} + \eta\right] = 0\ , \tag{5.44}$$

which constitutes the *optimality condition* of our problem.

Since the specific cost functions C_1 and C_2 in Eqs. (5.15) and (5.20) are independent of design α , the optimality condition (5.44) reduces when the functionals I_1 or I_2 are applied.

<u>5.4.4 Additional conditions</u>. The stationarity conditions $Re(\delta_\mu I^*) = 0$, $Re(\delta_\eta I^*) = 0$ and $Re(\delta_{\alpha_s} I^*) = 0$, respectively, are easily seen to re-establish the *constraint equations* (5.22) and (5.23), i.e.

$$V = \int_0^\ell f(\alpha\,,\,x)\,dx \tag{5.45}$$

$$\alpha - \alpha_{min} - \alpha_s^2 = 0 \tag{5.46}$$

and to produce the *switching condition*

$$\eta\,\alpha_s = 0\ . \tag{5.47}$$

It follows from Eqs. (5.46) and (5.47) that we have $\eta = 0$ in sub-intervals of inactive minimum constraint. Consequently, the optimality condition (5.44) reduces in such sub-intervals.

<u>5.4.5 The complete set of governing equations</u> for the general optimization problem consists of Eqs. (5.28), (5.35), (5.44)-(5.47), in addition to given boundary conditions for $u(x)$ and boundary conditions for $\lambda(x)$. The latter conditions are deduced from the requirement of vanishing boundary terms in Eq. (5.31).

5.5 EXAMPLE: OPTIMIZATION OF SOLID, COMPOSITE, VISCOELASTIC BEAMS SUBJECTED TO FORCED, TRANSVERSE VIBRATION

To exemplify the general theory for optimization of one-dimensional structures outlined in the preceeding sections, we now consider optimal design of a composite, viscoelastic beam that performs transverse vibra-

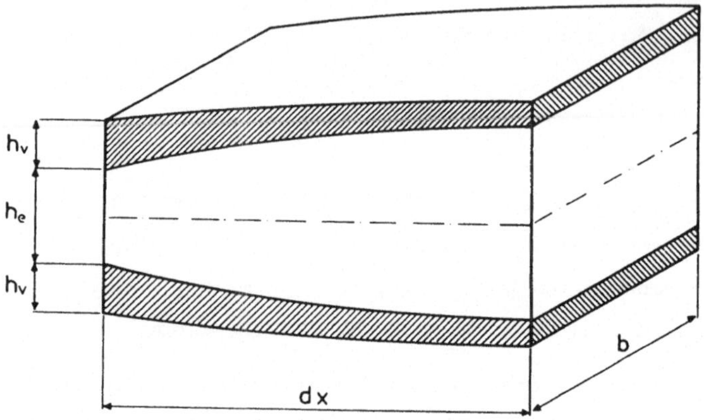

Fig. 15.

tions under a given external, harmonically varying, transverse load distribution $P(x, t)$ of the form (5.3).

The composite beam, an element of which is shown in Fig. 15, has given length ℓ and width b. The beam consists of a solid, viscoelastic (or purely elastic) core of given variable thickness $h_e(x)$, which is covered by two identical, solid layers of another viscoelastic material. Each of these layers have non-uniform thickness $h_v(x)$, which is taken as the *design variable* in our problem, and the total material volume

$$V = 2b \int_0^{\ell} h_v(x)\,dx \tag{5.48}$$

of the cover layers is prescribed. The cover thickness h_v may nowhere be less than a prescribed minimum thickness h_{min},

$$h_v(x) - h_{min} - h_s^2(x) = 0 \tag{5.49}$$

Eqs. (5.48) and (5.49) clearly correspond to Eqs. (5.22) and (5.23), respectively.

In fact, we may conceive the core of the beam as an originally independent (viscoelastic or elastic) structure, and the problem may interpreted as one of "optimal remodeling" [49]: namely to determine the optimal distribution of the viscoelastic material (damper material), such that the vibrational response of the remodeled, composite structure is minimized for a given amount of cover material. It is well known that viscoelastic materials with high dissipation of energy are often successfully used in practise for the purpose of damping vibration and noise.

In order to specialize the general, governing equations of Section 5.4 to our particular problem, we note that the second area moment of inertia for the cross-section of the core J_e is $J_e = b h_e^3/12$, and the complex bending stiffness $J_e K_e$ of the core is then given by

$$J_e K_e = J_e (k_{e1} + i k_{e2}) \ . \tag{5.50}$$

Similarly, the complex bending stiffness $J_v K_v$ of the two cover layers is

$$J_v K_v = J_v (k_{v1} + i k_{v2}) \ , \tag{5.51}$$

where the second area moment of inertia J_v is $J_v = b h_v^3/6 + b h_v (h_e + h_v)^2/2$. Note that h_v is not assumed to be small in comparison with h_e.

The complex state equation for the actual beam than has the form

$$\left[(J_e K_e + J_v K_v) u''(x) \right]'' - \omega^2 (\rho_e A_e + \rho_v A_v) u(x) = p(x) \ , \tag{5.52}$$

where $A_e = b h_e(x)$ and $A_v = 2b h_v(x)$ are the cross-sectional areas of the core and of the two cover layers, respectively, and where ρ_e and ρ_v are the mass densities of their materials. Eq. (5.52) corresponds to Eq. (5.28), and is equivalent to the two equations

$$\left[(J_e k_{e1} + J_v k_{v1}) u_1'' - (J_e k_{e2} + J_v k_{v2}) u_2'' \right]'' - \omega^2 (\rho_e A_e + \rho_v A_v) u_1 = p_1 \tag{5.52a}$$

$$\left[(J_e k_{e2} + J_v k_{v2}) u_1'' + (J_e k_{e1} + J_v k_{v1}) u_2'' \right]'' - \omega^2 (\rho_e A_e + \rho_v A_v) u_2 = p_2 \tag{5.52b}$$

By means of Eqs. (5.52) and (5.28), we are now able to identify the coefficients g_ν, $\nu = 0,\ldots,m$, $m = 2$, of the complex differential operator L defined in Eq. (5.13). We find

$$g_0(x) = -\omega^2(\rho_e A_e + \rho_v A_v)$$

$$g_1(x) = 0 \tag{5.53}$$

$$g_2(x) = (J_e K_e + J_v K_v) = (J_e k_{e1} + J_v k_{v1}) + i(J_e k_{e2} + J_v k_{v2}) \ .$$

With the form (5.36) of the adjoint operator L^a , Eqs. (5.53) give the following versions of the two state equations (5.35a-b) for the adjoint beam in our problem,

$$\left[(J_e k_{e1} + J_v k_{v1})\lambda_1'' + (J_e k_{e2} + J_v k_{v2})\lambda_2''\right]'' - \omega^2(\rho_e A_e + \rho_v A_v)\lambda_1 = -\mathrm{Re}\left(\frac{\partial C}{\partial u_1}\right) \tag{5.54a}$$

$$\left[-(J_e k_{e2} + J_v k_{v2})\lambda_1'' + (J_e k_{e1} + J_v k_{v1})\lambda_2''\right]'' - \omega^2(\rho_e A_e + \rho_v A_v)\lambda_2 = -\mathrm{Re}\left(\frac{\partial C}{\partial u_2}\right) \tag{5.54b}$$

On the basis of Eqs. (5.52a-b) and (5.54a-b), we may define real and imaginary parts of complex bending moments and shear forces in the actual and in the adjoint beam. Assuming that classical, homogeneous boundary conditions are specified for the actual beam, it is easily shown by means of Eqs. (5.31) that the adjoint beam will have the same boundary conditions as the actual beam [89].

According to Eqs. (5.43) and (5.54), the optimality condition (5.44) takes the following form for our example problem,

$$\frac{dJ_v}{dh_v}\left[\lambda_1''(k_{v1}u_1'' - k_{v2}u_2'') + \lambda_2''(k_{v2}u_1'' + k_{v1}u_2'')\right] - \omega^2\rho_v \frac{dA_v}{dh_v}(\lambda_1 u_1 + \lambda_2 u_2) + \mu \frac{dA_v}{dh_v} + \eta + \mathrm{Re}\left(\frac{\partial C}{\partial h_v}\right) = 0 \ . \tag{5.55}$$

Moreover, it is easily verified that Eq. (5.41) is satisfied, if classical boundary conditions are specified for the beam.

Finally, condition (5.47) translates into the following condition in the present case,

$$\eta h_s(x) = 0 \ . \tag{5.56}$$

We have now established the complete set of governing equations for our particular optimization problem. A numerical procedure based on successive iterations is developed in Ref. [89] for the solution of this complicated, strongly coupled and non-linear set of equations. Some numerical results from [89] will now be presented.

In the following examples, the given frequency ω of the loading is smaller than the fundamental transverse vibration frequency of the beam. The external, transverse load acting on the beams is chosen as $p \cos \omega t$, where p is a uniformly distributed or concentrated, force. The beam deflections $u(x, t)$ are then given by $u(x, t) = u_1(x) \cos \omega t - u_2(x) \sin \omega t$, cf. Rqs. (5.17) and (5.18). The load is easily identified in the following figures, and the spatial deflection components $u_1(x)$ and $u_2(x)$ are shown. Hatched areas in the figures indicate viscoelastic material. Zero thickness is allowed for the viscoelastic layers, for which Voight's material model, with $p_1 = 0$ in Eq. (5.1), is used. Note that the latter

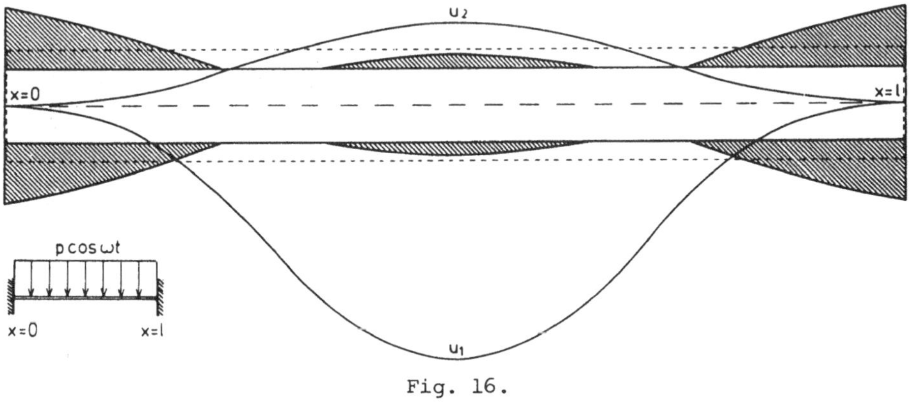

Fig. 16.

does not imply loss of generality, because both k_1 and k_2 given by Eqs. (5.7) will be different from zero in general.

Fig. 16 shows (hatched) the optimal distribution of *viscoelastic cover material* on a uniform, doubly clamped, *elastic* beam. The design is obtained on the basis of functional I_2, Eq. (5.20), i.e. the minimum dynamic compliance criterion is used. The angular frequency of the loading is $\omega = 100$ 1/sec, and the material volume of the cover layers is taken to be one-third of the total beam volume. The material data chosen for the cover layers are $k_{v1} = 0.206 \cdot 10^{12}$ N/m^2, $k_{v2}/k_{v1} = 0.5$ (high dissipation of energy), and for the elastic core, $k_{e1} = k_{v1}$ and $k_{e2} << k_{e1}$. The materials have equal mass densities.

The maximum value of the total deflection amplitude for the optimal beam in Fig. 16 is found to be 64% of the maximum total deflection amplitude for a corresponding, uniform beam (indicated by dashed lines in the figure), where the same amount of cover material is uniformly distributed over the core.

The cantilever beam in Fig. 17 is subjected to a concentrated, harmonically varying force acting at its free end, and the design is obtained on the basis of the functional I_1 defined by Eq. (5.15) with $n = 4$. The same materials as in Fig. 16 are used, but the volume of layer material is one-quarter of the total beam volume, and the angular frequency of the loading is taken to be $\omega = 50$ 1/sek. In this case, the maximum deflection is found to be 63% of the maximum deflection of the corresponding, uniform cantilever.

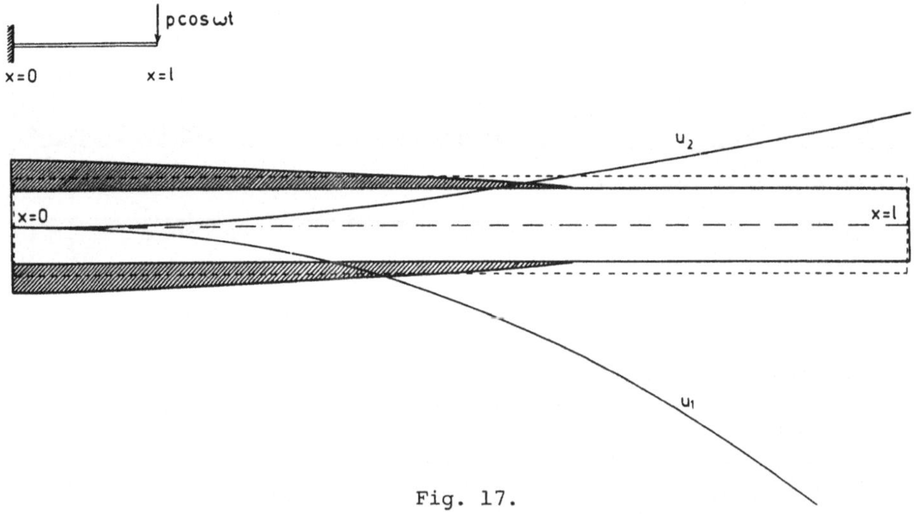

Fig. 17.

SECTION 6

OPTIMAL DESIGN OF SOLID, ELASTIC, AXISYMMETRIC PLATES

6.1 INTRODUCTION

This section gives an account of a new, regularized mathematical formulation of optimal design problems for solid, elastic, axisymmetric plates of prescribed material volume, plate domain and boundary conditions. As a typical design objective, we consider minimization of the compliance (maximization of the stiffness) for given static loading on the plate. This type of problem is, to some extent, less complicated than designing with respect to other objectives (e.g. maximum vibration frequency or buckling load), but it contains all the significant features that are inherent in optimal design of solid, elastic plates.

The new, regularized formulation, which alleviates some anomalies and difficulties encountered earlier in plate optimization problems, is based on a new compound plate model with two simultaneous design variables, namely, variable thickness of a solid part of the plate and variable concentration of a dense system of thin, integral stiffeners attached to the solid plate part. Necessary conditions of optimality are derived and numerical solutions are presented. The results are compared with results obtained from optimal design formulations applied heretofore, and substantiate the superiority of the new, regularized formulation.

The discussion in the following has implications for a number of similar two-dimensional optimization problems. It is based on the paper [94] by Cheng and the author.

6.2 THE PROBLEM

We consider the problem of finding the theoretically best plate model and mathematical formulation for optimal design of axisymmetric, thin, elastic, solid plates, whose thickness h is variable and identifies the distance between the upper and lower plate surface, which are assumed to be disposed symmetrically with respect to the plate mid-plane. The total plate volume is assumed to be specified, and in addition, maximum and minimum constraint values h_{max} and h_{min} for the plate thickness function, pertinent material properties, the inner and outer radii of the plate (which may be annular), and the boundary conditions, are assumed to be given.

For exemplification we consider minimum compliance, i.e. maximum integral stiffness, as the design objective. We adopt a polar coordinate system with origo in the plate centre, and assume for the minimum compliance problem that the distribution of static load $p(r, \theta)$ is of the special type

$$p(r, \theta) = f(r)\cos n\theta , \tag{6.1}$$

where f is a given, θ - independent function, and n a given integer. Eq. (6.1) thus models a rotationally symmetric load distribution for $n = 0$, and for $n \neq 0$, it models a load $p(r, \theta)$ that has the trace $f(r)$ for $\theta = 0$ and varies harmonically with θ in the circumferential direction. Assuming the boundary conditions to be homogeneous, the plate deflection $W(r, \theta)$ then attains the simple form

$$W(r, \theta) = w(r)\cos nt , \tag{6.2}$$

where w is independent of θ.

The assumptions introduced here offer the mathematical simplification that the a priori *partial* differential equations for our problem reduce to *ordinary* differential equations after a separation of variables, which implies that less computer space and -time are required for the numerical solution procedure. Note, however, that these mathematical simplifications do not impede a study of possible formation of integral plate stiffeners; although the plate is rotationally symmetric, it possesses the possibility of increasing its stiffness against circumferential varying loads by forming concentric, circumferential stiffeners that may effectively counteract the circumferential curvatures of the deflection function.

6.3 TRADITIONAL FORMULATION

In the traditional formulation for optimal design, the plate thickness $h(r)$ is used as the design variable, and the optimal solution is assumed to be a solid plate. In dimensionless form, where the inner and outer radii for an annular plate are R_i and 1, respectively, and we have $R_i = 0$ for a full plate, the problem is posed as follows

With $h(r)$ as the design variable, minimize

$$\Pi = \int_{\Omega} f(r)w(r)rdr \tag{6.3a}$$

subject to the constraints

$$\int_{\Omega} h(r)rdr = 1 , \tag{6.3b}$$

$$h_{min} \leq h(r) \leq h_{max} , \tag{6.3c}$$

where

$$r \in \Omega\left\{r \middle| R_i \leq r \leq 1\right\} . \tag{6.3d}$$

Within this formulation, the optimal thickness distribution $h(r)$ is sought in the class of continuous functions or piecewise continuous functions with a *finite* number of discontinuities, and the plate bending rigidity D is assumed to be isotropic, i.e. independent of orientation, and given by

$$D = \frac{Eh^3(r)}{12(1-\nu^2)} , \tag{6.4}$$

where E is Young's modulus and ν is Poisson's ratio of the plate material. The optimization problem has been considered in this form in several papers, see e.g. Refs. [95-100]. However, as has recently been shown by Cheng and the author [101], the optimal thickness distribution will generally not belong to the aforementioned class of functions. As a matter of fact, if the given constraint values h_{max} and h_{min} differ sufficiently from each other, the optimal thickness function will exhibit an *infinite* number of discontinuities (formation of infinitely thin stiffeners) in certain sub-regions of the plate domain Ω. Optimal thickness functions of this type can obviously not be determined on the basis of the traditional formulation, and it is therefore necessary to reformulate the optimization problem.

6.4 NEW, REGULARIZED FORMULATION WITH TWO SIMULTANEOUS DESIGN VARIABLES

In order to be able to determine a possible global optimal design subject to any given set of values of h_{max} and h_{min} for an axisymmetric plate, we follow [94] and expand our design space by constructing a new, generalized plate model. This model covers an axisymmetric, integrally stiffened plate consisting of a solid part of variable thickness $h_s(r)$, $0 < h_{min} \leq h_s \leq h_{max}$ that is equipped with a system of infinitely thin integral stiffeners of variable concentration $\mu(r)$. Fig. 18 shows a radial section through a small ring element of the plate. The element has the radial extent Δr and is equipped with a finite number of stiffeners. Each stiffener is circumferential, has rectangular cross-section of height $h_{max} - h_s$, and is placed symmetrically with respect to the plate mid-plane. The *concentration* $\mu(r)$ (or *density*) of the integral stiffeners is defined by

$$\mu(r) = \lim_{\Delta r \to 0} \frac{\sum_i \Delta c_i}{\Delta r} \quad , \quad 0 \leq \mu(r) \leq 1 \ , \tag{6.5}$$

where Δc_i is the width of the i'th stiffener of the element.

Note that this plate model, for $\mu(r) \equiv 0$ and $h_s \equiv h$, reduces to the solid plate of variable thickness $h(r)$ used in the traditional formulation for optimal design. On the other hand, for $\mu(r) \neq 0$ and $h_s \equiv h_{min}$, the model comprises another special case, namely the integrally stiffened plate model with $\mu(r)$ as the only design variable, which has very recently been used in Refs. [102,103] for determining a number of numerical results that are superior to results obtained by means of the traditional formulation.

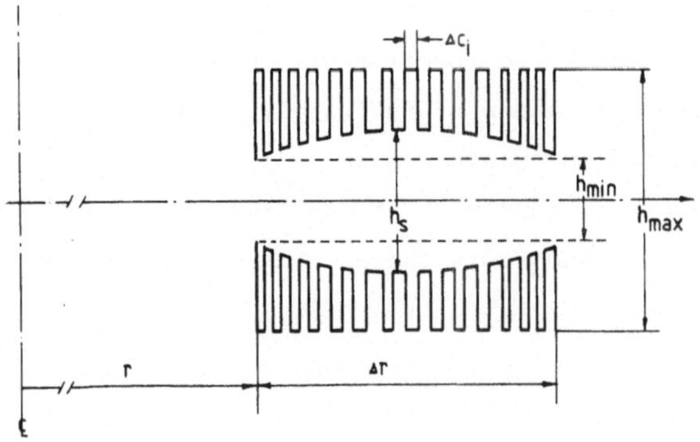

Fig. 18.

While integrally stiffened plate regions are excluded by the former special case plate model ($\mu \equiv 0$, $h_s \equiv h$) , and solid regions of intermediate thickness are excluded by the latter ($\mu \neq 0$, $h_s \equiv h_{min}$) , it will be shown in the following that the new generalized plate model is superior to those applied previously. In fact, it turns out that the optimal plate will, in general, be a compound plate that contains both sub-regions with integral stiffeners and sub-regions that are purely solid and of intermediate thickness.

By means of the new plate model with its two design variables $\mu(r)$ and $h_s(r)$, we now reformulate the plate optimization problem as follows,

With $h_s(r)$ and $\mu(r)$ as design variables, minimize

$$\Pi = \int_\Omega f(r)w(r)rdr \tag{6.6a}$$

subject to the constraints

$$\int_\Omega \Big[h_s(r) + \mu(r)(h_{max} - h_s(r))\Big]rdr = 1 \ , \tag{6.6b}$$

$$h_{min} \le h_s(r) \le h_{max} \ , \tag{6.6c}$$

$$0 \le \mu(r) \le 1 \ , \tag{6.6d}$$

where

$$r \in \Omega\Big\{r \,\Big|\, R_i \le r \le 1\Big\} \ . \tag{6.6e}$$

Now, since the optimal plate is geometrically anisotropic (cylindrically orthotropic) in possible sub-regions with $0 < \mu(r) < 1$, it is necessary to consider the plate bending rigidity as a tensor (i.e. to depend on orientation) in the moment-curvature relationships (Hooke's law) for such sub-regions of the plate. In Refs. [102,103] the components of the bending rigidity tensor are determined via two different paths, and we therefore just cite the results in what follows.

On the basis of Kirchhoff plate theory and with deflections in the form of Eq. (6.2), the θ-independent factorials κ_{rr}, $\kappa_{\theta\theta}$ and $\kappa_{r\theta}$ of the radial, circumferential and twisting curvatures are given by

$$\kappa_{rr} = w'' \ , \quad \kappa_{\theta\theta} = \frac{w'}{r} - \frac{n^2 w}{r^2} \ , \quad \kappa_{r\theta} = -\left(\frac{nw}{r}\right)' \ , \tag{6.7}$$

respectively, where primes denote differentiation with respect to r. The moment-curvature relations read [102]

$$m_{rr} = D_r(\kappa_{rr} + \nu\kappa_{\theta\theta}) \quad , \quad m_{\theta\theta} = D_\theta(\kappa_{\theta\theta} + \nu_r\kappa_{rr}) \quad ,$$

$$m_{r\theta} = D_{r\theta}(1-\nu)\kappa_{r\theta} \tag{6.8a}$$

or in inverse form,

$$\kappa_{rr} = \frac{\nu(m_{rr} - \nu_r m_{\theta\theta})}{\nu_r D_{r\theta}(1-\nu^2)} \quad , \quad \kappa_{\theta\theta} = \frac{m_{\theta\theta} - \nu m_{rr}}{D_{r\theta}(1-\nu^2)} \quad ,$$

$$\kappa_{r\theta} = \frac{m_{r\theta}}{D_{r\theta}(1-\nu)} \quad , \tag{6.8b}$$

where m_{rr} , $m_{\theta\theta}$ and $m_{r\theta}$ are the θ - independent factorials of the radial and circumferential bending moments and the twisting moment, respectively. In Eqs. (6.8a-b), ν is Poisson's ratio of the isotropic, linearly elastic plate material, and the components D_r , $D_{r\theta}$ and D_θ of the plate bending rigidity tensor, together with the symbol ν_r , are given by [102,103]

$$D_r = \frac{D_{max} D_s}{\mu D_s + (1-\mu) D_{max}} \quad , \quad D_{r\theta} = \mu D_{max} + (1-\mu) D_s$$

$$D_\theta = (1-\nu^2) D_{r\theta} + \nu^2 D_r \quad , \quad \nu_r = \nu \frac{D_r}{D_\theta} \quad , \tag{6.9}$$

respectively, where D_{max} and D_s are suitably non-dimensionalized scalar bending rigidities defined by

$$D_{max} = h^3_{max} \quad , \quad D_s = h^3_s \; . \tag{6.10}$$

Eqs. (6.8a-b) express by means of Eqs. (6.9) and (6.10) the moment-curvature relationships in possible cylindrically orthotropic sub-regions of the plate where $0 < \mu(r) < 1$ and $h_s < h_{max}$. However, the above equations reduce precisely to the corresponding well-known relationships for isotropic plates in those sub-regions in which the plate is purely solid, that is, where $\mu = 0$ (purely solid sub-region with plate thickness h_s , $h_{min} \le h_s \le h_{max}$) , or where $\mu = 1$ (purely solid sub-region with plate thickness h_{max}) . Consequently, we can apply Eqs. (6.8) - (6.10) throughout the plate in the following.

6.5 DERIVATION OF NECESSARY CONDITIONS FOR OPTIMALITY

The governing optimality equations for the regularized minimum compliance design problem (6.6a-e) can be derived by variational analysis. The compliance Π in (6.6a) can alternatively be written as [102]

$$\Pi = \int_\Omega \left\{ D_{r\theta}\left[(1-\nu^2)\kappa_{\theta\theta}^2 + 2(1-\nu)\kappa_{r\theta}^2 \right] + D_r(\kappa_{rr} + \nu\kappa_{\theta\theta})^2 \right\} rdr \ , \tag{6.11}$$

and we may use this expression in constructing an augmented functional Π^* defined by

$$\begin{aligned}\Pi^* = & \int_\Omega \left\{ D_{r\theta}\left[(1-\nu^2)\kappa_{\theta\theta}^2 + 2(1-\nu)\kappa_{r\theta}^2 \right] + D_r(\kappa_{rr} + \nu\kappa_{\theta\theta})^2 \right\} rdr \\ & - \Lambda\left\{ \int_\Omega \left[h_s + \mu(h_{max} - h_s) \right] rdr - 1 \right\} - \int_\Omega \lambda\left[h_s - h_{max} + \sigma^2 \right] rdr \\ & - \int_\Omega \beta\left[h_{min} - h_s + \tau^2 \right] rdr - \int_\Omega \gamma\left[\mu - 1 + \xi^2 \right] rdr - \int_\Omega \alpha\left[\eta^2 - \mu \right] rdr \ ,\end{aligned} \tag{6.12}$$

where the constraints (6.6b-d) are adjoined to the functional Π of Eq. (6.11) by means of the Lagrangian multipliers Λ , $\lambda(r)$, $\beta(r)$, $\gamma(r)$ and $\alpha(r)$, and where the real slack-variables $\sigma(r)$, $\tau(r)$, $\xi(r)$ and $\eta(r)$ are introduced for converting the inequality constraints on $h_s(r)$ and $\mu(r)$ to equality constraints.

The necessary condition for stationarity of Π^* with respect to the design variable $\mu(r)$ is now found to be

$$\begin{aligned}(D_{max} - D_s)\left[(1-\nu^2)\kappa_{\theta\theta}^2 + 2(1-\nu)\kappa_{r\theta}^2 \right] + D_r^2\left(\frac{1}{D_s} - \frac{1}{D_{max}} \right)(\kappa_{rr} + \nu\kappa_{\theta\theta})^2 \\ = \Lambda(h_{max} - h_s) + \gamma(r) - \alpha(r) \ ,\end{aligned} \tag{6.13}$$

and the stationarity condition with respect to the design variable $h_s(r)$ becomes

$$\begin{aligned}3h_s^2(1-\mu)\left[(1-\nu^2)\kappa_{\theta\theta}^2 + 2(1-\nu)\kappa_{r\theta}^2 \right] + \frac{3(1-\mu)}{h_s^4} D_r^2(\kappa_{rr} + \nu\kappa_{\theta\theta})^2 \\ = \Lambda(1-\mu) + \lambda(r) - \beta(r) \ .\end{aligned} \tag{6.14}$$

Conditions of stationarity of Π^* with respect to the Lagrangian multipliers Λ , λ , β , γ and α reestablish the plate volume constraint (6.6b) and the maximum and minimum constraints on $h_s(r)$ and $\mu(r)$ in (6.6c-d), and stationarity with respect to the slack variables σ , τ , ξ and η leads to switching conditions which, when combined with the constraints on h_s and μ , may be expressed together with appropriate Kuhn-Tecker conditions as follows,

$$\begin{aligned} &\gamma(r) = 0 \ , \quad \alpha(r) \geq 0 \quad \text{if} \quad \mu(r) = 0 \\ &\gamma(r) = 0 \ , \quad \alpha(r) = 0 \quad \text{if} \quad 0 < \mu(r) < 1 \\ &\gamma(r) \geq 0 \ , \quad \alpha(r) = 0 \quad \text{if} \quad \mu(r) = 0 \end{aligned} \tag{6.15}$$

and

$$\begin{aligned} &\lambda(r) = 0 \ , \quad \beta(r) \geq 0 \quad \text{if} \quad h_s = h_{min} \\ &\lambda(r) = 0 \ , \quad \beta(r) = 0 \quad \text{if} \quad h_{min} < h_s(r) < h_{max} \\ &\lambda(r) \geq 0 \ , \quad \beta(r) = 0 \quad \text{if} \quad h_s = h_{max} \end{aligned} \tag{6.16}$$

Eqs. (6.13) and (6.14) above constitute the two optimality conditions that are associated with the regularized formulation of the problem. The left-hand sides of these equations identify the gradients of the specific strain energy with respect to $\mu(r)$ and $h_s(r)$, respectively, and taking Eqs. (6.15) and (6.16) into account, Eqs. (6.13) and (6.14) show that in sub-regions where one of the design variables $\mu(r)$ and $h_s(r)$ is unconstrained, the gradient of the specific strain energy with respect to the particular design variable, should be constant.

By means of Eqs. (6.13)-(6.16), we are now able to derive specific conditions for the occurrence of sub-regions with integral stiffeners and of purely solid sub-regions of intermediate thickness in the optimal design.

6.5.1 Condition for sub-regions with integral stiffeners. An integrally stiffened sub-region is characterized by

$$0 < \mu(r) < 1 \ , \quad h_{min} \leq h_s(r) < h_{max} \ . \tag{6.17}$$

In view of Ineqs. (6.17) and the second of Eqs. (6.15), Eq. (6.13) reduces to

$$\frac{D_{max} - D_s}{h_{max} - h_s}\left[(1 - \nu^2)\kappa_{\theta\theta}^2 + 2(1 - \nu)\kappa_{r\theta}^2\right]$$

$$+ \frac{D_r^2}{h_{max} - h_s}\left(\frac{1}{D_s} - \frac{1}{D_{max}}\right)(\kappa_{rr} + \nu\kappa_{\theta\theta})^2 = \Lambda \ . \qquad (6.18)$$

Taking Ineqs. (6.17) and the first two of Eqs. (6.16) into account, we may express Eq. (6.14) as

$$3h_s^2\left[(1 - \nu^2)\kappa_{\theta\theta}^2 + 2(1 - \nu)\kappa_{r\theta}^2\right] + \frac{3}{h_s^4} D_r^2(\kappa_{rr} + \nu\kappa_{\theta\theta})^2 \leq \Lambda \ . \qquad (6.19)$$

Eliminating Λ between Eq. (6.18) and Ineq. (6.19), and expressing D_r, D_s and D_{max} in terms of μ, h_s and h_{max} by means of Eqs. (6.10) and the first Eqs. (6.9), we after dividing through by $h_{max} - h_s$ obtain the *necessary condition*

$$(2h_s + h_{max})\left[(1 - \nu^2)\kappa_{\theta\theta}^2 + 2(1 - \nu)\kappa_{r\theta}^2\right] \geq$$

$$\frac{h_{max}^3 h_s^2}{[\mu h_s^3 + (1 - \mu)h_{max}^3]^2}\left(h_s^2 + 2h_s h_{max} + 3h_{max}^2\right)(\kappa_{rr} + \nu\kappa_{\theta\theta})^2 \qquad (6.20)$$

for an integrally stiffened sub-region in the optimal design of a rotationally symmetric plate.

<u>6.5.2 Condition for purely solid sub-regions of intermediate thickness.</u>
A purely solid sub-region of intermediate thickness is associated with

$$\mu(r) = 0 \ , \quad h_{min} < h_s(r) < h_{max} \ . \qquad (6.21)$$

By Eqs. (6.21) and the first of Eqs. (6.15), and noting that $D_r = D_s$ for $\mu = 0$, the optimality condition (6.13) may thus be written as

$$\frac{D_{max} - D_s}{h_{max} - h_s}\ (1 - \nu^2)\kappa_{\theta\theta}^2 + 2(1 - \nu)\kappa_{r\theta}^2$$

$$+ \frac{D_s(D_{max} - D_s)}{D_{max}(h_{max} - h_s)}(\kappa_{rr} + \nu\kappa_{\theta\theta})^2 \leq \Lambda \ , \qquad (6.22)$$

while the optimality condition (6.14) reduces to

$$3h_s^2\left[(1-\nu^2)\kappa_{\theta\theta}^2 + 2(1-\nu)\kappa_{r\theta}^2\right] + 3h_s^2(\kappa_{rr} + \nu\kappa_{\theta\theta})^2 = \Lambda \tag{6.23}$$

since h_s is unconstrained and $\mu \neq 0$.

Combining Eqs. (6.22) and (6.23), using Eqs. (6.10), and dividing through by $h_{max} - h_s$, we obtain the inequality

$$(2h_s + h_{max})\left[(1-\nu^2)\kappa_{\theta\theta}^2 + 2(1-\nu)\kappa_{r\theta}^2\right] \leq$$
$$+ \frac{h_s^2}{h_{max}^3}\left(h_s^2 + 2h_s h_{max} + 3h_{max}^2\right)(\kappa_{rr} + \nu\kappa_{\theta\theta})^2 , \tag{6.24}$$

which constitutes a *necessary condition for a purely solid sub-region of intermediate thickness in the optimal design of a rotationally symmetric plate.*

Unfortunately, it does not seem possible to derive specific conditions for purely solid sub-regions of minimum thickness h_{min} or maximum thickness h_{max} on the present basis.

<u>6.5.3 Behaviour of the optimal design in the vicinity of plate edges</u>. At a simply supported or free plate edge, we have $m_{rr} = 0$, which is equivalent to $\kappa_{rr} + \nu\kappa_{\theta\theta} = 0$ since singular behaviour is excluded via the condition that h_s $(\geq h_{min}) > 0$. With $\kappa_{rr} + \nu\kappa_{\theta\theta} = 0$ and $(1-\nu^2)\kappa_{\theta\theta}^2 + 2(1-\nu)\kappa_{r\theta}^2 >$ 0 in general for a simply supported or free edge, and $2h_s + h_{max} > 0$, it is readily seen that the condition (6.24) is not satisfied.

Hence, *a purely solid sub-region of intermediate thickness will never appear at a simply supported or free edge of an optimally designed rotationally symmetric plate.* The optimal plate will either be integrally stiffened (note that condition (6.20) is satisfied) or solid with minimum or maximum thickness in the vicinity of a simply supported or free edge.

At a clamped edge $r = r^*$ of an axisymmetric plate, we have $w(r^*) = w'(r^*) = 0$, and hence $\kappa_{\theta\theta}(r^*) = \kappa_{r\theta}(r^*) = 0$. In view of the latter conditions, the fact that $h_s^2 + 2h_s h_{max} + 3h_{max}^2 > 0$ and that $(\kappa_{rr}(r^*) + \nu\kappa_{\theta\theta}(r^*))^2 > 0$ in general for a clamped edge, we see that condition (6.20) fails to be satisfied.

Thus, *a sub-region with integral stiffeners will not be found at a clamped edge of an optimally designed rotationally symmetric plate.* In the vicinity of such an edge, the plate will be purely solid, and its edge thickness will belong to the interval $h_{min} \leq h_s \leq h_{max}$.

The above results constitute a generalization of results obtained in [103] for a slightly different problem by means of the Kelley condition from optimal control theory.

It is shown in Ref. [94] that all the results and conclusions of Sections 6.5.1, 6.5.2 and 6.5.3 also hold good for the problem of maximizing a transverse vibration frequency for an axisymmetric, solid, elastic plate of given volume.

6.6 NUMERICAL RESULTS AND DISCUSSION

This section presents some numerical solutions obtained in Ref. [94] to the regularized formulation for minimization of plate compliance. The numerical procedure is outlined in [94]. The solutions are obtained by sub-dividing the plate into 100 ring-elements. In the following the compliance Π of each optimal plate will be stated in proportion to the corresponding compliance Π_u of a purely solid, uniform reference plate of thickness h_u that has the same total volume, plate radii and boundary conditions, and is made of the same material as the optimized plate. Poisson's ratio of the plate material is taken to be $\nu = 0.25$. The uniform reference plate is, of course, subjected to the same static loading as the optimal plate. Although we are able to cope with arbitrary static loading in the form of Eq. (6.1), we take $f(r) = \text{const.}$ in the examples.

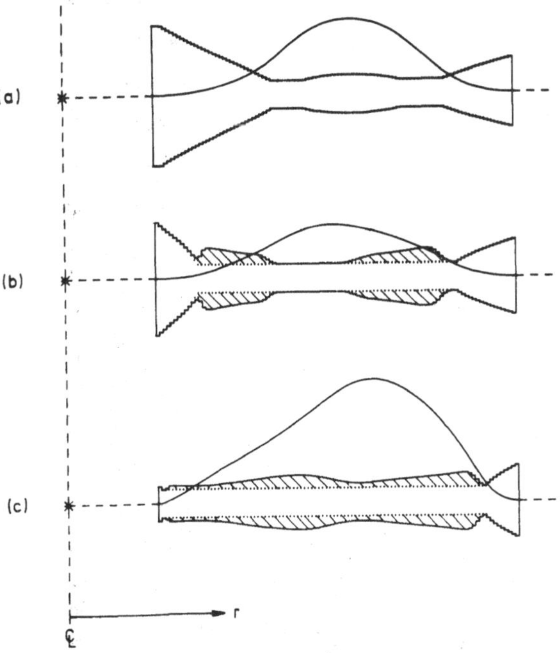

Fig. 19.

Fig. 19 shows radial sections through minimum compliance designs of axisymmetric, annular plates with clamped inner and outer edges. In each design, the unhatched area indicates the solid part of the plate, whose thickness is $h_s(r)$, and hatched areas indicate that integral stiffeners of total height $(h_{max} - h_s)$ are placed symmetrically with respect to the plate mid-plane. The sum of the extents of the upper and lower hatched areas in the normal direction at a specific value of the radial coordinate r , represents $\mu(r) \cdot (h_{max} - h_s(r))$ of the design. This function is plotted to the same scale as $h_s(r)$ in the figures, and illustrates the material consumption of the integral stiffeners by an equivalent thickness of purely solid material. The solid curve shown above each plate is the θ - independent factorial $w(r)$ of the deflection function.

The designs all have $h_u/h_{min} = 1.6579$, $h_{max}/h_{min} = 5$ and $R_i = 0.2$, and they serve to illustrate the influence on the optimal design of the circumferential wave number n of the external loading, Eq. (6.1). The optimal design in Fig. 19a corresponds to $n = 0$, i.e. axisymmetric load, and the designs in Fig. 19b and 19c correspond to $n = 2$ and $n = 4$, respectively, i.e. loads of the form $p(r, \theta) = \text{const} \cdot \cos 2\theta$ and $p(r, \theta) = \text{const} \cdot \cos 4\theta$. The compliances of the optimal designs are found to be $\Pi/\Pi_u = 0.463$, 0.491 and 0.357 , for $n = 0$, 2 and 4 , respectively.

The results in Fig. 19 clearly show that the significance of integral, circumferential stiffeners increases with increasing n . In the case of axisymmetric load, the optimal design, Fig. 19a, is almost entirely a purely solid plate, and only an examination of the numerical data reveals that a small sub-region with very low stiffener concentration is present. However, for $n = 4$ (Fig. 19c), most of the material volume, which is available for design in view of the minimum thickness constraint, is used for formation of stiffeners, and only small, solid sub-regions of intermediate thickness are found near the clamped edges of the plate.

	Π/Π_u		
Boundary conditions at inner and outer plate edge	Traditional formulation ($\mu \equiv 0$) , from [101]	Formulation with $h_s \equiv h_{min}$, from [102]	New, regularized formulation, Ref. [94]
clamped-clamped	0.536	0.415	0.357
simple supp.-clamped	0.564	0.407	0.351
free-clamped	0.617	0.404	0.349

Table 2 . Comparison of minimum compliances Π for axisymmetric annular plates determined via three different formulations for optimal design. The results correspond to $n = 4$, $R_i = 0.2$, $h_u/h_{min} = 1.6579$, $h_{max}/h_{min} = 5$ and $\nu = 0.25$.

The regularized formulation of a *general* plate optimization problem is quite complex and involves *four* functions as simultaneous design variables: the thickness of the solid part of the plate, the concentrations of two mutually orthogonal fields of integral stiffeners, and the orientation of the stiffeners. Results for this type of problem have only been obtained recently, see, e.g., Bendsøe [104,105] and Kohn and Vogelius [106], for elastic plates and Rozvany et al. [107-109] for plastic plates.

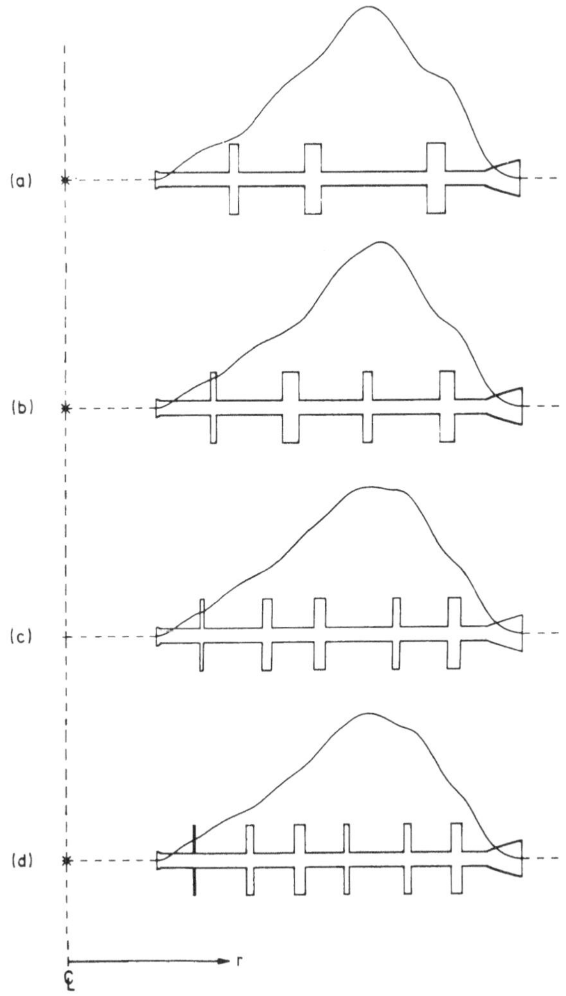

Fig. 20.

The designs obtained in [94] (some of which are presented here) are no doubt global optimal designs within the type of plate topology considered. However, these designs are obtained as solutions to a somewhat idealized mathematical formulation, and it is obvious that they must be regarded as limiting designs from the point of view of practical application.

In practice it is necessary to modify the design, which implies a less optimal value of the performance index. Let us therefore end this section with an assessment of the sensitivity of the current type of optimal designs with respect to their most necessary type of modification, namely lumping the system of infinitely many, infinitely thin integral stiffeners into a finite number of stiffeners of finite width (to meet non-buckling requirements, e.g.). As an example, consider the optimal design of Fig. 19c; sub-divide it into 3, 4, 5 and 6 sections, respectively; lump the infinitely thin stiffeners appropriately (which can be done in different ways), and obtain the series of modified designs shown in Figs. 20a-d. The compliances of these modified designs are given in Table 3 from [94] together with minimum compliance values determined via the two earlier and the new formulation for optimal design. We see that even the design with only three lumped stiffeners has a lower compliance than the minimum value determined via the traditional formulation for

	Π/Π_u	
Traditional formulation ($\mu \equiv 0$)	0.536	from [101]
Formulation with $h_s \equiv h_{min}$	0.415	from [102]
Regularized formulation, optimal design	0.357	Fig. 19c
Modified opt. design, 3 lumped stiffeners	0.444	Fig. 20a
Modified opt. design, 4 lumped stiffeners	0.438	Fig. 20b
Modified opt. design, 5 lumped stiffeners	0.413	Fig. 20c
Modified opt. design, 6 lumped stiffeners	0.401	Fig. 20d

Table 3. Compliances Π of optimized and modified designs of a clamped-clamped annular plate with $n = 4$, $R_i = 0.2$, $h_u/h_{min} = 1.6579$, $h_{max}/h_{min} = 5$ and $\nu = 0.25^1$.

optimal design. For the modified design with six lumped stiffeners, the compliance is already comparatively close to the compliance of the true optimal design. These results indicate clearly that optimal designs determined by means of the new, regularized formulation are rather insensitive with respect to reasonable modification of the type considered.

SECTION 7

REFERENCES

[1] Wasiutyński, Z. & Brandt, A.: Appl. Mech. Rev. 16, 344-350 (1963).

[2] Sheu, C.Y. & Prager, W.: Appl. Mech. Rev. 21, 985-992 (1968).

[3] Prager, W.: AGARD-Rept. No. 589 (1971).

[4] Niordson, F.I. & Pedersen, P.: Proc. 13th Int. Cong. Th. Appl. Mech. Moscow, 264-278, Springer-Verlag (1973).

[5] Rozvany, G.I.N. & Mróz, Z.: Appl. Mech. Rev. 30, 1461-1470 (1977).

[6] Venkayya, V.B.: Int. J. Num. Meth. in Engrg. 13, 203-228 (1978).

[7] Schmit, L.A.: AIAA J. 19, 1249-1263 (1981).

[8] Vanderplaats, G.N.: AIAA J. 20, 992-1000 (1982).

[9] Olhoff, N. & Taylor, J.E.: J. Appl. Mech. 50, 1139-1151 (1983).

[10] Prager, W.: Introduction to Structural Optimization, Springer-Verlag (1974).

[11] Hemp, W.S.: Optimum Structures, Clarendon, Oxford (1973).

[12] Gallagher, R.H. & Zienkiewicz, O.C.: Optimum Structural Design, John Wiley, New York (1973).

[13] Distefano, N.: Non-Linear Processes in Engineering, Academic Press, New York (1974).

[14] Rozvany, G.I.N.: Optimal Design of Flexural Systems, Pergamon, Oxford (1976). Russian version: Strojizdat, Moscow (1980).

[15] Reitman, M.I. & Shapiro, G.B.: Methods of Optimal Design of Deformable Bodies (in Russian), Nauka, Moscow (1976).

[16] Haug, E.J. & Arora, J.S.: Applied Optimal Design, John Wiley, New York (1979).

[17] Grinev, W.B. & Filippov, A.P.: Optimization of Beams Governed by Eigenvalue Problems (in Russian), Naukova Dumka, Kiev (1979).

[18] Haug, E.J. & Cea, J.: Optimization of Distributed Parameter Structures, Sijthoff & Nordhoff, Netherlands (1981).

[19] Kirsch, U.: Optimum Structural Design, McGraw-Hill, New York (1981).

[20] Carmichael, D.G.: Structural Modelling and Optimization, Ellis Horwood Ltd., Chichester (1981).

[21] Troitskij, W.A. & Petuchov, L.W.: Optimal Design of Elastic Bodies (in Russian), Nauka, Moscow (1982).

[22] Lepik, Ü.: Optimal Design of Inelastic Structures Under Dynamic Loading (in Russian; Extended summary in English), Walgus, Tallinn (1982).

[23] Banichuk, N.V.: Problems and Methods of Optimal Structural Design, Plenum Press, New York (1983).

[24] Brandt, A.M.: Criteria and Methods of Structural Optimization, Nijhoff Publishers, The Hague (1984).

[25] Haftka, R.T. & Kamat, M.P.: Elements of Structural Optimization, Nijhoff Publishers, The Hague (1985).

[26] Lurie, K.A.: Applied Optimal Control of Distributed Systems, Plenum Press, New York (to appear).

[27] Lagrange, J.L.: Miscellanea Taurinensia V, 123 (1770-1773).

[28] Clausen, T.: Mélanges Mathématiques et Astronomiques I, 279-294 (1849-1853).

[29] Keller, J.B.: Arch. Rat. Mech. Anal. 5, 275-285 (1960).

[30] Niordson, F.I.: Quart. Appl. Math. 23, 47-53 (1965).

[31] Ashley, H. & McIntosh, S.C. Jr.: Proc. 12th Int. Cong. Th. Appl. Mech. (Ed. M. Hetényi & W.G. Vincenti), 100-113, Stanford, Springer-Verlag (1969).

[32] Pierson, B.L.: Int. J. Num. Meth. Engrg. 4, 491-499 (1972).

[33] Reitman, M.I. & Shapiro, G.S.: All-Union Symp. "On the Problems of Optimization in Mech. of Solid Deformable Bodies", Vilnius, USSR, June (1974).

[34] Rao, S.S.: Shock Vib. Dig. 7, 61-70 (1975).

[35] Olhoff, N.: Shock Vib. Dig. 8, No. 8, 3-10 (Part I: Theory), 8, No. 9, 3-10 (Part II: Applications) (1976).

[36] Olhoff, N.: Proc. 15th Int. Cong. Th. Appl. Mech., Toronto, Canada, 133-149, North-Holland (1980).

[37] Keller, J.B. & Niordson, F.I.: J. Math. Mech. 16, 433-446 (1966).

[38] Taylor, J.E.: AIAA J. 6, 1379-1381 (1968).

[39] Masur, E.F.: J. Engrg. Mech. Div., ASCE, 96, 621-640 (1970).

[40] Prager, W. & Taylor, J.E.: J. Appl. Mech. 35, 102-106 (1968).

[41] Brach, R.M.: J. Opt. Th. & Appl. 11, 662-667 (1973).

[42] Taylor, J.E.: AIAA J. 5, 1911-1913 (1967).

[43] Vavrick, D.J. & Warner, W.H.: J. Struct. Mech. 6, 233-246 (1978).

[44] Seiranyan, A.P.: Int. J. Solids Struct. 15, 749-759 (1979).

[45] Olhoff, N. & Niordson, F.I.: ZAMM 59, T16-T26 (1979).

[46] Karihaloo, B.L. & Niordson, F.I.: J. Opt. Th. & Appl. 11, 638-654 (1973).

[47] Olhoff, N.: J. Struct. Mech. 4, 87-122 (1976).

[48] Turner, M.J.: AIAA J. 5, 406-412 (1967).

[49] Olhoff, N. & Taylor, J.E.: J. Opt. Th. & Appl. 27, 571-582 (1979).

[50] Taylor, J.E. & Liu, C.Y.: AIAA J. 6, 1497-1502 (1968).

[51] Tadjbakhsh, I. & Keller, J.B.: J. Appl. Mech. 29, 159-164 (1962).

[52] Masur, E.F.: J. Opt. Th. & Appl. 15, 69-84 (1975).

[53] Popelar, C.H.: J. Struct. Mech. 5, 45-66 (1977).

[54] Olhoff, N. & Rasmussen, S.H.: Int. J. Solids Struct. 13, 605-614 (1977).

[55] Brach, R.M.: Int. J. Solids Struct. 4, 667-674 (1968).

[56] Karihaloo, B.L. & Niordson, F.I.: Arch. of Mech. 24, 1029-1037 (1972).

[57] Huang, N.C. & Sheu, C.Y.: J. Appl. Mech. 35, 285-288 (1968).

[58] Frauenthal, J.C.: J. Struct. Mech. 1, 79-89 (1972).

[59] Adali, S.: Int. J. Solids Struct. 15, 935-949 (1979).

[60] Simitses, G.J., Kamat, M.P. & Smith, C.V. Jr.: AIAA J. 11, 1231-1232 (1973).

[61] Rasmussen, S.H.: J. Struct. Mech. 4, 307-320 (1976).

[62] Olhoff, N. & Taylor, J.E.: J. Struct. Mech. 6, 367-382 (1978).

[63] Taylor, J.E.: J. Appl. Mech. 34, 486-487 (1967).

[64] Banichuk, N.V. & Karihaloo, B.L.: Int. J. Solids Struct. 13, 725-733 (1977).

[65] Banichuk, N.V.: MTT (Mechanics of Solids), 9, 150-154 (1974).

[66] Mróz, Z. & Rozvany, G.I.N.: J. Struct. Mech. 5, 279-290 (1977).

[67] Szelag, D. & Mróz, Z.: ZAMM 58, 501-510 (1978).

[68] Masur, E.F.: Proc. IUTAM Symp. Optimization in Struct. Design (Eds. A. Sawczuk & Z. Mróz), 441-453, Warsaw, Springer-Verlag (1975).

[69] Bendsøe, M.P., Olhoff, N. & Taylor, J.E.: J. Struct. Mech. 11, 523-544 (1983-84).

[70] Taylor, J.E. & Bendsøe, M.P.: Int. J. Solids Struct. 20, 301-314 (1984).

[71] Masur, E.F. & Mróz, Z.: Proc. IUTAM Symp. on Variational Methods in the Mech. of Solids (Ed. S. Nemat-Nasser), Northwestern Univ., USA, Pergamon Press (1980)

[72] Masur, E.F.: Int. J. Solids Struct. 20, 211-231 (1984).

[73] Prager, S. & Prager, W.: Int. J. Mech. Sci. 21, 249-251 (1979).

[74] Gajewski, A.: Int. J. Mech. Sci. 23, 11-16 (1981).

[75] Blachut, J. & Gajewski, A.: Int. J. Solids Struct. 17, 653-667 (1981).

[76] Olhoff, N. & Plaut, R.H.: Int. J. Solids Struct. 19, 553-570 (1983).

[77] Rezaie-Keyvan, N. & Masur, E.F.: J. Struct. Mech. 13, 181-200 (1985).

[78] Olhoff, N.: J. Struct. Mech. 5, 107-134 (1977).

[79] Kamat, M.P. & Simitses, G.J.: Int. J. Solids Struct. 9, 415-429 (1973).

[80] Vepa, K.: Quart. Appl. Mech. 31, 329-341 (1973/74).

[81] Seiranyan, A.P.: MTT (Mechanics of Solids), 11, 147-152 (1976).

[82] Masur, E.F.: Proc. IUTAM Symp. on Optimization in Struct. Design, (Eds. A. Sawczuk & Z. Mróz), 99-102, Warsaw, Springer-Verlag (1975).

[83] Olhoff, N. & Parbery, R.: Int. J. Solids Struct. 20, 63-75 (1984).

[84] Bendsøe, M.P. & Olhoff, N.: Optimal Control Appl. & Meth. 6, 191-200 (1985).

[85] Troitskii, V.A.: MTT (Mechanics of Solids), 11, 145-152 (1976).

[86] Mróz, Z. & Rozvany, G.I.N.: J. Opt. Th. & Appl. 15, 85-101 (1975).

[87] Rozvany, G.I.N.: J. Struct. Mech. 3, 359-402 (1974/75).

[88] Pierson, B.L.: J. Struct. Mech. 5, 147-178 (1977).

[89] Lekszycki, T. & Olhoff, N.: Struct. Mech. 9, 363-387 (1981).

[90] Mróz, Z.: ZAMM 50, 303-309 (1970).

[91] Icerman, L.J.: Int. J. Solids Struct. 5, 473-489 (1969).

[92] Johnson, E.H., Rizzi, P., Ashley, H. & Segenreich, S.A.: AIAA J. 14, 1690-1698 (1976).

[93] Mróz, Z.: Proc. IUTAM Symp. on Struct. Control (Ed. H.H.E. Leipholz) Univ. of Waterloo, Torinto, Canada, North-Holland (1980).

[94] Cheng, K.-T. & Olhoff, N.: Int. J. Solids Struct. 18, 153-169 (1982).

[95] Banichuk, N.V. : MTT (Mechanics of Solids), 10, 180-188 (1975).

[96] Banichuk, N.V., Kartvelishvili, V.M. & Mironov, A.A.: MTT (Mechanics of Solids), 12, 68-78 (1977).

[97] Banichuk, N.V., Kartvelishvili, V.M. & Mironov, A.A.: MTT (Mechanics of Solids) 13, 124-131 (1978).

[98] Gura, N.M. & Seiranyan, A.P.: MTT (Mechanics of Solids), 12, 138-145 (1977).

[99] Armand, J.-L.: First Int. Conf. on Comp. Meth. in Nonlinear Mech., Austin, Texas, USA, Sept. (1974).

[100] Armand, J.-L. & Lodier, B.: Int. J. Num. Meth. Engrg. 13, 373-384 (1978).

[101] Cheng, K.-T. & Olhoff, N.: Int. J. Solids Struct. 17, 305-323 (1981).

[102] Cheng, K.-T.: Int. J. Solids Struct. 17, 795-810 (1981).

[103] Olhoff, N., Lurie, K.A., Cherkaev, A.V. & Fedorov, A.V.: Int. J. Solids Struct. 17, 931-948 (1981).

[104] Bendsøe, M.P.: Proc. Int. Symp. Optimum Structural Design, Oct. 1981, Univ. Ariz., Tucson, Arizona, pp. 13-29 to 13-34, Pineridge Press, Swansea (1982).

[105] Bendsøe, M.P.: Proc. Workshop on Homogenization and Effective Moduli of Materials and Media, Oct. 1984, Inst. for Mathematics and its Applications, Univ. of Minnesota.

[106] Kohn, R.V. & Vogelius, M.: Int. J. Solids Struct. 20, 333-350 (1984).

[107] Rozvany, G.I.N., Olhoff, N., Cheng, K.-T. & Taylor, J.E.: DCAMM Rept. No. 212, June 1981. J. Struct. Mech. 10, 1-32 (1982).

[108] Wang, C.-M., Rozvany, G.I.N. & Olhoff, N.: Computers and Struct. 18, 653-665 (1984).

[109] Wang, C.-M., Thevendran, V., Rozvany, G.I.N. & Olhoff, N.: Computers and Struct. 22, 519-528 (1986).

MINIMUM-WEIGHT PLATE DESIGN VIA PRAGER'S LAYOUT THEORY
(PRAGER MEMORIAL LECTURE)

G. I. N. Rozvany
FB.10 Essen University
4300 Essen 1,
West Germany.

Ong, T.-G.
Dept. of Civil Engrg.
Monash University,
Clayton, Vic. 3168, Australia.

During the last decade of his immensely creative life, Professor William Prager's research was directed at two central objectives, the derivation of a comprehensive set of *static-kinematic optimality criteria* and the development of an *optimal layout theory*. As the late Professor Prager's closest former associate, the first author will review briefly these fields in the first part of this memorial lecture.

Prager's intellectual heritage has found a number of useful applications since his tragic death in 1980; the most recent of these, a new approach to minimum-weight plate design, will be discussed in the second part of the lecture.

Static-Kinematic Optimality Criteria

This powerful approach to structural optimization was introduced by Prager and Shield [1] and extended considerably by the first author's research group (e.g. [2-4]). In formulating a *distributed parameter problem*, the Lagrangian function associated with the equilibrium condition can be regarded as a fictitious ("Pragerian" or adjoint) deflection field and then the Euler-Lagrange equations are interpreted as *generalised strain-stress relationships*. The latter, together with static and kinematic admissibility, constitute a necessary and sufficient condition of optimality (subject to existence) for convex problems and a necessary one for non-convex problems. Static-kinematic optimality criteria convert, in effect, a problem of structural optimization into a problem of (non-linear) structural analysis. While this analogy does not change the problem from a mathematical point of view, it provides a very useful insight into the nature of optimal solutions since for engineers it is easier to visualize (and anticipate) deflection or strain fields than abstract mathematical entities. Moreover, the above analogy enables us

NATO ASI Series, Vol. F27
Computer Aided Optimal Design: Structural and Mechanical Systems. Edited by C. A. Mota Soares

to employ in optimization existing numerical and analytical techniques for structural analysis, e.g. energy methods and finite element packages.

In applying the above approach to the optimal plastic design of structures with continuously varying cross sections [1,2], for example, we must find a statically admissible generalised stress field and a kinematically admissible generalised strain field such that the generalised strain components are given by the "G-gradients" [2,3,12] of the specific cost function with respect to the generalised stress components. The *specific cost function* is the relationship between the "cost" per unit length, area or volume and the generalised stresses (or stiffnesses) and its integral over the structural domain is the "total cost" which is to be minimised subject to design constraints. The *G-gradients* are simply *partial first derivatives* for differentiable functions and subgradients (having non-unique values given by a convex combination of the limiting gradients at discontinuities of the derivatives) for piece-wise differentiable functions. For discontinuous cost functions, the G-gradients contain *generalised functions* (e.g. Dirac distributions or impulses), see [3].

In the seventies, the Prager-Shield condition [1] for *optimal plastic design* was extended to discontinuous cost functions, multiple load conditions, allowance for the cost of reactions or unspecified forces (of non-zero cost) and connections (joints), prescribed cost distribution (segmentation), optimization of location of supports and segment boundaries and allowance for body forces (selfweight) [2,3,19]. Following a suggestion by Niordson, more recent optimality criteria [5] can handle limits on the spatial gradient of the cross-section distribution (="taper").

In extending static-kinematic optimality criteria to *optimal elastic design* [3], deflection, stress, buckling and natural frequency constraints [6], with additional optimization of support conditions or segmentation [7] and "Niordson-constraints" [8] have been considered. In optimal elastic design, two kinematically admissible deflection fields must be considered simultaneously: the elastic deflections caused by the external loads and the "Pragerian" (adjoint) displacement field which is the Lagrangian for the equilibrium condition. For prescribed deflection constraints, the latter becomes the elastic deflection field caused by "unit dummy loads" at the prescribed displacements.

An Illustrative Example: Optimal Plastic Design of a Beam with One Movable Support

Considering a beam (Fig. 1a) having a rectangular cross-section of given constant depth d, a variable width b and a yield stress s_y, the plastic moment capacity becomes $M_p = s_y bd^2/4$ and the cross-sectional area ψ = bd. Since in plastic design the yield condition $|M^s| \le M_p$ must be fulfilled where M^s denotes statically admissible bending moments, we obtain the specific cost function $\psi = k|M^s|$ with $k = 4/s_y d$ if we assume that all cross-sections are at yield in the optimal design ($|M^s| = M_p$). In this problem the generalised strain corresponding to the bending moment M is the beam curvature $\kappa = -du^2/dx^2$ where u is "Pragerian" (adjoint) beam deflection and x is the distance along the beam axis. The optimal kinematic admissible curvatures κ^k are given by the first derivatives of the specific cost function, $\kappa^k = G(\psi) = d\psi/dM^s$ with respect to the statically admissible moments. As the specific cost function ψ is non-differentiable at $M^s = 0$, the G-gradient becomes the subgradient, having a non-unique value of $-k \le \kappa = G(\psi) \le k$. The above specific cost function and the corresponding optimal curvature value are given in Fig.1g.

In addition to the cross-sections, the location of support B (Fig. 1a) is to be optimized. For the optimal location of a support of zero cost we have the optimality condition (e.g. [2], p.129), $u_B = u'_B = 0$ where $u'_B = du/dx|_B$.

A class of statically admissible moment fields M^s is shown in Fig. 1b and the corresponding "Pragerian" deflection field a in Fig. 1c. Kinematic admissibility of the latter (i.e. $u_E = 0$) implies $u_E = 0 = \int_B^C \kappa(x_E - x)dx = -kd(3L+a-d/2)+k(3L+a-d)^2/2 = 0$ or $d = (3L+a)(1-1/\sqrt{2})$. Similarly, $u_A = 0$

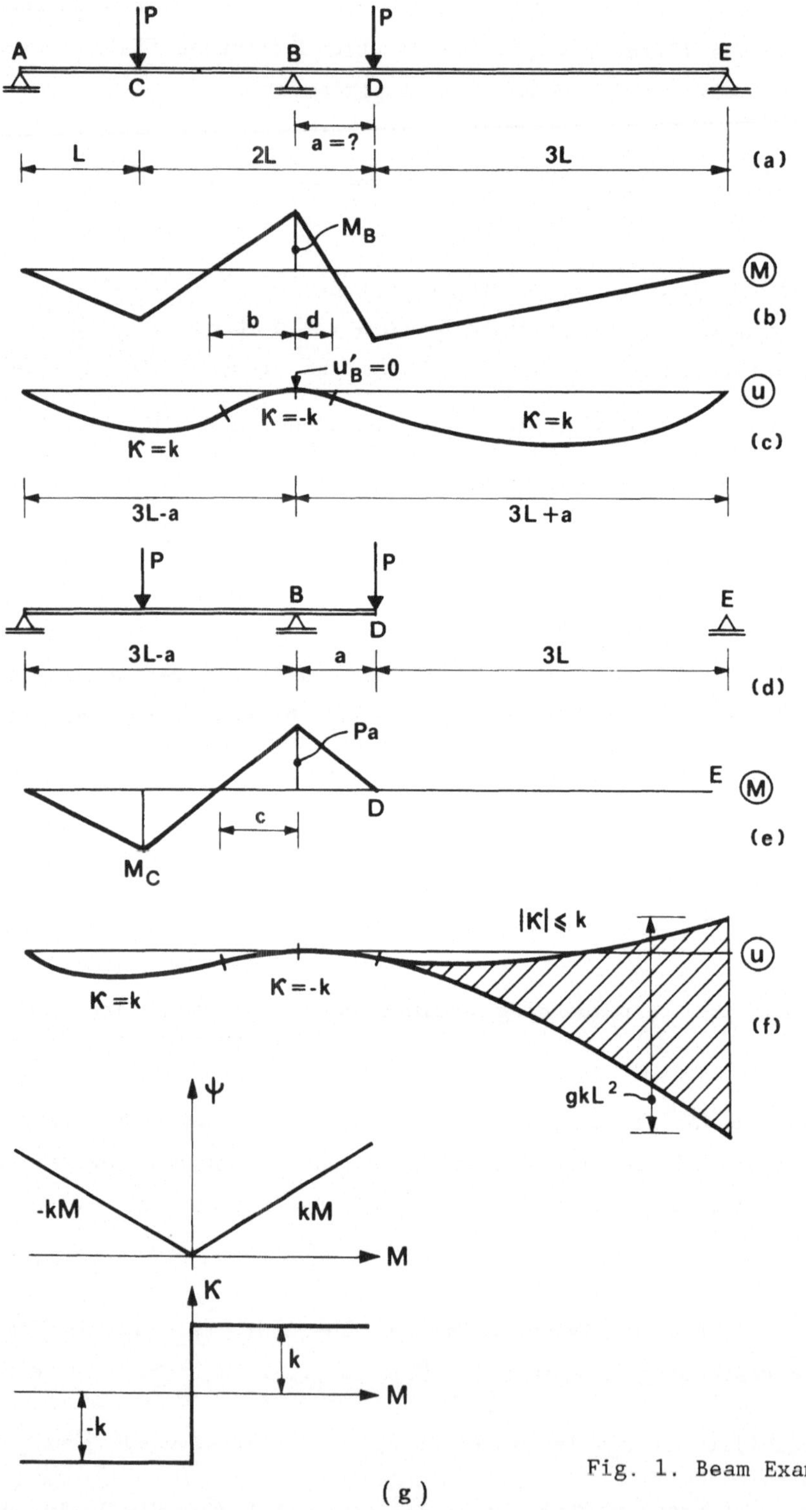

Fig. 1. Beam Example.

furnishes $b = (3L-a)(1-1/\sqrt{2})$ implying $b/d = (3L-a)/(3L+a)$. Moreover, statical considerations furnish $b = M_B/V_B^-$, $d = M_B/V_B^+$ where the shear forces to the left and right of the support B are $V_B^- = (M_B+PL)/(3L-a)$, $V_B^+ = (M_B+3PL)/(3L+a)$. Combining the kinematic and static requirements, we have $b = M_B(3L-a)/(M_B+pL) = M_B(3L-a)/(M_B+3pL)$ which is clearly infeasible. This means that the above class of solutions is non-optimal.

Assuming now that the cross-sectional area and moments between points D and E are zero (Fig. 1d and 1e), the Pragerian deflection field is shown in Fig. 1f. Static and kinematic consideration furnish again $c = Pa/V_B^-$ $= (3L-a)a/(L+a)$, $c = (3L-a)(1-1/\sqrt{2})$ yielding $a = L(\sqrt{2}-1)$. It is still necessary to show that the above solution admits $u_E = 0$. Since for $M = 0$ the curvature value is non-unique (Fig. 1g) it is easy to show that all u_E-values between $kL^2(2\sqrt{2}-6)$ and $kL^2(3+2\sqrt{2})$ are admissible (the non-unique Pragerian deflection field is represented by the shaded area in Fig. 1f). This confirms the optimality of the above solution (Figs. 1d and 1e).

Optimal Layout Theory

This theory [4, 9-14] is based on two underlying concepts, the above mentioned static-kinematic optimality criteria and the *structural universe*. The latter consist of all potential (or "feasible" or "candidate") members (centroidal axes or middle surfaces). Since a static-kinematic optimality condition gives a (usually non-unique) strain requirement also for vanishing members, its fulfilment for the entire structural universe constitutes a necessary and sufficient condition of optimality for convex specific cost functions (with linear subsidiary conditions). Other formulations of the same problem involving unknown geometrical parameters usually result in non-convexity (non-uniqueness).

The original (now termed *classical)* layout theory [2, 9-13], a generalisation of Michell's pioneering effort at the turn of the century [15], was used for low-density, grid-like structures in which at a given point members may run in any number of directions (Fig. 2a) but the effect of member intersections on both cost and strength (or stiffness) is neglected. Consequently the specific cost ψ is a sum of functions each of which depends only on the strength (e.g. moment capacity) or stiffness of an individual member: $\psi = \psi_1(s_1) + \ldots + \psi_n(s_n)$.

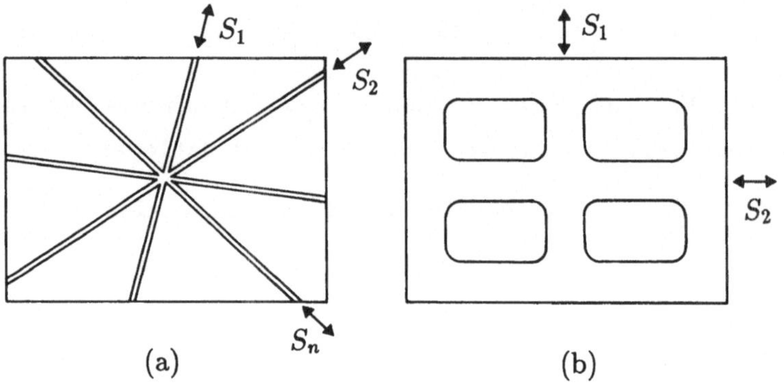

Fig. 2. Classical and advanced layout theory.

The development of *advanced* layout theory was prompted by the discoveries that optimized continua develop an infinite number of internal bounderies [16] and that least-weight plates contain a dense system of ribs [17]. Considering an elastic (or plastic) continuum, the microstructure is first optimized locally by minimizing the material volume ψ per unit area (or volume) for given stiffness (or strength) parameters s_1, s_2 (and possibly s_3) in the *principal* directions (Fig. 2b). It follows that the specific cost function $\psi(s_1, s_2)$ is a non-separable function of both (or all three) stiffnesses (or strength parameters). Advanced layout theory results in substantial extra savings if a high proportion of the feasible space is occupied by structural material but the optimal solution given by this theory tends to that of the "classical" layout theory if the material volume/feasible volume ratio tends to zero.

Classical layout theory has been applied to least-weight trusses or Michell frames (e.g. [9]) but it has been particularly successful in optimizing least-weight grillages (beam systems). This development was regarded as particularly important by Prager for the following reasons [10]:

(a) Grillages constitute the first class or truly two-dimensional optimization problems for which closed form analytical solutions are available for almost all possible boundary and loading conditions.

(b) Optimal grillages are more practical than Michell structures [15] because the latter are subject to instability which is ignored in their formulation.

(c) The optimal rib layout of least-weight plates has been found similar to that of minimum weight grillages.

(d) A computer algorithm (22,23) is available for generating analytically and plotting optimal beam layouts for a wide range boundary conditions.

Extensive reviews of the grillage theory [2, 9-14] show that analytical solutions are now available for any combination of simply supported and clamped boundaries, free edges and beam supported edges. Additional refinements include non-uniform depth, allowance for cost of supports, bending and shear dependent cost, upper constraint on beam density [18], partial discretization and allowance for selfweight (also for moment and shear dependent cost) [19,20]. Although it was shown earlier that plastically designed least-weight solutions are also valid for elastic grillages with prescribed stress, compliance or natural frequency [2,6], optimal grillage layouts for a prescribed elastic deflection were derived more recently [21].

Another application of the classical layout theory concerns archgrids (now termed "Prager structures") (24-27). A Prager structure can be defined as a surface structure consisting of intersecting arches or cables for which both the middle surface and member layout are optimized. Alternatively, a Prager-structure can be regarded as a special class of Michell space-frames for which (a) either the compressive or the tensile permissible stress tends to zero and (b) the vertical position of loads is unspecified and is to be optimized. This special class of Michell frames always reduces to a surface structure in 3D space (or a line structure in a plane). Closed form analytical solutions are now available for *any* axisymmetric load in space and for any parallel load system in a plane and also for additional selfweight. An interesting feature of Prager structures is the fact that the total structural weight is always proportional to the sum of the products of the loads and their elevations.

Optimal Plastic Design of Solid Plates

It was known already in the sixties that the weight minimization of solid plates without an upper constraint on the thickness is not a well-posed problem because the weight can be reduced to an arbitrarily small value by using a system of sufficiently high ribs [30,28]. More recently, Olhoff and Cheng [17] showed numerically that even with an upper constraint stiffener-like formations appear in the solution and their number increases with the number of finite elements used. Prager pointed out shortly before his death that the *layout* of such stiffeners is similar to that of least-weight grillages for the same boundary and loading conditions, obtained about a decade earlier by the first author and later confirmed by Prager (e.g. [19]). Subsequently the first author, Olhoff, Cheng, Taylor and Wang [28,29] obtained exact analytical minimum-weight solutions for plastically designed solid plates with a constraint on the maximum thickness. It was first established that considering given principal moments for a minimum weight solution ribs always run in the principal directions. Studying then the *local optimization* of solid Tresca-plates of variable thickness with ribs of variable depth and density in the principal directions, it was found that within a maximum thickness constraint and depending on the given values of the principal moments the solution may only consist of the following two types of regions [28]:

(a) a dense system of ribs in two principal directions with a plate of zero thickness in between the ribs; or

(b) ribs in one principal direction only and a solid plate whose moment capacity equals the smaller principal moment (M_2).

The above formulation has taken the weight saving at rib intersections into consideration when both *principal moments have the same sign*. For *principal moments of differing signs*, however, the Tresca condition requires a local widening of the ribs adjacent to rib intersections. It was shown that the weight of this extra widening approximately offsets the weight saving at the rib intersections. It was therefore assumed that in the latter (less frequent) case the effect of rib intersections on the structural weight can be ignored.

On the basis of the above optimal regions, a specific cost function $\psi(M_1, M_2)$ was established [28] and optimal solutions were obtained using the Prager-Shield condition [1] and optimal layout theory [4, 9-14].

It was shown subsequently by the first author, Wang and Olhoff [29] that the above optimal regions can be restricted further if, in addition to optimality criteria, static-kinematic admissibility is also taken into consideration. The latter investigation has established that in optimal region (a) above ribs can only run in one direction (with zero principal moment in the second direction, $M_2 = 0$) and in region (b) with non-maximum thickness the two principal moments must be equal $(M_1 = M_2)$ giving a plate without ribs. However, in minimum-weight solutions plates of maximum thickness may also occur with $M_1 = M_0$, $M_2 = M_0$ or $M_1 - M_2 = M_0$ (sgn $M_1 \neq$ sgn M_2) where M_0 is the maximum uniaxial moment capacity of a plate.

Optimal Elastic Design of Perforated and Composite Plates

A "perforated" plate may only have two thicknesses: a prescribed maximum thickness or zero thickness. The latter occurs over "perforations". Several mathematical studies (e.g. [31, 32]) came to the conclusion that elastic least-weight solutions for perforated disks and plates of given compliance consist of regions with two sets of intersecting ribs in the principal directions: one such set has a first order infinitesimal spacing and the other one a second order infinitesimal spacing.

A more recent investigation by the authors, Olhoff, Bendsoe, Szeto and Sandler [33] pointed out that on the basis of Saint Venant's principle the first/second order microstructure can only be optimal at lower rib densities. The above argument was followed up by detailed finite element/finite difference investigations by the first authors' research associates (Ong and Szeto at Monash University as well as Menkenhagen and Booz at Essen University) which exhibited a complete agreement of

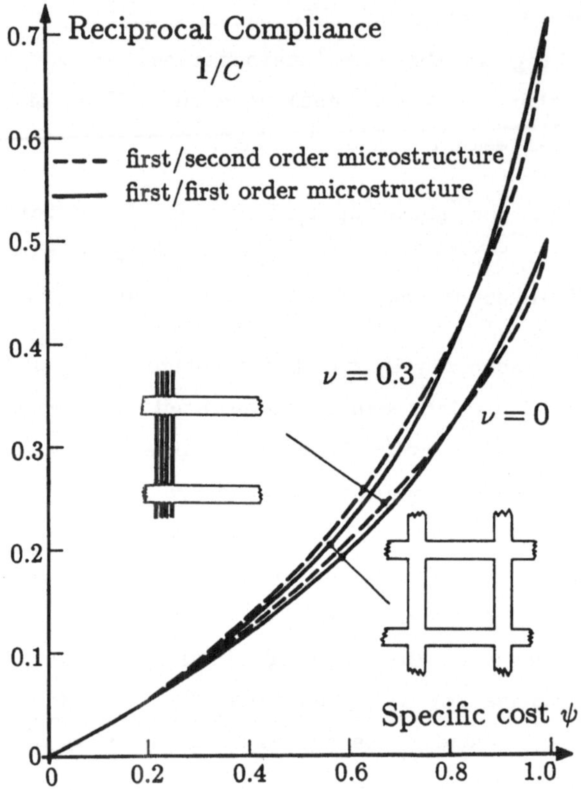

Fig. 3. Specific cost of various microstructures for perforated disks and plates

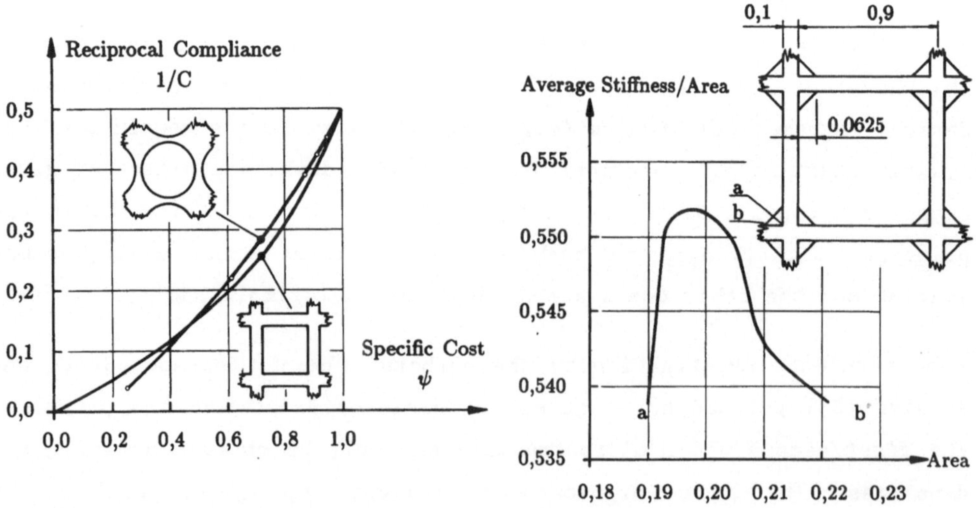

Figs. 4 and 5. Improvements on the prismatic first/first order microstructure.

various numerical solutions and with an analytical solution for first/second order microstructures. These investigations show (Fig. 3) that at higher rib densities a first/first order prismatic microstructure is more economical than a first second order one and hence the latter can not be optimal. Further preliminary finite element analyses by Booz and Menkenhagen have indicated that the prismatic first/first order microstructure can be improved further by using either corner fillets or circular holes (Figs. 4 and 5). Nevertheless a number of complete analytical solutions for axisymmetric plates have been determined [33] within the constraint that the microstructure is of first/second order (as suggested in [31, 32]). The specific cost function for this microstructure [33] is $\psi(s_1, s_2) = (s_1 - s_1 s_2 + s_2)/(1 - s_1 s_2)$ with $s_1 = d_1$, $s_2 = d_2/(1 + d_1 d_2 - d_1)$ where d_1 and d_2 are the first and second order rib densities, s_1 and s_2 are the specific stiffnesses for $\nu = 0$ where ν is Poisson's ratio (see Fig. 6a). For $\nu \neq 0$ the parameters s_1 and s_2 do *not* denote stiffnesses but for both cases the specific compliance is $c = M_1{}^2/s_1 + M_2{}^2/s_2 - 2\nu M_1 M_2$. For composite plates (in which the perforations are filled with a material of *lower stiffness and cost,* the respective rations for the two materials being $\alpha \leq 1$ and $\beta \leq 1$) the stiffnesses become $s_1 = d_1 + (1-d_1)/[d_2 - (1-d_2)/\alpha]$, $s_2 = [d_2 + \alpha(1-d_2)]/[1-d_1 + d_1 d_2 + \alpha d_1(1-d_2)]$ and the specific cost $\Psi = d_1 + \beta(1-d_1) + d_2(1-d_1)(1-\beta)$, see Fig. 6b for $\beta = 0.2$.

Using optimal layout theory for axisymmetric *perforated plates*, it was established that the optimum solution consists of (a) unperforated regions and (b) regions consisting of radial ribs only. On the basis of the above findings, analytical optimal solutions were obtained for simply supported and clamped circular plates with uniformly distributed full and partial loading as well as a central point load. The last case is particularly interesting because for a single point load the compliance constraint also implies a prescribed deflection constraint.

The (nondimensional) total weight Φ of various intuitive designs, for a circular perforated plate with $\nu = 0$ optimized within an assumed topography, is compared in Fig. 7 in which C is the total (nondimensional) compliance. It can be seen that, as predicted by the optimal layout theory, design D is optimal for all C values.

The optimal solution for clamped circular perforated plates may consist of either one or two unperforated regions and a region with radial ribs. The range of validity of these solutions for various ν-values is shown in Fig. 8.

It has also been found that for axisymmetric *composite plates* the solution may consist of (a) regions consisting entirely of the stiffer material (cross-hatched in Fig. 9); (b) regions consisting entirely of less stiff material (hatched in Fig. 9); and (c) regions with radial ribs made out of the stiffer material and the gaps filled with the less stiff material (unhatched in Fig. 9 which gives optimal region boundaries for $\alpha = \beta = 0.2$). Using again static/kinematic optimality criteria, optimal solutions for simply supported circular composite plates were determined by a semi- analytical method.

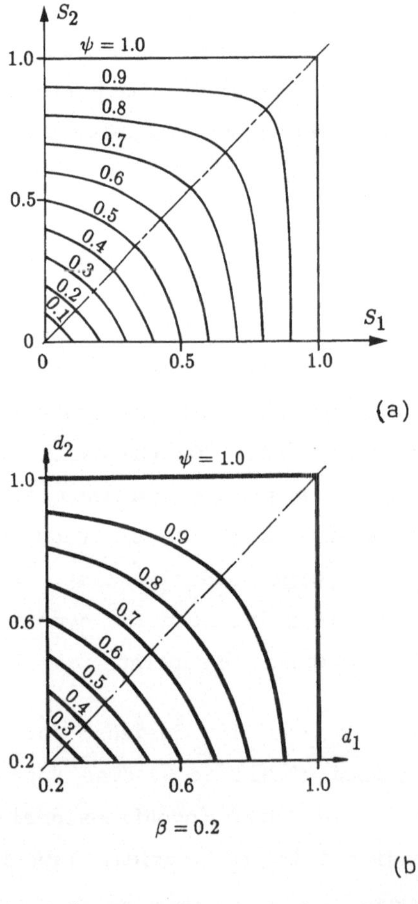

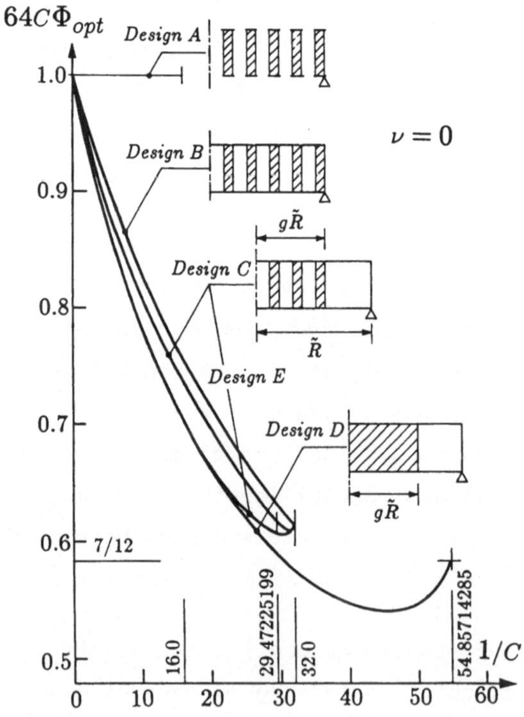

Fig. 6. Specific cost function for perforated and composite plates.

Fig. 7. Comparison of intuitive solutions for circular perforated plates.

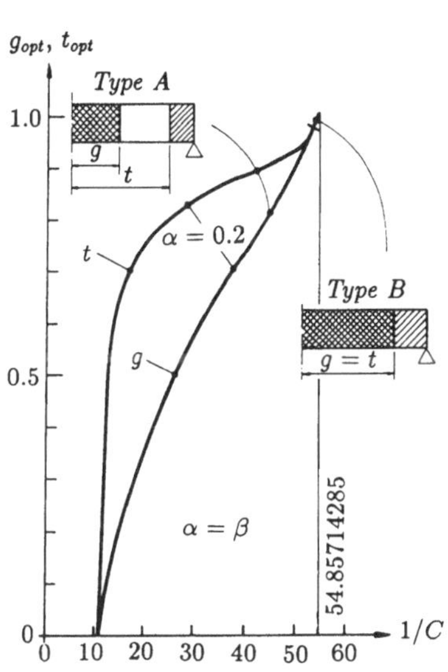

Fig. 8. Types of optimal solutions for clamped circular plates.

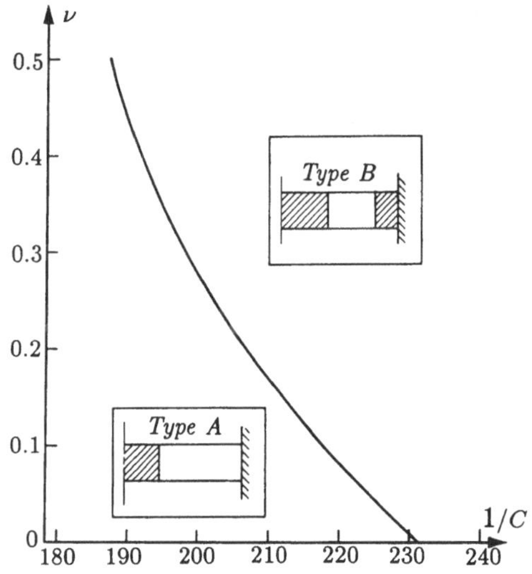

Fig. 9. Types of optimal solutions for composite circular plates.

Optimal Plastic Design of Perforated Plates

Extending the study of plastic solid plates [28, 29] to perforated plates, it has been found for plastic perforated axisymmetric plates the least-weight solution may only consist of two types of regions: (a) unperforated regions; and (b) ribs in the radial direction. On the above basis, the second author has obtained complete solutions for several loading/boundary conditions [14]. The above conclusions are being extended to *non-axisymmetric* plates. It is expected that the optimal solution for the latter consists of unperforated regions governed geometrically by the *yield-line theory for the Tresca yield condition* and regions with ribs in one direction only, governed by the *optimal grillage theory* (requiring a constant curvature in the rib direction). A combination of these two, now classical theories will furnish complete optimal solutions for plastic perforated plates.

REFERENCES

1. Prager, W.,Shield, R.T.: A general theory of optimal plastic design. *J. Appl. Mech.* 34 (1967) 184-186.

2. Rozvany, G.I.N.: *Optimal design of flexural systems*. Oxford: Pergamon Press (1976). Russian version: Moscow: Stroiizdat, (1980).

3. Rozvany, G.I.N.: Variational methods and optimality criteria, in Haug, E.J.; Cea, J. (eds.): *Optimization of distributed parameter structures, Proc. NATO ASI, Iowa, 1980*. Alphen aan Rhiijn: Sijthoff and Noordhoff (1981) 82-111.

4. Rozvany, G.I.N.; Ong, T.G.: Update to analytical methods in structural optimization in: *Applied Mechanics Update, ASME* (1986).

5. Rozvany, G.I.N.: Prager-Shield optimality criteria with bounded spatial gradients. *J. Engrg. Mech. ASCE* 110 (1984) 129-137.

6. Olhoff, N.; Rozvany, G.I.N.: Optimal grillage layout for given natural frequency. *J. Engrg. Mech. ASCE* 108 (1982) 971-974.

7. Rozvany, G.I.N.; Ong, T.G.; Karihaloo, B.L.: A general theory of optimal elastic design for structures with segmentation. *J. Appl. Mech, ASME*, 53 (1986) 242-248.

8. Rozvany, G.I.N.; Ong, T.G.; Yep, K.M.: Optimal elastic design with constrained taper: prescribed deflections. *J. Engrg. Mech. ASCE*, 112 (1986) 845-850.

9. Prager, W.; Rozvany, G.I.N.: Optimization of structural geometry, in: Bednarek, A.R., Cesari, L. (eds.): *Dynamical systems*. New York: Academic Press (1977) 265-294.

10. Prager, W.; Rozvany, G.I.N.: Optimal layout of grillages. *J. Struct. Mech.* 5 (1977) 1-18 (also: *14th IUTAM Congress*, Delft, (1976) paper No. 310).

11. Rozvany, G.I.N.: Optimality criteria for grids, shells and arches, in: Haug, E.J.; Cea, J. (eds.): *Optimization of distributed parameter structures, Proc. NATO ASI, Iowa, 1980*. Alphen aan Rhijn: Sijthoff and Noordhoff (1981) 112-151.

12. Rozvany, G.I.N.: A general theory of optimal structural layouts, *Proc. Int. Symp. on Optimum Structural Design*. Tucson, Arizona: University of Arizona (1981) 4.37-4.45. Extended version: Structural layout theory: the present state of knowledge. Chapter 7 in: Atrek, E.; Gallagher, R.H.; Ragsdell, K.M. and Zienkiewicz, O.C. (eds.): *New directions in optimum structural design*. Chichester, England: Wiley & Sons (1984) 167-196.

13. Rozvany, G.I.N.; Wang, C.M.: Extensions of Prager's layout theory, in: Eschenauer, H.; Olhoff, N. (eds.): *Optimization methods in structural design, Proc. Euromech. Colloquium*, Siegen (1982). Mannheim: Wissenschaftsverlag (1983) 103-110.

14. Rozvany, G.I.N.; Ong, T.G.: Optimal design of plates, shells and shellgrids, in: *Proc. IUTAM Symp. Inelastic Behaviour of Plates and Shells*, Rio de Janeiro, Aug. 1985, Berlin: Springer Verlag (1986) 357-384.

15. Michell, A.G.M.: The limits of economy of material in frame structures. *Phil Mag.*, 8 (1904) 589-597.

16. Kohn, R.V.; Strang, G.: Optimal design for torsional rigidity, in: Atluri, S.N.; Gallagher, R.H. and Zienkiewicz, O.C. (eds.): *Hybrid and mixed finite element methods*. Chichester, England: Wiley & Sons (1983) 281-288.

17. Cheng, K-T.; Olhoff, N.: An investigation concerning optimal design of solid elastic plates. *Int. J. Solids Struct.* 17 (1981) 305-323.

18. Rozvany, G.I.N.; Wang, C.M.: Constrained optimal layouts through Prager-Shield criteria. *J. Engrg. Mech. ASCE* 109 (1983) 648-653.

19. Rozvany, G.I.N.; Wang, C.M.: Optimal layout theory: allowance for selfweight. *J. Engrg. Mech. ASCE* 110 (1984) 66-83.

20. Rozvany, G.I.N.; Yep, K.M.; Sandler, R.: Optimal layout of long-span truss-grids, I - II, *Int. J. Solids Struct.* 22 (1986) 209-223, 225-238.

21. Rozvany, G.I.N.; Ong, T.G.: A general theory of optimal layouts for elastic structures. *J. Engrg. Mech. ASCE,* 112 (1986) 851-857.

22. Rozvany, G.I.N.; Hill, R.: A computer algorithm for deriving analytically and plotting optimal structural layout. *Proc. NASA Symp.: Future trends in computerised structural analysis and design*, Washington, (1978) 295-300. Oxford: Pergamon , 1978. Also *Computers and Struct.* 10 (1979) 295-300.

23. Hill, R.; Rozvany, G.I.N.: Prager's layout theory: a non-numeric computer method for generating optimal structural configurations and weight-influence surfaces. *Comp. Meth. Appl. Mech. Engrg.* 49 (1985) 131-148.

24. Rozvany, G.I.N.; Prager, W.: A new class of optimization problems: optimal archgrids. *Comp. Meth. Appl. Mech. Engrg.* 19 (1979) 127-150.

25. Rozvany, G.I.N.; Nakamura, H. Kuhnell, B.T.: Optimal archgrids: allowance for self-weight. *Comp. Meth. Appl. Mech. Engrg.* 24 (1980) 287-304.

26. Rozvany, G.I.N.; Wang, C.M.; Dow, M.: Prager-structures: archgrids and cable networks of optimal layout. *Comp. Meth. Appl. Mech. Engrg.* 31 (1982) 91-114.

27. Rozvany, G.I.N.; Wang, C.M.: On plane Prager-structures I-II. *Int. J. Mech. Sci.* 25 (1983) 519-527, 529-541.

28. Rozvany, G.I.N.; Olhoff, N.; Cheng, K.-T.; Taylor, J.: On the solid plate paradox in structural optimization. *J. Struct. Mech.* 10 (1982) 1-32.

29. Wang, C.-M.; Rozvany, G.I.N.; Olhoff, N.: Optimal plastic design of axisymmetric solid plates with a maximum thickness constraint. *Computers Struct.* 18 (1984) 653-665.

30. Kozlowski, W.; Mroz, Z.: Optimal design of solid plates. *Int. J. Solids Struct.* 5 (1969) 781-794.

31. Lurie, K.A.; Fedorov, A.V.; Cherkaev, A.V.: On the existence of solutions to some problems of optimal design for bars and plates. *J. Optimiz. Theory Appl.* 42 (1984) 247-281.

32. Lurie, K.A.; Fedorov, A.V.; Cherkaev, A.V.: Regularization of optimal design problems for bars and plates. I,II. *J. Optimiz. Theory Appl.* 37 (1982) 499-521, 523-543.

33. Rozvany, G.I.N.; Ong, T.G.; Olhoff, N.; Bendsoe, M.P.; Szeto, W.T.: Least-weight design of perforated elastic plates. *DCAMM-report.* No. 306, 1985. Extended version: Parts I and II, accepted. *Int. J. Solids Struct.*

DESIGN OF STRUCTURE AND CONTROLLERS FOR OPTIMAL PERFORMANCE

Martin Philip Bendsøe
Mathematical Institute
The Technical University of Denmark
Building 303, DK-2800 Lyngby
Denmark.

ABSTRACT. In this paper we consider the simultaneous optimal design of controls and structure for an active control of a flexible structure. Problem formulations related to mission control and to control of structural properties are identified and illustrated on a simple model problem, and the relationship to eigenfrequency optimization is discussed. We also discuss relevant design objectives for modal control of distributed structures or large scale structures and examples of optimal beam designs are presented.

1. INTRODUCTION.

In active control of flexible structures optimal control techniques are quite commonly employed, so that the best control strategy, in some sense, is obtained on a rational basis ([1]). Such techniques are usually used under the assumption of a , a priori, fixed design of the structure as well as given actuator and sensor positions for the active control system. However, considerable improvement in the performance of the system can be obtained by a simultaneous design of the structure, the control system and the control strategy. Optimal positioning of actuators and sensors has gained a lot of attention in the literature (see eg. [2] and references therein). Recently, the design of the structure has been included in studies of optimal control of a flexible structure ([3],[4],[5]) , and in this paper we will address the problem formulation and structural implications of a combined optimal design of structure, control strategy and control system.

In order to illustrate the basic properties of this type of problem we treat a simple mass-spring system where spring-stiffness, damping and controls are used to optimize certain energy criteria. We only employ certain functionals that allow the optimal control strategy to be easily identified so that implications for structural design can be emphasized, and from this example one can see the relation of this type of problem with the well-known eigenfrequency optimization problems.

NATO ASI Series, Vol. F27
Computer Aided Optimal Design: Structural and Mechanical Systems. Edited by C. A. Mota Soares

For distributed parameter structures or large scale structures control is often implemented via a reduced order model and the second part of the paper discusses optimal design of structure and controls for such systems. A commonly used reduced order model employs modal data for the structure and from this controls are designed in order to e.g. damp the vibrations of a few important modes ([6]). For such a modal control, optimal design can improve the performance of the controlled system and can also be used the reduce the undesirable effects of the use of a reduced order model. As an example we consider the so-called control spillover. This arises due to the local influence of actuators and results in a spillover of control energy into uncontrolled or unmodelled modes, and even though the system may remain stable the spillover degrades the system performance. However, via optimal design methods this spillover can be reduced considerably.

We will in this paper emphasize the structural design problems related to active control of structures and only very simple control techniques will be treated. For further information on the control aspects we refer to Ref. [7] and the list of references of that paper.

2. EXEMPLIFICATION.

One can illustrate the main features of problems of optimal design of structure and controls by considering the simple 2-degree of freedom system in Fig. 1.

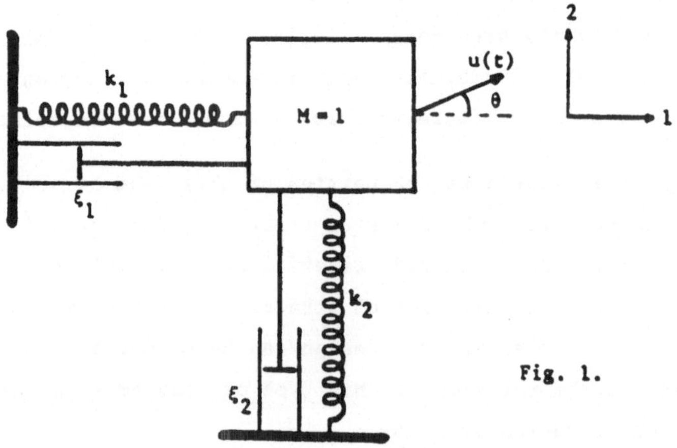

Fig. 1.

The system consists of one mass and two springs, with spring stiffnesses k_i , $i = 1,2$ and damping coefficients ξ_i , $i = 1,2$. The motion of the structure is controlled via the control force u which is designed in order to increase damping of the vibrations of spring no. 1 (mode 1). However, the control force also has a component that influences spring no. 2 (mode 2). If we use a simple velocity feedback for the control u so that u is proportional to the velocity v_1 of mode 1

$$u(t) = -\alpha \; v_1(t) \quad , \quad \alpha > 0 \; , \tag{1}$$

we can choose k_i, ξ_i, $i = 1,2$ and α so that the energy in the system is minimized, for any initial impulse in spring no. 1. The stateequations written using the statevector $\underline{x} = (x_1, v_1, x_2, v_2)$ are

$$\frac{d\underline{x}}{dt} = \underline{\underline{A}}\,\underline{x} \quad , \qquad \underline{\underline{A}} = \begin{pmatrix} 0 & 1 & 0 & 0 \\ -k_1 & -\alpha\cos\theta-\xi_1 & 0 & 0 \\ 0 & 0 & 0 & 1 \\ 0 & -\alpha\sin\theta & -k_2 & -\xi_2 \end{pmatrix} . \tag{2}$$

A measure of the effeciency of the damping is

$$E = \frac{1}{2}\int_0^\infty \underline{x}^T(t)\,\underline{\underline{Q}}\,\underline{x}(t)dt \tag{3}$$

where

$$Q = \begin{pmatrix} k_1 & 0 & 0 & 0 \\ 0 & 1 & 0 & 0 \\ 0 & 0 & k_2 & 0 \\ 0 & 0 & 0 & 1 \end{pmatrix} \tag{4}$$

and where $\underline{x}(t)$ is the solution to (2) , with initial conditions $\underline{x}(0) = \underline{x}_0$; E is thus the sum of the potential and kinetic energy in the system. A combined optimal design of structure and controller with respect to a minimization of E could thus take the form

$$\underset{k_i,\xi_i,\alpha}{\text{minimize}} \qquad E(\underline{x}_o) = \frac{1}{2}\int_o^\infty \underline{x}^T(t)\, \underline{\underline{Q}}\, \underline{x}(t)dt$$

$$\text{so :} \qquad \underline{x}(t) = e^{\underline{\underline{A}}t}\, \underline{x}_o \tag{5}$$

$$0 < k_{min} \le k_i \le k_{max} < \infty \;, \quad i = 1,2$$

$$0 < \xi_{min} \le \xi_i \le \xi_{max} < \infty \;, \quad i = 1,2$$

$$0 < \alpha_{min} \le \alpha \le \alpha_{max} < \infty$$

$$k_1 + k_2 \le V$$

$$\xi_1 + k_2 \le D$$

where V and D are resource constraints on stiffness and damping, respectively. The solution of (5) will depend on the choice of initial condition, but the problem can be given a suitable reformulation that removes this explicit dependence. To this end, we rewrite E as follows: let $\underline{\underline{P}}$ be a solution of the equation

$$\underline{\underline{A}}^T \underline{\underline{P}} + \underline{\underline{P}}\,\underline{\underline{A}} = -\underline{\underline{Q}} \;; \tag{6}$$

then $\underline{\underline{P}}$ is symmetric and positive definite (cf. [8]), and

$$\frac{d}{dt}(\underline{x}^T(t)\, \underline{\underline{P}}\, \underline{x}(t)) = -\underline{x}^T \underline{\underline{Q}}\, \underline{x} \tag{7}$$

so that

$$E = \frac{1}{2}\int_o^\infty \underline{x}^T \underline{\underline{Q}}\, \underline{x}\, dt = \frac{1}{2}[-\underline{x}^T \underline{\underline{P}}\, \underline{x}\,]_o^\infty = \frac{1}{2}\underline{x}_o^T \underline{\underline{P}}\, \underline{x}_o \;. \tag{8}$$

(The system is stable, so $\underline{x}(t) \to \underline{0}$ for $t \to \infty$).

The purpose of our optimal design is to minimize the total energy E for *any* impulse $\underline{x}_o$ with a given initial energy, and expression (8) allows us to write a functional that is independent of $\underline{x}_o$:

$$\max_{\underline{x}_o^T \underline{\underline{Q}}\, \underline{x}_o = 1} \quad \underline{x}_o^T \underline{\underline{P}}\, \underline{x}_o \tag{9}$$

which can be written as

$$\max_{\underline{x}_o^T \underline{x}_o = 1} \underline{x}_o^T \left(\sqrt{\underline{\underline{Q}}}^{-1} \underline{\underline{P}} \sqrt{\underline{\underline{Q}}}^{-1}\right) \underline{x}_o = \text{maximal eigenvalue of } \sqrt{\underline{\underline{Q}}}^{-1} \underline{\underline{P}} \sqrt{\underline{\underline{Q}}}^{-1} \,. \tag{10}$$

With this functional and the constraints on design from (5) we minimize the maximum of E , for any impulse with a unit initial energy. In the problem at hand, the control force u was designed so as to help damp out vibrations in spring no. 1, and we should thus only be considering impulses $\underline{x}_o$ in spring no. 1. With this in mind we can formulate the optimal design problem as:

$$\text{minimize} \quad \left(\text{maximal eigenvalue of} \begin{pmatrix} p_{11}/k_1 & p_{12}/\sqrt{k_1} \\ p_{12}/\sqrt{k_1} & p_{22} \end{pmatrix} \right) \tag{11}$$

$$\text{with:} \qquad 0 < k_{min} \le k_i \le k_{max} < \infty$$

etc.

where p_{ij} are the corresponding elements of the matrix $\underline{P}$ of (7) . The equation (6) can be solved analytically and then (11) can be solved using standard optimization algorithms. The results show that the spring for which the control is designed should be as stiff as possible and the coupling of the two modes caused by the feedback control makes it optimal to minimize the stiffness of the "uncontrolled" spring. In cases where the main concern is the coupling introduced via the control, this "control-spillover" can be reduced by solving (11) with an energy measure that only measures the energy in the second mode; in this case the α should be minimal.

Note that if the coupling of the two modes is not present (i.e. $\theta = 0$) , problem (11) corresponds to finding the optimal damping for a simple one mass/one spring system, for which the stiffness again should be as large as possible and for which the optimal value of $\xi_1 + \alpha$ of (2) is $1.572 \cdot \sqrt{k_1}$ (see Ref. [9] for details).

The problem formulated above optimizes the structure and controls in order to improve the damping characteristics of the system, i.e. the desired dynamics of the system is that the motion of the structure is minimal. In cases where specific movements are desired, such as placement of antennas, slewing of satelites etc., the functional of problem (5) should be modified to take this into account. If the desired dynamics are given as $\underline{z}(t)$, then a suitable functional is

$$\int_0^\infty (\underline{x}(t) - \underline{z}(t))^T \underline{Q} (\underline{x}(t) - \underline{z}(t)) dt \quad , \tag{12}$$

so that error is minimized. However, if precise positioning is to be achieved one should include requirements of the desired states at certain times as a constraint in (5) and more complicated types of control laws should be allowed in order to make the problem well posed. This type of problem is treated in Refs. [3] , [5] and in the preceeding chapters by Khot and Haftka.

3. ACTIVE CONTROL DAMPING OF A CANTILEVER.

In the control of large scale structures and distributed structures, interesting design criteria arise due to the difficulties in implementing distributed controls, and the commonly used technique of modal control gives rise to design problems that are very similar to the well-known problem of eigenfrequency optimization (cf. lectures by N. Olhoff). To illustrate this we consider an active control of a cantilevered beam by a finite number M of actuators, distributed along the length of the beam. The motion of the beam is governed by the equation

$$\rho a(x) \frac{\partial^2 w(x,t)}{\partial t^2} + \xi \rho a(x) \frac{\partial w(x,t)}{\partial t} + \frac{\partial^2}{\partial x^2}\left(Ea(x)^2 \frac{\partial^2}{\partial x^2} w(x,t)\right) = \sum_{i=1}^{M} b_i(x) f_i(t) \ . \tag{13}$$

Here ρ is the mass-density of the beam-material, ξ is the damping coefficient, $a(x)$ is the cross-sectional area of the geometrically similar cross-sections and E is Young's modulus for the beam-material. The actuator forces are $f_i(t)$, with actuator influence functions $b_i(x)$; the influence function can be modelled a δ-functions. We will seek to design the beam so that an active damping of the beam is optimal in some sense, and we use the varying cross-sections $a(x)$ of the beam as a distributed design variable.

A common technique for the design of controls for a distributed problem like (13) is to introduce a reduced order model, i.e. a finite dimensional approximation of (13) , and then use this model for the control design. Such reduced order models are conveniently constructed by introducing modal coordinates, and the controls are then designed from a model that takes a number of important modes into consideration (Refs. [6] and [10]-[13]).

If we introduce modal coordinates, we can write the solution $w(x,t)$ of (13) as

$$w(x,t) = \sum_{i=1}^{\infty} u_i(t)\varphi_i(x) \tag{14}$$

where the orthonormal modeshapes φ_i are given by

$$\frac{d^2}{dx^2}\left(E\,a(x)^2 \frac{d^2}{dx^2}\varphi_i(x)\right) = \lambda_i\,\rho\,a(x)\varphi_i(x) \tag{15}$$

$$\int_0^L \rho\,a(x)\varphi_i(x)\varphi_j(x)\,dx = \delta_{ij}$$

corresponding to the natural eigenfrequencies $\omega_i = \sqrt{\lambda_i}$, ordered according to $0 < \lambda_1 \le \lambda_2 \le \ldots$. The modal coordinates $u_i(t)$ then satisfy the infinite system of ordinary differential equations:

$$\ddot{u}_i(t) + \xi\,\dot{u}_i(t) + \lambda_i\,u_i(t) = \sum_{j=1}^{M} B_{ij}\,f_j(t) \tag{16}$$

with $B_{ij} = \int_0^L \varphi_i(x) b_j(x)\,dx$; the equations are coupled due to the control forces. If we choose to use the control forces to damp out the vibration of the N lower modes, using a velocity feedback, we can set

$$f_j(t) = G_{ji}\,\dot{u}_i \tag{17}$$

where $\underline{\underline{G}}$ is a $M \times N$ gain matrix. In this situation the dynamics of the lower modes will be governed by the equation (cf. Eq. (2)):

$$\underline{z} = (u_1,\ldots,u_N\,,\,\dot{u}_1,\ldots,\dot{u}_N)\ , \qquad \frac{d\underline{z}}{dt} = \underline{\underline{A}}\,\underline{z}$$

$$\underline{\underline{A}} = \begin{pmatrix} \underline{\underline{0}} & \underline{\underline{E}} \\ -\mathrm{diag}(\lambda_1,\ldots,\lambda_n) & -\mathrm{diag}(\xi_1,\ldots,\xi_n) + \underline{\underline{B}}\,\underline{\underline{G}} \end{pmatrix} . \tag{18}$$

We have assumed that the velocities of the lower modes can be observed directly, and this assumption makes the dynamics of the lower modes independent of the uncontrolled (residual) modes. The local influence of the actuators will, in general, make $B_{ij} \neq 0$ for almost all i and j , and thus the feedback law (17) will excite the higher modes and a spill-over of control energy into the higher modes will occur. Modal control has the following properties

i) With internal damping as in (13), a modal velocity feedback control can be constructed so that the stability margin of the system (18) is improved as compared to the uncontrolled system and at the same time the stability margin of the uncontrolled modes is unchanged ([6],[11]).

ii) With visco-elastic damping (i.e. damping proportional to stiffness), modal control can improve the stability margin of the full system (in this case the internal damping increases with modenumber) (cf. [13]).

iii) The control spill-over will, for a fixed number of actuators, decrease when the size of the model (18) is increased. (Refs. [11],[13],[14]).

For the sake of algebraic simplicity it is often convenient to consider cases where the number of actuators is chosen to be equal to the number of modes; then by choosing the feedback gain as

$$\underline{\underline{G}} = \underline{\underline{B}}^{-1} \operatorname{diag}(\alpha_1,\dots,\alpha_n) \tag{19}$$

the dynamics of the controlled modes decouple, and we have independent modal space control (cf. Ref. [12]):

$$\tilde{\underline{z}}_i = (u_i,\dot{u}_i) \quad , \quad \frac{d\tilde{\underline{z}}_i}{dt} = \underline{\underline{A}}_i \, \tilde{\underline{z}}_i$$

$$\underline{\underline{A}}_i = \begin{pmatrix} 0 & 1 \\ -\lambda_i & -\xi_i + \alpha_i \end{pmatrix} . \tag{20}$$

Note that for independent modal space control, the conclusion iii) above cannot be invoked.

From the remarks above we see that there are basically two types of design problems one should consider:

A. Design of structure and controllers with the purpose of optimizing the dynamics of the controlled modes.

B. Design of structure and controllers with the purpose of minimizing control spillover.

The system of section 2 can be considered as "the smallest large scale structure" and both design types described above were treated in that section. In Figs. 2 to 5 below and in table 1 we give results for the optimal design of a cantilevered beam, using the criteria of section 2 for a model with 2 modes. The beam is controlled with one actuator and the force of this actuator is proportional to the velocity of the first mode. The reduced order model consists of the first two modes and the design variables are the feedback gain, the position of the actuator and as a distributed design variable we use the cross-sectional area a(x) of the beam. The cases are:

Case 1. Coupling is ignored and we optimize the damping effect on mode 1. This problem reduces to maximizing the smallest eigenvalue, λ, of the beam (cf. Section 2). Fig. 2. See also chapters by N. Olhoff.

Case 2. Coupling is taken into account and we optimize the damping of the two first modes, for impulses in the first mode (cf. problem (11)). Fig. 3.

Case 3. Coupling is taken into account and we minimize the energy in the second mode, due to control damping of impulses in the first mode. Fig. 4.

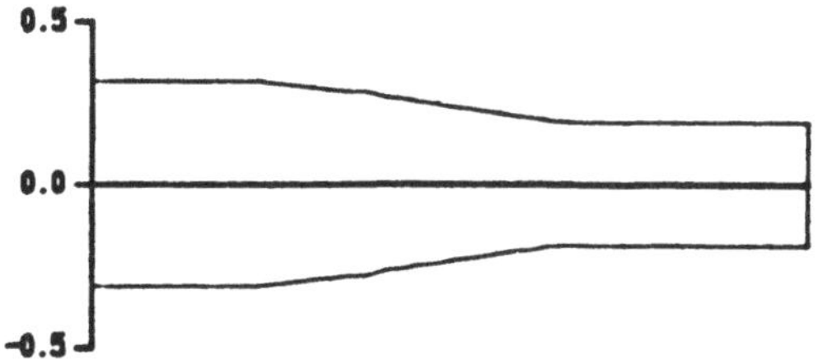

Fig. 2. Case 1. Maximization of first eigenvalue.

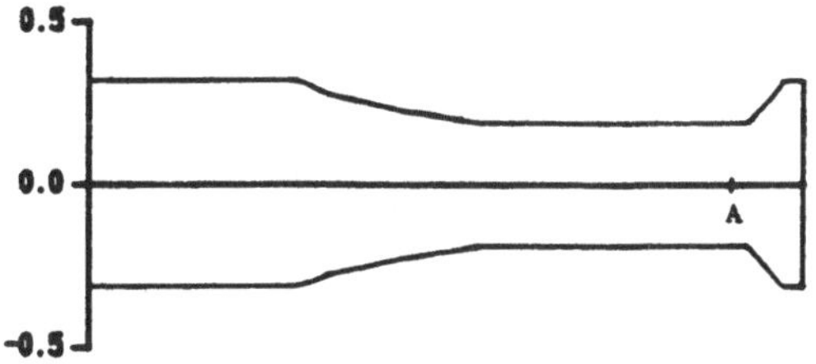

Fig. 3. Case 2. Minimization of energy in modes 1 and 2. One actuator, with optimal position A.

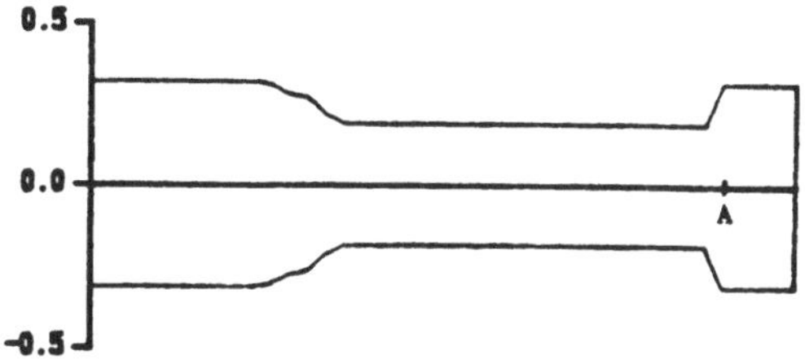

Fig. 4. Case 3: Minimization of energy in mode 2. Actuator at position A.

TABLE 1 Constraints: Volume $\leq$ 0.2 $0.113 \leq a(x) \leq 0.316$ Beam length L=1 ; ξ=0.01 ; ρ=1		Uniform beam of volume 0.2 Objective value π_u	Case 1 Max λ_1	Case 2 Min energy in mode 1+2	Case 3 Min energy in mode 2	Case 4 Min M 1 mode 1 actuator	Case 5 Min M 3 modes 3 actuators
Remarks: 1) $0.5 \leq \alpha \leq 2.0$	π / π_u	1.0	-	0.48	$4.4 \cdot 10^{-2}$	0.42 [4)]	0.43 [4)]
2) Actuator between 0.9 and 1.0 .	Vol	0.2	0.2	0.2	0.2	0.146	0.20
3) When actuator free to move, it will be at node for mode 2 and we have Case 1.	λ_1	2.47	7.50	4.61	2.46	1.27	2.83
	λ_2	97.1	101.8	85.1	90.4	59.6	63.7
	λ_3	762	723	574	513	476	428
4) Lower bounds on λ_i . Actuator positions fixed.	λ_4	2926	2666	2312	2091	1811	1773
	α	0.5	-	1.48 [1)]	0.5 [1)]	0.54 [1)]	0.50 [1)]
5) 3 actuators at 0.95, 0.70, 0.45.	Actuator position	0.95	-	0.9 [2)3)]	0.9 [2)]	0.95	[5)]

We note that the maximization of the damping results in an increase of the eigenfrequencies (Cases 1 and 2), while the minimization of spillover result in a decrease of the relevant eigenfrequencies.

The minimization of spillover considered above is only concerned with coupling of the first and second mode, and the criteria of spillover is the total energy in the second mode, due to an initial excitation of the first mode. It is, however, possible to give an estimate of the energy in all the uncontrolled modes due to initial impulses in the controlled modes and estimates of the peak energy or of the total energy can be obtained. In both cases an important measure of the coupling between the controlled and the uncontrolled modes is the factor (see [9],[14])

$$CP = \sum_{i=1}^{M} \sum_{j=N+1}^{\infty} \left(\int_0^L \rho\, a(x) b_i(x) \varphi_j(x) dx \right)^2 \tag{21}$$

which is the sum of the square of the norms of the projections of the force influence functions $b_i(x)$ on the uncontrolled modes; the orthonormality of the modes makes it possible to compute this factor from modal data of the controlled modes only:

$$CP = \sum_{i=1}^{M} \left[\int_0^L \rho\, a(x) b_i(x)^2 dx - \sum_{j=1}^{N} \left(\int_0^L \rho\, a(x) b_i(x) \varphi_j(x) dx \right)^2 \right] . \tag{22}$$

The spillover is, however, not only dependent on the coupling coefficient CP above, but it also depends on the dynamics of the controlled system and the size of the control forces needed for the damping of the controlled modes. Taking this into account one can write (cf. Ref. [9]):

$$\|\underline{z}_R(t)\|_E^2 \leq M \cdot e^{-\xi t} \cdot \|\underline{z}_o\|_E^2 \tag{23}$$

$$M = \frac{2\lambda_1+\xi}{2\lambda_1-\xi} \cdot \frac{1}{(\sigma-\xi/2)^2} \frac{1}{\mu} \cdot \rho(\underline{\underline{G}}^T\underline{\underline{G}}) \cdot K \cdot CP \tag{24}$$

for the energy $\|\underline{z}_R(t)\|_E$ in the residual modes due to an excitation $\underline{z}_o$ in the controlled modes. Here σ is the stability margin of the system (18), μ is given as

$$\mu = \min\{1, \lambda_1, \dots, \lambda_N\} \tag{25}$$

and $\rho(\cdot)$ denote the spectral radius of a matrix. Finally, the constant K is

$$K = \rho(\underline{\underline{D}}^T \underline{\underline{D}})\, \rho((\underline{\underline{D}}^T \underline{\underline{D}})^{-1}) \tag{26}$$

where $\underline{\underline{D}}$ is a matrix that transforms the coefficient matrix of (18) into Jordan normal form. The constant M in (23) depends only on the modal data of the controlled modes and a minimization of M will lead to a decrease in the eigenfrequencies of the controlled modes. Thus, in order not to degrade the performance of the active control damping, lower bounds on the eigenfrequencies should be imposed when spillover is minimized. Notice that M is a non-differentiable function of the eigenfrequences of the controlled modes, so that, in general, non-differentiable optimization techniques should be used. However, for independent modal space control, the controlled modes decouple and one can restate the problem (see [9]) as a differentiable problem by employing the technique of artificial bounds (cf. chapters by J.E. Taylor). Fig. 5 and 6 show optimal designs obtained for this special case.

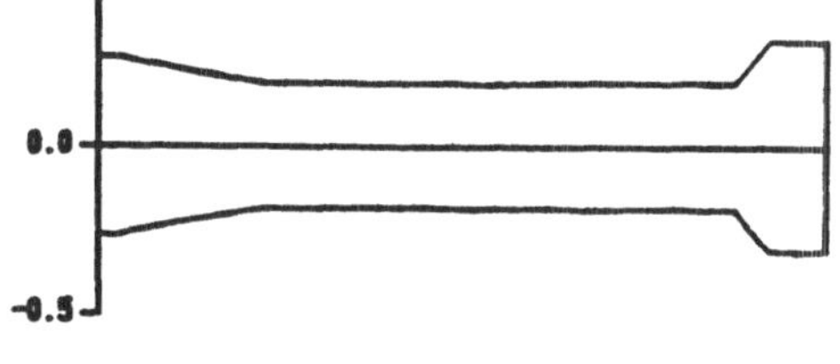

Fig. 5. Case 4: Minimization of spillover with one controlled mode.

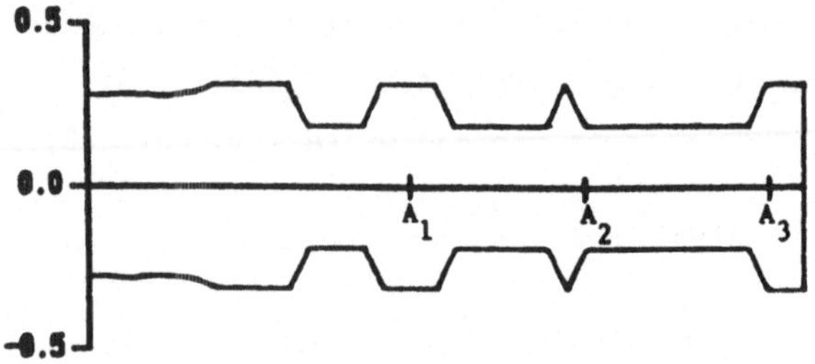

Fig. 6. Case 5: Minimization of spillover with 3 controlled modes. 3 actuators, at positions A_i.

REFERENCES

[1] M. Abdel - Rohman, H.H.E. Leipholz: Optimal Feedback Control of Elastic, Distributed Parameter Structures. Computers and Structures, 19 (1984), 801-805.

[2] G. Schultz, G. Heimbold: Dislocated Actuator/Sensor Positioning and Feedback Design for Flexible Structures. J. Guidance, Vol. 6 (1983), pp. 361-367.

[3] A.L. Hale, R.J. Lisowski, W.E. Dahl: Optimal Simultaneous Structural and Control Design of Maneuvering Flexible Spacecraft. Vol. 8 (1985), pp. 86-93.

[4] M.S. Khot, F.E. Eastep, V.B. Venkayya: Simultaneous Optimal Structural /Control Modifications to Enhance the Vibration Control of Large Flexible Structure. Paper presented at the AIAA Guidence and Control Conference, Snowmass, Colorado, April, 1985.

[5] A. Messac, J. Turner: Dual Structual - Control Optimization of Large Space Structures. Report, Charles Stark Draper Laboratory, Inc., Cambridge, Mass., USA.

[6] M.J. Balas: Modal Control of Certain Flexible Dynamic Systems. SIAM J. Control, Vol. 16 (1978), pp. 450-462.

[7] G.S. Murre, R.S. Ryan, H.M. Scofield, J.L. Sims: Dynamics and Control of Large Space Structures. J. Guidance, Control and Dynamics, Vol. 7 (1984), pp. 514-526.

[8] S. Barnett: Matrices in Control Theory. Van Nostrand Reinhold Company, London 1971.

[9] M.P. Bendsøe, N. Olhoff, J.E. Taylor: On the design of structure and controls for optimal performance of actively controlled flexible structures. Mechanics of Structures and Machines (to appear).

[10] M.J. Balas: Active Control of Flexible Systems. JOTA, Vol. 25 (1978), pp. 415-436.

[11] J.S. Gibson: An analysis of optimal modal regulation: convergence and stability. SIAM J. Control, Vol. 19 (1981), pp. 686-707.

[12] L. Meirovitch, H. Baruh, H.Öz: A Comparison of Control Techniques for Large Flexible Systems, J. Guidance, Vol. 6 (1983), pp. 302-310.

[13] Y. Sakawa: Feedback Control of Second Order Evolution Equations with Damping. SIAM J. Control and Optimization. 22 (1984), 343-361.

[14] M.J. Balas: Feedback Control of Dissipative Hyperbolic Distributed Parameter Systems with Finite Dimensional Controllers. J. Math. Anal. Appl. 98 (1984), pp. 1-24.

PART II

NUMERICAL METHODS IN OPTIMAL DESIGN

NUMERICAL OPTIMIZATION TECHNIQUES

Garret N. Vanderplaats
Department of Mechanical and Environmental Engineering
University of California
Santa Barbara, CA 93106

1.0 INTRODUCTION

Numerical optimization techniques provide a uniquely general and versatile tool for design automation. While these methods have been developed, to a large degree, by the operations research community, research in their application to engineering problems has been extensive as well. The first formal statement of nonlinear programming (numerical optimization) applied to structural design was offered by Schmit in 1960 [1]. Since that time, the field has evolved at an ever-increasing pace until it can now be considered to be reasonably mature.

Much of the work in recent years has dealt with efficient ways to use optimization on realistic problems where the underlying analysis is expensive. Optimization requires the repeated analysis of proposed designs and this can become unacceptably expensive. However, by careful formulation of the design problem, considerable efficiency can be gained so that, today, many large structures can be designed using as few as five detailed finite element analyses. Therefore, the proper study of structural optimization includes several aspects: efficient analysis methods, proper problem formulation, and efficient optimization algorithms. While the subject here is that of the optimization algorithms themselves, these other parts cannot be overlooked. Indeed, a clear understanding of the mathematical and physical nature of the design/ analysis problem directly leads to specialized algorithms for the optimization task.

The subject here will be limited, to a large degree, to general optimization concepts, noting that other parts of this lecture series dwell in detail on methods of gaining maximum overall efficiency and reliability. In order to use the range of capabilities available today, an understanding of basic optimization algorithms is essential as a building block. Also, for many applicatons in structural and mechanical design, some of the most

NATO ASI Series, Vol. F27
Computer Aided Optimal Design: Structural and Mechanical Systems. Edited by C. A. Mota Soares

powerful structural optimization methods do not apply, and so the fundamental tools must be used in order to achieve the desired solution.

Here, the general optimization problem statement will be given. With this, we outline a basic search strategy for optimization and will discuss the mathematical conditions that define an optimum design. We will then outline a variety of solution techniques. It will be seen here that, as opposed to structural analysis where the finite element method is almost universally used, there is seldom a clearly accepted design algorithm. This is not too surprising since the design objective, design variables, and the constraints are uniquely problem dependent. Thus, we might expect that the optimization algorithm will be different, depending on the nature of the problem and even the philosophy of the designer.

Having outlined a variety of solution techniques, some aspects relative to structural optimization will be identified. It will be seen that, for wide classes of design problems, the actual optimization algorithm is far less important than one might expect. An understanding of algorithms and how they fit into the overall design task provides the insight necessary to make the necessary choices.

Finally methods for obtaining the sensitivity of an optimized design to changes in problem parameters are discussed. It is shown that the sensitivity of the optimized design may be discontinuous with respect to the new parameter. The methods are demonstrated by a structural synthesis example.

2.0 THE BASIC PROBLEM STATEMENT

The general problem to be solved is: Find the set of design variables, $\underline{X}$, that will

Minimize $F(\underline{X})$ [1]

Subject to:

$g_j(\underline{X}) \leq 0 \qquad j=1,M$ [2]

$h_k(\underline{X}) = 0 \qquad k=1,L$ [3]

$X_i^\ell \leq X_i \leq X_i^u \qquad i=1,N$ [4]

Here, $\underline{X}$ contains the design variables, which can include member dimensions, geometric descriptions, and even material properties as examples. The function, $F(\underline{X})$ is the objective to be minimized, where it is noted that if $F(\underline{X})$ is to be maximized, we need only minimize its negative to have the problem in the form given here. The objective is most commonly taken as the weight of the structure since this is an easily definable parameter and because it is usually a good indication of cost.

The inequality conditions of Equations 2 are constraints that must be satisfied for the design to be considered feasible (acceptable). An obvious example is that the stress at each point in the structure must be less than a specified value under each loading condition. Such a constraint would be written

$$g = \sigma_{ij}/\bar{\sigma}_j - 1 \leq 0 \qquad [5]$$

where $\bar{\sigma}$ is the upper limit imposed on stress, and may be different for different materials in the structure and for different load conditions. Note that Equation 5 is always written in normalized form. Doing so for all constraints (including displacement, frequency and others) effectively nondimensionalizes the constraints so that if $g = -0.1$, for example, the constraint is satisfied by 10%.

Noting that a structure may be designed to support numerous independent loading conditions and that the stress may be calculated at hundreds or even thousands of points for each loading condition, it is clear that the set M of inequality constraints can become extremely large, even though the number of design variables, N, contained in $\underline{X}$ is relatively small.

The equality constraints, $h_k(\underline{X})$, impose precise conditions that must be satisfied at the optimum. Equality constraints are relatively uncommon in structural optimization, although they are sometimes included in multilevel optimization strategies. For example, it may be desirable to size the detailed dimensions of a stiffened panel as a sub-problem, while holding the overall stiffness of the panel fixed. These constraints are relatively difficult to deal with in practical design because they impose, in general, a set of nonlinear equations that must be dealt with simultaneous to the overall optimization task.

The conditions of Equation 4 are referred to as side constraints that simply limit the region of search for the optimum. These are themselves inequality constraints of the same form as Equation 2, but are best treated explicitly as bounds on the individual design variables. There are two reasons for this. First, side constraints can easily be incorporated into the optimization algorithm to gain efficiency. Second, and often more important, it may be necessary to insure that designs outside these limits are never considered. For example, if a minimum gage constraint is imposed on a member thickness in the structure, it should always be enforced. If the optimization algorithm is allowed to propose a design with a member of negative thickness, the analysis may proceed without error, but would obviously provide erroneous results.

At this point, no restrictions have been placed on the design task. The functions of Equations 1-3 may be highly nonlinear implicit functions of $\underline{X}$. Also, the design variables contained in $\underline{X}$ may be limited to a set of discrete values. Finally, no conditions have been imposed on the mathematical form of the problem, so the functions may even be discontinuous. Therefore, it is clear that the problem statement given here is extremely general and encompasses the majority of all engineering design. However, it will be seen that some limitations must be imposed in order to computationally solve the optimization task. Some of these are a statement of the present state of the art, while others represent fundamental assumptions in the mathematics of (present) optimization algorithms.

Among the present limitations to optimization methods are that the functions should be continuous in $\underline{X}$ and have continuous first (and sometimes second) derivatives. Also, the components of $\underline{X}$ should take on only continuous real values. Finally, there is an underlying assumption of mathematical convexity. This is simply the statement that we assume that there is only one unique optimum solution to the problem. These assumptions are, to a degree, more theoretical than practical. For example, algorithms that assume functional continuity often work quite well in the presence of function or derivative discontinuities, having noticeable difficulty only if such discontinuities exist quite near the optimum point. Algorithms do exist that do not assume this continuity, but these are usually considered to be too inefficient for application to realistic problems. For design tasks where the variables must take on discrete values, if the problem is solved assuming the design variables are continuous, the result can usually be rounded to the nearest required discrete value without major degradation of the design. Finally, for problems where more than one relative minimum (relative optimum) may exist, the probability of finding the best optimum is enhanced either by the choice of algorithm for solution or by beginning the optimization from several different starting points. It should be noted in this respect that if relative minima exist, they exist whether optimization is used or not. In this regard, optimization provides an efficient tool to find the best among the relative minima.

The optimization process usually begins with a proposed design, $\underline{X}^0$, provided by the engineer as input. This design may or may not satisfy all constraints. The design is then typically updated by modifying $\underline{X}$ as

$$\underline{X}^q = \underline{X}^{q-1} + \alpha \underline{S}^q \qquad [6]$$

where q is the iteration number. The vector, $\underline{S}$, is the search direction in n-dimensional space and α is a scalar move parameter. The product $\alpha\underline{S}$ is actually the change in $\underline{X}$ at this iteration. Thus, the optimization task is comprised of two parts. The first is to determine a search direction that will reduce the objective function without violating any constraints, and the second is to determine the single parameter, α, such that the design is improved as much as possible in this direction. This second task represents a problem in the single variable, α, and is called the line search, or the one-dimensional search.

The optimization process is continued until no search direction can be found that will further improve the design. Usually this is detected as a point of diminishing returns beyond which further improvement is not justified by the cost to achieve it. Ideally, however, a point will be reached where it can be shown to be an optimum (or at least a relative optimum) on a theoretical basis.

3.0 NECESSARY CONDITIONS FOR OPTIMALITY

The theoretical conditions that are satisfied at the optimum are referred to as the Kuhn-Tucker conditions. These conditions are derived from the fact that the Lagrangian function is stationary at the optimum, where the Lagrangian is defined as:

$$(\underline{X},\underline{\lambda}) = F(\underline{X}) + \sum^{M} \lambda_j g_j(\underline{X}) + \sum^{M+L} \lambda_k h_k(\underline{X}) \qquad [7]$$

The parameters, $\underline{\lambda}$, are referred to the dual variables. The Lagrangian is minimum with respect to $\underline{X}$ and maximum with respect to $\underline{\lambda}$ at the optimum. This is referred to as a saddle point and the stationary condition leads to the following three necessary conditions for optimality:

1. $\underline{X}^*$ is feasible.
2. $\lambda_j g_j(\underline{X}^*) = 0 \qquad \lambda_j \geq 0 \qquad j=1,M$
3. $\underline{\nabla}F(\underline{X}^*) + \sum_{j=1}^{M} \lambda_j \nabla g_j(\underline{X}^*) + \sum_{k=1}^{M+L} \lambda_k \underline{\nabla} h_k(\underline{X}^*) = \underline{0}$

   $\lambda_j \geq 0 \qquad \lambda_k$ unrestricted in sign.

These necessary conditions are also sufficient if the design space can be shown to be mathematically convex. While this is usually not possible to show, the Kuhn-Tucker conditions are still valuable to identify if a relative minimum has been found. Also, some of the more powerful modern algorithms use the Kuhn-Tucker conditions as a significant part of their derivation.

4.0 UNCONSTRAINED MINIMIZATION TECHNIQUES

While most engineering problems are constrained, unconstrained minimization techniques are still useful. First, for those problems that are actually unconstrained, these methods apply directly. Second, constrained problems can often be solved by converting them to a sequence of unconstrained problems. Finally, this provides a logical building block for the more general constrained optimization case. In the discussion here, we will assume that the gradient of the objective function is available, and so will omit discussion of methods that rely on function values only.

The Kuhn-Tucker conditions for unconstrained functions simply require that the gradient of the objective function vanishes at the optimum. The optimization process requires finding a search direction that will reduce the objective function, and then searching in that direction to gain as much improvement as possible.

4.1 Steepest Descent

A classical approach to this task is to set the search direction equal to the negative of the gradient of the objective function so

$$\underline{S} = -\underline{\nabla}F(\underline{X}) \qquad [8]$$

Now search in this direction, using Equation 6. A typical approach is to pick three values for α (including $\alpha = 0$) and evaluate the objective function for each of the resulting X-vectors. Then fit a quadratic polynomial to these points and determine the α for which $F(\alpha)$ is minimized. The objective function is evaluated for the proposed design and, if desired, the polynomial approximation is repeated until α is found to the desired precision. Alternative methods are available for solving the one-dimensional search problem (see ref. 2), but in general, this is a relatively standard numerical analysis task. However, it is important that it be done efficiently because the evaluation of $F(\underline{X})$ for each proposed α may be expensive. Once the minimum is found in direction $\underline{S}$, the gradient of the objective is calculated at this new design point and the search process is repeated until convergence is achieved.

While the steepest descent method is easily understood and implemented, it cannot be over-emphasized that this is a notoriously inefficient and unreliable algortihm. Its principal value is in providing background and because it is normally used as the first search direction for more modern algorithms.

4.2 Conjugate Directions

A simple modification to the steepest descent method can have a dramatic effect on the optimization efficiency. The conjugate direction of Fletcher and Reeves (3) is based on the use of H-conjugate directions, where H is the Hessian matrix. Using this method, it can be demonstrated that a quadratic function can be minimized in N or fewer iterations, whereas no proof of convergence is available for the steepest descent method.

Here, the first search direction is the steepest descent direction given by Equation 8. On subsequent iterations, the search direction is calculated from

$$\underline{S}^q = -\underline{\nabla}F(\underline{X}) + \beta\underline{S}^{q-1} \qquad [9]$$

where

$$\beta = |\underline{\nabla}F(X^q)|^2/|\underline{\nabla}F(\underline{X}^{q-1})|^2$$

Having determined the search direction, a one-dimensional search is then performed as before and the process is repeated until the solution has converged. Experience has shown that this simple modification dramatically improves the rate of convergence and the reliability of the optimization, even for the general case where the objective function is not quadratic.

4.3 Newton's Method

Newton's method is a classical second-order method for unconstrained minimization. The function is first approximated using a second-order Taylor series expansion:

$$\tilde{F} = F + \underline{\nabla}F\cdot\underline{\delta X} + \tfrac{1}{2}\,\underline{\partial X}^T H\underline{\delta X} \qquad [10]$$

where

$$\delta\underline{X} = \underline{X}^q - \underline{X}^{q-1}$$

H is the Hessian matrix (matrix of second derivatives) of F with respect to $\underline{X}$.

Taking the gradient of Equation 10 with respect to $\delta\underline{X}$ and equating it to zero gives the following estimate for the change in $\underline{X}$

$$\delta\underline{X} = -H^{-1}\underline{\nabla F} \qquad [11]$$

Noting that $F(\underline{X})$ may not be quadratic, $\underline{\partial X}$ may be used as the search direction, $\underline{S}$, in Equation 6 and a one-dimensional search performed. In this case, $\alpha = 1$ is a good initial estimate for the move parameter.

This method requires that the Hessian matrix be available and so is not usually an attractive algorithm. However, it is so powerful that it may be worthwhile to actually calculate H using finite difference methods. Also, in some classes of structural optimization problems where approximation techniques are used, the H matrix can be provided with minor effort, making this an attractive method.

4.4 Variable Metric Methods

Variable metric methods are algorithms which use gradient information to approximate the Hessian matrix (or its inverse) for use in Equation 11. The Davidon-Fletcher-Powell or DFP method (refs. 4,5) has been used for many years and is considered to be a powerful unconstrained minimization technique. A more recent algorithm known as the Broydon-Fletcher-Goldfarb-Shanno or BFGS method (refs. 6-9) is now considered to be the preferred method because it is less sensitive to numerical imprecision. In this method, the inverse of the Hessian matrix is initially set to the indentity matrix for use in the first iteration. On subsequent iterations, the inverse of the Hessian matrix is approximated as

$$H^{-1}_{new} = H^{-1} + \frac{\sigma+\tau}{\sigma^2}\,\underline{p}\underline{p}^T - \frac{1}{\sigma}\,[H^{-1}\underline{y}\underline{p}^T + \underline{p}(H^{-1}\underline{y})^T] \qquad [12]$$

where

$$\underline{p} = \underline{X}^q - \underline{X}^{q-1}$$

$$\underline{y} = \underline{\nabla}F(\underline{X}^q) - \underline{\nabla}F(\underline{X}^{q-1})$$

$$\sigma = \underline{p}\cdot\underline{y}$$

$$\tau = \underline{y}^T H^{-1}\underline{y}$$

Equation 12 provides an approximation to H^{-1} based on first-order information only. This is now used in Equation 11 to provide the desired search direction.

4.5 Example

Figure 1, taken from reference 2, shows a spring-mass system for which the equilibrium position is to be determined. The stiffness of spring i is given as

$$K_i = 500 + 200(5/3-i)^2 \quad N/m \tag{13}$$

and the weights are defined as

$$W_j - 50j \ N \tag{14}$$

The objective to be minimized is the total potential energy given by

$$PE = \sum_{i=1}^{6} \frac{1}{2} K_i \ (\delta L)^2 + \sum_{j=1}^{5} W_j Y_j \qquad N\text{-}m \tag{15}$$

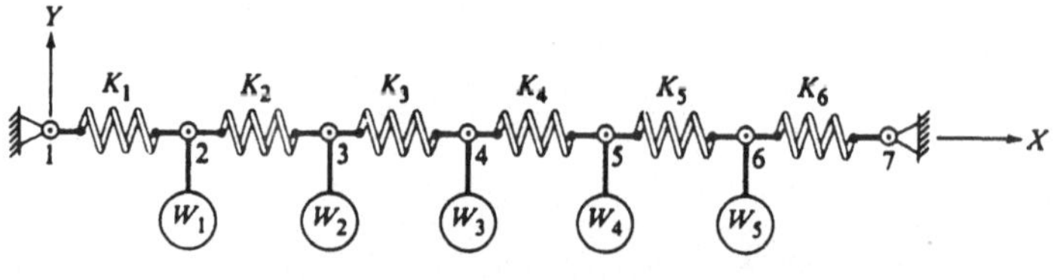

(a) Undeformed position

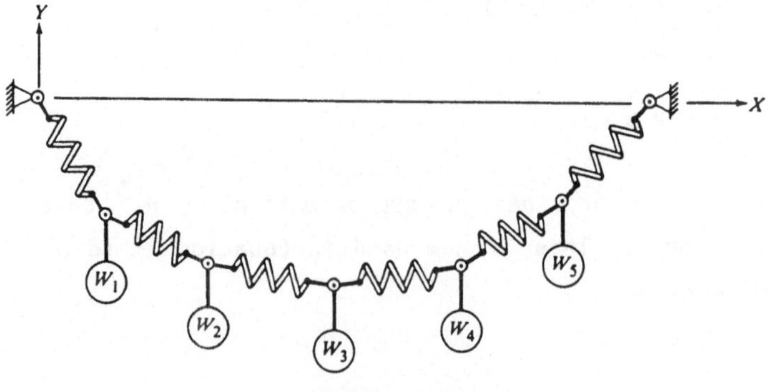

(b) Deformed position

Figure 1: Spring-Mass System

This problem was solved for using the following four algorithms and the results are given in tables 1 and 2.

1. Steepest Descent
2. Fletcher-Reeves conjugate direction method
3. Broydon-Fletcher-Goldfarb-Shanno variable metric method
4. Newton's method

In every case, all gradient information was calculated by finite differences and the function evaluations shown in Table 2 include this.

Table 1: Optimization Results

Design variable	Initial	Method 1	Method 2	Method 3	Method 4
X_1	10.0	10.3	10.4	10.2	10.4
X_2	20.0	20.7	21.1	20.8	21.1
X_3	30.0	31.0	31.6	31.4	31.7
X_4	40.0	41.3	42.0	41.7	42.1
X_5	50.0	51.1	51.6	51.4	51.8
Y_1	0.0	-2.65	-3.96	-4.64	-4.28
Y_2	0.0	-5.25	-7.77	-8.19	-7.90
Y_3	0.0	-7.35	-10.2	-10.0	-9.86
Y_4	0.0	-7.63	-9.52	-9.19	-9.40
Y_5	0.0	-4.97	-5.79	-5.43	-6.01

5.0 CONSTRAINED FUNCTION MINIMIZATION

The usual design task in engineering is to minimize some function such as weight or cost, subject to a variety of constraints. Therefore, the principal area of interest here is on this task, where the functions may be nonlinear and are usually implicit functions of the design variables. In this section, we discuss several algorithms for constrained optimization. We begin with the case where all functions are linear in the design variables because this provides a basis for further discussion. Also, many engineering problems can be cast as linear constrained optimization problems, such as limit analysis and design of structures. Finally, linear optimization can be used as a tool in solving the general nonlinear programming problem. Following the discussion of linear problems, the more

Table 2: Objective Function Versus Iteration Number

	Method			
Iteration	1	2	3	4
0	0.0	0.0	0.0	0.0
1	-60	-60	-60	-1256
2	-126	-292	-292	-1618
3	-151	-661	-666	-1987
4	-194	-1148	-941	-2175
5	-223	-1559	-1432	-2330
6	-318	-1895	-1812	-2912
7	-398	-2439	-2119	-3076
8	-1618	-2828	-2471	-3297
9	-1900	-3179	-2519	-3402
10	-1958	-3540	-2572	-3539
11	-2010	-3792	-3203	-4219
12	-2043	-4014	-3700	-4306
13	-2101	-4158	-3820	-4355
14	-2141	-4198	-3902	-4377
15	-2355	-4216	-4145	-4391
16	-2490	-4307	-4298	-4414
17	-2540	-4361	-4336	-4416
18	-2557	-4390	-4340	-4416
19	-2588	-4393	-4340	-4416
20	-2611	-4393	-4340	-
Final	-3964	-4393	-4378	-4416
Iterations	40†	22	26	19
Functions	587	331	383	378

† Maximum iterations allowed.

general nonlinear programming problem is addressed and a variety of algorithms are identified. Finally, a simple example is presented for comparison of the various methods.

5.1 Linear Constrained Minimization

Linear programming (refs. 10,11) is the oldest and the most completely developed of the modern numerical optimization techniques, having been first introduced by Dantzig in 1948 (10). In its basic form, this method solves the problem

Minimize $F = \underline{c}^T\underline{X}$ [16]

Subject to;

$$A\underline{X} = \underline{b} \qquad [17]$$

$$X_i \geq 0 \qquad i=1,N \qquad [18]$$

where matrix A contains the coefficients of the constraints and has dimensions MxN. In practice, this form of the problem is not really useful. For example, in engineering, the usual situation is to consider inequality constraints of the form of Equation 2. Also, it is often necessary to allow for negative values of the design variables, X_i.

Inequality constraints are handled by adding a non-negative "slack variable" so that this becomes, for the Qth inequality constraint;

$$\sum_{j=1}^{N} a_{ij}X_j + X_{N+Q} = 0 \qquad [19]$$

While this increases the number of design variables by the number of inequality constraints, this is not too critical an issue because very large linear problems can efficiently be solved.

To deal with the case where X_i may be negative, a simple transformation of variables can be used, where

$$X_i = X_i' - X_i'' \qquad [20]$$

where X' and X" are required to be positive so the difference in these positive variables can be negative. Substituting this into the general problem statement again increases the number of design variables by the number of negative X values to be allowed.

The details of solving this linear constrained optimization problem are beyond the scope of this discussion. However, it is sufficient to note that the linear programming problem is easily solved by the SIMPLEX (or preferably, the revised SIMPLEX) method and that computer codes for its solution are readily available on most major computing systems. Thus, for purposes of this discussion, the linear constrained optimization problem can be considered to be well in hand.

5.2 Sequential Linear Programming

Sequential Linear Programming (SLP), often referred to as Kelly's Cutting Plane Method (ref. 12), begins by creating a first order Taylor series expansion of the objective and constraint functions

$$\tilde{F} = F^{o} + \underline{\nabla F}(\underline{X}) \cdot \underline{\delta X} \qquad [21]$$

$$\tilde{g}_j = g_j^{o} + \underline{\nabla g}_j(\underline{X}) \cdot \underline{\delta X} \qquad j=1,M \qquad [22]$$

$$\delta X_i^{L} \leq \delta X_i \leq \delta X_i^{u} \qquad i=1,N \qquad [23]$$

This linearized problem is then solved using linear programming or other efficient optimization algorithm. The design is updated and a new linearization is performed, repeating the process until convergence is achieved. The move limits of Equation 23 are imposed to limit the design change to the region of applicability of the linearization and to prevent an unbounded solution. These move limits are reduced as the optimization progresses in order to effectively deal with the case where there may be fewer active constraints at the optimum than there are design variables (in this case, the linearization in the absence of move limits will cause oscillation or even be unbounded). This requirement to sequentially reduce the move limits is usually quoted as the principal disadvantage to the method. Also, for convex design problems, the solution will approach the optimum as a series of improving infeasible designs. This method is often considered a poor optimization algorithm. However, experience shows that it is often a very powerful and efficient method, and so should be considered as one of our design tools. If a sequential linear programming approach that creates a sequence of improving feasible designs is desired, the Method of Centers, developed by Baldur (ref. 13) provides this. The concept is to move to the center of the largest hypersphere bounded by the linearized constraints and the current value of the linearized objective function. This method is effective in providing a series of improving feasible designs, but usually converges to the optimum at a slower rate than the standard SLP method.

5.3 Sequential Unconstrained Minimization Techniques

These techniques use some form of penalty function to create an equivalent unconstrained optimization problem which is then solved by the methods of Section 4. Perhaps the best known is the Exterior Penalty Function method. Here, the following pseudo-objective function is minimized

$$\phi(\underline{X},r_p) = F(\underline{X}) + r_p \sum_{j=1}^{M} \mathrm{Max}[g_j(\underline{X}),0]^2 \qquad [24]$$

Here the scalar, r_p, is called the penalty parameter and is initially set to a small positive number. The pseudo-objective function is minimized with r_p held constant. Then r_p is increased, say by a factor of two, and the process is repeated, beginning with the optimum X-vector obtained in the previous unconstrained minimization. The concept is simply to penalize the design only when one or more constraints become violated. The reason for beginning with a small value for r_p and sequentially increasing it is to reduce the nonlinearities that would be introduced by a very large value of r_p. This method approaches the true constrained optimum as a sequence of improving, but infeasible designs.

An alternative approach is referred to as the interior penalty function method, where the pseudo-objective function is

$$\phi(\underline{X},r_p) = F(\underline{X}) - r_p \sum_{j-1}^{M} 1/g_j(\underline{X}) \qquad [25]$$

In this case, r_p begins as a large positive number and is sequentially reduced. It is assumed that the initial design is feasible and the penalty is imposed such that it grows as g_j approaches zero from the negative side. This method produces a sequence of improving feasible designs, but has the disadvantage that the pseudo-objective function is discontinuous at the constraint boundaries.

A third approach, which possesses the desired features of both the exterior and interior penalty function methods, is called the extended interior penalty function method. Here, the pseudo-objective function is defined as

$$\phi(\underline{X},r_p) = F(\underline{X}) + r_p \sum_{j=1}^{M} \tilde{g}_j(\underline{X}) \qquad [26]$$

where

$$\tilde{g}_j(\underline{X}) = -1/g_j(\underline{X}) \qquad \text{if } g_j(\underline{X}) \leq \varepsilon \qquad [27a]$$
$$\tilde{g}_j(\underline{X}) = [g_j(\underline{X}) - 2\varepsilon]/\varepsilon^2 \qquad \text{if } g_j(\underline{X}) > \varepsilon \qquad [27b]$$

and ε is a small negative number. Again, r_p is sequentially reduced during the optimization process. This method is referred to as the linear extended penalty function method (ref. 14). It has the advantage of producing a sequence of improving feasible designs, while maintaining function continuity across the constraint boundaries. The pseudo-objective function is continuous and has continuous first derivatives, but discontinuous second derivatives. If a second-order method such as Newton's method is to be used to solve the unconstrained subproblem, a quadratic form of this penalty function may be preferable (ref. 15).

A final sequential unconstrained minimization technique called the Augmented Lagrange Multiplier Method is considered to be a particularly powerful method (ref. 16). Here, the Lagrangian function, augmented by an exterior penalty term is minimized:

$$A(\underline{X},\underline{\lambda},r_p) = F(\underline{X}) + \sum_{j=1}^{M} [\lambda_j\psi_j + r_p\,\psi_j^2] \qquad [28]$$

where

$$\psi_j = \mathrm{Max}[g_j(\underline{X}), \frac{-\lambda_j}{2r_p}] \qquad [29]$$

The last term under the summation in Equation 28 is simply an exterior penalty function. The first term in the summation is the constraint term of the Lagrangian function. The basis of the method is that, if the optimum Lagrange multipliers were known, the constrained problem could be solved for almost any positive value for r_p, and r_p would not have to be sequentially updated. Since the values of the Lagrange multipliers are now known, an algorithm is needed to update them in a fashion that will converge to the optimum. Thus, the difference between this and the exterior penalty function method is that here both the penalty parameter, r_p, and the estimate of the Lagrange multipliers are updated. For the Lagrange multipliers the update formula is

$$\lambda_j^{p+1} = \lambda_j^p + 2r_p \psi_j \qquad j=1,M \qquad [30]$$

The attraction of this method is that it is possible to obtain a precise zero for active constraints, the penalty parameter r_p does not have to be made as large, and the basic algorithm is designed to drive the solution toward satisfaction of the Kuhn-Tucker condition.

In the Augmented Lagrange Multiplier method, as well as the exterior penalty function method, equality constraints can easily be included, and this is given in the references.

5.4 Method of Feasible Directions

This algorithm, attributable to Zoutendijk (ref. 17) is referred to as a direct method because it deals with the objective and constraints directly in order to determine a search direction for use in Equation 6. In this case, the solution for α in Equation 6 requires finding the best design such that the constraints are not violated. This one-dimensional search process can be solved using polynomial interpolation as for unconstrained functions except, now the α is found to minimize $F(\alpha)$ as well as the α_j which will drive $g_j(\alpha)$ to zero for j=1,M. The minimum positive value for α from all of these is the desired solution to the one-dimensional search.

It is assumed that the initial design is feasible. If no constraints are critical, the search direction is taken as the direction of steepest descent or (after the first iteration) a conjugate direction. Once one or more constraints become active $[g_j(\underline{X}) = 0,\ j \in J]$ the search direction is calculated from the following optimization sub-problem;

$$\text{Maximize } \beta \qquad [31]$$

Subject to;

$$\underline{\nabla}F(\underline{X}) \cdot \underline{S} + \beta \le 0 \qquad [32]$$

$$\underline{\nabla}g_j(\underline{X}) \cdot \underline{S} + \theta_j \beta \le 0 \qquad j \in J \qquad [33]$$

$$\underline{S} \cdot \underline{S} \le 1 \qquad [34]$$

The design variables for this sub-problem are the components of $\underline{S}$ as well as the value of the intermediate variable, β. Equation 32 is the usability requirement that states that the search direction must reduce the objective function and equation 33 is the feasibility requirement that requires no active constraints become violated for some small move in this direction. The parameter θ_j is referred to as a push-off factor that directs the search away from the critical constraints. This is because a search direction tangent to the constraint boundary will quickly go outside the feasible region if the constraint is nonlinear. This sub-problem is easily solved by converting it into a form similar to linear programming. Also, if the initial design is outside the feasible region a search direction can be efficiently found which will overcome the constraint violations (ref. 18). This method is contained in the CONMIN program (ref. 19) and has been widely applied to structural optimization problems.

The feasible directions method is often considered inefficient because it does not follow the critical constraint, but instead, pushes away from them. This "zig-zagging" can be alleviated somewhat by various numerical techniques, but remains a theoretical issue.

An alternative method is the Generalized Reduced Gradient method (ref. 20). This method converts the inquality constrained problem to an equality constrained problem by adding slack variables as is done in linear programming. Then a number of dependent variables is chosen equal to the total number of constraints. A search direction is then found in terms of the remaining independent design variables. During the one-dimensional search, the dependent variables are repeatedly updated using Newton's method applied to the constraints. This approach is often considered to be more efficient, but requires both the addition of many slack variables and the treatment of a sub-set of design variables as dependent.

A method that has many of the features of the Generalized Reduced Gradient method, but does not require the addition of slack variables or the separation of the design variables into independent and dependent sets, is referred to as the Modified Feasible Directions method (ref. 21). This method uses the direction-finding problem of Equations 31-34, but with the push-off factors, θ_j set to zero. The resulting search direction will be tangent to the critical constraints. During the one-dimensional search, these constraints will become violated and so it is necessary to move back to the constraint boundaries. Now, instead of treating some of the variables as dependent for this purpose, the shortest distance back to the constraint boundaries is sought in a least-squares sense. This leads to the perturbation vector

$$\underline{\delta X} = -A^T[AA^T]^{-1}\,\underline{G} \qquad [35]$$

where, the matrix A contains the gradients of the critical constraints and G contains their values. At each proposed $\underline{X}$ in the one-dimensional search, the constraints $\underline{G}$ are calculated and Equation 35 is applied to drive them to zero. This method has been found to be an effective general purpose algorithm which is a relatively simple modification to the feasible directions method, while having the essential features of the generalized reduced gradient method.

5.5 Sequential Quadratic Programming

Sequential Quadratic Programming (refs. 22-27) is a relatively new algorithm that is considered to be quite powerful for a wide class of problems. This method is often called a variable metric method for constrained optimization because of its formulation. The basic concept is to create a quadratic approximation to the Lagrangian function and linearize the constraints. This approximation is solved using Quadratic Programming (or any suitable optimizer) and the resulting perturbed design is treated as a search direction. A one-dimensional search is then performed in this direction as an unconstrained problem using a penalty function.

The problem solved for finding the search direction is

Minimize $$Q(\underline{S}) = F^\circ + \underline{\nabla}F(\underline{X}) \cdot \underline{S} + \frac{1}{2}\,\underline{S}^T B \underline{S} \qquad [36]$$

Subject to;

$$c_j g_j^\circ + \underline{\nabla} g_j(\underline{X}) \cdot \underline{S} \leq 0 \qquad j=1,M \qquad [37]$$

which c_j is an appropriately chosen constant, and B is an approximation to the Hessian matrix of the Lagrangian function. This is called a variable metric method because an update formula is used that is similar to that of the variable metric methods for unconstrained minimization. Here, however, B is an approximation to H instead of H^{-1}.

Numerous modifications can be made to improve the efficiency and reliability of this algorithm for engineering applications (ref. 28). Overall, it has been found to be a powerful method that contains many desirable features.

5.6 Sequential Convex Linearizations

This recent method has been shown to be particularly powerful for many structural optimization applications (ref. 29). The basic concept here is to first linearize the objective and constraint functions as is done in sequential linear programming. However, now instead of solving this linearized problem, reciprocal variables are used to create a conservative, convex, approximation. A typical constraint function is approximated as

$$\tilde{g}_j = g_j^o + \sum_{+} \frac{\partial g_i}{\partial X_i} (X_i - X_i^o) + \sum_{-} \frac{\partial g_i}{\partial X_i} \frac{X_i^o}{X_i} (X_i - X_i^o) \qquad [38]$$

where the + summation refers to those gradient terms that are positive and the - summation refers to those that are negative. Similarly, this mixed approximation is used for the objective function.

Now the approximate problem to be solved by the optimizer is not linear, but is still explicit and is easily solved. The key idea is that by using this mixed formulation, the essential nonlinearities of the original problem are retained. Thus, the move limits imposed during the approximate optimization are far less critical than for the Sequential Linear Programming method.

5.7 Example

Figure 2, from reference 2, shows a simple cantilevered beam made up of five rectangular cross-sections. The dimensions, b and h of each section are to be designed for minimum total material volume, so there are a total of ten design variables. The stress in each section is constrained and the deflection under the load is constrained. Also, the height was constrained to be less than twenty times the width for each section.

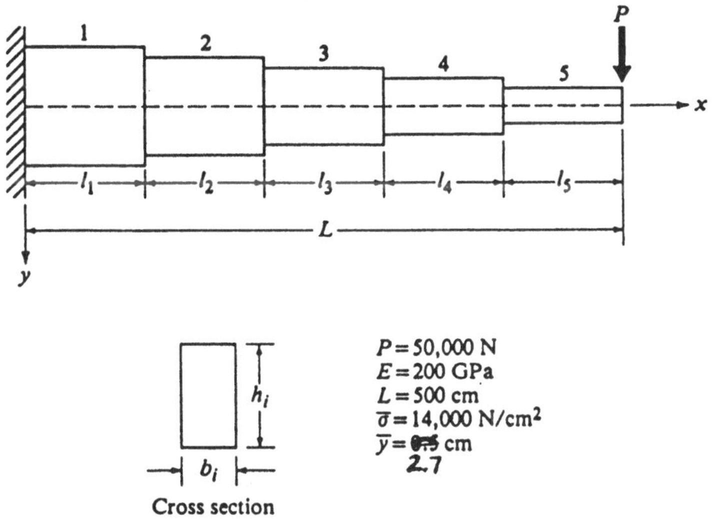

Figure 2: Cantilevered Beam

This problem was solved by the ADS program (ref. 30) using the following eight algorithms and the results are shown in Tables 3 and 4.

1. Sequential Linear Programming.
2. Exterior Penalty Function Method.
3. Linear Extended Interior Penalty Function Method.
4. Augmented Lagrange Multiplier Method.
5. Method of Feasible Directions.
6. Modified Method of Feasible Directions.
7. Sequential Quadratic Programming.
8. Sequential Convex Programming.

In each case, all gradient information was calculated by the finite difference method and the function evaluations shown in Table 4 includes this.

Table 3: Iteration History

Iteration	Method							
number	1	2	3	4	5	6	7	8
0	100,000	100,000	100,000	100,000	100,000	100,000	100,000	100,000
1	82,375	17,661	83,824	50,827	106,600	104,500	76,079	79,578
2	73,830	25,955	71,807	64,009	98,982	95,849	58,110	71,220
3	65,812	30,434	69,033	64,488	93,745	93,576	64,547	69,517
4	65,142	34,269	66,574	65,123	84,095	77,448	65,230	68,213
5	65,642	39,697	66,065	65,057	73,182	69,412	65,320	67,520
6	65,567	43,223	65,760	65,794	68,882	67,954	65,385	66,902
7	65,346	49,486	65,736	65,561	68,549	66,747	65,410	66,137
8	65,420	51,658	65,736	65,678	65,636	65,783	65,410	65,600
9	65,398	57,110	-	-	65,424	65,533	-	65,440
10	65,398	60,459	-	-	65,411	65,362	-	65,428
Iterations	10	16	8	8	11	11	8	
Optimum	65,398	66,159	65,736	65,678	65,411	65,362	65,410	65,428
Functions	110	989	492	533	140	130	106	111

Table 4: Design Variable Values

Variable	Initial	Method							
		1	2	3	4	5	6	7	8
b_1	5.0	2.19	2.24	2.17	2.17	2.19	2.19	2.19	2.19
b_2	5.0	2.20	2.26	2.24	2.27	2.21	2.20	2.20	2.21
b_3	5.0	2.52	2.52	2.54	2.52	2.52	2.52	2.52	2.52
b_4	5.0	2.78	2.90	2.79	2.79	2.78	2.78	2.78	2.78
b_5	5.0	2.99	3.24	3.00	3.00	2.99	2.99	2.99	2.99
h_1	40.0	43.73	41.71	43.35	42.94	43.82	43.72	43.82	43.78
h_2	40.0	44.11	44.87	44.67	44.99	44.14	44.09	44.09	44.10
h_3	40.0	50.47	50.30	50.61	50.47	50.47	50.47	50.47	50.47
h_4	40.0	55.55	53.81	55.63	55.42	55.55	55.55	55.55	55.55
h_5	40.0	59.84	56.77	59.93	59.77	59.84	59.84	59.84	59.83

6.0 SENSITIVITY OF OPTIMIZED DESIGNS TO PROBLEM PARAMETERS

Estimation of the sensitivity of an optimum design to some new problem parameter, P, (or perhaps multiple parameters) is an important issue in design. Two principal reasons for needing this information are first, that if the problem requirements are modified after the optimization is complete, this will provide the engineer with a measure of what effect such changes will have on the design, and second, this information is directly useful in formal multi-level and multi-discipline optimization. The discussion offered here is essentially a duplication of that given in reference 31.

The general optimization problem considered here is of the form given by equations 1-4.

Equality constraints are omitted in the followind discussion only for brevity and their addition is a direct extension of the methods presented here.

Assume the optimization problem has been solved, so the vector of optimum design variables, $\underline{X}^*$ is known. Now assume some new parameter, P, is to be changed, where P may be a load, allowable stress, material property, or any other problem parameter. It is desired to determine how the optimum design will change as a result of changing P. That is, we wish to find the total derivative $dF(\underline{X}^*)/dP$ as well as the rates of change of the optimum values of the design variables themselves, $\partial X_i/\partial P$.

In references 32-35, the Kuhn-Tucker conditions at the optimum are used to predict the required derivatives, based on the assumption that the Kuhn-Tucker conditions at $\underline{X}^*$ remain in force as P is changed. Normally, second derivatives of the objective and the binding constraints are required, as well as the Lagrange multipliers associated with the optimum design.

In reference 21, a method was introduced based on the concept of a feasible direction, for providing the optimum sensitivity information. This method requires only first derivatives of the objective function and binding constraints.

Here, the method of reference 21 is expanded to deal effectively with the possibility of a discontinuous derivative at $\underline{X}^*$. This first-order method provides the sensitivity information in the classical sense of a derivative. A second-order method is presented here for use in those design situations where second derivatives are economically available. In this case an approximate optimization problem is actually performed to estimate the design improvement which can be expected from a change in the new parameter.

The three available approaches to optimum design sensitivity are compared using simple examples to gain a conceptual understanding of the methods. The methods are then demonstrated by physical example.

6.1 Sensitivity Using the Kuhn-Tucker Conditions

At the optimum design, $\underline{X}^*$, some subset, J, of constraints will be critical. The Kuhn-Tucker conditions[36] which are satisfied at this point are

$$\nabla F(\underline{X}^*) + \sum_{j=1}^{m} \lambda_j \nabla g_j(\underline{X}^*) = 0 \qquad [39]$$

$$g_j(\underline{X}^*) = 0 \qquad j \in J \qquad [40]$$

$$\lambda_j \geqq 0 \qquad [41]$$

The needed sensitivity information comes from calculating the total derivative of $F(\underline{X}^*)$ with respect to the new parameter, P;

$$dF(\underline{X}^*) = \partial F(\underline{X}^*)/\partial P + \nabla F(\underline{X}^*) \cdot \underline{dX} \qquad [42]$$

where

$$dX_i = \partial X_i/\partial P \qquad [43]$$

If the Kuhn-Tucker conditions of equations (39)-(41) are to remain satisfied for some change δP in the independent parameter, P, then the rate of change of equation (39) with respect to P must vanish, where it is noted that equation (39) is actually n independent equations. This leads to a set of simultaneous equations which are solved for dX as well as $d\lambda$ (since the Lagrange multipliers are functions of P as well). This leads to the following set of simultaneous equations that are solved for $\underline{dX}$ and $\underline{d\lambda}$.

$$\begin{bmatrix} A_{n\times n} & B_{n\times J} \\ B^T_{J\times n} & 0_{J\times J} \end{bmatrix} \begin{Bmatrix} dX \\ d\lambda \end{Bmatrix} + \begin{Bmatrix} C_{n\times 1} \\ d_{j\times 1} \end{Bmatrix} = 0 \qquad [44]$$

where

$$A_{ik} = \frac{\partial^2 F(X^*)}{\partial X_i \partial X_k} + \sum_{j\in J} \lambda_j \frac{\partial^2 g_j(X^*)}{\partial X_i \partial X_k} \qquad [45a]$$

$$B_{ij} = \frac{\partial g_j(X^*)}{\partial X_i} \qquad [45b]$$

$$C_i = \frac{\partial^2 F(X^*)}{\partial X_i \partial P} + \sum_{j\in J} \frac{\partial^2 g_j(X^*)}{\partial X_i \partial P} \qquad [45c]$$

$$d_j = \frac{\partial g_j(X^*)}{\partial P} \qquad j\in J \qquad [45d]$$

The two significant features of this approach are that second-order information is required and that constraints which are critical at $\underline{X}^*$ remain critical when the independent parameter, P, is changed.

6.2 Sensitivity Using the Feasible Directions Concept

Conceptually, calculation of the sensitivity to parameter P can be viewed as seeking the greatest improvement in the expanded design space which includes the new "design variable" P. Thus, we mathematically seek the "constrained steepest descent direction." Assuming we are free to either increase or decrease P as necessary to improve the design, the information is obtained from the following sub-problem.

Treat P as an independent design variable and add it to the set of variables, $\underline{X}$, so

$$X_{n+1} = P \qquad [46]$$

Now solve the following direction-finding problem in the expanded space;

Find the components of S to

$$\text{Minimize} \quad \nabla F(\underline{X}^*) \cdot \underline{S} \qquad [47]$$

$$\text{Subject to;} \quad \nabla g_j(\underline{X}^*) \cdot \underline{S} \leqq 0 \qquad j \in J \qquad [48]$$

$$\underline{S} \cdot \underline{S} \leqq 1 \qquad [49]$$

Equation (47) represents the objective, which is to search in a direction as nearly possible to the steepest descent direction. This is constrained, however, by equation (48) which dictates that the search direction be tangent to, or away from, the boundaries of the critical constraints. Note that by virtue of the inequality condition of equation (48), the design may actually leave a constraint boundary if this will give maximum improvement. This is equivalent to saying that the Kuhn-Tucker conditions in force at $\underline{X}^*$ need not remain in force in the expanded design space. Equation (49) is required to insure a unique solution to the direction-finding process.

This geometric interpretation of a constrained steepest descent direction is also reached by considering a linear approximation to the problem in expanded space as; Find the perturbation of the design variables contained in the vector $\underline{S}$ which will

$$\text{Minimize} \quad \tilde{F}(\underline{X},P) = F(\underline{X}^*,P) + \underline{\nabla}F(\underline{X}^*,P) \cdot \underline{S} \qquad [50]$$

Subject to;

$$\tilde{g}_j(\underline{X},P) = g_j(\underline{X}^*,P) + \underline{\nabla}g_j(\underline{X}^*,P) \cdot \underline{S} \leqq 0 \qquad j \in J \qquad [51]$$

$$\underline{S} \cdot \underline{S} \leqq 1 \qquad [52]$$

where equation (52) arises from the need to bound the solution to the linearized problem. Noting that $F(\underline{X}^*,P)$ is constant and $g_j(\underline{X}^*,P) = 0$ for $j \in J$, it is clear that equations (50)-(52) are the same as equations (47)-(49).

Equations (47)-(49) represent a linear problem in S_i, i=1,n+1 subject to a single quadratic constraint. This is solved by conversion to a linear programming type problem of dimension J+1 and is easily solved numerically.[2]

Having solved the direction-finding problem of equations (47)-(49), the design can be updated by the common relationship

$$\underline{X} = \underline{X}^* + \alpha\underline{S} \qquad [53]$$

so $$X_i = X_i + \alpha\partial X_i/\partial\alpha \qquad [54]$$

and $$\partial X_i/\partial\alpha = S_i \qquad [55]$$

Considering the n+1 component, P,

$$P = P^* + \alpha\partial P/\partial a \qquad [56]$$
$$= P^* + \alpha S_{n+1}$$

For a specified change, $\delta P = P-P^*$, the move parameter, α, is

$$\alpha^* = \delta P/S_{n+1} \qquad [57]$$

In the event that $S_{n+1} = 0$, this indicates that the optimum design is not dependent on P and so P can be changed arbitrarily. In this case, $dF(\underline{X}^*) = 0$ and $\underline{dX} = 0$.

Here it is important that the sign on δP be the same as the sign on S_{n+1} since it was assumed that we will change P in the direction of maximum improvement.

The rate of change of the optimum objective is now found for a unit change in P as

$$dF(X^*)/dP = [\underline{\nabla}F(\underline{X}^*) \cdot \underline{S}]/S_{n+1} \qquad [58]$$

and the corresponding rates of change of the optimum design variables are

$$\underline{dX} = (1/S_{n+1})\underline{S} \qquad [59]$$

Multiplying equations (58) and (59) by δP gives the estimated change in $F(X^*)$ and X^* respectively. If δP is the opposite sign from S_{n+1}, equations (58) and (59) still apply if equation (48) is satisfied with strict equality. If this is not the case, the design dictated here will leave a constraint for maximum improvement. Thus a search in the opposite direction will violate one or more constraints.

Consider now the case where the sign of the change in parameter P is specified. Now it must be recognized that if P is decreased the optimum design may follow one constraint surface, but if P is increased, the design may follow a different constraint surface. In other words, the rate of change of the optimum design with respect to P may not be constant at $\underline{X}^*$. In this case, an additional constraint must be imposed on the sign of S_{n+1}. For example, if P is required to be positive, the direction-finding problem of equations (47)-(49) becomes

$$\text{Minimize} \quad \underline{\nabla}F(\underline{X}^*) \cdot \underline{S} - c\beta \qquad [60]$$

Subject to;

$$\underline{\nabla}g_j(\underline{X}^*) \cdot \underline{S} \leqq 0 \qquad [61]$$

$$-S_{n+1} + \beta \leqq 0 \qquad [62]$$

$$\underline{S} \cdot \underline{S} \leqq 1 \qquad [63]$$

The independent variables in this direction-finding problem are the components of $\underline{S}$ as well as the extra parameter, β. Here the constant, c, is a somewhat arbitrary positive number, say 10. The magnitude of c is not critical since it is used only to insure the resulting value of β will be positive and so S_{n+1} will in turn be positive. If δP is specified to be negative, equation (62) is replaced by

$$S_{n+1} + \beta \leqq 0 \qquad [64]$$

The result of the direction-finding problem is to find a search direction which will increase $F(\underline{X}^*)$ as little as necessary, decreasing $F(\underline{X}^*)$ if possible. Thus, if there is a specific reason to change P in a given direction, the optimum sensitivity is still found.

The method of equations (60)-(63) is actually the same as that used in the Method of Feasible Directions for overcoming constraint violations.(2)

The method of equations (47)-(49) and (60)-(63) is based on the assumption that P will be changed to gain maximum improvement (or minimum degradation) in the optimum objective. The only additional information required beyond that normally available is the gradient of the objective and critical constraints with respect to P. Also, it is important to note that the set J can include any near critical constraints which we do not wish to become violated for a small change in P. Finally, it should be noted that a search direction may be desired which moves away from the constraint boundaries to give a more conservative estimate of $dF(X^*)/dP$. This can be accomplished by adding a push-off factor to equation (48) or

(61) as is done in the conventional feasible directions algorithm.[2]

6.3 Sensitivity Using Second-Order Information

In cases where second derivatives are available, this information can be used to provide maximum guidance to the design process. Here, the information sought is not sensitivity in the mathematical sense, but rather is the best estimate of the new optimum based on a quadratic approximation to the objective and constraint functions about $\underline{X}^*$.

Consider a second-order Taylor series expansion about $\underline{X}^*$ of $F(\underline{X}^*)$ and $g_j(\underline{X}^*)$ $j \in K$, where K includes the set of critical and near critical constraints (could include the entire set m). Here also, we expand the set of design variables so $X_{n+1} = P$ as before.

The approximate optimization task now becomes, find the change in design variables, δX, to

$$\text{Minimize} \quad \tilde{F}(\underline{X}) = F(\underline{X}^*) + \underline{\nabla} F(\underline{X}^*) \cdot \underline{\delta X} + \frac{1}{2}\, \underline{\delta X}^T H_F \underline{\delta X} \qquad [65]$$

Subject to;

$$\tilde{g}_j(\underline{X}) = g_j(\underline{X}^*) + \underline{\nabla} g_j(\underline{X}^*) \cdot \underline{\delta X} + \frac{1}{2}\, \underline{\delta X}^T H_j \underline{\delta X} \leqq 0 \qquad [66]$$

$$\delta X_i^{\ell} \leqq \delta X_i \leqq \delta X_i^{u} \qquad i=1,n+1 \qquad [67]$$

where

$$\underline{\delta X} = \underline{X} - \underline{X}^* \qquad [68]$$

Here, P is treated as an independent variable and the move limits of equation (67) are imposed to limit the search to the region of validity of the quadratic approximation. In the event that δP (and therefore, the new value of P) is specified then the optimization is carried out with respect to the original n variables, with $\delta X_{n+1} = \delta P$.

The problem of equations (65)-(67) requires the same information as sensitivity based on the Kuhn-Tucker conditions (except the Lagrange multipliers are not needed here) but is complicated by the fact that a constrained optimization task is now required. However, the functions are now explicit and easily evaluated, along with their derivatives. The advantage here is that the approximate optimization task accounts for the complete set of constraints, K, and that no limiting assumptions are needed regarding the nature of the new optimum. Also, in the common situation in structural optimization using reciprocal variables, where the constraints are approximately linear, only the linear portion of equation (66) is needed, greatly simplifying the optimization task.

6.4 Comparison of Methods

In the previous sections, three distinct approaches were given for determining the sensitivity of an optimized design to some new parameter, P;

1. Based on the Kuhn-Tucker conditions at X^*.
2. Based on the Feasible Direction concept.
3. Based on second-order expansion of the problem.

These three approaches are compared here using simple examples to demonstrate, geometrically, the similarities and differences in the methods. Each method can be considered to produce a search direction, S^q, where q=1,2,3 corresponds to the particular method. The total derivative is proportional to S^q, so any differences in the methods are reflected in differing search directions.

Case 1 - Unconstrained Function

Consider the simple two-variable unconstrained problem;

$$\text{Minimize} \quad F = 2X_1^2 - 2X_1P + P^2 + 4X_1 - 4P \qquad [69]$$

While this is an explicit example function, it could as well represent the second-order approximation to a far more complicated problem.

The problem is first solved with respect to the single variable, X_1

and then sensitivity is calculated with respect to the new parameter, P.

Figure 3 is the two-variable function space for this problem, where initially P = 0, so the optimum with respect to X_1 is $F^* = -2$ at $X_1 = -1$. The search directions for the three methods are shown on the figure, and it is seen that methods 1 and 3, using second derivatives point to the solution of the quadratic problem. Method 2, being a first-order method, gives a steepest descent search direction.

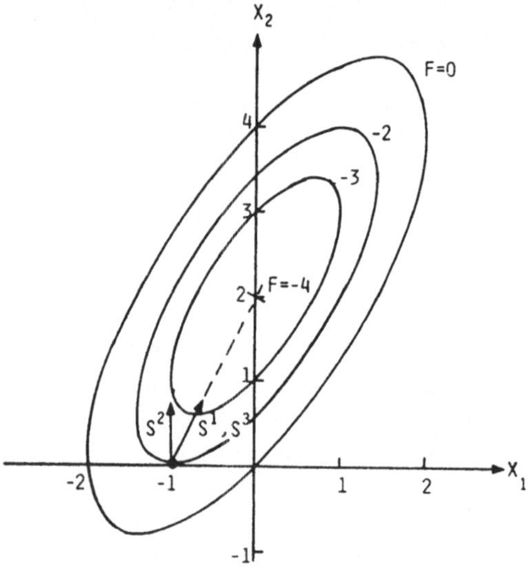

Figure 3. Unconstrained Problem.

<u>Case 2 - Constrained Problem with Discontinuous Derivative</u>

Figure 4 shows the results for the same problem with the addition of the linear inequality constraint,

$$g = -X_1 - 4P \leqq 0 \qquad [70]$$

Now minimization with respect to X_1 gives the constrained optimum, F=0, g=0 at $X_1 = 0$. Here the three approaches provide markedly different information. Method 1 produces a search direction which follows the constraint for either an increase or a decrease in P. If P is increased, method 2 gives a direction of steepest descent, while method 3 provides a direction toward the quadratic approximation to the minimum. If P is changed in the negative direction, each method results in a move along the constraint boundary. Clearly, because the optimum change direction

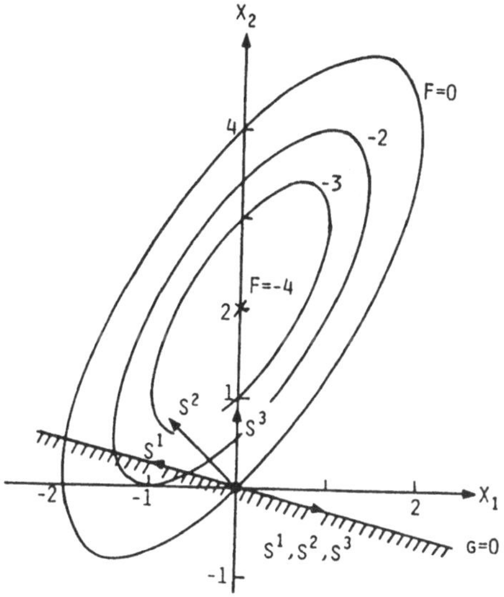

Figure 4. Discontinuous Sensitivities.

(whether first or second-order) is dependent on the sign of δP, the total derivative of the optimum objective is discontinuous at X^*.

Case 3 - Multiple Constraints

Here the constrained optimization problem to be solved is

$$\text{Minimize} \qquad F = X_1^2 + (P-1)^2 \qquad [71]$$

Subject to;

$$g_1 = -3X_1 - 2P + 10 \leqq 0 \qquad [72]$$

$$g_2 = -2X_1 - 3P + 10 \leqq 0 \qquad [73]$$

The problem is first solved with P held fixed at a value of 2. Then the sensitivity of the optimum to P is calculated. The two-variable function space and the calculated search directions are shown in Figure 5. This is a degenerate case which cannot be solved by method 1 due to matrix singularity. Here at X^*, two constraints are critical. However, because the problem is first a function of X_1 alone, one of these constraints is redundant. Methods 2 and 3 provide change vectors which follow the proper constraint boundary, depending on the sign of δP.

Case 4 - Dependence on the Magnitude of P

Here the problem to be solved is

Minimize $F = X_1^2 + P^2$ [74]

Subject to;

$$g = (X_1 - 5)^2 + (P - 5)^2 - 16 \leqq 0 \qquad [75]$$

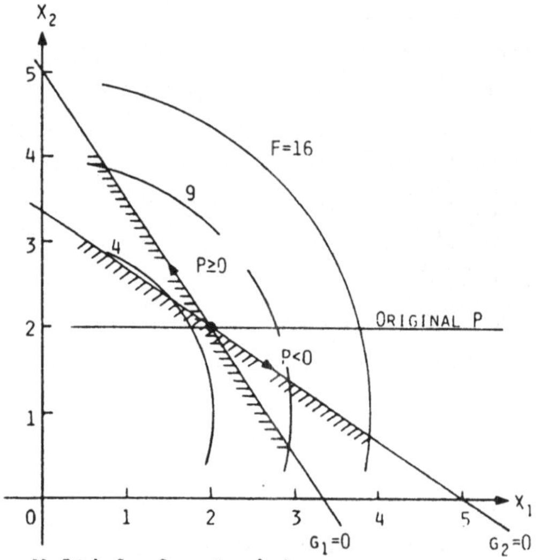

Figure 5. Multiple Constraints.

The two-variable function space and the calculated search directions are shown in Figure 6.

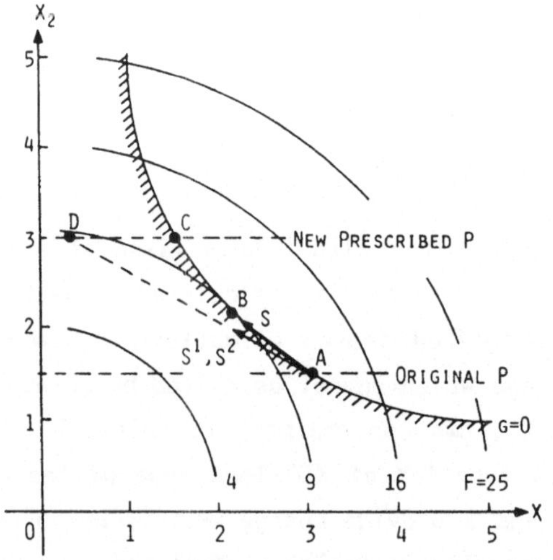

Figure 6. Specified Change in P.

Here, methods 1 and 2 provide the same search direction toward point D. Method 3 however, takes full advantage of the available second-order information to identify the new optimum based on the quadratic approximation. If P is allowed to change freely, the new optimum is at point B. This is the usual result for method 3 if no move limits are imposed on the design changes and the actual amount by which P is changed is not specified. However, if the change, δP, is specified, method 3 provides the best approximation to the new optimum with respect to the original design variable, X_1, at point C. Here, it is important to remember that this result is based on a quadratic approximation to the original problem, so in practice, move limits will be required to insure reasonable bounds on the solution. It is noteworthy that methods 1 and 2 provide the same search direction. This is because method 1, while requiring second derivatives of the constraints, does not actually use this information to update the approximation to the optimum design.

Consistency of the Sensitivity Calculations

Because the sensitivity is calculated about a numerically determined optimum, it is expected that the results will be dependent on the accuracy of $\underline{X}^*$. It is common that, while the objective function may be very near the theoretical optimum, the design vector, $\underline{X}^*$, is not this precise. Also, the gradients of the objective and constraint functions at the optimum are more variable in direction than in magnitude. These observations lead to the conclusion that the total derivative of $F(\underline{X}^*)$ with respect to P is reasonably stable, but that the rates of change of the independent design variables are more strongly dependent on the accuracy of the optimum.

This variability of the design vector sensitivities is seen from a simple example using linear sensitivity information.

Figure 7 shows the two-variable function space for the following problem;

Minimize $F = X_1^2 + X_2^2$ [76]

Subject to; $g = -X_1 - X_2 + P \leq 0$ [77]

where initially P=2. Points A, B and C are each near the optimum, with B being the precise optimum. The sensitivity of each "optimum" is shown by the search vectors and given numerically in Table 5.

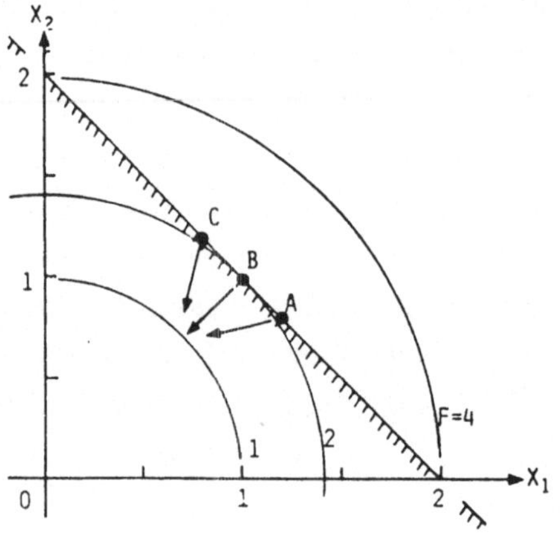

Figure 7. Sensitivity for Different Optima.

Table 5: Sensitivities for various optima

Var	X^A	S^A	X^B	S^B	X^C	S^C
X_1	1.2	-0.8	1.0	-0.5	0.8	-0.2
X_2	0.8	-0.2	1.0	-0.5	1.2	-0.8
F*	2.08	-	2.00	-	2.08	-
dF*/dP	-	-2.24	-	-2.00	-	-2.24

Thus, it is clear that the design sensitivity is dependent on the accuracy of $\underline{X}^*$. However, the information provided is still as useful, being the constrained steepest descent direction. This simply serves as a reminder, that if the design space is reasonably flat that, within numerical accuracy, the optimum design and its sensitivity is not unique.

6.5 Design Examples

Figure 8 is the 10-bar truss which is commonly used to demonstrate optimization procedures. The structure is loaded as shown by a single loading condition and is stress constrained. The allowable stress in each member is 25,000 psi, with the exception of member 9 which has a higher allowable stress. The cross-sectional areas of the members are the design variables and the total weight of the structure is to be minimized.

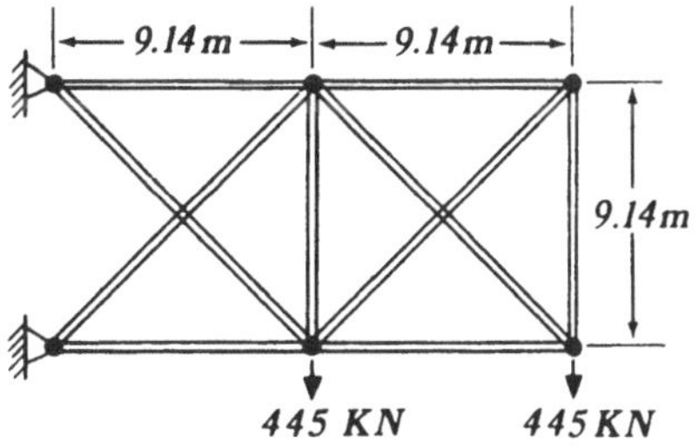

Figure 8. 10-Bar Truss.

Case 1: Comparison of Methods

Table 6 gives the optimum design and sensitivity results for a nominal design in which the allowable stress in member 9 is 30,000 psi, and will be changed in order to improve the optimum. The parameters of interest are listed in the first column and column 2 gives the initial optimum design. Columns 3 and 4 give the sensitivity based on the Kuhn-Tucker conditions and the present linear method, respectively. For this design, member 10 is very near its stress limit, but the actual binding constraint is the minimum size of the member. Thus, based on the Kuhn-Tucker conditions, the sensitivity of member 10 (as well as members 2, 5 and 6) is zero. In the linear method, this stress constraint is included in the active set and method recognizes that the size of member 10 must be increased to maintain feasibility.

The total derivative, $dF/d\bar{\sigma}_9$ is given for a 100% change in $\bar{\sigma}_9$, so the predicted change in F* is the total derivative times the fractional change to be made in $\bar{\sigma}_9$.

Columns 7-9 of Table 6 give the predicted design for each method for an allowable stress in member 9 of 40,000 psi (a 33% change from its nominal value), and column 10 gives the calculated optimum. The first two methods predicted essentially the same optimum except the present linear method accounted for the need to increase the size of member 10. The quadratic method provided the result nearest the calculated optimum, as

Table 6. Sensitivity of the optimum to the allowable stress in member 9.

(1) Var.	(2) X^*	(3) S^{K-T}	(4) S^L	(5) X^Q	(6) S^Q	(7) X^{K-T}	(8) X^L	(9) X^Q	(10) X^*
A_1	7.9292	-0.0492	-0.1195	7.902	-0.0991	7.913	7.889	7.896	7.900
A_2	0.1000	0.0000	0.0000	0.100	0.0000	0.100	0.100	0.100	0.100
A_3	8.0704	0.0499	0.1200	8.095	0.0883	8.087	8.110	8.100	8.100
A_4	3.9297	-0.0496	-0.1194	3.902	-0.0998	3.913	3.890	3.896	3.900
A_5	0.1000	0.0000	0.0000	0.100	0.0000	0.100	0.100	0.100	0.100
A_6	0.1000	0.0000	0.0000	0.100	0.0000	0.100	0.100	0.100	0.100
A_7	5.7566	0.0706	0.1699	5.794	0.1376	5.780	5.813	5.803	5.799
A_8	5.5566	-0.0698	-0.1691	5.545	-0.0418	5.533	5.500	5.543	5.515
A_9	4.6311	-4.6888	-4.7718	3.628	-3.6850	3.068	3.040	3.403	3.677
A_{10}	0.1000	0.0000	0.1708	0.144	0.1601	0.100	0.157	0.153	0.141
$\bar{\sigma}_9$	30,000	1.0000	1.0000	38,165	1.0000	40,000	40,000	40,000	40,000
F^*	1,545	-	-	1496.5	-	1,465	1,466	1,485	1,498
$dF/d\bar{\sigma}$	-	-240.5	-238.4	-	-178.6	-	-	-	-

would be expected.

Case 2: Discontinuity of the Sensitivity

Linear sensitivity of the 10 bar truss was calculated here for a nominal value of 37,500 psi for the allowable stress in member 9. Now the direction of change in the stress allowable was specified and the resulting problem was solved by the method of equations (60)-(64). If the direction of change in the allowable stress was not specified or was specified to be positive, the resulting sensitivities were zero, correctly indicating that the optimum weight cannot be reduced by changing $\bar{\sigma}_9$.

Table 7 gives the resulting sensitivities if the allowable stress is required to be reached. It is noteworthy that the projected optimum for a design allowable of 35,000 psi is very near the calculated optimum, even though the projected values of the design variables are not this precise.

Table 7. Discontinuity in sensitivity.

Var	X*	S^L	X^L	X*
A_1	7.9008	-0.0981	7.8943	7.9154
A_2	0.1000	0.2043	0.1136	0.1000
A_3	8.0998	0.1554	8.1102	8.0900
A_4	3.9000	-0.0963	3.8994	3.9099
A_5	0.1000	0.0000	0.1000	0.1000
A_6	0.1000	0.2299	0.1153	0.1000
A_7	5.7981	0.2076	5.8119	5.7848
A_8	5.5155	-0.1502	5.5055	5.5295
A_9	3.6791	3.6356	3.9215	3.9496
A_{10}	0.1413	0.2649	0.1590	0.1274
$\bar{\sigma}_9$	39,500	-1.0000	35,000	35,000
F*	1,498	-	1,512	1,511
$dF/d\bar{\sigma}_9$	-	215.7	-	-

This again underscores the somewhat non-unique nature of the optimum sensitivity calculations.

This example clearly demonstrates the need to account for the possibility that the design sensitivity may be discontinuous at X*. In practice, this discontinuity cannot be identified in any a priori fashion and so the mathematics of the procedure must be relied on to deal with that situation.

7.0 SUMMARY

The general optimization problem statement has been outlined and a variety of algorithms have been identified for its solution. The examples given here should not be considered as a definitive comparison of methods, but rather a general indication of algorithmic efficiencies. In the case of structural optimization, it is common to create an approximate problem to be solved by the optimizer. Therefore, since this is an explicit problem with a clear mathematical form, the actual optimization algorithm used for its solution becomes of lesser importance. The principal purpose here has been to familiarize the reader with the mathematical and algorithmic features of the optimization task and offer sufficient references for more detailed study of these and other methods.

Three procedures for calculating the sensitivity of an optimized design to some problem parameter have been presented and compared. The method based on the Kuhn-Tucker conditions assumes that the conditions in effect at the optimum remain in effect when P is changed. The linear method provides the sensitivity in the classical sense of a derivative, while accounting for the inequality constrained nature of design. The quadratic method is more appropriately considered as an improved estimate of the optimum using available second-order information.

In practice, careful problem formulation can be expected to improve the quality of the projected optimum. For example, for many structural optimization problems, the use of reciprocal variables will allow much larger perturbations in the design without significant constraint violations.

As noted in the introduction, sensitivity information is valuable in its own right for estimating the effect that design changes will have on the optimum, wihout re-optimization. Additionally, this capability provides a convenient tool for use in multi-level and multi-discipline design, particularly where distributed computing is desirable. This general area of study is expected to be the direction of future research using these techniques.

8.0 REFERENCES

1. Schmit, L.A., Structural Design by Systematic Synthesis, Proceedings, 2nd Conference on Electronic Computations, ASCE, New ork, pp. 105-122, 1960.

2. Vanderplaats, G.N., Numerical Optimization Techniques for Engineering Design: with Applications, McGraw-Hill, 1984.

3. Fletcher, R. and C.M. Reeves, Function Minimization by Conjugate Gradients, Br. Computer J., Vol. 7, No. 2, pp. 149-154, 1964.

4. Davidon, W.C., "Variable Metric Method for Minimization," Argonne National Laboratory, ANL-5990 Rev., University of Chicago, 1959.

5. Fletcher, R. and M.J.D. Powell, A Rapidly Convergent Method for Minimization, Computer J., Vol. 6, No. 2, pp. 163-168, 1963.

6. Broydon, C.G., The Convergence of a Class of Double Rank Minimization Algorithms, parts I and II, J. Inst. Math. Appl., Vol. 6, pp. 76-90, pp. 222-231, 1970.

7. Fletcher, R., A New Approach to Variable Metric Algorithms, Computer J., Vol. 13, pp. 317-322, 1970.

8. Goldfarb, D., A Family of Variable Metric Methods Derived by Variational Means, Math. Comput., Vol. 24, pp. 23-36, 1970.

9. Shanno, D.F., Conditioning of Quasi-Newton Methods for Function Minimization, Math. Comput., Vol. 24, pp. 647-656, 1970.

10. Dantzig, G.B., "Programming in a Linear Structure," Comptroller, USAF, Washington, DC, February 1948.

11. Dantzig, G.B., Linear Programming and Extensions, Princeton University Press, Princeton, NJ, 1963.

12. Kelley, J.E., The Cutting Plane Method for Solving Convex Programs, J. SIAM, pp. 703-713, 1960.

13. Baldur, R., Structural Optimization by Inscribed Hyperspheres, J. Engin. Mech., ASCE, Vol. 98, No. EM3, pp. 503-508, June 1972.

14. Cassis, J.H. and L.A. Schmit, On Implementation of the Extended Interior Penalty Function, Int. J. Num. Meth. Engin., Vol. 10, No. 1, pp. 3-23, 1976.

15. Haftka, R.T. and J.H. Starnes, Applications of a Quadratic Extended Interior Penalty Function for Structural Optimization, AIAA J., Vol. 14, No. 6, pp. 718-724, June 1976.

16. Pierre, D.A. and M.J. Lowe, "Mathematical Programming via Augmented Lagrangians," Applied Mathematics and Computation Series, Addison-Wesley, Reading, Mass., 1975.

17. Zoutendijk, G., "Methods for Feasible Directions," Elsevier, Amsterdam, 1960.

18. Vanderplaats, G.N. and F. Moses, Structural Optimization by Methods of Feasible Directions, J. Computers Struct., Vol. 3, pp. 739-755, July 1973.

19. Vanderplaats, G.N., CONMIN - A FORTRAN Program for Constrained Function Minimization; User's Manual, NASA TM X-62,282, August 1973.

20. Gabriele, G.A. and K.M. Ragsdell, The Generalized Reduced Gradient Method: A Reliable Tool for Optimal Design, ASME J. Engin. Ind., Series B, Vol. 99, No. 2, pp. 394-400, May 1977.

21. Vanderplaats, G.N., "An Efficient Feasible Direction Algorithm for Design Synthesis," AIAA Journal, Vol. 22, No. 11, Nov. 1984.

22. Powell, M.J.D., A Fast Algorithm for Nonlinearly Constrained Optimization Calculations, No. DAMPTP77/NA 2, University of Cambridge, England, 1977.

23. Powell, M.J.D., Algorithms for Nonlinear Constraints that use Lagrangian Functions, Math. Prog., Vol. 14, No. 2, 1978.

24. Biggs, M.C., Constrained Minimization Using Recursive Equality Quadratic Programming in "Numerical Methods for Nonlinear Optimization," F.A. Lootsma (ed.), Academic Press, London, pp. 411-428, 1972.

25. Biggs, M.C., Constrained Minimization Using Recursive Quadratic Programming: Some Alternative Subproblem Formulations, in "Towards Global Optimization," L.C.W. Dixon and G.P. Szego (eds.), North-Holland Publishing Co., pp. 341-349, 1975.

26. Han, S.P., "Superlinearly Convergent Variable Metric Algorithms for General Nonlinear Programming Problems," Math. Prog., Vol. 11, pp. 263-282, 1976.

27. Han, S.P., "A Globally Convergent Method for Nonlinear Programming," J. Optim. Theory Appl., Vol. 22, No. 3, pp. 297-309, 1977.

28. Vanderplaats, G.N. and H. Sugimoto, "Application of Variable Metric Methods to Structural Synthesis," Engineering Computations, Vol. 2, No. 2, pp. 96-100, June 1985.

29. Fleury, C. and V. Braibant, "Structural Optimization. A New Dual Method Using Mixed Variables," Int. J. for Numerical Methods in Engineering (to appear).

30. Vanderplaats, G.N. and H. Sugimoto, "A General-Purpose Optimization Program for Engineering Design," Computers and Structures (to appear).

31. Vanderplaats, G.N. and Yoshida, N., "Efficient Calculation of Optimum Design Sensitivity," AIAA Journal, Vol. 23, No. 11, Nov. 1985, pp. 1798-1803.

32. Sobieszczanski-Sobieski, J., Barthelemy, J.F., and Riley, K.M., "Sensitivity of Optimum Solutions to Problem Parameters," AIAA Journal, Vol. 20, Sept. 1982, pp. 1291-1299.

33. Barthelemy, J.F. and Sobieszczanski-Sobieski, J., "Extrapolation of Optimum Design Based on Sensitivity Derivatives," AIAA Journal, Vol. 21, May 1983, pp. 707-709.

34. Barthelemy, J.F. and Sobieszczanski-Sobieski, J., "Optimum Sensitivity Derivatives of Objective Functions in Nonlinear Programming," AIAA Journal, Vol. 21, June 1983, pp. 913-915.

35. Schmit, L.A. and Cheng, K.J., "Optimum Design Sensitivity Based on Approximation Concepts and Dual Methods," Proc. AIAA/ASME/ASCE/AHS 23rd Structures, Structural Dynamics and Materials Conference, New Orleans, LA., May 1982, AIAA Paper No. 82-0713-CP.

36. Zangwill, W.I., Nonlinear Programming: A Unified Approach, Prentice-Hall, Englewood Cliffs, NJ, 1969.

FINITE ELEMENTS IN OPTIMAL STRUCTURAL DESIGN

Raphael T. Haftka
Department of Aerospace and Ocean Engineering
Virginia Polytechnic Institute and State University
Blacksburg, Virginia 24061, USA

and

Manohar P. Kamat
School of Engineering Science and Mechanics
Georgia Institute of Technology
Atlanta, Georgia 30332, USA

I. INTRODUCTION

The optimization of a structure modeled by finite elements can proceed in two diametrically opposed directions. The first direction is that of interfacing a finite element software package with an optimization package where both packages are treated primarily as black boxes. The second direction is the intimate integration of the finite element analysis and optimization processes. Many research structural optimization programs followed the second path for reasons of efficiency and convenience to the researchers who wrote these programs. However, in production codes the tendency is to follow the more modular first direction. The integrated approach is probably justified only when it reflects algorithmic integration of the analysis and optimization processes. This type of integration is presently at the research stage.

The next four sections of this chapter are concerned with the interfacing of black-box finite element and optimization codes. The last chapter is devoted to a discussion of the integrated approach.

II. THE FINITE-ELEMENT AND OPTIMIZATION INTERFACE

The interface required to connect an optimization package with a finite element package is typically composed of three elements: (i) design variable expression; (ii) constraint and objective function calculation; and (iii) sensitivity calculations.

The first element, design variable expression, is concerned with expressing the values of the design variables supplied by the optimization program as a finite element model. This part of the interface can be implemented as a preprocessor to the finite element program. It requires a strategy for selecting structural parameters as design variables and relating them to the finite element model. It is important not to attempt

NATO ASI Series, Vol. F27
Computer Aided Optimal Design: Structural and Mechanical Systems. Edited by C. A. Mota Soares

a one-to-one correspondence between design variables and finite elements except for skeletal structures (trusses and frames). For structures containing two and three dimensional elements, the accuracy of the finite element analysis can be destroyed if such one-to-one correspondence is enforced. As an example consider the square plate design (Fig. 1) obtained by the author (Ref. 1) by associating a design variable with almost each individual element thickness. The 7 x 7 mesh was reasonable for analyzing a uniform thickness plate, but it is inappropriate for analyzing the final design shown in Fig. 1. Similar results are obtained by associating a design variable with each node of a boundary in shape optimization as shown in Fig. 2 (taken from Ref. 2).

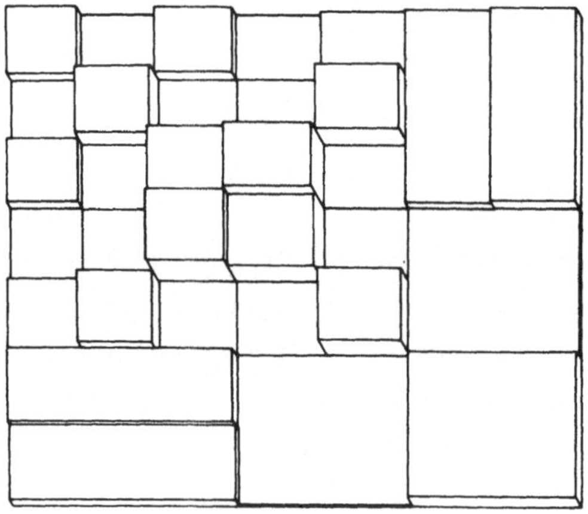

Figure 1: Plate design from Reference 1

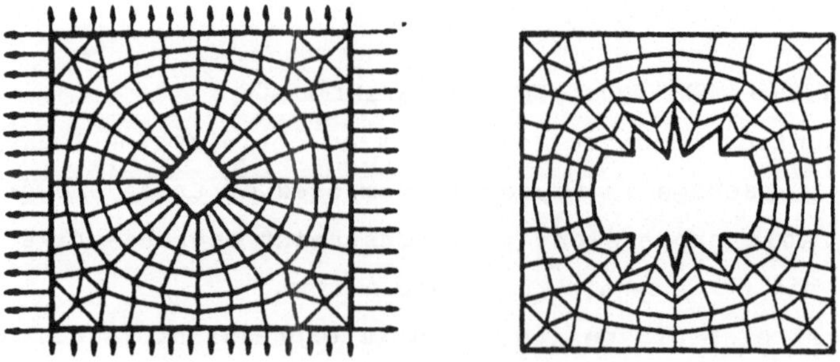

Figure 2: Optimal hole shape from Reference 2

The second element of the interface, the calculation of objective function and constraints, typically requires a post processor which uses the structural response obtained by the finite element program to obtain the desired quantity.

The third element of the interface, the calculation of sensitivities is the most troublesome, and is the subject of discussion of the following sections. Because the techniques associated with sensitivity calculation are substantially different for static analysis, eigenproblems, and transient response, each one of these classes of problem is dealt with in a separate section.

III. CALCULATION OF SENSITIVITIES FOR STATIC LOADS

The equations of equilibrium in terms of the nodal displacement vector U are generated from a finite element model in the form

$$KU = F \tag{1}$$

where K is the stiffness matrix and F a load vector. A typical constraint, such as a limit on a displacement or stress component may be written as

$$g(U,x) \geqq 0 \tag{2}$$

where, for the sake of simplified notation, it is assumed that g depends only on a single design variable x. Using the differentiation chain rule, we obtain

$$dg/dx = \partial g/\partial x + Z^T dU/dx \tag{3}$$

where Z is a vector with components $z_i = \partial g/\partial u_i$. The first term in Eq. (3) is usually zero or very easy to obtain so we discuss only the computation of the second term. Differentiating Eq. (1) with respect to x we obtain

$$KdU/dx = dF/dx - (dK/dx)U \tag{4}$$

Premultiplying Eq. (4) by $Z^T K^{-1}$ obtain

$$Z^T(dU/dx) = Z^T K^{-1}[dF/dx - (dK/dx)U] \tag{5}$$

Numerically, the calculation of $Z^T dU/dx$ may be performed in two different ways. The first, called the direct method (sometimes referred to as the behavior space approach), consists of solving Eq. (4) for dU/dx and then taking the scalar product with Z. The second approach, called the adjoint variable method, defines an adjoint variable vector Λ which is the solution of the system

$$K\Lambda = Z \tag{6}$$

and then Eq. (3) is rewritten as

$$dg/dx = \partial g/\partial x + \Lambda^T[dF/dx - (dK/dx)U] \tag{7}$$

where use has been made of the symmetry of K.

The adjoint variable method is also known as the dummy load method because Z is often described as a dummy load. When the g in Eq. (2) is an upper limit on a single displacement component, the dummy load also has a single non-zero component corresponding to the constrained displacement component. Similarly, when g is an upper limit on the stress in a truss member, the dummy load is composed of a pair of equal and opposite forces acting on the two nodes of the member.

Both the direct and adjoint variable methods require the solution of a system of equations as the major component of the computational effort. We should note, however, that the matrix K of the equations is usually available in a factored form from the solution of Eq. (1) for the displacements. The solution for dU/dx or Λ is therefore much cheaper than the original solution of Eq. (1). The difference between the computational effort associated with the direct method and with the adjoint variable method depends on the relative number of constraints and design variables. The direct method requires the solution of Eq. (4) once for each design variable, while the adjoint variable method requires the solution of Eq. (6) once for each constraint. Consequently, the direct method is more efficient than the adjoint variable method when the number of design variables is smaller than the number of constraints. The adjoint variable method is more efficient than the direct method when the number of design variables is larger than the number of constraints.

In practical design situations we usually have to consider several load cases. The effort associated with the direct method is approximately proportional to the number of load cases. The number of critical constraints, on the other hand, does not change significantly with the number of load cases, and is usually of the same order as the number of design variables. Therefore, in a multiple-load-case situation the adjoint method is preferable.

Both the direct and adjoint methods require the derivatives of the stiffness matrix with respect to the design variable. When it is difficult to obtain this derivative it is possible to fall back on a finite difference approach. For example, using the forward difference formula we can approximate dU/dx as

$$\frac{dU}{dx} \cong \frac{U(x + \Delta x) - U(x)}{\Delta x} \tag{8}$$

The evaluation of dU/dx from Eq. (8) is more expensive than from Eq. (4) because the calculation of $U(x + \Delta x)$ requires assembling a new stiffness matrix $K(x + \Delta x)$ and solving a new system

$$K(x + \Delta x)U(x + \Delta x) = F(x + \Delta x) \tag{9}$$

The major difference in computational expense is due to the fact that Eq. (1) and Eq. (4) have the same matrix K while Eq. (9) has a new matrix which requires an extra factorization. Beside the computational expense of the finite difference approach it is also associated with numerical problems. The error in Eq. (8) is

$$\frac{dU}{dx} - \frac{U(x + \Delta x) - U(x)}{\Delta x} = \frac{\Delta x}{2} \frac{d^2U}{dx^2} (x + \xi\Delta x) \qquad 0 \leq \xi \leq 1 \qquad (10)$$

and is called the truncation error. The selection of a large Δx is associated with high value of the truncation error. On the other hand, too small values of Δx can cause the round-off error to become excessive. Occasionally it is difficult to find a good value of Δx which is a good compromise (see Refs. 3 and 4 for discussion of optimal Δx).

A method which combines the simplicity of the finite difference approach with the computational efficiency of the analytical approach is the semi-analytical approach (Ref. 5). The finite difference method is used to calculate dF/dx and dK/dx and then the direct or adjoint methods are used for obtaining the constraint derivatives. For example, using the forward difference approximation

$$\frac{dK}{dx} \cong \frac{K(x + \Delta x) - K(x)}{\Delta x} \qquad (11)$$

where the error is $0.5\ \Delta x d^2K(x + \xi\Delta x)/dx^2$ for some $0 \leq \xi \leq 1$. The semi-analytical approach, however, can occasionally suffer from excessive truncation errors as shown in Ref. 6 and the following example.

Example - Cantilever beam subject to a displacement constraint

Figure 3 shows a two-element cantilever beam under an end load. We want to find the derivative of the displacement at the tip with respect to the moment of inertia I_1 and the length ℓ_1. The problem is simple enough so that an analytical solution based on elementary beam theory is easily obtained:

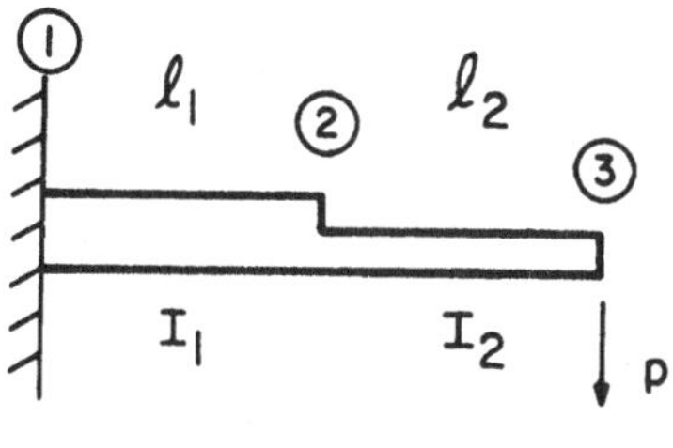

Figure 3: Beam example for derivatives of static response

$$w_{tip} = \frac{p}{3EI_1}(\ell_1^3 + 3\ell_1^2\ell_2 + 3\ell_1\ell_2^2) + \frac{p\ell_2^3}{3EI_2} \tag{12}$$

so that

$$\frac{\partial w_{tip}}{\partial I_1} = \frac{-p}{3EI_1^2}(\ell_1^3 + 3\ell_1^2\ell_2 + 3\ell_1\ell_2^2) \tag{13}$$

$$\frac{\partial w_{tip}}{\partial \ell_1} = \frac{p}{3EI_1}(3\ell_1^2 + 6\ell_1\ell_2 + 3\ell_2^2) = \frac{p}{EI_1}(\ell_1 + \ell_2)^2 \tag{14}$$

The finite element solution is based on a standard cubic beam element. We denote the displacement and rotation at the i-th node as w_i and θ_i, respectively. The element stiffness matrix is

$$[K^E] = \frac{EI}{\ell^3}\begin{bmatrix} 12 & 6\ell & -12 & 6\ell \\ 6\ell & 4\ell^2 & -6\ell & 2\ell^2 \\ -12 & -6\ell & 12 & -6\ell \\ 6\ell & 2\ell^2 & -6\ell & 4\ell^2 \end{bmatrix} \tag{15}$$

so that the global stiffness matrix, corresponding to degrees of freedom w_2, θ_2, w_3, θ_3 is

$$K = E\begin{bmatrix} 12\left(\frac{I_1}{\ell_1^3} + \frac{I_2}{\ell_2^3}\right) & -6\left(\frac{I_1}{\ell_1^2} - \frac{I_2}{\ell_2^2}\right) & -12\frac{I_2}{\ell_2^3} & 6\frac{I_2}{\ell_2^2} \\ & 4\left(\frac{I_1}{\ell_1} + \frac{I_2}{\ell_2}\right) & -6\frac{I_2}{\ell_2^2} & 2\frac{I_2}{\ell^2} \\ & & 12\frac{I_2}{\ell_2^3} & -6\frac{I_2}{\ell_2^2} \\ \text{sym} & & & 4\frac{I_2}{\ell_2} \end{bmatrix} \tag{16}$$

The load vector F is $F^T = [0,0,p,0]$ and the solution of Eq. (1) is

$$U = \begin{Bmatrix} w_2 \\ \theta_2 \\ w_3 \\ \theta_3 \end{Bmatrix} = \left(\frac{p}{E}\right)\begin{Bmatrix} \ell_1^3/3I_1 + \ell_2\ell_1^2/2I_1 \\ \ell_1^2/2I_1 + \ell_2\ell_1/I_1 \\ (\ell_1^3 + 3\ell_1^2\ell_2 + 3\ell_1\ell_2^2)/3I_1 + \ell_2^3/3I_2 \\ \ell_1^2/2I_1 + \ell_2\ell_1/I_1 + \ell_2^2/I_2 \end{Bmatrix} \tag{17}$$

We first use analytical methods for the derivative calculation so that we need $(\partial K/\partial I_1)U$ and $(\partial K/\partial \ell_1)U$

$$\frac{\partial K}{\partial I_1} U = \left(\frac{E}{\ell_1^3}\right) \begin{bmatrix} 12 & -6\ell_1 & 0 & 0 \\ & 4\ell_1^2 & 0 & 0 \\ & & 0 & 0 \\ \text{sym} & & & 0 \end{bmatrix} \begin{Bmatrix} w_2 \\ \theta_2 \\ w_3 \\ \theta_3 \end{Bmatrix}$$

$$= \left(\frac{E}{\ell_1^3}\right) \begin{Bmatrix} 12w_2 - 6\ell_1\theta_2 \\ -6\ell_1 w_2 + 4\ell_1^2\theta_2 \\ 0 \\ 0 \end{Bmatrix} = \left(\frac{P}{I_1}\right) \begin{Bmatrix} 1 \\ \ell_2 \\ 0 \\ 0 \end{Bmatrix} \tag{18}$$

where the expression for w_2 and θ_2 from Eq. (17) was used. Similarly

$$\frac{\partial K}{\partial \ell_1} U = \left(\frac{EI_1}{\ell_1^4}\right) \begin{bmatrix} -36 & 12\ell_1 & 0 & 0 \\ & -4\ell_1^2 & 0 & 0 \\ & & 0 & 0 \\ \text{sym} & & & 0 \end{bmatrix} \begin{Bmatrix} w_2 \\ \theta_2 \\ w_3 \\ \theta_3 \end{Bmatrix}$$

$$= \left(\frac{4EI_1}{\ell_1^4}\right) \begin{Bmatrix} -9w_2 + 3\ell_1\theta_2 \\ 3\ell_1 w_2 - \ell_1^2\theta_2 \\ 0 \\ 0 \end{Bmatrix} = \left(\frac{p}{\ell_1}\right) \begin{Bmatrix} -6(1 + \ell_2/\ell_1) \\ 2(\ell_1 + \ell_2) \\ 0 \\ 0 \end{Bmatrix} \tag{19}$$

Using the direct method

$$\frac{\partial U}{\partial I_1} = K^{-1}\left[\frac{\partial F}{\partial I_1} - \frac{\partial K}{\partial I_1} U\right]$$

or

$$\frac{\partial}{\partial I_1} \begin{Bmatrix} w_2 \\ \theta_2 \\ w_3 \\ \theta_3 \end{Bmatrix} = -K^{-1} \begin{Bmatrix} p/I_1 \\ p\ell_2/I_1 \\ 0 \\ 0 \end{Bmatrix} = -\frac{p}{EI_1^2} \begin{Bmatrix} \ell_1^2\ell_2/2 + \ell_1^3/3 \\ \ell_1\ell_2 + \ell_1^2/2 \\ \ell_1^2\ell_2 + \ell_1\ell_2^2 + \ell_1^3/3 \\ \ell_1\ell_2 + \ell_1^2/2 \end{Bmatrix} \tag{20}$$

so that $\partial w_{tip}/\partial I_1 = \partial w_3/\partial I_1$ agrees with the result from Eq. (13).

Similarly

$$\frac{\partial U}{\partial \ell_1} = K^{-1}\left[\frac{\partial E}{\partial \ell_1} - \frac{\partial K}{\partial \ell_1} U\right]$$

or

$$\frac{\partial}{\partial \ell_1} \begin{Bmatrix} w_2 \\ \theta_2 \\ w_3 \\ \theta_3 \end{Bmatrix} = -K^{-1} \begin{Bmatrix} -(6p/\ell_1)(1 + \ell_2/\ell_1) \\ (2p/\ell_1)(\ell_1 + \ell_2) \\ 0 \\ 0 \end{Bmatrix} = \left(\frac{p}{E}\right) \begin{Bmatrix} (\ell_1^2 + \ell_1\ell_2)/I_1 \\ (\ell_1 + \ell_2)/I_1 \\ (\ell_1 + \ell_2)^2/I_1 \\ (\ell_1 + \ell_2)I_1 \end{Bmatrix} \tag{21}$$

So that $\partial w_{tip}/\partial \ell_1 = \partial w_3/\partial \ell_1$ agrees with the result from Eq. (14).

Using the adjoint method, $Z^T = \partial w_{tip}/\partial U = [0,0,1,0]$ so,

$$\Lambda = K^{-1}Z = K^{-1}\begin{Bmatrix} 0 \\ 0 \\ 1 \\ 0 \end{Bmatrix} = (\frac{1}{E}) \begin{Bmatrix} \ell_1^3/3I_1 + \ell_2\ell_1^2/2I_1 \\ \ell_1^2/2I_1 + \ell_2\ell_1/I_1 \\ (\ell_1^3 + 3\ell_1^2\ell_2 + 3\ell_1\ell_2^2)/3I_1 + \ell_2^3/3I_2 \\ \ell_1^2/2I_1 + \ell_2\ell_1/I_1 + \ell_2^2/I_2 \end{Bmatrix} \tag{22}$$

so that, from Eq. (7)

$$\frac{\partial w_{tip}}{\partial I_1} = -\Lambda^T(\partial K/\partial I_1)U = -\frac{p}{EI_1}(\ell_1^3/3I_1 + \ell_2\ell_1^2/2I_1 + \ell_1^2\ell_2/2I_1 + \ell_2^2\ell_1/I_1)$$

$$= -\frac{p}{EI_1^2}(\ell_1^2\ell_2 + \ell_2^2\ell_1 + \ell_1^3/3) \tag{23}$$

and

$$\frac{\partial w_{tip}}{\partial \ell_1} = -\Lambda^T(\partial K/\partial \ell_1)U = -(\frac{p}{E_1\ell_1})(\ell_1 + \ell_2)(-2\ell_1^2/I_1 - 3\ell_2\ell_1/I_1$$

$$+ \ell_1^2/I_1 + 2\ell_2\ell_1/I_1) = \frac{p}{I_1E_1}(\ell_1 + \ell_2)^2 \tag{24}$$

Next we compare the exact derivatives to those obtained by finite differences. Using the forward difference approach

$$\frac{\partial w_{tip}}{\partial I_1} \cong \frac{w_{tip}(I_1 + \Delta I_1) - w_{tip}(I_1)}{\Delta I_1} \tag{25}$$

and the truncation error e_t is approximately

$$e_t = \frac{\partial^2 w_{tip}}{\partial I_1^2}\frac{\Delta I_1}{2} = \frac{p}{3EI_1^3}(\ell_1^3 + 3\ell_1\ell_2) + 3\ell_1\ell_2^2)\Delta I_1 \tag{26}$$

the relative error

$$\frac{e_t}{\frac{\partial w_{tip}}{\partial I_1}} = -\frac{\Delta I_1}{I_1} \tag{27}$$

so that it is enough to take $\Delta I_1/I_1 = 10^{-3}$ to get a negligible truncation error. Similarly, the truncation error for the derivative with respect to ℓ_1 is approximately

$$e_t = \frac{\partial^2 w_{tip}}{\partial \ell_1^2}\frac{\Delta \ell_1}{2} = \frac{p}{EI_1}(\ell_1 + \ell_2)\Delta \ell_1 \tag{28}$$

and the relative error is

$$\frac{e_t}{\dfrac{\partial w_{tip}}{\partial \ell_1}} = \frac{\Delta \ell_1}{\ell_1 + \ell_2} \tag{29}$$

and again it is enough to take $\Delta\ell_1/(\ell_1 + \ell_2) = 10^{-3}$. The error analysis for the semi-analytical method is more complicated. Using Eq. (23)

$$\frac{\partial w_{tip}}{\partial I_1} \simeq -\Lambda^T \frac{K(I_1 + \Delta I_1) - K(I_1)}{\Delta I_1} U \tag{30}$$

and the truncation error is

$$e_t = -\frac{\Delta I_1}{2} \Lambda^T \frac{\partial^2 K}{\partial I_1^2} U = 0 \tag{31}$$

because K is a linear function of I_1. The situation is not as good for the truncation error of $\partial w_{tip}/\partial \ell_1$ which is approximately

$$e_t = -\frac{\Delta \ell_1}{2} \Lambda^T \frac{\partial^2 K}{\partial \ell_1^2} U = -\frac{p\Delta\ell_1}{EI_1\ell_1}(3\ell_1^2 + 7\ell_1\ell_2 + 4\ell_2^2) \tag{32}$$

and the relative error is

$$\frac{e_t}{\dfrac{\partial w_{tip}}{\partial \ell_1}} = -\frac{3\ell_1^2 + 7\ell_1\ell_2 + 4\ell_2^2}{(\ell_1 + \ell_2)^2} \frac{\Delta\ell_1}{\ell_1} \tag{33}$$

Comparing the semi-analytical error to the one obtained by the finite-difference approach, Eq. (29) we note that it is seven times larger when $\ell_1 = \ell_2$. As shown in Ref. 6, this larger error for the semi-analytical method becomes very serious with increasing number of elements.

IV. SENSITIVITY OF VIBRATION AND BUCKLING CONSTRAINTS

Vibration and buckling problems lead to an eigenvalue problem of the type

$$KU - \mu MU = 0 \tag{34}$$

where K is the stiffness matrix, M is the mass matrix (vibration) or the geometric stiffness matrix (buckling) and U is the mode shape. For vibration problems μ is the square of the frequency of free vibration and for buckling problems it is the buckling load factor. Both K and M are symmetric. The mode shape is often normalized, with a symmetric, positive definite matrix W as

$$U^T W U = 1 \tag{35}$$

where, in vibration problems, the normalizing matrix W is often equal to the mass matrix M. Equations (34) and (35) hold for all eigen-pairs (μ_k, U_k). Differentiating these equations with respect to a design variable x for a particular eigen-pair we obtain

$$(K - \mu_k M)(dU_k/dx) - (d\mu_k/dx)MU_k = -(dK/dx - \mu_k dM/dx)U_k \tag{36}$$

and

$$U_k^T W(dU_k/dx) = - 0.5U_k^T(dW/dx)U_k \tag{37}$$

where use has been made of the symmetry of the matrix W. Equations (36) and (37) are valid only for the case of distinct eigenvalues. In the case of repeated eigenvalues the repeated eigenvalues may not even be differentiable, and only directional derivatives may be obtained (Ref. 7). If only the derivatives of the eigenvalues are required they may be obtained by premultiplying Eq. (36) by U_k^T to obtain

$$d\mu_k/dx = U_k^T(dK/dx - \mu_k dM/dx)U_k/U_k^T M U_k \tag{38}$$

When eigenvector derivatives are required Eqs. (36) and (37) are combined as

$$\begin{bmatrix} K - \mu_k M & - MU_k \\ - U_k^T W & 0 \end{bmatrix} \begin{Bmatrix} dU_k/dx \\ d\mu_k/dx \end{Bmatrix} = \begin{Bmatrix} - (\frac{dK}{dx} - \mu_k \frac{dM}{dx})U_k \\ - \frac{1}{2} U_k^T \frac{dW}{dx} U_k \end{Bmatrix} \tag{39}$$

The system (39) may be solved for the derivatives of the eigenvalue and the eigenvector. Care must be taken in the solution process because the principal minor $K - \mu_k M$ is singular. Cardani and Mantegazza [Ref. 8] discuss several solution strategies.

One of the more popular solution techniques is due to Nelson (Ref. 9). Nelson's method temporarily replaces the normalization equation, Eq. (35), by the requirement that the largest component of the eigenvector be equal to one. Denoting this re-normalized vector $\bar{U}$, and assuming that its largest component is the m-th one, Eq. (35) is replaced by

$$\bar{u}_m = 1 \tag{40}$$

and Eq. (37) by

$$(d\bar{U}_k/dx)_m = 0 \tag{41}$$

Equation (36) is applicable also to $\bar{U}_k$ and is written as

$$(K - \mu_k M)(d\bar{U}_k/dx) = R \tag{42}$$

where

$$R = (d\mu_k/dx)M\bar{U}_k - (dK/dx - \mu_k dM/dx)\bar{U}_k \tag{43}$$

Equation (41) is now used to reduce the order of Eq. (42) by deleting the m-th row and m-th column. When the eigenvalue μ_k is distinct the reduced system is not singular and may be solved by standard techniques. Once $\bar{U}_k$ and $d\bar{U}_k/dx$ are found, U_k and dU_k/dx are easy to obtain. In fact, it is easy to check that

$$U_k = \bar{U}_k/(\bar{U}_k^T W \bar{U}_k) \tag{44}$$

and

$$\frac{dU_k}{dx} = \frac{d\bar{U}_k/dx}{\bar{U}_k^T W \bar{U}_k} - \frac{\bar{U}_k \left[\bar{U}_k^T \frac{dW}{dx} \bar{U}_k + 2\bar{U}_k^T W \frac{d\bar{U}_k}{dx} \right]}{(\bar{U}_k^T W \bar{U}_k)^2} \tag{45}$$

Example

For the beam shown in Fig. 4 initially $I_1 = I_2$, and we want to

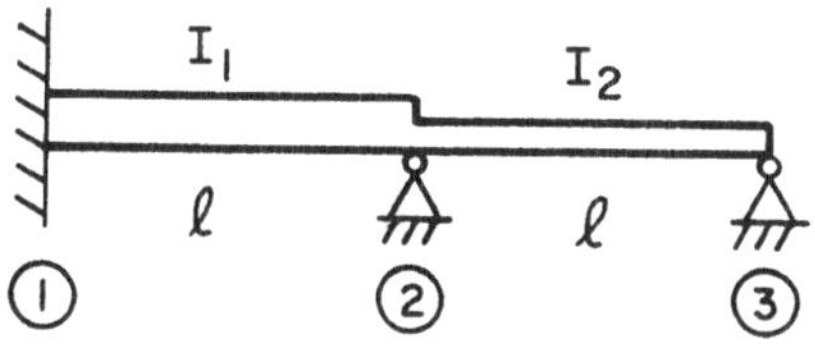

Figure 4: Beam example for eigenvalue derivatives

stiffen the beam for the purpose of increasing the fundamental frequency with the least additional mass. We assume that the mass is proportional to the square root of the moment of inertia

$$m = \alpha \sqrt{I}$$

where α is a given constant. We need the derivatives of the fundamental frequency of the beam with respect to the moments of inertia I_1 and I_2. We employ a finite element analysis of the beam using a 2 element representation. The only non-zero degrees of freedom are the rotations θ_2, θ_3 at nodes 2 and 3. The element stiffness matrix is given in Eq. (15) and the global stiffness matrix K is found to be

$$K = \frac{E}{\ell}\begin{bmatrix} 4(I_1 + I_2) & 2I_2 \\ 2I_2 & 4I_2 \end{bmatrix} \tag{46}$$

where E is Young's modulus. The element consistent mass matrix is

$$[M^E] = \frac{m}{420}\begin{bmatrix} 156 & 22\ell & 54 & -13\ell \\ & 4\ell^2 & 13\ell & -3\ell^2 \\ & & 156 & -22\ell \\ \text{sym} & & & 4\ell^2 \end{bmatrix} \tag{47}$$

The global mass matrix is found to be

$$M = \frac{\ell^2}{420}\begin{bmatrix} 4(m_1 + m_2) & -3m_2 \\ -3m_2 & 4m_2 \end{bmatrix} \tag{48}$$

where m_1, m_2 are the masses of the two beam segments

$$m_1 = \alpha\sqrt{I_1} \qquad m_2 = \alpha\sqrt{I_2} \tag{49}$$

The displacement vector U is

$$U = \begin{Bmatrix} \theta_2 \\ \theta_3 \end{Bmatrix} \tag{50}$$

The eigenvalue μ in Eq. (34) is the square of the frequency, and we take the mass matrix as the normalizing matrix W in Eq. (35). Initially $I_1 = I_2 = I$ and we can find the eigenvalues from the equation

$$|K - \mu M| = 0 \tag{51}$$

or

$$\frac{EI}{\ell}\begin{bmatrix} 8 & 2 \\ 2 & 4 \end{bmatrix} - \frac{\mu\alpha\sqrt{I}\ell^2}{420}\begin{bmatrix} 8 & -3 \\ -3 & 4 \end{bmatrix} = 0 \tag{52}$$

Define

$$\bar{\mu} = \frac{\mu\alpha\ell^3}{420E\sqrt{I}}$$

then

$$0 = \begin{vmatrix} 8 - 8\bar{\mu} & 2 + 3\bar{\mu} \\ 2 + 3\bar{\mu} & 4 - 4\bar{\mu} \end{vmatrix} = 23\bar{\mu}^2 - 76\bar{\mu} + 28$$

$$\mu_1 = 0.4222, \; \bar{\mu}_2 = 2.88$$

so that

$$\mu_1 = \frac{177E\sqrt{I}}{\alpha\ell^3} \tag{53}$$

The corresponding eigenvector satisfies Eq. (34) or

$$\begin{bmatrix} 4.62 & 3.27 \\ 3.27 & 2.31 \end{bmatrix} \begin{Bmatrix} \theta_2 \\ \theta_3 \end{Bmatrix} \quad \begin{Bmatrix} 0 \\ 0 \end{Bmatrix} \tag{54}$$

and Eq. (35) with W = M

$$\frac{\partial\ell^2\sqrt{I}}{420} (8\theta_2^2 - 6\theta_2\theta_3 + 4\theta_3^2) = 1 \tag{55}$$

The solution of the above equations gives

$$\begin{Bmatrix} \theta_2 \\ \theta_3 \end{Bmatrix} = \frac{5.85}{(I\alpha^2)^{1/4}\ell} (-\ 0.707,\ 1.00)^T \tag{56}$$

Now we can go to the derivative calculation

$$\frac{\partial K}{\partial I_1} - \mu_1 \frac{\mu M}{\partial I_1} = \frac{E}{\ell}\begin{bmatrix} 4 & 0 \\ 0 & 0 \end{bmatrix} - \left(\frac{177E\sqrt{I}}{\alpha\ell^3}\right)\left(\frac{\ell^2\alpha}{420\sqrt{I}}\right)\begin{bmatrix} 2 & 0 \\ 0 & 0 \end{bmatrix} = \frac{E}{\ell}\begin{bmatrix} 3.16 & 0 \\ 0 & 0 \end{bmatrix}$$

$$\frac{\partial K}{\partial I_2} - \mu_1 \frac{\partial M}{\partial I_2} = \frac{E}{\ell}\begin{bmatrix} 4 & 2 \\ 2 & 4 \end{bmatrix} - \left(\frac{177E\sqrt{I}}{\alpha\ell^3}\right)\left(\frac{\ell^2\alpha}{420\sqrt{I}}\right)\begin{bmatrix} 2 & -\ 1.5 \\ -\ 1.5 & 2 \end{bmatrix}$$

$$= \frac{E}{\ell}\begin{bmatrix} 3.16 & 2.63 \\ 2.63 & 3.16 \end{bmatrix} \tag{57}$$

From Eq. (38) we obtain

$$\frac{\partial\mu_1}{\partial I_2} = [\theta_2\theta_3] \left[\frac{\partial K}{\partial I_1} - \mu_1 \frac{\partial M}{\partial I_1}\right] \begin{Bmatrix} \theta_3 \\ \theta_3 \end{Bmatrix} = \left(\frac{177E}{\alpha\sqrt{I\ell^3}}\right) = \frac{\mu_1}{I}$$

$$\frac{\partial\mu_1}{\partial I_2} = [\theta_2\theta_3] \left[\frac{\partial K}{\partial I_1} - \mu_1 \frac{\partial M}{\partial I_2}\right] \begin{Bmatrix} \theta_2 \\ \theta_3 \end{Bmatrix} = \frac{122E}{\alpha\sqrt{I\ell^3}} = \frac{0.64\mu_1}{I} \tag{58}$$

It is therefore more efficient to stiffen beam 1 than to stiffen beam 2 in terms of minimum weight increase. This result holds at the design point where $I_1 = I_2$. As I_1 is increased, we may expect that $\partial\mu_1/\partial I_2$ will become larger and eventually it will become more effective to increase I_2.

It should be noted that vibration and buckling constraints do not always require the solution of an eigenvalue problem. For example, one of the most common buckling or vibration constraints is a lower limit on the lowest eigenvalue, that is

$$\mu_1 \geq \mu_{min} \tag{59}$$

This constraint can usually be reformulated as

$$K_T = K - \mu_{min}M \text{ is positive definite} \tag{60}$$

The advantage of this formulation is that the constraint may be evaluated by factoring K_T

$$K_T = LDL^T \tag{61}$$

where L is a lower triangular matrix with ones on its diagonal and D is a diagonal matrix. The condition that K_T is positive definite now reduces to the condition that all the terms in the matrix D are positive. This approach has the advantage that matrix factorization is usually much cheaper than matrix eigensolution. The disadvantage of this alternative approach is that the derivatives of D are expensive to evaluate. They are calculated best by finite differences (Ref. 10) so that each derivative is as expensive to calculate as the matrix D itself. The derivatives of the eigenvalues (Eq. (38)), on the other hand, are very inexpensive. Thus, the alternate approach is recommended only if the number of design variables is very small, or if the derivatives of K and M are not readily available.

V. SENSITIVITY OF CONSTRAINTS ON TRANSIENT BEHAVIOR

Constraints on Transient Behavior

Compared to constraints on steady-state behavior, constraints on transient behavior depend on one additional parameter-time. That is, a typical constraint may be written as

$$g(x,U,t) \geq 0 \qquad 0 \leq t \leq t_f \tag{62}$$

where for simplicity we assume that the constraint is enforced during a period of time beginning at zero and ending at some time t_f. For actual computation the constraint has to be discretized at a series of time points t_i

$$g_i = g(x,U,t_i) \geq 0 \qquad i = 1,\ldots,n_t \tag{63}$$

The distribution of time points t_i has to be dense enough to minify the chance of significant constraint violation between time points. This type of constraint discretization has the adverse effect that it greatly increases the number of constraints. Because the number of constraints is an important factor in the cost of an optimization problem, the approach represented by Eq. (63) is not very satisfactory. There are several alternative ways of removing the time dependence of a constraint that do not replace it by many equivalent constraints. We discuss three alternatives:

(i) Equivalent exterior constraints. This approach replaces $g(x,U,t)$ by a single constraint $\bar{g}(x,U)$. One form is

$$\bar{g}(x,U) = \frac{1}{t_f} \int_0^{t_f} [g(x,U,t) - |g(x,U,t)|]dt \geq 0 \tag{64}$$

Note that if the constraint $g(x,U,t) \geq 0$ is violated over any finite period of time $\bar{g}(x,U) \geq 0$ is also violated. If, however, $g(x,U,t)$ is not violated anywhere $\bar{g}(x,U)$ is zero. The disadvantage of this formulation is that in the feasible domain there is no indication of how near to being critical the constraint is. Therefore, this form of equivalent constraint is not suitable for optimization methods which tend to operate in the interior of the feasible domain (such as the method of feasible directions or the interior penalty function method). It is mostly used with the gradient projection method which operates on the outside boundary of the feasible domain. For use with an exterior penalty function formulation, $\bar{g}(x,U)$ of Eq. (64) has the disadvantage of having a discontinuous derivative. A continuously differentiable form which is more compatible with the exterior penalty function is

$$\bar{g}(x,U) = [\frac{1}{t_f} \int_0^{t_f} < -g(x,U,t)>^2 dt]^{1/2} \tag{65}$$

where $\langle a \rangle$ denotes max (a,o).

The savings obtained by replacing the discretized constraint, Eq. (63) by an integrated one may seem illusory because the integral usually requires the evaluation of $g(x,U,t)$ at many time points. The savings however are in the computations of constraint derivatives and search directions. The savings in derivative calculations are discussed later in this section.

(ii) Equivalent interior constraint. This is another form of an integrated constraint

$$\bar{g}(x,U) = [\frac{1}{t_f} \int_0^{t_f} dt/g(x,U,t)]^{-1} \geq 0 \tag{66}$$

where the equivalent constraint $\bar{g}$ is defined only when $g(x,U,t) > 0$. Thus this form is of use only with optimization techniques which are restricted to the interior of the feasible domain.

(iii) Critical point constraint. Unlike the two previous constraint forms this form is not an integrated constraint. Rather, the constraint is replaced by

$$g(x,U,t_{mi}) \geq 0 \qquad i = 1,2,\ldots \tag{67}$$

where t_{mi} are points where the constraint has a local minimum. Figure 5 shows a typical situation. The constraint function has two local minima; an interior one t_{m1} and a boundary minimum at $t_{m2} = t_f$.

These local minima are critical points in the sense that they represent time points that are likely to be involved first in constraint violation. The advantage of this form of equivalent constraint over the other two is that there is no blurring due to the integrated effect. For example, consider a change in design which moves the constraint g from the solid to the dashed line in Fig. 5. An integrated constraint $\bar{g}$ may become more positive indicating a beneficial effect, while the critical point constraint would indicate the increased danger of constraint violation. The disadvantage of the critical point formulation is that the number of critical points could conceivably be large.

One attractive feature of the critical point constraint is that for the purpose of obtaining first derivatives the location of the critical point may be assumed to be fixed. This is shown by differentiating Eq. (67) with respect to a design variable x

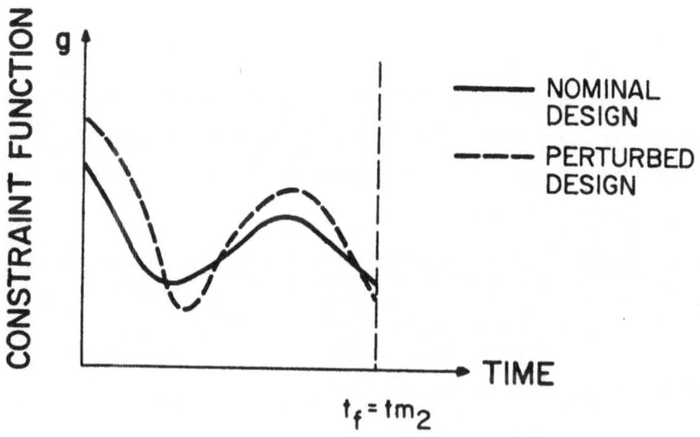

Figure 5: Critical points

$$\frac{dg(t_{mi})}{dx} = \frac{\partial g}{\partial x} + \frac{\partial g}{\partial U}\frac{dU}{dx} + \frac{\partial g}{\partial t}\frac{dt_{mi}}{dx} \tag{68}$$

The last term in Eq. (68) is always zero. At an interior minimum such as t_{mi} in Fig. 5 $\partial g/\partial t$ is zero. At a boundary minimum, if $\partial g/\partial t$ is not zero (it must be negative) then dt_{mi}/dx is zero, because an infinitesimal change in the design variable cannot change the slope by a finite amount to make it positive (which is necessary if the minimum location has moved away from the boundary).

In the following we assume that the equivalent constraint is given as

$$\bar{g}(x,U) = \int_0^{t_f} p(x,U,t)dt \geq 0 \tag{69}$$

This form obviously represents the equivalent exterior constraint of Eq. (64). It also represents the critical point constraint with

$$p(x,U,t) = g(x,U,t)\delta(t - t_{mi}) \tag{70}$$

The form of Eq. (69) does not include the equivalent interior constraint, but it corresponds to the form of the reciprocal of that constraint. The methods discussed in the remainder of this section for obtaining derivatives of $\bar{g}(x,U)$ can be adapted to the interior constraint by noting that

$$\frac{d}{dx}\left[\frac{1}{\bar{g}(x,U)}\right] = \frac{-1}{\bar{g}^2(x,U)}\frac{d\bar{g}}{dx} \tag{71}$$

Similarly the form of Eq. (69) corresponds to the square of the constraint of Eq. (65), and it is easy to obtain the derivative of $\bar{g}$ from that of $\bar{g}^2$.

Derivatives of Constraints

The derivative of the equivalent constraint, Eq. (69) with respect to a design variable x is

$$\frac{d\bar{g}}{dx} = \int_0^{t_f} \left(\frac{\partial p}{\partial x} + \frac{\partial p}{\partial U}\frac{dU}{dx}\right)dt \tag{72}$$

To evaluate the integral we need to differentiate the equations of motion with respect to x. These equations are written in the general form

$$A\dot{U} = F(U,x,t) \qquad U(0) = U_0 \tag{73}$$

where U is a vector of generalized degrees of freedom and F is a vector which includes contributions of external and internal loads.

We now discuss several methods for calculating the constraint derivative Eq. (72) using Eq. (73). We start with the simplest - the direct method.

As in the steady-state case the direct method proceeds by differentiating Eq. (73) to obtain an equation for dU/dx.

$$A \frac{d\dot{U}}{dx} = J \frac{dU}{dx} - \frac{dA}{dx} \dot{U} + \frac{\partial F}{\partial x} , \qquad \frac{dU}{dx}(0) = 0 \tag{74}$$

where J is the Jacobian matrix

$$j_{ij} = \frac{\partial f_i}{\partial u_j} \tag{75}$$

The direct method consists of solving for dU/dx from Eq. (74), and then substituting into Eq. (72). The disadvantage of this method is that each design variable requires the solution of a system of differential equations, Eq. (74). When we have many design variables and few constraint functions we can, as in the steady-state case, use a vector of adjoint variables which depends only on the constraint function and not on the design variables. This procedure is again called the adjoint variable method. The first step is to premultiply Eq. (74) by the transpose of the adjoint vector Λ and integrate

$$\int_0^{t_f} \Lambda^T [A \frac{d\dot{U}}{dx} - J \frac{dU}{dx}]dt = \int_0^{t_f} \Lambda^T[\frac{\partial F}{\partial x} - \frac{dA}{dx} \dot{U}]dt \tag{76}$$

Integrating by parts we obtain

$$\Lambda^T A \frac{dU}{dx} \Big|_0^{t_f} - \int_0^{t_f} [\dot{\Lambda}^T A + \Lambda^T \dot{A} + \Lambda^T J] \frac{dU}{dx} dt = \int_0^{t_f} \Lambda^T[\frac{\partial F}{\partial x} - \frac{dA}{dx} \dot{U}]dt \tag{77}$$

we now require Λ to satisfy (recall that p is defined by Eq. (69))

$$A^T \dot{\Lambda} + (J^T + \dot{A}^T)\Lambda = (\frac{\partial p}{\partial U})^T , \qquad \Lambda(t_f) = 0 \tag{78}$$

then

$$\int_0^{t_f} \frac{\partial p}{\partial U} \frac{dU}{dx} dt = - \int_0^{t_f} \Lambda^T [\frac{\partial F}{\partial x} - \frac{dA}{dx} \dot{U}]dt \tag{79}$$

where use has been made of the fact that dU/dx is zero at t = 0. Combining Eq. (79) with (72) we get

$$\frac{d\bar{g}}{dx} = \int_0^{t_f} [\frac{\partial p}{\partial x} - \Lambda^T (\frac{\partial F}{\partial x} - \frac{dA}{dx} \dot{U})]dt \tag{80}$$

Equation (78) is a system of ordinary differential equations for Λ which are integrated backwards (from t_f to t = 0). This system has to be solved once for each constraint rather than once for each design variable. As in the steady-state case the direct method is preferable when the number of design variables is smaller than the number of constraints, and the adjoint variable method is preferable when the number of design variables is larger than the number of constraints.

Equation (78) takes a simpler form in the case of a single critical point constraint

$$A^T\dot{\Lambda} + (J^T + \dot{A}^T)\Lambda = (\frac{\partial g}{\partial U})^T \delta(t - t_{mi}) , \qquad \Lambda(t_f) = 0 \tag{81}$$

If there is only one critical point, by integrating Eq. (81) from $t_{mi} - \varepsilon$ to $t_{mi} + \varepsilon$ for an infinitesimal ε it is easily shown that this is equivalent to

$$A^T\dot{\Lambda} + (J^T + \dot{A}^T)\Lambda = 0, \qquad \Lambda^T(t_{mi}) = - \frac{\partial g}{\partial U} (t_{mi})A^{-1} \tag{82}$$

A third method which is available for derivative calculation is the Green's function approach [11]. This method is useful when the number of degrees of freedom in Eq. (73) is smaller than the number of design variables or constraints. This can happen when the order of Eq. (73) has been reduced by employing modal analysis. The Green's function method will be discussed for the csse of A = I in Eq. (73) so that Eq. (79) becomes

$$\frac{d\dot{U}}{dx} = J \frac{dU}{dx} + \frac{\partial F}{\partial x} , \qquad \frac{dU}{dx} (0) = 0 \tag{83}$$

The solution of Eq. (83) may be written [11] in terms of a Green's function $K(t,\tau)$ as

$$\frac{dU}{dx} = \int_0^{t_f} K(t,\tau) \frac{\partial F}{\partial x} (\tau)d\tau \tag{84}$$

where $K(t,\tau)$ satisfies

$$\dot{K}(t,\tau) - J(t)K(t,\tau) = \delta(t - \tau)I \tag{85}$$
$$K(0,\tau) = 0$$

and where $\delta(t - \tau)$ is the Dirac delta function. It is easy to check, by direct substitution, that dU/dx defined by Eq. (84), indeed satisfies Eq. (83).

If the elements of J are bounded then it is easy to show that Eq. (85) is equivalent to

$$\begin{aligned} &K(t,\tau) = 0 && t < \tau \\ &K(\tau,\tau) = I && \\ &\dot{K}(t,\tau) - J(t)K(t,\tau) = 0 , && t > \tau \end{aligned} \tag{86}$$

so that the integration in Eq. (84) needs to be carried out only up to $\tau = t$. To see how dU/dx is evaluated with the aid of Eq. (86) assume that we divide the interval $0 \leq t \leq t_f$ into n subintervals with end points at $\tau_o = 0 < t_1 < \ldots < t_n = t_f$. The points τ_i are dense enough to evaluate Eq. (89) by numerical integration and to interpolate dU/dx to other time points of interest with sufficient accuracy. We now define the initial value problems

$$\dot{K}(t,\tau_k) - J(t)K(t,\tau_k) = 0$$

$$K(\tau_k,\tau_k) = I \qquad k = 0,1,\ldots,n-1 \tag{87}$$

Each of the equations in (87) is integrated from τ_k to τ_{k+} to yield $K(\tau_{k+}, \tau_k)$. The value of K for any other pairs of points is given by (see Ref. 11 for proof)

$$K(\tau_j,\tau_k) = K(\tau_j,\tau_{j-})K(\tau_{j-},\tau_{j-})\ldots K(\tau_{k+},\tau_k) \qquad j > k \tag{88}$$

The solution for K is equivalent to solving n_m systems of the type of (74) of (81) where n_m is the number of degrees of freedom of the vector U. Therefore, the method would become competitive when the number of design variables and constraints both reach or exceed n_m.

Example

We have a nonlinear single degree of freedom system governed by the differential equation

$$a\dot{u} = (u - b)^2, \qquad u(0) = 0 \tag{89}$$

and a constraint on the response u in the form

$$g(u) = c - u(t) \geq 0, \qquad 0 \leq t \leq t_f \tag{90}$$

The response has been calculated and found to be monotonically increasing, so that using the critical point constraint

$$\bar{g}(u) = g[u(t_f)] = c - u(t_f)$$

We want to use the direct, adjoint variable and Green's function methods to calculate the derivatives of $\bar{g}$ with respect to a and b.

The problem is simple enough so that u may be integrated directly to yield

$$u = \frac{b^2 t}{bt + a} \tag{91}$$

In the notation adopted in this section

$$A = a$$

$$J = \frac{\partial F}{\partial u} = 2(u - b) \tag{92}$$

Direct Method. The direct method requires us to write Eq. (74) for x = a and x = b. For x = a we obtain

$$a \frac{d\dot{u}}{da} = 2(u - b) \frac{du}{da} - \dot{u} \quad , \qquad \frac{du}{da}(0) = 0 \tag{93}$$

In general the values for u and $\dot{u}$ would be available only numerically and Eq. (93) will also be integrated numerically. Here, however, we have the closed-form solution for u, so that we can substitute

$$a \frac{d\dot{u}}{da} = \frac{2ab}{bt + a} \frac{du}{da} - \frac{ab^2}{(bt + a)^2} \qquad \frac{du}{da}(0) = 0 \tag{94}$$

and solve analytically to obtain

$$\frac{du}{da} = \frac{-b^2 t}{(bt + a)^2} \tag{95}$$

then

$$\frac{d\bar{g}}{da} = \frac{-du}{da}(t_f) = \frac{b^2 t_f}{(bt_f + a)^2} \tag{96}$$

We repeat the process for x = b. Equation (74) becomes

$$a \frac{d\dot{u}}{db} = 2(u - b) \frac{du}{db} - 2(u - b) \quad , \qquad \frac{du}{db}(0) = 0 \tag{97}$$

Solving for du/db we obtain

$$\frac{du}{db} = \frac{b^2 t^2 + 2abt}{(bt + a)^2} \tag{98}$$

and then

$$\frac{d\bar{g}}{db} = \frac{-du}{db}(t_f) = -\frac{b^2 t_f^2 + 2bat_f}{(bt_f + a)^2} \tag{99}$$

Adjoint Method. The adjoint method requires the solution of Eq. (82) which becomes

$$a\dot{\lambda} + 2(u - b)\lambda = 0 \quad , \qquad \lambda(t_f) = \frac{-1}{a} \frac{\partial g}{\partial u}(t_f) = \frac{1}{a} \tag{100}$$

or

$$a\dot{\lambda} - \frac{2ab}{bt + a} \lambda = 0 \quad , \qquad \lambda(t_f) = \frac{1}{a} \tag{101}$$

which can be integrated to yield

$$\lambda = \frac{1}{a} \left(\frac{bt + a}{bt_f + a}\right)^2 \tag{102}$$

Then $d\bar{g}/da$ is obtained from Eq. (80) which becomes

$$\frac{d\bar{g}}{da} = \int_{0}^{t_f} \lambda \dot{u}\, dt = \int_{0}^{t_f} \frac{1}{a} \left(\frac{bt + a}{bt_f + a}\right)^2 \frac{ab^2}{(bt + a)^2}\, dt = \frac{b^2 t_f}{(bt_f + a)^2} \tag{103}$$

Similarly, from Eq. (80) $d\bar{g}/db$ is

$$\frac{d\bar{g}}{db} = \int_{0}^{t_f} 2\lambda(u - b)dt = -\frac{2}{a} \int_{0}^{t_f} \left(\frac{bt + a}{bt_f + a}\right)^2 \frac{ab}{bt + a}\, dt = -\frac{b^2 t_f^2 + 2bat_f}{(bt_f + a)^2} \tag{104}$$

<u>Green's Function Method.</u> We recast the problem as

$$\dot{u} = (u - b)^2/a \tag{105}$$

so that the Jacobian J is

$$J = 2(u - b)/a \tag{106}$$

and Eq. (86) becomes

$$\dot{k}(t,\tau) - [2(u - b)/a]k(t,\tau) = 0 \qquad k(t,t) = 1 \tag{107}$$

or

$$\dot{k}(t,\tau) + \frac{2b}{bt + a} k(t,\tau) = 0 \tag{108}$$

The solution of Eq. (108) with the initial condition $K(\tau,\tau)$ is

$$k = \left(\frac{b\tau + a}{bt + a}\right)^2 \qquad t \geq \tau \tag{109}$$

so that from Eq. (84)

$$\frac{du}{da} = \int_{0}^{t_f} \frac{\partial F}{\partial a} k d\tau = -\int_{0}^{t_f} \left(\frac{b\tau + a}{bt + a}\right) \frac{(u - b)^2}{a^2} d\tau = \frac{-b^2 t}{(bt + a)^2} \text{ and}$$

similarly

$$\frac{du}{db} = \int_{0}^{t_f} \frac{\partial F}{\partial b} kdt = -\int_{0}^{t_f} 2\left(\frac{b\tau + a}{bt + a}\right)^2 \left(\frac{u - b}{a}\right) d\tau = -\frac{b^2t^2 + 2abt}{(bt + a)^2} \tag{110}$$

VI. SIMULTANEOUS ANALYSIS AND DESIGN

In its early days structural optimization employed the calculus of variations to obtain the Euler-Lagrange optimality differential equations and these were solved simultaneously with the differential equations of structural response. This approach is still used today for the optimization of individual structural elements such as beam-columns (Ref. 12); however, for built-up structures modeled by finite elements, a nested approach is typical. Resizing rules based on optimality criteria require that the structural response be calculated repeatedly for each set of trial structural design variables (see, for example, Ref. 13). This preference for the nested over the simultaneous approach is probably due to the simplicity of the structural resizing rules which are possible when the structural response is known. This simplicity contrasts with the difficulty of solving the large systems of nonlinear algebraic equations which are obtained from a simultaneous formulation.

In the last twenty-five years direct search methods have been gaining ground as the standard for structural optimization. These techniques are commonly used in a nested approach, with the structural analysis equations repeatedly solved during each design iteration. Part of the reason for the popularity of the nested approach is that the structural analysis equations are solved by techniques which are quite different than those used for the design optimization. An exception is the design of a structure subject to constraints on its collapse load. There, the analysis problem ("limit analysis") is often approximated as a linear program and solved by the simplex method. The structural design problem in that case ("limit design") is easily formulated as a single linear program with the element forces and structural parameters both treated as design variables (Ref. 14, for example).

In the late sixties, Schmit, Fox and their coworkers (Refs. 15-19) tried to integrate structural analysis and design by employing conjugate gradient (CG) minimization techniques for solving linear structural analysis problems. They found that CG methods were not competitive with the traditional direct Gaussian elimination techniques. More recently techniques for unconstrained minimization have become more efficient and their application to structural analysis has become more feasible (e.g., Ref. 20). The emergence of the preconditioned CG techniques (e.g., Ref. 21) and the element-by-element (EBE) formulations of Hughes and coworkers (Ref. 22) make CG methods particularly attractive for structural analysis.

In view of the increasing use of optimization methods for structural analysis, there is merit in considering again the simultaneous approach to analysis and design. This section (based on Refs. 23, 24) describes the use of the simultaneous approach for the design of structures subject to stress constraints both in the linear and nonlinear range.

Nested and Simultaneous Formulations of the Structural Design Problem

The traditional-nested-formulation of the structural design problem may be written as find X to minimize m(X) subject to

$$g_j(X,U) \geq 0 \qquad j = 1,\ldots,n \tag{111}$$

where m is an objective function, X is a vector of design parameters, U is the displacement vector and g_j are constraint functions such as stress and displacement constraints. The displacement vector U is the solution to a linear or nonlinear system of algebraic equations (the equations of equilibrium)

$$F(U,P,X) = 0 \tag{112}$$

where P is a load vector. The optimization problem is usually solved by repeatedly calculating U and its derivatives with respect to the components of X, x_j. U is calculated from Eq. (112) and $\frac{dU}{dx_j}$ is calculated either by differentiating Eq. (112) or by finite differences. Based on U and its derivatives the constraint functions g_j and their derivatives can be evaluated and a numerical optimization technique can use this information to improve X.

A simultaneous approach treats X and U equally as design variables and solves the following expanded problem

find X,U to minimize m(X) subject to

$$g_j(X,U) \geq 0 \qquad j = 1,\ldots,n \tag{113}$$

and

$$F(U,P,X) = 0.$$

The equations of equilibrium are treated here as nonlinear equality constraints. In comparing the simultaneous and nested approaches two points should be noted. First, the simultaneous approach trades the solution of a bigger problem against the repeated solution of smaller problems (the response solution). Second, sensitivity calculations for the simultaneous approach are very inexpensive, because g_j and F are usually easily differentiated with respect to both U and X.

Penalty Function Solution Technique

Problem (113) is replaced by

$$\text{minimize } \phi(X,U,r) = cm(X) + r\sum_{j=1}^{m} p[g_j(X,U)] + \frac{c_1}{\sqrt{r}} F^T B^{-1} F \tag{114}$$

for $r = r_1, r_2, \ldots$ where $r_i \to 0$

where

$$p(g) = \begin{cases} 1/g \text{ if } g > g_0 \\ \frac{1}{g_0}\,[(g/g_0)^2 - 3g/g_0 + 3] \text{ if } g \leq g_0 \end{cases} \qquad (115)$$

is an extended interior penalty function (Ref. 25) with g_0 being a transition parameter. The constants c and c_1 are chosen to balance the contribution of the objective function, inequality constraints and equality constraints to ϕ. The matrix B is used to improve the conditioning of the last term. In the present work it was taken to be the element-by-element approximation to the inverse of the small-displacement stiffness matrix (see Refs. 22, 23) of the initial structural design. Because it is an element-by-element approximation, the calculation of $B^{-1}F$ may be performed without forming any large system stiffness matrices.

The matrix B^{-1} is intended to improve the conditioning of the minimization problem, and its role is explained here for the linear case where

$$F(U,P,X) = P - KU \qquad (116)$$

where K is the system stiffness matrix. The term containing B^{-1} may be written as

$$(P - KU)^T B^{-1} (P - KU) \qquad (117)$$

Without the B^{-1} the second-derivative matrix of Eq. (117) with respect to the displacement unknowns would be K^TK which has the square of the condition number of K, and so is very ill conditioned. Because B^{-1} is an approximation to K^{-1} the second-derivative matrix now has a condition number similar to K.

The sequence of unconstrained problems was solved by a CG package based on Beale's restarted CG algorithm (Ref. 26). An initial structural configuration was chosen and the displacement field based on a linear analysis was calculated by Gaussian elimination and used for starting the CG algorithm. For cases when the behavior was highly nonlinear this initialization resulted in slow convergence, therefore the residual vector F(U,P,X) was replaced by $\bar{F}(r,U,P,X)$ where

$$\bar{F}(r,U,P,X) = \frac{1}{\lambda(r)} F(U,\lambda(r)P,X) \qquad (118)$$

where $\lambda(r)$ is a scaling factor which was set to a small value for the initial penalty multiplier r_1 and was gradually increased to 1 as r was decreased. This is equivalent to increasing the nonlinearity of the problem gradually as the optimization proceeds.

Seventy-Two-Bar Truss Example

The first example is a 72-bar truss shown in Fig. 6. The truss was designed subject to a maximum stress constraint of 25,000 psi, and a minimum area of 0.1 in². Two loading cases were considered. The first loading case consisted of 4 downward 5000 lb loads at nodes 1-4 plus a 5000 lb in the X-direction and 5000 lb in the Y-direction at node 1. The second loading condition was with the same loads scaled up by a factor of 10. The traditional nested formulation has 72 design variables and was solved by Powell's projected Lagrangian technique (Ref. 27) which is a sequential and quadratic programming algorithm recognized as one of the best techniques for constrained optimization. The simultaneous formulation has 120 variables including 48 displacement variables, and was solved with the penalty technique described in the previous section.

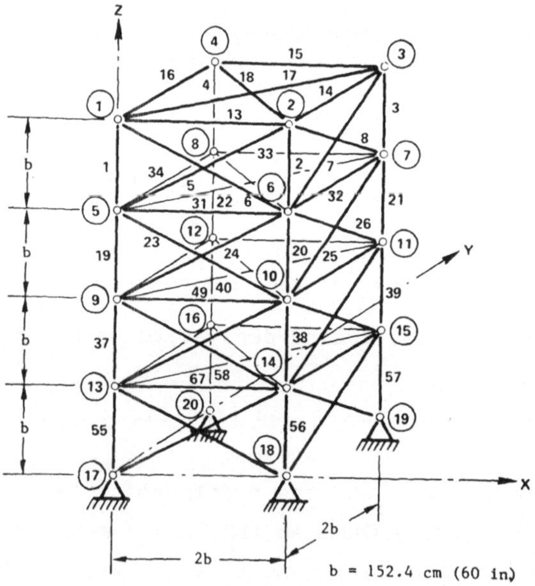

Figure 6: Seventy bar truss

The results of the three optimization algorithms are compared in Table 1. It is seen that both solution techniques produced similar weight with differences of the order of 1 percent. For both load cases the simultaneous approach was significantly better in terms of CPU time than the nested approach.

Table 1

Comparison of results of nested and simultaneous optimization procedures for 72-bar truss.

	Penalty-function Formulation (simultaneous)	Projected Lagrangian (nested)	
mass (lb)	96.5	95.7	nominal loads
CPU* (sec)	70.9 (20 r values) 52.4 (10 r values)	90.2	
mass (lb)	493.7	498.9	higher (x 10) loads
CPU* (sec)	222.6 (20 r values) 258.4 (10 r values)	288.3	

*IBM 3084

Antenna Example

The second example is a 55 meter antenna-reflector structure shown in Fig. 7. The structure was supported at the six vertices of its lower surface and subjected to a 10000 lb tensile load at the center of its upper surface. The structure was designed subject to the same stress and minimum area constraints as the 72-bar truss. The nested approach has 420 truss elements as design variables while the simultaneous approach has 730 design variables including 310 displacement variables.

The simultaneous solution procedure required 49 minutes of IBM 3084 CPU time with 30 r values and yielded an optimum weight of 1529.0 lb. The nested approach based on Ref. 27 failed to complete even a single iteration because of convergence difficulties. As these types of difficulties were encountered by other researchers for large problems with sequential quadratic programming algorithms, we switched to the generalized reduced gradient algorithm (Ref. 28). Because of availability of computer resources the simultaneous and nested procedures were run on an IBM 4341. The simultaneous design procedure required 502 minutes of CPU time and yielded an optimum weight of 1527.9 lb. The nested approach based on Ref. 28 required 1961 minutes and yielded a final weight of 1519.6 lb. The two designs were almost identical. It is interesting to note the small differences between the two designs obtained by the simultaneous approach on the two computers. These small differences are thought to be due to differences in the implementation of double-precision accuracy on the two machines magnified by the more than 10,000 CG iterations required for convergence. This example clearly demonstrates the advantage of the simultaneous approach for large problems.

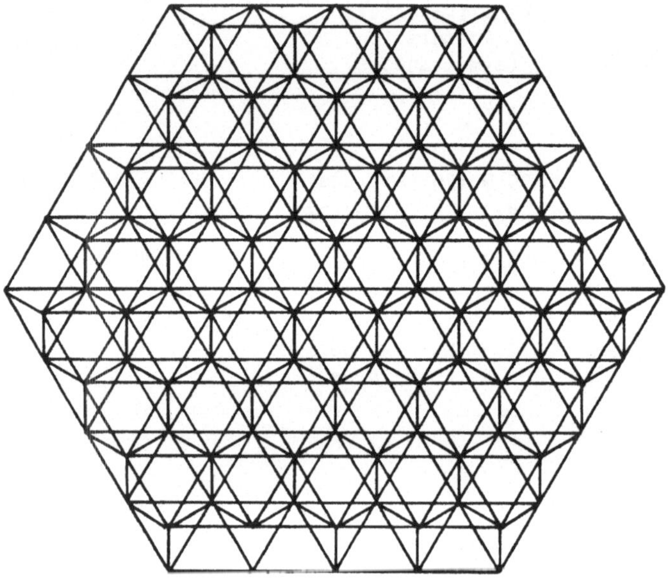

Figure 7: Antenna reflector structure

REFERENCES

1. Prasad, B. and Haftka, R.T., "Structural Optimization with Plate Finite Elements," Journal of the Structural Division, ASCE, Vol. 105, No. ST11, pp. 2367-2382, 1979.

2. Braibant, V., Fleury, C. and Beckers, P., "Shape Optimal Design. An Approach Matching CAD and Optimization Concepts," Report SA-109, Aerospace Laboratory of the University of Liege, Belgium, 1983.

3. Gill, P.E., Murray, W., Saunders, M.A. and Wright, M.H., "Computing Forward-Difference Intervals for Numerical Optimization," SIAM J. Sci. and Stat. Comput., Vol. 4, No. 2, pp. 310-321, June 1983.

4. Iott, J., Haftka, R.T. and Adelman, H.M., "Selecting Step Sizes in Sensitivity Analysis by Finite Differences," NASA TM 86382, 1985.

5. Camarda, C.J. and Adelman, H.M., "Static and Dynamic Structural-Sensitivity Derivative Calculation in the Finite-Element-Based Engineering Analysis Language (EAL) System," NASA TM-85743, 1984.

6. Barthelemy, B.M., Chon, C.T. and Haftka, R.T., "Accuracy of Finite-Difference Approximations to Sensitivity Derivatives of Static Structural Response," paper presented at the First World Congress on Computational Mechanics, Austin, Texas, S.

7. Haug, E.J., Komkov, V. and Choi, K.K., Design Sensitivity Analysis of Structural Systems, Academic Press, 1986.

8. Cardani, C. and Mantegazza, P., "Calculation of Eigenvalue and Eigenvector Derivatives for Algebraic Flutter and Divergence Eigenproblems," AIAA Journal, Vol. 17, pp. 408-412, 1979.

9. Nelson, R.B., "Simplified Calculation of Eigenvector Derivatives," AIAA Journal, Vol. 14, pp. 1201-1205, 1976.

10. Haftka, R.T., "Design for Temperature and Thermal Buckling Constraints Employing a Noneigenvalue Formulation," Journal of Spacecraft, Vol. 20, pp. 363-367, 1983.

11. Hwang, J.T., Dougherty, E.P., Rabitz, S. and Rabitz, H., "The Green's Function Method of Sensitivity Analysis in Chemical Kinetics," J. Chem. Phys., Vol. 69, pp. 5180-5191, 1978.

12. Plaut, R.H., Johnson, L.W. and Olhoff, N., "Bimodal Optimization of Compressed Columns of Elastic Foundations," Journal of Applied Mechanics, 1986.

13. Khot, N.S., Berke, L. and Venkayya, V.B., "Comparison of Optimality Criteria Algorithms for Minimum Weight Design of Structures," AIAA Journal, Vol. 17, No. 2, pp. 182-189, 1979.

14. Haftka, R.T. and Kamat, M.P., Elements of Structural Optimization, Martinus Nijhoff, The Netherlands, 1985.

15. Fox, R.L. and Schmit, L.A., "An Integrated Approach to Structural Synthesis and Analysis," AIAA Journal, Vol. 3, pp. 1104-1112, June 1965.

16. Fox, R.L. and Schmit, L.A., "Advances in the Integrated Approach to Structural Synthesis," Journal of Spacecraft and Rockets, Vol. 3, pp. 858-866, June 1966.

17. Schmit, L.A., Bogner, F.K. and Fox, R.L., "Finite Deflection Discrete Element Analysis Using Plate and Shell Discrete Elements," AIAA Journal, Vol. 6, No. 5, pp. 781-791, 1968.

18. Fox, R.L. and Stanton, E.L., "Developments in Structural Analysis by Direct Energy Minimization," AIAA Journal, Vol. 6, No. 6, pp. 1036-1042, 1968.

19. Fox, R.L. and Kapoor, M.P., "A Minimization Method for the Solution of Eigenproblem Arising in Structural Dynamics," Proceedings of the Second Conference on Matrix Methods in Structural Mechanics, Wright-Patterson AFB, Ohio, AFFDL-TR-68-150, 1968.

20. Kamat, M.P. and Hayduk, R.J., "Recent Developments in Quasi-Newton Methods for Structural Analysis and Synthesis," AIAA Journal, Vol. 20, No. 5, pp. 672-679, 1982.

21. Johnson, O.B., Micchelli, C.A. and Paul, G., "Polynomial Preconditioners for Conjugate Gradient Calculations," SIAM Journal for Numerical Analysis, Vol. 20, pp. 362-376, 1983.

22. Hughes, T.J.R., Winget, J., Levit, I. and Tezduyar, T.E., "New Alternating Direction Procedures in Finite Element Analysis Based on EBE Approximate Factorization," Computer Methods for Nonlinear Solids and Structural Mechanics, (S. Atluri and N. Perrone, editors), AMD, Vol. 54, pp. 75-109, 1983.

23. Haftka, R.T., "Simultaneous Analysis and Design," AIAA Journal, Vol. 23, No. 7, pp. 1099-1103, 1985.

24. Haftka, R.T. and Kamat, M.P., "Simultaneous Nonlinear Analysis and Design," presented at the ASME Design Automation Conference, Cincinnati, Ohio, September 1985.

25. Haftka, R.T. and Starnes, J.H., Jr., "Application of a Quadratic Extended Interior Penalty Function for Structural Optimization," AIAA Journal, Vol. 14, pp. 718-724, 1976.

26. Powell, M.J.D., "Restart Procedures for the Conjugate Gradient Methods," Mathematical Progress, Vol. 11, pp. 42-49, 1976.

27. Powell, M.J.D., "A Fast Algorithm for Nonlinearity Constrained Optimization Calculations," Proceedings of the 1977 Dundee Conference on Numerical Analysis, Lecture Notes in Mathematics, Vol. 630, pp. 144-157, Springer-Verlag, Berlin, 1978.

28. Lasdon, L.S. and Warren, A.D., "Generalized Reduced Gradient Software for Linearly and Nonlinearly Constrained Problems," Design and Implementation of Optimization Software, (H. Greenberg, ed.) Sijthoff and Nordhoff Pub., 1979.

STRUCTURAL OPTIMIZATION USING OPTIMALITY CRITERIA

L. Berke
National Aeronautics and Space Administration
Lewis Research Center
Cleveland, Ohio 44135

and

N.S. Khot
Air Force Wright Aeronautical Laboratories
Wright-Patterson Air Force Base, OH 45433

ABSTRACT

First, the historical background leading to the optimality criteria approach is discussed pointing out the role of the traditional design methods on one hand, and Prager's work based on variational principles on the other hand as the two motivating influences. This is followed by the formal development of the method utilizing the separability properties of discretized structures or models. The importance of the single constraint case is pointed out and the associated particularly simple yet powerful optimality criteria is presented followed by extension to multiple constraints. Examples are used to illustrate the approach for displacement, stress and eigenvalue related constraints.

INTRODUCTION

Optimality Criteria (OC) methods of structural optimization, the subject of this discussion, trace their origin to intuitive traditional approaches to the problem of strength design, most directly to the Fully Stressed Design (FSD) criteria with its associated stress ratio resizing algorithm. The FSD criteria states that a structure is of minimum weight if every member is at its maximum allowable stress, or at minimum size at least under one of the loading conditions. This is correct for statically determinate and most indeterminate structures. This criteria, and its relative, the Simultaneous Failure Modes (SFM) criteria, are based on the intuitive, but incomplete assumption that if a structure is

NATO ASI Series, Vol. F27
Computer Aided Optimal Design: Structural and
Mechanical Systems. Edited by C. A. Mota Soares

sized just not to fail locally and globally under its critical loads, then it must be an optimum structure, because one could not reduce the sizes and therefore the weight, any further. What is missing is the important influence of internal force distribution, also the function of the size variables in indeterminate structures.

This idea of characterizing an optimum structure through conditions that are believed to exist at optimum, and then apply a resizing procedure that directly satisfies those conditions, is the fundamental approach that designers always used and became formalized as the Optimality Criteria methods during the late 60's. Because of this ancestry in essentially intuitive methods, confusion persisted as to the exactness and validity even of the formally derived OC methods that followed. It is perhaps beneficial to briefly glance over the history of the development of analysis and optimization methods to better appreciate why and how optimality criteria methods were developed and how they fit into the current optimization technology.

HISTORICAL BACKGROUND

Structural engineers were among the first to start to utilize computers as soon as they became available. Solutions to specific design problems were the first targets both in civil and in aeronautical engineering. At the same time developments were under way in general purpose structural analysis technology that soon resulted in structural engineers to become one of the major users of computers. The resulting general purpose automated analysis capability, that after a few other names became known as the Finite Element Method (FEM), enjoyed rapid development.

Recognizing the possibilities offered by computers the idea of developing general purpose structural optimization methods also emerged in the 50's. The proper framework was found to be the Mathematical Programming (MP) methods of operations research. Consequently, structural optimization was recast as a problem in nonlinear mathematical programming

(Refs. 1 and 2) opening up vast possibilities for further developments. The new method quickly proved very successful for problems of a moderate number of design variables, such as the ones formerly treated by SFM. Hopes were very high, but as it turned out there was a price to pay for generality. While finite element analysis capability rose from a few hundred to thousands of elements, the capability of MP methods rose only from a few times ten to a few hundred design variables. MP methods clearly were not able to keep up to optimize what now could be routinely analyzed. There was a need towards the end of the 60's to increase optimization capabilities to better match finite element practices, at least for the simple sizing problem. In the course of these developments the time honored FSD and SFM methods were shown to be correct most of the time but not all of the time. SFM was easily replaced by MP methods because of the small number of design variables that were involved. This was not the case with FSD, usually applied to the large finite element models to perform a few resizing iterations to meet stress constraints. It was known to converge very rapidly for conventional metal structures and was accepted as a practical approach (Refs. 3 to 5). A method was needed that like FSD also exhibited weak or no dependence on the number of design variables, and was applicable not only to stress constraints, but to other important stiffness constraints as well, and was theoretically respectable. Formal methods of the calculus of variation, applied by Prager and co-workers to specific distributed parameter problems, showed great promise, suggesting that interpretation to discretized (e.g., FEM) models could lead to a new class of general purpose optimization methods.

The first discretized OC methods were based on strain energy distributions characterizing optimum structural designs for stress constraints. For example as Prager showed (Ref. 6), if the work of the applied loads is limited as an equality constraint, the optimum structure with uniform material properties has uniform energy density distribution. A similar criterion was proposed (Ref. 7) for discretized structures stating that at optimum the ratio of strain energy to strain energy capacity is constant for every member. The optimality criteria of Refs. 6 and 7 were valid, and equivalent, for uniform material properties and stress allowables, and a single loading condition. They were both heuristic outside of these conditions, yet very efficient, and apparently yielded correct results, within a practical range of deviation from these limiting conditions.

The more or less heuristic discretized OC methods up until this point were focused on the strength design problem. At the same time many interesting special problems were addressed by Prager and co-workers (Refs. 8 to 13) using variational approaches. The optimality conditions were stated in terms of differential equations with their solutions describing the shape of the optimum structure. Many references could be listed; the ones above are only typical examples. To combine the power and exactness of these special variational methods with the generality of finite element methods became a tempting proposition. It meant simply to change the integrals into summations, the differential equations into algebraic equations, and serendipitously exploit the separability properties of discretized formulations. Turning attention to certain stiffness constraints theoretically valid optimality criteria were derived for discretized structures with displacement constraints (Ref. 14) employing classical Lagrangian multiplier methods of mathematical optimization. To satisfy the optimality criteria an algorithm was proposed based on the attitude that if it provides an exact direct formula for statically determinate structures, same as FSD, then it will converge in a few iterations for most practical structures, again same as FSD. The difference, however, is an important one; this criteria, unlike FSD, was theoretically correct guaranteeing at least a local optimum when satisfied. This new and somewhat controversial direction soon started to enjoy major developments and clarification by interested researchers (Refs. 15 to 22). Essentially complete developments were available by the mid 70's (Refs. 23 to 29), followed by broadening applications (Refs. 30 to 45).

Ironically, as will be seen, the new OC methods that were introduced were as efficient numerically as FSD but were not directly applicable to stress constraints, and could not replace the discredited FSD for that purpose. FSD remains the work-horse for strength design in most design offices despite its shortcomings.

Optimality Criteria Methods are based on radically different thinking from the one that went into the development of the Mathematical Programming Methods. Most MP methods concentrate on obtaining information from conditions around the current design point in design space in order to find the answer to two questions; in what direction to go, and how far, to best reduce the weight or cost directly. This is repeated until no more reduction is produced between iterations within some selected tolerance. Optimality Criteria methods on the other hand, exact or heuristic, derive or state conditions which characterize the optimum design, and then find or change the design to satisfy those conditions while indirectly optimizing the structure. The OC approach results in finding the close neighborhood of the optimum usually very quickly. This is the basic difference in the philosophy of the two approaches as observed by Haug in Ref. 46. The variational methods applied by Prager produced differential equations with their solutions defining the optimum structure. The derivable discretized OC methods produce uncoupled algebraic equations to be solved to define the optimum structure. These equations are nonlinear and require iterative solutions, but structural behavior, usually well understood by structural engineers, has a strong, and usually beneficial effect resulting in fast convergence to a practical solution.

Optimality criteria methods had a brief but vigorous development in the seventies including the ASOP/FASTOP production level capabilities for strength and aeroelastic constraints (Refs. 30,40,44,45). These two codes are in production use by a number of organizations engaged in aircraft design. Not only have OC methods provided an efficient capability closely connected with structural behavior, but they also presented a severe challenge to MP methods. As pointed out earlier, OC methods, unlike MP methods, are not limited by the number of design variables. Convergence depends on structural behavior, usually well understood by structural engineers, inheriting the behavior of the stress ratio algorithm for FSD in this respect. A few iterations are usually sufficient for a structural model with thousands of finite elements. In response to this challenge the computational aspects of MP methods have been vastly improved, while accompanied unfortunately by some belligerence towards OC methods. The common computational elements of OC and MP methods were investigated and utilized. As a result, researchers of MP methods turned to such useful concepts as variable linking, sensitivity analyses, rapid reanalysis techniques, substructuring, dual programming, or anything else to slay the OC dragon, or at least to attempt to catch up with it. However, it is generality that remains the best feature of MP methods versus efficiency of any specialized method. At the present time there is a very simple and powerful OC approach for a single constraint. It is an important capability because in many practical cases there is a single overriding problem to be alleviated. For multiple constraints the simplicity is somewhat compromized by the need to evaluate the Lagrange multipliers. In that case the dividing lines are fast fading (Refs. 47 to 49) and Multiplier Methods (MM) would probably be a better common designation for a number of efficient OC and MP approaches.

OPTIMALITY CRITERIA METHODS FOR DISCRETIZED STRUCTURES WITH STIFFNESS CONSTRAINTS

Single Constraint

The single constraint case is important in practice. Often there is an overriding single requirement that dictates the material distribution. For example, the forward swept wing of the X-29 experimental aircraft (Fig. 1) had to be designed, in addition to strength, for divergence not to occur up to a certain velocity. An optimality criteria method (Refs. 50 and 51) was used in the aeroelastic tailoring studies to stiffen the composite wing for this requirement. The finite element model of the wing with about two thousand elements is shown on Fig. 2. Only around half a dozen iterations were usually needed to adjust the basic FSD strength design to increase divergence velocity to the desired value.

The basic ideas will be introduced by first considering multiple constraints, then by specializing the results for single constraints. Further specialization for trusses with a single displacement constraint will be used as the simplest way to illustrate the method. Finite element formulations, and problems associated with evaluation of the Lagrangian multipliers in case of multiple constraints will complete the presentation.

The success of numerical optimality criteria methods hinges on the concept of separability that results in uncoupled optimality criteria equations. This in turn allows the formulation of simple recursion relations to calculate the next values of the size variables in the iterative process. In our case an additional condition concerning the derivatives of the constraint functions $G_j = C_j(X) - \overline{C}_j$ also have to be satisfied.

A function of $f(X)$, where $X = (x_1, x_2, \ldots, x_n)$ is a vector of the variables x_i, is separable if

$$f(X) = \sum_{i=1}^{n} f_i(x_i) \qquad [1a]$$

that is, if the function can be written as the sum of contributions, each a function of only one of the variables. In the simple derivation that follows both the merit function and the contributions $C_j(X)$ in the constraint functions $G_j(X)$ are required to be separable explicitly. Furthermore, the functions $C_j(X)$ contain internal forces that in case of indeterminate structures are implicit functions of all the variables x_i.

For the uncoupled optimality criteria equations that will be derived to be rigorously valid, in addition to separability, the following condition for the derivatives of $C_j(x)$ also has to hold

$$\frac{\partial C_j(X)}{\partial x_i} = \frac{dc_{ij}(x_i)}{dx_i} \quad i=1,\ldots, n \qquad [1b]$$

Equation (1b) expresses a stronger condition than separability when implicit dependence on all the variables may be involved. This condition expresses the requirement for the sum of the implicit derivatives to vanish so that the optimality criteria equations uncouple leading to the simple recursion relations to be derived next. These conditions are satisfied for the stiffness related constraints (Ref. 17) considered here, but not for stress constraints, frustrating attempts to develop a rigorous yet simple method for that most important case.

The problem of optimum sizing of a discretized structure can be stated as

$$\text{minimize} \qquad W = \sum_{i=1}^{n} W_i(x_i) \qquad [2a]$$

$$\text{subject to} \qquad G_j = C_j - \bar{C}_j \_ 0, \quad j=1,\ldots, m \qquad [2b]$$

$$\text{where} \qquad C_j = \sum_{i=1}^{n} c_{ij}(x_i) \qquad [3]$$

and $\overline{C}_j$ are the limiting values of the constraints. To derive the optimality criteria equations we form the Lagrangian

$$L(x,\lambda) = \sum_{i=1}^{n} W_i(x_i) + \sum_{j=1}^{m} \lambda_j \sum_{i=1}^{n} c_{ij}(x_i) - \bar{C}_j \qquad [4]$$

and require the derivatives relative to x_i to vanish. Using the shorthand notation $,_i$ for the operator d/dx_i, the optimality conditions are:

$$\frac{\partial L}{\partial x_i} = W_{i,i} + \sum_{j=1}^{m} \lambda_j C_{ij,i} = 0 \quad i=1,\ldots, n \qquad [5]$$

and

$$\lambda_j G_j = 0; \qquad \lambda_j \geq 0 \qquad j=1,\ldots, m \qquad [6]$$

where λ_j are the Lagrangian multipliers. It is also required usually that the sizes x_i satisfy some practical constraints on minimum and maximum sizes, that is $x_i^o \leq x_i \leq x_i^1$, where x_i^o and x_i^1 are the minimum and maximum acceptable sizes, respectively.

We now specialize Eqs. (5) and (6) to a single constraint by dropping the index j and the summation on j. Equations (5) and (6) become

$$W_{i,i} + \lambda C_{i,i} = 0 \quad i=1,\ldots, n \qquad [7]$$

and either

$$\lambda > 0; \; G = 0 \qquad [8]$$

or

$$\lambda = 0; \; G < 0 \qquad [9]$$

If the condition $\lambda G = 0$ is satisfied through Eq. (9), we do not have a problem, constraint G is satisfied a priori because of some stronger requirement on the sizes. Equation (7) can be rewritten in a more enlightening form by simply dividing through by $w_{i,i}$ and λ to obtain

$$\frac{C_{i,i}}{w_{i,i}} = \frac{1}{\lambda} = \text{constant} \quad i=1,\ldots, n \qquad [10]$$

or in words:

$$\frac{\text{CHANGE IN CONSTRAINT}}{\text{CHANGE IN MERIT FUNCTION}} = \text{"COST OF IMPROVEMENT"} = \text{CONSTANT}$$

for all variables in an optimum structure. It should be remarked that this word statement does not require Eq. (1b) to be satisfied; it requires only that the derivatives can be calculated. Unlike Eq. (10), it applies for example also to stress constraints. In that case, however, no simple expression can be obtained because Eq. (1b) is not satisfied.

This statement that in an optimum structure the cost of improvement is the same for all design variables is of general validity (Ref. 17) as opposed to the earlier heuristic statements relating stress or energy distributions to totally exploit the material. It also has to be realized that for this statement to have meaning the variables have to represent a "diminishing return of investment," a condition satisfied by size variables against stiffness or stress constraints. To produce a useful expression for computational procedures, further specialization is required representing this last condition on diminishing return.

We denoted the contribution of each variable to the merit function as $W_i(x_i)$ and the contribution to satisfy the constraint $\overline{C}$ by $C_i(x_i)$. The presence of x_i in the functions $W_i(x_i)$ and $C_i(x_i)$ represents a cross section property contributing to weight and stiffness, such as a bar area or plate thickness. In the case of bars and plates with only in-plane load carrying capacity the size variable appears to the first power. If bending is involved then in $C_i(x_i)$ the inertia of the cross section appears and has to be expressed in terms of x_i. The well-known simple relationship $I = ax^p$ covers many practical cases where "a" and "p" are constants and x is a bar area or plate thickness. The specialization involves simply factoring out x_i^p by requiring the following relationships to hold:

$$W_i(x_i) = w_i x_i \qquad [11]$$

$$C_i(x_i) = c_i/px_i^p) \qquad [12]$$

where w_i and c_i now are simply the expressions "left over" after x_i and px_i^p have been factored out respectively.

Substituting Eqs. (11) and (12) into Eq. (7) the following simple expression is obtained:

$$w_i - \lambda(c_i/x_i^{p+1}) = 0, \qquad i=1,\ldots,n \tag{13}$$

This equation can also be rewritten as

$$1 = \lambda c_i/(w_i x_i^{p+1}) \qquad [14]$$

and in the more useful forms

$$x_i = \lambda^{1/p+1}(c_i/w_i)^{1/p+1} \qquad [15]$$

or

$$x_i = \lambda(c_i/w_i x_i^p) \qquad [16]$$

Equations (15) and (16) represent the simplest way to obtain useful expressions for the size variables using optimality criteria methods. In Eqs. (14 to 16) w_i is constant, but c_i, the contribution to the constraint, contains the internal forces in the i-th member. Equation (15) would be a direct sizing formula if c_i also would be a constant. However, except in statically determinate structures and displacement constraint, the internal forces are functions of the relative values of all x_i, therefore Eq. (15) is nonlinear. The c_i have to be recomputed each time new x_i have been obtained suggesting the recursive relations:

$$x_i^{k+1} = \lambda_k^{1/p+1}\ (c_i/w_i)_k^{1/p+1} \qquad [17]$$

or

$$x_i^{k+1} = \lambda_k\ c_i/(w_i x_i^p)_k \qquad [18]$$

where k indicates the k-th iteration. These equations contain the essence of numerical optimality criteria methods.

Equation (17) has been proposed first for displacement constraints (Ref. 14). In that case it mimics the stress ratio algorithm in the important aspect that both give the correct optimum design in a single sizing step for statically determinate structures where the internal forces are independent of the size variables. The experience of generation of engineers showed that in practical structures the sensitivity of internal forces to reasonably small changes in size variables is also small resulting in small changes in the internal forces and convergence in a few iterations. Static indeterminacy is a question of degree, as has been discussed in detail by Hardy Cross of moment distribution fame half a century ago (Ref. 52). Cross distinguished between "normal" and "hybrid" behavior in this respect. Internal load distribution in frames for example is quite insensitive to redesign, while very sensitive in X-laced trusses. That is why the well-known ten bar truss is a good test problem. The consequence is that the convergence of optimality criteria methods depends on this hard-to-specify quality of the structure rather than on the number of design variables like most mathematical programming methods. Essentially cantilever beam wing boxes with thousands of finite element sizes in their model might converge in half a dozen iterations while a ten bar truss problem can take dozens of reanalyses and resizing.

Before discussing other recursion relations that were found successful in practice the role of the Lagrangian multiplier has to be examined. For a single constraint, the term $\lambda^{1/p+1}$ in Eq. (17), or λ in Eq. (18), can be viewed as a simple scaler. For any choice of λ the relative sizes x_i represent an optimum design. The choice of λ will determine a value of the constrained quantity associated with that relative design. There is a particular value of the multiplier that will result in a design with the desired value of the constraint, thus satisfying it in an equality sense.

For the purpose of discussing the various forms of the recursion relations it can be assumed that the value of λ is known. The methods to evaluate λ will be briefly introduced after the recursion relations are discussed, and will be examined in some detail in conjunction with the case of multiple constraints where they pose a more difficult problem.

To simplify the presentation the notation

$$D_i = \lambda c_i / w_i x_i^{p+1} \qquad [19]$$

is introduced for the right hand side of Eq. (14) which then becomes

$$1 = D_i \qquad i=1,\ldots,n. \qquad [20]$$

Equation (20) is valid at the optimum, and has been manipulated in a number of ways to obtain expressions that include an arbitrary relaxation parameter to control the step size between iterations. Multiplying Eq. (20) by x_i^q on both sides and then taking the q-th root one obtains the recursive relation

$$x_i^{k+1} = x_i^k (D_i)_k^{1/q} \qquad [21]$$

where q is the arbitrary parameter to control step size. There are a number of ways to view this expression. One is that with $q = p + 1$ it is equivalent to Eq. (17) and with $q = 1$ to Eq. (18), capturing both. Another way is that for small values of q the deviation of D_i from unity is accentuated causing larger differences in x_i between iterations. The opposite is true of course for large values of q. Examining Eqs. (3), (4), (12), and (19) it can be seen that for example, if x_i is smaller than it should be in order to satisfy the constraint, then D_i will be larger than unity, therefore Eq. (21) will have a tendency to correct that. When the optimum is reached D_i becomes unity and Eq. (21) stops modifying the variables x_i.

A linearized form of Eq. (21) can also be derived by first rewriting it as

$$x_i^{k+1} = x_i^k \, (1 + (D_i - 1))_k^{1/q}$$

where $(D_i - 1)$ vanishes as the optimum is approached. Then, expanding the term with the exponent $1/q$ in a binomial series and retaining only the linear term the following linearized form of Eq. (21) is obtained

$$x_i^{k+1} = x_i^k \, (1 + (1/q)(D_i - 1))_k \qquad [22a]$$

or

$$x_i^{k+1} = x_i^k/(1 - (1/q)(D_i - 1))_k \qquad [22b]$$

if reciprocal variables are used in Eq. (21).

Another form of Eq. (22a) can be derived by multiplying both sides of Eq. (19) by $x_i(1-a)$ and rearranging terms to obtain

$$x_i^{k+1} = x_i^k(a+(1-a)D_i)_k \qquad [23]$$

where a is a relaxation parameter. This form was proposed to improve convergence (Ref. 20), but it is easy to see that with $q = 1/(1-a)$ it is equivalent to Eq. (22a).

The performance of the various proposed recursion relations was critically examined (Ref. 53), including expressions using reciprocal variables $z_i = 1/x_i$. One can say that there are only three useful independent forms, Eqs. (21) and (22a) and (b). The first one is exponential and modifies the variables by multiplying them with a factor that becomes unity at optimum. The other two are linearized and modify the variables by adding a factor that becomes zero at optimum. It is a sensitive measure of convergence to observe how the D_i approach unity for each variable during iterations.

To use Eqs. (21) or (22a) and (b) a method is needed to evaluate the multiplier λ in each iteration. For the case of a single constraint the design obtained by the current value of x_i can be analyzed, and then scaled simply by the formula

$$x_i^{k+1} = x_i^k \, (C^k/\bar{C})_k \qquad [24]$$

where C^k is the current and $\bar{C}$ is the desired value of the constraint. Instead of using Eq. (24), λ can also be evaluated by back-substituting x_i from the optimality criteria expression in the form of Eq. (15) into the constraint expression and solve for λ. Omitting the algebra, the expression that can be then back-substituted into Eq. (15) is

$$\lambda^{1/p+1} = \frac{1}{\bar{C}^{1/p}} \sum_{i=1}^{n} C_i^{1/p(p+1)} w_i^{1/p+1} \qquad [25]$$

Equations (15) and (25) are equivalent to solving the "dual programming" problem where x_i are the primal, and λ is the dual variable. In case of multiple constraints no simple equivalent to Eqs. (24) and (25) can be derived. As opposed to the case of single constraint, were Eq. (24) is the best approach, formal dual programming (Refs. 47 and 48) may provide a view with potential for smooth transition from OC to MP techniques as specifics of the problem require.

There is one more consideration that has to be discussed before the method is illustrated with a simple example. In most practical cases not all the design variables are free to take on values that the equations presented so far would assign to them. There are many reasons for this. One might want to simply assign certain values to some of the variables for fabrication or whatever reasons and exclude them from the group of variables that are free to change. Usually there are limitations on minimum, sometimes on maximum sizes. Even in research problems, involving an automated analysis module, minimum sizes should be stipulated to avoid unwanted mechanisms and therefore singularity to develop. Most frequently there is a design satisfying stress constraints, obtained for example by FSD, in which case added stiffness to limit displacements or to increase system buckling loads is the optimization problem at hand. There is a need, therefore, to recognize a grouping of the design variables into active and passive groups, possibly changing within each iteration. To accomplish this, the equations presented so far need only a minor correction of the target constraint value $\overline{C}$. Consider that in addition to the n active x_i there are p passive x_i whose contribution to achieve the target value $\overline{C}$ is constant, therefore can be subtracted from $\overline{C}$ as a quantity already "payed". This contribution then is computed to be

$$C^o = \sum_{i=n+1}^{p} C_i(x_i) \qquad [26]$$

and all the equations presented are valid by simply replacing $\overline{C}$ with C^* where

$$C^* = \overline{C} - C^o \qquad [27]$$

In the remainder of this presentation C^* will be used for this modified target constraint value to signify the presence of possible passive variables.

Application to Trusses

To illustrate how the equations presented so far can be specialized, the simplest case, trusses under displacement constraints, will be considered. For this case the following definitions are valid with $x_i = A_i$ to signify that the cross section areas are the size variables

$$W_i(x_i) = L_i \rho_i A_i = w_i A_i \qquad [31]$$

$$C_i(x_i) = (S_i^p S_i^v L_i)/(E_i A_i) = c_i/A_i \qquad [32]$$

$$D_i = \lambda(S_i^p S_i^V)/(E_i \rho_i A_i^2) \qquad [33]$$

$$\lambda^{1/2} = \frac{1}{C^*} \sum_{i=1}^{n} (S_i^p S_i^v L_i^2 \rho_i/E_i)^{1/2} \qquad [34]$$

where for the i-th bar

L_i length

ρ_i density

E_i Young's modulus

S_i^p member force due to actual loads P

S_i^v member force due to virtual loads V

If one combines the above expressions, Eq. (15) can be written as

$$A_i^{k+1} = \left(\frac{1}{C^*}\sum_{\ell=1}^{n} S_\ell^p S_\ell^v L_\ell \rho_\ell / E_\ell\right)_k \left(S_i^p S_i^v / E_i \rho_i\right)_k^{1/2} \qquad [35]$$

providing a formal solution to the special dual programming problem of trusses with generalized displacement constraints. However, because of its simplicity Eq. (24) is recommended for computation of λ instead of Eq. (34). Because most computations are performed with finite element methods the recursive relations will be derived next using the conventional matrix formulation of the displacement method. Generalized displacement, and typical eigen value constraints will be considered.

FINITE ELEMENT FORMULATION

Displacement Constraints

All the former expressions can be written in terms of expressions used in finite element computations. In deriving Eq. (7) c_i was taken as constant, although it contains internal forces that are functions of all the x_i, and a sum of additional implicit derivatives should have been considered. These terms vanish exactly, thus the derived optimality criteria are exact due to satisfying the condition expressed in Eqs. (1a) and (1b).

Equation (3) can be written for a single constraint as

$$C = \sum_{i=1}^{n} c_i/x_i = r'^P K r^V \tag{36}$$

where

r^P system displacements due to loads P

r^V system displacements due to loads V

K system stiffness matrix

r' transpose of r

Because r^P and r^V are solutions we have

$$Kr^P = P \tag{37}$$

$$Kr^V = V \tag{38}$$

$$K_i r_i^P = S_i^P \tag{39}$$

$$K_i r_i^V = S_i^V \tag{40}$$

where K_i, r_i, and S_i are element stiffness matrix, displacements and member nodal forces respectively. The Lagrangian can be written as

$$L = \sum_{i=1}^{n} w_i x_i + \lambda(T'^P K r^V - C^*) \tag{41}$$

and the optimality criteria becomes

$$\frac{\partial L}{\partial x_i} = w_i + \lambda \left(\frac{\partial r'^P}{\partial x_i} K r^V + r'^P \frac{\partial K}{M\chi_i} r^V + r'^P K \frac{\partial r^V}{\partial x_i} \right) = 0 \tag{42}$$

This expression can be simplified by considering Eq. (37) as

$$r'^{P}K = P \qquad [43]$$

and differentiate with respect to X_i to obtain

$$\frac{\partial r'^{P}}{\partial x_i} K + r'^{P} \frac{\partial K}{\partial x_i} = 0 \qquad [44]$$

from which

$$\frac{\partial r'^{P}}{\partial x_i} = -r'^{P} \frac{\partial K}{\partial x_i} K^{-1} \qquad [45]$$

Substituting Eq. (45), and a similar one for the case of r^V into Eq. (42), the first two terms in the parentheses cancel and the third one transforms to yield

$$w_i - \lambda \left(r'^{P} \frac{\partial K}{\partial x_i} r^V \right) = 0 \qquad [46]$$

as the optimality criteria.

The derivative term in Eq. (46) can be simplified considering the definition of K and, as an example, finite elements with their stiffness matrix depending linearly on x_i. Then we have

$$\frac{\partial K}{\partial x_i} = \frac{1}{x_i} K_i \qquad [47]$$

resulting in the expression

$$w_i - \lambda \frac{1}{x_i} r_i'^{P} K_i r_i^V = 0 \qquad [48]$$

or

$$1 = \lambda \frac{r_i'^P K_i r_i^V}{w_i x_i} = D_i \qquad [49]$$

as the optimality criteria.

Equation (49) can be used to derive recursion relations that are equivalent to Eqs. (20) and (22a) and (b). It is easy to see that for example for trusses the product $r_i'^P k_i r_i^V$ is equivalent to $s_i^P s_i^V L_i/A_i$ and that the right hand side of Eq. (49) is equivalent to D_i of Eqs. (20) and (22a) and (b).

The method of derivation presented here for the finite element formulation showing the satisfaction of Eqs. (1a) and (1b) was introduced (Ref. 28) during the early development of discretized optimality criteria methods. It can be viewed as the forerunner of what later became a separate area of optimization research, now referred to as "sensitivity analysis".

System Buckling Constraint

The derivation of optimality criteria for constraint on system instability is outlined next. Details can be found in Refs. 25, 27 and 28. The formulation presented here is only one of the many possible ways to prescribe a target value μ^* for load intensity μ at which instability occurs. The constraint equation may be given in the form

$$y'Ky - \mu^* y'K_G y = 0 \qquad [50]$$

where we require the buckling displacements y, and the load intensity μ^* to be the solution of the eigen value problem

$$Ky - \mu K_G y = 0 \qquad [51]$$

Here K_G is the geometric stiffness matrix. We can again write the Lagrangian as

$$L = \sum_{i=1}^{n} w_i x_i + \lambda(y'Ky - \mu^* y'K_G y) \qquad [52]$$

to yield the optimality criteria equations

$$1 = \lambda \frac{y'K_i y}{w_i x_i} = D_i \qquad i=1,\ldots,n \tag{53}$$

The derivation of Eq. (53) follows the same line as displacement constraints making use of Eq. (51) and of the fact that K_G is not a function of the x_i size variables.

Vibration Frequency Constraint

Derivation for constraints on frequency proceeds along the same lines as for constraints on system buckling.

The constraint equation may be given in the form

$$y'Ky - \omega^{*2} y'My = 0 \tag{54}$$

where ω^* is the target value for the frequency ω, and M is the mass matrix. It is required that the mode shape y and the eigenvalue ω^2 be the solution of the eigen value problem

$$Ky - \omega^2 My \tag{55}$$

The Lagrangian can be written as

$$L = \sum_{i=1}^{n} w_i x_i + \lambda(y'Ky - \omega^{*2} y'My) \tag{56}$$

to yield the optimality criteria

$$1 = \lambda \frac{y'K_i y - \omega^2 y'M_i y}{w_i x_i} = D_i = 0 \qquad i=1,\ldots,n \tag{57}$$

with the use of Eq. (55). The derivation follows the same steps as earlier, but one remark has to be made concerning scaling of the variables to achieve the target value ω^* once the relative values of the design have been obtained. Because unlike K_G, the mass matrix M is the function of the design variables x_i, without nonstructural mass scaling is ineffective to change the frequency ω. The scaling procedure has to be well thought out for a given situation involving nonstructural mass.

MULTIPLE CONSTRAINTS

The modification to multiple constraints of the expressions, developed for single constraints is trivial on the surface, but it eliminates simple scaling as the method to evaluate the Lagrangian multipliers λ_j that appear in Eq. (5). The modifications necessary are to replace $C_i(x_i)$ with $C_{ij}(x_i)$, λ with λ_j and retain the summation as indicated in Eq. (5). To follow the forms of the expressions presented for single constraints, Eq. (5) is rewritten as

$$1 = \sum_{j=1}^{n} \lambda_j \frac{C_{ij,i}}{W_{i,i}} \qquad [58]$$

If the requirements for the structure of the functions $W_i(x_i)$ and $C_{ij}(x_i)$ expressed in Eqs. (11) and (12) are again invoked, Eq. (58) becomes

$$1 = \sum_{j=1}^{m} \lambda_j \frac{c_{ij}}{w_i x_i^{P+1}} \qquad [59]$$

The definition for D_i to be substituted in the recursion relations for multiple constraints then is given as

$$D_i = \sum_{j=1}^{m} \lambda_j \frac{c_{ij}}{w_i x_i^{P+1}} \qquad [60]$$

In case of further specialization for trusses as a simple example, Eq. (33) becomes

$$D_i = \sum_{j=1}^{m} \lambda_j \left(S_{ij}^P S_{ij}^V\right) / \left(\rho_i A_i^2\right) \qquad [61]$$

where S_{ij}^P and S_{ij}^V are again the member forces, but for the j-th combination of the actual load condition P_j and virtual load V_j associated with the j-th constraint condition.

With the above modification there is no difference in the recursion relations for single or multiple constraints. As will be discussed next, there is a major difference associated with the evaluation of the Lagrangian multipliers.

EVALUATION OF THE LAGRANGIAN MULTIPLIERS IN CASE OF MULTIPLE CONSTRAINTS

Unless relative values of λ_i are known, the simple scaling procedure recommended in case of single constraints is not applicable. Essentially there are two classes of approaches to obtain the λ_j - s. One class considers the coupling effects among the active λ_j - s, and the other class of approaches does not.

To facilitate the discussion we consider the case when Eqs. (11) and (12) are valid and $p = 1$. In that case the Lagrangian is written as

$$L(x,\lambda) = \sum_{i=1}^{n} w_i x_i + \sum_{j=1}^{m} \lambda_j G_j \qquad [62]$$

with

$$G_j = C_j - C_j^* < 0 \qquad j=1,\ldots,m \qquad [63]$$

where

$$C_j = \sum_{i=1}^{n} \frac{c_{ij}}{x_i} \qquad [64]$$

is the value of the j-th constrained behavior variable that has the target value C^*_j. The optimality criteria becomes

$$1 = \sum_{j=1}^{m} \lambda_j \frac{c_{ij}}{w_i x_i^2} \qquad i=1,\dots,n \qquad [65]$$

It is instructive to write out Eq. (65) in the following form to illustrate the dual nature of the formulation:

$$\lambda_1 \begin{bmatrix} c_{11}/x_1 \\ c_{2i}/x_2 \\ \vdots \\ c_{n1}/x_n \end{bmatrix} + \lambda_2 \begin{bmatrix} c_{12}/x_1 \\ c_{22}/x_2 \\ \vdots \\ c_{n2}/x_n \end{bmatrix} + \dots + \lambda_m \begin{bmatrix} c_{1m}/x_1 \\ c_{2m}/x_2 \\ \vdots \\ c_{mn}/x_n \end{bmatrix} = \begin{bmatrix} w_1 x_1 \\ w_2 x_2 \\ \vdots \\ w_n x_n \end{bmatrix} \qquad [66a]$$

$$\lambda_1 \quad C^*_1 + \lambda_2 \quad C^*_2 + \dots + \lambda_m \quad C^*_m = W \qquad [66b]$$

The equations for the "weighted" design variables represent the optimality conditions for any choice of λ_j. The vector of the design variables is a sum of design vectors for each constraint C^*_j separately, multiplied by a coefficient λ_j. Viewing the equations in the column direction the sum of the elements in each vector is the value C_j of the j-th constraint to be less than or equal to the target value C^*_j. If a constraint, say the k-th is oversatisfied, then it is not active and $\lambda_k = 0$. If the k-th constraint is not satisfied, one would want to increase the participation of the k-th vector as the most cost effective component to improve constraint satisfaction. This is achieved by increasing λ_k through some formula involving $G_j = C_j - C^*_j$ as the measure of the lack of satisfaction. This suggests the two updating formulas

$$\lambda_j^{k+1} = \lambda_j^k + p_k G_j^k \quad [67a]$$

and

$$\lambda_j^{k+1} = \lambda_j^k (C_j/C_j^*)_k^{p_k} \quad [67b]$$

Where p_k is a parameter to control convergence and possibly change as $G_j \to 0$. Reference 54 provides formal presentation of the first-order multiplier iteration indicated in Eq. (67a) to solve the dual programming problem. The exponential form in Eq. (67b) was suggested in Refs. 18 and 20 somewhat heuristically. These are two simple uncoupled algorithms that can be used with success. Initial values are needed, but inactive constraints are automatically elliminated.

Historically the first uncoupled approach suggested for structures (Ref. 15) was very much along the heuristic lines of FSD. It became known as the "envelope method" because it suggested to obtain the optimum design for each constraint separately, then select for each variable the largest value among the separate designs. This rather practical approach gave surprisingly good designs that were, of course, approximations violating the optimality criteria equations. It can certainly serve as an approach to obtain initial estimates of x_i and λ_j.

The next class of approaches considers the coupling effect among the λ_j leading to a set of nonlinear simultaneous equations to solve. There are three obvious ways to derive essentially the same expressions for our special problem. The one given here was suggested in Ref. 42 and is the simplest in terms of the algebra. The constraint and the optimality criteria equations can be written in the following matrix forms with typical elements as indicated.

$$\left[\frac{c_{ij}}{x_i}\right]_{m\times n} \begin{bmatrix} 1 \\ 1 \\ \cdot \\ \cdot \\ \cdot \\ 1 \end{bmatrix}_{n\times 1} = \left[C_j^*\right]_{m\times 1} \qquad [68]$$

$$\left[\frac{c_{ik}}{w_i x_i^2}\right]_{n\times m} \left[\lambda_k\right]_{m\times 1} = \begin{bmatrix} 1 \\ 1 \\ \cdot \\ \cdot \\ \cdot \\ 1 \end{bmatrix}_{n\times 1} \qquad [69]$$

Substituting Eq. (69) into Eq. (68) one obtains the linear equations for the λ (with summation on repeated i)

$$\left[\frac{c_{ij}\, c_{ik}}{w_i x_i^3}\right] \left[\lambda_k\right] = \left[C_j^*\right] \qquad [70]$$

valid at the solution of the dual programming problem involving the previous specializations.

Equation (70) has been used with success in the iterative form

$$\left[\lambda_j\right]_{\nu+1} = \left[E_{jk}\right]_{\nu}^{-1} \left[(P + 1)Cj - PC_j^*\right] \qquad [71]$$

where

$$\left[E_{jk}\right]_{\nu} = \left[\frac{C_{ij} C_{ik}}{w_i x_i^3}\right]_{\nu} \qquad [72]$$

with summation on the repeated index i. Here p_ν is again a step size parameter and ν represent the iteration number. It is to be noted that no starting values are needed if Eq. (71) is used.

There are two other ways to derive coupled equations for the λ_j - s. The most obvious way is to do the same substitution of the x_i in terms of the λ_j as has been done in case of the single constraint. That is to solve the optimality conditions for the x_i then eliminate the x_i from the constraint equations in terms of the λ_j. This results in a set of nonlinear equations for the λ_j - s. The Newton-Raphson approach requires the matrix of derivatives which turns out to consist of the same terms as E_{jk} defined above. Another way to obtain the λ_j is by formal application of the dual programming approach (Ref. 47). This consists of again expressing the x_i from the optimality conditions in terms of the λ_j - s and back-substitute these expressions into the Lagrangian $L(x,\lambda)$ in terms of the λ_j with x_i eliminated to obtain the dual Lagrangian $L(\lambda,\lambda)$. From the problem statement in terms of the Lagrangian multipliers the derivative of $L(x,\lambda)$ relative to λ_j set to zero expresses the conditions for the satisfaction of the constraints. The derivatives of $L(\lambda,\lambda)$ relative to λ_j - s set to zero then provides equations in λ_j since the x_i - s have been eliminated. The equations so obtained are again contain the same terms as E_{jk} and seemingly nothing new has been gained. However, in the general case when the special conditions employed to obtain Eq. (70) do not hold, a direct MP method can be still used to achieve the condition

$$\frac{\partial L(\lambda,\lambda)}{\partial \lambda_j} = 0 \qquad [73]$$

thus obtain values for the λ_j - s. The formal dual programming approach (Refs. 47 and 48) in case of separable problems provides flexibility to deal with problems where the specializations utilized in the traditional O.C. methods are not applicable. On the other hand to view O.C. methods through dual programming confuses the basic philosophy, that led to O.C. methods, with the resulting computational aspects, that can be quite similar among various approaches. Different roads can lead to the same place (Ref. 49) each providing a different view along the way, some simpler to travel than others. Classical O.C. methods strive for simplicity and extreme efficiency. In exchange they solve only restricted special problems which, however, are the ones encountered the most frequently in practice.

STRESS CONSTRAINTS

Finally, a few remarks are perhaps appropriate concerning stress constraints, the most fundamental problem in structural design. Unfortunately, if directly stated in terms of member forces and allowable stresses, stress constraints do not satisfy Eq. (1b), and the resulting optimality criteria equations do not uncouple. The traditional FSD with stress ratio algorithm is equivalent to neglecting the coupling terms as if Eq (1b) were satisfied.

When stress constraints are stated in terms of relative displacements, the equations presented here apply, but two difficulties arise. One is that this is a case of large number of constraints and requires the evaluation of as many Lagrangian multipliers as there are design variables. The other difficulty is that only in case of trusses is there a clear definition of relative displacements equivalent to stress constraints. Consequently only the case of trusses was studied with Eq. (67b) being simple and effective to update the large number of active Lagrangian multipliers. A special case (Ref. 42) can be created assuming that the optimum design is fully stressed even in some cases of nonuniform stress allowables. Placing opposing unit virtual loads at both ends of the bars and equilibrating them with a unit virtual internal bar force diagonalizes Eq. (71) and the simple formula

$$\lambda_i = S_i^p \; (E_i \rho_i / \sigma_{oi}^2) \qquad [74]$$

is obtained. It is, however, not simpler than Eq. (67), valid only if the optimum structure is indeed fully stressed, and it has not been explored for multiple load conditions. However, it has interesting theoretical implications for fully stressed optimum designs in case of nonuniform material properties.

It can be pointed out that although Refs. 55 to 57 present an efficient approach, it is based on the unpopular Force Method. This leaves us with FSD as the most practical, best understood approach (Ref. 58) that serves designers well if caution is exercised. In most practical applications (Refs. 50 and 51) an FSD design is obtained and the resulting sizes treated as minimum values for the design variables during optimization for other constraints. While this is an approach that is satisfactory in practice, theoretically it is clearly an approximation. MP methods with efficient use of computational shortcuts may provide an approach that allows thousands of finite element member sizes to be treated, perhaps with creative "variable linking" (Ref. 59) that is beneficial, in any case, to smooth the design.

Finally, particularly in research problems, puzzling behavior may occur, as pointed out by Prager (Ref. 60) for a few cases. Fortunately, problems encountered in practice are usually safe from pitfalls.

EXAMPLES

A few simple applications will be used to illustrate convergence behavior for various combinations of the three recurrence relations and the three suggested methods to evaluate the Lagrangian multipliers. Equations (21), (22a) and (22b) will be referred to as options 1, 2, and 3 for the recurrence relations and Eqs. (67a), (67b) and (71) as options A, B, and C for the evaluation of the Lagrangian multipliers. The step size parameters p_k and q_k can be modified in each iteration simply by a factor slightly different from unity. It is a good practice for small test problems to cautiously start with $p = 0.5$ and $q = 2$ and increase their effects by 10 percent in each iteration. In production applications when the method is expected to be applied to the same type of structures repeatedly then a best combination of options with "tuned" parameters can be developed.

Example 1. Three-Bar Truss Problem

The simple structure shown in Fig. 3 is probably the one most studied in human history. The "variable linking" of $A_1 = A_3$ is used to create a two-dimensional problem. The properties chosen here lead to particularly simple expressions. With material densities $\rho_1 = \rho_3 = 1/(2\sqrt{2})$ lbs/in^3 and $\rho_2 = 1/10$ lbs/in^3. the merit function becomes

$$W = 100A_1 + 10A_2 \qquad [75]$$

The constraint chosen is one inch elongation of Bar number 1. This can be viewed either as a displacement or as a stress constraint of 70.710678 KSI in Bar number 1. Simple algebra leads to the constraint equation

$$G = (L/E)(P_1/A_1+P_2/(A_1 + \sqrt{2}A_2)) - \overline{C} = 0 \qquad [76]$$

With $L = 100$ in., $P_1 = P_2 = 100$ K and $E = 10^4$ ksi Eq. (76) simplifies to

$$A_2 = (A_1/\sqrt{2})(\overline{C}A_1 - 2)/(1 - \overline{C}A_1) \qquad [77]$$

where in our case $\overline{C} = 1$. (Note that the dimension of $\overline{C}$ is work).

Equations (75) and (77) are plotted on Fig. 4 with the optimum solution indicated graphically. Substituting Eq. (77) into Eq. (75) a quadratic equation is obtained providing the analytical solution: $A_1 = 1.275846$ and $A_2 = 2.368357$ given here with probably sufficient accuracy. The merit function at optimum takes on the value: $W = 151.268188$ lbs.

All nine combinations of the three recursion relations and the three approaches to evaluate the Lagrangian multiplier were tested on this problem, and without scaling first. In this case only the converged design satisfies the constraint.

In case of single constraints scaling camouflages the effect of any approach to update the multiplier because it is sufficient by itself. The initial values for all runs were $A_1 = A_2 = 1$, $\lambda = 100$, $q = 2$, and $p = 0.5$. A multiplier of 1.1 was used to decrease q and to increase p each iteration. Table I shows the typical results using the combination 2B according to our earlier definition. Table II shows the results with the same 2B combination, but also using scaling. Wide variations of convergence behavior can be obtained when experimenting with the various options.

Example 2. Five-Bar Truss Problem

To show the problem dependency of the convergence behavior the four problems indicated in Fig. 5 were solved with the previously defined combinations 1C to 3C and scaling. The value of the merit function versus iterations is plotted in Fig. 6 for all cases. Designs are given in Table III.

Problem 1 is an elementary check on the methods. Problem 2 is more difficult with both constraints active at the optimum and with tendency for criticality to switch during iterations. Problem 3 is again a trivial check but with numerical sensitivity if started with unit bar areas as done here with all problems. Problem 4 is almost a statically determinate problem explaining the quick solution by combination 1C that would size a determinate structure in one step.

As can be seen, an option with good convergence can be found for any of the problems and perhaps tuned for better performance if repeated optimization runs are expected. Problems 2 and 3 were sensitive to computational accuracy and required double precision. This cautions against indiscriminate use of approximation concepts in optimization.

Example 3. Multiple Frequency Constraints.

An application of the optimality criterion algorithm to design a structure with multiple frequency constraints (Ref. 61) is illustrated by designing the 38 member truss shown in

Fig. 7(a). The elastic modulus and weight density of the material were 10^7 psi and 0.1 lbs/in.^{3}. A nonstructural mass of 0.1 lb-s^2/in., was attached at the top eight node points 4 to 18. The relative cross-sectional area of all the members for the first iteration was equal to unity. The minimum gauge constraint was equal to 0.005 in.^{2}. The truss was designed to satisfy three constraint conditions. These were: (1) $\omega_1^2 = \omega_2^2 = 2500$ (Case A); (2) $\omega_1^2 = 2500$, $\omega_2^2 = 5000$ (Case B); (3) $\omega_1^2 = 2500$, $\omega_2^2 = 5000$ (Case C). It was also required that for Case B, the vibration Mode 1 (Fig. 7(b)) and Mode 2 (Fig. 7(c)) be associated with ω_1^2 and ω_2^2 respectively. However, in Case C, ω_1^2 was required to be associated with Mode 2 and ω_2^2 with Mode 1. The iteration history for the three constraints is given in Table IV. The designs satisfying these constraints were obtained in sequence. The distribution of cross-sectional areas of the members for the three designs are given in Table V. Details of the design procedure are given in Ref. 61.

Example 4. Design of a Dome Structure for Limit Load

The algorithm used in this example is based on the optimality criteria (Ref. 62) that the ratio of geometric nonlinear strain energy density to mass density is equal for all elements. This statement is a special case of Eq. (10).

The 30 member three-dimensional dome structure shown in Fig. 8 was optimized for a concentrated load of 2000 lbs applied in the vertically downward direction at node 1 and with a minimum size constraint of 0.1 in.^{2}. The iteration history for this dome structure is given in Table VI. The weight of the structure after ten iterations was 779.4680 lbs. With an additional twenty iterations the weight of the structure was reduced to 766.1880 lbs. The cross-sectional areas of the members for the optimum design are given in Table VII. The analysis method used is based on the direct minimization of the total potential. The procedure for scaling to limit load and the effect of the step size parameters is discussed in Ref. 62.

Example 5. Large Trusses

A series of trusses are being shown on Fig. 9 with the same overall geometry but with the number of members increasing from 106 to 1027. The loading consists of two concentrated loads as shown on the first truss. This loading, with the other end supported by symmetry conditions, would tend to twist and bend the flat truss-slab. Two displacement constraints were imposed at the two loaded points, 10 in. at the higher load and 20 in. at the lower load. Both constraints were active allowing the truss-slab to twist in an opposite manner from the twist that would be induced by the loads in a structure with uniform member areas. This could be viewed as a "poor man's" aeroelastic tailoring approach. Figure 10 shows the convergence curves for the trusses as a function of the number of truss members. This set of examples was created to show that the number of analyses is indeed a weak function of the number of independent design variables if an optimality criteria algorithm is used. This allows direct application to large finite element models without artificial approximation concepts or variable linking that are needed by MP methods. One should remark that while approximation concepts might be dangerous in cases sensitive to accuracy (see Example 2), variable linking can be employed beneficially if certain smoothness of the design is required.

Figure 1. Grumman/Darpa X-29A Advanced Technology Demonstrator

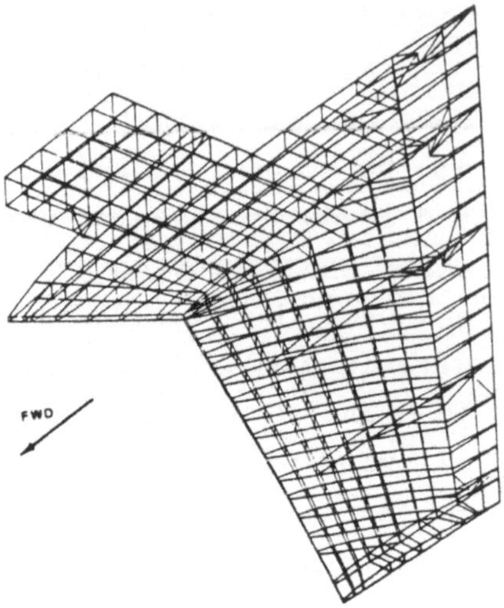

Figure 2. Isometric View of Wing Finite Element Model for Preliminary Design

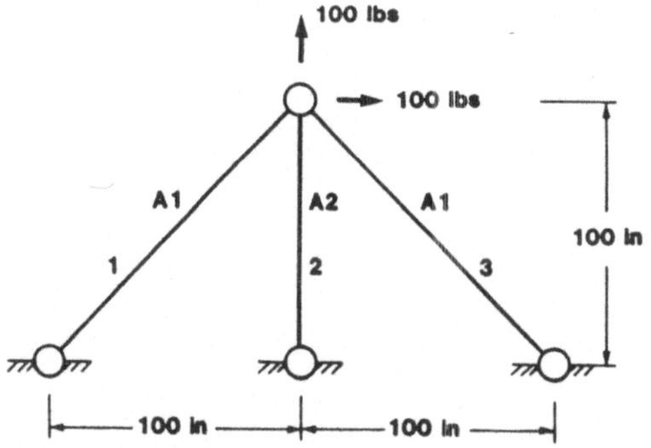

Figure 3. Three-Bar Truss

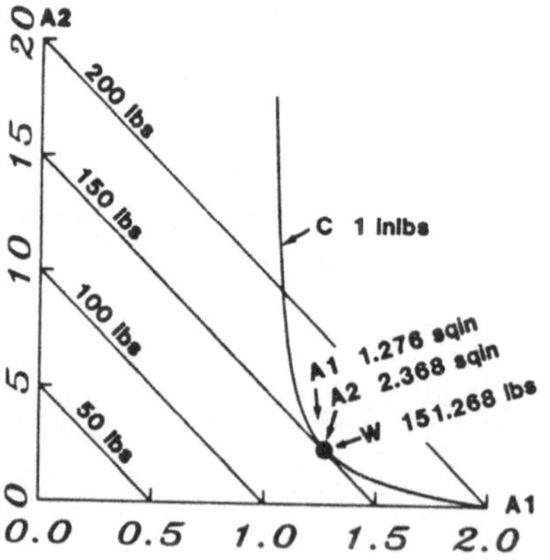

Figure 4. Design Space for Three-Bar Truss

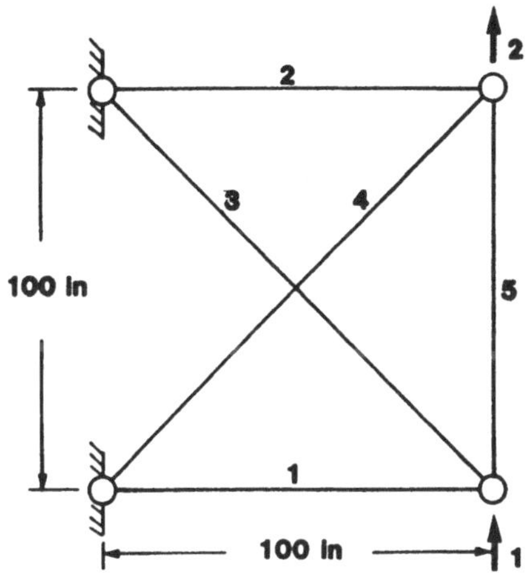

DEFINITION OF FOUR PROBLEMS

PRM NO.	P1 (kips)	P2 (kips)	C1 (in kips)	C2 (in kips)
1	—	100.	—	2.
2	100.	—	—	2.
3	-100.	100.	-0.1	0.1
4	100.	100.	2.	2.

E = 10,000 ksi

ρ = 0.1 lbs/cuin

Figure 5. Five-Bar Truss and Problem Definitions

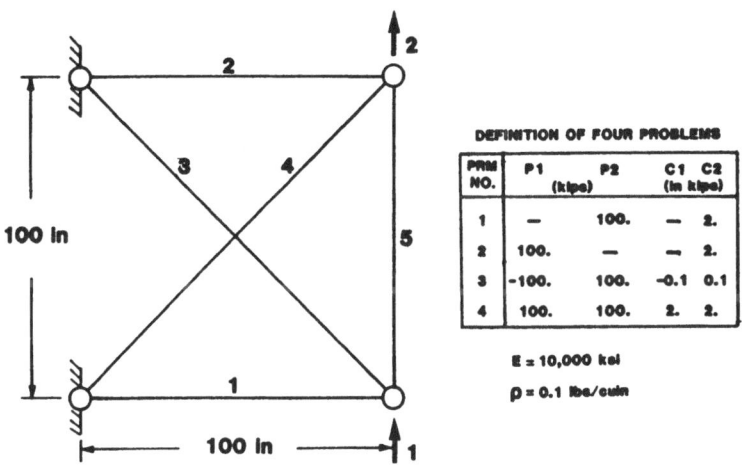

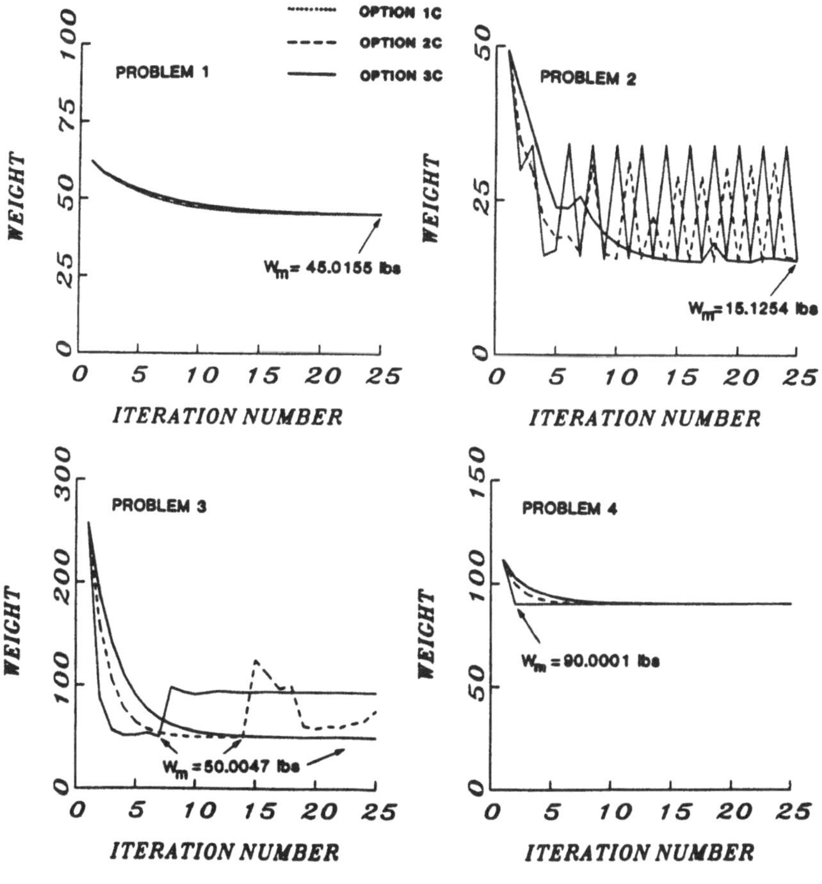

Figure 6. Convergence Curves for the Five-Bar Truss problems

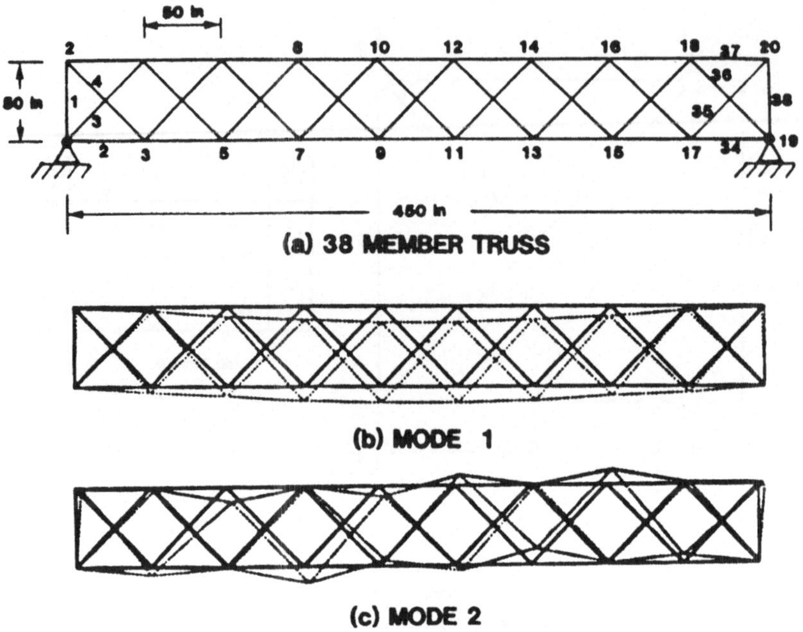

Figure 7. 38 Member Truss and Vibration Modes

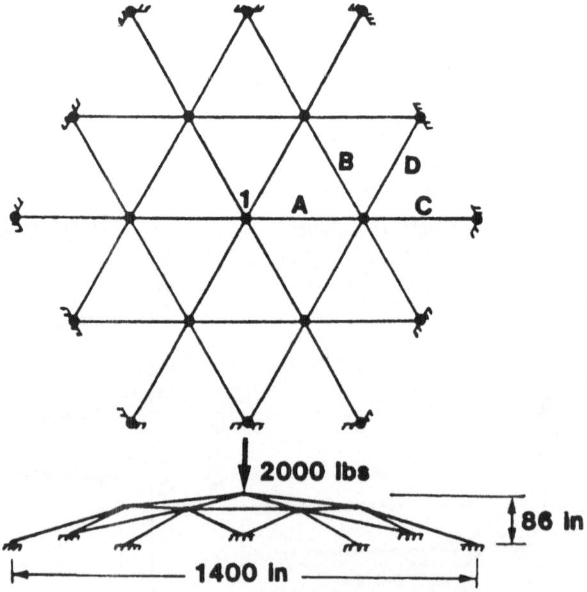

Figure 8. Shallow Truss-Dome Structure

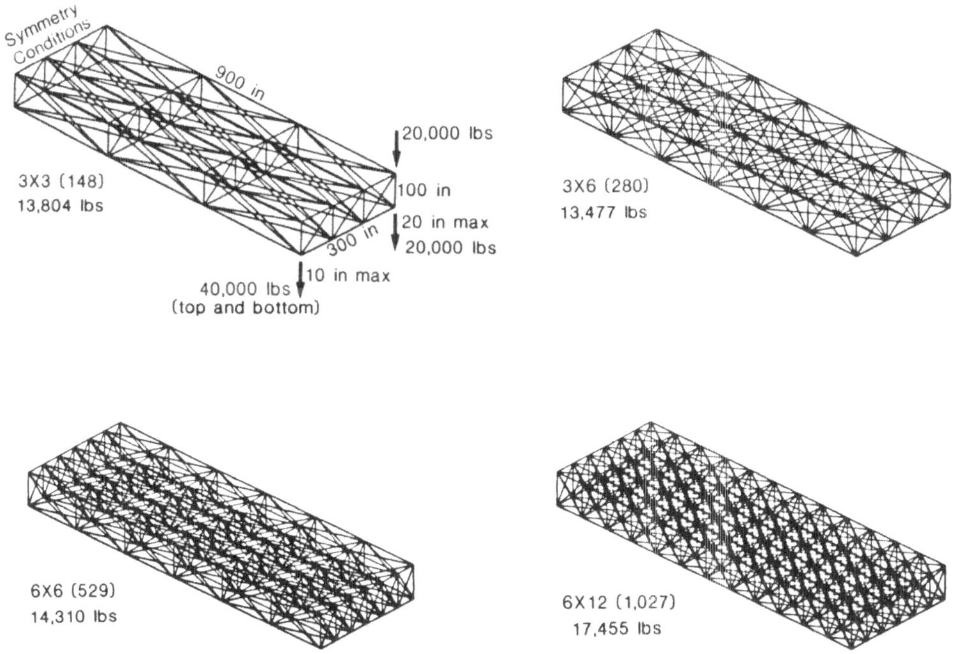

Figure 9. 3-D Truss Problems

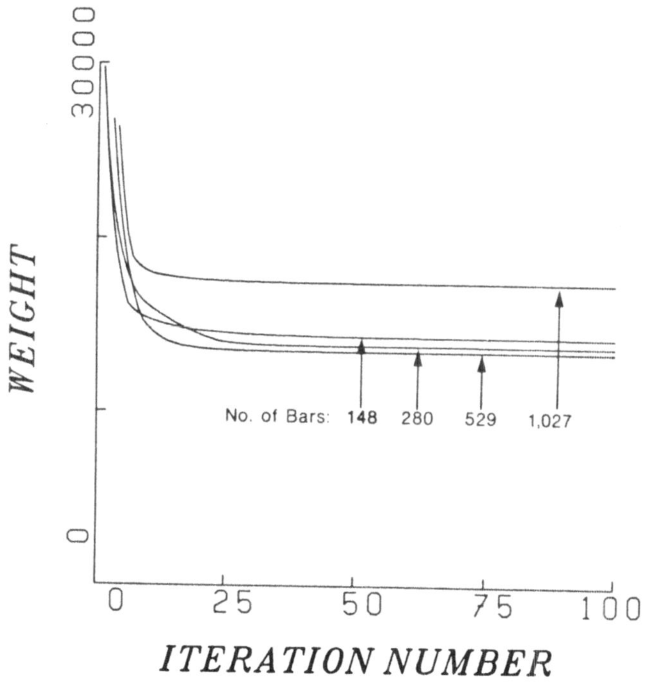

Figure 10. Convergence Curves for 3-D Truss Problems

Table 1. Iterations for the Three-Bar Truss Problem with No Scaling

IT	A1	A2	STRESS1	W	LAMBDA	D1	D2
0	1.000000	1.000000	99.999985	110.000000	100.000000	0.482843	1.000042
1	1.283121	2.297330	70.710754	151.285400	86.746979	0.952849	1.130454
2	1.276890	2.357945	70.710693	151.268433	308.154541	0.993002	1.427426
3	1.276171	2.365108	70.710678	151.268127	72.997498	0.997812	0.694675
4	1.275991	2.366898	70.710678	151.268112	274.773193	0.999017	1.347764
5	1.275928	2.367541	70.710678	151.268143	102.296478	0.999451	0.822350
6	1.275897	2.367827	70.710739	151.267960	185.702133	0.999645	1.107987
7	1.275884	2.367981	70.710678	151.268173	138.799835	0.999748	0.957903
8	1.275874	2.368065	70.710678	151.268051	155.697968	0.999804	1.014537
9	1.275869	2.368118	70.710678	151.268112	150.100693	0.999840	0.996135
10	1.275865	2.368152	70.710739	151.267960	151.519913	0.999863	1.000833
11	1.275864	2.368178	70.710678	151.268143	151.224945	0.999880	0.999859
12	1.275861	2.368193	70.710739	151.267990	151.273911	0.999892	1.000019
13	1.275861	2.368208	70.710678	151.268143	151.267487	0.999901	0.999999
14	1.275859	2.368217	70.710678	151.268036	151.268127	0.999906	1.000000
15	1.275859	2.368225	70.710678	151.268127	151.268112	0.999913	1.000001
16	1.275858	2.368229	70.710678	151.268066	151.268112	0.999915	1.000000
17	1.275858	2.368235	70.710678	151.268127	151.268173	0.999920	1.000001
18	1.275857	2.368237	70.710678	151.268051	151.268051	0.999920	1.000000
19	1.275857	2.368240	70.710678	151.268097	151.268112	0.999923	1.000001
20	1.275857	2.368244	70.710678	151.268127	151.268234	0.999925	1.000001

Table 2. Iterations for the Three-Bar Truss Problem with Scaling.

IT	A1	A2	STRESS1	W	LAMBDA	D1/D2	SCALER
0	1.000000	1.000000	99.999985	110.000000	100.000000	0.482843	1.000000
1	1.268580	2.442827	70.710678	151.286301	166.133148	1.050949	1.100064
2	1.276562	2.361213	70.710678	151.268295	142.728683	0.995195	0.926889
3	1.275859	2.368215	70.710754	151.268021	155.347916	0.999906	1.043271
4	1.275847	2.368340	70.710678	151.268127	150.016769	0.999990	0.982692
5	1.275846	2.368351	70.710678	151.268051	151.495651	0.999997	1.004917
6	1.275846	2.368358	70.710678	151.268112	151.254410	1.000001	0.999204
7	1.275846	2.368356	70.710678	151.268097	151.267670	1.000000	1.000044
8	1.275846	2.368357	70.710678	151.268112	151.268234	1.000000	1.000002
9	1.275846	2.368356	70.710678	151.268097	151.268234	1.000000	1.000000
10	1.275846	2.368356	70.710678	151.268097	151.268234	1.000000	1.000000

Table 3. Minimum Weight Designs for the Five-Bar Problems

PR	OPT	A1	A2	A3	A4	A5	W-min
	1C	0.0010	1.4994	0.0010	2.1204	0.0010	45.0155
1	2C	0.0010	1.4990	0.0010	2.1199	0.0010	45.0239
	3C	0.0010	1.4994	0.0010	2.1204	0.0010	45.0155
	1C	0.4944	0.0233	0.6992	0.0329	0.0024	15.5577
2	2C	0.4981	0.0058	0.7044	0.0082	0.0010	15.1254
	3C	0.4979	0.0066	0.7042	0.0091	0.0010	15.1430
	1C	0.0010	0.0010	0.0010	0.0010	5.0000	50.0047
3	2C	0.0010	0.0010	0.0010	0.0010	4.9999	50.0047
	3C	0.0010	0.0010	0.0010	0.0010	4.9999	50.0047
	1C	1.5000	1.5000	2.1213	2.1213	0.0010	90.0001
4	2C	1.5000	1.5000	2.1213	2.1213	0.0010	90.0010
	3C	1.5000	1.5000	2.1212	2.1213	0.0010	90.0001

Table 4. Iterations for the 38 Member Truss Problems

ITER. NO.	CASE (A)	CASE (B)	CASE (C)
1	29.09	29.09	29.09
2	21.76	21.76	21.76
3	17.98	17.98	17.98
4	16.44	16.44	16.44
5	15.89	15.89	15.89
6	15.59	15.59	15.59
7	15.38	15.38	15.38
8	15.24	15.33	15.24
9	15.14	15.32	15.14
10	15.11	15.32	15.11
11	15.09		15.09
12	15.08		15.08
13			22.96 (1.5)*
14			24.57 (1.6)*
15			26.20 (1.7)*
16			27.84 (1.8)*
17			29.49 (1.9)*
18			31.16 (2.0)*

* ratio of the second and first eigenvalues

Table 5. Minimum Weight Designs for the 38 Member Truss Problems

ELEMENTS	CASE (A)	CASE (B)	CASE (C)
1-38	0.037927	0.038338	0.079492
2-34	0.037865	0.038268	0.079232
3-36	0.053478	0.054060	0.111750
4-35	0.066035	0.066830	0.140917
5-37	0.138371	0.132426	0.296488
6-30	0.130099	0.130045	0.273760
7-32	0.065329	0.066069	0.137912
8-31	0.037743	0.038862	0.078599
9-33	0.073028	0.66194	0.158215
10-26	0.180685	0.179983	0.373786
11-28	0.036597	0.037747	0.073884
12-27	0.041226	0.040699	0.087490
13-29	0.005571	0.017975	0.005358
14-22	0.235310	0.233954	0.481237
15-24	0.039558	0.039012	0.080676
16-23	0.00500	0.000714	0.00500
17-25	0.028882	0.035034	0.055150
18	0.234603	0.233762	0.474714
19-20	0.00500	0.009498	0.00500
21	0.056495	0.061631	0.112273
WEIGHT (LBS)	15.08	15.32	31.16

Table 6. Iterations for the Shallow Truss-Dome Structure

ITER. NO.	WEIGHT (LBS)
1	1660.7960
2	983.1602
3	875.2096
4	840.2418
5	818.9048
6	804.9048
7	795.1926
8	788.2850
9	783.2366
10	779.4680
11	776.6074
15	770.3064
20	767.4552
25	766.5488
30	766.1880

Table 7. Designs for the Shallow Truss-Dome Structure

ELEMENT LOCATION	AREA (SQIN)
A	1.6926
B	1.3754
C	0.2693
D	0.1000

REFERENCES

1. Klein, B.: Direct Use of Extremal Principles in Solving Certain Optimizing Problems Involving Inequalities. J. Oper. Res. Soc. Am., vol. 3, 1955, pp. 168-175, 548.

2. Schmidt, L.A.: Structural Design by Systematic Synthesis. Proceedings of the Second Conference on Electronic Computation, ASCE, 1960, pp. 105-132.

3. Razani, R.: Behavior of Fully-Stressed Design of Structures and its Relationship to Minimum Weight Design. AIAA J. vol. 3, no. 12, Dec. 1965, pp. 2262-2268.

4. Kicher, T.P.: Optimum Design-Minimum Weight Versus Fully-Stressed. J. Struct. Div. Am. Soc. Civ. Eng., vol. 92, no. ST6, Dec. 1966, pp. 265-279.

5. Melosh, R.J.: Convergence in Fully-Stressed Designing. Symposium on Structural Optimization, AGARD CP-36, 1969, pp. 7-1 to 7-15.

6. Prager, W.; and Taylor, J.E.: Problems of Optimal Structural Design, J. Appl. Mech., vol. 90, no. 1, Mar. 1968, pp. 102-106.

7. Venkayya, V.B.; Khot, N.S.; and Reddy, V.S.: Energy Distribution in an Optimum Structural Design. AFFDL-TR-68-156, Sept. 1968.

8. Taylor, J.E.: The Strongest Column: An Energy Approach. J. Appl. Mech., vol. 89, no. 2, June 1967, pp. 486-487.

9. Sheu, C.Y.; and Prager, W.: Recent Developments in Optimal Structural Design. Appl. Mech. Rev., vol. 21, no. 10, Oct. 1968, pp. 985-992.

10. Prager, W.; and Shield, R.T.: Optimal Design of Multi-purpose Structures, Int. J. Solids Struct., vol. 4, no. 4, 1968, pp. 469-475.

11. Sheu, C.Y.; and Prager, W.: Minimum-Weight Design with Piecewise Constant Specific Stiffness, Journal of Optimization Theory and Application, vol. 2, no. 3, May 1968, pp. 179-186.

12. Prager, W.: Optimization of Structural Design. Journal of Optimization Theory and Applications, vol. 6, no. 1, July 1970, pp. 1-21.

13. Shield, R.T.; and Prager, W.: Optimal Structural Design for Given Deflection, A. Angew. Math. Phys., vol. 21, no. 4, 1970, pp. 513-523.

14. Berke, L: An Efficient Approach to the Minimum Weight Design of Deflection Limited Structures. AFFDL-TM-70-4, 1970.

15. Gellatly, R.A.; and Berke, L.: Optimum Structural Design. AFFDL-TR-70-165, Apr. 1971.

16. Prager, W.; and Marcal, P.V.: Optimality Criteria in Structural Design, AFFDL-TR-70-166, May 1971.

17. Venkayya, V.B., et al.: Design of Optimum Structures for Dynamic Loads. Proceedings of the Third Conference on Matrix Methods in Structural Mechanics, R.M. Bader, et al., eds., Air Force Flight Dynamics Lab, Oct. 1971.

18. Berke, L.: Convergence Behavior of Iterative Resizing Procedures Based on Optimality Criteria. AFFDL-TM-72-1-FBR, 1972.

19. Gellatly, R.A.; and Berke, L.: Optimality-Criterion Based Algorithm. Optimum Structural Design, R.H. Gallagher, and O.C. Zienkiewicz, eds., John Wiley & Sons, 1972, pp. 33-49.

20. Kiusalaas, J.: Minimum Weight Design of Structures Via Optimality Criteria. NASA TN D-7115, 1972.

21. Nagtegaal, J.C.: A New Approach to Optimal Design of Elastic Structures. Comput. Methods Appl. Mech. Eng., vol. 2, no. 3, July-Aug. 1973, pp. 255-264.

22. Khot, N.S., et al.: Application of Optimality Criterion to Fiber Reinforced Composites. AFFDL-TR-73-6, May 1973.

23. Venkayya, V.B.; Khot, N.S.; and Berke, L.: Application of Optimality Criteria Approaches to Automated Design of Large Practical Structures. Second Symposium on Structural Optimization, AGARD-CP-123, 1973, pp. 3-1 to 3-19.

24. Khot, N.S., et al.: Optimization of Fiber Reinforced Composite Structures. Int. J. Solids Struct. vol. 9, no. 10, 1973, pp. 1225-1236.

25. Khot, N.S.; Venkayya, V.B.; and Berke, L.: Optimization and Structures for Strength and Stability Requirements. AFFDL-TR-73-98, Dec. 1973.

26. Prager, W.: Minimum Weight Design of Statically Determinate Truss Subject to Constraints on Compliance, Stress, and Cross-Sectional Area. J. Appl. Mech., vol. 40, no. 1, Mar. 1973, pp. 313-314.

27. Kiusalaas, J.: Optimal Design of Structures with Buckling Constraints. Int. J. Solids Struct., vol. 9, no. 7, 1973, pp. 863-878.

28. Berke, L.; and Khot, N.S.: Use of Optimality Criteria Methods for Large Scale Systems. Structural Optimization, AGARD-LS-70, 1974, pp. 1-1 to 1-29.

29. Berke, L.; and Venkayya, V.B.: Review of Optimality Criteria Approaches to Structural Optimization. Structural Optimization Symposium, L.A. Schmit, ed., ASME, New York, 1974, pp. 23-34.

30. Wilkinson, K.; Lerner, E.; and Taylor, R.F.: Practical Design of Minimum Weight Aircraft Structures for Strength and Flutter Requirements. AIAA Paper 74-986, Aug. 1974.

31. Venkayya, V.B.; and Khot, N.S.: Design of Optimum Structures to Impulse Type Loading. AIAA J., vol. 13, no. 8, Aug. 1975, pp. 989-994.

32. Khot, N.S.; Venkayya, V.B.; and Berke, L.: Optimum Structural Design with Stability Constraints. Int. J. Numer Methods Eng., vol. 10, no. 5, 1976, pp. 1097-1114.

33. Khot, N.S.; Venkayya, V.B.; and Berke, L.: Optimum Design of Composite Structures with Stress and Displacement Constraints. AIAA J., vol. 14, no. 2, Feb. 1976, pp. 131-132.

34. Rizzi, P.: Optimization of Multi-Constrained Structures Based on Optimality Criteria. 17th Structures, Structural Dynamics, and Materials Conference, AIAA, 1976, pp. 448-462.

35. Segenreich, S.A.; and McIntosh, S.C., Jr.: Weight Optimization Under Multiple Equality Constraints Using an Optimality Criteria. 17th Structures, Structural Dynamics, and Materials Conference, AIAA, 1976, Additional Paper No. 3.

36. Dobbs, M.W.; and Nelson, R.B.: Application of Optimality Criteria to Automated Structural Design. AIAA J., vol. 14, no. 10, Oct. 1976, pp. 1436-1443.

37. Taylor, J.E.; and Rossow, M.P.: Optimal Structural Design Algorithm Using Optimality Criteria. Advances in Engineering Science, vol. 2, NASA CP-2001-vol-2, 1976, pp. 521-530.

38. Khot, N.S., et al.: Optimum Design of Composite Wing Structures With Twist Constraint for Aeroelastic Tailoring. AFFDL-TR-76-117, Dec. 1976.

39. Khot, N.S.; Venkayya, V.B.; and Berke, L.: Experiences with Minimum Weight Design of Structures Using Optimality Criteria Methods. 2nd International Conference on Vehicle Structural Mechanics, SAE-P-71, Society of Automotive Engineers, 1977, pp. 191-201.

40. Wilkinson, K., et al: FASTOP: A Flutter and Strength Optimization Program for Lifting-Surface Structures, J. Aircr., vol. 4, no. 6, June 1977, pp. 581-587.

41. Sanders, G.; and Fleury, C.: A Mixed Method in Structural Optimization. Int. J. Numer. Methods Eng., vol. 13, no. 2, 1978, pp. 385-404.

42. Berke, L.; and Khot, N.S.: A Simple Virtual Strain Energy Method to Fully Stress Design Structures With Dissimilar Stress Allowables and Material Properties. AFFDL-TM-77-28-FBR, Dec. 1977.

43. Khan, M.R.; Willmert, K.D.; and Thornton, W.A.: A New Optimality Criterion Method for Large Scale Structures. 19th AIAA/ASME/SAE Structures, Structural Dynamics and Materials Conference, AIAA, 1978, pp. 47-58.

44. Isakson, G., et al.: ASOP-3: A Program for Optimum Structural Design to Satisfy Strength and Deflection Constraints, J. Aircr., vol. 15, no. 7, July 1978, pp. 422-428.

45. Markowitz, J.; and Isakson, G.: FASTOP-3: A Strength, Deflection and Flutter Optimization Program for Metallic and Composite Structures, Vol. I Theory and Application, Vol. II - Program User's Manual, AFFDL-TR-78-50, May 1978.

46. Haug, E.J.; and Arora, J.S.: Applied Optimal Design, Wiley-Interscience, 1979, p. 215.

47. Schmidt, L.A., Jr.; and Fleury, C.: An Improved Analysis/Synthesis Capability Based on Dual Methods - ACCESS 3. 20th Structures, Structural Dynamic and Materials Conference, AIAA, 1979, pp. 23-50.

48. Fleury, C.; and Schmit, L.A. Jr.: Dual Methods and Approximation Concepts in Structural Synthesis. NASA CR-3226, 1980.

49. Khot, N.S.; Berke, L.; and Venkayya, V.B.: Minimum Weight Design of Structures by the Optimality Criterion and Projection Method. 20th Structures, Structural Dynamics and Materials Conference, AIAA, 1979, pp. 11-22.

0. Lerner, E.: The Application of Practical Optimization Techniques in the Preliminary Structural Design of a Forward-Swept Wing. Second International Symposium on Aeroelasticity and Structural Dynamics, Deutsche Gesellschaft fuer Luft- und Raumfahrt, Bonn, Germany, 1985, pp. 381-392.

1. Lerner, E.; and Markowitz, J.: An Efficient Structural Resizing Procedure for Meeting Static Aeroeleastic Design Objectives. J. Aircr., vol. 16, no. 2, Feb. 1979, pp. 65-71.

2. Cross, H.: The Relation of Analysis to Structural Design, Am. Soc. Civ. Eng. Proc., vol. 61, no. 8, Oct. 1935, pp. 1119-1130; and vol. 61, no. 10, pt. 1, Dec. 1935, pp. 1551-1557.

3. Khot, N.S.; Berke, L.; and Venkayya, V.B.: Comparison of Optimality Criteria Algorithms for Minimum Weight Design of Structures, AIAA J., Vol. 17, no. 2, Feb. 1979, pp. 182-190.

4. Bertsekas, D.P.: Constrained Optimization and Lagrange Multiplier Methods. Academic Press, 1982, p. 104.

5. Gellatly, R.A.; and Berke, L.: A Preliminary Study of a New Approach to the Optimization of Strength Limited Structures, AFFDL-TM-75-162-FBR, Sept. 1975..

56. Gellatly, R.A.; and Thom, R.D.: Force Method Optimization, AFWAL-TR-80-3006, Feb. 1980.

57. Gellatly, R.A.; and Thom, R.D.: Optimality Criteria Using a Force Method of Analysis Approach. Foundations of Structural Optimization: A Unified Approach. A.J. Morris, ed., Wiley-Interscience, 1982, pp. 237-272.

58. Adelman, H.M.; Haftka, R.T.; and Tsach, U.: Application of Fully Stressed Design Procedures to Redundant and Non-Isotropic Structures, NASA TM-81842, 1980.

59. Walsh, J.: Application of Mathematical Optimization Procedures to a Structural Model of a Large Finite-Element Wing, NASA TM-87597, 1986.

60. Prager, W.: Unexpected Results in Structural Optimization, J. Struct. Mech., vol. 9, no. 1, 1981, pp. 71-90.

61. Khot, N.S.: Optimization of Structures with Multiple Frequency Constraints. Comput Struct., vol. 20, no. 5, 1985, pp. 869-876.

62. Khot, N.S.; and Kamat, M.P.: Minimum Weight Design of Truss Structures with Geometric Nonlinear Behavior. AIAA J., vol. 23, no. 1, Jan. 1985, pp. 139-144.

OPTIMAL DESIGN OF ELASTIC-PLASTIC STRUCTURES

C.Cinquini
Department of Structural Mechanics
University of Pavia
Via Abbiategrasso 209
I 27100 Pavia - Italy

R.Contro
Department of Structural Engineering
Politecnico of Milan
P.zza L. da Vinci, 32
I 20131 Milano - Italy

1. INTRODUCTION

The optimal design problem for linear elastic structures has been the subject of abundant literature, and various behavioural constraints have been taken into account that very often concern structural deformation. On the other hand, the structural behaviour beyond the elastic limit has been considered in many papers dealing with optimal design for prescribed plastic collapse load (rigid-plastic models). Comprehensive surveys can be found, for istance, in Ref.s 1 to 5.

It can be noticed that the above mentioned problems are completely different: elastic approaches usually refer to service load conditions while plastic formulations deal with ultimate conditions, and suggest design solutions for which the actual behaviour in service conditions are to be studied in a separate way. In practice, the application fields of the proposed optimal design approaches are very different.

NATO ASI Series, Vol. F27
Computer Aided Optimal Design: Structural and Mechanical Systems. Edited by C. A. Mota Soares

Variational formulations have been proposed for both kinds of problems. For plastic problems, static and kinematic formulations are emphasized in this way and the optimality criterion is found to be dependent on the specific dissipated power. For optimal design problems in linear elastic field, as well known, variational formulations very frequently involve an adjoint problem, and the optimality criterion shows an elastic energy which depends on both actual and adjoint problems.
Anyway , on the basis of variational formulations, optimal design problems can be completely described in an analytic way and closed-form solutions are allowed for simple examples. As numerical approaches are concerned, mathematical programming methods and optimality criteria methods can be used, once problems are formulated in a discrete way.
The limits shown by rigid-plastic models can be overcome by considering suitable models of elastic-plastic (or elastic-hardening) constitutive laws. Particular formulations of such kind of problems are to be studied if the aim of preserving the general form of linear elastic optimization problems is pursued. In particular, holonomic laws are to be assumed and direct formulations of analysis problems are to be investigated.
In the present paper, this kind of problems is studied by using variational formulations, with reference to beams in bending. Holonomic nonlinear constitutive laws are firstly considered, in a general way, for optimal design problems, and the special form of optimality criterion is discussed. Analogous approaches are proposed for piecewise linear constitutive laws (see also Ref.6) and some analytical solutions for simple examples are discussed as well.
Starting from such continuous approaches, a finite element formulation is also shown, which leads to numerical methods of solution. An iterative procedure is conceived which makes use of a sort of scaling factor method and of an optimal criteria method. Some solutions were presented in Ref.7, where a simple F.E. model for beams in bending was proposed.

In the present paper a more general numerical tool is discussed with reference to beam and portal frames.

2. GENERAL CONSTITUTIVE LAW

2.1. General formulation

Structural optimization problems in presence of a material exhibiting a general non linear constitutive law are the subject of the present study.
In particular, for the sake of simplicity, problems of beams in bending are dealt with. The general layout of the structure is given and the usual hypotheses of small depths and deflections are adopted. The behaviour of material is described by a holonomic constitutive law defined in terms of moment function M and curvature function χ (see Fig.1). The design variable is denoted by h; it represents a local geometric dimension. A linear form is assumed for the cost function

$$\Gamma(h) = \int_{\Omega} h d\,\Omega \tag{2.1}$$

where Ω denotes the domain defined by the structure geometry. The local bending stiffness $s = M,_{\chi}$ depends on the design variable h; therefore

$$M = M(h,\chi) \tag{2.2}$$

If P and u denote given external load and deflection function respectively, the equilibrium condition reads

$$P + (M\,(h,\chi))'' = 0 \tag{2.3}$$

and the compatibility condition is

$$\chi + u'' = 0 \tag{2.4}$$

where a prime denotes differentiation with respect to abscissa x ($x \in \Omega$). The set of Rel.s 2.2 to 2.4 together with appropriate boundary conditions on Ω, defines the governing relations of the optimal design problem.
Upper and lower bounds on deflection function u are assumed as behavioural constraints

$$- u^- \leq u \leq u^+ \tag{2.5}$$

and side (or technological) constraints are prescribed on design variable h

$$h_{min} \leq h \leq h_{max} \tag{2.6}$$

2.2. Optimality criterion

Applying the Lagrangian multipliers η and ν to the equality constraints of Rel.s 2.3 an 2.4 respectively and the non-negative Lagrangian multipliers μ^+, μ^-, α and β to the inequality constraints of Rel.s 2.5 and 2.6, the following functional can be obtained

$$L_1 = \int_\Omega \{h + \eta(P + M'') + \nu(\chi + u'') + \mu^+(u - u^+) + \mu^-(-u - u^-) + \alpha(h - h_{max}) + \beta(h_{min} - h)\}\, d\Omega \tag{2.7}$$

The stationarity conditions of L_1 are necessary conditions for the optimality of design. The stationarity conditions with respect to Lagrangian multipliers provide the constraints of Rel.s 2.3 to 2.6, together with the orthogonality constraints

$$\mu^+(u - u^+) = 0$$
$$\mu^-(-u - u^-) = 0 \qquad (2.8)$$

$$\alpha(h - h_{max}) = 0$$
$$\beta(h_{min} - h) = 0 \qquad (2.9)$$

The stationarity conditions with respect to h, χ and u provide respectively

$$1 + \eta'' M,_h + \alpha - \beta = 0 \qquad (2.10)$$

$$\eta'' M,_\chi + \nu = 0 \qquad (2.11)$$

$$\nu'' + \mu^+ - \mu^- = 0 \qquad (2.12)$$

In order to write Rel.s 2.10 to 2.12, Green's formula is taken into account; accordingly, natural boundary conditions on $\partial\Omega$ are

$$\eta\delta M' = 0 \qquad \eta'\delta M = 0$$
$$\nu\delta u' = 0 \qquad \nu'\delta u = 0 \qquad (2.13)$$

In a similar way as in linear elastic optimization problems, an adjont problem can be defined through the equations presented so far. The adjont external load is furnished by $P^* = \mu^+ - \mu^-$ and Rel. 2.12 provides the equilibrium conditions, where $\nu = M^*$ represents the moment function. The adjont deflection function is given by $\eta = u^*$ and in Rel. 2.11 the constitutive law of the adjoint problem can be recognized; an obvious compatibility condition is given by assuming the adjoint curvature function as

$$\chi^{*} = -\eta'' \qquad (2.14)$$

So, Rel.s 2.11 and 2.12 read respectively

$$M^{*} - M,_{\chi}\ \chi^{*} = 0 \qquad (2.15)$$
$$P^{*} + M^{*''} = 0 \qquad (2.16)$$

The adjoint problem shows an elastic behaviour, as the adjoint bending stiffness $s^{*}=M,_{\chi}$ (see Rel. 2.15) does not depend on the curvature χ^{*} and it is defined by the stress state of the actual problem.
The nature of the boundary conditions of Rel.s 2.13 is now clear

$$u^{*}\ \delta M' = 0 \qquad u^{*'}\delta M = 0$$
$$M^{*}\ \delta u' = 0 \qquad M^{*'}\delta u = 0 \qquad (2.17)$$

The optimality criterion is given by Rel. 2.10, that reads

$$1 - M,_{h}\ \chi^{*} + \alpha - \beta = 0 \qquad (2.18)$$

3. PIECEWISE LINEAR CONSTITUTIVE LAW

3.1. General formulation

If the behaviour of beams in bending is described by means of a piecewise linear constitutive law, a modified form of the optimal design problem can be shown. The approach proposed hereinafter lies in minimizing a functional, derived from an extremum principle of Maier /8/, with respect to deflection function and non negative plastic multipliers.
The moment-curvature relationship shows a linear elastic branch and a set of subsequent "hardening" branches (in both

positive an negative sides), the total number of which will be denoted by b (Fig.2). Let $s_o = s_o(h)$ and $s_i = s_i(h)$ ($i = 1 \dots b$) denote elastic and hardening stiffness respectively, $R_i = R_i(h)$ ($i=1 \dots b$) being positive parameters representing yield limits of moment.
The assumed holonomic constitutive law can be described as follows

$$M = s_o(\chi - \sum_{i=1}^{b} \theta_i \lambda_i) \qquad (3.1)$$

$$M = \theta_i(R_i + s_i\lambda_i + \phi_i) \qquad (3.2)$$

$$\lambda_i\phi_i = 0 \qquad (3.3)$$

$$i = 1 \dots b$$

$$\lambda_i \geq 0 \qquad \phi_i \leq 0 \qquad (3.4)$$

$$\theta_i = \text{SIGN } (M) \qquad (3.5)$$

The bending moment and the elastic curvature are related by Rel. 3.1, where the elastic curvature is expressed as a difference between total curvature and plastic curvature χ_p. Plastic multipliers λ_i ($i=1 \dots b$) define the plastic part of deformation ($\chi_p = \theta_1\lambda_1 + \theta_2\lambda_2 + \dots + \theta_b\lambda_b$). Functions ϕ_i($i=1 \dots b$), so-called load functions, once set equal to zero, given the equations of hardening branches. Clearly, the case $\lambda_i=0$ for all $i=1 \dots b$ corresponds to a linear elastic behaviour. Moreover, if the last stiffness s_k in the positive (negative) side is supposed to vanishe, a perfectly-plastic behaviour is defined for $M = R_k$($M=-R_k$).
Now, the optimal design problem can be defined for beams in bending which exhibit such a piecewise linear constitutive law. The same formulation as seen in Sect. 2.1 is proposed, to within Rel. 2.2, which is substituted by the set of Rel.s 3.1 to 3.5.

3.2. Optimality criterion

Looking for the optimality criterion referred to optimal design problems in presence of a piecewise linear constitutive law, a Lagrangian functional can be defined.

$$L = L_1 + L_2 \tag{3.6}$$

where L_1 is the functional of Rel. 2.7 and L_2 takes into account moment-curvature relationships

$$L_2 = \int_\Omega \{\zeta[M - s_o(\chi - \sum_{i=1}^{b} \theta_i\lambda_i)] + \sum_{i=1}^{b} [\gamma_i(M - \theta_i(R_i + s_i\lambda_i + \phi_i)) + \psi_i\lambda_i\phi_i]\} d\Omega \tag{3.7}$$

From the stationarity conditions of L, othogonality conditions of Rel.s 2.8 and 2.9 and boundary conditions of Rel.2.13 are found, while Rel.s 2.10 and 2.11 are to be respectively modified in the form

$$1 - \zeta s_{o,h}(\chi - \sum_{i=1}^{b} \theta_i\lambda_i) - \sum_{i=1}^{b} \gamma_i\theta_i(R_{i,h} + s_{i,h}\lambda_i) + \alpha - \beta = 0 \tag{3.8}$$

$$\nu - s_o\zeta = 0 \tag{3.9}$$

The stationarity condition of L with respect to deflection function u is given by Rel.2.12 and the stationarity conditions with respect to M, λ_i and ϕ_i are to be added

$$\eta'' + \zeta + \sum_{i=1}^{b} \gamma_i = 0 \tag{3.10}$$

$$\left.\begin{aligned} &\theta_i(s_o\zeta - s_i\gamma_i) + \psi_i\phi_i \geq 0 \\ &\lambda_i[\theta_i(s_o\zeta - s_i\gamma_i) + \psi_i\phi_i] = 0 \end{aligned}\right\} \quad i=1\ldots b \tag{3.11}$$

$$-\theta_i\gamma_i + \psi_i\lambda_i \leq 0 \qquad i=1\ldots b \qquad (3.12)$$
$$\phi_i(-\theta_i\gamma_i + \psi_i\lambda_i) = 0$$

The orthogonality constraints of Rel.3.12, accounting for the assumption $\lambda_i\phi_i=0$, involve

$$\phi_i\gamma_i = 0 \qquad i=1\ldots b \qquad (3.13)$$

For the adjoint problem, defined in Sect. 2.2, the constitutive law is furnished by Rel.s 3.9 to 3.11; from Rel.s 3.9 and 3.10 one has

$$M^* = s_o(\chi^* - \sum_{i=1}^{b} \gamma_i) \qquad (3.14)$$

while for any $\lambda > 0$ Rel.s 3.11 read

$$M^* = s_i\,\gamma_i \qquad (3.15)$$

Any way a linear elastic behaviour is found for the adjoint problem, for which the constitutive law is meaningfully expressed either by Rel. 3.14, if the actual problem shows an elastic behaviour ($\gamma_i=0$ for i=1...b) or by Rel. 3.15 if in the actual problem plasticity is active.
The optimality criterion of Rel.3.8 reads

$$1 - (\chi^* - \sum_{i=1}^{b} \gamma_i)\,.\,s_{o,h}(\chi - \sum_{i=1}^{b} \theta_i\lambda_i)$$
$$-\sum_{i=1}^{b} \gamma_i\theta_i(R_{i,h} + s_{i,h}\,\lambda_i) + \alpha - \beta = 0 \qquad (3.16)$$

If λ_i and γ_i vanishe for any i=1...b (linear elastic behaviour for both actual and adjoint problems), the optimality criterion shows the classic form

$$1 - \chi^*\,s_{o,h}\,\chi + \alpha - \beta = 0 \qquad (3.17)$$

By assuming, without loss of generality,

$$s_o = A_o h^p \qquad (3.18)$$

$$s_i = A_i h^p \qquad i=1...b \qquad (3.19)$$

$$R_i = B_i h^q \qquad (3.20)$$

the optimality criterion of Rel.3.16 can be simplified in form

$$1 - \chi^* A_o p h^{p-1}(\chi - \sum_{i=1}^{b} \theta_i \lambda_i) + (p-q) \sum_{i=1}^{b} \gamma_i \theta_i B_i h^{q-1} + \alpha - \beta = 0 \qquad (3.21)$$

or alternatively

$$1-(\chi^* - \sum_{i=1}^{b} \gamma_i) A_o p h^{p-1} \chi - \sum_{i=1}^{b} \gamma_i \theta_i B_i h^{q-1} + \alpha - \beta = 0 \qquad (3.22)$$

4. EXAMPLES

4.1. Operating remarks

In order to find analytical or numerical solutions, problems exhibiting the previous seen piecewise linear constitutive law can be described in terms of deflection function u, plastic multipliers λ_i and load functions ϕ_i. Accordingly, the corresponding optimal design problem requires design variable h and adjoint parameters $\eta(=u^*)$ and γ_i to be added. In the adjoint problem also the external load $\mu^+ - \mu^-$ is a unknown function.
Thus, as the actual problem is concerned, the governing equations presented so far can be summarized as follows

$$P - [A_o h^p (u'' + \sum_{i=1}^{b} \theta_i \lambda_i)]'' = 0 \qquad (4.1)$$

$$-A_o h^p (u'' + \sum_{j=1}^{b} \theta_j \lambda_j) = \theta_i (B_i h^q + A_i h^p \lambda_i \phi_i) \qquad (4.2)$$

$$\lambda_i \phi_i = 0 \qquad i=1...b \qquad (4.3)$$

$$\lambda_i \geq 0 \qquad \phi_i \leq 0 \qquad (4.4)$$

$$\theta_i = \text{SIGN}\ (M) \tag{4.5}$$

and, for the adjoint problem, one has

$$\mu^+ - \mu^- - A_o h^p(\eta'' + \sum_{i=1}^{b} \gamma_i)'' = 0 \tag{4.6}$$

$$-\theta_i[A_o h^p(\eta'' + \sum_{j=1}^{b} \gamma_j) + A_i h^p \gamma_i] + \psi_i \phi_i \geq 0$$

$$i=1\ldots b$$

$$\lambda_i\{-\theta_i|A_o h^p(\eta'' + \sum_{j=1}^{b} \gamma_j) + A_i h^p \gamma_i| + \psi_i \phi_i\} = 0 \tag{4.7}$$

The orthogonality constraint of Rel.3.13 is to be considered, and the optimality criterion can be employed in the form given by Rel.3.21

$$1 - \eta'' A_o p h^{p-1}(u'' + \sum_{i=1}^{b} \theta_i \lambda_i) + (p-q) \sum_{i=1}^{b} \gamma_i \theta_i B_i h^{q-1} + \alpha - \beta = 0 \tag{4.8}$$

In the following, the form of moment curvature relationship is assumed as shown in Fig.3.The same behaviour is considered in both positive and negative sides and for the total number of hardening branches b=4 is assumed. In particular, the assumptions

$$\theta_1 = \theta_2 = +1$$

$$\theta_3 = \theta_4 = -1$$

$$s_3 = s_4 = 0$$

are made. An elastic-hardening-perfectly plastic behaviour is described in such a way.

4.2. Looking for closed form solutions

Some simple examples will be discussed hereinafter, in order to give some elements about the way to obtaine closed-form solutions, as in analogous elastic optimization problems. For the sake of simplicity assume $h_{min}=0$ and the upper bound h_{max} not to be active. Let a statically determinate beam be considered. Depending on the load, moment funtion $M=M(x)$ can be calculated and adjoint moment function $M^*=M^*(x)$ can be found to within the actual value F of the adjoint load. If the abscissa $\bar{x}$ where the deflection constraints of Rel.2.5 are active cannot be found a-priori, also $\bar{x}$ is a unknown parameter in function $M^*(x)$. Anyway, if the beam is supposed to remain in elastic domaine, one has

$$u'' = \frac{-M}{A_o h^p} \tag{4.9}$$

$$\eta'' = \frac{-M^*}{A_o h^p} \tag{4.10}$$

and from the optimality criterion

$$1 - \eta'' A_o \, p h^{p-1} \, u'' = 0 \tag{4.11}$$

it follows

$$h^{p+1} = p \, \frac{MM^*}{A_o} \tag{4.12}$$

Then

$$u''(x) = \frac{-M}{A_o^{\frac{1}{p+1}} \, (MM^* p)^{\frac{p}{p+1}}} \tag{4.13}$$

By twice integrating Rel.4.13, deflection function u is found to within parameters F and $\bar{x}$ as well as 2m integration constants (where m represents the number of integration fields). These parameters can be found through 2m geometric boundary conditions (statically determinate beams) and by prescribing the d deflection constraints of Rel.2.5 to be active.
Finally, the admissibility condition for the elastic solution will be investigated

$$-B_1 h^q \leq M \leq B_1 h^q \tag{4.14}$$

The elastic-hardening behaviour can be investigated in a similar way. If $\lambda_1>0$ is assumed ($\lambda_2=\lambda_3=\lambda_4=0$), one has

$$u'' + \lambda_1 = - \frac{M}{A_o h^p} \tag{4.15}$$

$$\lambda_1 = \frac{M}{A_1 h^p} - \frac{B_1}{A_1} h^{q-p} \tag{4.16}$$

$$\eta'' + \gamma_1 = - \frac{M^*}{A_o h^p} \tag{4.17}$$

$$\gamma_1 = \frac{M^*}{A_1 h^p} \tag{4.18}$$

The optimality criterion reads

$$1 - \eta'' \; A_o p h^{p-1}(u''+\lambda_1)+(p-q)\gamma_1 B_1 h^{q-1} = 0 \tag{4.19}$$

By elimitating u'', η'', λ_1 and γ_1 from Rel.s 4.15 to 4.19, it follows

$$1 - (\frac{1}{A_o} + \frac{1}{A_1})p \frac{MM^*}{h^{p+1}} + (p-q)\gamma_1 \, B \, h^{q-1} = 0 \tag{4.20}$$

If the function h=h(x) can be calculated from Rel.4.20, an analytical form of curvature function u", to be integrated, can be found by Rel.s 4.15 and 4.16 and the optimization problem is solved as in elastic cases. The admissibility condition for such a solution reads

$$B_1 \, h^q \leq M \leq B_2 h^q \tag{4.21}$$

Generally speaking, an optimal solution for a beam can involve both elastic and plastic behaviours. So, temptative hypotheses are to be done in order to define elastic and plastic regions along the beam.
Moreover, for statically undeterminate problems, the same method could be employed, if moment functions are calculated referring to a statically determinate scheme and by founding a-posteriori redundancy unknown parameters.

4.3. Cantilever beam

Let the statically determinate beam of Fig. 4 be considered for which

$$M = - Q \frac{x^2}{2} \qquad\qquad x \in (0, l) \qquad\qquad (4.22)$$
$$M = -F\, x$$

according to the assumption $u(0)=u^+$

In the elastic hypotesis, from Rel. 4.12 and 4.13, it follows

$$\left.\begin{aligned} u &= u^+\left[\frac{2}{3}\left(\frac{x}{l}\right)^{5/2} - \frac{5}{3}\,\frac{x}{l} + 1\right] \\ h &= \frac{1}{5}\,\frac{Q}{A_o}\,\frac{l^{5/2}}{u^+}\,x^{3/2} \end{aligned}\right\} \text{for } p=1 \qquad (4.23)$$

$$\left.\begin{aligned} u &= u^+\left(\frac{x}{l} - 1\right)^2 \\ h &= \frac{1}{2}\left(\frac{Q}{A_o u^+}\right)^{1/2} l\, x \end{aligned}\right\} \text{for } p=2 \qquad (4.24)$$

The admissibility condition for p=q=1 (sandwich beam) requires

$$\frac{u^+}{l} \leq \frac{5}{2}\,\frac{B_1 l}{A_o} \qquad (4.25)$$

It can be noticed that the same condition for a uniform design solution reads

$$\frac{u^+}{l} \leq \frac{1}{4}\,\frac{B_1 l}{A_o} \qquad (4.26)$$

If Rel. 4.25 is not fulfilled, the behaviour beyond the elastic limit is to be investigated. Thus, having assumed $\lambda_3 > 0$ along all the beam, the solution ca be calculated in the form

$$u = \left(u^+ + \frac{B_1}{A_1}\frac{l^2}{2}\right)\left[\frac{2}{3}\left(\frac{x}{l}\right)^{5/2} - \frac{5}{3}\frac{x}{l} + 1\right] - \frac{B_1}{A_1}\frac{(x-l)^2}{2}$$

$$h = \frac{A_o + A_1}{2u^+ A_1 + B_1 l^2}\ \frac{2}{5}\, Q\, l^{5/2}\, x^{3/2} \tag{4.27}$$

The admissibility condition

$$B_1\ h(x) \le \frac{Q\ x^2}{2} \tag{4.28}$$

cannot be fulfilled at least in a suitable neighbourhood of $x=0$. Then, if the admissibility condition of the elastic solution (Rel.4.25) is not fulfilled, the optimal solution involves an elastic region for $x \in (0,x_1)$ and a plastic in $x \in (x_1,l)$. By using Rel.s 4.12 and 4.13 (elastic behaviour) and 4.20 (plastic behaviour), the following solution can be found

$$h = \left(\frac{QF}{2A_o}\right)^{1/2} x^{3/2}$$

$$u = \left(\frac{Q}{2A_oF}\right)^{1/2}\left(\frac{4}{15}\, x^{5/2} + C_1 x\right) + u^+ \qquad x \in (0,x_1) \tag{4.29}$$

$$h = \left[\left(\frac{1}{A_o} + \frac{1}{A_1}\right)\frac{QF}{2}\right]^{1/2} x^{3/2}$$

$$u = \left[\left(\frac{1}{A_o} + \frac{1}{A_1}\right)\frac{Ql}{2F}\right]^{1/2}\frac{2}{3}\, l^2 \left[\frac{2}{5}\left(\frac{x}{l}\right)^{5/2} - \frac{x}{l} + \frac{3}{5}\right] - \frac{B_1}{A_1}\frac{(x-l)^2}{2} \qquad x \in (x_1,l) \tag{4.30}$$

Unknown parameters x_1, F and C_1 can be calculated through continuity conditions for u and u' in $x=x_1$ and by prescribing the yield condition in $x=x_1$

$$B_1 \left(\frac{QF}{2A_o}\right)^{1/2} x_1^{3/2} = \frac{Q\ x_1^2}{2} \tag{4.31}$$

If $p=2$ and $q=1$ is assumed , the admissibility condition for the elastic solution of Rel.4.24 reads

$$Q \le \frac{B_1^2}{A_o\ u^+} \tag{4.32}$$

In this case, looking for plastic solutions, from Rel.4.20 it follows

$$1 - \left(\frac{1}{A_o} + \frac{1}{A_1}\right)\frac{QFx^3}{h^3} + \frac{B_1F_x}{A_1h^2} = 0 \qquad (4.33)$$

Let the structure of Fig.5 as another example be now considered for the optimal solution of which different plasticity conditions will be found. In the elastic hypotèsis, from Rel. 4.12 and 4.13, it follows

$$\left.\begin{aligned} u &= u^+(1 - \frac{x}{l})^2 \\ h &= \frac{p\, l^2}{2A_o u^+}\, x \end{aligned}\right\} \quad \text{for } p=1 \qquad (4.34)$$

$$\left.\begin{aligned} u &= u^+\left[\frac{3}{4}\left(\frac{x}{l}\right)^{7/3} - \frac{7}{4}\frac{x}{l} + 1\right] \\ h &= \frac{3P}{7A_o}\,\frac{l^{7/3}}{u^+}\, x^{2/3} \end{aligned}\right\} \quad \text{for } p=2 \qquad (4.35)$$

In the case of sandwich beam (p=q=1), the admissibility condition for elastic solution implies

$$\frac{u^+}{l} \leq \frac{B_1\, l}{2\, A_o} \qquad (4.36)$$

while the corresponding condition for a uniform design solution is

$$\frac{u^+}{l} \leq \frac{B_1\, l}{3\, A_o} \qquad (4.37)$$

If $M \leq -B_1\, h(x)$ is assumed along all the beam, the following form can be obtained for the design function

$$h = \frac{A_o + A_1}{2u^+A_1 + B_1 l^2}\;\frac{P\, l^2}{A_o}\; x \qquad (4.38)$$

and the admissibility condition provides

$$\frac{B_1 l}{2A_o} \leq \frac{u^+}{l} \leq \frac{B_2 l}{2A_o} + \frac{(B_2 - B_1)\, l}{2A_1} \qquad (4.39)$$

Then, depending on the behavioural constraint on deflection function, the optimal solution can show either an elastic or a plastic behaviour along all the beam. Analogously the collapse condition ($M=-B_1 h$) is attained for any $x \in (0,l)$

5. FINITE ELEMENT APPROACH

5.1. General formulation

Finite element models which allow for formulations similar to the approaches seen in the previous sections are proposed in Ref.s 9 and 10. Operating remarks on such approaches will be presented in next section. In the present section the finite element formulation is proposed having as fundamental idea the continuous approach presented in Sect.s 3 and 4.
The holonomic, piecewise linear, elastic-plastic problem is formulated via finite element method by suitably modelling displacement function u and non negative plastic multipliers λ_i. Generally speaking, such a model allows for spreading plasticity along the elements, overcoming in this way the classic methods based on plastic hinges.
For the sake of simplicity the beam F.E. proposed in Ref.9 is considered in this section. A model implying more general applications will be accounted for in next section.
The set of governing equations represents a discrete form of Rel.s 4.1 to 4.5. Let $\underline{P}$ and $\underline{u}$ denote nodal external load vector and nodal displacement vector respectively. Similarly, $\underline{\lambda}$ represents the vector of plastic multiplier and $\underline{\phi}$ the vector of load functions, the physical meaning of which follows from the continuous approach. Thus, in the assembled form, one has

$$\underline{P} - \underline{K}_{uu}\, \underline{u} + \underline{K}_{u\lambda}\, \underline{\lambda} = \underline{0} \tag{5.1}$$

$$\underline{K}_{\lambda u}\, \underline{u} - \underline{K}_{\lambda\lambda}\, \underline{\lambda} - \underline{R} - \underline{\phi} = \underline{0} \tag{5.2}$$

$$\underline{\phi}_d\, \underline{\lambda} = \underline{0} \tag{5.3}$$

$$\underline{\lambda} \geq \underline{0} \quad ; \quad \underline{\phi} \leq \underline{0} \tag{5.4}$$

Special notation $\underline{\phi}_d$ denotes a diagonal matrix defined in such a way that $\underline{\phi}$=diag $\underline{\phi}_d$; analogous notations will be used in the following In Rel.5.1 matrix $\underline{K}_{uu}$ represents the elastic stiffness matrix and the contribution of deformations to the equilibrium equations is taken into account through matrix $\underline{K}_{u\lambda}$. Rel.5.2 derives from the assemblage after having modelled plastic multipliers; vector $\underline{R}$ collects yelding limit moments.

Now, with the view to optimal design problems, let $h_j (j=1...n_h)$ denote the set of design variables and let the set of finite elements be subdivided into n_h subsets, each one constituted by the elements to which the same design variable h is attributed. The geometric domaine (total length) defined by the j-th subset will be denoted by l_j and the total number of elements belonging to this subset will be denoted by $m_j (j=1...n_h)$. In order to point out the role played by design variables in Rel.s 5.1 and 5.2, the assemblage operations are to be accounted for. If $\underline{u}_i$, $\underline{\lambda}_i$ and $\underline{\phi}_i$ are vectors related to the i-th finite element, the following classic relations can be proposed

$$\begin{aligned} u_i &= \underline{L}_i \, \underline{u} \\ \underline{\lambda}_i &= \underline{Z}_i \, \underline{\lambda} \qquad\qquad i = 1 \ldots n_e \\ \underline{\phi}_i &= \underline{Z}_i \, \underline{\phi} \end{aligned} \tag{5.5}$$

being L_i and Z_i costumary connectivity matrices and n_e the total number of finite elements. Now, with obvious notations and in accordance with the assumptions of Rel.s 3.18 to 3.20.

$$\begin{aligned} \underline{K}_{uu} &= \sum_{j=1}^{n_h} h_j^p \sum_{j=1}^{m_j} \underline{L}_i^T \, \underline{k}_{uu}^i \, \underline{L}_i \\ \underline{K}_{u\lambda} &= \sum_{j=1}^{n_h} h_j^p \sum_{i=1}^{m_j} \underline{L}_i^T \, \underline{k}_{u\lambda}^i \, \underline{Z}_i = \underline{K}_{\lambda u}^T \\ \underline{K}_{\lambda\lambda} &= \sum_{j=1}^{n_h} h_j^p \sum_{i=1}^{m_j} \underline{Z}_i^T \, \underline{k}_{\lambda\lambda}^i \, \underline{Z}_i \\ \underline{R} &= \sum_{j=1}^{n_h} h_j^q \sum_{i=1}^{m_j} \underline{Z}_i^T \, \underline{r}^i \end{aligned} \tag{5.6}$$

and

$$\underline{P} = \sum_{i=1}^{n_e} \underline{L}_i^T \underline{P}_i \tag{5.7}$$

Finally, in discrete form, the cost function of Rel.2.1 and behavioural and side constraints of Rel.2.5 and 2.6 can be respectively formulated as

$$\Gamma = \sum_{j=1}^{n_h} \ell_j h_j \tag{5.8}$$

$$\underline{T}\,\underline{u} - \underline{\hat{u}} \leq \underline{0} \tag{5.9}$$

$$h_{min} \leq h_j \leq h_{max} \qquad j=1...n_h \tag{5.10}$$

5.2. Optimality criterion

Looking for the optimality criterion, the discrete form of Lagrangian functional L of Rel.3.6 is

$$\begin{aligned} \bar{L} = & \sum_{j=1}^{n_h} \ell_j h_j + \underline{\eta}^T(\underline{P} - \underline{K}_{uu}\,\underline{u} + \underline{K}_{u\lambda}\,\underline{\lambda}) \\ & + \underline{\gamma}^T(\underline{K}_{\lambda u}\,\underline{u} - \underline{K}_{\lambda\lambda}\,\underline{\lambda} - \underline{R} - \underline{\phi}) + \underline{\psi}^T \underline{\Phi}_d\,\underline{\lambda} \\ & + \underline{\mu}^T(\underline{T}\,\underline{u} - \underline{\hat{u}}) + \sum_{j=1}^{n_h} \{\alpha_j(h_j - h_{max}) + \beta_j(h_{min} - h_j)\} \end{aligned} \tag{5.11}$$

where $\underline{\mu} \geq \underline{0}$ and $\alpha_j, \beta_j \geq 0$. The orthogonality constraints read

$$\underline{\mu}_d(\underline{T}\,\underline{u} - \underline{\hat{u}}) = \underline{0} \tag{5.12}$$

$$\alpha_j(h_j - h_{max}) = 0 \qquad j=1...n_h$$

$$\beta_j(h_{min} - h_j) = 0 \tag{5.13}$$

and the stationarity conditions with respect to vectors $\underline{u}$, $\underline{\lambda}$ and $\underline{\phi}$ provide respectively

$$-\underline{K}_{uu}\,\underline{\eta} + \underline{K}_{u\lambda}\,\underline{\gamma} + \underline{T}\,\underline{\mu} = \underline{0} \tag{5.14}$$

$$\underline{K}_{\lambda u}\,\underline{\eta} - K_{\lambda\lambda}\,\underline{\gamma} + \underline{\Phi}_d\,\underline{\psi} > \underline{0} \tag{5.15}$$

$$\lambda_d(\underline{K}_{\lambda u}\,\underline{\eta} - K_{\lambda\lambda}\,\underline{\gamma} + \underline{\Phi}_d\,\underline{\psi}) = \underline{0}$$

$$-\underline{\gamma} + \underline{\lambda}_d \, \underline{\psi} \leq \underline{0} \tag{5.16}$$
$$\underline{\phi}_d(-\underline{\gamma} + \underline{\lambda}_d \, \underline{\psi}) = \underline{0}$$

The optimality criterion is given by the stationarity conditions with respect to the design variables, that read

$$l_j + p\, h_j^{p-1} \sum_{j=1}^{m_j} \{\underline{n}^T \underline{L}_i^T(-\underline{k}_{uu}^i \underline{L}_i \underline{u} + \underline{k}_{u\lambda}^i \underline{Z}_i \underline{\lambda})$$
$$+ \underline{\lambda}^T \underline{Z}_i^T(\underline{k}_{\lambda u}^i \underline{L}_i \underline{u} - \underline{k}_{\lambda\lambda}^i \underline{Z}_i \underline{\lambda})\} \tag{5.17}$$
$$- q\, \underline{h}_j^{q-1} \sum_{j=1}^{m_j} \underline{\gamma}^T \underline{Z}_i^T \underline{r}^i + \alpha_j - \beta_j = 0 \qquad j=1\ldots n_h$$

From Rel.s 5.3 and 5.16 the following orthogonality constraint is derived

$$\underline{\phi}_d \, \underline{\gamma} = 0 \tag{5.18}$$

and the optimality criterion can be simplified in the form

$$l_j + p\, h_j^{p-1} \sum_{i=1}^{m_j} \underline{n}^T \underline{L}_i^T(-\underline{k}_{uu}^i \underline{L}_i \underline{u} + \underline{k}_{u\lambda}^i \underline{Z}_i \underline{\lambda})$$
$$+(p-q)h_j^{q-1} \sum_{i=1}^{m_j} \underline{\gamma}^T \underline{Z}_i^T \underline{r}^i + \alpha_j - \beta_j = 0 \qquad j=1\ldots n_h \tag{5.19}$$

5.3. Numerical procedure

The solution of the optimal design problem previous formulated can be obtained by solving the system of simultaneous equations and inequalities of Rel.s 5.1 to 5.4, 5.9, 5.10 and 5.14 to 5.17. Two different (actual and adjoint) analysis problems can be focussed, which are related by optimality criterion of Rel. 5.17 (or 5.19). Thus, a suitable numerical iterative procedure can be proposed.

For the sake of simplicity, assume technological contraints of Rel.s 5.10 not to be active and just one of contraints of Rel.s 5.9 to be **active**. So, Rel.s 5.13 involve $\alpha_j = \beta_j = 0$ $(j=1\ldots n_h)$ and Rel.5.12 implies just one nonvanishing term in the dual load vector $\underline{F} = \underline{T}^T\underline{\mu}$. The iterative procedure can be summarized as follows

a) for a given design solution $\bar{h}_j (j=1...n_h)$ solve the analysis problem of Rel.s 5.1 to 5.4;
b) modify the solution by defining a design factor f, in such a way the design $h'_j = f\bar{h}_j$ implies one (the k-th) of the behavioural constraints to be active;
c) solve the adjoint problem, assuming a unitary load, corresponding to the k-th term of vector F;
d) by using the optimality criterion , define a set of n_h factors s_j for having a new design $\bar{h}_j = s_j \, h'_j \; (j=1...n_h)$ and return to step a).

Such a procedure is summarized in Table 1.

6. AN OUTLINE OF THE NONLINEAR FINITE ELEMENT BEAM MODEL AND ITS IMPLICATIONS ON THE OPTIMUM DESIGN FORMULATION

Having in mind to take into account the nonlinear material behaviour throughout the optimum design process, a suitable finite element model has to be considered. Selection of such a model is essentially led by the purpose of modifying as little as possible the optimal design procedure developed for elastic materials and based on optimality criteria. A model which has been thought to correspond to this dictate and has been firstly proposed in Ref.9 and then refined in Ref.11, in the framework of theory of plasticity, will be now re-considered in order to clarify peculiar features of the optimal design formulation based on finite elements and previously presented.

6.1 Basic concepts

For the sake of brevity and in so far as understanding of this paper needs, only the main ideas are redrawn from the

referenced works 9,10,11 to which the Reader may turn for deeper informations.
Note that adopting such a model involves some notations to be modified with respect to the constitutive relations seen in previous Section, whereas the basic concepts are preserved. Mean axial strain $\delta(x)$ and curvature $\chi(x)$ govern the deformability of any cross section, supposed to be symmetric about z-axis and placed on a point x on centroidal axis of a straight beam element. According to slender beam theory only the axial strain ε is considered to be significant; because of the Bernoulli hypothesis ε is related to δ and χ by the following relation

$$\varepsilon(x,z) = \delta(x) - z\chi(x) \tag{6.1}$$

In turn, δ and χ depend on axial and transversal displacement u and v, according to the compatibility relations

$$\begin{aligned} \delta(x) &= du/dx \\ \chi(x) &= -d^2v/dx^2 \end{aligned} \tag{6.2a,b}$$

If δ and χ are considered as generalized (global) strains, the generalized stresses are defined through the virtual work equation

$$\underline{Q}^T \, \underline{q} \, \underline{\underline{\Delta}} \, [N \; M]\begin{Bmatrix}\delta\\ \chi\end{Bmatrix} = \int_A \sigma \, \varepsilon \, dA = \int_A \sigma(\delta - z\chi) \, dA \tag{6.3}$$

so that it is

$$\begin{aligned} N(x) &= \int_A \sigma(x,z) \, dA \\ M(x) &= -\int_A \sigma(x,z) \, z \, dA \end{aligned} \tag{6.4a,b}$$

Neglecting the influence of shear stress on yielding, material behaviour at a point B(z) of the cross section x is conceived to be governed by the linearized constitutive curve of Fig.6a, that is described by the following relations

$$\left.\begin{aligned} &\sigma_B(z) = E(\varepsilon_B(z) - p_B(z)) \quad ; \quad p_B(z) = \underline{n}_B\ \underline{\lambda}_B(z) \\ &\underline{n}_B = [1 \ldots 1_Y + 1_{Y+1} \ldots - 1_{Y^+ + Y^-}] \end{aligned}\right\} \quad (6.5a,b,c)$$

$$\left.\begin{aligned} &\underline{\phi}_B(z) = \underline{n}_B^T\ \sigma_B(z) - \underline{h}_B\ \underline{\lambda}(z) - \underline{r}_B \\ &\underline{r}_B^T = [r_1 \ldots r_{Y^+} \quad -r_{Y+1} \ldots -r_{Y^+ + Y^-}] \end{aligned}\right\} \quad (6.5d,e)$$

$$\underline{\lambda}_B(z) \geq \underline{0} \quad , \quad \underline{\phi}_B(z) \leq \underline{0} \quad , \quad \underline{\phi}_B^T(z)\ \underline{\lambda}_B(z) = 0 \qquad (6.5f\text{-}h)$$

where p represents the plastic part of strain $\varepsilon_B(z)$, which is expressed as summation of variables $\underline{\lambda}_B(z)$. In the framework of the theory of plasticity $\underline{\lambda}_B(z)$ are measures of plastic deformations. In the context of the present design formulation they should be better considered as simple deviations from linearity, since finite (no incremental) form of Rel.s 6.5 implies holonomic behaviour.

Hardening matrix $\underline{h}_B$ colets the slopes of branches in Fig.6b and, for the optimization model, might be simply $\underline{h}_B$= diag$[h_y]$, $y = 1 \ldots Y$ with $Y = Y^+ + Y^-$(no interaction between branches).

Note that $\underline{n}_B$, $\underline{h}_B$, $\underline{r}_B$ are constants and the variables of Rel.s 6.5 are functions of z only.

In short, the element model is based on independent interpolation of displacements u(x), v(x) and of multipliers $\underline{\lambda}(x,z)$. As displacement model $\underline{\psi}(x)$, ensuring continuity for u, v and dv/dx after assemblage, a linear distribution for u(x) and a cubic distribution for v(x), integrated by a fourth order bubble function, is assumed. Hence the vector $\underline{u}$ of the nodal degrees of freedom will collect 7 parameters

$$\underline{u}(x) = \begin{Bmatrix} u(x) \\ v(x) \end{Bmatrix} = \underline{\psi}(x)\ \underline{u}$$

with (6.6a,b)

$$\underline{u}^T = \{u_1\ u_2\ v_1\ v_2\ \theta_1\ \theta_2\ \alpha\}$$

where u_i, v_i, θ_i(i=1,2) are the axial, transversal displacements and rotations at the ends of the element; α is an additional parameter homogeneous with a displacement. Parameter α concerns the bubble function and shows that, by virtue of the independence between displacement and multiplier fields, hierarchic models and relevant procedure can be used when it needs. The assumed displacement model $\underline{\psi}(x)$ determines, by means of eq.s 6.1 and 6.2 the strain distribution: $\delta(x)$ is constant and $\chi(x)$ parabolically along the element. Multipliers $\underline{\lambda}$ are independently interpolated as follows

$$\lambda_y(x,z) = \Lambda_s(z)\ \Lambda_{\bar{s}}(x)\ \Lambda_y^{\bar{s}s} \quad , \quad y = 1...Y \qquad (6.7)$$

where

$$\Lambda_y^{\bar{s}s} = \lambda_y(x_{\bar{s}}\ ,\ z_s) \qquad (6.8)$$

are parameters which define the distribution of multipliers according to the interpolation polynomials $\Lambda_s(z)$ and $\Lambda_{\bar{s}}(x)$. Rel.6.8 shows that $\Lambda_y^{\bar{s}s}$ are the values that the multipliers assume at points z_s ($s = 1...S$) placed on cross section $x_{\bar{s}}$ ($\bar{s} = 1...\bar{S}$)(see Fig.9c). In other words, spreading of non-linearity (plasticity) is controlled by $\bar{S}$ cross sections and within them by S suitably chosen control points. The total number of parameters Λ for a beam element is Y x S x $\bar{S}$.
The crucial point to be discussed is the mutual independence of the interpolation models $\underline{\psi}(x)$ and $\Lambda_s(z)\ \Lambda_{\bar{s}}(x)$ for displacement and multiplier fields, respectively.
In fact, any arbitrarily chosen interpolation for p, determined through the Rel.6.5b by Rel.s 6.7,8, might be not compatible with the total strains defined by Rel.s 6.1,2 and by the displacement model $\underline{\psi}(x)$. This lack of compatibility would produce non vanishing distributions of the generalized stresses N(x), M(x) even in an isolated element subject to plastic strains only, which is unacceptable because of its static determinacy. This inconvenience is avoided by enforcing that N(x) be constant and M(x) vary parabolically with x, as consistency with the assumed displacement model requires. The above action, not constraining the order of polynomials $\Lambda_s(z)\ \Lambda_{\bar{s}}(x)$, is an operative interpretation for the beam element and derives from a general procedure illustrated in Ref.11.

Good description of the structural behaviour, given by the model outline, permits to reduce the number of the discretizing elements and hence the number of the nodal variables.

6.2. Explicit expression for the constitutive matrices

Application of the optimality criterion requires that the matrices formally introduced by Rel.s 5.6 be derived with respect to the design variables. As consequence, calculus of such derivatives needs that $\underline{K}_{uu}$, $\underline{K}_{u\lambda}$, $\underline{K}_{\lambda\lambda}$ and $\underline{R}$ be made in an explicit form, also suitable for an efficient organization of computations and for a rational employement of computer devices. The sectional behaviour will be firstly considered and then the elemental model will be derived, having in mind to evidentiate which components depend on the design variables and which do not.

A consistency condition, similar to 6.3, is written for yield function and plastic multipliers distributed along a cross section, thus defining the generalized yield function $\underline{f}$ and the generalized multipliers $\underline{l}$

$$\underline{f}^T \, \underline{l} = \int_A \phi_B^T(z) \; \lambda_B(z) \; dA \qquad (6.9)$$

The sectional behaviour is assumed to be governed by relations quite analogous to Rel.6.5 but expressed in terms of sectional generalized variables

$$\underline{Q} = \underline{d}(\underline{q} - \underline{n} \; \underline{l})$$

$$\underline{f} = \underline{n}^T \; \underline{Q} - \underline{h} \; \underline{l} - \underline{r} \qquad (6.10a\text{-}e)$$

$$\underline{l} \geq \underline{0} \quad , \quad \underline{f} \leq \underline{0} \quad , \quad \underline{f}^T \; \underline{l} = 0$$

Besides the elastic matrix $\underline{d}$, the arrays $\underline{n}$, $\underline{h}$ and $\underline{r}$ contain geometric properties of the cross sections as it will be shown in the following.

Supposing that for a section s $\varepsilon_B(z)$ and $\lambda_B(z)$ be expressed as functions of these generalized quantities $\underline{q}$ and $\underline{l}$ through the interpolation matrices $\underline{\beta}(z)$ and $\underline{\Lambda}(z)$

$$\varepsilon_B(z) = \underline{\beta}(z)\underline{q} \quad , \quad \lambda_B(z) = \underline{\Lambda}(z)\underline{1} \tag{6.11a,b}$$

from Rel.s 6.3 and 6.9 we have

$$\underline{Q} = \int_A \underline{\beta}^T(z)\ \sigma_B(z)\ dA, \underline{f} = \int_A \underline{\Lambda}^T(z)\ \phi_B(z) dA \tag{6.12a,b}$$

Comparing expressions derived from Rel.s 6.10a,b with those obtained by introducing Rel.s 6.5a,b and 6.11 in Rel.6.12a and Rel.s 6.5a,c,d in Rel.6.12b it results

$$\begin{aligned} &\underline{d} = E\,\underline{\Gamma}_{qq} \quad ; \quad \underline{n} = \underline{\Gamma}_{qq}^{-1}\,\underline{\Gamma}_{q1} \\ &\underline{h} = E(\underline{\Gamma}_{11} - \underline{\Gamma}_{1q}\,\underline{\Gamma}_{qq}^{-1}\,\underline{\Gamma}_{q1}) \quad ; \quad \underline{r} = \int_A \underline{\Lambda}^T\,\underline{r}_B\ dA \end{aligned} \tag{6.13a-d}$$

where

$$\begin{aligned} \underline{\Gamma}_{qq} &= \int_A \underline{\beta}^T(z)\ \underline{\beta}(z)\ dA \\ \underline{\Gamma}_{q1} &= \underline{\Gamma}_{1q} = \int_A \underline{\beta}^T(z)\ \underline{n}_B\ \underline{\Lambda}(z)\ dA \\ \underline{\Gamma}_{11} &= \int_A \underline{\Lambda}^T(z)\ \underline{n}_B^T\ \underline{n}_B\ \underline{\Lambda}(z) + \frac{1}{E}\int_A \underline{\Lambda}^T(z)\ \underline{h}_B\ \underline{\Lambda}(z)\ dA \\ \underline{r} &= \int_A \underline{\Lambda}^T(z)\ \underline{r}_B\ dA \end{aligned} \tag{6.14a-d}$$

Definition of matrices $\underline{n}$ and $\underline{h}$ requires that $\underline{\Gamma}_{qq}$ be non singular which is true if the entries of $\underline{\beta}(z)$ are linearly independent. In this case $\underline{\Gamma}_{qq}$ is symmetric and positive--defined.

Moreover, $\underline{h}$ is symmetric and positive semidefinite, as it can be proved, when $\underline{h}_B = \underline{0}$ or is symmetric and positive semidefinite.

Expression of matrices $\underline{\Gamma}$ are completely determined once interpolation functions $\underline{\beta}(z)$ and $\underline{\Lambda}(z)$ have been fixed.

According to Bernoulli hypothesis we assume

$$\underline{\beta}(z) \triangleq [1 - z] \tag{6.15}$$

On this basis, $\underline{\Gamma}_{qq}$ can be calculated and produces the well--known matrix $\underline{d} = E\begin{vmatrix} A & 0 \\ 0 & I \end{vmatrix}$ where A is the area and I is the moment of inertia of a cross section.

For $\underline{\Lambda}(z)$ a Lagrangian polynomial is considered and as zero points (control points) S points are selected along z-axis. Their coordinates z_s(s=1...S) can be suitably chosen in such a way that any $\int_A P(z)dA$, where P(z) is a polynomial of degree up to 2S-1, is exactly calculated by a weighted summation

$$\int_A P(z)\ dA = \sum_{1}^{S}{}_s\ w_s\ P(z_s) \qquad (6.16)$$

provided that coefficients w_s be correctly evaluated in dependence on the shape of the cross section (see Ref.11). It has been shown that w_s and z_s can be expressed as

$$\begin{aligned} w_s &= A\ \bar{w}_s \\ z_s &= \sqrt{I/A}\ \bar{z}_s \end{aligned} \qquad (6.17a,b)$$

where $\bar{w}_s$ and $\bar{z}_s$ are numerical values tabulated for certain total numbers S of control points. In order to express the integrals as funtion of cross section areas only, the following relation between the cross section area A and the moment of inertia I is assumed.

$$I = \alpha\ A^2 \qquad (6.18)$$

where α is a coefficient which can be calculated for different types of profiles (for IPE beams $\alpha \simeq 3$).
It is worth-while noting that the only arrays depending on A, among those introduced in this section, are $\underline{d}$, $\underline{n}$, $\underline{h}$ and $\underline{r}$. It is also evident that using such a model leads to assume as design variable the cross section area of the straight beam element. Now the beam element model can be formulated. Again the starting point is given by constitutive relations of the same form as Rel.s 6.5 and 6.10, but in terms of generalized variables for the element

$$\begin{aligned} &\underline{\sigma} = \underline{D}(\underline{\varepsilon} - \underline{N}\ \underline{\lambda}) \\ &\underline{\phi} = \underline{N}^T\ \underline{\sigma} - \underline{H}\ \underline{\lambda} - \underline{R} \\ &\underline{\lambda} \geq \underline{0}\ ,\quad \underline{\phi} \leq \underline{0}\ ,\quad \underline{\phi}^T\ \underline{\lambda} = 0 \end{aligned} \qquad (6.19a-e)$$

in which generalized stresses and strains $\underline{\sigma}$ and $\underline{\varepsilon}$ have to be redefined for the element through consistent relations involving generalized stresses $\underline{Q}$ and generalized strains $\underline{q}$ introduced for the cross section

$$\underline{\sigma}^T \ \underline{\varepsilon} = \int_0^L \underline{Q}^T(x) \ q(x) \ dx \tag{6.20}$$

Similarly for generalized yield functions $\underline{\phi}$ and generalized multipliers $\underline{\lambda}$ one obtains

$$\underline{\phi}^T \ \underline{\lambda} = \int_0^L \underline{f}^T(x) \ \underline{l}(x) \ dx \tag{6.21}$$

L is the length of the finite element.
Henceforth the procedure is very similar to the previous one. Combining the elemental interpolation relations

$$\begin{aligned} \underline{q}(x) &= \underline{b}(x) \ \underline{\varepsilon} \\ \underline{l}(x) &= \underline{\Lambda}(x) \ \underline{\lambda} \end{aligned} \tag{6.22a,b}$$

with the consistency conditions of Rel.s 6.20,21, after some algebra, one has

$$\begin{aligned} \underline{N} &= \int_0^L \underline{s}^T(x) \ \underline{n} \ \underline{\Lambda}(x) \ dx \\ \underline{H} &= \int_0^L \underline{\Lambda}^T(x) \ \underline{h} \ \underline{\Lambda}(x) \ dx \\ \underline{R} &= \int_0^L \underline{\Lambda}^T(x) \ \underline{r} \ dx \end{aligned} \tag{6.23a-c}$$

where

$$\underline{s}(x) = \underline{d} \ \underline{b}(x) \ \underline{D}^{-1} \tag{6.24}$$

is the stress interpolation function ($\underline{Q}(x) = s(x)\underline{\sigma}$) and $\underline{D}$ is the generalized elastic matrix

$$\underline{D} = \int_0^L \underline{b}^T(x) \ \underline{d} \ \underline{b}(x) \ dx \tag{6.25}$$

that is symmetric and positive definite since $\underline{d}$ is so and $\underline{b}(x)$ consists of functions linearly independent.
It remain to define $\underline{b}(x)$ and $\underline{\Lambda}(x)$.
Once the displacement model $\underline{\psi}(x)$ has been assumed, one can define

$$\underline{q}(x) = \begin{Bmatrix} \delta(x) \\ \chi(x) \end{Bmatrix} = \begin{Bmatrix} du/dx \\ d^2v/dx^2 \end{Bmatrix} = \underline{B}(x)\ \underline{u} \qquad (6.26)$$

where $\underline{u}$ collects the nodal displacement, in a generalized sense, as, for istance, in Rel.6.6b.
Interpolation matrix $\underline{b}(x)$ is defined by identifying as generalized strains $\underline{\varepsilon}$ quantities which represent purely straining modes. So, if $\underline{u}$ is partitioned into rigid modes $\underline{\rho}$ and straining modes $\underline{\varepsilon}$, we can write

$$\underline{u} = [\underline{A}_\rho\ \underline{A}_\varepsilon] \begin{Bmatrix} \underline{\rho} \\ \underline{\varepsilon} \end{Bmatrix}$$

and since $\underline{B}(x)\ \underline{A}_\rho = \underline{0}$ (6.27a-c)

it results

$$\underline{b}(x) = \underline{B}(x)\ A_\varepsilon$$

In the case of the previously considered displacement model, $\underline{\varepsilon}$ is a vector of 4 elements to be chosen among the 7 entries of $\underline{u}$, being R=3 the number of rigid motions for a plane beam element. In this way it is also possible to avoid that an isolated unloaded element be subject to a stress state when only a plastic deformation distribution exists.
As for the beam section model, $\underline{\Lambda}(x)$ is assumed to be a Lagrangian polynomial such that the integral expression of $\underline{N}$,$\underline{H}$ and $\underline{R}$ can be substituted by summation like Rel.6.16. From Rel.6.7 it can be argued that $\underline{\Lambda}(x)$ and $\underline{\Lambda}(z)$ in general comprise Lagrangian polynomial of different order, since the control points number s for $\underline{\Lambda}(z)$ may be different from the control sections number $\bar{S}$ concerning $\underline{\Lambda}(x)$.
Constitutive equations combined with the compatibility equation $\underline{\varepsilon} = \underline{C}\ \underline{u}$ and equilibrium conditions give the following expressions for the constitutive matrices

$$\underline{k}_{uu} = \underline{C}^T \underline{D}\,\underline{C}, \quad \underline{k}_{u\lambda} = \underline{k}_{u\lambda}^T = \underline{C}^T \underline{D}\,\underline{N}, \quad \underline{k}_{\lambda\lambda} = \underline{H} + \underline{N}^T \underline{D}\,\underline{N} \qquad (6.28a\text{-}c)$$

in which only $\underline{D}$ $\underline{N}$ and $\underline{H}$ are functions of the design variables A_i. Also the vector

$$\underline{R} = \int_0^L \underline{\Lambda}^T(x)\ \underline{r}\ dx \text{ with } \underline{r} = \int_A \underline{\Lambda}^T(z)\ \underline{r}_B\ dA \qquad (6.29)$$

depends on the design variables A_i.
For the outlined model Rel.s 5.6 have to be modified since the dependence of the constitutive matrices on the design variables is different. Rel.s 5.6 preserve their symbolic validity in general and an operative validity only for the less refined model proposed in Ref. 9.
We have now sound reasons to emphasize the noticeable merits of the adopted nonlinear beam model for the optimal design procedure. In fact it makes possible to evidentiate the design variables A_i in all the arrays which must be derived with respect to them in order to check the optimality condition

$$\frac{1}{L_i}\left(\underline{u}^* \frac{\partial \underline{K}_{uu}}{\partial A_i} \underline{u} - \underline{u} \frac{\partial \underline{K}_{u\lambda}}{\partial A_i} \underline{\lambda} - \underline{\lambda}^* \frac{\partial \underline{K}_{\lambda u}}{\partial A_i} \underline{u} - \underline{\lambda}^* \frac{\partial \underline{K}_{\lambda u}}{\partial A_i} \underline{\lambda} + \underline{\lambda}^* \frac{\partial \underline{R}}{\partial A_i}\right) = 1 \tag{6.30}$$

(for i = 1... number of design variables) where $\underline{K}_{uu}$, $\underline{K}_{u\lambda} = \underline{K}_{\lambda u}^T$ and $\underline{R}$ are obtained by assembling the homonymous arrays in Rel.s 6.28a-c and 6.29, (see Rel.s 5.5,6). Starred quantities derive from the adjoint problem, and are rather easy to calculate, as it is shown in Appendix.
In Rel.s 6.28a-c and 6.29, not only we meet again

$$\underline{d} = E\, \underline{\Gamma}_{qq} = E \int_A \underline{\beta}^T(z)\, \underline{\beta}(z)\, dA = E \int_A \begin{Bmatrix} 1 \\ -z \end{Bmatrix} [1 - z]\, dA = E \begin{bmatrix} A & 0 \\ 0 & I \end{bmatrix} \tag{6.31}$$

but again, the use of Lagrangian polynomials for $\underline{\Lambda}(z)$ and Gaussian points as control (zero) points, permits to consider algebraic summations instead of the integrals defining $\underline{n}$, $\underline{h}$ and $\underline{r}$, and hence allows to evidentiate the design variables. Thus we can write

$$\underline{d} = \underline{d}_A \underline{\bar{d}} \quad , \quad \underline{n} = \underline{n}_A \underline{\bar{n}} \quad , \quad \underline{h} = \underline{h}_A \underline{\bar{h}} \quad , \quad \underline{r} = \underline{r}_A \underline{\bar{r}} \tag{6.32a-d}$$

in which the overlined quantities do not depend to the cross section areas and on moments of inertia.
For the same reasons, not only $\underline{D}$ but also $\underline{N}$, $\underline{H}$ and $\underline{R}$ can be expressed as algebraic summations from which the design variables can be again drawn out.
So, it is

$$\underline{D} = \underline{D}_A \, \bar{\underline{D}} \quad , \quad \underline{N} = \underline{N}_A \, \bar{\underline{N}} \quad , \quad \underline{H} = \underline{H}_A \, \bar{\underline{H}} \quad , \underline{R} = \underline{R}_A \, \bar{\underline{R}} \qquad (6.33a\text{-}d)$$

where overline as the same meaning as above.

Matrices defined by Rel.s 6.28a-c become

$$\underline{k}_{uu} = \underline{C}^T \, \underline{D}_A \, \bar{\underline{D}} \, \underline{C} \quad , \quad \underline{k}_{u\lambda} = \underline{k}_{\lambda u}^T = \underline{C}^T \, \underline{D}_A \, \bar{\underline{D}} \, \underline{N}_A \, \bar{\underline{N}} \quad ,$$
$$\qquad (6.34a\text{-}c)$$
$$\underline{k}_{\lambda\lambda} = \underline{H}_A \, \bar{\underline{H}} + \bar{\underline{N}}^T \, \underline{N}_A^T \, \underline{D}_A \, \bar{\underline{D}} \, \underline{N}_A \, \bar{\underline{N}}$$

Hence the required derivatives are simply

$$\frac{\partial \underline{k}_{uu}}{\partial A_i} = \underline{C}^T \frac{\partial \underline{D}_A}{\partial A_i} \bar{\underline{D}} \, \underline{C}$$

$$\frac{\partial \underline{k}_{u\lambda}}{\partial A_i} = \frac{\partial \underline{k}_{\lambda u}^T}{\partial A_i} = \underline{C}^T \frac{\partial \underline{D}_A}{\partial A_i} \bar{\underline{D}} \, \underline{N}_A \, \bar{\underline{N}} + \underline{C}^T \, \underline{D}_A \, \bar{\underline{D}} \frac{\partial \underline{N}_A}{\partial A_i} \bar{\underline{N}} \qquad (6.35a\text{-}c)$$

$$\frac{\partial \underline{k}_{\lambda\lambda}}{\partial A_i} = \bar{\underline{N}}^T \frac{\partial \underline{N}_A^T}{\partial A_i} \underline{D}_A \, \bar{\underline{D}} \, \underline{N}_A \, \bar{\underline{N}} + \bar{\underline{N}}^T \, \underline{N}_A^T \frac{\partial \underline{D}_A}{\partial A_i} \bar{\underline{D}} \, \underline{N}_A \, \underline{N}$$
$$+ \bar{\underline{N}}^T \, \underline{N}_A^T \, \underline{D}_A \, \bar{\underline{D}} \frac{\partial \underline{N}_A}{\partial A_i} \bar{\underline{N}} + \frac{\partial \underline{H}_A}{\partial A_i} \bar{\underline{H}}$$

Similarly, derivatives of the assembled matrices in the optimal conditions of Rel.6.30 can be calculated.

6.3. Concluding remarks on merits of the adopted model in view of the optimal design procedure

The model taken into consideration in the present finite element optimal design approach allows to describe the inelastic (nonlinear) behaviour of members subject to both axial force and bending moment. Measures of deviation from linearity are modelled over the cross section and along the element length independently of the displacement field. In this way it is possible a good description of material nonlinearity with an accuracy degree which can be selected separately from that suggested by load conditions for displacements.

What above summarized has the following impact on the proposed optimal design approach:

a) Description needs for frames in nonlinear range can be very well satisfied even if the structure is discretized with only one element for any beam column member;

b) As consequence, more refined subdivisions into finite elements are involved by need to have more than one design variable for any member;

c) In spite of a certain complexity of formulation, computational burden is considerably reduced because of a suitable interpolation of parameters that represent nonlinearity;

d) In addition, the above interpolation models allow that the design variables be easily evidentiated;

e) For this reason optimality criterion can be made explicit and numerically checked on the basis of quantities which can be efficiently handled by a properly devised design--synthesis integrated system;

f) Finally, slight modifications are to be given to similarly founded elastic optimum design procedure, which is a purpose basically postulated.

REFERENCE

1. A.SAWCZUK,Z.MROZ (Eds.). Optimization in Structural Design. Proc. IUTAM Symp., Warsaw 1973, Springer-Verlag, Berlin (1975)
2. E.J.HAUG, J.CEA (Eds.). Optimization of Distributed Parameter Structures. Vol.1 and 2, Proc.Nato ASI, Iowa City 1980, Noordhoff, The Netherlands (1981)
3. A.J.MORRIS (Ed.). Foundations of Structural Optimization: A Unified Apprach. Proc. Nato ASI, Liege, Belgium 1980, Chichester (1982)
4. R.H.GALLAGHER (Ed.). Proc.Int.Symp. Optimum Structural Design, Univ. of Arizona, Tucson, Arizona (1981)
5. H.ESCHENAUER, N.OLHOFF (Eds.). Optimization Methods in Structural Design. Proc. Euromech Coll. 164, Siegen, FRG, 1982, Bibliographisches Institut, Mannheim (1983)
6. C.CINQUINI. Optimality Criteria for Non-Linear Behaviour Material: Application to Beam in Bending, Eng.Struct., Vol.6, pp.61-64 (1984)
7. C.CINQUINI, R.CONTRO. Optimal Design of Beams Discretized by Elastic Plastic Finite Element, Comp.Struct., Vol.20, N.1-3, pp.475-485 (1985)
8. G.MAIER. Teoremi di Minimo in Termini Finiti per continui Elasto-plastici con Leggi Costitutive Linearizzate a

Tratti, Rend. Ist. Lomb.Sci.Lett., A103, 1066 (1969)
9. L.CORRADI. A Displacement Formulation for the Finite Element Elastic Plastic Problem, Meccanica, 18, 77-91 (1983)
10. L.CORRADI, C.POGGI. An Analysis Procedure for Non-Linear Elastic-Plastic Frames Accounting for the Spreading of Local Plasticity. Costr.Met., N.1, pp.2-15, (1985)
11. L.CORRADI, C.POGGI. A Refined Finite Elemet Model for the Analysis of Elastic-Plastic Frames, Int.J.Num.Meth.Eng., Vol.20, 2155-2174, (1984)

APPENDIX

Starred quantities of the adjoint problem, needed to complete the check of the optimality criterion of Rel.6.30, can be calculated by 'shrewdly' handling the analysis procedure. Let the adjoint problem be written as follows (+)

$$\begin{aligned} &\underline{K}_{uu}\,\underline{u}^* - \underline{K}_{u\lambda}\,\underline{\lambda}^* = \underline{P}^* \\ &\underline{K}_{\lambda u}\,\underline{u}^* - \underline{K}_{\lambda\lambda}\,\underline{\lambda}^* + \underline{\phi}_d\,\underline{v}^* \geq \underline{0} \\ &\underline{\lambda}_d(\underline{K}_{\lambda u}\,\underline{u}^* - \underline{K}_{\lambda\lambda}\,\underline{\lambda}^* + \underline{\phi}_d\,\underline{v}^*) = \underline{0} \\ &-\underline{\lambda}^* + \underline{\lambda}_d\,\underline{v}^* \geq \underline{0} \\ &\underline{\phi}_d(-\underline{\lambda}^* + \underline{\lambda}_d\,\underline{v}^*) = \underline{0} \end{aligned} \qquad (1a\text{-}e)$$

Since $\underline{\phi}_d\,\underline{\lambda} = \underline{0}$, from Rel.1.e it derives

$$\underline{\phi}_d\,\underline{\lambda}^* = \underline{0} \qquad (2)$$

Moreover, Rel.1.a solved with respect to $\underline{u}^*$ gives

$$\underline{u}^* = \underline{K}_{uu}^{-1}\,(\underline{P}^* + \underline{K}_{u\lambda}\,\underline{\lambda}^*) \qquad (3)$$

which, substituted in (1.b), yields

$$\underline{\phi}_d\,\underline{v}^* + \underline{K}_{\lambda u}\,\underline{K}_{uu}^{-1}\,P^* - (\underline{K}_{\lambda\lambda} - \underline{K}_{\lambda u}\,\underline{K}_{uu}^{-1}\,\underline{K}_{u\lambda})\underline{\lambda}^* \geq \underline{0} \qquad (4)$$

This inequality can be expressed in terms of $\underline{C}$, $\underline{D} = \underline{D}^T$, $\underline{N}$ and

(+) The quantities of the adjoint problem have the same meaning as in the previous sections but here they are indicated with different symbols to underline their mechanic interpretation.

$\underline{H}$ as

$$\underline{N}^T \underline{D}\,\underline{C}\,\underline{K}_{uu}^{-1}\,\underline{P}^* - (\underline{H} + \underline{N}^T\underline{D}\,\underline{N} - \underline{N}^T\underline{D}\,\underline{C}\,\underline{K}_{uu}^{-1}\,\underline{C}^T\,\underline{D}\,\underline{N})\underline{\lambda}^* + \underline{\phi}_d\,\underline{v}^*$$

$$= \underline{N}^T\,\underline{D}\,\underline{C}\,\underline{K}_{uu}^{-1}\,\underline{P}^* - \underline{A}\,\underline{\lambda}^* + \underline{\phi}_d\,\underline{v}^* \geq 0 \qquad (5)$$

where $\underline{A} = -(\underline{N}^T\,\underline{Z}\,\underline{N} - \underline{H})$ (5')

and $\underline{Z} = \underline{D}\,\underline{C}\,\underline{K}_{uu}^{-1}\,\underline{C}^T\,\underline{D} - \underline{D}$ (5")

Since

$$\underline{K}_{uu}^{-1}\,\underline{P}^* = \underline{u}_o^* \qquad (6)$$

and consequently

$$\underline{N}^T\,\underline{D}\,\underline{C}\,\underline{u}_o^* = \underline{N}^T\,\underline{D}(\underline{C}\,\underline{u}_o^*) = \underline{N}^T\,\underline{D}\,\underline{\varepsilon}_o^* = \underline{N}^T\,\underline{\sigma}_{el}^* \qquad (7)$$

Rel.5 becomes

$$\underline{N}^T\,\underline{\sigma}_{el}^* - \underline{A}\,\underline{\lambda}^* + \underline{\phi}_d\,\underline{v}^* \geq \underline{0} \qquad (8)$$

Now let Rel. 1c be considered. Since the left-hand side term of Rel.8 is equal to the expression multiplied by $\underline{\lambda}_d$ in Rel. 1c, this last can be rewritten as

$$\lambda_d(\underline{N}^T\,\underline{\sigma}_{el}^*\,\underline{A}\,\underline{\lambda}^* + \underline{\phi}_d\,\underline{v}^*) = \underline{0} \qquad (9a)$$

Matrix $\underline{A}$ is partitioned on the basis of the 3 subvectors $\underline{\lambda}_i^*$, $\underline{\lambda}_j^*$ and $\underline{\lambda}_k^*$, with i, j, k $\in$ I, J, K.

Sets I,J,K collect the indices which denote, in the adopted inelastic analysis procedure by holonomic steps, the plasticity conditions that are active, no active, no active in the current and active in the previous step, respectively. One obtains

$$\underline{\lambda}_d(\underline{N}^T\,\underline{\sigma}_{el}^* - \begin{bmatrix} \underline{A}_{ii} & \underline{A}_{ij} & \underline{A}_{ik} \\ \underline{A}_{ji} & \underline{A}_{jj} & \underline{A}_{jk} \\ \underline{A}_{ki} & \underline{A}_{kj} & \underline{A}_{kk} \end{bmatrix} \begin{matrix} \underline{\lambda}_i^* \\ \underline{\lambda}_j^* \\ \underline{\lambda}_k^* \end{matrix} + \underline{\phi}_d\,\underline{v}^*) = \underline{0} \qquad (9b)$$

and in an expanded form

$$\underline{\lambda}_d \begin{pmatrix} \underline{N}_i^T\,\underline{\sigma}_{el}^* - (\underline{A}_{ii}\,\underline{\lambda}_i^* + \underline{A}_{ij}\,\underline{\lambda}_j^* + \underline{A}_{ik}\,\underline{\lambda}_k^*) + \underline{\phi}_i^T\,\underline{v}_i^* \\ \underline{N}_j^T\,\underline{\sigma}_{el}^* - (\underline{A}_{ji}\,\underline{\lambda}_i^* + \underline{A}_{jj}\,\underline{\lambda}_j^* + \underline{A}_{jk}\,\underline{\lambda}_k^*) + \underline{\phi}_j^T\,\underline{v}_j^* \\ \underline{N}_k^T\,\underline{\sigma}_{el}^* - (\underline{A}_{ki}\,\underline{\lambda}_i^* + \underline{A}_{kj}\,\underline{\lambda}_j^* + \underline{A}_{kk}\,\underline{\lambda}_k^*) + \underline{\phi}_k^T\,\underline{v}_k^* \end{pmatrix} = \underline{0} \qquad (9c)$$

where $\underline{\phi}_i^T = \underline{0}$, $\underline{\phi}_j^T < 0$, $\underline{\phi}_k^T < 0$ (10a,b,c)

The entries of the diagonal matrix $\underline{\lambda}_d$ can be rearranged to form the vectors $\underline{\lambda}_i, \underline{\lambda}_j$ and $\underline{\lambda}_k$ constrained as follows

$$\underline{\lambda}_i \geq \underline{0} \quad , \quad \underline{\lambda}_j = \underline{0} \quad , \quad \underline{\lambda}_k = \underline{0} \tag{11a,b,c}$$

Taking into account Rel.s 10a and 11a, the first row in Rel.s 9c becomes

$$\underline{N}_i^T \underline{\sigma}_{el}^* - (\underline{A}_{ii} \underline{\lambda}_i^* + \underline{A}_{ij} \underline{\lambda}_j^* + A_{ik}^* \underline{\lambda}_k^*) = \underline{0} \tag{12}$$

Because of Rel.2 and ineq.s of Rel.s 10b,c it results

$$\underline{\lambda}_j^* = \underline{0} \quad , \quad \underline{\lambda}_k^* = \underline{0} \tag{13}$$

Hence Rel.12 reduces to

$$\underline{N}_i^T \underline{\sigma}_{el}^* - \underline{A}_{ii} \underline{\lambda}_i^* = \underline{0} \tag{14}$$

from which

$$\underline{\lambda}_i^* = \underline{A}_{ii}^{-1}(\underline{N}_i^T \underline{\sigma}_{el}^*) \tag{15}$$

Substitution of vector $\underline{\lambda}_i^*$ in Rel.3 yields $\underline{u}^*$.
So , computation of the adjoint quantities $\underline{u}^*$ and $\underline{\lambda}^*$ simply requires that the arrays appearing in the right-hand sides of Rel.s 3 and 15 be suitably saved during the analysis procedure.

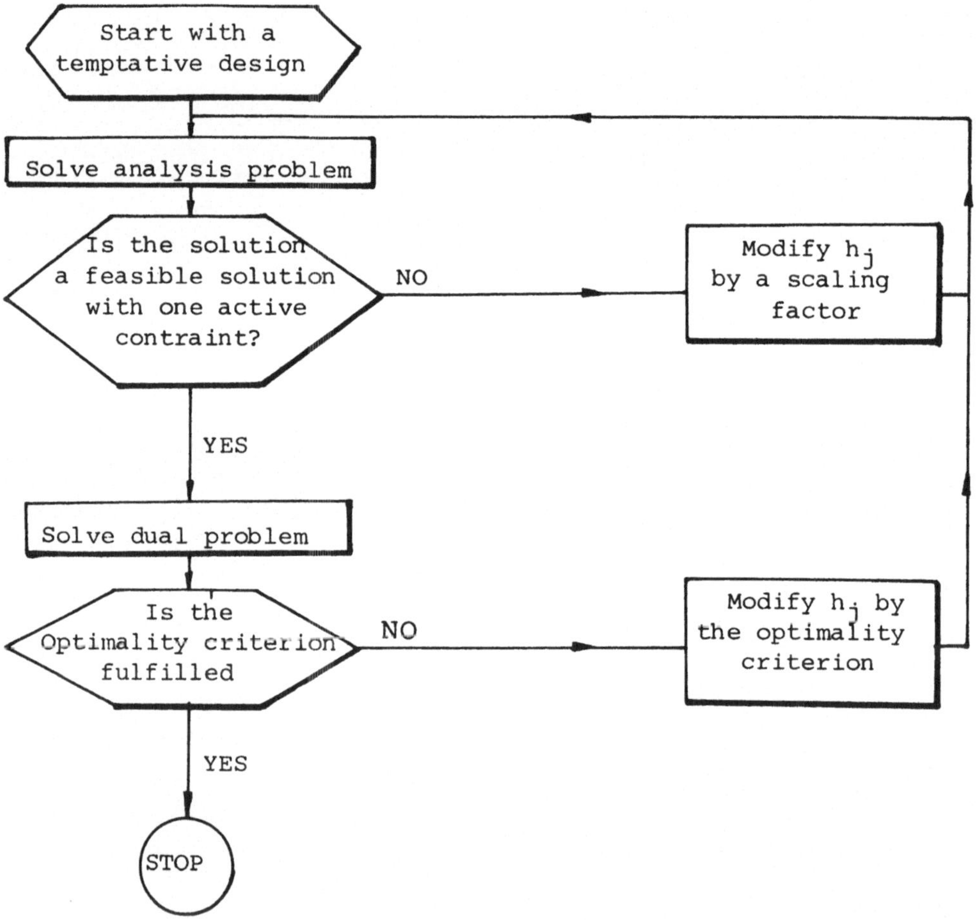
Start with a temptative design
Solve analysis problem
Is the solution a feasible solution with one active contraint?
NO
Modify h_j by a scaling factor
YES
Solve dual problem
Is the Optimality criterion fulfilled
NO
Modify h_j by the optimality criterion
YES
STOP

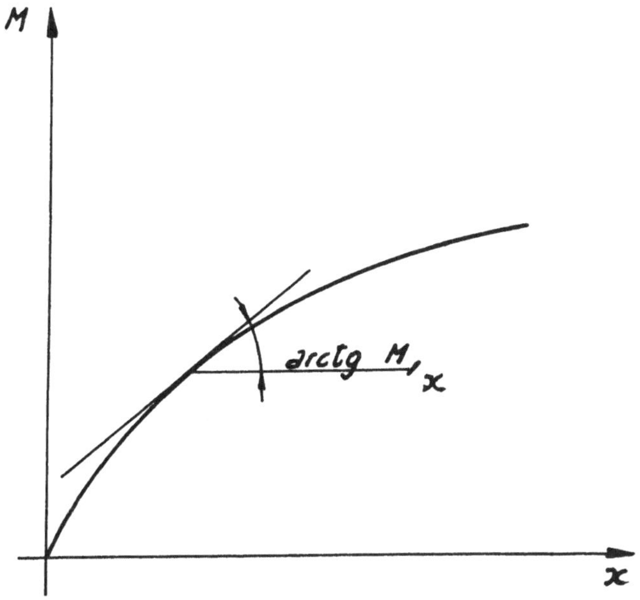

FIG. 1

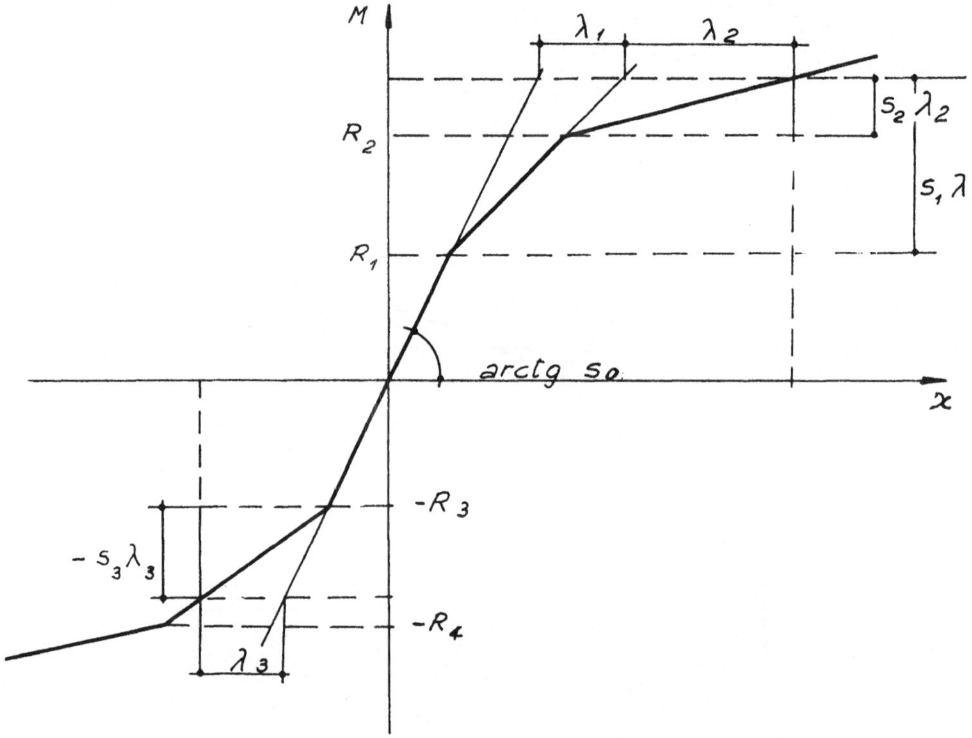

FIG. 2

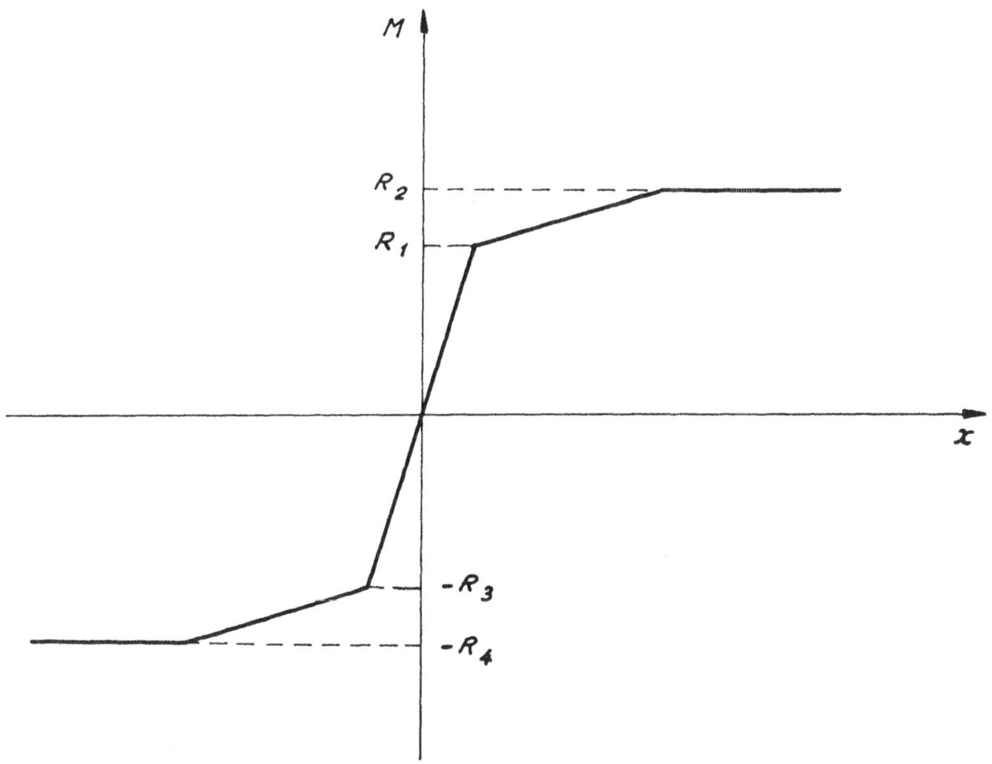

FIG. 3

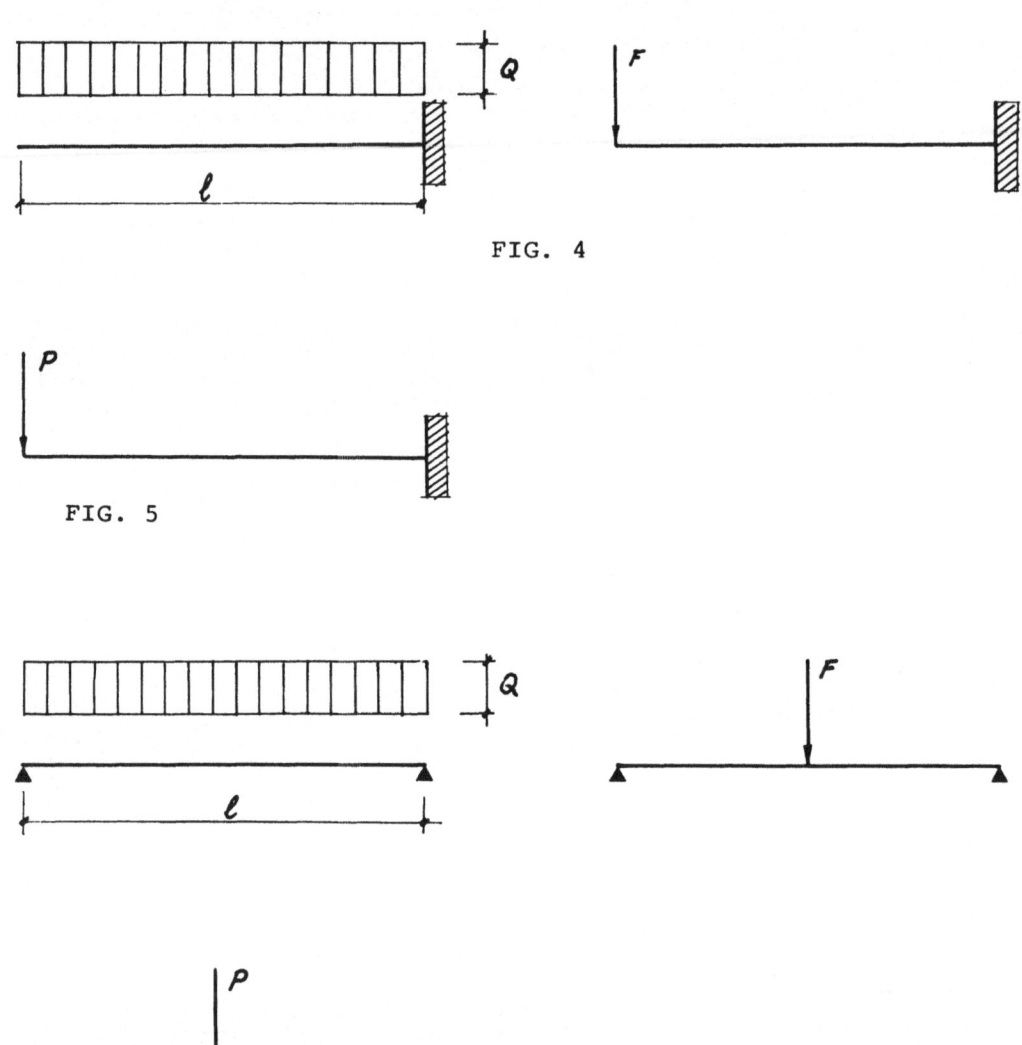

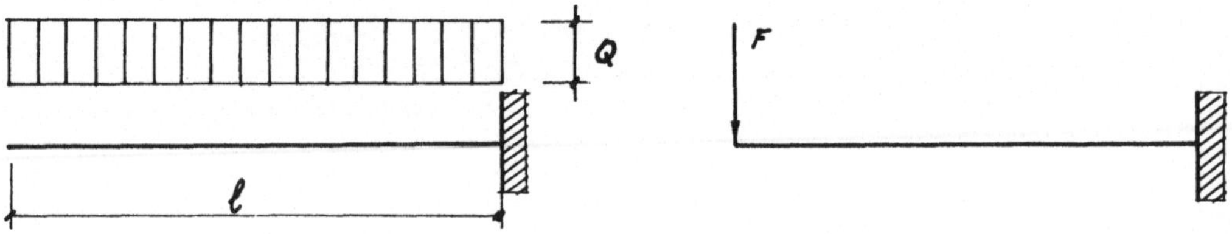

FIG. 4

FIG. 5

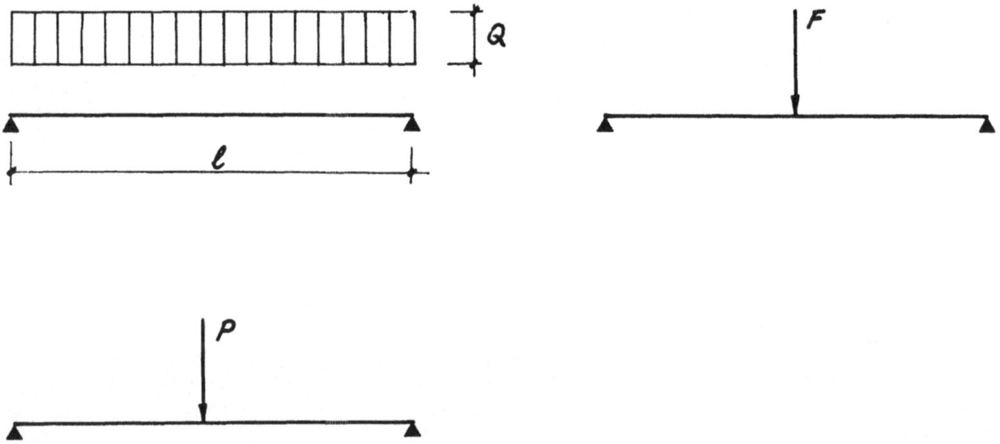

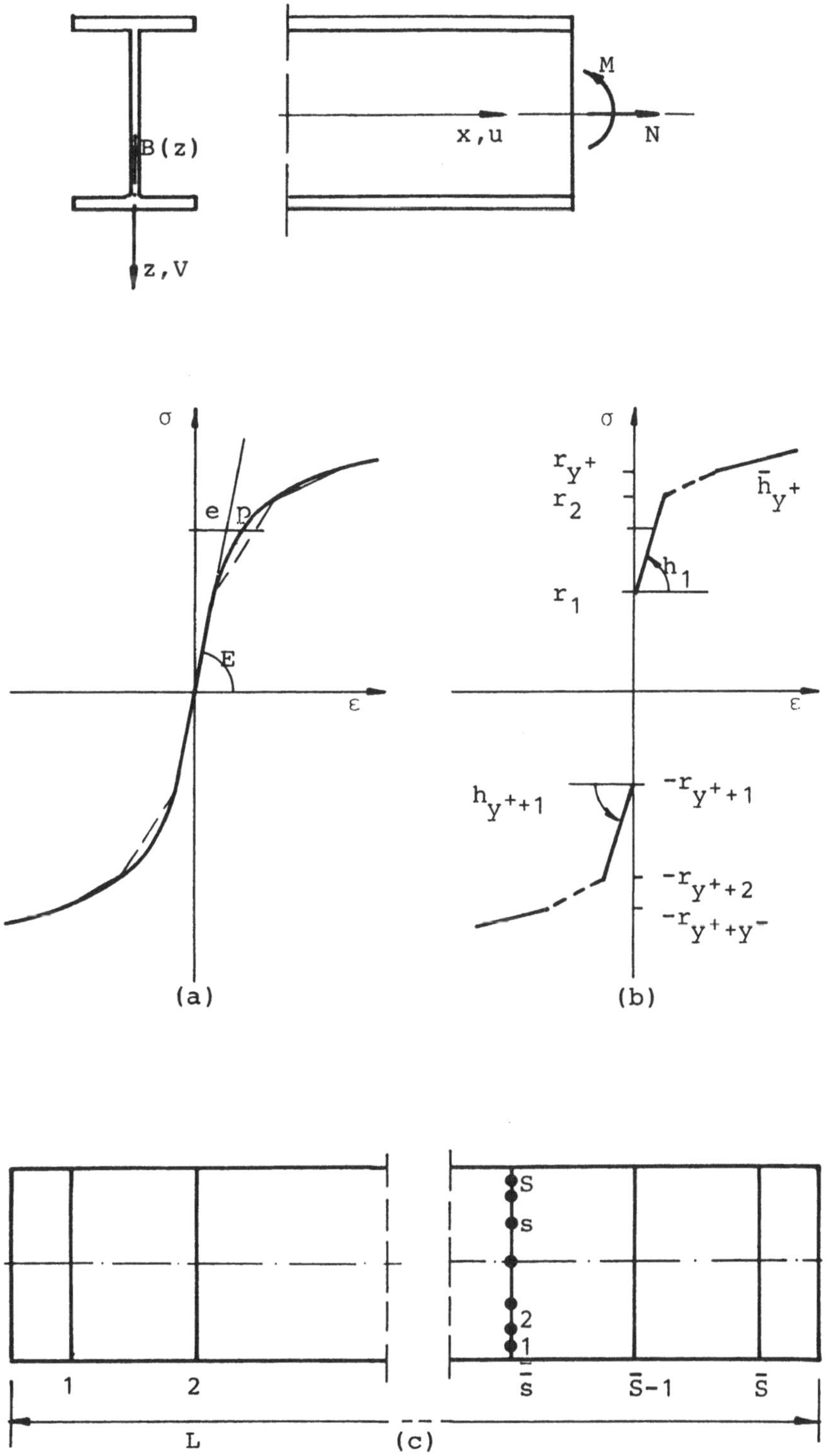

B(z)
z,V
M
x,u
N
σ
e
p
E
ε
(a)
σ
r_{y^+}
r_2
$\bar{h}_{y^+}$
h_1
r_1
ε
h_{y^++1}
$-r_{y^++1}$
$-r_{y^++2}$
$-r_{y^++y^-}$
(b)
S
s
2
1
1
2
$\bar{s}$
$\bar{S}-1$
$\bar{S}$
L
(c)

Fig.6

COMPUTER-AUTOMATED DESIGN OF BUILDING FRAMEWORKS UNDER VARIOUS PERFORMANCE CONDITIONS

Donald E. Grierson
Solid Mechanics Division
University of Waterloo, Canada

INTRODUCTION

This study addresses the task faced by designers of structures for which "limit-states" criteria must be satisfied at one or more distinct loading levels. For example, the specified limit states may concern acceptable elastic displacements under service loads, acceptable elastic stresses under factored service loads, and adequate strength reserve under ultimate loads against buckling instability and/or plastic collapse of the structure. Ideally, while satisfying the various performance criteria, the most economical design of the structure is sought.

The conventional approach to such design is to separately proportion the structure to satisfy one set of performance criteria (e.g., stress limits), and to then modify the structure to satisfy the one or more other sets of criteria that are of concern to the design (e.g., displacement limits, failure limits, etc.). A drawback of this approach, however, is that decisions taken at any one time to satisfy some of the performance criteria are usually made in the absence of explicit information as to their consequences for the other criteria. As such, they may result in the violation of criteria that were otherwise satisfied at a previous design stage. Moreover, such an approach makes it difficult to have explicit concern for design economy. At best, a somewhat cumbersome trial-and-error process is required to achieve a reasonably efficient design.

An earlier study by Grierson and Schmit [1] considered thin-walled structures comprised of bar, membrane and/or shear panel elements under uniaxial stress, and developed a computer-automated synthesis capability whereby a minimum weight structure is found while satisfying any combination of limit states criteria con-

NATO ASI Series, Vol. F27
Computer Aided Optimal Design: Structural and Mechanical Systems. Edited by C.A. Mota Soares

cerning acceptable elastic stresses and displacements under service loads and adequate post-elastic strength reserve of the structure under ultimate loads. As well, fabrication conditions can be imposed to ensure member-continuity and structure-symmetry requirements.

Further studies by Chiu [2] and by Grierson and Chiu [3,4] extended the design method to planar frameworks comprised of beam and column members under combined axial and bending stresses. These studies took simultaneous account of service and ultimate performance criteria, but considered the sizes of the cross-sections for the members of the structure as continuous variables to the synthesis process (see Fig. 1, where different types of member sizing variables are graphically illustrated). In theory, this design approach tacitly assumes the availability of custom-fabricated sections that have the exact size, stiffness and strength properties required for the members of the minimum weight structure.

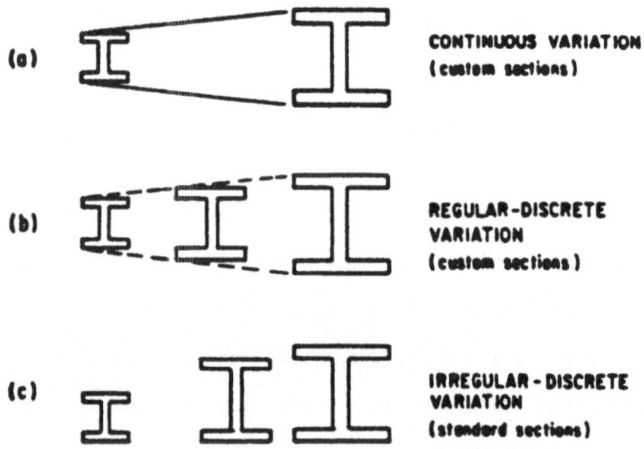

Figure 1 : Different Types of Member Sizing Variables for Design

Still further studies by Lee [5] and by Grierson and Lee [6,7] extended the design method such that the sizes of the cross-sections for the members of planar frameworks can be taken as discrete variables to the design. For the case when both elastic and plastic performance criteria must be satisfied, the design cross-section for each member is selected from among a specified set of regular-discrete sections (specifically, a set of discrete cross-sections of different sizes but of the same shape; see Fig. 1b). For the case when elastic stress and/or displacement criteria alone are to be satisfied, the design cross-section for each member can instead be selected from among available commercial-standard sections (which do not obey a constant-shape rule as their size varies; see Fig. 1c).

The strategy employed by the synthesis technique to classify the databank of standard steel sections is quite general and independent of the units of measurement adopted for the design, [7]. This implies that the design method may be directly applied for a variety of different steel codes (Canadian, American, Japanese, European, etc.). Studies by Cameron [8] and by Grierson and Cameron [9] have thus far extended the synthesis capability to allow computer-automated design for the standard steel sections specified by both the Canadian Institute of Steel Construction [10] and the American Institute of Steel Construction [11]; stress criteria specified by Canadian limits states design [12] and by American working stress design [13] and load and resistance factor design [14] are directly accounted for.

Further studies by Hall [15] and by Grierson and Hall [16] have extended the design method to account for structure instability in the presence of large (second-order) displacements. A modified Newton-Raphsen technique enables the second-order analysis and the structural design process to be conducted simultaneously.

Still further on-going studies by Kramer [17] and by Grierson and Kramer [18] are extending the design method to account for frequency, stress and displacement limit states under dynamic loads.

The essential details of the design method are first described in the following, and then the designs of two frameworks are presented to illustrate the method.

THE DESIGN PROBLEM

For the purposes of this study, all loads are taken as being static and the distinct loading levels associated with the various limit states are proportionally related to each other. The framework is discretized into an assemblage of n prismatic members, which may be of a variety of types (wide-flange beams, hollow-box columns, double-angle bracing struts, etc.). The design variable for each member i is its cross-section area a_i. In its general form, the minimum weight design problem is:

$$\text{Minimize:} \quad \sum_{i=1}^{n} w_i a_i \tag{1a}$$

$$\text{Subject to:} \quad \underset{\wedge}{\delta}_j \leq \delta_j \leq \hat{\delta}_j \qquad (j = 1,2,\ldots,d) \tag{1b}$$

$$\underset{\wedge}{\sigma}_k \leq \sigma_k \leq \hat{\sigma}_k \qquad (k = 1,2,\ldots,s) \tag{1c}$$

$$\underset{\wedge}{\alpha}_m \leq \alpha_m \leq \hat{\alpha}_m \qquad (m = 1,2,\ldots,p) \tag{1d}$$

$$a_i \in A_i \qquad (i = 1,2,\ldots,n) \tag{1e}$$

Equation (1a) defines the weight of the structure (w_i is the weight coefficient for member i); eqs. (1b) define the d service-load constraints on displacements δ_j (quantities with under- and super-imposed 'hat' ^ denote specified lower- and upper-bounds, respectively); eqs. (1c) define the s service-load constraints on stresses σ_k; eqs. (1d) define the p ultimate-load constraints on plastic collapse-load factors α_m; eqs. (1e) require each member cross-section area a_i to belong to a specified set of discrete section areas $A_i \equiv \{a_1, a_2, ..\}_i$. Depending on the design, eqs. (1b) and (1c) may account for small (first-order) or large (second-order) displacement effects for the structure. The lower and upper bounds on displacements and stresses account for the two possible senses of response action (e.g., left or right sway, tensile or compressive stress). The compressive stress bounds, which serve to guard against local buckling of members and of flange and web elements of cross-sections, are progressively updated to reflect the member and section properties that prevail at each stage of the design history [8].

FORMULATION OF THE DESIGN

In their present form, the performance constraints eqs. (1b), (1c) and (1d) are but implicit functions of the sizing variables for the design. To facilitate the numerical implementation of the synthesis process, these constraints are formulated as explicit functions of the sizing variables through the use of sensitivity analysis techniques.

Service-Load Constraints. The (global-axis) stiffness matrix $\underline{K}^i$ for a planar member i can be expressed as:

$$\underline{K}_i = \underline{K}_i^A a_i + \underline{K}_i^B I_i + \underline{K}_i^G N_i \tag{2}$$

where, for the member, $\underline{K}_i^A$ and $\underline{K}_i^B$ are constant matrices that correspond to the first-order axial and bending stiffness properties, the constant matrix $\underline{K}_i^G$ corresponds to the second-order geometric stiffness properties, a_i and I_i are the cross-section area and moment of inertia, and N_i is the member axial force (i.e., that prevailing at any one stage of iterative design process). The geometric stiffness term in eq. (2) is omitted if the design only has concern for first-order displacement and stress effects.

To facilitate formulation of the service-load constraints eqs. (1c) and (1d) at each design stage, the stiffness matrix $\underline{K}^i$ for each member i is expressed as a linear function of the cross-section area a_i as follows

$$\underline{K}_i = \underline{K}_i^A a_i + \underline{K}_i^B k_i a_i + \underline{K}_i^G \sigma_i a_i = \underline{K}_i^* a_i \tag{3}$$

where $\underline{K}_i^* = \underline{K}_i^A + k_i \underline{K}_i^B + \sigma_i \underline{K}_i^G$ is taken as a constant matrix for the current design stage, in which k_i is a constant that depends on the shape of the cross-section and σ_i is the prevailing axial stress for the cross-section. The matrix $\underline{K}_i^*$ is updated for each design stage.

Recognizing that elastic displacements and stresses vary inversely with the member cross-section areas a_i, corresponding 'good' quality performance constraints are achieved by formulating them as explicit-linear functions of the 'reciprocal' sizing variables

$$x_i = 1/a_i \qquad (i = 1,2,\ldots,n) \qquad (4)$$

To this end, first-order Taylor's series expansions are employed to formulate the displacement constraints eqs. (1b) as

$$\underset{\wedge}{\delta}_j \leq \delta_j^\circ + \sum_{i=1}^{n} \left(\frac{\partial \delta_j}{\partial x_i}\right)^\circ (x_i - x_i^\circ) \leq \hat{\delta}_j \qquad (j = 1,2,\ldots,d) \qquad (5)$$

and the stress constraints eqs. (1c) as

$$\underset{\wedge}{\sigma}_k \leq \sigma_k^\circ + \sum_{i=1}^{n} \left(\frac{\partial \sigma_k}{\partial x_i}\right)^\circ (x_i - x_i^\circ) \leq \hat{\sigma}_k \qquad (k = 1,2,\ldots,s) \qquad (6)$$

where the superscript $^\circ$ indicates known quantities evaluated for the current design stage (e.g., the initial 'trial' design), and the x_i are the sizing variables to the next weight optimization.

The displacement and stress gradients in eqs. (5) and (6) are evaluated using 'virtual-load' sensitivity analysis techniques in conjunction with elastic structural analysis [19]. Each displacement gradient is found as

$$\left(\frac{\partial \delta_j}{\partial x_i}\right)^\circ = \underline{b}_j^T \left(\frac{\partial \underline{u}}{\partial x_i}\right)^\circ = -\underline{b}_j^T \left(\underline{K}^{-1} \frac{\partial \underline{K}}{\partial x_i} \underline{u}\right)^\circ = \frac{1}{x_i^\circ} \left(\underline{u}_j^T \underline{K}_i \underline{u}\right)^\circ \qquad (7)$$

and each stress gradient is found as

$$\left(\frac{\partial \sigma_k}{\partial x_i}\right)^\circ = \underline{t}_k^T \left(\frac{\partial \underline{u}}{\partial x_i}\right)^\circ = -\underline{t}_k^T \left(\underline{K}^{-1} \frac{\partial \underline{K}}{\partial x_i} \underline{u}\right) = \frac{1}{x_i^\circ} \left(\underline{u}_k^T \underline{K}_i \underline{u}\right)^\circ \qquad (8)$$

where, from eqs. (3) and (4), $\underline{K} = \Sigma\, \underline{K}_i = \Sigma\, \underline{K}_i^\star / x_i$ is the stiffness matrix for the structure, $\underline{u}$ is the vector of nodal displacements due to the applied load vector $\underline{P}$, and $\underline{u}_j$ and $\underline{u}_k$ are vectors of 'virtual' nodal displacements due to 'virtual' load vectors $\underline{b}_j$ and $\underline{t}_k$. For designs where large (second-order) displacement effects are to be accounted for, the various vectors of displacements are found using a two-cycle secant-stiffness second-order analysis procedure, [15].

The specified vector $\underline{b}_j$ identifies the particular displacement δ_j of concern to the design (e.g., if $\underline{b}_j^T = [0,1,0,\ldots,0]$ then $\delta_j = u_2$, the displacement corresponding to degree-of-freedom 2), and $\underline{t}_k$ is row k of the global-axis stress matrix for the member associated with the stress σ_k. Equations (7) and (8) are based on the assumption that the vectors $\underline{P}$, $\underline{b}_j$ and $\underline{t}_k$ are all invariant with changes in the design; i.e., $\partial\underline{P}/\partial x_i = \partial\underline{b}_j/\partial x_i = \partial\underline{t}_k/\partial x_i = 0$. This recognizes that the applied load vector $\underline{P}$ excludes structure self-weight, that $\underline{b}_j$ is always a constant vector, and that $\underline{t}_k$ is a constant vector for truss structures; at the final stage of the synthesis history, there is negligible error inherent in this assumption for flexural structures, for which vectors $\underline{t}_k$ are functions of variable member neutral-axis positions [2].

The displacement and stress values δ_j° and σ_k° for the current design that appear in eqs. (5) and (6) can be expressed as

$$\delta_j^\circ = \sum_{i=1}^{n} \left(\frac{\partial \delta_j}{\partial x_i}\right)^\circ x_i^\circ \quad ; \qquad \sigma_k^\circ = \sum_{i=1}^{n} \left(\frac{\partial \sigma_k}{\partial x_i}\right)^\circ x_i^\circ \tag{9}$$

Therefore, from eqs. (5), (6) and (9), and adopting the concise notation $d_{ij}^\circ = (\partial\delta_j/\partial x_i)^\circ$ and $s_{ik}^\circ = (\partial\sigma_k/\partial x_i)^\circ$, the 'explicit-linear' performance constraints on elastic displacements are

$$\underset{\wedge}{\delta}_j \leq \sum_{i=1}^{n} d_{ij}^\circ \, x_i \leq \hat{\delta}_j \qquad (j = 1,2,\ldots,d) \tag{10}$$

and on elastic stresses,

$$\underset{\wedge}{\sigma}_k \leq \sum_{i=1}^{n} s_{ik}^\circ \, x_i \leq \hat{\sigma}_k \qquad (k = 1,2,\ldots,s) \tag{11}$$

<u>Ultimate-Load Constraints</u>. Consistent with the service-load constraints, the ultimate-load constraints are also formulated in terms of the reciprocal sizing variables x_i defined by eq. (4). To this end, first-order Taylor's series expansions and sensitivity analysis techniques are employed to formulate the load-factor constraints eqs. (1d) as the 'explicit-linear' constraints [1]

$$\underset{\wedge}{\alpha}_m - 2\alpha_m^\circ \leq \sum_{i=1}^{n} \left(\frac{\partial \alpha_m}{\partial x_i}\right)^\circ x_i \leq \hat{\alpha}_m - 2\alpha_m^\circ \qquad (m = 1,2,\ldots,p) \tag{12}$$

where, again, the superscript ° denotes quantities evaluated for the current design stage, and the x_i are the variables to the next weight optimization.

The load factor α_m° in eq. (12) defines the load level at which plastic-collapse mechanism m forms for the current design. It is evaluated using a 'finite-incremental' plastic analysis technique due to Franchi [21] as

$$\alpha_m^\circ = (\underline{R}^T \dot{\underline{\lambda}}_m)^\circ = \sum_{i=1}^{n} (\underline{R}_i^T \dot{\underline{\lambda}}_{im})^\circ \tag{13}$$

where $\underline{R}$ is the vector of plastic capacities for all members of the structure (the subvector $\underline{R}_i$ refers to that for member i) and $\dot{\underline{\lambda}}_m$ is the vector of member plastic deformation-rates associated with collapse mechanism m (the subvector $\dot{\underline{\lambda}}_{im}$ refers to that for member i). Moreover, the load-factor sensitivity gradient in eq. (12) is evaluated as [1]

$$\left(\frac{\partial \alpha_m}{\partial x_i}\right)^\circ = -\frac{1}{x_i^\circ} (\underline{R}_i^T \dot{\underline{\lambda}}_{im})^\circ \tag{14}$$

Equation (14) derives, in part, from the fact that for fixed structure topology the vector $\dot{\underline{\lambda}}_m$ of plastic deformation-rates characterizing collapse mechanism m is invariant with changes in the design (i.e., $\partial \dot{\underline{\lambda}}_m / \partial x_i = 0$).

To facilitate efficient plastic analysis for each design stage, a piecewiselinear (PWL) yield condition is adopted to govern plastic behaviour at the two end-sections j and k of each member i. As such, the vector $\underline{R}_i$ of plastic capacities for each member i can be expressed as [2,3]

$$\underline{R}_i^T = [\underline{R}_j^T ; \underline{R}_k^T] \tag{15}$$

where the components of the subvectors $\underline{R}_j = \underline{R}_k$ are functions of the 'principal' axial and/or bending plastic capacities N_p and M_p for the member. For a member under axial or bending stress alone, $\underline{R}_i^T = [N_p ; N_p]$ or $\underline{R}_i^T = [M_p, M_p; M_p, M_p]$, respectively. For a member under combined axial and bending stresses, $\underline{R}_j^T = \underline{R}_k^T = [r_1, r_2, \ldots]$, where each component r_ℓ is the orthogonal distance from the origin of the stress space to a particular linear yield surface of the PWL yield condition for the cross-section.

The principal axial plastic capacity N_{pi} of each member i is linearly related to the cross-section area a_i as

$$N_{pi} = \sigma_y a_i \tag{16}$$

where σ_y is the material yield stress. For all members i having the same cross-section shape, the principal bending plastic capacity M_{pi} is related to a_i as

$$M_{pi} = \sigma_y m_i a_i^{1.5} \tag{17}$$

where the constant m_i depends only on the shape of the cross-section. Having the member cross-section areas $a_i^\circ = a_i$ (i=1,2, ...,n) from the previous weight optimization, the vector $\underline{R}_i$ of plastic capacities for each member i, eqs. (15), is updated through eqs. (16) and (17). Then, plastic sensitivity analysis is conducted to formulate the explicit ultimate-load constraints eqs. (12) for the next weight optimization.

The number p of ultimate-load constraints comprising eqs. (12) progressively increases from stage to stage of the synthesis process as different collapse mechanisms become 'critical' to the design (i.e., the constraint set for each load case is progressively augmented to account for any 'new' mechanism that forms at the lowest load level for any given design stage). As such, each collapse mechanism m referred to by eqs. (12) is either a new 'critical' mechanism for the current design, or it is a mechanism that has been accounted for by the weight optimization of the previous design stage. In the former case, the load factor α_m and the vector $\dot{\underline{\lambda}}_m$ of plastic deformation-rates for the mechanism are determined directly by the plastic analysis for the current design stage. The corresponding ultimate-load constraint eq. (12) is then formulated through eq. (14) and added to the constraint set for the next weight optimization. For the latter case, however, the vector $\dot{\underline{\lambda}}_m$ is already known from a plastic analysis conducted at a previous design stage (recall that $\dot{\underline{\lambda}}_m$ is invariant with changes in the design). As such, the load factor α_m is updated through eq. (13) directly (i.e., without need for further plastic analysis). Thereafter, the corresponding ultimate load constraint eq. (12) is updated through eq. (14) for the next weight optimization.

<u>The Explicit Design Problem</u>. From eqs. (4), (10), (11) and (12), and adopting the concise notation $p^\circ_{im} = (\partial\alpha_m/\partial x_i)^\circ$, the minimum weight design problem eqs. (1) is expressed explicitly in terms of the reciprocal sizing variables x_i as

$$\text{Minimize:} \quad \sum_{i=1}^{n} w_i/x_i \tag{18a}$$

$$\text{Subject to:} \quad \underset{\wedge}{\delta}_j \leq \sum_{i=1}^{n} d^\circ_{ij}x_i \leq \hat{\delta}_j \qquad (j=1,2,\ldots,d) \tag{18b}$$

$$\underset{\wedge}{\sigma}_k \leq \sum_{i=1}^{n} s^\circ_{ik}x_i \leq \hat{\sigma}_k \qquad (k=1,2,\ldots,s) \tag{18c}$$

$$\underset{\wedge}{\alpha}_m - 2\alpha^\circ_m \leq \sum_{i=1}^{n} p^\circ_{im}x_i \leq \hat{\alpha}_m - 2\alpha^\circ_m \quad (m=1,2,\ldots,p) \tag{18d}$$

$$x_i \in X_i \tag{18e}$$

where the components of each discrete set $X_i \equiv \{x_1,x_2,\ldots\}_i$ in eqs. (18e) are the reciprocals of the cross-section areas comprising the corresponding discrete set A_i in eqs. (1e).

THE SYNTHESIS PROCESS

The discrete weight optimization problem eqs. (18) is solved for each design stage using a 'generalized optimality criteria' technique due to Fleury [22]. After each design stage, the sensitivity coefficients d°_{ij}, s°_{ik} and p°_{im} are updated, if necessary new ultimate-load constraints are added to the constraint set, and the weight optimization is repeated to find an improved (lower weight) design. The iterative process terminates with the final design when structure weight convergence occurs with no constraint infeasibilities.

The flowchart in Fig. 2 presents the general strategy of the design method for computer implementation. The central activity

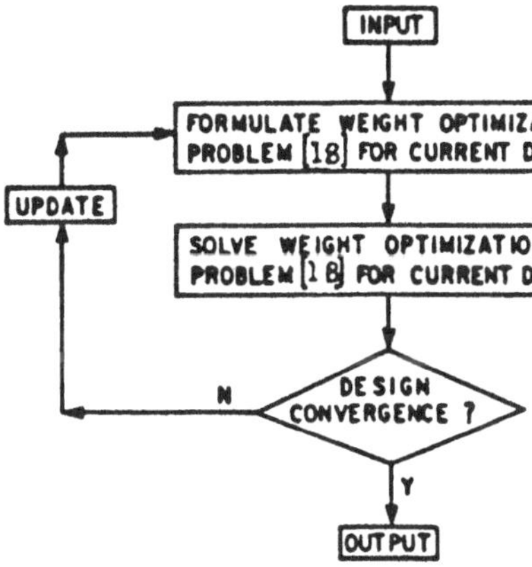

Figure 2 : Structural Synthesis Flowchart

of the synthesis process involves the formulation and solution of the weight optimization problem eqs. (18) for each design stage.

For designs that account for standard steel sections (Fig. 1(c)), the synthesis process is conducted in two distinct phases. The first phase involves a specified number (e.g., three) of design stages for which the member sizes are taken to be continuous variables so as to establish a reasonably well-proportioned structure as the starting basis for the second design phase involving discrete standard sections. Thereafter, a 'section-selection' technique [5] is employed to identify the discrete sets of reciprocal areas X_i in eq. (18e) from the databank of standard sections specified for the design.

Designs that account for large (second-order) displacement effects are also conducted in two distinct phases. The first phase involves a specified number (e.g., three) of design stages for which first-order effects are alone considered so as to establish a reasonably well-proportioned structure as the starting basis for the second phase of the design involving both first-order and second-order effects.

Referring to Fig. 2, the basic computer input concerns typical structural analysis data plus the performance and fabrication conditions for the design. For the case of design using standard sections, the standard-section category for each member is specified (i.e., wide-flange or tee, etc.). For each design stage (e.g., the initial 'trial' design), the weight optimization problem eqs. (18) is formulated using the previously described sensitivity analysis techniques. To ensure the fabrication requirements relating to member continuity and structure symmetry, the formulation is done in terms of a corresponding reduced set of independent sizing variables through the use of a 'design-variable-linking' technique [20]. Recognizing, with a view to numerical efficiency, that but a limited number of the performance conditions will actually control the design, a 'constraint-selection' technique [5] is applied to retain only those conditions in eqs. (18) that are potentially 'active' for the current design stage. To avoid oscillating or divergent behaviour of the synthesis history, which may occur at intermediate design stages as a consequence of the approximate nature of the performance conditions, a 'move-limit' technique [5] is applied that takes a weighted average of the design points existing before and after solving eqs. (18) as the basis for updating the design for the next stage.

EXAMPLE APPLICATIONS

Two steel frameworks are considered in the following. In each case, several comparative designs are conducted for illustration purposes. Designs are conducted using a STRUctural SYnthesis computer program named STRUSY developed at the University of Waterloo, Canada [2,5,8,15].

Two-Storey Frame

A variety of different designs of the pin-support frame in Figure 3(a) are presented in the following to illustrate the synthesis technique and associated computer implementation, [9]. Specifically, for the five independent loading schemes indicated in Figure 3(b), the following three design cases are considered:

Case 1: Limit-states design using Canadian standard steel sections [10,12]

Case 2: Working-stress design using American standard steel sections [11,13]

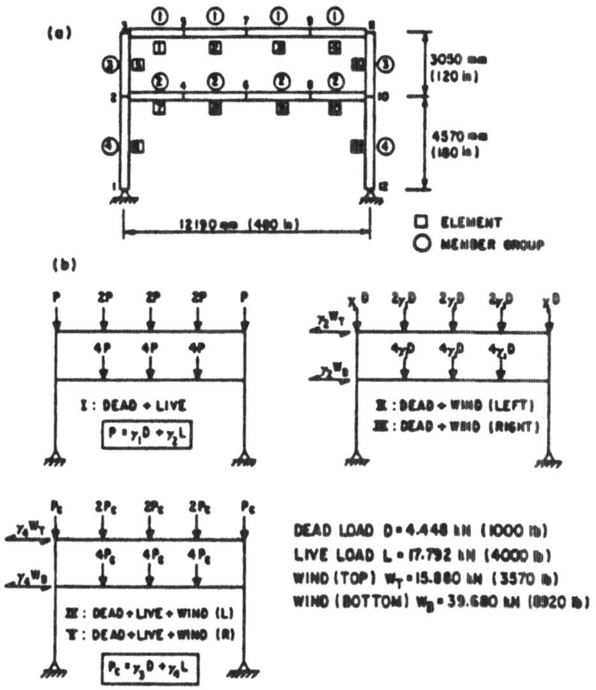

Figure 3 : Two-Storey Frame (a) Geometry; (b) Loading

Case 3: Design under both service and plastic-collapse performance constraints using discrete steel sections [6]

It is assumed that shear stresses and second-order geometric (P-Δ) effects are negligible and that members are braced against lateral buckling. Note that the plastic collapse constraints eqs. (18d) are eliminated from the weight optimization problem for design cases 1 and 2. For each case, for comparison purposes, a corresponding design where member sizes are taken as continuous variables is also presented, for which eqs. (18e) are replaced in the weight optimization problem by the side-constraints

$$\underset{\wedge}{x}_i \le x_i \le \hat{x}_i \qquad (i=1,2,\ldots,n) \tag{19}$$

where, from eq. (4), $\underset{\wedge}{x}_i = 1/\hat{a}_i$ and $\hat{x}_i = 1/\underset{\wedge}{a}_i$, in which $\underset{\wedge}{a}_i$ and $\hat{a}_i$ are the specified lower- and upper-bounds on the cross-section area for member i, respectively. The load factors γ_i (i=1,2,3,4) in Fig. 3(b) take on different values depending on the loading levels at which the various performance conditions are imposed.

To satisfy conventional fabrication requirements concerning member continuity and structure symmetry, the twelve elements for the frame (denoted by □ in Fig. 3(a)) are linked together into four independent member groups (denoted by 0 in Fig. 3(a)); i.e., the girder for each storey is required to be a prismatic member and the two columns for each storey are required to be identical prismatic members.

Stress and displacement performance conditions are imposed for each design case. Member stresses under the five loading schemes are constrained at all joint connections and load points, for a total of eighty stress conditions. The midspan vertical deflection of each girder under loading scheme I is limited to 1/360 of its span length; i.e., 34 mm (1.33 in). The lateral deflection of each storey under loading schemes II and III is limited to 1/400 of the height of the frame at that level; i.e., 11.5 mm (0.45 in) for the bottom storey and 19 mm (0.75 in) for the top storey. The displacement conditions are imposed at the service load level for all design cases, in which case $\gamma_1 = \gamma_2 = 1.0$ for loading schemes I, II and III in Figure 3(b).

CASE 1: CISC Limit-States Design, [12] In keeping with the limit-states design philosophy of ref. [12], the stress and displacement performance conditions are imposed at two different loading levels. Namely, stresses are constrained under factored service loads while displacements are constrained under unfactored service loads. For the stress conditions, the design loads are defined by setting $\gamma_1 = \gamma_3 = 1.25$, $\gamma_2 = 1.50$ and $\gamma_4 = 0.7 \times 1.5 = 1.05$ for the five loading schemes in Fig. 3(b).

All members of the frame are specified to be of the same steel grade: density = 7.85×10^{-6} kg/mm^3 (0.284 lb/in^3); Young's modulus $E = 2 \times 10^5$ MPa (29×10^6 psi); yield stress $F_y = 300$ MPa (43.5×10^3 psi). Allowable stresses for the members are limited to 90% of the steel yield stress; i.e., 270 MPa (39.2×10^3 psi). The design cross-section for each member is to be selected from among the Canadian standard wide-flange (W or WWF) sections specified by CISC [10].

Upon applying the STRUSY computer program [2,5,8,15] to

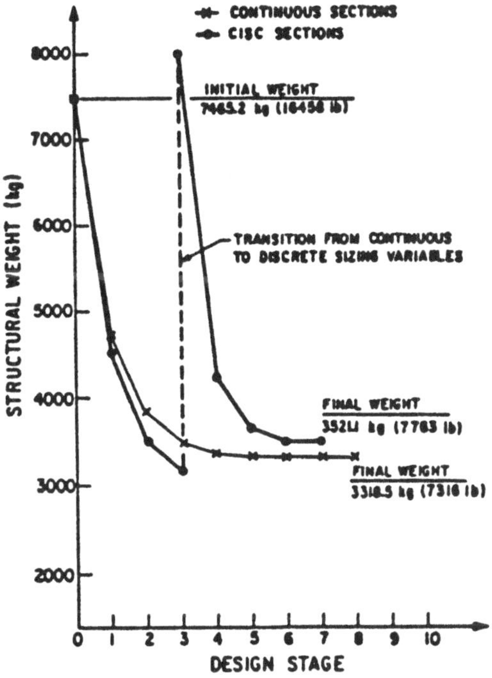

Figure 4 : Two-Storey Frame Synthesis History for Design Case 1

conduct the design process (Fig. 2), the minimum-weight structure is found after seven design stages requiring a total CPU time of 12.1 seconds on an IBM 4341 computer (University of Waterloo, Canada). The CISC design standard sections found for the members are given in Table 1. The history of the synthesis process is illustrated in Fig. 4. The convergence to the minimum-weight structure is monotonic, except at the transition between the first design phase involving continuous sizing variables and the second design phase involving discrete standard sections as sizing variables. A corresponding synthesis history for which member sizes were considered as continuous variables to the design is also illustrated in Fig. 4; to provide a basis for comparison, the fundamental shape and stiffness properties of the individual member cross-sections were taken to be those of the minimum-weight CISC standard section design. As expected, the final structure weight for the latter design is less than that for the final design using standard sections.

CASE 2: AISC Working-Stress Design, [13] For this design case, the stress and displacement performance conditions are both imposed at the specified service-load level in keeping with the working-stress design philosophy of ref. [13]. For the stress conditions, the design loads are defined by setting $\gamma_1 = \gamma_2 = 1.0$ and $\gamma_3 = \gamma_4 = 0.75$ for the five loading schemes in Fig. 3(b).

For the steel grade adopted for the design, the density and Young's modulus are the same as for design case 1 while the yield stress $F_y = 248$ MPa (36×10^3 psi). Allowable stresses are limited to 60% of the steel yield stress; i.e., 149 MPa (21.6×10^3 psi). The design cross-section for each member is to be selected from among the American standard wide-flange (W) sections specified by AISC [11].

Upon applying the STRUSY computer program [2,5,8,15], the minimum-weight structure is found after seven design stages requiring a total CPU time of 14.4 seconds on an IBM 4341 computer. The AISC design standard sections found for the members are given in Table 1. The history of the synthesis process is illustrated in Fig. 5. As for design case 1, the convergence to the minimum-weight structure is essentially monotonic. The synthesis history for a corresponding continuous-variable design is also illustrated in Fig. 5. The fundamental shape and stiffness properties of the member cross-sections for this design were taken to be those of the minimum-weight AISC standard section design.

CASE 3: Service and Plastic Collapse Performance Conditions. For this design case, the stress and displacement performance conditions are imposed exactly as for design case 1. In addition, ultimate-load constraints are imposed such that the plastic collapse load factor for the frame under loading schemes I, IV and V is limited to be not less than 2.35 (i.e., the frame is required to be capable of withstanding a 135% overload beyond the service load level without failure occurring in any plastic mechanism mode), for which $\gamma_1 = \gamma_2 = \gamma_3 = 2.35$ and $\gamma_4 = 0.7 \times 2.35 = 1.65$ for loading schemes I, IV and V in Figure 3(b).

All members of the frame are specified to have the same steel grade as for design case 1. Each member is specified to have a wide-flange cross-section having the 'constant-shape'

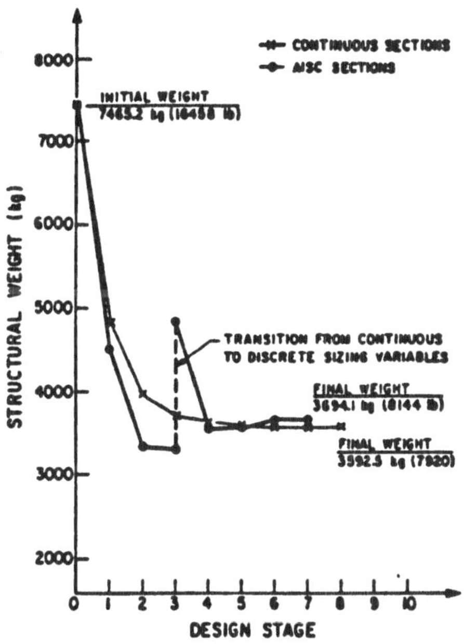

Figure 5 : Two-Storey Frame Synthesis History for Design Case 2

characteristics indicated in Figure 6(a); for the given values of the three constants C_1, C_2 and C_3, and any cross-section area a, the parameters k, m and n directly define the stiffness, strength and neutral-axis position of the corresponding section. These section parameters remain constant throughout the synthesis history.

For the plastic analyses that are conducted over the synthesis history to establish the load-factor performance conditions governing the design, ideal plastic behaviour is assumed confined to the element end-sections defined by joint connections and load points. The eight-sided piecewiselinear yield surface in Figure 6(b) is adopted to govern plastic behaviour at each such section (the orthogonal distances r_1, r_2, ..., r_8 from the origin of the stress space to the linear yield modes are fixed functions of the principal flexural and axial plastic capacities M_p and N_p for any section that maintains a constant shape regardless of its size), [2].

The design cross-section areas of the members are to be selected from among an available set of 180 discrete area values;

(a)

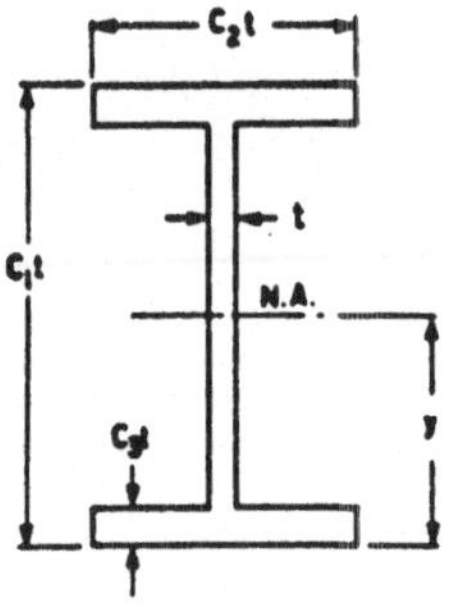

MEMBERS	C_1	C_2	C_3
1	53.73	23.60	1.45
2	57.43	21.71	1.42
3	53.73	23.60	1.45
4	77.78	33.33	2.22

$$I = ka^2;\ M_p = \sigma_y m a^{1.5};\ y = na^{0.5}$$

$$k = \frac{(C_1-2C_3)^3 + 2C_2C_3 + 6C_2C_3(C_1-C_3)^2}{12(C_1+2C_2C_3-2C_3)^2}$$

$$m = \frac{C_2C_1^2-(C_2-1)(C_1-2C_3)^2}{4(C_1+2C_2C_3-2C_3)^{1.5}}$$

$$n = \frac{C_1}{2(C_1+2C_2C_3-2C_3)^{0.5}}$$

(b)

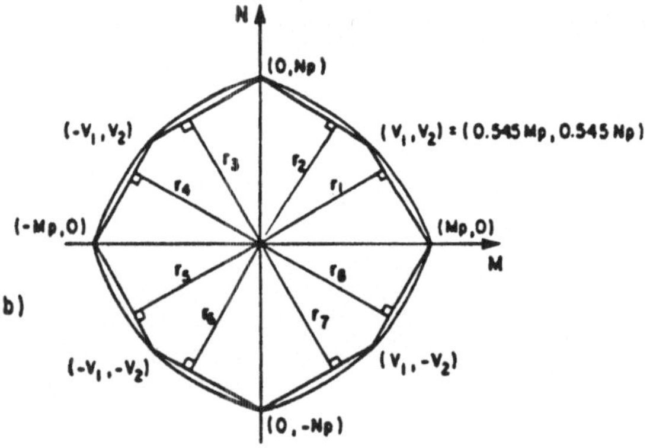

Figure 6 : Two-Storey Frame Design Case 3
(a) Wide-Flange Cross-Section
(b) Piecewiselinear Yield Condition

commencing with a lower-bound area of 2000 mm^2, the discrete area values increment by 500 mm^2 up to an upper-bound area of 92000 mm^2.

Upon applying the STRUSY computer program [2,5,8,15], the minimum-weight structure was found after fourteen design stages requiring a total CPU time of 42.6 seconds on an IBM 4341 computer. The discrete design cross-section areas found for the members are given in Table 1. The history of the synthesis process is illustrated in Figure 7 (the non-monotonic convergence behaviour at intermediate design stages occurs as a consequence of new plastic collapse load-factor constraints becoming active in the design). The five mechanisms shown in Figure 8 were identified as the most 'critical' failure modes for the frame at various stages of the synthesis history; the particular stages

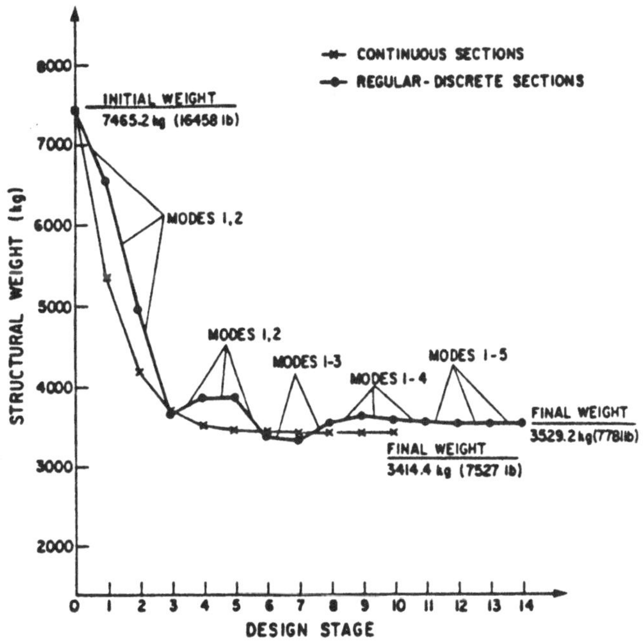

Figure 7 : Two-Storey Frame Synthesis History for Design Case 3

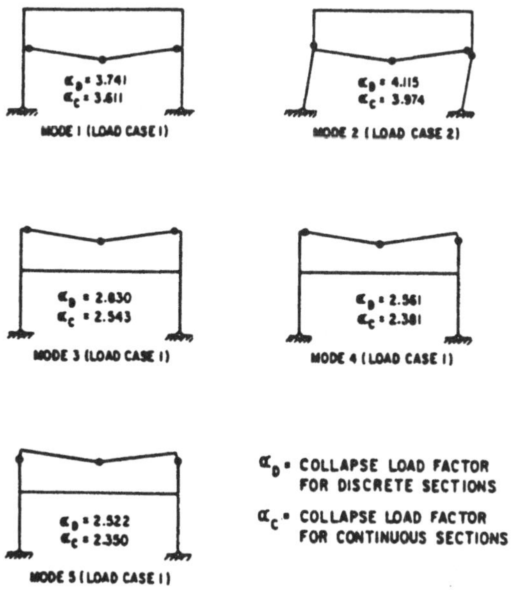

Figure 8 : Critical Plastic Collapse Modes for Two-Storey Frame Design Case 3

at which they became critical are indicated in Figure 7. The collapse-load factors indicated in Figure 8 are those prevailing for the various mechanism modes at the final design stage. Plastic collapse mode 5 is alone active at the final design stage under loading scheme I (the corresponding load factor is not exactly equal to the limit value of 2.35 because of the discrete nature of the member sizing variables).

The synthesis history for a corresponding continuous-variable design is also illustrated in Figure 7. The fundamental section properties for this design were taken to be the same as those specified for the discrete section design.

Table 1 : Design Cross-Sections for Minimum-Weight Two-Storey Frame

Member Group	Element No.	Design Case 1		Design Case 2		Case 3
		Designation (CISC, metric)	Area (mm^2)	Designation (AISC, imperial)	Area (mm^2)	Area (mm^2)
1	1,2,3,4	W410 × 54	6810	W 21 × 44	8387	8000
2	7,8,9,10	W610 × 101	13000	W 24 × 76	14452	14500
3	5,6	W410 × 54	6810	W 18 × 40	7613	7000
4	11,12	WWF700 × 141	18100	W 24 × 84	15935	14500

<u>Ten-Storey Frame</u>, [15]

The ten-storey rigid frame shown in Figure 9(a) is to be designed accounting for large (second-order) displacement effects under the single loading scheme shown in Figure 9(b). For comparison purposes, a corresponding first-order (small displacement) design is also conducted. For both designs, member sizes are taken as continuous variables, in which case eqs. (18e) are replaced by eqs. (19) in the weight optimization problem. The limits put on the sizing variable for each member i are taken to be $\check{a}_i$ = 2850 mm² and $\hat{a}_i$ = 57000 mm² (which represent the section areas of the smallest and largest W-shapes available in Canada). All members are specified to have the same steel grade as for the previous two-storey frame example design case 1,

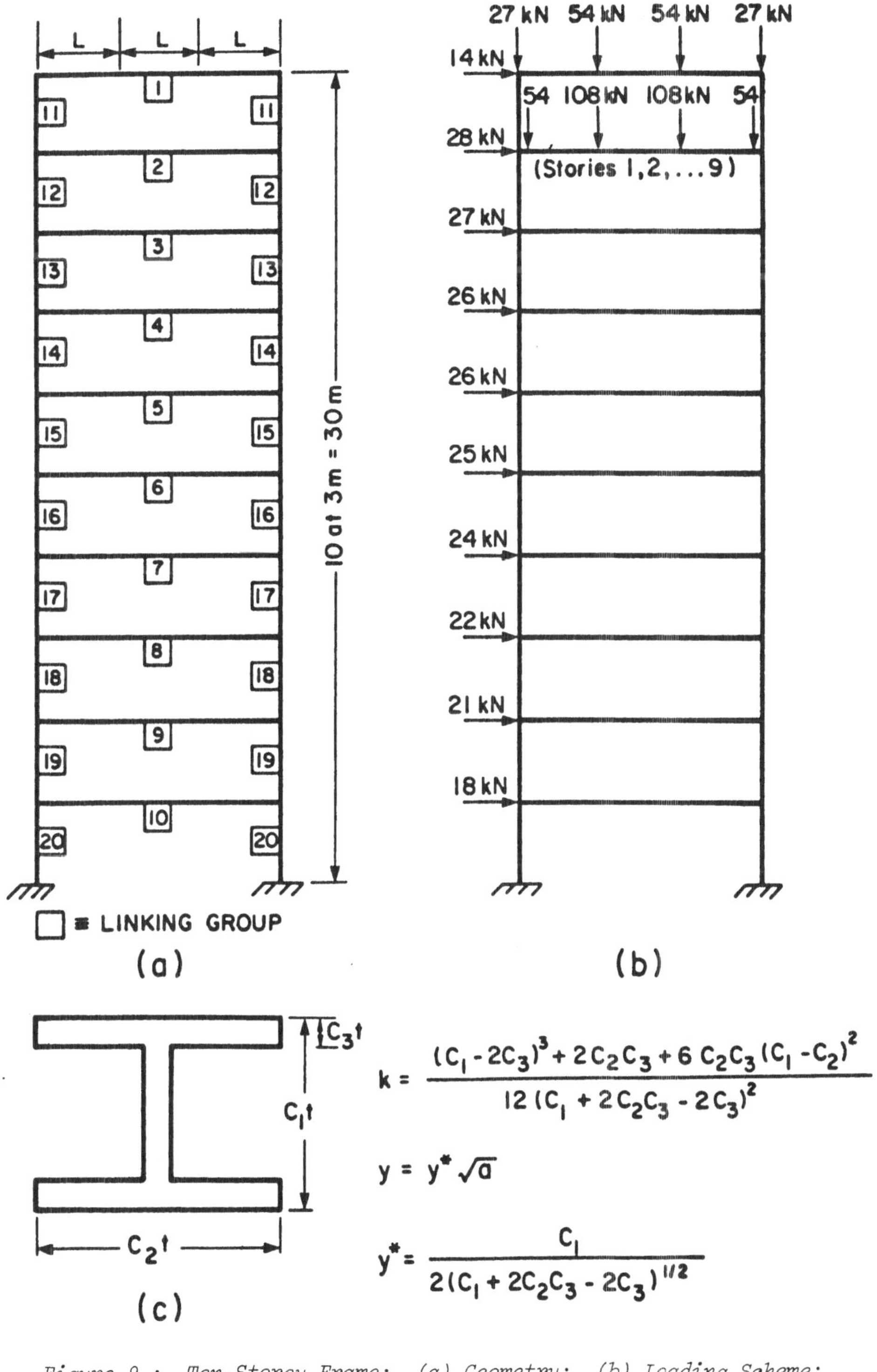

$$k = \frac{(C_1 - 2C_3)^3 + 2C_2C_3 + 6\,C_2C_3\,(C_1 - C_2)^2}{12\,(C_1 + 2C_2C_3 - 2C_3)^2}$$

$$y = y^* \sqrt{a}$$

$$y^* = \frac{C_1}{2(C_1 + 2C_2C_3 - 2C_3)^{1/2}}$$

Figure 9 : Ten-Storey Frame: (a) Geometry; (b) Loading Scheme; (c) Generic Cross-Section (Constant-Shape W-Section) as well as the 'constant-shape' wide-flange cross-section characteristics indicated in Figure 9(c).

To satisfy conventional fabrication requirements concerning member continuity and structure symmetry, the fifty elements of the frame (i.e., the 30 girder elements and 20 column elements defined by joints and load points in Figure 9(b)) are linked together into twenty independent sizing groups (denoted by □ in Figure 9(a)); i.e., the girder for each storey is required to be a prismatic member and the two columns for each storey are required to be identical prismatic members.

Stress and displacement performance conditions are imposed for both the first-order and second-order designs of the frame. In the latter case, the sensitivity coefficients d°_{ij} and s°_{ik} in eqs. (18b) and (18c) are calculated accounting for second-order (P-Δ) displacement effects. Horizontal sway displacement constraints are imposed at each storey, for a total of ten displacement conditions. The maximum horizontal sway permitted at any storey level is H/150, where H is the storey height measured from the supports (sway displacements of substantial magnitude have been arbitrarily permitted so that pronounced second-order effects will occur). Member stresses are constrained at all joints and load points, for a total of eighty stress conditions. Allowable stresses for the members are limited to the steel yield stress, i.e., 300 MPa (while not shown herein, it is readily possible to consider stress bounds that progressively change over the design history so as to account for the changing slenderness properties of the member cross-sections as the design evolves towards the minimum-weight structure, [8].

Upon applying the STRUSY program [2,5,8,15] to conduct the first-order design of the frame, the minimum-weight structure is found after six design stages requiring 49.3 seconds of CPU time (IBM 4341). The corresponding second-order design requires eight design stages and 70.0 seconds of CPU time. The results for the two designs are given in Table 2 and illustrated in Figure 10. Note from Figure 10 that the two designs are identical for the first three stages, after which the second-order design becomes heavier as a consequence of the large displacement (P-Δ) effects. Even then, the optimal weight shown in Figure 10 for the second-

Table 2 : Final Designs for Ten-Storey Frame

MEMBER	LINKING GROUP	FIRST-ORDER DESIGN		SECOND-ORDER DESIGN		
		AREA (mm^2)	MOMENT OF INERTIA ($\times 10^6$ mm^4)	AREA (mm^2)	MOMENT OF INERTIA ($\times 10^6$ mm^4)	% DIFFERENCE IN AREA w.r.t. FIRST-ORDER DESIGN
GIRDERS	1	3 667	59.1	3 694	59.9	0.74
	2	5 745	145	5 780	147	0.61
	3	6 511	194	6 595	199	1.29
	4	7 301	225	7 412	232	1.52
	5	7 903	264	8 433	300	6.71
	6	9 024	318	9 577	358	6.13
	7	9 679	366	10 190	405	5.28
	8	10 178	404	10 691	446	5.04
	9	11 061	432	11 950	505	8.04
	10	9 279	364	9 439	376	1.72
COLUMNS	11	4 722	39.8	4 744	40.1	0.47
	12	7 639	38.6	7 704	39.2	0.85
	13	8 076	75.4	8 168	77.2	1.14
	14	8 795	81.7	8 901	83.7	1.21
	15	10 157	104	10 327	107	1.67
	16	11 024	128	11 268	134	2.21
	17	12 262	146	12 572	153	2.53
	18	12 328	205	12 568	213	1.95
	19	13 474	244	13 668	251	1.44
	20	15 480	277	15 633	291	0.98

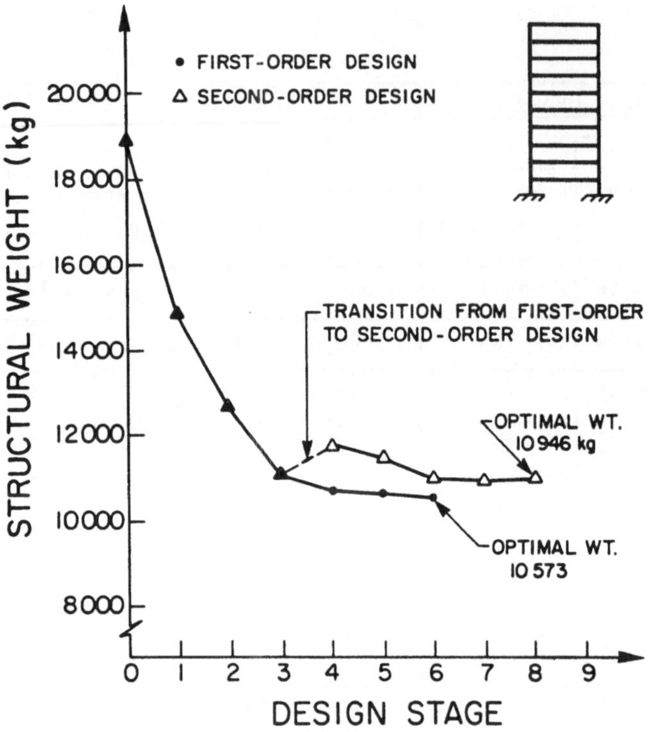

Figure 10 : Synthesis History for Ten-Storey Frame

order design represents only a 3.5% increase over that for the first-order design. This small weight increase is a direct consequence of the design synthesis optimization capability, which optimally redistributes the structural weight between the girders and columns to account for second-order effects. Such optimal weight redistribution is implied by the last column of Table 2, from which it can be determined that the total weight of the girders increased by 4.2% while that for the columns increased by only 1.5% (i.e., 76% of the additional structural weight required to accomodate the second-order effects is taken up by the girders).

ACKNOWLEDGEMENT

This study was sponsored by the Natural Sciences and Engineering Research Council of Canada under grant A5306.

REFERENCES

1. Grierson, D.E. and Schmit, L.A., "Synthesis under Service and Ultimate Performance Constraints", Journal of Computers and Structures, 15(4), pp. 405-417, 1982.
2. Chiu, T.C.W., "Structural Synthesis of Skeletal Frameworks under Service and Ultimate Performance Constraints", M.A.Sc. Thesis, University of Waterloo, Canada, 1982.
3. Grierson, D.E. and Chiu, T.C.W., "Synthesis of Frameworks under Multilevel Performance Constraints", NASA Conference Proceeding 2245, 1982.
4. Grierson, D.E. and Chiu, T.C.W., "Optimal Synthesis of Frameworks under Multilevel Performance Constraints", Journal of Computers and Structures, 18(5), pp. 889-898, 1984.
5. Lee, W.H., "Optimal Structural Synthesis of Frameworks using Discrete and Commercially Available Standard Sections", M.A.Sc. Thesis, University of Waterloo, Canada, 1983.
6. Grierson, D.E. and Lee, W.H., "Discrete Optimization of Frameworks under Elastic and Plastic Performance Constraints", Proceedings of the Michael R. Horne Conference on Instability and Plastic Collapse of Steel Structures", University of Manchester, England, pp. 130-139, 1983.
7. Grierson, D.E. and Lee, W.H., "Optimal Synthesis of Steel Frameworks using Standard Sections", Journal of Structural Mechanics, 12(3), pp. 335-370, 1984.
8. Cameron, G.E., "Optimal Synthesis of Steel Frameworks Accounting for all Building Code Requirements and using Available Sections", M.A.Sc. Thesis, University of Waterloo, Canada, 1984.
9. Grierson, D.E. and Cameron, G.E., "Computer-Automated Synthesis of Building Frameworks", Canadian Journal of Civil Engineering, 12(4), pp. 863-874, 1984.
10. Canadian Institute of Steel Construction (CISC), *Handbook of Steel Construction*, Third Edition, 1980.
11. American Institute of Steel Construction (AISC), *Manual of Steel Construction*, Eighth Edition, 1980.
12. Canadian Standards Association, CSA Standard CAN3-S16.1-M78, "Steel Structures for Buildings - Limit States Design", Rexdale, Canada, 1978.
13. AISC Specifications for the Design, Fabrication and Erection of Structural Steel for Buildings, November, 1978.
14. Proposed "Load and Resistance Factor Design (LRFD) Specification for Structural Steel Buildings", AISC, 1983.
15. Hall, S.K., "Synthesis of Frameworks under Large Displacements", M.A.Sc. Thesis, University of Waterloo, Canada, 1986.
16. Grierson, D.E. and Hall, S.K., "Automated Synthesis of Frameworks under Large Displacements", First World Congress on Computational Mechanics, The University of Texas at Austin, September, 1986.
17. Kramer, G.J.E., "Synthesis of Frameworks Accounting for Stresses and Displacements due to Dynamic Loading", M.A.Sc. Thesis, University of Waterloo, Canada, in progress, 1986.
18. Grierson, D.E. and Kramer, G.J.E., "Optimal Synthesis of Frameworks under Dynamic Loads", 10th CANCAM, University of Western Ontario, Canada, June, 1985.

19. Fleury, C. and Sander, G., "Structural Optimization by Finite Elements", LTAS Report SA-58, University of Liege, Belgium, 1978.
20. Schmit, L.A., Jr. and Miura, H., "An Advanced Structural Analysis/Synthesis Capability - ACCESS 2", International Journal of Numerical Methods in Engineering, 12, pp. 353-377, 1978.
21. Franchi, A., "STRUPL-ANALYSIS: Fundamentals for a General Software System", Ph.D. Thesis, Univ. of Waterloo, Canada, 1977.
22. Fleury, C., "Structural Weight Optimization by Dual Methods of Convex Programming", Int. J. Num. Meth. in Engng., 14, pp. 1761-1783, 1979.

OPTIMUM CONTROL OF STRUCTURES

Raphael T. Haftka
Virginia Polytechnic Institute and State University
Blacksburg, Virginia 24061 USA

Introduction

In the past decade the interest in using active control systems to improve structural performance has increased dramatically. The basic idea is to build structures which use an active control system as a substitute for strength, stiffness or damping. An early application of the concept was the gust-alleviation system designed to prolong the fatigue life of the aging fleet of B-52 bombers. In that application, sensors were used to anticipate gust loads on the wing, and control surface on the wing were deflected to cancel part of these gust loads. Today, these types of systems are already considered in the preliminary design stage (e.g. Ref. 1). The same idea may be used to improve the ride quality in an airplane. Other aeronautical applications such as active flutter suppression are discussed in Ref. 2.

There has been also a lot of interest in active control of buildings (e.g. Ref. 3) subjected to seismic excitation, and in active control of automotive suspensions. However, at present it seems that the major challenge in the field of active control of structures is the design of control systems for very large space structures. Because of the high cost of lifting mass to orbit, there is a great incentive to make these structures light (and therefore flexible). On the other hand, many of these structures, especially antenna structures, have very stringent requirements on their shape accuracy. This combination of light weight and required shape accuracy is expected to necessitate the extensive use of active control systems. The perceived challenge of the problem has led to intensive research and many publications devoted to active control of large space structures.

The first part of this paper presents the theory of a popular method for optimizing the control system for a given structure. Problems associated with structural modelling are also discussed. The second part of the paper is concerned with current attempts to integrate the design of the structure with the design of the control system. The major issue in

NATO ASI Series, Vol. F27
Computer Aided Optimal Design: Structural and Mechanical Systems. Edited by C. A. Mota Soares

this context seems to be the formulation of the combined problems in terms of objective function and constraints.

Linear Quadratic Optimal Control of Structures

The equations of motion of the structure are usually written in a descretized form as

$$M\ddot{q} + C\dot{q} + Kq = L \tag{1}$$

where M, C and K are the mass, damping and stiffness matrices, respectively, L is the load vector, q is the structural response vector and a dot denotes differentiation with respect to time. Equation (1) can be reduced to a more general first order form

$$\dot{x} = Ax + f \tag{2}$$

One way of accomplishing this reduction to a first order system is to define $x^T = \lfloor q, \dot{q} \rfloor$ and then

$$A = \begin{bmatrix} 0 & I \\ -M^{-1}C & -M^{-1}K \end{bmatrix} \qquad f = \begin{Bmatrix} 0 \\ M^{-1}L \end{Bmatrix} \tag{3}$$

where I denotes the unit matrix. In the consideration of the control problem we limit ourselves to linear control so that f is given as

$$f = Bu + w \tag{4}$$

where u is the control vector, and w is a vector of noise in the control commands assumed to be zero-mean, Gaussian stationary process. The control law is assumed to be linear and time-invariant, that is

$$u = -Gx \tag{5}$$

where G is a matrix of constant gains which is to be found by solving an optimal control problem. The solution for the matrix G is fairly simple if we define the objective function to be minimized as

$$J = E(x^T Qx + u^T Ru) \tag{6}$$

where E is the expected-value operator and Q and R are positive definite matrices. J is usually referred to as the quadratic performance index and the terms including the matrices Q and R measure the magnitude of the response and control effort, respectively. The matrices Q and R may be selected to make these measures of response and control effort physically meaningful. For example, if the matrix Q is selected to be

$$Q = \frac{1}{2}\begin{bmatrix} K & 0 \\ 0 & M \end{bmatrix} \tag{7}$$

then with $x^T = [q,\dot{q}]$, the first term in Eq. (6) is the expected value of the total energy (kinetic plus elastic) in the system. The gain matrix G which minimizes the quadratic performance index is given as (see Ref. 4 for additional details)

$$G = R^{-1}B^TS \tag{8}$$

where S is the solution of a matrix Ricatti equation

$$A^TS + SA - SBR^{-1}B^TS + Q = 0 \tag{9}$$

Several software packages are available for the solution of matrix Ricatti equations (e.g. Ref. 5).

Equation (5) embodies the unrealistic assumption that we can measure the entire response vector (called also "state" vector, in control jargon). Instead we should expect to measure a smaller vector z given as

$$z = Mx + v \tag{10}$$

where v is measurement noise. From z we need to estimate x, and one popular approach to estimating x is called the Kalman filter. If the measurement noise is also a stationary, zero-mean Gaussian process then the optimal estimate $\hat{x}$ is found by solving

$$\dot{\hat{x}} = Ax + Bu + F(z-M\hat{x}) \tag{11}$$

where F is the filter gain matrix and is given by

$$F = TMV^{-1} \tag{12}$$

where V is the covariance matrix (or intensity) of the noise v and T is found from the solution of another Ricatti equation

$$AT + TA^T - TM^TV^{-1}MT + W \tag{13}$$

and W is the covariance matrix of w. The matrices V and W play an analogous role to the matrices Q and R. They are usually selected not on the basis of known physical estimates of the noise, but so as to generate an observer with desired properties. Similarly, the matrices Q and R that design the control gains are often selected to produce desirable properties of the controller such as sufficient stability margins.

The major difficulty associated with the use of the above approach to structural control is that it results in a high order controller, (same order as the order of x), which is difficult to implement in hardware. Therefore, control specialists typically seek a reduced-order model of the structure, usually one based on a small number of vibration modes. Sophisticated schemes have been developed to select the vibration modes which should be used in the reduced model (e.g. Ref. 6). With a reduced structural model there is always the danger that the higher-order modes will be excited by the controller and will become unstable, a phenomenon called spillover (Ref. 7). At present the issue of how to obtain a satisfactory low-order model of the structure seems to be the only point of interaction between the structural analyst and the control specialist.

Combined Control-Structure Design

The need for active control systems is rooted in the requirement for light-weight structures and the attendant flexibility. However, control systems and their power supplies also have mass, and therefore it makes sense to optimize the total mass of the system, designing both the structure and control system simultaneously.

One of the obstacles to simultaneous structural-control design is that very few control specialists are also familiar with structural design, and very few structural analysts are familiar with control design. For this reason, some researchers preferred to attack the simultaneous design problem indirectly. For example, some structural analysts (e.g. Refs. 8, 9) attempted to enhance the performance of the control system by changing the stiffness properties of the structure. The control system and the structure were still designed separately, but the structural optimization was governed by control-related stiffness requirements. Similarly, it is possible (see Ref. 10) to use the design of the control system to infer about required structural changes, because some of the gain matrices obtained in control design are equivalent to modifications in the structural stiffness matrix.

More truly integrated structure-control design procedures were proposed which operate simultaneously on a set of structural and control design variables. The mass of the structure can be combined with the quadratic performance index used in optimum control to obtain a composite objective function (e.g. Refs. 11, 12), or the quadratic performance index is optimized for a given structural mass (Ref. 13). Alternatively, the structure and control system are optimized with constraints placed on the eigenvalues of the closed-loop system (e.g. Refs. 14, 15).

The present treatment of the combined design problem is based on Ref. 16. It is based on the assumption that the mass m_c of the control system and its power supply depends on the control effort c_E as

$$m_c = \alpha c_E^{\beta} \tag{14}$$

where α and β are constants and

$$c_E = E(u^T R u) \tag{15}$$

The objective function to be minimized is the total mass, m_T

$$m_T + m_s + m_c \tag{16}$$

where m_s is the mass of the structure. A constraint is placed on the magnitude of the response as

$$r = E(x^T Q x) \leq \sigma_1 \tag{17}$$

Minimizing m_T subject to the constraint on r makes physical sense, but it does not allow us to use the easy solution (via Ricatti equation) of the linear-quadratic optimal control methodology. So instead of solving the optimization problem by varying the control and structural parameters simultaneously we define a nested problem which requires the solution of an optimal control problem for each set of structural parameters as follows

find structural parameters
and k to minimize $m_T = m_s + m_c$
such that $r \leq \sigma_1$ (18)

where the control system is designed by minimizing

$$r + kc_E \tag{19}$$

The additional parameter k allows us to control the value of the structural response r because obviously r is a montonically increasing function of k (the higher the cost of control effort the higher the optimized response). We need to show that the reformulated nested problem has the same optimal solution as the original problem.

Clearly the optimum of the nested problem (18) cannot be better than the original optimum because it satisfies the constraint (17), so we need to show that the nested problem does not yield a larger mass. Given the optimum of the original problem we take the optimum structure and design a k and a control system for it to minimize $r + kc_E$ such that $r = \sigma_1$. The value of the c_E that we obtain must be lower or equal to the optimal c_E obtained in the original problem. Otherwise the control system obtained in the original optimization produces a better value of $r + kc_E$, which is a contradiction.

Beam Example

Reference 16 presented a free-free beam (see Fig. 1) example under white-noise disturbance. The beam supports a uniform payload and is controlled by two attitude-control units with torque actuators symmetrically installed on the structure. The beam was modeled by finite elements and its cross-sectional thickness used as structural design variables as well as the actuator location. Five elements were used to model half of the beam. In the results presented here it is assumed that the control mass m_c is proportional to the control effort c_E (i.e., β = 1), that the beam stiffness is proportional to the thickness, and that the payload is equal to the structural mass. Even though only two actuators are used, it was assumed that the entire state vector is sensed so that there is no need for a Kalman filter.

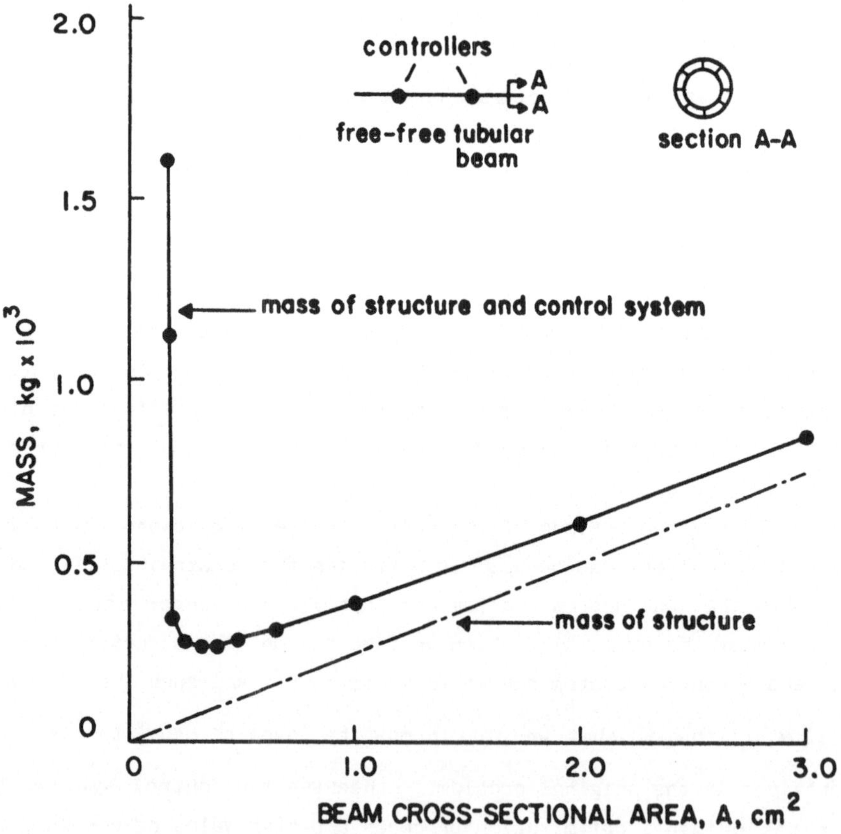

Fig. 1: Combined structure/control optimization of beam structure

Table 1 compares four cases with varying response requirements (σ_1) and control cost (α). As can be seen from the table, the average thickness (proportional to m_s) is very sensitive to the mass cost of the control. When that cost is reduced by a factor of 10 the mass is reduced by a factor of 2-3. Also, both control effort and mass are sensitive to the required response limit σ_1. It thus appears to be quite important to do the combined optimization. Another demonstration of this importance is shown in Fig. 1 which shows the total mass as a function of the structural mass for a uniform beam. The figure indicates that even with an optimal control system the control effort is sensitive to the structural stiffness. In fact if the structure is made too flexible the total mass becomes very high.

Table 1

Effect of control cost and response limit on combined optimum design for controlled beam

response limit	σ_1	0.01	0.001	0.01	0.001
control cost	α	1.0	1.0	0.1	0.1
normalized beam thickness	$\xi 1$	0.262	0.822	0.120	0.342
	$\xi 2$	0.620	1.785	0.184	0.506
	$\xi 3$	0.811	2.23	0.156	0.408
	$\xi 4$	0.639	1.14	0.159	0.430
	$\xi 5$	0.114	0.713	0.242	1.27
actuator location	$\bar{x}_c$	0.786	0.622	0.386	0.391
average thickness	$\bar{\xi}$	0.489	1.338	0.172	0.591
total normalized mass	m_T	1.063	2.179	0.268	0.603

References

1. Gould, J.D., "Effect of Active Control System Nonlinearities on the L=1011-3(ACS) Design Gust Loads", AIAA Paper 85-0755, presented at the AIAA/ASME/ASCE/AHS Structures, Structural Dynamics and Material Conference, Orlando, Florida, April 1985.
2. Hitch, H.P.Y., "Active Controls for Civil Aircraft", Aeronautical Journal, October 1979, pp. 389-398.
3. Meirovitch, L. and Silverberg, L.M., "Control of Structures Subjected to Seismic Excitation", ASCE Journal of Engineering Mechanics, Vol. 109, No. 2, pp. 604-618, April 1983.
4. Kwakernaak, H. and Sivan, R., "Linear Optimal Control Systems", Wiley-Interscience, 1972.
5. Bingulac, S.P., "Recent Advances in the L-A-S Software used in CAD of Control Systems", IFAC Symposium on CAD, Copenhagen, July 1985.
6. Skelton, R.E. and Hughes, P.C., "Model Cost Analysis for Linear Matrix-Second-Order Systems", Journal of Dynamic Systems Measurement and Control, Vol. 102, September 1980.
7. Balas, M.J., "Active Control of Flexible Systems", Journal of Optimization Theory and Applications, Vol. 25, No. 3, pp. 415-436, July 1978.
8. Khot, N.S., Venkayya, V.B. and Eastep, F.E., "Structural Modifications to Reduce the LOS-Error in Large Space Structures", AIAA Paper 84-099TCP, presented at the AIAA/ASME/ASCE/AHS 25th, Structures, Structural Dynamics and Material Conference, Palm Springs, California, May 1984.
9. Venkayya, V.B. and Tischler, V.A., "Frequency Control and its Effects on the Dynamic Response of Flexible Structures", AIAA Journal, Vol. 22, No. 9, pp. 1293-1298, 1984.
10. Hanks, B.R. and Skelton, R.E., "Designing Structures for Reduced Response by Modern Control Theory", Paper presented at the AIAA/ASME/ASCE/AHS 24th Structures, Structural Dynamics and Material Conference, Lake Tahoe, Nevada, May 1983.
11. Salama, M., Hamidi, M. and Demestz, L., "Optimization of Controlled Structures", Proceedings of the JPL Workshop on Identification and Control of Flexible Space Structures, JPL Publication 85-29, Vol. II, April 1985, pp. 311-327.
12. Hale, A.L., Lisowski, R.J. and Dahl, W.G., "Optimum Simultaneous Structural and Control Design of Maneuvering Flexible Spacecraft", Journal of Guidance and Control, Vol. 8, No. 1, pp. 86-93, 1985.
13. Messac, A., Turner, J. and Soosaar, K., "An Integrated Control and Minimum Mass Structural Optimization Algorithm for Large Space Structures", Proceedings of the JPL Workshop on Identification and Control of Flexible Space Structures, JPL Publication 85-29, Vol. II, April 1985, pp. 231-236.
14. Bodden, D.S. and Junkins, J.L., "Eigenvalue Optimization Algorithms for Structure/Controller Design Iterations", Journal of Guidance and Control, Vol. 8, No. 6, pp. 697-706, 1985.
15. Haftka, R.T., Martinovic, Z.N. and Hallauer, W.L., "Enhanced Vibration Controllability by Minor Structural Modification", AIAA Journal, Vol. 23, No. 8, pp. 1260-1266, 1985.
16. Onoda, T. and Haftka, R.T., "An Approach to Structure-Control Simultaneous Optimization for Large Flexible Spacecraft", to be published.

MINIMUM WEIGHT AND OPTIMAL CONTROL DESIGN OF SPACE STRUCTURES

N. S. Khot
Structures and Dynamics Division
Flight Dynamics Laboratory, AFWAL/FIBRA
Wright-Patterson Air Force Base, Ohio 45433-6553, USA

ABSTRACT

Algorithms are presented to design a minimum weight structure and to improve the dynamic response of a closed-loop control system. Constraints are imposed either on the structural response quantities or on the complex eigenvalue distribution of the closed-loop system. Use of the algorithms is illustrated by solving different problems.

I. INTRODUCTION

The design requirements on space structures are stringent, because of the environment in which they are used and the methods used to launch them. A space structure has to be minimum weight and also satisfy all the design requirements. One of the important design requirements is to minimize the line-of-sight (LOS) error.

The optimum design of space structures has recently been investigated by using two approaches. In the first approach, an optimum structure is initially designed to satisfy constraints on weight, displacements, frequency distribution etc., and then an optimum control system is designed to improve the dynamic response of the structure to initial disturbances[1]. An integrated approach using finite element analysis with weight as the objective function and constraints on the damping parameters and the eigenvalue distribution of the closed-loop system is proposed in Refs. 2-3. In Ref. 4, a structure/control optimization problem has been formulated with weight as the objective function and constraints on the closed-loop eigenvalue distribution and the minimum Frobenious norm of the required control gains. This paper contains a summary of some of the results obtained by the author by using both approaches in structure and control optimization.

NATO ASI Series, Vol. F27
Computer Aided Optimal Design: Structural and Mechanical Systems. Edited by C. A. Mota Soares

II. BASIC EQUATIONS OF ANALYSIS

The equations of motion for a Large Space Structure with no external disturbance are given by

$$[\mathrm{M}]\{\ddot{\mathrm{U}}\} + [\mathrm{E}]\{\dot{\mathrm{U}}\} + [\mathrm{K}]\{\mathrm{U}\} = [\mathrm{D}]\{\mathrm{F}\} \tag{1}$$

where [M] is the mass matrix, [E] is the damping matrix, and [K] is the total stiffness matrix. These matrices are nxn where n is the number of degrees of freedom of the structure. In Eq. 1, [D] is the nxp applied load distribution matrix relating the control input vector $\{F\}$ to the coordinate system. The number of elements in $\{F\}$ is assumed to be p. $\{U\}$ in Eq. 1 is a vector defining the amplitudes of motion. Using the coordinate transformation,

$$\{\mathrm{U}\} = [\phi]\{\eta\} \tag{2}$$

where $[\phi]$ is the modal matrix whose columns are eigenvectors $\{U\}_j$ (normalized with respect to the mass matrix [M]), Eq. 1 can be transformed into an n uncoupled system of differential equations as

$$[\overline{\mathrm{M}}]\{\ddot{\eta}\} + [\overline{\mathrm{E}}]\{\dot{\eta}\} + [\overline{\mathrm{K}}]\{\eta\} = [\phi]^{\mathrm{T}}[\mathrm{D}]\{\mathrm{F}\} \tag{3}$$

where

$$[\overline{\mathrm{M}}] = \begin{pmatrix} \nwarrow & & \\ & \mathrm{I} & \\ & & \searrow \end{pmatrix}; \quad [\overline{\mathrm{E}}] = \begin{pmatrix} \nwarrow & & \\ & 2\varsigma\omega & \\ & & \searrow \end{pmatrix}; \quad [\overline{\mathrm{K}}] = \begin{pmatrix} \nwarrow & & \\ & \omega^2 & \\ & & \searrow \end{pmatrix} \tag{4}$$

In Eq. 4, $\{\varsigma\}$ and $\{\omega\}$ are the vectors of modal damping factors and circular frequencies respectively. The second order Eq. 3 can be reduced to a first order equation by using the state variable vector $\{x\}$ given by

$$\{x\}_{2n} = \frac{\eta}{\dot{\eta}} \tag{5}$$

The state space representation of Eq. 3 using Eq. 5 can be written as

$$\{\dot{x}\} = [\mathrm{A}]\{\mathrm{x}\} + [\mathrm{B}]\{\mathrm{f}\} \tag{6}$$

where [A] is a $2n \times 2n$ plant matrix, [B] is $2n \times p$ input matrix and $\{f\}$ is an $p \times 2n$ control input vector. The plant matrix and the input matrix are of the form

$$[\mathrm{A}] = \left(\begin{array}{ccc|ccc} \nwarrow & & & \nwarrow & & \\ & 0 & & & \mathrm{I} & \\ & & \searrow & & & \searrow \\ \hline \nwarrow & & & \nwarrow & & \\ & -\omega^2 & & & -2\varsigma\omega & \\ & & \searrow & & & \searrow \end{array}\right) \quad [\mathrm{B}] = \left(\frac{0}{\phi^{\mathrm{T}}\mathrm{D}}\right) \tag{7}$$

Eq. 6 is known as the state input equation and the state output equation is given by

$$\{y\} = [\ c\ ]\{x\} \tag{8}$$

where $\{y\}$ is a $q \times 1$ output vector and [C] is a $q \times 2n$ output matrix. If the sensors and the actuators are colocated, then q=p and $[c] = [B]^T$. In order to design a controller using a linear quadratic regulator, a performance index (PI) can be defined as

$$\mathrm{PI} = \int_0^t \left(\{x\}^{\mathrm{T}}[\mathrm{Q}]\{\mathrm{x}\} + \{\mathrm{f}\}^{\mathrm{T}}[\mathrm{R}]\{\mathrm{f}\}\right)\, \mathrm{dt} \tag{9}$$

where [Q] is the state weighting matrix which has to be positive semi-definite, and [R] is the control weighting matrix which has to be positive definite. The result of minimizing the quadratic performance index and satisfying the state equation gives the state feedback control law

$$\{f\} = -[\tilde{\mathrm{G}}]\{x\} \tag{10}$$

where $[\tilde{G}]$ is the optimum gain matrix given by

$$[\tilde{\mathrm{G}}] = [\mathrm{R}]^{-1}[\mathrm{B}]^{\mathrm{T}}[\tilde{\mathrm{P}}] \tag{11}$$

and $[\tilde{P}]$ is a symmetric positive definite matrix called the Riccati matrix. Substituting Eq. 10 in Eq. 6 gives the governing equations for the optimal closed loop system in the form

$$\{\dot{x}\} = [\overline{\mathrm{A}}]\{x\} \tag{12}$$

where

$$[\overline{\mathrm{A}}] = [\mathrm{A}] - [\mathrm{B}][\tilde{\mathrm{G}}] \tag{13}$$

For an initial given condition x(o) the solution to Eq. 12 with no forcing function, is given by

$$x(t) = \mathrm{e}^{[\overline{\mathrm{A}}]t}x(0) \tag{14}$$

The eigenvalues of the closed-loop matrix, $[\tilde{A}]$, are a set of complex conjugate pairs written as

$$\lambda_i = \tilde{\sigma}_i \pm j\tilde{\omega}_i \tag{15}$$

The damping factors ξ_i and the damped frequencies $\tilde{\omega}_i$ are related to the complex eigenvalues through

$$\xi_i = -\frac{\tilde{\sigma}_i}{(\tilde{\sigma}_i^2 + \tilde{\omega}_i^2)^{1/2}} \tag{16}$$

III. DESCRIPTION OF THE NOMINAL STRUCTURAL MODEL

The structural model selected for the investigation is a tetrahedral truss (Fig. 1).[5] This structure, ACOSS-FOUR, in spite of its simplicity, models the feed tower in a generic class of large antenna applications. The apex of the structure, node 1, represents the antenna feed, and its motion in the X-Y plane has to be actively controlled to improve the performance of the structure's line-of-sight (LOS). The structure has twelve degrees of freedom and four masses of 2 units each are attached at nodes 1 through 4. The coordinates of the node points are given in Table 1. The actuators and sensors are collocated in six bipods. The dimensions of the structure are given in unspecified consistent units. The elastic modulus of the members was assumed to be 1.0, and the density of the structural material was assumed to be 0.001. The square of the LOS is equal to $(\text{LOS-X})^2 + (\text{LOS-Y})^2$ where LOS-X and LOS-Y are the components of the LOS in the X and Y direction. The nominal design is denoted by Design A with cross-sectional areas of the members equal to those assigned by the Charles Stark Draper Lab.[5] This design was used as the initial design for the optimization prolems and was considered a basic design for comparison with the optimum designs. Comparison of the different designs was made by comparing weight, structural frequencies, closed-loop eigenvalues, damping parameters and the dynamic response to an initial disturbance. The weighting matrices [Q] and [R] in Eq. 9 were assumed to be identity matrices for all the designs and the elements of the vector of modal damping factor $\{\varsigma\}$ in Eq. 7 was set to zero.

Using Design A as the initial design, four optimum designs with constraints as given below were obtained.

Design B: Minimum static displacements associated with the LOS and the weight is the same as Design A.

Design C: Minimum weight with constraints on the structural frequencies.

Design D: Minimum weight with constraints on the damping parameters of the closed-loop system.

Design E: Minimum weight with constraints on the closed-loop frequency distribution and the damping parameters.

The cross-sectional areas of the members for all the designs are given in Table 2. Table 3 contains the square of the structural frequencies. The closed-loop eigenvalues and the damping parameters for all the designs are shown in Tables 4 and 5, respectively. In order to compare the dynamic response, all the designs were subjected to the same initial condition. A unit displacement was imposed at node 2 in the X direction at $t = o$. The transient response was simulated by finding the solution to Eq. 16 for the period $t = o$ to $t = 25$ seconds at the time interval of $t = 0.05$ seconds. The magnitude of the LOS and the integrand of the performance index, PI, in Eq. 9 was calculated at each time interval. The time histories for all the designs are shown in Fig. 2.

IV. DESIGN WITH MINIMUM STATIC DISPLACEMENTS ASSOCIATED WITH THE LOS ERROR (DESIGN B)

This design was obtained by using optimality criterion approach to minimize $(LOS)^2$ while keeping the weight of the structure the same as Design A.[1] The nonstructural masses attached at nodes 1 through 4 were assumed to be the static loads acting in the three coordinate directions. In addition to this, a load of 29.695 units was applied at node 2 in the X direction to simulate a unit displacement which was used to initiate the dynamic response of the structure.

V. MINIMUM WEIGHT DESIGN WITH CONSTRAINTS ON STRUCTURAL FREQUENCIES (DESIGN C)

This design was obtained by using optimality criterion approach with constraints on the square of the first two frequencies, ω_1^2 and ω_2^2. It was specified that the values for these frequencies should be the same as that for Design A, *i.e.*, $\omega_1^2 = 1.80$ and $\omega_2^2 = 2.77$. The initial weight was 43.7 units and the optimum design weighed 15.22. The optimum design was obtained with twenty iterations.

VI. MINIMUM WEIGHT DESIGN WITH CONSTRAINTS ON THE EIGENVALUES AND DAMPING FACTORS OF THE CLOSED-LOOP SYSTEM (DESIGNS D AND E)

The optimization problem for the integrated structure/control design can be stated as:

find $\{A\}$

$$\textit{to minimize the weight } \mathrm{W} = \sum_{i=1}^{m} \rho_i \mathrm{A}_i l_i \tag{17}$$

$$\textit{subject to:} \quad g_j(\xi_i) = \xi_i - \bar{\xi}_i = 0 \tag{18}$$

$$g_j(\tilde{\omega}_i) = \tilde{\omega}_i - \bar{\tilde{\omega}}_i = 0 \tag{19}$$

where $g_j(\xi_i)$ and $g_j(\tilde{\omega}_i)$ represent equality constraints on the damping factors and on the imaginary part of the closed-loop eigenvalues. In Eqs. 18 and 19 $\bar{\xi}_i$ and $\bar{\tilde{\omega}}_i$ are desired values of the damping parameters and imaginary parts of the eigenvalues. The optimization problem was solved by using the VMCON optimization subroutine which is based on Powell's algorithm for nonlinear constraints that uses Lagrangian functions. This subroutine needs the sensitivities of the objective function and the constraints with respect to the design variables. The sensitivity of the objective function, W, can be explicitly written by differentiating Eq. 17 with respect to A_i. The sensitivity of the closed-loop eigenvalues and the damping parameters can be obtained by using finite differences [2] or using analytical approach[3]. The sensitivity of the closed-loop eigenvalues with respect to the design variable A_l is given by

$$\frac{\partial \lambda}{\partial \mathrm{A}_l} = \{\beta\}_i^{\mathrm{T}} [\bar{\mathrm{A}}]_{,l} \{\alpha\}_i \tag{20}$$

where $\{\alpha\}_i$ and $\{\beta\}_i$ are the right-hand and left-hand eigenvectors of the closed-loop plant matrix $[\bar{A}]$. The gradient of $[\bar{A}]$ with respect to the design variable A, can be written by differentiating the right side of Eq. 13.[3]

Design D was obtained with the constraint that the damping parameter ξ_1 of the closed-loop system be increased to 0.3 from 0.05464 of the initial design (Design A). The weight of the optimized structure was 47.46 which was higher than that of the initial design. The final design was obtained after eleven iterations. A finite-difference scheme was used to determine the gradients which was very time consuming.

The optimum Design E was obtained by imposing the following constraints on the closed-loop frequencies and the damping parameters:

1) $\tilde{\omega}_1$ must be equal to 1.341 *i.e.*, the same as that for Design A.

2) $\tilde{\omega}_2$ must be equal or greater than 1.5.

3) The damping parameters $\bar{\xi}_i$, associated with the four lowest frequencies of the closed-loop system, must be equal to 0.1093 *i.e.*, 100% more than that for Design A for the first mode.

The optimum Design E weighed 24.01 units. For this problem it was necersary to impose equality constraints on the damping parameters in steps in order to avoid divergence in the iterative procedure. Initially, equality constraints were imposed only on $\tilde{\omega}_1$ and $\tilde{\xi}_1$. After obtaining a design satisfying these constraints, constraints on the other three damping parameters were imposed. The first step needed about 29 iterations, and an additional 60 iterations were required to obtain the minimum weight design. The problem of slow convergence may be due to the VMCON subroutine.

VII. COMPARISON OF ACOSS-FOUR DESIGNS A THROUGH E

Some of the observations which can be made by comparing the different optimum designs were as follows: The lowest and highest values of ω_j^2 associated with the 1st mode and the 12th mode were for Design B. ω_1^2 for Designs C and E was nearly equal to that of Design A since it was constrained, while for Designs B and D it decreased. The structural frequency bands for Designs C and E were narrower than that for Design A. The damping parameter associated with the 1st mode of the closed-loop system was a maximum for Design B while a minimum for Design C. For Design C most of the damping parameters associated with the closed-loop system were smaller than those for the other designs. Comparing the dynamic response of all the designs in Fig. 2, it was seen that the amplitudes of the LOS for Designs B, D and E damped out quicker than that for Design A. But in the case of Design C the amplitudes of LOS were larger than that for Design A. Values of the total performance index, PI, over a period of 25 seconds for Deigns A, B, C, D and E were 764.4, 372.1, 125.8, 797.6 and 320.3 respectively.

VIII. MINIMUM WEIGHT DESIGN WITH CONSTRAINTS ON THE NORM OF THE GAIN MATRIX

In this optimization problem, instead of using the linear quadratic regulator to determine the gain matrix $[\tilde{G}]$, it was computed such that Eq. 12 satisfies the eigenvalue constraints, and the Frobenious norm of the gain matrix $S_G = [\tilde{G}]^T[\bar{R}][\tilde{G}]$ was minimized, where $[\bar{R}]$ is the control gain weighting matrix. The optimization problem can be stated as,(4)

$$\textit{minimize the weight } \mathrm{W} = \sum_{i=1}^{m} \rho_i \mathrm{A}_i l_i \tag{21}$$

$$\begin{aligned} \textit{subject to}: \quad & \mathrm{S_G} = \bar{\mathrm{S}}_\mathrm{G} \\ & \tilde{\sigma}_i = \bar{\tilde{\sigma}}_i \\ & \tilde{\omega}_i = \tilde{\omega}_i^* \end{aligned} \tag{22}$$

where S_G is the desired value of the gain norm and $\bar{\tilde{\sigma}}_i$, $\bar{\tilde{\omega}}_i^*$ are the desired values of the real and imaginary parts of the eigenvalues of the closed-loop matrix $[\tilde{A}]$. This is a nested optimization problem. The solution to the optimization problem needs sensitivities of the objective function and the constraint functions with respect to the elements of the gain matrix $[\tilde{G}]$ and the design variables A_i. The sensitivity of the gain norm, S_G, and the eigenvalues, λ_j, of the closed-loop system with respect to the elements of the gain matrix are given by

$$\frac{\partial S_G}{\partial g_{rs}} = \sum_{j=1} 2\bar{R}_{rj} G_{js} \tag{23}$$

$$\frac{\partial \lambda}{\partial g_{rs}} = \{\beta\}_j^T \{b_r\} \{\alpha\}_j \tag{24}$$

where $\{\alpha\}_j$ and $\{\beta\}_j$ are the right-hand and left-hand eigenvectors of the closed-loop matrix $[\bar{A}]$ and $\{b_r\}$ is the rth column vector of [B]. The sensitivity of the gain norm, S_G, with respect to the design variable, A_i, has to be obtained numerically, since closed-form solutions for control gains in terms of structural design variables are not available except for simple problems with single input. A finite difference scheme for the general approximation will have to be used.

In order to present the optimization concept discussed above, a simple two bar truss[4] shown in Fig. 3 was considered. A control force F(t) was located at the apex. A nonstructural mass of 2 units was placed at the vertex. The elastic modulus was E = 1 and the mass density was $\rho_i = 0.001$. A_i (min) was 10 units. The structure and control objective functions were

$$W = \rho_1 A_1 l_1 + \rho_2 A_2 l_2 \qquad S_G = \sum_{i=1}^{4} g_i^2 \tag{25}$$

with specified equality constraints on the closed-loop eigenvalues as

$$\tilde{\sigma}_1 = \bar{\tilde{\sigma}}_1 = -0.228 \; ; \qquad \tilde{\sigma}_2 = \bar{\tilde{\sigma}}_2 = -0.361$$

$$\tilde{\omega}_1 = \tilde{\omega}_1^* = 1.17 \; ; \qquad \tilde{\omega}_2 = \tilde{\omega}_2^* = 4.81 \tag{26}$$

and the minimum gain norm $S_G = S_G = 1500$. The optimization was done by using the mathematical optimization program, IDESIGN. The gradients of the objective function and the gain norm were expressed as formulas which was possible for this simple problem. The results of the optimization are given in Table 6. For this problem, a single optimal design was not obtained. Corresponding to five different initial designs, three optimum designs were obtained satisfying all the constraints. The existence of a multiple minimum for this case suggests that the problem is under constrained and additional design freedom exists. In order to obtain a unique design, it may be necessary to impose additional constraints on quantities such as the structural frequencies.

IX. SUMMARY

It has been shown that by changing the cross-sectional areas of the members of a structure, a minimum weight design and improvement in the dynamic characteristics of a closed-loop system can be achieved by using the tools of optimization. Two approaches were presented. In one case, the structure was optimized with constraints on the structural response, while in the other case the structural design and control design were integrated. In the second case, two approaches were proposed. In the first approach the objective function was the weight of the structure, and the constraints were on the eigenvalues and damping parameters of the linear quadratic regulator closed-loop system. The second approach consisted of minimizing the weight of the structure with constraints on the distribution of the closed-loop eigenvalues for a specified value of the Fobenious norm of the gain matrix. The feasibility of the integrated approaches was illustrated on sample problems.

X. REFERENCES

1. Khot, N. S., Venkayya, V. B. and Eastep, F. E., "Structural Modification of Large Flexible Structures to Improve Controllability," (84-1906) Proceedings of the AIAA Guidance and Control Conference, Seattle, WA, August, 1984.
2. Khot, N. S., Venkayya, V. B. and Eastep, F. E., "Optimal Structural Modifications to Enhance the Active vibration Control of Flexible Structure," AIAA J., Vol. 24, No. 8, 1986.
3. Khot, N. S., "Structure/Control Optimization to Improve the Dynamic Response of Space Structures," Presented at the International Conference on Computational Mechanics, Tokyo, Japan, May 25-29, 1986, and to be published in Computational Mechanics, an International Journal.
4. Khot, N. S., Oz, H., Eastep, F. E. and Venkayya, V. B., "Optimal Structural Designs to Modify the Vibration Control Gain Norm of Flexible Structures," (86-0840-CP) Presented at AIAA/ASME/ASCE/AHS 27th Structures, Structural Dynamics and Materials Conference, San Antonio, Texas, May 19-21, 1986.
5. Strunce, R. R. et. al., "ACOSS FOUR (Active Control of Space Structures) Theory Appendix," RADC-TR-80-78, Vol. II, 1980.

Table 1. Node Point Coordinates for ACOSS-FOUR

Node	X	Y	Z
1	0.0	0.0	10.165
2	-5.0	-2.887	2.00
3	5.0	-2.887	2.00
4	0.0	5.7735	2.00
5	-6.0	-1.1547	0.0
6	-4.0	-4.6188	0.0
7	4.0	-4.6188	0.0
8	6.0	-1.1547	0.0
9	-2.0	5.7735	0.0
10	2.0	5.7735	0.0

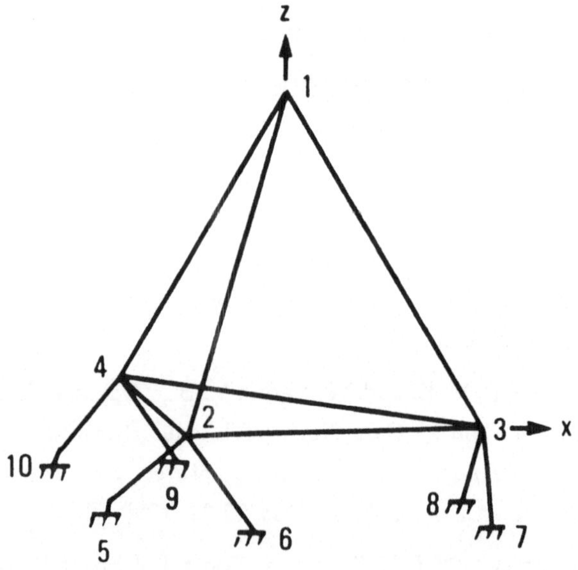

Fig. 1. ACOSS-FOUR Finite Element Model

Table 2. Cross-Sectional Areas of the Members of ACOSS-FOUR

Element No.	Design A	Design B	Design C	Design D	Design E
1(1-2)	1000.	642.1	124.70	998.22	477.82
2(2-3)	1000.	693.0	245.53	998.62	479.06
3(1-3)	100.	42.89	229.37	354.69	162.39
4(1-4)	100.	244.60	229.40	354.79	181.78
5(2-4)	1000.	812.40	245.49	998.62	479.48
6(3-4)	1000.	1142.00	134.72	998.22	484.08
7(2-5)	100.	891.1	197.05	14.74	131.44
8(2-6)	100.	10.0	197.07	14.73	62.47
9(3-7)	100.	10.0	229.95	15.34	75.65
10(3-8)	100.	978.9	125.49	46.65	125.18
11(4-9)	100.	10.0	229.93	15.34	41.28
12(4-10)	100.	902.8	125.55	46.74	45.91
Weight	43.69	43.69	15.22	47.56	24.01

Table 3. Natural Frequencies (ω_j^2) of ACOSS-FOUR

Mode	Design A	Design B	Design C	Design D	Design E
1	1.80	.4847	1.80	.7283	1.804
2	2.77	.9673	2.77	1.074	2.259
3	8.35	1.130	7.54	1.734	5.713
4	8.74	2.880	9.90	2.147	5.905
5	11.54	11.65	13.62	3.754	11.37
6	17.67	29.72	25.71	6.101	13.44
7	21.73	44.20	28.71	25.04	18.91
8	22.61	66.47	38.15	34.76	27.30
9	72.92	139.40	43.85	68.91	40.70
10	85.57	171.0	44.05	81.05	46.07
11	105.77	186.8	48.12	101.0	54.93
12	166.54	198.4	57.04	171.2	83.25

Table 4. The Closed-Loop Eigenvalues for ACOSS-FOUR

Design A		Design B		Design C		Design D		Design E	
Real Part	Imaginary Part	Real Part	Imaginary Part	Real Part	Imaginary Part	Real Part	Imaginary Part	Real Part	Imaginary Part
-0.073	± 1.34	-0.277	± 0.717	-0.053	± 1.34	-0.270	± 0.859	-0.147	± 1.34
-0.109	± 1.66	-0.335	± 0.992	-0.052	± 1.66	-0.286	± 1.040	-0.165	± 1.50
-0.213	± 2.88	-0.364	± 1.05	-0.080	± 2.75	-0.352	± 1.30	-0.263	± 2.39
-0.237	± 2.95	-0.287	± 1.67	-0.165	± 3.14	-0.375	± 1.45	-0.265	± 2.41
-0.285	± 3.39	-0.199	± 3.41	-0.145	± 3.70	-0.302	± 1.92	-0.267	± 3.37
-0.363	± 4.19	-0.163	± 5.41	-0.273	± 5.06	-0.337	± 2.45	-0.326	± 3.65
-0.355	± 4.65	-0.222	± 6.65	-0.342	± 5.34	-0.301	± 5.00	-0.315	± 4.34
-0.344	± 4.74	-0.207	± 8.15	-0.324	± 6.17	-0.278	± 5.89	-0.316	± 5.22
-0.292	± 8.53	-0.103	±11.80	-0.337	± 6.61	-0.244	± 8.30	-0.301	± 6.37
-0.276	± 9.25	-0.347	±13.10	-0.254	± 6.63	-0.245	± 9.00	-0.272	± 6.78
-0.214	±10.30	-0.348	±13.70	-0.355	± 6.93	-0.197	±10.00	-0.249	± 7.41
-0.083	±12.90	-0.342	±14.10	-0.320	± 7.20	-0.074	±13.10	-0.098	± 9.12

DESIGN A

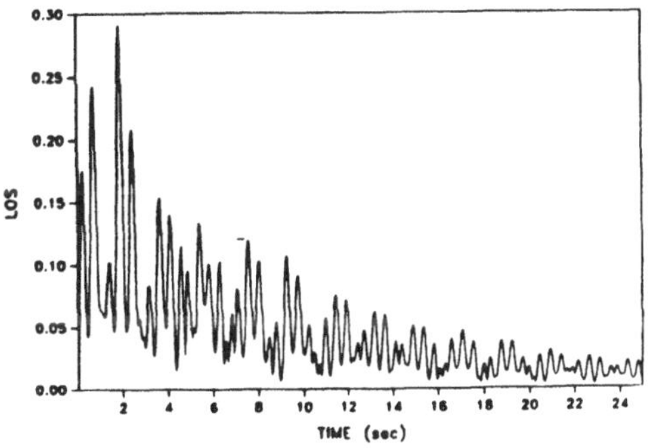

DESIGN B

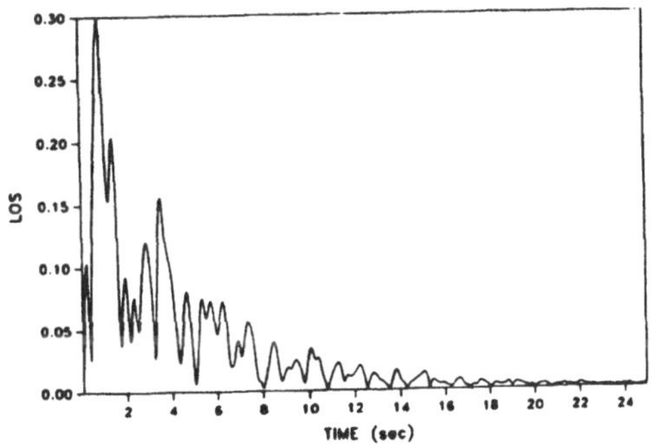

DESIGN C

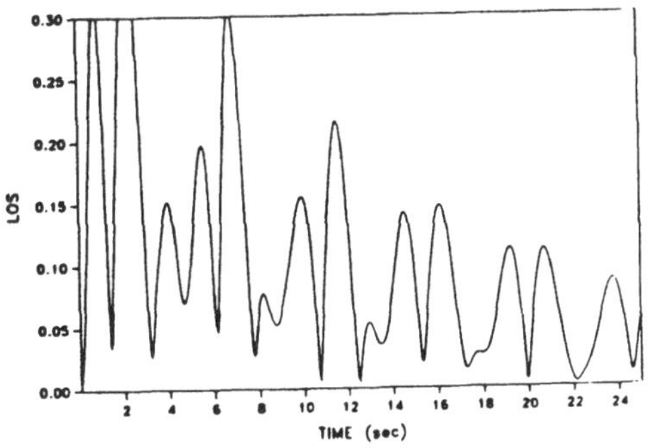

DESIGN D

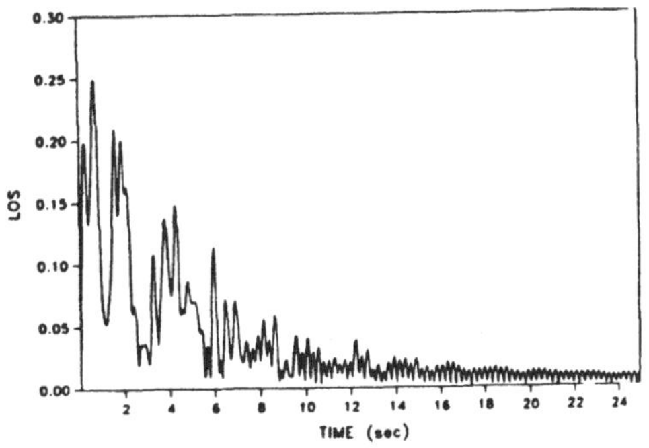

DESIGN E

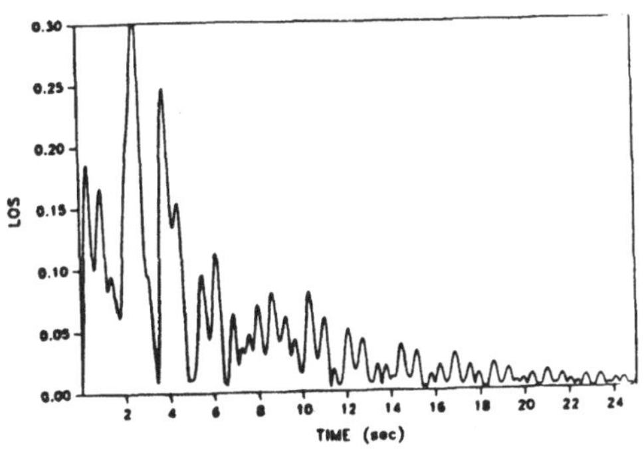

Fig. 2. LOS Transient Response of ACOSS-FOUR Design

Table 5. The Modal Damping Parameters of the Closed-Loop Eigenvalues for ACOSS-FOUR

Mode	Design A	Design B	Design C	Design D	Design E
1	0.0546	0.3630	0.0394	0.3000	0.1093
2	0.0653	0.3202	0.0311	0.2660	0.1093
3	0.0737	0.3263	0.0292	0.2604	0.1093
4	0.0801	0.1693	0.0525	0.2506	0.1093
5	0.0839	0.0582	0.0392	0.1551	0.07906
6	0.0864	0.0298	0.0538	0.1363	0.08921
7	0.0760	0.0333	0.0638	0.0602	0.07230
8	0.0723	0.0253	0.0524	0.0471	0.06052
9	0.0341	0.0086	0.0509	0.0294	0.04724
10	0.0298	0.0265	0.0384	0.0271	0.04004
11	0.0207	0.0254	0.0511	0.0196	0.03355
12	0.0064	0.0243	0.0444	0.0057	0.01080

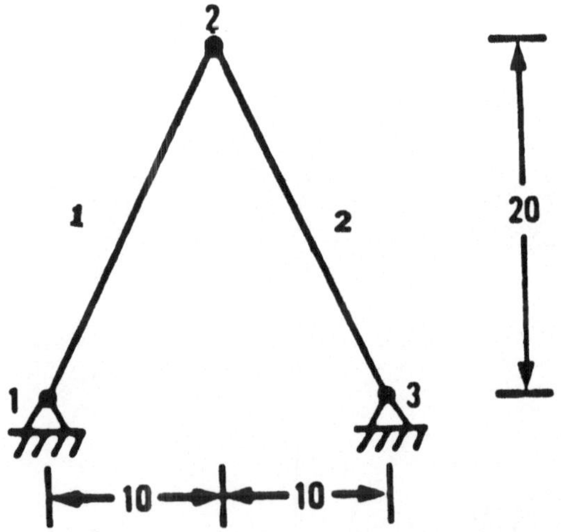

Fig. 3. Two Bar Truss

Table 6. 2-Bar Truss Designs

DESIGN		A_1	A_2	W	g_1	g_2	g_3	g_4	S_G	ω_1	ω_2	N-ITRS
B1	INIT. DESIGN	1000.	500.	33.54	-15.221	-9.8735	0.0468	2.1243	333.68	2.3997	5.2708	10
B2*	OPT. DESIGN	10.	1324.9	29.85	18.874	-33.533	-1.8257	3.9968	1500*	0.3777	5.4505	
B3	INIT. DESIGN	1000.	1000.	44.72	-11.467	-45.251	-0.5077	2.678	2186.5	2.9906	5.9813	12
B4*	OPT. DESIGN	10.	739.8	16.76	-13.66	36.014	-1.5562	3.7273	1500*	0.3777	4.0771	
B5	INIT. DESIGN	500.	500.	22.36	-12.325	18.852	-0.2915	2.4627	513	0.6031	4.6900	13
B6	INIT. DESIGN	100.	500.	13.41	-19.181	51.007	-0.9685	3.1396	2980	1.1502	3.4775	21
B7	INIT. DESIGN	100.	100.	4.72	11.170	45.954	1.4368	0.7343	2239	0.9457	1.8914	25
B8*	OPT. DESIGN	184.66	53.59	5.33	27.221	27.485	1.8863	0.2848	1500*	0.8256	2.1553	

$\{\tilde{\sigma}_1 = -0.0228,\ \tilde{\sigma}_2 = -0.361,\ \tilde{\omega}_1 = 1.17,\ \tilde{\omega}_2 = 4.81\}$ For All Design Iterations;

ω_j = Uncontrolled Natural Frequency

$S = \int_0^{t_f} F^2(t)dt = \int_0^{t_f} x^T G^T G\, x\, dt$ = Control Effort; N-ITRS: Number of Iterations

PART III

SHAPE OPTIMAL DESIGN

SENSITIVITY ANALYSIS AND OPTIMAL DESIGN WITH ACCOUNT FOR VARYING SHAPE AND SUPPORT CONDITIONS

Zenon Mróz
Institute of Fundamental Technological Research
Polish Academy of Sciences
00-049 Warsaw, Poland

ABSTRACT

The present article deals with an extended class of design problems when besides material or dimensional variables, also structure shape, external support conditions, action through initial distortions is considered. In particular, a problem of mixed boundary conditions is discussed when imposed strength and stiffness requirements may lead to conflicting design decisions. The sensitivity analysis for both linear and non-linear structures provides a uniform variational approach to such variety of problems by generating expressions of functional gradients explicitly in terms of state fields of primary and adjoint structures.

1. INTRODUCTION

The optimal design methods in structural mechanics have developed in several directions and different classes of problems have been considered. Most frequently, the optimization procedure is aimed at determining the cross sectional dimensions of a structural member, such as beam, plate or more generally stiffness parameters related to both material and geometric properties. The dimension or stiffness variables occur only in the stiffness or compliance matrices of a structure and do not affect equilibrium, compatibility or boundary conditions. A more difficult class of problems is specified by considering shape or more generally, configuration variation of a structure. Whereas for a disk the shape variation is performed on its plane boundaries, for a shell it may be referred to its edges or to median

NATO ASI Series, Vol. F27
Computer Aided Optimal Design: Structural and Mechanical Systems. Edited by C. A. Mota Soares

surface, and for a truss the configuration variation occurs due to position variation of its joints. In general, equilibrium, compatibility and boundary conditions are affected by such structure variations. A different class of problems is generated by considering variations of structure topology, such as connectivity in bar structures, number of joints or connections between substructures, etc. This class of problems is not well investigated and does not fit into variational formulation of structure modification.

Variation of support conditions, that is support position or orientation provides also variation of structure configuration. However, this class of problems has its specific characteristic. The optimality conditions express the concept of optimal reaction of supports in terms of local deflections and deflection gradients and have a clear mechanical interpretation. A more general formulation of this problem involves optimal structure-foundation interaction, distribution of line or surface stiffeners and application of prestressing fields. A closely related class of problems is associated with external loading or displacement distribution. The concept of optimal action of loads has again a clear mechanical meaning that can be used in analysing the problem of optimal interaction between structural elements. In this paper we shall briefly discuss these classes of problems which are so far much less investigated.

In Section 2, we shall discuss the sensitivity analysis for both linear and non-linear structures, whereas in Section 3 the problem of optimal support reaction on beam structures with boundary conditions expressed by both lateral loads and displacements or initial distortions is discussed. In Section 4, variation of shape of a structure is briefly discussed.

2. SENSITIVITY ANALYSIS FOR LINEAR AND NON-LINEAR STRUCTURES

To provide a uniform treatment for a variety of problems of optimal design, let us briefly discuss first the sensitivity analysis with respect to design variables. As stated in the introduction, these design variables can represent various structure modifications, namely

- i) dimensional or material variables,
- ii) configuration or shape variables,

iii) support or loading variables,

iv) stiffener or hinge variables.

It should be noted that in structural mechanics the sensitivity analysis is important since any redesign process requires assessment of variation of local or global structural response characteristic due to structure modification.

2.1. Variation of material variables in linear elasticity

Let us consider first the variation of dimensional or material variables in a linear elastic structure. Assume the stiffness or compliance matrices $\underset{\sim}{D}$ and $\underset{\sim}{E}$ to depend on a set of design functions $s_k(x)$, so that Hookes law takes the form

$$\underset{\sim}{\sigma} = \underset{\sim}{D}(s_k)\underset{\sim}{\varepsilon}\ , \qquad \underset{\sim}{\varepsilon} = \underset{\sim}{C}(s_k)\underset{\sim}{\sigma} \tag{1}$$

where $\underset{\sim}{D}\underset{\sim}{\varepsilon} = D_{ijkl}\varepsilon_{kl}$ denotes the matrix product of tensors of different orders and the scalar product of tensors or vectors of the same order is denoted by dot, thus $\underset{\sim}{\sigma}\cdot\underset{\sim}{\varepsilon} = \sigma_{ij}\varepsilon_{ij}$. For small variations δs_k of design functions, we have

$$\begin{aligned}\delta\underset{\sim}{\sigma} &= \underset{\sim}{D}\delta\underset{\sim}{\varepsilon} + \delta\underset{\sim}{D}\underset{\sim}{\varepsilon} = \underset{\sim}{D}\delta\underset{\sim}{\varepsilon} + \frac{\partial \underset{\sim}{D}}{\partial s_k}\delta s_k\underset{\sim}{\varepsilon} = \delta\underset{\sim}{\sigma}' + \delta\underset{\sim}{\sigma}'' \\ \delta\underset{\sim}{\varepsilon} &= \underset{\sim}{E}\delta\underset{\sim}{\sigma} + \delta\underset{\sim}{E}\underset{\sim}{\sigma} = \underset{\sim}{E}\delta\underset{\sim}{\sigma} + \frac{\partial \underset{\sim}{E}}{\partial s_k}\delta s_k\underset{\sim}{\sigma} = \delta\underset{\sim}{\varepsilon}' + \delta\underset{\sim}{\varepsilon}''\end{aligned} \tag{2}$$

The equations of equilibrium and compatibility are expressed through virtual work equations

$$\int \underset{\sim}{\sigma}\cdot\delta\underset{\sim}{\varepsilon}\, dV = \int \underset{\sim}{T}^{o}\cdot\delta\underset{\sim}{u}\, dS_T + \int \underset{\sim}{f}\cdot\delta\underset{\sim}{u}\ dV \tag{3}$$

$$\int \delta\underset{\sim}{\sigma}\cdot\underset{\sim}{\varepsilon}\, dV = \int \delta\underset{\sim}{T}\cdot\underset{\sim}{u}^{o} dS_u + \int \delta\underset{\sim}{f}\cdot\underset{\sim}{u}\ dV \tag{4}$$

where $\underset{\sim}{T}^{o}$ and $\underset{\sim}{u}^{o}$ are the specified tractions and displacements on S_T and S_u whereas $\underset{\sim}{f}$ denotes the body force vector, depending in general on the design functions, $\underset{\sim}{f} = \underset{\sim}{f}(s_k)$, for instance in the case of disks or plates of varying thickness with account for self-weight.

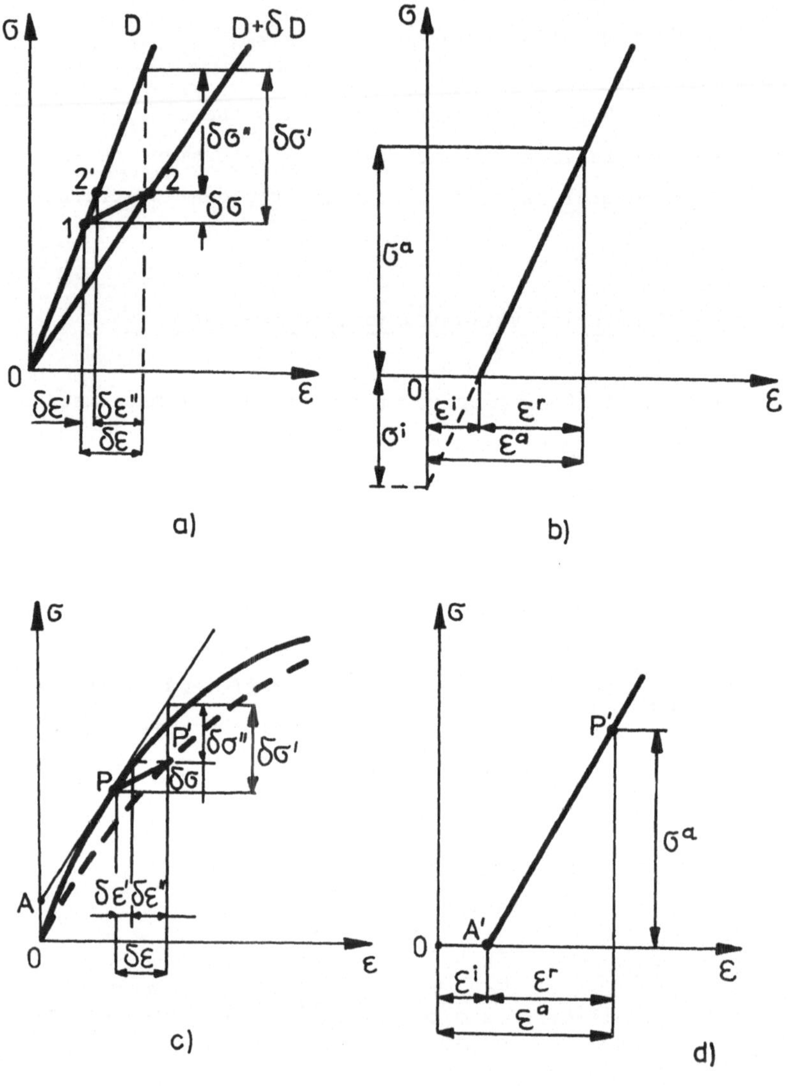

Fig.1. a) Variation of stress and strain due to stiffness modulus variation, b) stress and strain in the adjoint structure, c) non-linear stress-strain relation, d) stress-strain relation in the adjoint structure.

Consider a general functional of the form

$$G = \int \psi(\underset{\sim}{\sigma},\underset{\sim}{u},s_k)\,dV + \int h(\underset{\sim}{u},\underset{\sim}{T})\,dS \tag{5}$$

where the integrands are continuous and differentiable functions of their arguments. Since for any statically admissible stress field $\underset{\sim}{\sigma}^s$ in equilibrium with body forces $\underset{\sim}{f}^s$ and surface tractions $\underset{\sim}{T}^s$, and kinematically admissible fields $\underset{\sim}{u}^k,\underset{\sim}{\varepsilon}^k$ the following virtual work equation holds

$$\int \underset{\sim}{\sigma}^s\cdot\underset{\sim}{\varepsilon}^k\,dV - \int \underset{\sim}{f}^s\cdot\underset{\sim}{u}^k\,dV - \int \underset{\sim}{T}^s\cdot\underset{\sim}{u}^k\,dS = 0 \tag{6}$$

this equation can be added to (5) as the constraint condition. Let the stress field $\underset{\sim}{\sigma}^s$ be identified with the actual field $\underset{\sim}{\sigma}$ whereas $\underset{\sim}{u}^k$ and $\underset{\sim}{\varepsilon}^k$ be identified as fields associated with an adjoint structure of the same shape and elastic stiffness moduli but with different boundary conditions and initial strain or stress fields within the structure. Denoting the state fields within the adjoint structure by $\underset{\sim}{\sigma}^a,\underset{\sim}{\varepsilon}^a,\underset{\sim}{u}^a$, we obtain the augmented functional

$$\bar{G} = \int \psi\,dV + \int h\,dS - \int \underset{\sim}{\sigma}\cdot\underset{\sim}{\varepsilon}^a\,dV + \int \underset{\sim}{f}\cdot\underset{\sim}{u}^a\,dV + \int \underset{\sim}{T}\cdot\underset{\sim}{u}^a\,dS \tag{7}$$

where for simplicity the Lagrange multiplier was assumed as unity. In what follows, we shall show that considering the first variation of (7) and requiring the stationarity with respect to the fields $\underset{\sim}{u}$ and $\underset{\sim}{u}^a$, we obtain the equilibrium conditions for both primary and adjoint structures together with respective boundary conditions. The variation of (7) with respect to s_k then provides the first order sensitivity of the structure. Consider the variation of (7)

$$\begin{aligned}\delta\bar{G} = &\int \Big(\frac{\partial\psi}{\partial\underset{\sim}{\sigma}}\cdot\delta\underset{\sim}{\sigma} + \frac{\partial\psi}{\partial\underset{\sim}{u}}\cdot\delta\underset{\sim}{u} + \frac{\partial\psi}{\partial s_k}\cdot\delta s_k\Big)dV + \int \Big(\frac{\partial h}{\partial\underset{\sim}{u}}\cdot\delta\underset{\sim}{u} + \frac{\partial h}{\partial\underset{\sim}{T}}\cdot\delta\underset{\sim}{T}\Big)dS - \\ &- \int (\underset{\sim}{\sigma}\cdot\delta\underset{\sim}{\varepsilon}^a + \delta\underset{\sim}{\sigma}\cdot\underset{\sim}{\varepsilon}^a)dV + \int (\delta\underset{\sim}{f}\cdot\underset{\sim}{u}^a + \underset{\sim}{f}\cdot\delta\underset{\sim}{u}^a)dV + \int (\delta\underset{\sim}{T}\cdot\underset{\sim}{u}^a + \underset{\sim}{T}\cdot\delta\underset{\sim}{u}^a)dS\end{aligned} \tag{8}$$

and assume that the constitutive relations for the adjoint structure take the form, cf. Fig.1

$$\underset{\sim}{\sigma}^a = \underset{\sim}{D}(\underset{\sim}{\varepsilon}^a - \underset{\sim}{\varepsilon}^{ai})$$

where $\underset{\sim}{\varepsilon}^{ai}$ is the initial strain field to be specified later. Assuming that $\underset{\sim}{u}^{o} = 0$ on S_u and $\underset{\sim}{T}^{o} = 0$ on S_T, the expression (8) takes the form

$$\delta\bar{G} = \int \left(\frac{\partial\psi}{\partial s_k} \delta s_k + \delta \underset{\sim}{f} \cdot \underset{\sim}{u}^{a} \right) dV - \left\{ \int (\underset{\sim}{\sigma} \cdot \delta \underset{\sim}{\varepsilon}^{a} - \underset{\sim}{f} \cdot \delta \underset{\sim}{u}^{a}) dV - \int \underset{\sim}{T} \cdot \delta \underset{\sim}{u}^{a} \, dS \right\} -$$
$$- \left\{ \int [(\underset{\sim}{\varepsilon}^{a} - \underset{\sim}{\varepsilon}^{ai}) \cdot \delta \underset{\sim}{\sigma} - \underset{\sim}{f}^{a} \cdot \delta \underset{\sim}{u}] dV - \int \underset{\sim}{T}^{a} \cdot \delta \underset{\sim}{u} \, dS_T \right\} - \left\{ \int [(\underset{\sim}{\varepsilon}^{ai} - \frac{\partial\psi}{\partial \underset{\sim}{\sigma}}) \cdot \delta \underset{\sim}{\sigma} + \right.$$
$$\left. + (\underset{\sim}{f}^{a} - \frac{\partial\psi}{\partial \underset{\sim}{u}}) \cdot \delta \underset{\sim}{u}] dV + \int (\underset{\sim}{T}^{ao} - \frac{\partial h}{\partial \underset{\sim}{u}}) \cdot \delta \underset{\sim}{u} \, dS_T - \int (\underset{\sim}{u}^{ao} + \frac{\partial h}{\partial \underset{\sim}{T}}) \cdot \delta \underset{\sim}{T} \, dS_u \right\} \tag{10}$$

Let us note that the expression within the first pair of braces can be regarded as the equilibrium equation for the primary structure expressed through the virtual work equation

$$\int \underset{\sim}{\sigma} \cdot \delta \underset{\sim}{\varepsilon}^{a} \, dV = \int \underset{\sim}{f} \cdot \delta \underset{\sim}{u}^{a} \, dV + \int \underset{\sim}{T} \cdot \delta \underset{\sim}{u}^{a} \, dS \tag{11}$$

The expression in the second pair of braces by virtue of (2) and (9) can be presented as follows

$$\int [(\underset{\sim}{\varepsilon}^{a} - \underset{\sim}{\varepsilon}^{ai}) \cdot \delta \underset{\sim}{\sigma} - \underset{\sim}{f}^{a} \cdot \delta \underset{\sim}{u}] dV - \int \underset{\sim}{T}^{a} \cdot \delta \underset{\sim}{u} \, dS_T = \left\{ \int (\underset{\sim}{\sigma}^{a} \cdot \delta \underset{\sim}{\varepsilon} - \underset{\sim}{f}^{a} \cdot \delta \underset{\sim}{u}) dV - \right.$$
$$\left. - \int \underset{\sim}{T}^{a} \cdot \delta \underset{\sim}{u} \, dS_T \right\} + \int (\underset{\sim}{\varepsilon}^{a} - \underset{\sim}{\varepsilon}^{ai}) \frac{\partial \underset{\sim}{\dot{D}}}{\partial s_k} \cdot \underset{\sim}{\varepsilon} \delta s_k \, dV \tag{12}$$

Since the equilibrium conditions for the adjoint structure are expressed by the virtual work equation

$$\int \underset{\sim}{\sigma}^{a} \cdot d\underset{\sim}{\varepsilon} \, dV = \int \underset{\sim}{f}^{a} \cdot \delta \underset{\sim}{u} \, dV + \int \underset{\sim}{T}^{a} \cdot \delta \underset{\sim}{u} \, dS_T \tag{13}$$

only the last term of (12) does not vanish. The expression in the third pair of braces of (10) vanishes when the following loading and boundary conditions are satisfied for the adjoint structure

$$\underset{\sim}{\varepsilon}^{ai} = \frac{\partial\psi}{\partial \underset{\sim}{\sigma}} \qquad \underset{\sim}{f}^{a} = \frac{\partial\psi}{\partial \underset{\sim}{u}} \quad \text{within } V$$
$$\underset{\sim}{T}^{ao} = \frac{\partial h}{\partial \underset{\sim}{u}} \quad \text{on } S_T \qquad \underset{\sim}{u}^{ao} = - \frac{\partial h}{\partial \underset{\sim}{T}} \quad \text{on } S_u \tag{14}$$

In view of (11)-(14), the variation G is the expressed as follows

$$\delta G = \delta\bar{G} = \int \left[\frac{\partial\psi}{\partial s_k} + \frac{\partial \underset{\sim}{f}}{\partial s_k} \cdot \underset{\sim}{u}^a - (\underset{\sim}{\varepsilon}^a - \underset{\sim}{\varepsilon}^{ai}) \cdot \frac{\partial \underset{\sim}{D}}{\partial s_k} \cdot \underset{\sim}{\varepsilon} \right] \delta s_k \, dV \tag{15}$$

that is explicity in terms of the variations δs_k of the design functions. Noting that

$$\underset{\sim\sim}{DE} = \underset{\sim}{I} , \qquad \frac{\partial \underset{\sim}{D}}{\partial s_k} \underset{\sim}{E} = -\underset{\sim}{D} \frac{\partial \underset{\sim}{E}}{\partial s_k} \tag{16}$$

the expression (15) can be presented in the equivalent form

$$\delta G = \int \left[\frac{\partial\psi}{\partial s_k} + \frac{\partial \underset{\sim}{f}}{\partial s_k} \cdot \underset{\sim}{u}^a + \underset{\sim}{\sigma}^a \cdot \frac{\partial \underset{\sim}{E}}{\partial s_k} \cdot \underset{\sim}{\sigma} \right] \delta s_k \, dV \tag{17}$$

It is seen that augmenting the functional G by the bilinear form (6), we generate both equilibrium and boundary conditions for both structures and also the sensitivity expression with respect to the design functions.

The presented approach (which will be called the adjoint structure method is based on the concept of an adjoint structure and its state fields. Having determined solutions for both primary and adjoint structures, the first variations or derivatives with respect to design functions or parameters are explicitly expressed. The other alternative approach (which will be called the direct method is possible by direct calculation of variations of stress, strain and displacement fields due to variation of design parameters. Consider the decomposition (2) where the stress or strain variations $\delta\underset{\sim}{\sigma}''$ or $\delta\underset{\sim}{\varepsilon}''$ can be regarded as initial fields within the elastic body of constant stiffness or compliance matrices $\underset{\sim}{D}$ and $\underset{\sim}{E}$. The constitutive relations, equilibrium equations, and boundary conditions now are

$$\begin{aligned} &\delta\underset{\sim}{\sigma} = \underset{\sim}{D}(\delta\underset{\sim}{\varepsilon} - \delta\underset{\sim}{\varepsilon}'') \\ &\delta\sigma_{ij,j} = 0 , \quad \delta\sigma_{ij}n_j = 0 \text{ on } S_T , \quad \delta\underset{\sim}{u} = 0 \text{ on } S_u \end{aligned} \tag{18}$$

or for the field $\delta\sigma'_{ij}$

$$\begin{aligned} &\delta\sigma'_{ij,j} = -\delta\sigma''_{ij,j} \\ &\delta\underset{\sim}{\sigma}' = \underset{\sim}{D}\delta\underset{\sim}{\varepsilon} \qquad \delta\underset{\sim}{\sigma}'' = \delta\underset{\sim\sim}{D\varepsilon} \\ &\delta\sigma'_{ij}\, n_j = -\delta\sigma''_{ij}\, n_j \text{ on } S_T , \quad \delta\underset{\sim}{u} = 0 \text{ on } S_u \end{aligned} \tag{19}$$

When the stiffness matrix D depends on n parameters, the direct approach requires n+1 solutions whereas the adjoint structure approach requires only two solutions and is more economical. However, when there are m functionals and n stiffness parameters, then the direct approach would require n+1 solutions and the adjoint structure would need m+1 solutions in order to generate all variations. The choice between the two approaches depends on the ratio of m to n and the convenience of generating solutions associated with one or other approach.

2.2. Variation of material variables in non-linear structures

The sensitivity analysis for a linear elastic structure can now be generalized for a physically non-linear material within small strain theory or to geometrically non-linear theory. Consider first a general non-linear stress--strain relation for the small strain theory following from the elastic potential, so that

$$\underset{\sim}{\sigma} = \underset{\sim}{S}(\underset{\sim}{\varepsilon}, s_k) = \frac{\partial U}{\partial \underset{\sim}{\varepsilon}} \tag{20}$$

where U is the specific strain energy per unit volume. The variation of stress is expressed as follows

$$\delta\underset{\sim}{\sigma} = \frac{\partial \underset{\sim}{S}}{\partial \underset{\sim}{\varepsilon}} \delta\underset{\sim}{\varepsilon} + \frac{\partial \underset{\sim}{S}}{\partial s_k} \delta s_k = \frac{\partial^2 U}{\partial \underset{\sim}{\varepsilon} \partial \underset{\sim}{\varepsilon}} \delta\underset{\sim}{\varepsilon} + \frac{\partial^2 U}{\partial \underset{\sim}{\varepsilon} \partial s_k} \delta s_k = \delta\underset{\sim}{\sigma}' + \delta\underset{\sim}{\sigma}'' \tag{21}$$

The decomposition (21) is identical to (2) provided the tangent stiffness matrix $\underset{\sim}{D}_t$ is specified by the relation

$$\underset{\sim}{D}_t = \frac{\partial \underset{\sim}{S}}{\partial \underset{\sim}{\varepsilon}} = \frac{\partial^2 U}{\partial \underset{\sim}{\varepsilon} \partial \underset{\sim}{\varepsilon}} \tag{22}$$

The adjoint structure is now described by a linear relationship (9), that is

$$\underset{\sim}{\sigma}^a = \underset{\sim}{D}_t(\underset{\sim}{\varepsilon}^a - \underset{\sim}{\varepsilon}^{ai}) \tag{23}$$

The sensitivity expressions (15) or (17) are valid in this case. Figs.1c,d illustrate the constitutive relations (20) and (23) for both primary and adjoint structures.

The case of non-linear geometric theory can be treated similarly. Assume a fixed Cartesian coordinate system to describe particle location during a finite deformation process. A typical particle at the point $x_i (i=1,2,3)$ is displaced to the point y_i in the deformed configuration. Introducing the displacement field $u_i(x)$, we can specify the deformation gradient

$$x_i \to y_i = x_i + u_i , \qquad F_{ij} = \frac{\partial y_i}{\partial x_j} = \delta_{ij} + u_{i,j} \tag{24}$$

where δ_{ij} denotes the Kronecker symbol. The material is assumed to be hyper-elastic obeying the constitutive law of the form

$$t_{ij} = \frac{\partial U(\underset{\sim}{F}, s_k)}{\partial F_{ij}} = t_{ij}(\underset{\sim}{F}, s_k) \tag{25}$$

where $\underset{\sim}{t}$ is the first (non-symmetric) Piola-Kirchhoff stress tensor related to the symmetric Cauchy stress $\underset{\sim}{\sigma}$ by the formula

$$t_{ij}F_{kj} = \det |\underset{\sim}{F}| \sigma_{ik} \tag{26}$$

and $U(\underset{\sim}{F}, s_k)$ is the specific strain energy per unit volume in the initial configuration B_o. The equilibrium equations and boundary conditions now are

$$t_{ij,i} = f_i , \qquad t_{ij}n_i = T_i^o \text{ on } S_T , \qquad u_i = u_i^o \text{ on } S_u \tag{27}$$

and the virtual work equation (3) now takes the form

$$\int t_{ij} \delta u_{j,i} \, dV_o = \int \underset{\sim}{T}^o \cdot \delta \underset{\sim}{u} \, dS_T + \int \underset{\sim}{f} \cdot \delta \underset{\sim}{u} \, dV_o \tag{28}$$

where V_o denotes the structure volume in the initial configuration with $S = S_T \cup S_u$ specified in this configuration. Instead of (21), we now have

$$\delta \underset{\sim}{t}_{ij} = \frac{\partial^2 U}{\partial F_{ij} \partial F_{kl}} \delta F_{kl} + \frac{\partial^2 U}{\partial F_{ij} \partial s_k} \delta s_k = \delta t'_{ij} + \delta t''_{ij} \tag{29}$$

The functional (5) is now expressed in the underformed configuration in terms of $\underset{\sim}{t}$, $\underset{\sim}{u}$, s_k and boundary tractions or displacements. The adjoint structure is linear with the constitutive relation and the initial displacement gradient specified as follows

$$t^a_{ij} = \frac{\partial^2 U}{\partial F_{ij} \partial F_{kl}} (u^a_{k,l} - u^i_{k,l}) = D^t_{ijkl}(u^a_{k,l} - u^i_{k,l})$$
$$u^i_{k,l} = \frac{\partial \psi}{\partial t_{kl}} \tag{30}$$

whereas the body forces $\underset{\sim}{f}^a$, surface tractions and displacements $\underset{\sim}{T}^{ao}$, $\underset{\sim}{u}^{ao}$ are specified by (14). The formulae (15) remain valid provided $\underset{\sim}{\varepsilon}^a$ and $\underset{\sim}{\varepsilon}^{ai}$ are replaced by $u^a_{i,j}$ and $u^{ai}_{i,j}$.

Now, let us express these relations in terms of the symmetric Piola Kirchhoff stress tensor s_{ij} and the conjugate Green strain tensor e_{ij} . We have

$$s_{kl} = \frac{\partial U(\underset{\sim}{e}, s_k)}{\partial e_{kl}} \tag{31}$$

where

$$e_{kl} = \tfrac{1}{2}(u_{k,l} + u_{l,k} + u_{m,k}\, u_{m,l}) = \varepsilon_{kl} + \tfrac{1}{2}\, u_{m,k}\, u_{m,l}$$
$$t_{ij} = s_{ik}\, F_{jk} = s_{ik} \frac{\partial y_j}{\partial x_k} = s_{ij}(\delta_{jk} + u_{j,k}) \tag{32}$$

and the virtual work equation takes the form

$$\int s_{kl}\, \delta e_{kl}\, dV_o = \int T^o_i\, \delta u_i\, dS_T + \int f_i\, \delta u_i\, dV_o \tag{33}$$

Consider now the increments of t_{kl} and s_{kl} and the respective equilibrium equations

$$\delta t_{ij} = \delta s_{ik}(\delta_{jk} + u_{j,k}) + s_{ik}\, \delta u_{j,k}$$
$$\delta t_{ij,i} = [\delta s_{ik}(\delta_{jk} + u_{j,k})]_{,j} + (s_{ik}\, \delta u_{j,k})_{,j} + f_i = 0 \tag{34}$$

and the incremental constitutive equations

$$\delta s_{kl} = \frac{\partial^2 U(e_{mn})}{\partial e_{kl} \partial e_{ij}}\, \delta e_{ij}\, , \qquad \delta e_{ij} = \frac{\partial^2 W(s_{mn})}{\partial s_{ij} \partial s_{kl}}\, \delta s_{kl} \tag{35}$$

These relations indicate the respective relations for the adjoint problem. Let us write

$$e^a_{ij} = \frac{\partial^2 W(s^s_{mn})}{\partial s^s_{ij}\partial s^s_{kl}} s^s_{kl} \qquad s^s_{ij} = \frac{\partial^2 U(e_{mn})}{\partial e_{ij}\partial e_{kl}} e^a_{kl}$$

$$s^n_{ij} = s_{ik} u^a_{j,k} \qquad s^a_{ij} = s^s_{ij} + s^n_{ij} \tag{36}$$

so that the equilibrium and virtual work equations take the form

$$[s^s_{ik}(\delta_{jk} + u_{j,k})]_{,i} + (s_{ik} u^a_{j,k})_{,i} + f^a_j = 0 \tag{37}$$

and

$$\int \underset{\sim}{T}\cdot\delta\underset{\sim}{u}^a\, dS + \int \underset{\sim}{f}\cdot\delta\underset{\sim}{u}^a\, dV = \int s^s_{ij}\delta e^a_{ij}\, dV + \int s^n_{ij}\delta u^a_{j,i}\, dV$$

$$\int \delta\underset{\sim}{T}\cdot\delta\underset{\sim}{u}^a\, dS + \int \delta\underset{\sim}{f}\cdot\delta\underset{\sim}{u}^a\, dV = \int \delta s^s_{ij}\delta e^a_{ij}\, dV + \int s_{ik}\delta u_{j,k}\delta u^a_{j,i}\, dV \tag{38}$$

Here the total stress tensor s^a_{ij} in the adjoint system is decomposed into symmetric and non-symmetric parts s^s_{ij} and s^n_{ij} satisfying the constitutive relations (36) and the equilibrium equations (37). The strain e^a_{ij} is specified as follows

$$e^a_{ij} = \tfrac{1}{2}(u^a_{i,j} + u^a_{j,i} + u_{,mi}\, u^a_{,mj} + u^a_{m,i}\, u_{m,j}) \tag{39}$$

Consider, for example, the functional

$$I = f(\underset{\sim}{u}, s_k)dV \tag{40}$$

Following as previously, we have

$$\delta I = \int \left(\frac{\partial f}{\partial \underset{\sim}{u}} \cdot\delta\underset{\sim}{u} + \frac{\partial f}{\partial s_k}\delta s_k\right)dV = \int \left(\underset{\sim}{f}^a\cdot\delta\underset{\sim}{u} + \frac{\partial f}{\partial s_k}\delta s_k\right)dV =$$

$$= \int s^s_{ij}\, \delta e_{ij}\, dV + \int s^n_{ml}\, \delta u_{l,m}\, dV + \int \frac{\partial f}{\partial s_k}\delta s_k\, dV \tag{41}$$

and since

$$\delta e_{ij} = \frac{\partial^2 W}{\partial s^s_{ij}\partial s^s_{kl}}\delta s^s_{kl} + \frac{\partial^2 W}{\partial s^s_{ij}\partial s_k}\delta s_k \tag{42}$$

in view of the virtual work equation (38), we obtain

$$\delta I = \int \left(s^{s}_{ij} \frac{\partial^2 W}{\partial s^{s}_{ij} \partial s_k} + \frac{\partial f}{\partial s_k} \right) \delta s_k \, dV \tag{43}$$

2.3. Variation of the buckling load for a plate

The general discussion of sensitivity analysis for non-linear structures can be applied to a case of plate loaded by the in-plane boundary tractions λT^{o}_{ij}. The onset of buckling is specified by the following equality

$$\int U^B(k_{ij},h)dA + \tfrac{1}{2}\lambda \int N^{o}_{ij}\, w_{,i}\, w_{,j}\, dA = 0 \tag{44}$$

where U^B is the specific flexural energy of the buckling mode associated with the lateral deflection w and the curvature k_{ij} . The plane forces are $N_{ij} = \lambda N^{o}_{ij}$ and the in-plane displacements are u_i, so that the generalized strains are specified as follows

$$\begin{aligned} e_{ij} &= \tfrac{1}{2}(u_{i,j} + u_{j,i}) + \tfrac{1}{2}\, w_{,i}\, w_{,j} = \varepsilon_{ij} + \gamma_{ij} \\ k_{ij} &= -\, w_{,ij} \end{aligned} \tag{45}$$

and the equilibrium equations are

$$N_{ij,j} = 0\,, \qquad M_{ij,ij} = -\, N_{ij}\, w_{,ij} \tag{46}$$

The constitutive relations for a physically linear material are formulated as follows

$$\begin{aligned} M_{ij} &= D^{B}_{ijkl}(h)k_{kl} = \frac{\partial U^B}{\partial k_{kl}} \\ N_{ij} &= D^{A}_{ijkl}(h)e_{kl} = \frac{\partial U^A}{\partial e_{ij}} = D^{A}_{ijkl}(\varepsilon_{kl} + \gamma_{kl}) = N^{A}_{ij} + N^{B}_{ij} \end{aligned} \tag{47}$$

where D^{B}_{ijkl} , and D^{A}_{ijkl} are flexural and in-plane stiffness matrices of the plate generated by the specific strain energies $U^B(\underset{\sim}{k},h)$ and $U^A(\underset{\sim}{e},h)$. The virtual work equation associated with (45)-(46) has the form

$$\int M_{ij}\delta k_{ij}\, dA + \int N_{ij}\, w_{,i}\, \delta w_{,j}\, dA = 0 \tag{48}$$

Consider the perturbation of Eq.(44) associated with the plate thickness variation h. We have

$$\int \frac{\partial U^B}{\partial k_{ij}} \delta k_{ij}\, dA + \int \frac{\partial U^B}{\partial h} \delta h\, dA + \tfrac{1}{2}\lambda \int N^o_{ij}(\delta w_{,i}\, w_{,j} + w_{,i}\, \delta w_{,j})dA + $$
$$+ \tfrac{1}{2}\lambda \int \delta N^o_{ij}\, w_{,i}\, w_{,j}\, dA + \tfrac{1}{2}\delta\lambda \int N^o_{ij}\, w_{,i}\, w_{,j}\, dA = 0 \tag{49}$$

In view of the virtual work equation (48), the first and the third terms of (49) cancel. In order to eliminate the variation N_{ij}, let us formulate an adjoint problem from which we determine plane forces N^a_{ij}, strains ε^a_{ij}, and displacements u^a_i. Assume that there is an initial strain field applied to the adjoint plate

$$\varepsilon^i_{ij} = w_{,i}\, w_{,j} \tag{50}$$

so that

$$\tfrac{1}{2}(u^a_{i,j} + u^a_{j,i}) = \varepsilon^a_{ij} = \varepsilon^i_{ij} + \varepsilon^r_{ij} \tag{51}$$

and

$$N^a_{ij} = D^A_{ijkl}(\varepsilon^a_{kl} - \varepsilon^i_{kl}) = D^A_{ijkl}\, \varepsilon^r_{kl}$$
$$N^a_{ij,j} = 0\ , \qquad u^a_i = 0 \text{ on } S_u\ , \qquad N^a_{ij}\, n_j = 0 \text{ on } S_T \tag{52}$$

where S_T and S_u now denote loaded and supported plate boundaries. In view of (52), we can write

$$\int \delta N^o_{ij}\, w_{,i}\, w_{,j}\, dA = \int \delta N^o_{ij}\, \varepsilon^i_{ij}\, dA = \int \delta N^o_{ij}(\varepsilon^a_{ij} - \varepsilon^r_{ij})dA =$$
$$= - \int \delta N^o_{ij}\, \varepsilon^r_{ij}\, dA \tag{53}$$

since δN^o_{ij} is a self-equilibrated state with vanishing tractions on the loaded boundary S_T. We have

$$\int \delta N^o_{ij} w_{,i} w_{,j} \, dA = - \int D^A_{ijkl} \delta\varepsilon_{kl} \varepsilon^r_{ij} \, dA - \int \varepsilon_{kl} \frac{\partial D^A_{ijkl}}{\partial h} \varepsilon^r_{ij} \delta h \, dA =$$

$$= - \int \varepsilon_{kl} \frac{\partial D_{ijkl}}{\partial h} \varepsilon^r_{ij} \delta h \, dA \tag{54}$$

since

$$\int D^A_{ijkl} \delta\varepsilon_{kl} \varepsilon^r_{ij} \, dA = \int N^a_{kl} \delta\varepsilon_{kl} \, dA = 0 \tag{55}$$

In view of (54), from (48) it follows that

$$\delta\lambda = \frac{- 2 \int \frac{\partial U^B}{\partial h} \delta h \, dA + 2\lambda \int \frac{\partial U^A}{\partial h} \delta h \, dA}{\int N^o_{ij} w_{,i} w_{,j} \, dA} =$$

$$= \frac{- 2 \int \frac{\partial U^B}{\partial h} \delta h \, dA + \lambda \int \frac{\partial D^A_{ijkl}}{\partial h} \varepsilon^r_{ij} \varepsilon_{kl} \delta h \, dA}{\int N^o_{ij} w_{,i} w_{,j} \, dA} \tag{56}$$

where

$$2U^B = \frac{Eh^3}{12(1-\nu^2)} [k^2_{xx} + k^2_{yy} + 2\nu k_{xx} k_{yy} + 2(1-\nu) k^2_{xy}] =$$

$$= \frac{Eh^3}{12(1-\nu^2)} [(w_{,xx} + w_{,yy})^2 - 2(1-\nu)(w_{,xx} w_{,yy} - w^2_{,xy})] \tag{57}$$

and

$$2U^A = \frac{Eh}{1-\nu^2} [\varepsilon_{xx} \varepsilon^r_{xx} + \varepsilon_{yy} \varepsilon^r_{yy} + \nu(\varepsilon_y \varepsilon^r_x + \varepsilon_x \varepsilon^r_y) + 2(1-\nu)\varepsilon_{xy} \varepsilon^r_{xy}] \tag{58}$$

is the mutual energy of two states ε_{ij} and $\varepsilon^r_{ij} = \varepsilon^a_{ij} - w_{,i} w_{,j}$. In deriving the formulae (56)-(58) it was assumed that the stress state in the plate before buckling is specified by linear geometric relations. A more general case when non-linear theory is used before buckling can be treated similarly.

3. OPTIMAL DESIGN OF SUPPORTS OF STRUCTURES SUBJECTED TO LOADS AND INITIAL DISTORTIONS

In formulating optimal design problems for structures, it is usually assumed that support conditions are specified. However, a more extended class of problems where support stiffness, position, orientation or prestress are also to be determined, offers more realistic solutions since static or dynamic response is sensitive to support conditions. Moreover, the optimality conditions for support action can easily be introduced into practical design and also utilized in active control of structures.

The research on this class of problems was concerned with the optimal location and orientation of point supports assuming static behaviour [4-6], next stability or vibration control was discussed [5,8]. In these papers the structures were assumed to be subjected to external loads with support attached to rigid boundaries. The optimal supporting action then corresponds to the stiffest structural response. However, the problem becomes different when initial distortions or displacements are imposed on the structure, for instance, due to foundation settlement, assemblage errors or temperature fields. Then, the optimal design should correspond to a more flexible structure which could deform at a low stress level. Conflicting design requirements will occur when external loads or distortions are imposed simultaneously on a structure. This case will be briefly discussed in this section.

3.1. Conditions for optimal support reaction in the case of applied loads

Consider a plane beam structure of specified configuration, cross-sectional dimensions, loading conditions and some of supports. Let the positions and directions of some additional supports be at the designer choice together with magnitudes of support reactions. The optimal support reaction now corresponds to minimization of global static compliance or local deflections and stresses.

Denote the generalized stresses and strains by $\tilde{Q}$ and $\tilde{q}$ and the specific stress and strain energies by

$$W(\tilde{Q}) = \int_0^{Q} \tilde{q}\cdot d\tilde{Q}\ , \qquad U(q) = \int_0^{q} \tilde{Q}\cdot d\tilde{q} \tag{59}$$

For a linear elastic material both $W(\underset{\sim}{Q})$ and $U(\underset{\sim}{q})$ are quadratic functions and

$$\underset{\sim}{q} = \frac{\partial W}{\partial \underset{\sim}{Q}} = \underset{\sim}{L}\underset{\sim}{Q} , \qquad \underset{\sim}{Q} = \frac{\partial U}{\partial \underset{\sim}{q}} = \underset{\sim}{M}\underset{\sim}{q} , \qquad \underset{\sim}{L} = \underset{\sim}{M}^{-1} \tag{60}$$

where $\underset{\sim}{L}$ and $\underset{\sim}{M}$ are the constant compliance and stiffness matrices. For a non--linear material, the tangent stiffness and compliance matrices are

$$\underset{\sim}{L}^t = \frac{\partial^2 W}{\partial \underset{\sim}{Q} \partial \underset{\sim}{Q}} , \qquad \underset{\sim}{M}^t = \frac{\partial^2 U}{\partial \underset{\sim}{q} \partial \underset{\sim}{q}} \tag{61}$$

Consider the stress functional

$$G = \int \phi(\underset{\sim}{Q})\, dx \tag{62}$$

which is to be minimized by support reaction. Following previous derivations, introduce an adjoint structure submitted to an initial strain field $\underset{\sim}{q}^i$ specified by the potential law

$$\underset{\sim}{q}^i = \frac{\partial \phi}{\partial \underset{\sim}{Q}} \tag{63}$$

The initial strain field induces the residual stress state $\underset{\sim}{Q}^a$ and the associated strain $\underset{\sim}{q}^a$, so that

$$\underset{\sim}{q}^a = \underset{\sim}{q}^i + \underset{\sim}{q}^r , \qquad \underset{\sim}{Q}^a = \underset{\sim}{M}^t(\underset{\sim}{q}^a - \underset{\sim}{q}^i) = \underset{\sim}{M}^t \underset{\sim}{q}^r \tag{64}$$

Let the position, direction and magnitude of support reaction be varied. Assume that from the position A the support is translated through a distance $\dot{s}$ along the beam axis, rotated through the angle $\delta\phi$ in the structure plane and its magnitude be varied from R to R + δR. Let the stress strain and displacement fields in the primary structure before and after variation be $\underset{\sim}{Q}$, $\underset{\sim}{q}$, $\underset{\sim}{u}$, and $\underset{\sim}{Q}' = \underset{\sim}{Q} + \delta\underset{\sim}{Q}$, $\underset{\sim}{q}' = \underset{\sim}{q} + \delta\underset{\sim}{q}$, $\underset{\sim}{u}' = \underset{\sim}{u} + \delta\underset{\sim}{u}$, whereas the states in the adjoint structure be $\underset{\sim}{Q}^a$, $\underset{\sim}{q}^a$, and $\underset{\sim}{u}^a$. By the principle of virtual work, there is

$$\begin{aligned} &\int \underset{\sim}{p}\cdot\underset{\sim}{u}^a\, dx - R u^a_R = \int \underset{\sim}{Q}\cdot\underset{\sim}{q}^a\, dx \\ &\int \underset{\sim}{p}\cdot\underset{\sim}{u}^a\, dx - (R + \delta R)\cos\phi(u^a_R + u^a_{R,s}\,\delta s) - (R + \delta R)\sin(\delta\phi)u^a_s = \\ &= \int \underset{\sim}{Q}'\cdot\underset{\sim}{q}^a\, dx \end{aligned} \tag{65}$$

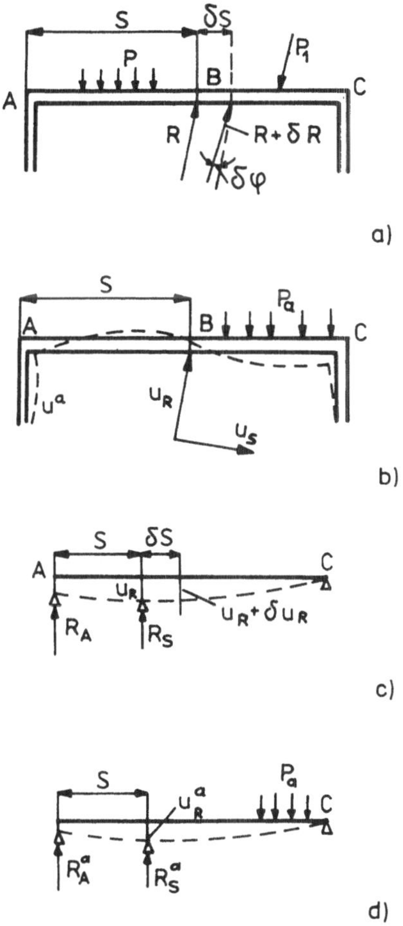

Fig.2. Primary and adjoint structure in the case of (a,b) support reaction control and (c,d) support displacement control.

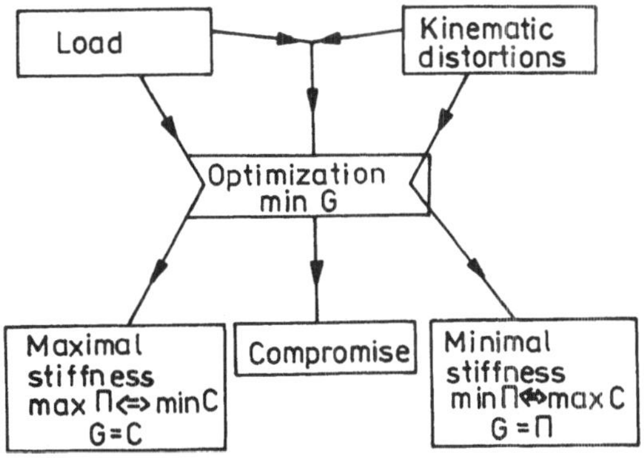

Fig.3. Conflicting design situation in the case of acting loads and displacements or distorsions.

where the external loading is denoted by $\underset{\sim}{p}$ and u_R^a , u_s^a are the displacement components along the force R and in the perpendicular direction. The derivative

$$u_{R,s}^a = \left(\frac{\partial u_R^a}{\partial x} \right)_{x=s} \tag{66}$$

is assumed to be continuous within the adjoint structure. For small variations δR, δs and $\delta\phi$, from (65) it follows that

$$\int \underset{\sim}{q}^a \cdot \delta \underset{\sim}{Q} \, dx = - R u_R^a - R u_{R,s}^a \, \delta s - R u_s^a \, \delta\phi \tag{67}$$

Note that the reaction expends a negative work, so we have put a negative sign before the term Ru_R^a in (65). Also note that there is no force R^a applied at A to the adjoint structure Figs.2a,b.

Consider now the variation of the functional (62). In view of (64), we have

$$\delta G = \int \frac{\partial \phi}{\partial \underset{\sim}{Q}} \cdot \delta \underset{\sim}{Q} \, dx = \int \underset{\sim}{q}^i \cdot \underset{\sim}{Q} \, dx = \int (\underset{\sim}{q}^a - \underset{\sim}{q}^r) \cdot \delta \underset{\sim}{Q} \, dx = \int \underset{\sim}{q}^a \cdot \delta \underset{\sim}{Q} \, dx \tag{68}$$

since

$$\int \underset{\sim}{q}^r \cdot \delta \underset{\sim}{Q} \, dx = \int \underset{\sim}{q}^r \cdot \underset{\sim}{M}^t \delta \underset{\sim}{q} \, dx = \int \delta \underset{\sim}{q} \cdot \underset{\sim}{M}^t \underset{\sim}{q}^r \, dx = \int \delta \underset{\sim}{q} \cdot \underset{\sim}{Q}^a \, dx = 0 \tag{69}$$

Using (67), we obtain the sensitivity expression associated with support variation

$$\delta G = - u_R^a \, \delta R - R u_{R,s}^a \, \delta s - R u_s^a \, \delta\phi \tag{70}$$

and since δR, δs, and $\delta\phi$ are independent variations, the stationarity condition $\delta G = 0$ implies that

$$u_R^a = R u_{R,s}^a = R u_s^a = 0 \tag{71}$$

Thus, the optimal support conditions (71) are expressed directly in terms of the displacement field of the adjoint structure: for $R \neq 0$, the two displacement components u_R^a and u_s^a should vanish at the point A of the adjoint structure together with the derivarive $u_{R,s}^a$. When the displacement functional

$$H = \int \psi(\underset{\sim}{u})dx \tag{72}$$

the adjoint structure is loaded by

$$\underset{\sim}{p}^a = \frac{\partial H}{\partial \underset{\sim}{u}} \tag{73}$$

and the optimality conditions (71) apply.

3.2. Action of loads and distortions or displacements

Consider now the case when both surface loads, initial distortions, and boundary displacements are imposed on the structure. These distortions can be distributed within the structure and specified as a field $\underset{\sim}{q}^i(x)$, distributed on its boundary portion S_u as an initial displacement field $u_b^i(x)$, or concentrated at point supports u_j^i (here j denotes the consecutive support number). The initial distortion field is kinematically inadmissible and it induces the elastic strain fields $\underset{\sim}{q}^r(x)$ and boundary or support displacements $u_b^r(x)$, u_j^r, so that

$$\underset{\sim}{q} = \underset{\sim}{q}^i + \underset{\sim}{q}^r, \qquad u_b^o - u_b = u_b^i + u_b^r, \qquad u_j^o - u_j = u_j^i + u_j^r \tag{74}$$

where u_b, u_j are the displacements on the structure boundary at interaction points between distributed or concentrated supports and the structure, whereas u_b^o, u_j^o are the displacements of the supporting boundary and of points of interaction of supports with foundation. The difference $u_b^o - u_b$ or $u_j^o - u_j$ represents the elastic and initial displacements of distributed and concentrated supports which can be conceived as distributed elastic springs on the boundary S_u and at points j.

Consider first the case when the supporting boundary is rigid with no initial displacement, $u_b = u_b^o = u_b^i = u_b^r = 0$, and when the rigid point supports posses imposed displacements, $u_j^r = u_j^i = 0$, $u_j = u_j^o$. Moreover, we assume that there is an initial distortion field $q^i(x)$ within the structure.

The elastic stress field $\underset{\sim}{Q}$ is related to $\underset{\sim}{q}^r$ by the constitutive relation

$$\underset{\sim}{Q} = \underset{\sim}{Q}(\underset{\sim}{q}^r) = \frac{\partial U(q^r)}{\partial \underset{\sim}{q}^r} \tag{75}$$

where $\underset{\sim}{q}^r = \underset{\sim}{q} - \underset{\sim}{q}^i$. Since $\underset{\sim}{Q}$ is in equilibrium with the applied load $\underset{\sim}{p}$ and the support reactions R_j, the virtual work equations take the form

$$\int \underset{\sim}{Q} \cdot d\underset{\sim}{q}\, dx - \int \underset{\sim}{p} \cdot \delta \underset{\sim}{u}\, dx = 0 \tag{76}$$

and this equation can be regarded as the stationarity condition of the potential energy

$$\Pi_u = \int U(\underset{\sim}{q}^r) dx - \int \underset{\sim}{p} \cdot \underset{\sim}{u}\, dx \tag{77}$$

Considering a statically admissible stress variation $\delta \underset{\sim}{Q}$ and the associated support reaction δR_j, we can write

$$\int (\underset{\sim}{q}^i + \underset{\sim}{q}^r) \cdot \delta \underset{\sim}{Q}\, dx - u_j^o\, \delta R_j = 0 \tag{78}$$

and this equation can be regarded as the stationarity condition of the complementary energy

$$\Pi_Q = \int [W(\underset{\sim}{Q}) + \underset{\sim}{Q} \cdot \underset{\sim}{q}^i] dx - R_j(\underset{\sim}{Q}) \cdot u_j^o \tag{79}$$

When $\underset{\sim}{p} = 0$ and there is only distortion field $\underset{\sim}{q}^i(x)$, for a linear elastic material we have

$$- \Pi_Q = \Pi_u = \int U(\underset{\sim}{q}^r) dx = \tfrac{1}{2} \int \underset{\sim}{Q}^r \cdot \underset{\sim}{q}^r\, dx = \int W(\underset{\sim}{Q}^r) dx \tag{80}$$

and $\Pi_Q < 0$, $\Pi_u > 0$. The following equivalence of various optimal design formulations now occurs

$$\max \Pi_Q \rightarrow \min \Pi_u \rightarrow \min \int U\, dx \rightarrow \min \int W\, dx \tag{81}$$

On the other hand, when there are no distortions and a loaded structure is rigidly supported, we have

$$\Pi_u = \int [U(\underset{\sim}{q}) - \underset{\sim}{p} \cdot \underset{\sim}{u}] dx\,, \qquad \Pi_Q = \int W(\underset{\sim}{Q}) dx \tag{82}$$

and for a linear structure there is

$$\Pi_Q = -\Pi_u = \tfrac{1}{2}\int \underset{\sim}{Q}\cdot\underset{\sim}{q}\,dx = \tfrac{1}{2}\int \underset{\sim}{p}\cdot\underset{\sim}{u}\,dx = \tfrac{1}{2}\int U\,dx \tag{83}$$

thus $\Pi_Q > 0$ and $\Pi_u < 0$. The following formulations of optimal design are equivalent

$$\min \Pi_Q \rightarrow \max \Pi_u \rightarrow \min \int \underset{\sim}{p}\cdot\underset{\sim}{u}\,dx \rightarrow \min \int W\,dx \rightarrow \min \int U\,dx \tag{84}$$

We may regard therefore in this case Π_Q as a measure of mean structure compliance whereas Π_u is regarded as a measure of mean structure stiffness as both are proportional to work of surface loads on induced displacements. Comparing (81) and (84) it is seen that there is a conflicting situation when both loads and distortions or initial displacements are imposed simultaneously on the structure. In fact, when the global elastic energy is to be minimized, the structure attains maximal stiffness for the case of applied loads and minimal stiffness (maximal compliance) for the case of imposed distortions and support displacements. This situation is schematically illustrated in Fig.3.

Instead of the complementary or potential energies, we can now use the total elastic energy stored in the structure as a measure of structure quality and the objective function to be minimized. For a linear structure, we have

$$\Omega(\underset{\sim}{Q}) = \int W(\underset{\sim}{Q})dx = \int W(\underset{\sim}{Q}^p)dx + \int W(\underset{\sim}{Q}^d)dx \tag{85}$$

where $\underset{\sim}{Q} = \underset{\sim}{Q}^p + \underset{\sim}{Q}^d$ and $\underset{\sim}{Q}^p$, $\underset{\sim}{Q}^d$ are the stress fields due to separate action of loads and distortions plus displacements with vanishing loads.

Let us derive the optimal support conditions assuming that only the support position $x = s$ and the imposed support displacement u_R are varied. Consider a beam structure subjected to load $\underset{\sim}{p}(x)$, initial distortion field $\underset{\sim}{q}^i(x)$ and with specified displacements at some fixed supports. Assume the displacement u_R of the support to be subject to variation together with the support position $x = s$. Consider the functional

$$G = \int F(\underset{\sim}{Q},\underset{\sim}{u})dx \tag{86}$$

and introduce the adjoint structure subjected to initial distortions $\underset{\sim}{q}^{ai}$ and load $\underset{\sim}{p}^{a}$

$$\underset{\sim}{q}^{ai} = \frac{\partial F}{\partial \underset{\sim}{Q}} , \qquad \underset{\sim}{p}^{a} = \frac{\partial F}{\partial \underset{\sim}{u}} \tag{87}$$

and with vanishing displacements at all supports, $u_j^a = 0$. It is assumed that at $x = s$ there is also support in the adjoint structure with the respective reaction R_s^a. Following previous derivation of (70), the sensitivity expression is now derived in the form

$$\delta\bar{G} = - R_s^a\, \delta u_R + (R_s^a\, u_{R,s} + R_s\, u_{R,s}^a)\delta s \tag{88}$$

and hence the stationarity conditions for the functional $\bar{G}$ take the form

$$R_s^a = 0 , \qquad R_s\, u_{R,s}^a = 0 \tag{89}$$

Consider now the case when the functional G coincides with the elastic stress energy $\Omega(\underset{\sim}{Q})$, specified by Eq.(85). When distortions and support displacements vanish, then also $u_R = 0$ and $\min\Omega \rightarrow \max\Pi_u \rightarrow \min\Pi_Q$. Equations (87) then yield

$$\underset{\sim}{p}^{a} = 0 , \qquad \underset{\sim}{q}^{ai} = \frac{\partial W}{\partial \underset{\sim}{Q}} = \underset{\sim}{q} \tag{90}$$

that is the initial distortions of the adjoint structure are equal to the total strains of the primary structure and are kinematically admissible, thus $\underset{\sim}{q}^{ar} = \underset{\sim}{q}^{a} - \underset{\sim}{q}^{ai} = 0$, $u^a = u$, $R^a = 0$. On the other hand, when only initial distortions and support displacements occur within the structure, then $\min\Omega(\underset{\sim}{Q}) \rightarrow \min \Pi_u \rightarrow \max \Pi_Q$, and

$$\begin{aligned} &\underset{\sim}{q}^{ai} = \frac{\partial W}{\partial \underset{\sim}{Q}} = \underset{\sim}{q}^{r}, \qquad \underset{\sim}{q}^{ar} = - \underset{\sim}{q}^{ai}, \qquad \underset{\sim}{q}^{a} = 0, \\ &\underset{\sim}{Q}^{a} = - \underset{\sim}{Q}, \qquad \underset{\sim}{R}_s^a = - \underset{\sim}{R}_s , \qquad \underset{\sim}{u}^{a} = 0 \end{aligned} \tag{91}$$

In the case of combined loading, the sensitivity expressions are

$$\frac{\partial\Omega}{\partial u_R} = R_s^d , \qquad \frac{\partial\Omega}{\partial s} = R_s^p\, u_{R,s}^p - R_s^d\, u_{R,s}^d \tag{92}$$

and are specified by the solution of the primary structure.

3.3. Example

To illustrate the general theory, consider a simple beam structure shown in Fig.4, for which the sensitivity derivatives can be expressed in analytical forms. Assume the objective functional to be equal to the total elastic stress energy, $G = \Omega$. Assume the beam to be loaded by the lateral pressure p, initial curvature $q^i(x) = k^i(x)$ and with the imposed support displacements u_R at B. Introduce the following non-dimensional quantities

$$P = \frac{3l^3}{EI} p, \qquad k = 100\, q^i l = -100\, l \frac{d^2w}{dx^2}, \qquad \Delta = 100 \frac{u_R}{l} \tag{93}$$

and assume that P and k may vary within the interval (-1, +1). The support reaction R and the deflection slope $w_{,x}$ due to loading and initial distortion field are

$$\begin{aligned} R^P &= \frac{1}{24} \frac{EI}{l^2} \frac{-6 + 4\eta - \eta^2}{\eta} P \\ w^P_{,s} &= \frac{1}{144} \eta(6 - 12\eta + 5\eta^2)P \\ R^{k\Delta} &= \frac{3}{200} \frac{EI}{l^2} \left(\frac{k}{\eta} + \frac{\Delta}{\eta^3} \right) \\ w^{k\Delta}_{,s} &= \frac{1}{400} \left(-\eta k + \frac{3\Delta}{\eta} \right) \end{aligned} \tag{94}$$

The sensitivity derivatives of the functional Ω with respect to w_R and s are

$$\begin{aligned} \frac{\partial \Omega}{\partial u_R} &= R^{k\Delta} = \frac{3}{200} \frac{EI}{l^2} \left(\frac{k}{\eta} + \frac{\Delta}{\eta^3} \right) \\ \frac{\partial \Omega}{\partial s} &= R^P w^P_{,s} - R^{k\Delta} w^k_{,s} = \frac{1}{3456} (-6 + 4\eta - \eta^3)(6 - 12\eta + 5\eta^2)P^2 - \\ &- \frac{3}{80000} \left(\frac{k}{\eta} + \frac{\Delta}{\eta^3}\right) \left(-\eta k + \frac{3\Delta}{\eta} \right) \end{aligned} \tag{95}$$

in the expressions (94)-(95) the symbol Δ denotes the total non-dimensional displacement of the support at B. It is equal to initial displacement, $\Delta = \Delta^i$, when there is no control displacement and equals $\Delta = \Delta^i + \Delta^s$, when the control displacement Δ^s is superposed. The sensitivity derivatives (95) contain the effect of load P, support displacement Δ and initial curvature k. Using

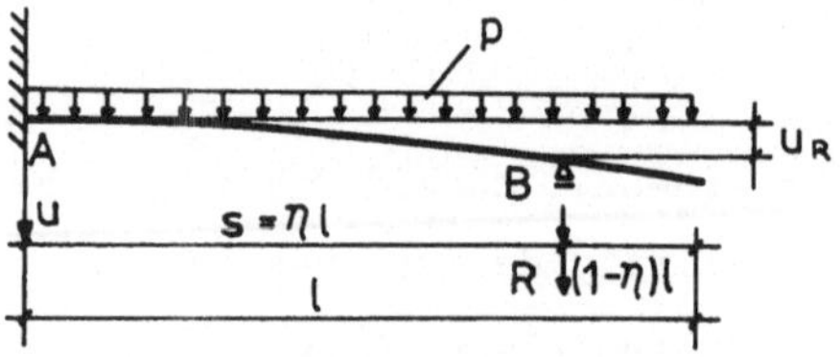

Fig.4. Beam supported at B and subjected to lateral loading, initial distortion field and initial support displacement Δ^i.

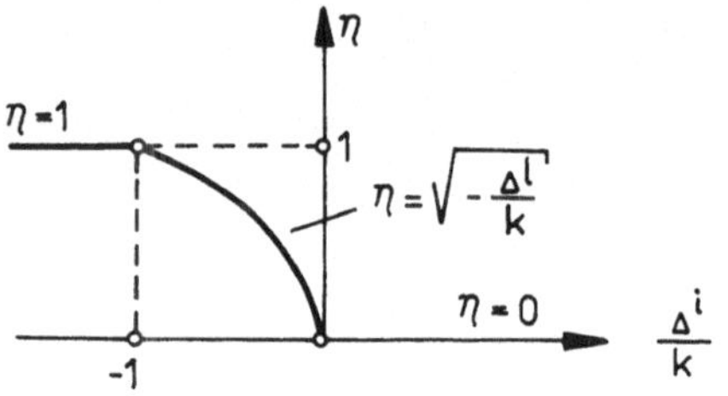

Fig.5. Variation of η depending on initial support displacement.

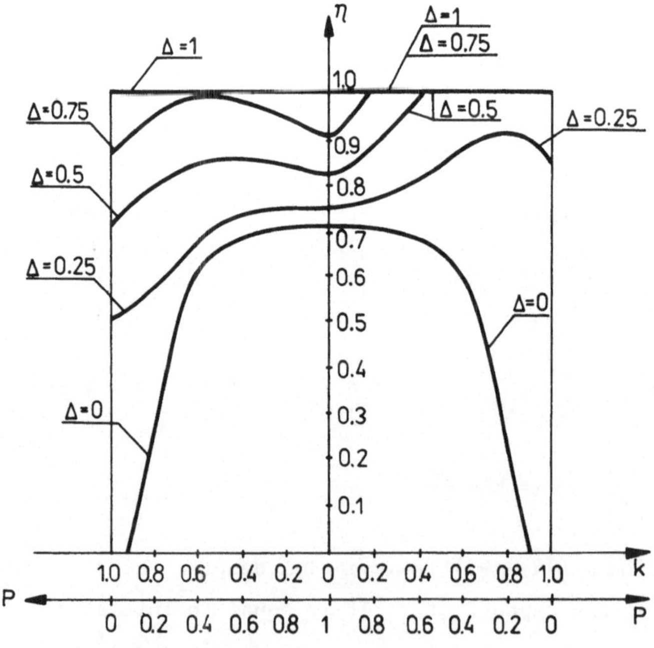

Fig.6. Design diagram η, k, P.

these derivatives the optimal solutions can be generated for various combinations of P, k, and Δ.

i) Beam acted on by the lateral load P.

The optimal support reaction corresponds to $\Delta_{opt} = 0$, $\eta_{opt} = 0.71$ that is to rigid support with no initial displacement, acting at the point at zero deflection slope.

ii) Beam subjected to initial constant curvature $k^i = k =$ const. and to support displacement Δ^i.

When $\Delta = \Delta^i$ is fixed and η is varying, we have

$$\begin{aligned} \eta_{opt} &= 1 && \text{for} \quad \frac{\Delta^i}{k} \leqq -1 \\ \eta_{opt} &= \sqrt{-\frac{\Delta^i}{k}} && \text{for} \quad -1 \leqq \frac{\Delta^i}{k} \leqq 0 \\ \eta_{opt} &= 0 && \text{for} \quad \frac{\Delta^i}{k} \geqq 0 \end{aligned} \tag{96}$$

The optimal solution is presented in Fig.5 on the diagram $(\eta, \Delta^i/k)$. When both Δ and η are varied, from the sensitivity derivatives (95) it follows that

$$\Delta_{opt} = -\eta^2 k \qquad \text{for arbitraty } \eta \tag{97}$$

iii) Beam subjected to the lateral load P and the initial curvature k^i.

When there is no support displacement, $\Delta = 0$, and η is varied, the optimal solution is presented in Fig.6 where on the abscissa axis the values of load P and of curvature k are presented, such that $P + |k| = 1$, $0 \leqq P \leqq 1$, $-1 \leqq k \leqq 1$. When Δ and k are varied, then the optimal solution corresponds to $\eta_{opt} = 0.71$, $\Delta_{opt} = -\eta^2 k = -0.504\ k$.

iv) Beam subjected to load P and initial displacement Δ^i.

When $\Delta = \Delta^i$ is fixed, the optimal values of η are plotted in Fig.6 as intersection points of curves corresponding to specified values of Δ with the η - axis (P = 1). When both Δ and η are varied, the optimal solution is that obtained for i).

v) Beam is subjected to simultaneous action of P, k, and Δ.

When $\Delta = \Delta^i$ is fixed and η varied, the optimal values of η are presented in Fig.6 for different values of P, k, and Δ. When Δ and η are subject to variation, the solution of the case iii) applies.

4. STRUCTURE SHAPE VARIATION

In this section we shall briefly discuss the case when the shape of external boundary is not specified in advance but can vary in order to attain the desired properties of a structure. Besides the external boundary, we can also investigate the variation of shape of interfaces in a composite structure, shape of reinforcing layers or of discontinuity surfaces in displacement components. The concept of shape variation can therefore be considered in a broader context, not necessarily related with the shape variation of external boundaries.

In discussing shape variation, several classes of problems can be distinguished. A most typical problem is associated with regular boundary variation of a structure for which state fields (stress, strain, displacement) are specified within the structure domain, so that normal gradients of these fields are known on the boundary subject to variation. This class of problems will be discussed in this chapter. Second class of problems arises when governing equations and state fields are specified on a surface or line (e.g. shell, arch, bar), so that only gradients of state fields within tangent plane are available. This class of problems is more difficult and needs a separate treatment.

The problems of shape variation can also be classified according to their regularity. A regular boundary variation does not involve any singularities and gradients of state variables are bounded. A singular boundary variation introduces crack-like or notch-like shapes for which local gradients of displacements or stresses are infinite. Sensitivity analysis with singular boundary variation is typical in fracture mechanics where crack growth process can be treated as shape variation. A quasi-singular boundary variation is referred to such cases where local gradients of state fields at the boundary reach great values as compared to respective values within the structure domain (e.g. notch root stresses, interface displacement gradients of a composite). Both singular and quasi-singular cases require special treatment, both analytical and numerical.

Consider an elastic body occupying of volume V and undergoing a deformation process specified by the displacement field $\underset{\sim}{u}(\underset{\sim}{x},t)$ where t is a time--like parameter. In a finite deformation process the initial configuration C_o is deformed into C_d, $C_o \rightarrow C_d$, $\underset{\sim}{x}^d = \underset{\sim}{x} + \underset{\sim}{u}$. Consider now the transformation process resulting in variation of boundary shape of the initial configuration. To describe boundary variation, the transformation vector function $\underset{\sim}{\varphi}(\underset{\sim}{x},t)$ should be specified on the boundary surface S, so that $\underset{\sim}{x}^t =$ $= \underset{\sim}{x} + \underset{\sim}{\varphi}$, $C_o \rightarrow C_t$. However, it is more convenient to specify the transformation field not only on S but also in the interior and exterior of the body as the space field. In this way, for any instantaneous shape of the structure, its further modification is specified. In a continuing modification process, we can consider the transformation rate field $\dot{\underset{\sim}{\varphi}}(\underset{\sim}{x},t)$ or the infinitesimal variation field $\delta\underset{\sim}{\varphi}(\underset{\sim}{x},t) = \dot{\underset{\sim}{\varphi}}\,dt$ specified within a domain enclosing the instantaneous structure configuration. Obviously, the continuation of the transformation field and its rate from surface into enclosing domain is quite arbitrary provided proper continuity conditions are satisfied. In fact, first variations of functionals can be expressed as surface integrals in terms of surface data only.

Consider an infinitesimal transformation field $\delta\underset{\sim}{\varphi}(\underset{\sim}{x})$ from the assumed configuration, so that a typical point P, initially placed at $\underset{\sim}{x}$ passes to the position $\underset{\sim}{x}^*$, so that

$$P \rightarrow P^*: \quad \underset{\sim}{x}^* = \underset{\sim}{x} + \delta\underset{\sim}{\varphi}(\underset{\sim}{x}) \tag{98}$$

and $\delta\underset{\sim}{\varphi}$ is assumed as the differentiable field. The volume and surface elements of a structure and the unit normal vector to the boundary surface are then transformed as follows [2]

$$\delta(dV) = \delta\varphi_{k,k}\, dV$$

$$\delta(dS) = (\delta_{kl} - n_k\, n_l)\delta\varphi_{k,l}\, dS \qquad (99)$$

$$\delta n_j = (n_j n_l - \delta_{jl})n_k\, \delta\varphi_{k,l}$$

where comma preceding an index denotes partial differentiation. The variations of state fields within a structure are now expressed in a fixed reference system as follows

$$\delta u = \delta\bar{u} + u_{,k}\, \delta\varphi_k\, , \qquad \delta\underset{\sim}{\varepsilon} = \delta\bar{\underset{\sim}{\varepsilon}} + \underset{\sim}{\varepsilon}_{,k}\, \delta\varphi_k\, , \qquad \delta\underset{\sim}{\sigma} = \delta\bar{\underset{\sim}{\sigma}} + \underset{\sim}{\sigma}_{,k}\, \delta\varphi_k \qquad (100)$$

and the variations of surface tractions and body forces are

$$\delta T_i = \delta(\sigma_{ij}n_j) = \delta\sigma_{ij}n_j + \sigma_{ij}\delta n_j = \delta\bar{\sigma}_{ij}n_j + \sigma_{ij,k}\delta\varphi_k n_j +$$

$$+ \sigma_{ij}(n_j n_l - \delta_{jl})n_k\delta_{kl} \qquad (101)$$

$$\delta f = \delta\bar{f} + f_{,k}\, \delta\varphi_k$$

Consider now the first variations or derivatives of volume and surface integrals. Consider the volume integral

$$I_v = \int f\, dV \qquad (102)$$

and its variation

$$\delta I_v = \int \delta f\, dV + \int f\delta(dV) = \int (\delta f + f\delta\varphi_{k,k})dV = \int \delta\bar{f}\, dV + \int f\delta\varphi_n dS \qquad (103)$$

where $\delta\varphi_n = \delta\varphi_k n_k$ is the normal component of $\delta\underset{\sim}{\varphi}$. The time derivative of I_v is expressed similarly

$$\frac{d}{dt}(I_v) = \int \left(\frac{Df}{Dt} + f\dot{\varphi}_{k,k}\right)dV = \int \left[\frac{\partial f}{\partial t} + (f\varphi_k)_{,k}\right]dV =$$

$$= \int \frac{\partial f}{\partial t}\, dV + \int f\dot{\varphi}_n\, dS \qquad (104)$$

and is familiar in continuum mechanics. Here Df/Dt denotes the material time derivative and $\partial f/\partial t$ is the local derivative with the position of material point fixed.

Consider now the surface integral

$$I_s = \int f \, dS \tag{105}$$

over the regular surface S. In view of (99) its variation can be expressed as follows

$$\begin{aligned} \delta I_s &= \int \delta f \, dS + \int f\delta(dS) = \int \delta f + f(\delta_{kl} - n_k n_l)\delta\varphi_{k,l} = \\ &= \int [\delta f + f(\delta\varphi_{\alpha,\alpha} - 2K_m \delta\varphi_n)]dS \end{aligned} \tag{106}$$

where $\delta\varphi_\alpha$ $(\alpha = 1,2)$ are the tangential components of $\delta\underset{\sim}{\varphi}$ referred to a curvilinear reference system within the surface and K_m is the mean surface curvature (now comma denoting the covariant derivative). The last expression can next be transformed as follows

$$\begin{aligned} \delta I_s &= \int (\delta f - f_{,\alpha}\delta\varphi_\alpha - 2fK_m\delta\varphi_n)dS + \oint f \, \delta\varphi_\mu \, dl = \\ &= \int (\delta_n f - 2fK_m\delta\varphi_n)dS + \oint f \, \delta\varphi_\mu \, dl \end{aligned} \tag{107}$$

where we applied the Green theorem for a continuous and differentiable vector field $\underset{\sim}{v}$ specified on a regular surface S bounded by a piecewise smooth closed curve

$$\int v_{\alpha,\alpha} \, dS = \int_\Gamma v_\alpha \mu_\alpha \, dl = \int_\Gamma v_\mu \, dl \tag{108}$$

where l is the arc length and $\underset{\sim}{\mu}$ denotes the unit vector normal to Γ and tangent to S, pointing towards the outside of S. The symbol $\delta_n f$ denotes the variation along the direction normal to S, thus

$$\delta_n f = \delta\bar{f} + f_{,n} \, \delta\varphi_n \tag{109}$$

and $\delta\varphi_\mu = \delta\varphi_\alpha\mu_\alpha = \delta\varphi_i\mu_i$ is the component of the transformation vector $\delta\underset{\sim}{\varphi}$ in the direction of $\underset{\sim}{\mu}$. To calculate the variation of surface integral over a

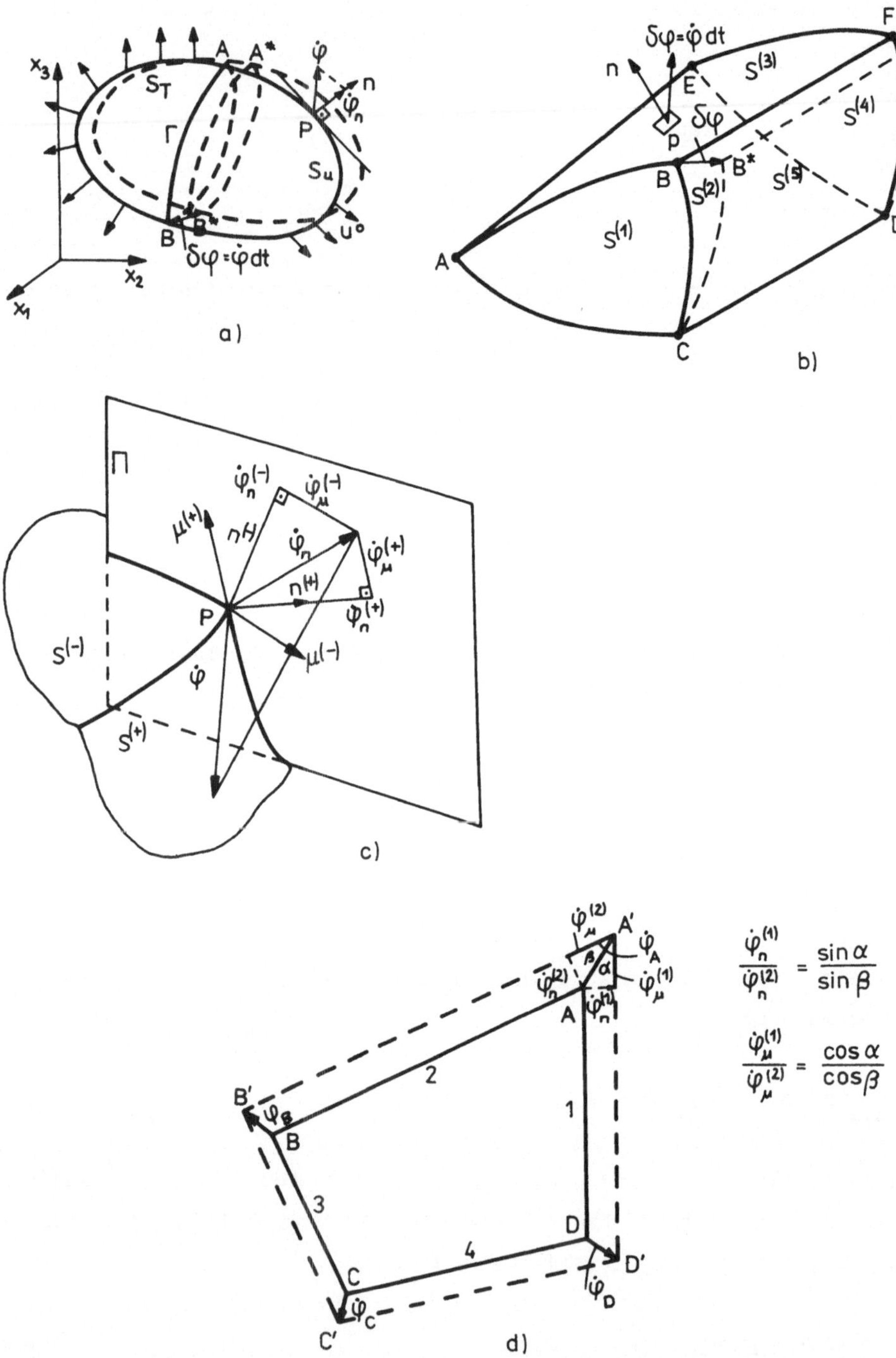

Fig.7. Boundary shape variation, a) regular boundary, b) piecewise regular boundary, c) decomposition of the transformation vector at the edge of intersection of regular boundaries, d) polygonal boundary.

piecewise regular surface, the formula (107) can be directly applied over each regular surface with subsequent addition of consecutive variations. In particular, the variation of the surface integral over a closed piecewise regular surface S takes the form

$$\delta I_S = \int (\delta_n f - 2fK_m \delta\varphi_n) dS + \Sigma \int (f^+ \delta\varphi_\mu^+ + f^- \delta\varphi_\mu^-) dl \tag{110}$$

where the sum of the line integrals is taken over all edges of the surface S, and (+) and (-) signs refer to quantities evaluated on the two regular surface sections intersecting along the edge Γ.

Consider the augmented functional (7), and its variation

$$\begin{aligned}\delta\bar{G} = &\int \delta\psi\, dV + \int \psi\delta(dV) + \int \delta h\, dS + \int h\delta(dS) - \int \delta(\underset{\sim}{\sigma}\cdot\underset{\sim}{\varepsilon}^a) dV - \\ &- \int (\underset{\sim}{\sigma}\cdot\underset{\sim}{\varepsilon}^a)\delta(dV) + \int \delta(\underset{\sim}{f}\cdot\underset{\sim}{u}^a) dV + \int (\underset{\sim}{f}\cdot\underset{\sim}{u}^a)\delta(dV) + \int \delta(\underset{\sim}{T}\cdot\underset{\sim}{u}^a) dS + \\ &+ \int (\underset{\sim}{T}\cdot\underset{\sim}{u}^a)\delta(dS)\end{aligned} \tag{111}$$

Omitting details of derivation, expression (111) can be presented in a form containing only boundary integrals

$$\begin{aligned}\delta G = &\int [\psi - \underset{\sim}{\sigma}\cdot\underset{\sim}{\varepsilon}^a + \underset{\sim}{f}\cdot\underset{\sim}{u}^a + (h + \underset{\sim}{T}\cdot\underset{\sim}{u}^a)_{,n} - (h + \underset{\sim}{T}\cdot\underset{\sim}{u}^a)2K_m]\delta\varphi_n\, dS + \\ &+ \int \Big(\frac{\partial h}{\partial \underset{\sim}{T}} + \underset{\sim}{u}^a\Big)\cdot(\delta\underset{\sim}{T}^o - \underset{\sim}{T}^o_{,k}\delta\varphi_k)\, dS_T + \int \Big(\frac{\partial h}{\partial \underset{\sim}{u}} - \underset{\sim}{T}^a\Big)\cdot(\delta\underset{\sim}{u}^o - \underset{\sim}{u}^o_{,k}\delta\varphi_k) dS_u + \\ &+ \int \delta\bar{\underset{\sim}{f}}\cdot\underset{\sim}{u}^a\, dV + \oint_\Gamma [h + \underset{\sim}{T}\cdot\underset{\sim}{u}^a]\delta\varphi_\mu\, dl\end{aligned} \tag{112}$$

where the line integral can be selected along the curve Γ separating the boundary portions S_T and S_u, Fig.7a. Since $\delta\varphi_\mu^+ = -\delta\varphi_\mu^- = \delta\varphi_\mu$, the integrand $[h + T\cdot u^a] = (h + \underset{\sim}{T}\cdot\underset{\sim}{u}^a)^+ - (h + \underset{\sim}{T}\cdot\underset{\sim a}{u}^a)^-$ represents the discontinuity of $h + \underset{\sim}{T}\cdot\underset{\sim}{u}^a$ along Γ. This line integral describe therefore the interaction between boundaries with discontinuous boundary conditions.

Consider now a case when the boundary is formed by a set of regular surfaces intersecting at edges L_k, Fig.7b. Following (110), the line integrals along L_k in (112) will appear in the form

$$\sum_k \int_{\Gamma_k} [(h + \underset{\sim}{T}\cdot\underset{\sim}{u}^a)^+ \delta\varphi_\mu^+ + (h + \underset{\sim}{T}\cdot\underset{\sim}{u}^a)^- \delta\varphi_\mu^-] dl_k \qquad (113)$$

which should be added to (112) as the contribution along the edges of intersection of regular surface sections.

REFERENCES

1. K.Dems and Z.Mróz, "Variational approach by means of adjoint systems to structural optimization and sensitivity analysis. Part I. Variation of material parameters within fixed domain", Int.J.Solids Struct., 19 , 677-- 692, 1983.

2. K.Dems and Z.Mróz, "Variational approach by means of adjoint systems to structural optimization and sensitivity analysis. Part II. Structure shape variation", Int.J.Solids Struct., 20, 527-552, 1984.

3. Z.Mróz, M.P.Kamat and R.H.Plaut, "Sensitivity analysis and optimal design of non-linear beams and plates", J.Struct.Mech., 13, 245-266, 1985.

4. Z.Mróz and G.I.N.Rozvany, "Optimal design of structures with variable support conditions",J.Opt.Theory Appl., 15, 85-101, 1975.

5. D.Szelag and Z.Mróz, "Optimal design of elastic beams with upspecified support conditions", Z.Angew.Math.Mech., 58, 501-510, 1978.

6. Z.Mróz and T.Lekszycki, "Optimal support reaction in elastic frame structures", Comp.Struct., 14, 179-185, 1982.

7. A.Garstecki and Z.Mróz, "Optimal design of elastic structures subjected to loads and initial distortions", J.Struct.Mech., 1986, (in print).

8. N.Olhoff and J.E.Taylor, "Designing continuous columns for minimum total cost and interior supports", J.Struct.Mech., 6, 367-382, 1978.

9. E.J.Haug, V.Komkov and K.K.Choi, "Design sensitivity analysis of structural systems", Acad.Press, 1985.

10. G.Szefer, Z.Mróz and L.Demkowicz, "Variational approach to sensitivity analysis in non-linear elasticity", Arch.Mech. 1987 (in print).

SHAPE DESIGN SENSITIVITY ANALYSIS AND OPTIMAL DESIGN OF STRUCTURAL SYSTEMS*

Kyung K. Choi
Department of Mechanical Engineering
and
Center for Computer Aided Design
College of Engineering
The University of Iowa
Iowa City, Iowa 52242

ABSTRACT

The material derivative concept of continuum mechanics and an adjoint variable method of design sensitivity analysis are used to relate variations in structural shape to measures of structural performance. A domain method of shape design sensitivity analysis is used to best utilize the basic character of the finite element method that gives accurate information not on the boundary but in the domain. Implementation of shape design sensitivity analysis using finite element computer codes is discussed. Recent numerical results are used to demonstrate accuracy that can be obtained using the method. Result of design sensitivity analysis is used to carry out design optimization of a built-up structure.

1. INTRODUCTION

A substantial literature has been developed in the field of shape design sensitivity analysis and optimization of structural components [1-3] over the past few years. Contributions to this field have been made using two fundamentally different approaches to structural modeling and analysis. The first approach uses a discretized structural model, based on finite element analysis, and proceeds to carry out shape design sensitivity analysis by controlling finite element node movement and differentiating the algebraic finite element equations [4-6]. The second approach to shape design sensitivity analysis uses an elasticity model of

*Research supported by NASA-Langley Grant NAG-1-215.

NATO ASI Series, Vol. F27
Computer Aided Optimal Design: Structural and Mechanical Systems. Edited by C. A. Mota Soares

the structure and the material derivative method of continuum mechanics to account for changes in shape of the structure [7-13]. Using this approach, expressions for design sensitivity in terms of domain shape change are derived in the continuous setting and evaluated using any available method of structural analysis; e.g., finite element analysis, boundary element analysis, photoelasticity, etc.

Shape design sensitivity analysis for several structural components has been treated in Refs. 2, 9, and 10 where sensitivity information is explicitly expressed as integrals, using integration by parts and boundary and/or interface conditions to obtain identities for transformation of domain integrals to boundary integrals. Numerical calculation of design sensitivity information in terms of the resulting boundary integrals thus requires stresses, strains, and/or normal derivatives of state and adjoint variables on the boundary. However, when the finite element method is used for analysis of built-up structures, the accuracy of numerical results for state and adjoint variables on interface boundaries may not be good [14].

To overcome this difficulty, a domain method of shape design sensitivity analysis is developed in Ref. 15, in which design sensitivity information is expressed as domain integrals, instead of boundary integrals (boundary method). The domain and the boundary methods are analytically equivalent. However, when one uses an approximate numerical method such as finite element analysis, the resulting design sensitivity approximations may give quite different numerical values. Moreover, the domain method offers a remarkable simplification in derivation of shape design sensitivity formulas for built-up structures since interface conditions are not required to obtain shape design sensitivity formulas. In the domain method, numerical evaluation of the sensitivity information is more complicated and inefficient than the result of the boundary method, since the domain method requires integration over the entire domain, whereas the boundary method requires integration over only the variable boundary. To alleviate this problem, a boundary layer of finite elements that vary during the perturbation of the shape of a structural component is introduced in Ref. 16.

In shape design problems, nodal points of the finite element model move as shape changes. In Ref. 17, a method of automatic regridding to account for shape change has been developed using a velocity field in the domain that obeys the governing deformation equations of the elastic solid.

Using the domain method of Ref. 15 and results of conventional (sizing) design sensitivity analysis theory of Ref. 2, a design component method is developed in Ref. 18 for unified and systematic organization of design sensitivity analysis of built-up structures, with both conventional and shape design variables. That is, conventional and shape design sensitivity formulas for each standard component type can be derived. The result is standard formulas that can be used for design sensitivity analysis of built-up structures, by simply adding contributions from each component. The method gives a systematic organization of computations for design sensitivity analysis that is similar to the way in which computations are organized within a finite element code.

A numerical method has been developed in Ref. 19 to implement the results of the design component method, using the versatility and convenience of existing finite element codes. It is shown in Ref. 19 that calculations can be carried out outside existing finite element codes, using postprocessing data only. Thus, design sensitivity analysis software does not have to be imbedded in an existing finite element code.

The purpose of this paper is to combine these developments to present a unified method of shape design sensitivity analysis and numerical implementation of the method with existing finite element codes. Even though only static response is considered here, the method is also applicable for eigenvalue design sensitivity analysis as shown in Ref. 2.

The design sensitivity analysis method presented here supports optimality criteria method for structural optimization and serves as the foundation for iterative methods of structural optimization using nonlinear programming. At a more practical level, the design sensitivity analysis method can be used to develop an interactive computer-aided design system [2]. A large scale built-up structure is optimized to demonstrate capability of the method.

2. VARIATIONAL FORM OF GOVERNING EQUATIONS

While a substantial library of structural components must be considered to implement the design sensitivity analysis for broad classes of applications, the component library to be considered in this paper is limited to truss, beam, plate, plane elastic solid, and three dimensional solid components. Even though this is a somewhat restricted class of components, it is general enough that significant applications can be made and practicality of the method can be demonstrated.

In the actual formulation, the truss and beam components, including both bending and torsion of the beam, are incorporated into a single component. Similarly, plate and plane elastic solid components are combined as a single component. To be more specific, the following formation of beam/truss, three dimensional elastic solid, and plate/plane elastic solid components are employed:

A. BEAM/TRUSS

Consider the beam/truss component of Fig. 2.1. The energy bilinear form (internal virtual work) [2] of the component is

$$a_{u,\Omega}(z,\bar{z}) = \int_0^{\ell} EI^1 w^1_{xx}\bar{w}^1_{xx}dx + \int_0^{\ell} EI^2 w^2_{xx}\bar{w}^2_{xx}dx + \int_0^{\ell} GJ\theta_x\bar{\theta}_x dx + \int_0^{\ell} hEv_x\bar{v}_x dx \qquad [2.1]$$

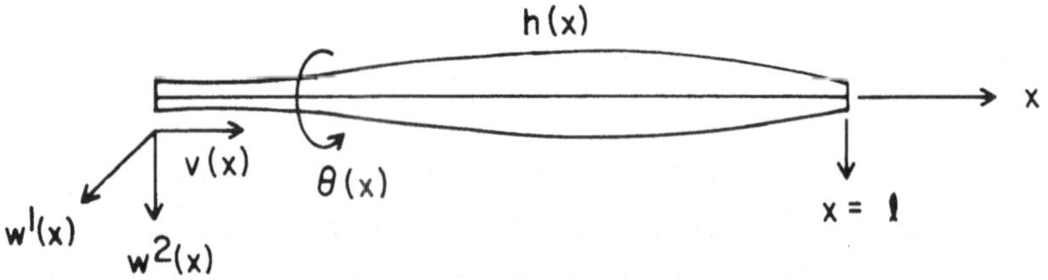

Figure 2.1 Beam/Truss Component

where w^1, w^2, θ, and v are two orthogonal lateral displacements, angle of twist, and axial displacement, respectively, and $z = [w^1, w^2, \theta, v]^T$. Throughout this paper, an overbar; e.g., $\bar{z}$, denotes a virtual displacement. Subscript x in Eq. 2.1 denotes derivative with respect to x. In Eq. 2.1, E, G, I^1, I^2, J, and h are Young's modulus, shear modulus, two moments of inertia, torsional moment of inertia, and cross-sectional area of the component, respectively. The conventional design variable is u = h(x) and the shape design variable is the length of the domain $\Omega = [0,\ell]$. The load linear form (external virtual work) [2] of the component is

$$\ell_{u,\Omega}(\bar{z}) = \int_0^\ell q^1\bar{w}^1 dx + \int_0^\ell q^2\bar{w}^2 dx + \int_0^\ell r\bar{\theta} dx + \int_0^\ell f\bar{v} dx \qquad [2.2]$$

where q^1, q^2, r, and f are two orthogonal lateral loads, twisting moment, and axial load, respectively, as shown in Fig. 2.2 [2]. If there are point loads, a Dirac delta measure can be used for q^1, q^2, r, and f in Eq. 2.2 [2].

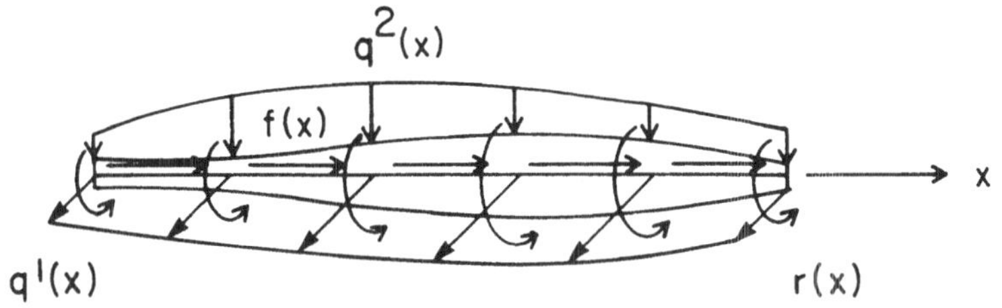

Figure 2.2 External Loads For Beam/Truss

The variational equation of the beam/truss component is [2]

$$a_{u,\Omega}(z,\bar{z}) = \ell_{u,\Omega}(\bar{z}), \qquad \text{for all } \bar{z} \in Z \qquad [2.3]$$

where Z is the space of kinematically admissible displacement. That is, $Z \subset [H^2(0,\ell)]^2 \times [H^1(0,\ell)]^2$ and elements of Z satisfy kinematic boundary conditions where $H^i(0,\ell)$ is the Sobolev space of order i [2]. As possible boundary conditions, the beam/truss component can be simply supported, clamped, cantilevered, or clamped-simply supported. It is shown in Ref. 2 that the variational Eq. 2.3 is applicable for all boundary conditions mentioned.

B. THREE DIMENSIONAL ELASTIC SOLID

Consider the three dimensional elastic solid of Fig. 2.3. For plane elastic solid, results of the three dimensional elastic solid may be reduced.

The strain tensor is defined as

$$\varepsilon^{ij}(z) = \frac{1}{2}\left(z^i_j + z^j_i\right), \quad i,j = 1,2,3, \quad x \in \Omega \qquad [2.4]$$

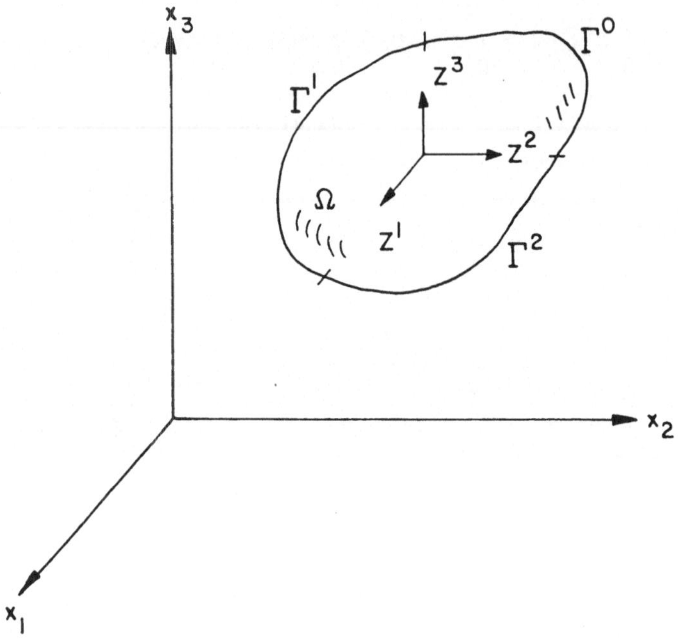

Figure 2.3 Three Dimensional Elastic Solid

where $z = [z^1, z^2, z^3]^T$ is displacement field and subscript i, i = 1,2,3, denotes derivatives with respect to variable x_i. The stress-strain relation (generalized Hooke's Law) is

$$\sigma^{ij}(z) = \sum_{k,\ell=1}^{3} C^{ijk\ell}\varepsilon^{k\ell}(z), \quad i,j,k,\ell = 1,2,3, \quad x \in \Omega \qquad [2.5]$$

where C is the elastic modulus tensor, satisfying symmetry relations $C^{ijk\ell} = C^{jik\ell}$ and $C^{ijk\ell} = C^{ij\ell k}$, $i,j,k,\ell = 1,2,3$. The energy bilinear form [2] of the three dimensional elastic solid is

$$a_{u,\Omega}(z,\bar{z}) = \iiint_\Omega \left[\sum_{i,j=1}^{3} \sigma^{ij}(z)\varepsilon^{ij}(\bar{z})\right]d\Omega \qquad [2.6]$$

Even though shape design variable, which is the shape of the domain Ω, is the only design variable in this case, subscript u is left for general treatment. The load linear form [2] of the three dimensional elastic solid is

$$\ell_{u,\Omega}(\bar{z}) = \iiint_\Omega [\sum_{i=1}^{3} f^i \bar{z}^i] d\Omega + \iint_{\Gamma^2} [\sum_{i=1}^{3} T^i \bar{z}^i] d\Gamma \qquad [2.7]$$

where Γ^0, Γ^1, and Γ^2 are clamped, traction free, and loaded boundaries, respectively, $f = [f^1, f^2, f^3]^T$ is the body force, and $T = [T^1, T^2, T^3]^T$ is the traction force.

The variational equation of the three dimensional elastic solid is [2]

$$a_{u,\Omega}(z,\bar{z}) = \ell_{u,\Omega}(\bar{z}), \quad \text{for all } \bar{z} \in Z \qquad [2.8]$$

where Z is the space of kinematically admissible displacements; i.e.,

$$Z = \{z \in [H^1(\Omega)]^3 : z^i(x) = 0, \ i = 1, 2, 3, \quad x \in \Gamma^0\} \qquad [2.9]$$

For plane elasticity problems in which either all components of stress in the x_3-direction are zero or all components of strain in the x_3-direction are zero, Eq. 2.8 remains valid, with limits of summation running from 1 to 2 and an appropriate modification of the generalized Hooke's Law of Eq. 2.5.

C. PLATE/PLANE LEASTIC SOLID

Consider the plate/plane elastic solid component of Fig. 2.4. The energy bilinear form [2] of the component is

$$a_{u,\Omega}(z,\bar{z}) = \iint_\Omega \hat{D}(t)[(w_{11} + \nu w_{22})\bar{w}_{11} + (w_{22} + \nu w_{11})\bar{w}_{22} + 2(1-\nu)w_{12}\bar{w}_{12}]d\Omega + \iint_\Omega t[\sum_{i,j=1}^{2} \sigma^{ij}(v)\varepsilon^{ij}(\bar{v})]d\Omega \qquad [2.10]$$

where $z = [w, v^1, v^2]^T$ is the displacement field. In Eq. 2.10, $\hat{D}(t) = Et^3/[12(1-\nu^2)]$, ν, and t are flexural rigidity, Poisson's ratio, and thickness of the component, respectively. Also, $\sigma^{ij}(v)$ and $\varepsilon^{ij}(v)$ are stress and strain due to an in-plane displacement field $v = [v^1, v^2]^T$, respectively. For this component, the conventional design variable is $u = t(x)$ and the shape design variable is the shape of the domain Ω. The load linear form [2] of the component is

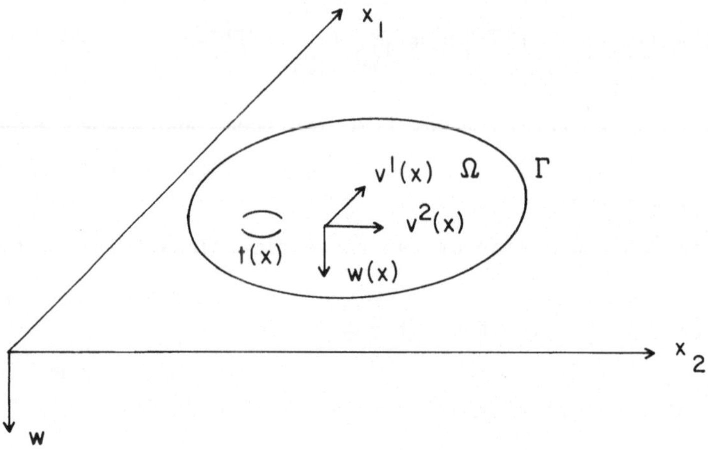

Figure 2.4 Plate/Plane Elastic Solid Component

$$\ell_{u,\Omega}(\bar{z}) = \iint_{\Omega} q\bar{w}d\Omega + \iint_{\Omega} [\sum_{i=1}^{2} f^i\bar{v}^i]d\Omega + \int_{\Gamma^2} [\sum_{i=1}^{2} T^i\bar{v}^i]d\Gamma \qquad [2.11]$$

where q, $f = [f^1, f^2]^T$ and $T = [T^1, T^2]^T$ are lateral load, body force, and traction force, respectively, as shown in Fig. 2.5. As in the beam/truss component, if there are point loads, a Dirac delta measure [2] can be used.

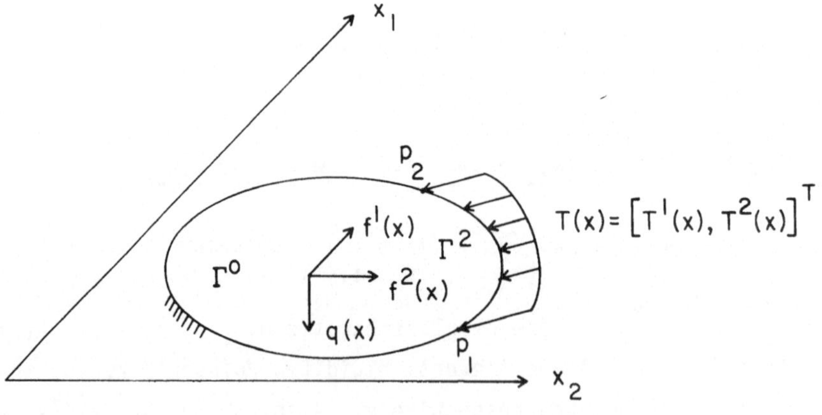

Figure 2.5 External Loads For Plate/Plane Elastic Solid

The variational equation of the plate/plane elastic solid component is [2]

$$a_{u,\Omega}(z,\bar{z}) = \ell_{u,\Omega}(\bar{z}), \qquad \text{for all } \bar{z} \in Z \qquad [2.12]$$

where $Z \subset H^2(\Omega) \times [H^1(\Omega)]^2$ and elements of Z satisfy kinematic boundary conditions. For plane leastic solid, kinematic boundary condition is

$$v^i(x) = 0, \quad i = 1,2, \quad x \in \Gamma^0 \qquad [2.13]$$

For plate, the boundary can be clamped, simply supported, or free edge. While the calculation may not be as simple as in the case of beam, the variational Eq. 2.12 is valid for all boundary conditions considered [2].

Note that Eqs. 2.3, 2.8, and 2.12, the variational equations for different structural components are all in the same form.

3. MATERIAL DERIVATIVE FOR SHAPE DESIGN SENSITIVITY ANALYSIS

The first step in shape design sensitivity analysis is development of relationships between a variation in shape of a structural component and the resulting variations in functionals that may arise in the shape design problems. Since the shape of domain a structural component occupies is treated as the design variable, it is convenient to think of Ω as a continuous medium and utilize the material derivative idea of continuum mechanics. In this section, the definition of material derivative is introduced and several material derivative formulas that will be used in later sections are derived.

Consider a domain Ω in one, two, or three dimensions, shown schematically in Fig. 3.1. Suppose that only one parameter τ defines the transformation T, as shown in Fig. 3.1. The mapping $T : x \to x_\tau(x)$, $x \in \Omega$, is given by

$$\left.\begin{aligned} x_\tau &= T(x,\tau) \\ \Omega_\tau &\equiv T(\Omega,\tau) \end{aligned}\right\} \qquad [3.1]$$

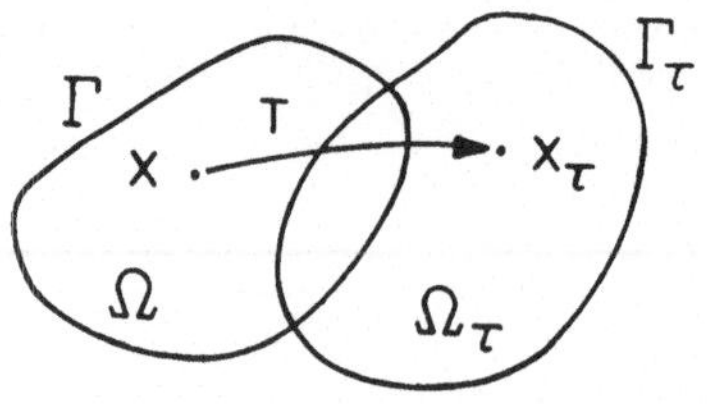

Figure 3.1 One Parameter Family of Mappings

The process of deforming Ω to Ω_τ by the mapping of Eq. 3.1 may be viewed as a dynamic process of deforming a continuum, with τ playing the role of time. At the initial time $\tau = 0$, the domain is Ω. Trajectories of points $x \in \Omega$, beginning at $\tau = 0$, can now be followed. The initial point moves to $x_\tau = T(x,\tau)$. Thinking of τ as time, a design velocity can be defined as

$$V(x_\tau,\tau) \equiv \frac{dx_\tau}{d\tau} = \frac{\partial T(x,\tau)}{\partial \tau} \qquad [3.2]$$

In a neighborhood of $\tau = 0$, under reasonable regularity hypotheses [2],

$$T(x,\tau) = T(x,0) + \tau \frac{\partial T}{\partial \tau}(x,0) + 0(\tau^2)$$

$$= x + \tau V(x,0) + 0(\tau^2)$$

Ignoring higher order terms,

$$T(x,\tau) = x + \tau V(x) \qquad [3.3]$$

where $V(x) \equiv V(x,0)$. In this paper, only the transformation T of Eq. 3.2 will be considered, the geometry of which is shown in Fig. 3.2. Variations of the domain Ω by the design velocity field V(x) are denoted as $\Omega_\tau = T(\Omega,\tau)$ and the boundary of Ω_τ is denoted as Γ_τ. Henceforth in the paper, the term "design velocity" will be referred to simply as "velocity".

Let Ω be a C^k-regular open set; i.e., its boundary Γ is closed and bounded and can be locally represented by a C^k-function. Let $V(x) \in R^n$ in Eq. 3.2 be a vector defined on a neighborhood U of the closure $\bar{\Omega}$ of Ω

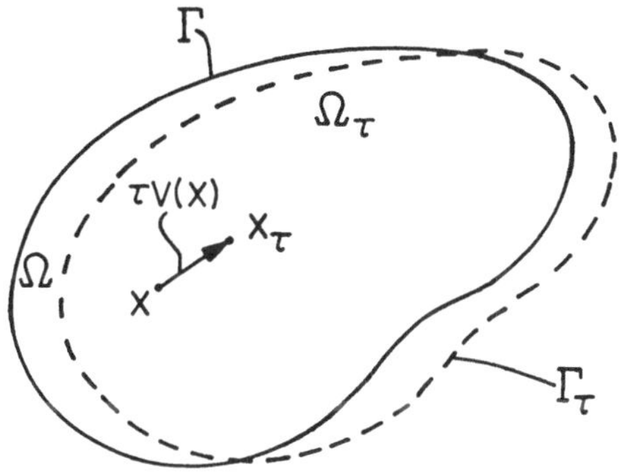

Figure 3.2 Variation of Domain

and V(x) and its derivatives up to order $k \geq 1$ be continuous. With these hypotheses, it has been shown [20] that for small τ, $T(x,\tau)$ is a homeomorphism (a one-to-one, continuous map with a continuous inverse) from U to $U_\tau \equiv T(U,\tau)$ and that $T(x,\tau)$ and its inverse mapping $T^{-1}(x_\tau,\tau)$ have C^k-regularity and Ω_τ has C^k-regularity.

Suppose $z_\tau(x_\tau)$ is a smooth solution of the elasiticity equations. Then the mapping $z_\tau(x_\tau) \equiv z_\tau(x + \tau V(x))$ is defined on Ω and $z_\tau(x_\tau)$ depends on τ in two ways. First, it is the solution of the boundary-value problem on Ω_τ. Second, it is evaluated at a point x_τ that moves with τ. The pointwise material derivative (which is shown to exist in Ref. 2) at $x \in \Omega$ is defined as

$$\dot{z}(x) = \frac{d}{d\tau} z_\tau(x+\tau V(x))\Big|_{\tau=0} = \lim_{\tau\to 0} \frac{z_\tau(x + \tau V(x)) - z(x)}{\tau} \qquad [3.4]$$

If z_τ has a regular extension to a neighborhood U_τ of $\bar{\Omega}_\tau$, then

$$\dot{z}(x) = z'(x) + \nabla z^T V(x) \qquad [3.5]$$

where

$$z'(x) \equiv \lim_{\tau\to 0} \frac{z_\tau(x) - z(x)}{\tau} \qquad [3.6]$$

is the partial derivative of z.

One attractive feature of the partial derivative is that, with reasonable smoothness assumptions, it commutes with the derivatives with respect to x_i [2]; i.e.,

$$\left(\frac{\partial z}{\partial x_i}\right)' = \frac{\partial}{\partial x_i}(z') , \quad i = 1,2,3 \qquad [3.7]$$

A pair of technical material derivative formulas that are used throughout the remainder of the paper are summarized in this section. Their proofs are presented in Ref. 2.

Lemma 3.1: Let ψ_1 be a domain functional, defined as an integral over Ω_τ,

$$\psi_1 = \iint_{\Omega_\tau} f_\tau(x_\tau)\, d\Omega_\tau \qquad [3.8]$$

where f_τ is a regular function defined on Ω_τ. If Ω has C^k-regularity, then the material derivative of ψ_1 at Ω is

$$\psi_1' = \iint_\Omega f'(x) d\Omega + \int_\Gamma f(x)\ (V^T n)\ d\Gamma \qquad [3.9]$$

or, equivalently,

$$\psi_1' = \iint_\Omega [f'(x) + \nabla f(x)^T V(x) + f(x) \mathrm{div}\ V(x)] d\Omega \qquad [3.10]$$

It is interesting and important to note that only the normal component $(V^T n)$ of the boundary velocity appearing in Eq. 3.9 is needed to account for the effect of domain variation. In fact, it is shown by Theorem 3.5.2 of Ref. 2 that if a general domain functional ψ has a gradient at Ω and if Ω has C^{k+1}-regularity, then only the normal component $(V^T n)$ of the velocity field on the boundary is needed for derivative calculations.

In contrast to Eq. 3.9, use of the mathematically equivalent result given in Eq. 3.10 requires that the velocity field $V(x)$ be defined throughout the domain Ω. Of course, it must be consistent with $(V^T n)$ on Γ. Nevertheless, there are an infinite number of velocity fields that satisfy this condition, for each of which the result of Eqs. 3.9 and 3.10 must be the same.

Next, consider a functional defined as an integration over Γ_τ,

$$\psi_2 = \int_{\Gamma_\tau} g_\tau(x_\tau)\, d\Gamma_\tau \qquad [3.11]$$

Lemma 3.2: Suppose g_τ in Eq. 3.11 is a regular function defined on Γ_τ. If Ω is C^{k+1} regular, the material derivative of ψ_2 is

$$\psi_2' = \int_\Gamma [g'(x) + (\nabla g^T n + H\, g(x))\,(V^T n)]\, d\Gamma \qquad [3.12]$$

where H is the curvatue of Γ in R^2 and twice the mean curvature in R^3.

4. ADJOINT VARIABLE FORMULATION OF SHAPE DESIGN SENSITIVITY ANALYSIS

As seen in Section 3, the static response of a structure depends on the shape of the domain. Existence of the material derivative $\dot{z}$, which is proved in Ref. 2, and material derivative formulas presented in Section 3 are used in this section to derive an adjoint variable method for design sensitivity analysis of several functionals. Since the finite element method is used for numerical analysis of the structural systems in this paper, only the domain method of shape design sensitivity analysis is presented in this section.

The variational equations of several structural components of Eqs. 2.3, 2.8, and 2.12 on a deformed domain, is of the form

$$a_{u,\Omega_\tau}(z_\tau, \bar{z}_\tau) \equiv \iint_{\Omega_\tau} c(z_\tau, \bar{z}_\tau)\, d\Omega_\tau$$
$$= \iint_{\Omega_\tau} f^T \bar{z}_\tau d\Omega_\tau + \int_{\Gamma_\tau^2} T^T \bar{z}_\tau d\Gamma_\tau \equiv \ell_{u,\Omega_\tau}(\bar{z}_\tau), \quad \text{for all } \bar{z}_\tau \in Z_\tau \qquad [4.1]$$

where Z_τ is the space of kinematically admissible displacements on Ω_τ and $c(\cdot,\cdot)$ is a bilinear mapping that is defined by the integrand of Eqs. 2.1, 2.6, and 2.10.

Taking the material derivative of both sides of Eq. 4.1, using Eqs. 3.10 and 3.11 and noting that the partial derivatives with respect to τ and x commute,

$$[a_{u,\Omega}(z,\bar{z})]' \equiv a'_{u,V}(z,\bar{z}) + a_{u,\Omega}(\dot{z},\bar{z}) = \ell'_{u,V}(\bar{z}), \quad \text{for all } \bar{z} \in Z \qquad [4.2]$$

where, using Eq. 3.5,

$$[a_{u,\Omega}(z,\bar{z})]' = \iint_\Omega [c(z,\bar{z}') + c(z',\bar{z}) + \nabla c(z,\bar{z})^T V + c(z,\bar{z}) \mathrm{div}\, V]\, d\Omega$$

$$= \iint_\Omega [c(z,\dot{\bar{z}} - \nabla\bar{z}^T V) + c(\dot{z} - \nabla z^T V, \bar{z}) + \nabla c(z,\bar{z})^T V$$

$$+ c(z,\bar{z}) \mathrm{div}\, V]\, d\Omega \qquad [4.3]$$

and

$$\ell'_{u,V}(\bar{z}) = \iint_\Omega [f^T \bar{z}' + \nabla(f^T \bar{z})^T V + f^T \bar{z} \mathrm{div}\, V] d\Omega$$

$$+ \int_{\Gamma^2} [T^T \bar{z}' + (\nabla(T^T \bar{z})^T n + H T^T \bar{z})(V^T n)] d\Gamma$$

$$= \iint_\Omega [f^T(\dot{\bar{z}} - \nabla\bar{z}^T V) + \nabla(f^T \bar{z})^T V + f^T \bar{z} \mathrm{div}\, V] d\Omega$$

$$+ \int_{\Gamma^2} [T^T(\dot{\bar{z}} - \nabla\bar{z}^T V) + (\nabla(T^T \bar{z})^T n + H T^T \bar{z})(V^T n)] d\Gamma \qquad [4.4]$$

The fact that the partial derivatives of the coefficients, which depend on cross-sectional area and thickness, in the bilinear mapping $c(\cdot,\cdot)$ are zero has been used in Eq. 4.3 and $f' = T' = 0$ has been used in Eq. 4.4. For boundary variations, it is supposed that the boundary $\Gamma = \Gamma^0 \cup \Gamma^1 \cup \Gamma^2$ is varied, except that the curve $\partial\Gamma^2$ that bounds the loaded surface Γ^2 is fixed for three dimensional elastic solid, so the velocity field V at $\partial\Gamma^2$ is zero. For the case in which $\partial\Gamma^2$ is not fixed, variation of the traction term in Eq. 4.1 (given as an integral over Γ^2) gives an additional term that was not discussed in lemmas. For this case, the interested reader is referred to Ref. 21. For plane elastic solid component case, these additional terms will be given in Section 5.

For $\bar{z}_\tau$, select $\bar{z}_\tau(x + \tau V(x)) = \bar{z}(x)$; i.e., choose $\bar{z}$ as constant on the line $x_\tau = x + \tau V(x)$. Then, since $H^m(\Omega)$ is preserved by $T(x,\tau)$ (homeomorphism property noted in Section 3), if $\bar{z}$ is an arbitrary element of $H^m(\Omega)$ that satisfies kinematic boundary conditions on Γ, $\bar{z}_\tau$ is an arbitrary element of $H^m(\Omega_\tau)$ that satisfies kinematic boundary conditions on Γ_τ. In this case, using Eq. 3.5,

$$\dot{\bar{z}} = \bar{z}' + \nabla\bar{z}^T V = 0 \qquad [4.5]$$

From Eqs. 4.2, 4.3, and 4.4, using Eq. 4.5,

$$a'_{u,V}(z,\bar{z}) = \iint_\Omega [-c(z,\nabla\bar{z}^T V) - c(\nabla z^T V,\bar{z}) + \nabla c(z,\bar{z})^T V + c(z,\bar{z})\,\mathrm{div}\, V]\,d\Omega \qquad [4.6]$$

and

$$\ell'_{u,V}(\bar{z}) = \iint_\Omega [\bar{z}^T(\nabla f^T V) + f^T\bar{z}\,\mathrm{div}\, V]\,d\Omega + \int_{\Gamma^2} [-T^T(\nabla\bar{z}^T V) + (\nabla(T^T\bar{z})^T n + HT^T\bar{z})(V^T n)]\,d\Gamma \qquad [4.7]$$

Then, Eq. 4.2 can be rewritten to provide the result

$$a_{u,\Omega}(\dot{z},\bar{z}) = \ell'_{u,V}(\bar{z}) - a'_{u,V}(z,\bar{z}), \qquad \text{for all } \bar{z} \in Z \qquad [4.8]$$

Consider a displacement functional that defines the displacement at nodal point $\hat{x} \in \Omega$

$$\psi_1 = z(\hat{x}) = \iint_\Omega \hat{\delta}(x-\hat{x})z(x)\,d\Omega \qquad [4.9]$$

where $\hat{\delta}(x)$ is the Dirac delta measure at the origin. Taking the first variation of Eq. 4.9, using the material derivative,

$$\psi_1' = \dot{z}(\hat{x}) = \iint_\Omega \hat{\delta}(x-\hat{x})\dot{z}(x)\,d\Omega \qquad [4.10]$$

The objective now is to obtain an explicit expression for ψ_1' in terms of the velocity field V, which requires eliminating $\dot{z}$. An adjoint equation is introduced by replacing $\dot{z} \in Z$ in Eq. 4.10 by a virtual displacement $\bar{\lambda} \in Z$ and equating terms involving $\bar{\lambda}$ to the energy bilinear form, yielding the adjoint equation for the adjoint variable λ,

$$a_{u,\Omega}(\lambda,\bar{\lambda}) = \iint_\Omega \hat{\delta}(x-\hat{x})\bar{\lambda}(x)\,d\Omega, \qquad \text{for all } \bar{\lambda} \in Z \qquad [4.11]$$

Denote the solution of Eq. 4.11 as $\lambda^{(1)}$.

To take advantage of the adjoint equation, evaluate Eq. 4.11 at $\bar{\lambda} = \dot{z}$, since $\dot{z} \in Z$ [2], to obtain the expression

$$a_{u,\Omega}(\lambda^{(1)},\dot{z}) = \iint_{\Omega} \hat{\delta}(x-\hat{x})\dot{z}(x)d\Omega \qquad [4.12]$$

Similarly, evaluate the identity of Eq. 5.8 at $\bar{z} = \lambda^{(1)}$, since both are in Z, to obtain

$$a_{u,\Omega}(\dot{z},\lambda^{(1)}) = \ell'_{u,V}(\lambda^{(1)}) - a'_{u,V}(z,\lambda^{(1)}) \qquad [4.13]$$

Recalling that the energy bilinear form $a_{u,\Omega}(\cdot,\cdot)$ is symmetric in its arguments, the left sides of Eqs. 4.12 and 4.13 are equal, so

$$\iint_{\Omega} \hat{\delta}(x-\hat{x})\dot{z}(x)d\Omega = \ell'_{u,V}(\lambda^{(1)}) - a'_{u,V}(z,\lambda^{(1)}) \qquad [4.14]$$

Using Eqs. 4.14, Eq. 4.10 yields

$$\psi_1' = \ell'_{u,V}(\lambda^{(1)}) - a'_{u,V}(z,\lambda^{(1)}) \qquad [4.15]$$

Explicit expressions of the terms in Eq. 4.15, for each structural component can be obtained using Eqs. 4.6 and 4.7. These explicit expressions will be derived in Section 5. This order of presentation was chosen to show basic idea of the adjoint variable method without complicate derivation of expressions.

Note that evaluation of the design sensitivity formula of Eq. 4.15 requires solution of Eq. 4.1 for z. Similarly, Eq. 4.11 must be solved for the adjoint variable $\lambda^{(1)}$. This is an efficient calculation, using finite element analysis, if the boundary-value problem for z has already been solved, requiring only evaluation of the solution of the same set of finite element equations with different right side, called an adjoint load.

Next, consider a locally averaged stress functional over a test volume Ω_p of the three dimensional elastic solid,

$$\psi_2 = \iiint_{\Omega} g(\sigma(z))\, m_p \, d\Omega = \frac{\iiint_{\Omega_p} g(\sigma(z))d\Omega}{\iint_{\Omega_p} d\Omega} \qquad [4.16]$$

where σ denotes the stress tensor, Ω_p is an open set, and m_p is a characteristic function that is constant on Ω_p, zero outside of Ω_p, and whose integral is 1. Here, g is assumed to be continuously differentiable with respect to its arguments. Note that $g(\sigma(z))$ might involve principal stresses, von Mises failure criterion, or some other material failure criteria. Taking the first variation of Eq. 4.16, using Eq. 3.10 [10],

$$\psi_2' = [\iiint_{\Omega_p} (g' + \nabla g^T V + g \operatorname{div} V) d\Omega \iiint_{\Omega_p} d\Omega$$

$$- \iiint_{\Omega_p} g \, d\Omega \iiint_{\Omega_p} \operatorname{div} V d\Omega] / (\iiint_{\Omega_p} d\Omega)^2$$

$$= \iiint_{\Omega} \sum_{i,j=1}^{3} g_{\sigma^{ij}}(z) [\sigma^{ij}(\dot{z}) - \sigma^{ij}(\nabla z^T V)] m_p d\Omega$$

$$+ \iiint_{\Omega} \sum_{k=1}^{3} [\sum_{i,j=1}^{3} g_{\sigma^{ij}}(z) \sigma_k^{ij}(z) V^k] m_p d\Omega + \iiint_{\Omega} g \operatorname{div} V m_p d\Omega$$

$$- \iiint_{\Omega} g \, m_p d\Omega \iiint_{\Omega} m_p \operatorname{div} V d\Omega \qquad [4.17]$$

It can be shown that

$$\sigma^{ij}(\nabla z^T V) = \sum_{k,\ell=1}^{3} C^{ijk\ell} (\nabla z_\ell^{k^T} V + \nabla z^{k^T} V_\ell) \qquad [4.18]$$

and

$$\sum_{k=1}^{3} \sigma_k^{ij}(z) V^k = \sum_{k,\ell=1}^{3} C^{ijk\ell} (\nabla z_\ell^{k^T} V) \qquad [4.19]$$

Using these results, Eq. 4.17 becomes

$$\psi_2' = \iiint_{\Omega} [\sum_{i,j=1}^{3} g_{\sigma^{ij}}(z) \sigma^{ij}(\dot{z})] m_p d\Omega$$

$$- \iiint_{\Omega} \sum_{i,j=1}^{3} [\sum_{k,\ell=1}^{3} g_{\sigma^{ij}}(z) C^{ijk\ell} (\nabla z^{k^T} V_\ell)] m_p d\Omega$$

$$+ \iiint_{\Omega} g \operatorname{div} V m_p d\Omega - \iiint_{\Omega} g m_p d\Omega \iiint_{\Omega} m_p \operatorname{div} V d\Omega \qquad [4.20]$$

As in the displacement functional case, an adjoint equation is introduced by replacing $\dot{z} \in Z$ in the term on the right of Eq. 4.20 by a virtual displacement $\bar{\lambda} \in Z$ and equate the result to the energy bilinear form,

$$a_{u,\Omega}(\lambda,\bar{\lambda}) = \iint_{\Omega} \left[\sum_{i,j=1}^{3} g_{\sigma^{ij}}(z)\sigma^{ij}(\bar{\lambda}) \right] m_p d\Omega, \qquad \text{for all } \bar{\lambda} \in Z \qquad [4.21]$$

Denote the solution of Eq. 4.21 as $\lambda^{(2)}$. By the same method used for the displacement functional, the sensitivity formula is obtained as

$$\psi_2' = \ell'_{u,V}(\lambda^{(2)}), - a'_{u,V}(z,\lambda^{(2)})$$
$$- \iiint_{\Omega} \sum_{i,j=1}^{3} \left[\sum_{k,\ell=1}^{3} g_{\sigma^{ij}}(z) C^{ijk\ell}(\nabla z^{k^T} V_\ell) \right] m_p d\Omega$$
$$+ \iiint_{\Omega} g \operatorname{div} V m_p d\Omega - \iiint_{\Omega} g m_p d\Omega \iiint_{\Omega} m_p \operatorname{div} V d\Omega \qquad [4.22]$$

where explicit expressions of the first two terms in Eq. 4.22 for the three dimensional elastic solid can be obtained using Eq. 2.6, 2.7, 4.6, and 4.7. These explicit expressions will be derived in Section 5. Note that these terms have the same form as those of Eq. 4.15 for the three dimensional elastic solid. The difference is that terms in Eq. 4.15 are evaluated at $\lambda^{(1)}$ and terms in Eq. 4.22 are evaluated at $\lambda^{(2)}$. That is, once the expressions for terms in Eq. 4.15 are derived, they can be used for different functionals.

Finally consider a locally averaged stress functional over a test area Ω_p in a plate/plane elastic solid component,

$$\psi_3 = \iint_{\Omega} [g^1(t,w_{ij}) + g^2(\sigma(v))] m_p \, d\Omega \qquad [4.23]$$

where $g^1(t,w_{ij})$ and $g^2(\sigma(v))$ are principal stress, von Mises yield stress, or some other stress measures due to lateral displacement w and in-plane displacement field v, respectively. Here, $g^1(t,w_{ij})$ is measured at the extreme fiber and m_p is a characteristic function on that is constant on Ω_p, zero outside Ω_p, and whose integral is 1. Taking the first variation of Eq. 4.23, using Eq. 3.10,

$$\psi_3' = \iint_{\Omega} \sum_{i,j=1}^{2} g^1_{w_{ij}} [\dot{w}_{ij} - (\nabla w^T V)_{ij}] \, m_p d\Omega$$
$$+ \iint_{\Omega} \left[\sum_{i,j=1}^{2} g^2_{\sigma^{ij}} \, \sigma^{ij}(\dot{v}) \right] m_p d\Omega + \iint_{\Omega} \operatorname{div}(g^1 V) \, m_p d\Omega$$

$$- \iint_\Omega \sum_{i,j=1}^{2} \left[\sum_{k,\ell=1}^{2} g^2_{\sigma^{ij}}(v) C^{ijk\ell} (\nabla v^{k^T} V_\ell) \right] m_p d\Omega$$

$$+ \iint_\Omega g^2 \text{div } V \, m_p d\Omega - \iint_\Omega (g^1 + g^2) \, m_p d\Omega \iint_\Omega m_p \text{div } V \, d\Omega \qquad [4.24]$$

Define an adjoint equation by replacing $\dot{w}$ and $\dot{v}$ in Eq. 4.24 by virtual displacements $\bar{\eta}$ and $\bar{\xi}$, respectively, and equate terms involving $\bar{\eta}$ and $\bar{\xi}$ in Eq. 4.24 to the energy bilinear form,

$$a_{u,\Omega}(\lambda,\bar{\lambda}) = \iint_\Omega \left[\sum_{i,j=1}^{2} g^1_{w_{ij}} \bar{\eta}_{ij} \right] m_p d\Omega$$

$$+ \iint_\Omega \left[\sum_{i,j=1}^{2} g^2_{\sigma^{ij}} \sigma^{ij}(\bar{\xi}) \right] m_p d\Omega, \qquad \text{for all } \bar{\lambda} \in Z \qquad [4.25]$$

where $\lambda = [\eta, \xi^1, \xi^2]^T$ is an adjoint variable. Denote the solution of Eq. 4.25 as $\lambda^{(3)}$. By the same method used for the displacement functional, the sensitivity formula is obtained as

$$\psi_3' = \ell'_{u,V}(\lambda^{(3)}) - a'_{u,V}(z,\lambda^{(3)}) - \iint_\Omega \left[\sum_{i,j=1}^{2} g^1_{w_{ij}} (\nabla w^T V) \right] m_p d\Omega$$

$$+ \iint_\Omega \text{div}(g^1 V) m_p d\Omega - \iint_\Omega \sum_{i,j=1}^{2} \left[\sum_{k,\ell=1}^{2} g^2_{\sigma^{ij}}(v) C^{ijk\ell} (\nabla v^{k^T} V_\ell) \right] m_p d\Omega$$

$$+ \iint_\Omega g^2 \text{div } V \, m_p d\Omega - \iint_\Omega (g^1 + g^2) m_p d\Omega \iint_\Omega m_p \text{div } V \, d\Omega \qquad [4.26]$$

where explicit expressions of the first two terms in Eq. 4.26 for the plate/plane elastic solid component can be obtained using Eqs. 2.10, 2.11, 4.6 and 4.7. These explicit expressions will be derived in Section 5. As in the displacement functional case, evaluation of the design sensitivity formula of Eq. 4.26 requires solutions z and $\lambda^{(3)}$ of Eqs. 4.1 and 4.25. Design sensitivity information for locally averaged stress functional over a test length Ω_p in a beam/truss component can be derived using the same procedure.

5. SHAPE DESIGN SENSITIVITY ANALYSIS OF STRUCTURAL COMPONENTS

In this section, explicit expressions for terms in Eqs. 4.6 and 4.7 are derived for each structural components by identifying bilinear mapping c(•,•) and loading terms. The result is standard expressions that can be used for design sensitivity analysis of different functionals. These results can also be used for design sensitivity analysis of built-up structures which will be shown in Section 9.

A. BEAM/TRUSS

Using energy bilinear and load linear forms of Eqs. 2.1 and 2.2 for beam/truss component, Eqs. 4.6 and 4.7 become

$$\begin{aligned} a'_{u,V}(z,\bar{z}) = & \int_0^{\ell} \{-EI^1[3w^1_{xx}\bar{w}^1_{xx}V_x + (w^1_{xx}\bar{w}^1_x + w^1_x\bar{w}^1_{xx})V_{xx}] \\ & + EI^1_x w^1_{xx}\bar{w}^1_{xx}V\}\, dx + \int_0^{\ell} \{-EI^2[3w^2_{xx}\bar{w}^2_{xx}V_x \\ & + (w^2_{xx}\bar{w}^2_x + w^2_x\bar{w}^2_{xx})V_{xx}] + EI^2_x w^2_{xx}\bar{w}^2_{xx}V\}\, dx \\ & + \int_0^{\ell} (-GJ\theta_x\bar{\theta}_x V_x + GJ_x\theta_x\bar{\theta}_x V)\, dx \\ & + \int_0^{\ell} (-hEv_x\bar{v}_x V_x + h_x E v_x\bar{v}_x V)\, dx \end{aligned} \qquad [5.1]$$

and

$$\begin{aligned} \ell'_{u,V}(\bar{z}) = & \int_0^{\ell} (q^1_x\bar{w}^1 V + q^1\bar{w}^1 V_x)\, dx + \int_0^{\ell} (q^2_x\bar{w}^2 V + q^2\bar{w}^2 V_x)\, dx \\ & + \int_0^{\ell} (r_x\bar{\theta}V + r\bar{\theta}V_x)\, dx + \int_0^{\ell} (f_x\bar{v}V + f\bar{v}V_x)\, dx \end{aligned} \qquad [5.2]$$

B. THREE DIMENSIONAL ELASTIC SOLID

Using energy bilinear and load linear forms of Eqs. 2.6 and 2.7 for three dimensional elastic solid, Eqs. 4.6 and 4.7 become

$$a'_{u,V}(z,\bar{z}) = -\iiint_\Omega \sum_{i,j=1}^{3} [\sigma^{ij}(z)\varepsilon^{ij}(\nabla\bar{z}^T V) + \sigma^{ij}(\bar{z})\varepsilon^{ij}(\nabla z^T V)]d\Omega$$

$$+ \iiint_\Omega \nabla[\sum_{i,j=1}^{3} \sigma^{ij}(z)\varepsilon^{ij}(\bar{z})]^T V d\Omega + \iiint_\Omega [\sum_{i,j=1}^{3} \sigma^{ij}(z)\varepsilon^{ij}(\bar{z})]\text{div } V d\Omega \qquad [5.3]$$

and

$$\ell'_{u,V}(\bar{z}) = \iiint_\Omega \sum_{i=1}^{3} \bar{z}^i(\nabla f^{i^T} V)d\Omega + \iiint_\Omega [\sum_{i=1}^{3} f^i\bar{z}^i] \text{ div } V d\Omega$$

$$+ \iint_{\Gamma^2} \{-\sum_{i=1}^{3} T^i(\nabla\bar{z}^{i^T} V) + (\nabla[\sum_{i=1}^{3} T^i\bar{z}^i]^T n + H[\sum_{i=1}^{3} T^i\bar{z}^i])(V^T n)\}d\Gamma \qquad [5.4]$$

It can be verified that

$$\sum_{i,j=1}^{3} \sigma^{ij}(z)\varepsilon^{ij}(\nabla\bar{z}^T V) = \sum_{i,j=1}^{3} \sigma^{ij}(z)(\nabla\bar{z}_j^{i^T} V + \nabla\bar{z}^{i^T} V_j) \qquad [5.5]$$

and

$$\nabla[\sum_{i,j=1}^{3} \sigma^{ij}(z)\varepsilon^{ij}(\bar{z})]^T V = \sum_{i,j=1}^{3} [\sigma^{ij}(z)(\nabla\bar{z}_j^{i^T} V) + \sigma^{ij}(\bar{z})(\nabla z_j^{i^T} V)] \qquad [5.6]$$

where $V_j = [V_j^1, V_j^2, V_j^3]^T$. Using the above results, Eqs. 5.3 becomes

$$a'_{u,V}(z,\bar{z}) = -\iiint_\Omega \sum_{i,j=1}^{3} [\sigma^{ij}(z)(\nabla\bar{z}^{i^T} V_j) + \sigma^{ij}(\bar{z})(\nabla z^{i^T} V_j)]d\Omega$$

$$+ \iiint_\Omega [\sum_{i,j=1}^{3} \sigma^{ij}(z)\varepsilon^{ij}(\bar{z})] \text{ div } V d\Omega \qquad [5.7]$$

C. PLATE/PLANE ELASTIC SOLID

As in the beam/truss and three dimensional elastic solid components, explicit expressions for terms in Eqs. 4.6 and 4.7 can be obtained using energy bilinear and load linear forms of Eqs. 2.10 and 2.11 for plate/plane elastic solid component. For plane elastic solid component, Eqs. 5.7 and 5.4 remain valid, with limits of summation running from 1 to 2 and an appropriate modification of the generalized Hooke's Law of Eq. 2.5. Equations 4.6 and 4.7 become, for plate/plane elastic solid component,

$$a'_{u,V}(z,\bar{z}) = \iint_\Omega -\hat{D}(t)\{4(w_{11}\bar{w}_{11}V^1_1 + w_{22}\bar{w}_{22}V^2_2) + [\nu(w_{11}\bar{w}_{22} + w_{22}\bar{w}_{11})$$

$$- (w_{11}\bar{w}_{11} + w_{22}\bar{w}_{22}) + 2(1-\nu)w_{12}\bar{w}_{12}]\mathrm{div}\, V$$

$$+ 2(w_{11}\bar{w}_{12} + w_{12}\bar{w}_{11} + w_{22}\bar{w}_{12} + w_{12}\bar{w}_{22})(V^1_2 + V^2_1)$$

$$+ [w_1\bar{w}_{11} + w_{11}\bar{w}_1 + \nu\,(w_1\bar{w}_{22} + w_{22}\bar{w}_1)]V^1_{11}$$

$$+ [w_1\bar{w}_{22} + w_{22}\bar{w}_1 + \nu(w_1\bar{w}_{11} + w_{11}\bar{w}_1)]V^1_{22}$$

$$+ [w_2\bar{w}_{11} + w_{11}\bar{w}_2 + \nu(w_2\bar{w}_{22} + w_{22}\bar{w}_2)]V^2_{11}$$

$$+ [w_2\bar{w}_{22} + w_{22}\bar{w}_2 + \nu\,(w_2\bar{w}_{11} + w_{11}\bar{w}_2)]V^2_{22}$$

$$+ 2(1-\nu)[(w_1\bar{w}_{12} + w_{12}\bar{w}_1)V^1_{12} + (w_2\bar{w}_{12} + w_{12}\bar{w}_2)V^2_{12}]\}$$

$$+ \frac{Et^2}{4(1-\nu^2)}[w_{11}\bar{w}_{11} + \nu(w_{11}\bar{w}_{22} + w_{22}\bar{w}_{11}) + w_{22}\bar{w}_{22}$$

$$+ 2(1-\nu)w_{12}\bar{w}_{12}]\nabla t^T V\, d\Omega$$

$$+ \iint_\Omega \{-t \sum_{i,j=1}^{2} [\sigma^{ij}(v)(\nabla\bar{v}^{i\,T} V_j) + \sigma^{ij}(\bar{v})(\nabla v^{i\,T} V_j)$$

$$- \sigma^{ij}(v)\varepsilon^{ij}(\bar{v})\mathrm{div} V] + \sum_{i,j=1}^{2} \sigma^{ij}(v)\varepsilon^{ij}(\bar{v})\nabla t^T V\}d\Omega \qquad [5.8]$$

and

$$\ell'_{u,V}(\bar{z}) = \iint_\Omega [\bar{w}(\nabla q^T V) + q\bar{w}\ \mathrm{div}\ V]d\Omega$$

$$+ \iint_\Omega \sum_{i=1}^{2} [\bar{v}^i(\nabla f^{i\,T} V) + f^i\bar{v}^i\ \mathrm{div}\ V]d\Omega$$

$$+ \int_\Gamma \sum_{i=1}^{2} \{-T^i(\nabla\bar{v}^{i\,T} V) + [\nabla(T^i\bar{v}^i)^T n + HT^i\bar{v}^i](V^T n)\}d\Gamma$$

$$+ \sum_{i=1}^{2} [T^i\bar{v}^i V_T|_{p_2} - T^i\bar{v}^i V_T|_{p_1}] \qquad [5.9]$$

In Eq. 5.9, H is the curvature of the loaded boundary Γ^2 and the last two terms on the right account for corner effects due to movement of points p_1

and p_2 in Fig. 2.5 [21]. In these two terms, the notation $\big|_{p_i}$ indicates that the terms are evaluated at point p_i and V_T is the component of velocity V tangent to Γ, which is positive if it is in a counter-clockwise direction in Fig. 2.5.

Results given in Eqs. 5.1, 5.2, 5.4, and 5.7 - 5.9 can be used in Eqs. 4.15, 4.22, and 4.26 for each structural component and each functional. This allows one to develop a modular computer program that will carry out numerical integrations of terms in Eqs. 5.1, 5.2, 5.4, and 5.7 - 5.9 using the same shape functions that are employed in finite element analysis codes. The result will then be a general algorithm and numerical method for design sensitivity analysis that can be implemented with existing finite element codes which will be discussed in Section 6.

6. IMPLEMENTATION OF DESIGN SENSITIVITY ANALYSIS WITH EXISTING FINITE ELEMENT CODES

To obtain design sensitivity information, Eqs. 4.1 and 4.11 must be solved for displacement functionals and Eqs. 4.1, 4.21 and 4.25 must be solved for stress functionals. Once the original and adjoint structures are solved, one can integrate Eqs. 4.15, 4.22, and 4.26 numerically to obtain the desired sensitivity information. The finite element method can be viewed as an application of the Galerkin method to Eqs. 4.1, 4.11, 4.21, and 4.25 for an approximate solution of the boundary-value problem. Note that the energy bilinear forms for Eqs. 4.1, 4.11, 4.21, and 4.25 are same. Hence, the adjoint structures of Eqs. 4.11, 4.21, and 4.25 are the same as that of Eq. 4.1, with different adjoint loads. The adjoint load of Eq. 4.11, is a simple unit load at the point $\hat{x}$ in the positive direction of $z(\hat{x})$. To calculate the adjoint load using the load functional on the right side of Eqs. 4.21 and 4.25, one should use the same shape functions that are used in the finite element code. Since m_p is a characteristic function defined on finite element Ω_p, numerical integration of the load functional is done on Ω_p only and the adjoint equivalent nodal force acts only on the nodal points of Ω_p.

For numerical implementation with existing finite element codes, one can proceed as in the flow chart of Fig. 6.1. In the beginning, the model is defined by identifying the finite element model, original structural load, design variables, and constraint functionals. In the next step, an

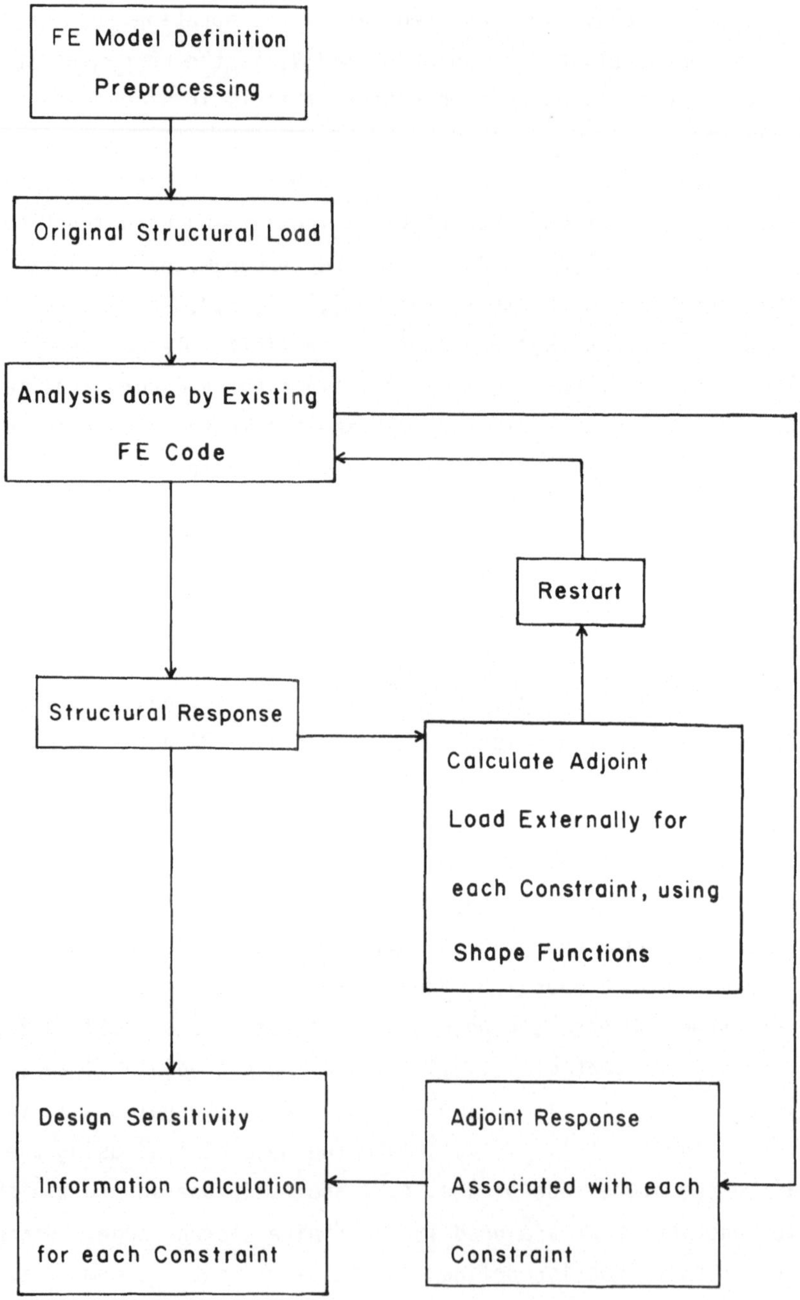

Figure 6.1 Flowchart of Design Sensitivity Calculation Procedure

existing finite element code is called to obtain structural response. With the structural response obtained, one calculates an adjoint load, external to the finite element code, using the shape functions of the code. The adjoint load is then input to the finite element code, to obtain an adjoint response for each constraint functional. For adjoint analysis, one can use the multiloading (restart) option of the finite element code, so that only forward and backward substitutions are performed to obtain each adjoint response. Using the original and adjoint structural responses, design sensitivity information is calculated for each constraint functional, by carrying out only numerical integration. This procedure allows one to carry out calculations outside finite element codes, using postprocessing data only. That is, the design sensitivity software does not have to be imbedded into finite element codes. Moreover, the method does not require differentiation of stiffness and mass matrices and the uncertainty of numerical accuracy associated with selection of a finite difference perturbation can be eliminated.

7. NUMERICAL EXAMPLES

Substantial numerical experimentation has been carried out using the material derivative shape design sensitivity analysis formulation, with the boundary method. Good results have been reported [2,23] for a variety of single structural components. These studies have shown that great care must be taken in projecting stress information to the boundary to achieve acceptable design sensitivity accuracy. Higher order elements and extrapolation from Gauss points have been shown to be essential in achieving acceptable accuracy. Substantially inaccurate results have been observed when low order elements are used and elementary boundary projection approaches are employed.

Numerical experimentation with the domain method [15,16,18,22] has indicated consistently good results for structural components, without the requirement for sophisticated elements, clever boundary projection methods, or drastically refined grids. In order to be more quantitative, two examples are discussed to permit numerical comparison.

Consider a plane elastic solid that is composed of two materials of substantially different modulus of elasticity ($E^2/E^1 = 7.65$) and subjected to simple tension, as shown in Fig. 7.1. The finite element configuration,

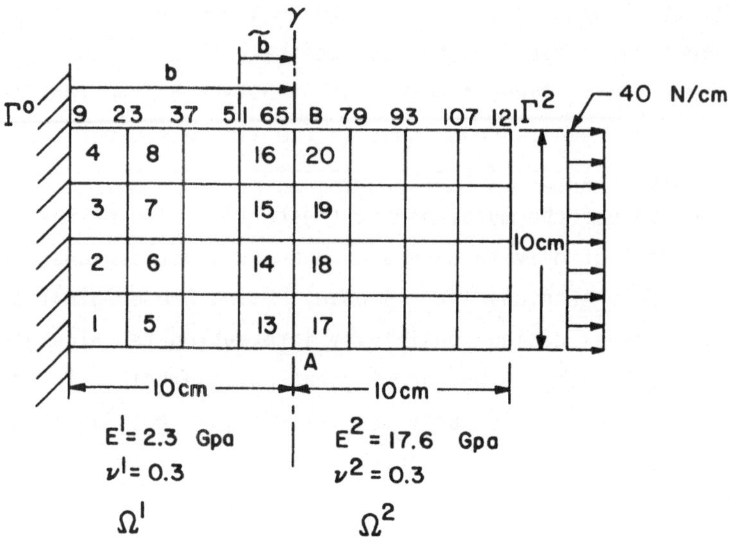

Figure 7.1 Interface Problem

dimensions, material properties of each body, and loading conditions are shown in Fig. 7.1. Body i occupies domain Ω^i, i=1,2, AB is the interface boundary γ, and Γ^0 and Γ^2 are the clamped and loaded boundaries, respectively. The design variable b controls the position of the interface boundary γ, while the overall dimensions of the structure are fixed.

The expression for design sensitivity of the von Mises yield stress functional associated with interface boundary movement with the domain method is obtained by simply adding results of Eq. 4.26 for both segments of the structure. For the plane stress interface problem, terms in Eq. 4.26 due to plate bending must be dropped. For the boundary method, design sensitivity computations are carried out in Ref. 10 (Eq. 42) that is analytically equivalent to the result of the domain method.

For numerical computation, the finite element method is used to approximate the state and adjoint equations of Eqs. 4.1 and 4.25, respectively. In order to compute the design sensitivity expressions of Eq. 4.26 one must define a design velocity field V that satisfies regularity properties defined in Refs. 2 and 9, in terms of variations in the design variable b. To have a continuous design velocity field, one may define

$$\left.\begin{aligned} V^1 &= \frac{x_1}{b}\,\delta b \\ V^2 &= 0 \end{aligned}\right\}, \text{ on } \Omega^1 \qquad [7.1]$$

and

$$\left.\begin{aligned} V^1 &= \frac{20 - x_1}{20 - b} \\ V^2 &= 0 \end{aligned}\right\}, \text{ on } \Omega^2 \qquad [7.2]$$

The finite element model shown in Fig. 7.1 contains 32 elements, 121 nodal points, and 224 degrees-of-freedom. The 8-noded isoparametric element is employed for design sensitivity analysis. For the boundary method, stresses and strains are obtained at Gauss points and extrapolated to the boundary to obtain accurate results on the boundary [23]. Define ψ^1 and ψ^2 as the functional values for the initial design b and modified design b + δb, respectively. Let $\Delta\psi = \psi^2 - \psi^1$ and let ψ' be the predicted difference from sensitivity analysis. The ratio $\psi'/\Delta\psi$ times 100 is used as a measure of accuracy; i.e., 100% means that the predicted change is exactly the same as the actual change. Notice that this accuracy measure will not give meaningful information when $\Delta\psi$ is very small compared to ψ^1, because the difference $\Delta\psi$ may lose precision due to the subtraction $\psi^2 - \psi^1$.

Numerical results with a 3% design change; i.e., $\delta b = 0.03b$, are shown in Table 7.1 for the boundary method and in Table 7.2 for the domain method. Due to symmetry, sensitivity results for only the lower half of the structure are given. These results indicate that the domain method gives excellent results, whereas accuracy of the boundary method is not acceptable. For elements 22 and 29, the predicted values are less accurate than others. However, the magnitude of actual differences $\Delta\psi$ for those elements are smaller than others, so $\Delta\psi$ may lose precision.

A disadvantage of the domain method is that a velocity field must be defined in the domain and satisfy regularity properties. There is no unique way of defining domain velocity fields for a given normal velocity field (V^Tn) on the boundary. Also, numerical evaluation of the sensitivity result of Eq. 4.26 is more complicated than evaluation of Eq. 42 of Ref. 10, since Eq. 4.26 requires integration over the entire domain, whereas Eq. 42 of Ref. 10 requires integration over only the variable boundary. This problem can be alleviated by introducing a boundary layer [16] of finite elements that vary during the perturbation of the shape of

Table 7.1. Boundary Method for Interface Problem

El. No.	ψ^1	ψ^2	$\Delta\psi$	ψ'	$(\psi'/\Delta\psi \times 100)\%$
1	393.01304	393.17922	0.16618	0.20403	122.8
2	364.37867	364.76664	0.38796	0.67218	173.3
5	388.07514	388.36215	0.28701	0.56684	197.5
6	402.26903	402.83406	0.56503	0.42080	74.5
9	386.43461	386.84976	0.41515	-0.08520	-20.5
10	407.14612	407.48249	0.33637	0.14159	42.1
13	388.59634	388.95414	0.35780	-0.53089	-148.4
14	379.04276	379.25247	0.20971	-1.90134	-906.6
17	441.68524	442.25032	0.56507	-13.85905	-2452.6
18	424.05820	425.22910	1.17089	-13.63066	-1164.1
21	424.19015	424.70840	0.51825	-0.21408	-41.3
22	378.85433	378.97497	0.12064	0.76770	636.4
25	407.71528	408.23368	0.51840	0.49780	96.2
26	387.87307	387.32342	-0.54962	-0.48837	88.9
29	400.61014	400.60112	-0.00903	0.01423	-157.7
30	394.61705	394.00702	-0.61003	-0.57794	94.7

Table 7.2. Domain Method for Interface Problem

El. No.	ψ^1	ψ^2	$\Delta\psi$	ψ'	$(\psi'/\Delta\psi \times 100)\%$
1	393.01304	393.17922	0.16618	0.17954	108.0
2	364.37867	364.76664	0.38796	0.37840	97.5
5	388.07514	388.36215	0.28701	0.28671	99.9
6	402.26903	402.83406	0.56503	0.59634	105.5
9	386.43461	386.84976	0.41515	0.41515	100.6
10	407.14612	407.48249	0.33637	0.33637	109.6
13	388.59634	388.95414	0.35780	0.37548	104.9
14	379.04276	379.25247	0.20971	0.20159	96.1
17	441.68524	442.25032	0.56507	0.57069	101.0
18	424.05820	425.22910	1.17089	1.12871	96.4
21	424.19015	424.70840	0.51825	0.53919	104.0
22	378.85433	378.97497	0.12064	0.06396	53.0
25	407.71528	408.23368	0.51840	0.51710	99.7
26	387.87307	387.32342	-0.54962	-0.56083	102.0
29	400.61014	400.60112	-0.00903	-0.00298	33.0
30	394.61705	394.00702	-0.61003	-0.58529	95.9

a structural component. This approach is illustrated schematically in Fig. 7.2. The domain Ω is divided into subdomains Ω_1 and Ω_2, with inner

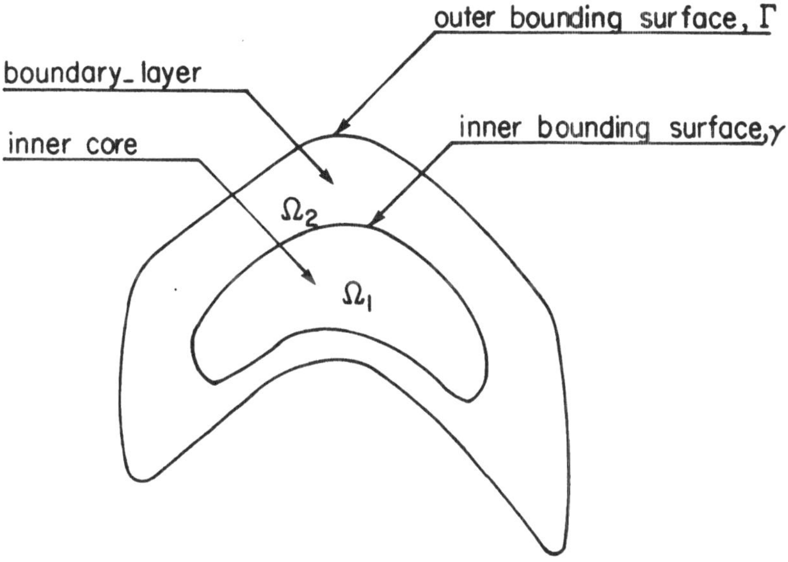

Figure 7.2 Boundary Layer

core Ω_1 held fixed and only boundary layer Ω_2 modified. In this way, the velocity field need be defined only on Ω_2. The thickness of the boundary layer Ω_2 will depend on trade-offs between numerical accuracy and numerical efficiency. In practice, Ω_1 can be a substructure of the finite element model.

To demonstrate feasibility of the boundary layer approach, two examples are solved by the boundary layer approach. The first example is the plane stress interface problem discussed in this section. For a body of given geometry there is a large number of possible boundary-layers, some of which are better than others, from the viewpoint of accuracy and efficiency. It is difficult to estimate the size and location of the best boundary-layers in advance. They can be determined by analyzing the structure and measuring the strain energy density [24].

The boundary-layer is chosen to include elements 13 thru 20 in Fig. 7.1. The design variable $\tilde{b}$ for this case is distance between node 51 and node 65 in Fig. 7.1. Consequently, regions outside the boundary-layer remain unchanged. Numerical results with a 3% design change are shown in Table 7.3 for the boundary-layer approach. Due to symmetry, the shape design sensitivity analysis results of the lower half of the structure are shown. Shape design sensitivity analysis results obtained with the boundary-layer approach are excellent, as shown in Table 7.3.

Table 7.3. Boundary Layer Approach for Interface Problem ($E^2/E^1 = 7.65$)

El. No.	ψ^1	ψ^2	$\Delta\psi$	ψ'	$(\psi'/\Delta\psi x100)\%$
1	393.01304	393.07967	0.06663	0.06770	101.6
2	364.37867	364.29542	-0.08325	-0.08412	101.0
5	388.07514	388.20344	0.12830	0.12916	100.7
6	402.26903	402.19633	0.07270	-0.07126	98.0
9	386.43461	386.47687	0.04227	0.04103	97.1
10	407.14612	407.46571	0.31960	0.32954	103.1
13	388.59634	388.70300	0.10666	0.10834	101.6
14	379.04276	379.52487	0.48210	0.48590	100.8
17	441.68524	442.17008	0.48484	0.47615	98.2
18	424.05820	425.31717	1.25897	1.23636	98.2
21	424.19015	424.44270	0.25254	0.25680	101.7
22	378.85433	378.99459	0.14025	0.11902	84.9
25	407.71528	407.99105	0.27577	0.27248	98.8
26	387.87304	387.59778	-0.27526	-0.27540	100.0
29	400.61014	400.62571	0.01557	0.01543	99.1
30	394.61705	394.45461	-0.16244	-0.16104	99.1

Next, to test validity of the boundary-layer approach, Young's modulus is changed to $E^1 = 0.2$ MPa and $E^2 = 100$ MPa for Ω^1 and Ω^2, respectively. In other words, the ratio between E^2 and E^1 is raised to 500, from 7.65, to check a more severe condition. Design sensitivity results are given in Table 7.4. Accuracy of design sensitivity is excellent. For elements 9 and 22, the magnitude of actual change are small, so finite differences may not be accurate. Numerical results obtained with the boundary method given in Ref. 16, indicates that worse results arise if the ratio E^2/E^1 is increased.

Results for the plane stress interface problem clearly indicate that the boundary approach may have considerable difficulty in handling problems with singular characteristics. Accuracy of the boundary approach rapidly deteriorates in the vicinity of a singularity. On the other hand, the boundary-layer approach can give good sensitivity results throughout the domain. Also, in this interface problem, 56% of cpu time is saved by using the boundary-layer approach instead of the domain approach, without sacrificing accuracy of design sensitivity.

Next, the classical fillet shown in Fig. 7.3 is used to study accuracy of the boundary layer approach. The design for this problem is the shape

Table 7.4. Boundary Layer Approach for Interface Problem ($E^2/E^1 = 500$)

El. No.	ψ^1	ψ^2	$\Delta\psi$	ψ'	$(\psi'/\Delta\psi x100)\%$
1	392.30557	392.39359	0.08802	0.08926	101.4
2	365.24352	365.14384	-0.09967	-0.10047	100.8
5	386.63113	386.77199	0.14086	0.14138	100.4
6	402.93637	402.89886	-0.03752	-0.03477	92.7
9	386.58037	386.56330	-0.01706	-0.02016	118.2
10	403.05338	403.58214	0.52876	0.54150	102.4
13	392.16507	392.15462	-0.01046	-0.01033	98.8
14	365.55409	366.17638	0.62229	0.62552	100.5
17	477.04122	478.18542	1.14420	1.11943	97.8
18	440.11698	442.56025	2.44327	2.39258	97.9
21	442.63898	443.09215	0.45317	0.46452	102.5
22	362.56664	362.67559	0.10895	0.07117	65.3
25	412.83466	413.32142	0.48676	0.48000	98.6
26	379.90505	379.41355	-0.49149	-0.49033	99.8
29	401.08818	401.11877	0.03059	0.03031	99.1
30	391.19616	390.92213	-0.27403	-0.27158	99.1

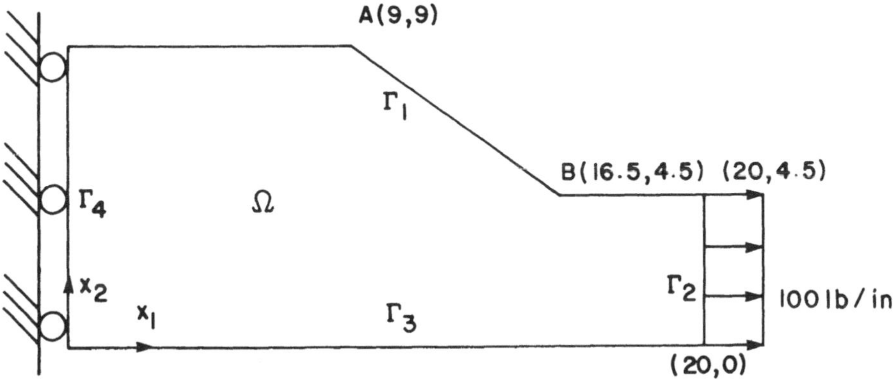

Figure 7.3 Fillet

of the varying boundary Γ_1 between points A and B, without moving these two points. B-spline representation is used for the varying boundary Γ_1. Due to symmetry, only the upper half of a fillet is analyzed. Dimensions of the structure and applied loads are given in Fig. 7.3. For material property, Young's modulus and Poisson's ratio are 3.0 x 10^7 psi and ν = 0.293, respectively. The segment Γ_3 is the center-line of the fillet and Γ_2 is the uniformly loaded edge. Sensitivity of von Mises stress averaged over individual finite elements is employed to test accuracy of the boundary-layer approach. The expression for design sensitivity is obtained from Eq. 4.26 with nonzero velocity field on the boundary layer.

The boundary-layer (27% of the total area) shown in Fig. 7.4 is chosen after analyzing the structure and measuring the strain energy density. In Fig. 7.5, a finite element model with optimized boundary profile Γ_1 and 319 elements and 1994 active degrees-of-freedom is shown. The element type used is an 8-node isoparametric element.

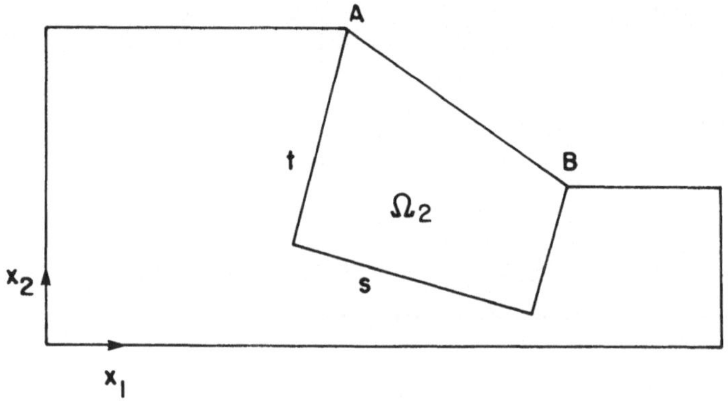

Figure 7.4 Boundary-Layer of Fillet

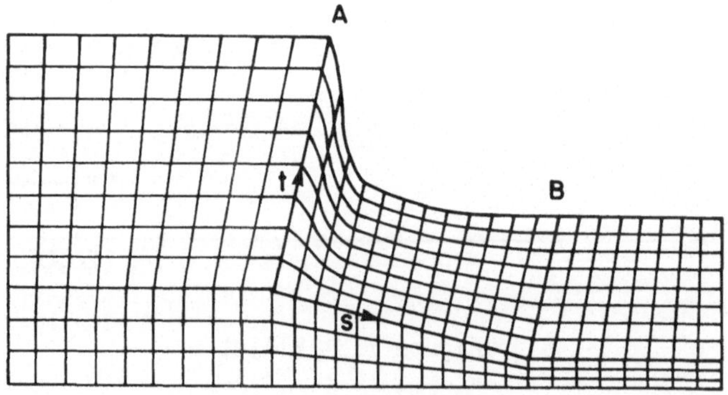

Figure 7.5 Finite Element Mesh of Fillet

In Table 7.5, shape design sensitivity results for a fillet with optimized boundary profile Γ_1 (see Fig. 7.5) are given, obtained with 0.1% design perturbation. From Table 7.5, it can be seen that this approach can yield excellent shape design sensitivity results.

Table 7.5. Boundary Layer Approach for Fillet

El. No.	ψ^1	ψ^2	$\Delta\psi$	ψ'	$(\psi'/\Delta\psi x100)\%$
3	306.91341	307.18632	0.27291	0.27802	101.9
13	326.43889	326.65762	0.21873	0.22298	101.9
23	386.17664	386.33965	0.16301	0.16693	102.4
33	471.26543	471.38415	0.11872	0.12219	102.9
43	570.91518	570.95727	0.04209	0.04294	102.0
53	669.26309	669.16035	-0.10274	-0.10693	104.1
63	736.39264	736.12922	-0.26341	-0.27187	103.2
73	682.49507	682.26122	-0.23385	-0.23826	101.9
83	761.77304	761.44391	-0.32913	-0.33921	103.1
93	911.47561	911.14248	-0.33314	-0.33869	101.7
103	764.32105	764.00138	-0.31967	-0.32424	101.4
113	884.95173	884.66270	-0.28902	-0.28788	99.6
123	768.06301	767.74016	-0.32285	-0.32715	101.3
133	857.37113	857.10506	-0.26607	-0.26249	98.7
143	999.87828	999.87875	0.00047	0.00045	95.4
153	1009.58379	1009.52243	-0.06137	-0.05071	82.6
163	1000.99060	1000.92960	-0.06100	-0.05863	96.1
173	999.42406	999.33430	-0.08976	-0.08873	98.8
183	1001.15587	1001.04706	-0.10881	-0.10766	98.9
193	958.70044	958.26124	-0.43919	-0.44416	101.1
203	980.40747	980.03216	-0.37531	-0.37607	100.2
213	993.57091	993.33304	-0.23787	-0.23708	99.7
223	1000.47920	1000.29798	-0.18122	-0.17969	99.2
233	762.37599	762.01522	-0.36076	-0.40403	112.0
243	778.19389	777.74914	-0.44475	-0.45837	103.1
253	881.29226	880.57387	-0.71838	-0.69488	96.7
263	1220.22663	1219.47520	-0.75143	-0.72158	96.0
273	835.72273	835.22544	-0.49729	-0.48555	97.6
283	922.89834	922.29381	-0.60453	-0.58846	97.3
293	1033.88104	1033.07352	-0.80752	-0.77133	95.5
303	1093.40867	1090.74702	-2.66165	-2.67483	100.5
313	936.38542	936.01573	-0.36969	-0.37920	102.6

8. AUTOMATIC REGRIDDING FOR SHAPE DESIGN

For numerical implementation of shape design sensitivity analysis, one must parameterize the boundary Γ of the domain Ω. For this purpose, one may use Bezier curves or surfaces [25]. The next step is to develop a general method of defining and computing a velocity field in the domain, in terms of the perturbation of the boundary Γ. Moreover, the velocity field must satisfy certain regularity conditions. It is shown in Refs. 2 and 9 that C^1-regular and C^2-regular velocity fields are sufficient for shape design sensitivity analysis of truss and elastic solid problems and beam and plate problems, respectively. However, observing Eqs. 5.1, 5.2, 5.4, and 5.7 - 5.9, one may relax these regularity conditions. That is, for truss and elastic solid problems, the highest order derivative of the velocity field that appears in Eqs. 5.1, 5.2, 5.4, and 5.7 - 5.9 is one. Thus, one may use a C^0-regular velocity field with an integrable first derivative. Similarly, one may use a C^1-regular velocity field with an integrable second derivative for beam and plate problems. Therefore, regularity of the velocity field must be at least at the level of regularity of the displacement field of the structural component considered. This suggests use of displacement shape functions to systematically define the velocity field in the domain. Moreover, one can select a velocity field that obeys the governing equation of the structure. That is, the perturbation of the boundary can be considered as a displacement at the boundary. With no additional external forces and a given displacement at the boundary, one can use the finite element code to find the displacement (domain velocity) field that satisfies the required regularity conditions. Thus

$$[K]\{V\} = \{f\} \qquad [8.1]$$

where $[K]$ is the reduced stiffness matrix, $\{V\}$ is the velocity vector of the nodes of varying domain, and $\{f\}$ is the unknown ficticious boundary force that produces a perturbation of the boundary. In segmented form, Eq. 8.1 becomes

$$\begin{bmatrix} K_{bb} & K_{bd}^T \\ K_{bd} & K_{dd} \end{bmatrix} \begin{Bmatrix} V_b \\ V_d \end{Bmatrix} = \begin{Bmatrix} f_b \\ 0 \end{Bmatrix} \qquad [8.2]$$

where $\{V_b\}$ is the given perturbation of nodes on the boundary, $\{V_d\}$ is the node velocity vector in the interior of the domain and $\{f_b\}$ is the ficticious boundary force acting on the varying boundary. Equation for the unknown interior node velocity vector can be obtained from Eq. 8.2 as

$$[K_{dd}]\ \{V_d\} = -\ [K_{bd}]\ \{V_b\} \qquad [8.3]$$

If Bezier curves or surfaces [17] are used for boundary representation, positions of control points are selected as design parameter b_i, i = 1,2,...,k. To use Eqs. 4.15, 4.22, and 4.26 for sensitivity computation, interior node velcity vector $\{V_d\}$ should be expressed in terms of variations of design parameter δb_i, i = 1,2,...,k. To obtain this expression, boundary perturbation $\{V_b\}$ should be written in terms of variation δb of the design parameter. Once the inverse matrix $[K_{dd}]^{-1}$ is obtained, Eq. 8.3 can be used to express $\{V_d\}$ in terms of the variation δb of design parameter. However, this requires large computational effort. To gain computational efficiency, following method is used: first by pertubing a design parameter b_i a unit magnitude, boundary perturbation $\{V_b\}$ can be obtained. Then Eq. 8.3 can be solved to obtain $\{V_d\}$. Using $\{V_d\}$ and displacement shape functions, Eqs. 4.15, 4.22, or 4.26 can be evaluated which gives $\frac{\partial \psi}{\partial b_i}$. This method requires to solve Eq. 8.3 k times. However, much as in the adjoint analysis, this is an efficient calculation, if Eq. 8.3 has already been solved, requiring only evaluation of the solution of the same set of finite element equations with different right side for each unit perturbation of b_i, i = 1,2,...,k. Once design change has been determined using iterative design process, regridding of interior grid points can be carried out using $\{V_d\}$ of Eq. 8.3.

The automatic regridding method presented here can be used with the boundary-layer approach very effectively. That is, for the fixed domain Ω_1, V_d can be set equal to zero and thus reduce the dimension of $[K_{dd}]$ in Eq. 8.3.

To demonstrate feasibility of the method, an engine bearing cap shown in Fig. 8.1 is treated. The engine bearing cap is modeled as a three dimensional elastic solid. Due to symmetry, only the right half of the cap is analyzed. The finite element configuration and loading conditions are shown in Fig. 8.1. The material used is steel with Young's modulus and Poisson's ratio of $E = 1.0 \times 10^7$ psi and $\nu = 0.3$, respectively. The finite element model shown in Fig. 8.1 contains 82 elements, 768 nodal points, and 2111 degrees-of-freedom. For analysis, ANSYS finite element STIF95 [26], which is a 20-noded isoparametric element, is used.

The design variables for this problem are the shape of the varying surface Γ_1, distance C_5 of clamping bolt center line AB, and distance C_6 of edge from cap centerline (Fig. 8.2). For surface Γ_1, a Bezier surface

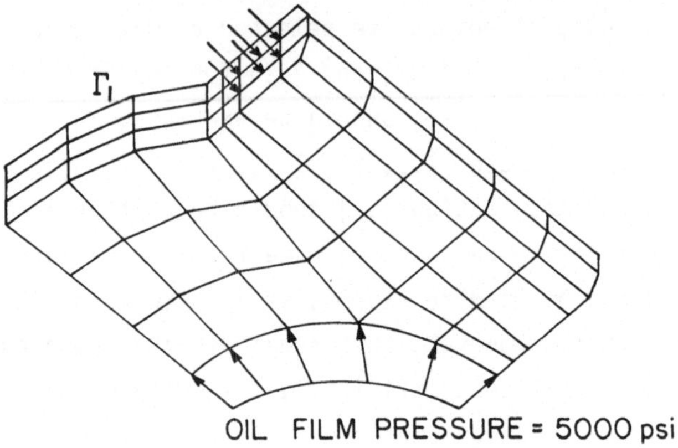

Figure 8.1 Engine Bearing Cap

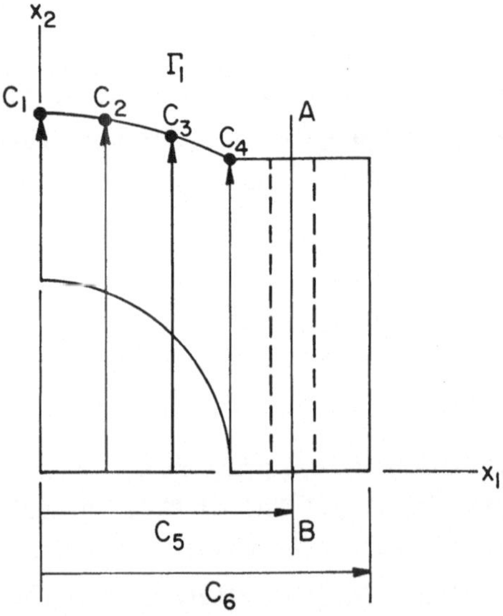

Figure 8.2 Shape Design Parameters of Engine Bearing Cap

with 4 x 4 control points is used. For simplicity, only x_2-coordinates of four control points C_1 thru C_4 are allowed to be varying. That is, surface Γ_1 has curvature in the x_1-direction only.

The expression for design sensitivity of von Mises stress averaged over individual finite element is obtained from Eq. 4.22. Numerical computation of design sensitivity information has been carried out using ANSYS finite element code [26] and the computational procedure of Section 6. For computation of domain velocity vector, Eq. 8.2 is solved using ANSYS finite element code.

Numerical results with a 1% design change are shown in Table 8.1 for randomly selected finite elements. Accuracy of design sensitivity is excellent except for elements 5, 22, 36, 56, and 57 where the magnitudes of actual change are small.

Table 8.1. Shape Sensitivity of Engine Bearing Cap

El. No.	ψ^1	ψ^2	$\Delta\psi$	ψ'	$(\psi'/\Delta\psi x100)\%$
1	9829.4564	9727.3229	-102.1335	-109.7298	107.4
2	9631.2028	9565.8641	-65.3387	-70.6936	108.2
4	10620.3320	10648.0140	27.6820	25.7748	93.1
5	11444.4800	11448.0190	3.5390	0.4482	12.7
7	13584.5710	13604.0520	19.4810	17.5377	90.0
8	13641.4950	13674.1020	32.6070	31.2587	95.9
10	17933.5910	17964.5170	30.9260	29.8750	96.6
11	18498.6300	18526.7900	28.1600	27.5316	97.8
13	21202.2630	21222.8600	20.5970	20.559	100.8
14	34270.5140	34294.7650	24.2510	23.7614	98.0
16	8367.4820	8152.6800	-214.8020	-230.0584	107.1
18	9686.1116	9652.6069	-33.5047	-38.2462	114.2
20	12670.2480	12634.3500	-35.8980	-38.4216	107.0
22	16248.4050	16256.2990	7.8940	6.3254	80.1
24	30901.3690	30862.5960	-38.7730	-36.4113	93.9
26	7311.4083	6999.4094	-311.9989	-321.7022	103.1
27	6857.7422	6519.3747	-338.3675	-344.1961	101.7
29	7917.4211	7774.3107	-143.1104	-148.4639	103.7
30	7234.2502	7081.2085	-153.0417	-159.7947	104.4
32	9528.8157	9395.4689	-133.3468	-137.7809	103.3
33	8521.5076	8387.1466	-134.3610	-141.4600	105.3
35	13328.4650	13264.9790	-63.4860	-59.4243	93.6
36	12091.2310	12159.1600	67.9290	87.2240	128.4
38	24661.4990	23849.7610	-811.7380	-842.1029	103.7
39	44231.0680	42109.0220	-2122.0460	-2222.5504	104.7
41	7349.5330	6999.9927	-349.5403	360.6076	103.2
42	6920.5279	6531.4016	-389.1263	-404.7290	104.0
44	5998.6512	5844.9335	-153.7177	-165.1199	107.4
45	5762.5105	5630.1473	-132.3632	-142.4745	107.6
47	7016.8980	6905.9260	-110.9720	-115.6099	104.2
48	6822.9614	6736.9477	-86.0137	-90.5011	105.2
50	9706.9951	9449.8267	-257.1684	-262.1729	102.0
51	9639.7495	9411.8005	-227.9490	-228.3732	100.2
53	13634.1000	12964.2560	-669.8440	-701.6882	104.8

Table 8.1 Continued

54	19874.8650	18390.3600	-1484.5050	-1586.2730	106.9
56	6080.3933	6078.1016	-2.2917	-3.6525	159.4
57	6121.4120	6114.6667	-6.7453	-8.1242	120.4
59	5832.8322	5697.4132	-135.4190	-139.1451	102.8
60	6266.5615	6098.1882	-168.3733	-172.9635	102.7
62	7041.7283	6971.4204	-70.3079	-79.6051	113.2
63	8230.6127	8059.7990	-170.8137	-188.3943	110.3
65	4816.1908	4793.3159	-22.8749	-24.1961	105.8
66	4787.5653	4761.5085	-26.0568	-27.6278	106.0
68	3537.2024	3578.9546	41.7522	40.3578	96.7
69	3692.9881	3725.9600	32.9719	31.0383	94.1
71	6541.8233	6585.9308	44.1075	45.1422	102.4
72	6643.3182	6680.1317	36.8135	38.0789	103.4
74	3872.7605	3898.8240	26.0635	25.4267	97.6
75	3820.6962	3843.9362	23.2400	22.5210	96.9
77	3918.9608	4017.5877	98.6269	100.0753	101.5
78	3932.8001	4024.4881	91.6880	92.9458	101.4
80	6240.3854	6285.3485	44.9631	46.3209	103.0
81	6158.4620	6202.3951	43.9331	45.2521	103.0

9. DESIGN COMPONENT METHOD FOR BUILT-UP STRUCTURES

In this section, design sensitivity analysis method for built-up structures is presented. Both shape and conventional (sizing) design variables for components of built-up structures are considered. For conventional design sensitivity analysis, distributed parameter structural design sensitivity analysis theory of Ref. 2 is used.

Consider a built-up structure that is made up of m > 1 structural components that are interconnected by kinematic constraints at their interfaces. Using the principle of virtual work for built-up structures [2], one can obtain the variational formulation of the governing equations,

$$a_{u,\Omega}(z,\bar{z}) = \ell_{u,\Omega}(\bar{z}), \qquad \text{for all } \bar{z} \in Z \tag{9.1}$$

where

$$a_{u,\Omega}(z,\bar{z}) = \sum_{i=1}^{m} a_{u^i,\Omega^i}(z,\bar{z}) \tag{9.2}$$

$$\ell_{u,\Omega}(\bar{z}) = \sum_{i=1}^{m} \ell_{u^i,\Omega^i}(\bar{z}) \tag{9.3}$$

and Z is the space of kinematically admissible displacements [2], which is defined as the set of displacement fields that satisfy homogeneous boundary conditions and kinematic interface conditions between components. In Eqs. 9.2 and 9.3, $a_{u^i,\Omega^i}(z,\bar{z})$ and $\ell_{u^i,\Omega^i}(\bar{z})$ are energy bilinear and load linear forms of component i with domain Ω^i. Note from Eqs. 9.2 and 9.3, the energy bilinear and load linear forms of Eq. 9.1 are simply summations of corresponding terms from each component. Thus, as will be seen later, the design sensitivity analysis of the built-up structure is a simple additive process.

In this section, design sensitivity information for displacement functional is derived for general built-up structures. Once this is done, extention to locally averaged stress functional can be carried out easily.

Define $\dot{z}$ as the total variation of z, due to both conventional and shape design changes [2],

$$\begin{aligned}\dot{z} &= \frac{d}{d\tau} z_\tau(x+\tau V(x),\ u+\tau\delta u)\Big|_{\tau=0} \\ &= \frac{d}{d\tau} z(x,u+\tau\delta u)\Big|_{\tau=0} + \frac{d}{d\tau} z_\tau(x+\tau V(x),u)\Big|_{\tau=0}\end{aligned} \quad [9.4]$$

The first variation of Eq. 9.1 is [2]

$$a'_{\delta u,\Omega}(z,\bar{z}) + a'_{u,V}(z,\bar{z}) + a_{u,\Omega}(\dot{z},\bar{z}) = \ell'_{\delta u,\Omega}(\bar{z}) + \ell'_{u,V}(\bar{z}),\ \text{for all}\ \bar{z}\in Z \quad [9.5]$$

where $\dot{z} = [\dot{w}^1,\ \dot{w}^2,\ \dot{\theta},\ \dot{v}]^T$ for beam/truss component, $\dot{z} = [\dot{z}^1,\ \dot{z}^2,\ \dot{z}^3]^T$ for, three dimensional elastic solid, and $\dot{z} = [\dot{w},\ \dot{v}^1,\ \dot{v}^2]^T$ for plate/plane elastic solid component. The notation of Eq. 9.5 is chosen to clearly display which variables are held fixed and which vary.

Consider a displacement functional that defines the displacement z at nodal point $\hat{x} \in \Omega^r$

$$\psi = \iint_{\Omega^r} \hat{\delta}(x-\hat{x})z(x)\ d\Omega \quad [9.6]$$

where $\hat{\delta}(x)$ is the Dirac delta measure at the origin. Taking the first variation of Eq. 9.6, one obtains [2]

$$\psi' = \iint_{\Omega^r} \hat{\delta}(x-\hat{x})\dot{z}(x)\ d\Omega \quad [9.7]$$

Define a variational adjoint equation by replacing $\dot{z}$ in the term on the right of Eq. 9.7 by a virtual displacement $\overline{\lambda}$ and equate the result to the energy bilinear form evaluated at the adjoint variable λ; i.e.,

$$a_{u,\Omega}(\lambda,\overline{\lambda}) = \iint_{\Omega^r} \hat{\delta}(x-\hat{x})\overline{\lambda}(x)\, d\Omega, \qquad \text{for all } \overline{\lambda} \in Z \qquad [9.8]$$

Denote the solution of Eq. 9.8 as λ. Since $\dot{z}$ satisfies kinematic boundary and interface conditions [2], Eq. 9.8 can be evaluated at $\overline{\lambda} = \dot{z}$ and Eq. 9.5 can be evaluated at $\overline{z} = \lambda$, to obtain

$$\begin{aligned} \psi' = &\sum_{i=1}^{m} [\ell'_{\delta u^i,\Omega^i}(\lambda) - a'_{\delta u^i,\Omega^i}(z,\lambda)] \\ &+ \sum_{i=1}^{m} [\ell'_{u^i,V^i}(\lambda) - a'_{u^i,V^i}(z,\lambda)] \end{aligned} \qquad [9.9]$$

where the first term on the right is due to conventional design variation and the second term is due to shape design variation. Note that Eq. 9.9 is valid for general built-up structures that are composed of $m \geqslant 1$ structural components. For explicit expressions of the second term on the right of Eq. 9.9, results of Eqs. 5.1, 5.2, 5.4, and 5.7 - 5.9 can be used. For the first term on the right of Eq. 9.9, the interested reader is referred to Refs. 2 and 18.

As seen in this derivation, one can systematically organize design sensitivity expressions for built-up structures, using the design component method. Moreover, one can develop a modular computer program that will carry out numerical integration of terms in Eqs. 5.1, 5.2, 5.4, and 5.7 - 5.9, using the same shape functions that are employed by the finite element analysis of the original structure. The result will then be a general algorithm and numerical method for design sensitivity analysis that can be implemented with existing finite element codes as shown in Section 6.

To demonstrate accuracy of the design component method, a truss-beam-plate built-up structure is treated in this section. Consider the truss-beam-plate built-up structure shown in Fig. 9.1. A distributed vertical load f(x) is applied to the plates. The points supported by the trusses are at the intersections of two crossing beams nearest to the free edges of the structure. No external loads are applied to the truss and beam components. The plates and beams are assumed to be welded together.

Coordinates of intersection points of beams and plates are supposed to be in the mid-planes of the plates and neutral axes of the beams. Beam components have rectangular cross-sections.

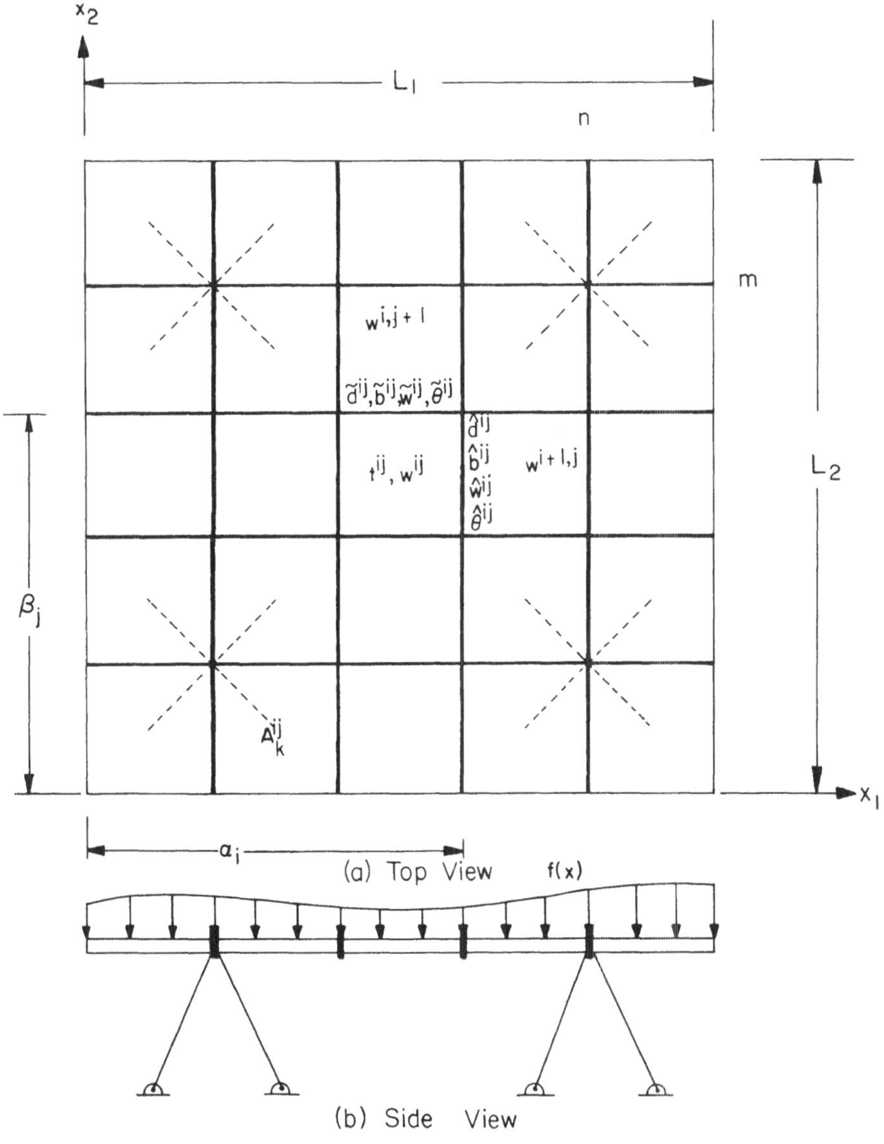

Figure 9.1 Truss-Beam-Plate Built-up Structure

The design variables for this built-up structure are thickness $t^{ij}(x)$ of each plate component, width $\tilde{d}^{ij}(x_1)$ and height $\tilde{b}^{ij}(x_1)$ of each longitudinal beam compoenent, width $d^{ij}(x_2)$ and height $b^{ij}(x_2)$ of each transverse beam component, cross-sectional areas h^k (k=1,16) of the four 4-bar truss components, and positions α_i (i=1,4) and β_j (j=1,4) of transverse and longitudinal beam components, respectively. The lengths of the trusses are fixed, but they may change their ground positions, and the outside boundary of the entire structure is fixed; i.e., only the locations α_i and β_j, i,j=1,4, of beams are shape variables. Dimensions of the structure and the numbering and spacing of beams in both directions are shown in Fig. 9.1.

For numerical calculations, conventional and shape design sensitivity calculations are carried out seperately. For plate components, 12 degree-of-freedom non-conforming rectangular elements [27] are used. For beam components, Hermite cubic shape functions are used. The finite element model used for design sensitivity analysis is shown in Fig. 9.2. Only one quarter of the entire structure is analyzed, due to symmetry. A total of 484 elements, with 1281 degrees-of-freedom, are used to model the built-up structure, including 400 rectangular plate elements, 80 beam elements and 4 truss elements.

For numerical data, Young's modulus and Poisson's ratio are 3.0×10^7 psi and 0.3, respectively. The overall dimensions are $L_1 \times L_2$ = 15 in. x 15 in. At the nominal design, beam components are located 3 in. apart. Other dimensions of the built-up structure at the nominal design are; uniform thickness t = 0.1 in. for plate components, uniform height h = 0.5 in. and width d = 0.15 in. for beam components, and length ℓ = 5.364 in. and cross-sectional area h = 0.1 in.^{2} for truss components. A uniform distributed load f = 0.1 lb/in.^{2} is applied on the plate components.

In Table 9.1, design sensitivity accuracy results are given for several functionals, with 1% uniform change in all conventional design variables except the cross-sectional areas of truss components. Design sensitivity results for displacements, bending stresses σ^{11} and σ^{22} at the extreme fiber of longitudinal and transverse beam components, and von Mises yield stress

$$g(\sigma) = ({\sigma^{11}}^2 + {\sigma^{22}}^2 + 3{\sigma^{12}}^2 - \sigma^{11}\sigma^{22})^{1/2} \qquad [9.10]$$

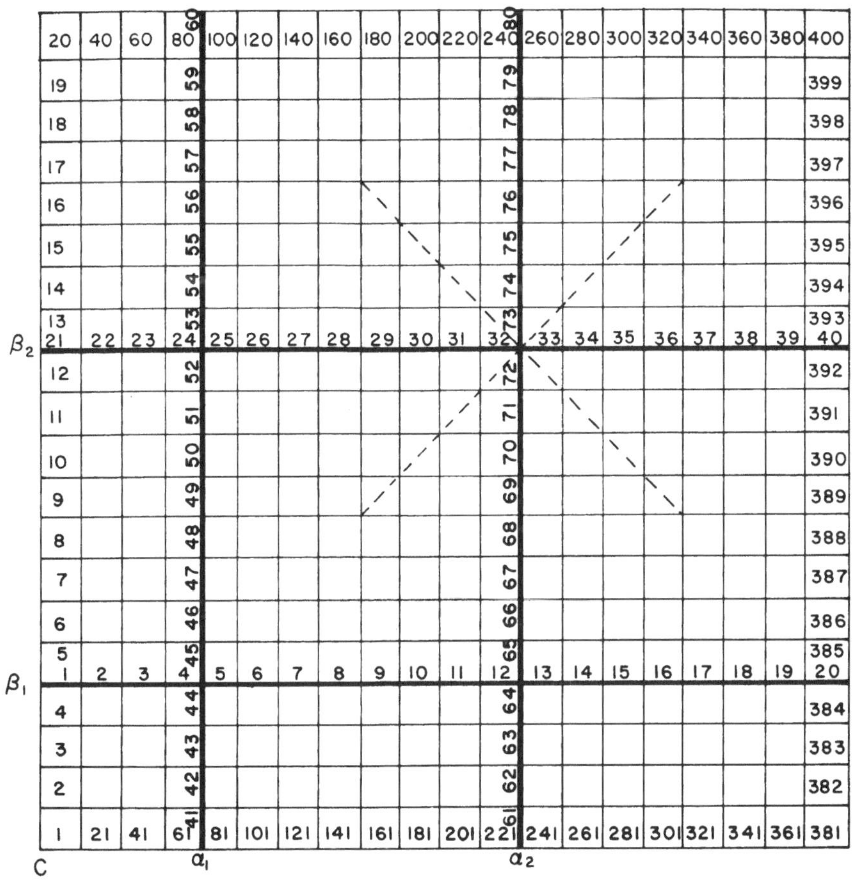

Figure 9.2 Finite Element Model of A Truss-Beam-Plate Built-up Structure

at the extreme fiber of plate components are given in Table 9.1. Results given in Table 9.1 show good agreement between predictions ψ' and finite differences $\Delta\psi$ except for von Mises yield stresses on plate elements 177, 358, 380 and 400 which are acceptable but not good. However, note that these elements have low von Mises yield stress and $\Delta\psi$ is small, compared to others, and may not be accurate.

For shape design sensitivity calculations, since the built-up structure is symmetric with respect to the center C, the locations α_i and β_j, i,j=1,2, of transverse and longitudinal beams, measured from the center C, are taken as design variables.

Table 9.1. Conventional Design Sensitivity of Truss-Beam-Plate Built-Up Structure

(a) Displacement

Node No.	ψ^1	ψ^2	$\Delta\psi$	ψ'	$(\psi'/\Delta\psi x100)\%$
53	-2.9755E-04	-2.8723E-04	1.0315E-05	9.5761E-06	92.8
95	-2.5890E-04	-2.4993E-04	8.9703E-06	8.3204E-06	92.8
113	-2.2837E-04	-2.2048E-04	7.8934E-06	7.3284E-06	92.8
137	-4.2898E-04	-4.1320E-04	1.5781E-05	1.6171E-05	102.5
179	-4.0427E-04	-3.8949E-04	1.4788E-05	1.5042E-05	101.7
221	-3.7551E-04	-3.6186E-04	1.3649E-05	1.3747E-05	100.7
268	-1.2025E-04	-1.1601E-04	4.2443E-06	4.1932E-06	98.8
310	1.4419E-05	1.3799E-05	-6.1962E-07	-6.1013E-07	98.5
335	1.5392E-04	1.4824E-04	-5.6883E-06	-5.3932E-06	94.8
352	5.5382E-05	5.3329E-05	-2.0534E-06	-1.9995E-06	97.4

(b) Bending Stress on Beam Element

El. No.	ψ^1	ψ^2	$\Delta\psi$	ψ'	$(\psi'/\Delta\psi x100)\%$
1	160.415	155.757	-4.658	-5.250	112.7
4	128.072	124.347	-3.725	-4.271	114.7
21	365.447	355.610	-9.837	-9.608	97.7
25	305.855	297.648	-8.207	-8.191	99.8
30	-69.361	-67.350	2.011	2.233	111.0
45	106.547	103.409	-3.138	-3.383	107.8
49	24.570	23.801	-0.769	-0.963	125.2
54	-74.026	-72.026	2.000	2.141	107.1
60	-4.789	-4.624	0.165	0.162	98.0
80	2.141	2.063	-0.079	-0.080	102.1

(c) von Mises Stress on Plate Element

El. No.	ψ^1	ψ^2	$\Delta\psi$	ψ'	$(\psi'/\Delta\psi x100)\%$
1	49.128	47.808	-1.320	-1.560	118.2
5	34.279	33.323	-0.956	-1.059	110.8
9	62.143	60.507	-1.636	-1.695	103.6
13	77.076	75.067	-2.009	-1.982	98.6
17	92.755	90.404	-2.351	-2.157	91.7
22	44.262	43.040	-1.222	-1.459	119.4

Table 9.1(c) Continued

26	42.663	41.471	-1.192	-1.247	104.6
30	63.522	61.831	-1.691	-1.729	102.2
34	77.930	75.877	-2.053	-1.946	94.8
38	91.868	89.533	-2.335	-2.105	90.1
44	33.332	32.332	-1.000	-1.189	118.9
48	52.498	51.074	-1.424	-1.477	103.8
52	67.527	65.707	-1.820	-1.761	96.7
56	79.045	76.974	-2.071	-1.895	91.0
65	38.362	37.273	-1.090	-1.150	105.5
69	46.061	44.733	-1.328	-1.297	97.7
73	70.749	68.874	-1.875	-1.784	95.2
77	67.127	65.280	-1.847	-1.609	87.1
85	38.300	37.201	-1.099	-1.104	100.5
89	47.855	46.506	-1.348	-1.288	95.6
93	67.866	66.083	-1.783	-1.759	98.7
97	62.719	61.070	-1.649	-1.602	97.2
106	43.048	41.825	-1.223	-1.191	97.4
112	62.228	60.640	-1.588	-1.477	93.0
118	59.270	57.835	-1.435	-1.385	96.5
128	55.122	53.691	-1.431	-1.362	95.1
134	48.012	46.800	-1.211	-1.119	92.4
140	56.823	55.539	-1.284	-1.214	94.5
151	53.218	51.887	-1.331	-1.183	88.9
155	37.003	36.148	-0.855	-0.735	85.9
159	37.897	37.097	-0.800	-0.691	86.4
169	52.077	50.777	-1.300	-1.163	89.5
173	56.294	54.890	-1.404	-1.292	92.0
177	23.383	22.944	-0.439	-0.276	62.8
192	61.326	59.720	-1.605	-1.568	97.7
198	27.041	26.479	-0.562	-0.451	80.3
212	67.683	65.843	-1.840	-1.854	100.8
216	53.350	52.077	-1.273	-1.264	99.3
235	82.704	80.664	-2.041	-2.127	104.2
239	100.980	98.872	-2.107	-2.104	99.8
255	79.509	77.476	-2.033	-1.968	96.8
259	101.657	99.539	-2.117	-2.040	96.4
274	67.361	65.559	-1.802	-1.765	98.0
278	64.853	63.399	-1.454	-1.403	96.5
288	37.003	36.148	-0.855	-0.735	85.9
319	27.894	27.305	-0.590	-0.586	99.4
337	21.848	21.459	-0.389	-0.419	107.6
358	14.556	14.427	-0.129	-0.190	146.8
380	12.189	12.077	-0.113	-0.161	142.8
400	8.471	8.381	-0.090	-0.119	132.8

As mentioned in Section 8, for shape design sensitivity calculations, one must define a velocity field that has C^1-regularity with its second derivative integrable. The beam components are allowed to move in transverse directions only. Hence, V^1 is a function of x_1 only and V^2 is a function of x_2 only. The velocity field in each plate component is represented by Hermite cubic functions in each direction. That is, $V^1(x_1)$ and $V^2(x_2)$ are represented by Hermite cubic functions. To see the velocity field representation graphically, consider Fig. 9.3, in which the shape functions for $V^1(x_1)$ are plotted. In Fig. 9.3, $\delta\alpha_1$ and $\delta\alpha_2$ denote perturbations of locations of transverse beams. From Fig. 9.3, one obtains $V^1(x_1) = \phi^1(x_1) + \phi^2(x_1)$. That is,

$$V^1(x_1) = \begin{cases} -\dfrac{2x^2}{\alpha_1^3}\left(x - \dfrac{3\alpha_1}{2}\right)\delta\alpha_1, & 0 \leq x \leq \alpha_1 \\[2ex] \dfrac{2(x-\alpha_1)^2}{(\alpha_2-\alpha_1)^3}\left[(x-\alpha_1) - \dfrac{3(\alpha_2-\alpha_1)}{2}\right](\delta\alpha_1-\delta\alpha_2) + \delta\alpha_1, & \alpha_1 \leq x \leq \alpha_2 \\[2ex] \dfrac{2(x-\alpha_2)^2}{\left(\dfrac{L_1}{2}-\alpha_2\right)^3}\left[(x-\alpha_2) - \dfrac{3\left(\dfrac{L_1}{2}-\alpha_2\right)}{2}\right]\delta\alpha_2 + \delta\alpha_2, & \alpha_2 \leq x \leq \dfrac{L_1}{2} \end{cases} \qquad [9.11]$$

and a similar expression for $V^2(x_2)$.

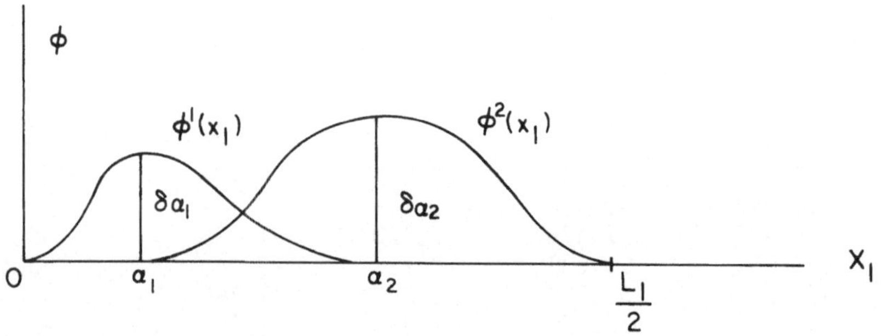

Figure 9.3 Shape Functions For The Velocity $V^1(x_1)$

In Table 9.2, design sensitivity accuracy results are given for several functionals, with a 0.25% uniform change in shape design parameters. Sensitivity results for displacements, bending stress for beam components, and von Mises yield stress for plate components, are given in Table 9.2. Results given in Table 9.2 show excellent agreement between predictions ψ' and the finite differences $\Delta\psi$.

Table 9.2. Shape Design Sensitivity of Truss-Beam-Plate Built-Up Structure

(a) Displacement

Node No.	ψ^1	ψ^2	$\Delta\psi$	ψ'	$(\psi'/\Delta\psi x100)\%$
53	-2.9755E-04	-3.0123E-04	-3.6843E-06	-3.6386E-06	98.8
95	-2.5890E-04	-2.6186E-04	-2.9640E-06	-2.9212E-06	98.6
113	-2.2837E-04	-2.3080E-04	-2.4214E-06	-2.3798E-06	98.3
137	-4.2898E-04	-4.3951E-04	-1.0535E-05	-1.0489E-05	99.6
179	-4.0427E-04	-4.1394E-04	-9.6675E-06	-9.6264E-06	99.6
221	-3.7551E-04	-3.8411E-04	-8.5986E-06	-8.5631E-06	99.6
268	-1.2025E-04	-1.2319E-04	-2.9391E-06	-2.9273E-06	99.6
310	1.4419E-05	1.5210E-05	7.9096E-07	7.9419E-07	100.4
335	1.5392E-04	1.5978E-04	5.8589E-06	5.8686E-06	100.2
352	5.5382E-05	5.7951E-05	2.5690E-06	2.5786E-06	100.4

(b) Bending Stress on Beam Element

El. No.	ψ^1	ψ^2	$\Delta\psi$	ψ'	$(\psi'/\Delta\psi x100)\%$
1	160.415	163.634	3.219	3.095	96.2
4	128.072	131.041	2.968	2.871	96.7
21	365.447	370.311	4.864	4.586	94.3
25	305.855	310.385	4.530	4.399	97.1
30	-69.361	-66.654	2.708	2.820	104.2
45	106.547	109.385	2.838	2.794	98.4
49	24.570	26.757	2.187	2.143	98.0
54	-74.026	-73.243	0.783	0.663	84.7
60	-4.789	-4.853	-0.064	-0.065	101.8
80	2.141	2.113	-0.028	-0.026	91.2

(c) von Mises Stress on Plate Element

El. No.	ψ^1	ψ^2	$\Delta\psi$	ψ'	$(\psi'/\Delta\psi x100)\%$
1	49.128	50.043	0.915	0.915	100.1
5	31.279	34.881	0.602	0.600	99.6
9	62.143	63.067	0.925	0.924	99.9
13	77.076	77.949	0.873	0.871	99.8
17	92.755	93.843	1.088	1.090	100.2
22	44.262	45.153	0.891	0.891	100.0

Table 9.2(c) Continued

26	42.663	43.440	0.778	0.776	99.8
30	63.522	64.437	0.914	0.913	99.8
34	77.930	78.891	0.961	0.960	99.9
38	91.868	93.039	1.171	1.172	100.1
44	33.332	34.143	0.811	0.810	99.9
48	52.498	53.329	0.831	0.830	99.9
52	67.527	68.410	0.883	0.881	99.7
56	79.045	80.149	1.104	1.104	100.0
65	38.362	39.116	0.754	0.749	99.4
69	46.061	46.804	0.743	0.742	99.9
73	70.749	71.663	0.914	0.911	99.7
77	67.127	68.334	1.207	1.209	100.1
85	38.300	39.039	0.740	0.738	99.8
89	47.855	48.520	0.665	0.663	99.7
93	67.866	68.758	0.892	0.889	99.6
97	62.719	64.012	1.293	1.294	100.1
106	43.048	43.773	0.726	0.725	99.9
112	62.228	62.978	0.749	0.745	99.3
118	59.270	60.553	1.283	1.284	100.1
128	55.122	55.881	0.759	0.756	99.7
134	48.012	48.846	0.834	0.830	99.4
140	56.823	58.228	1.405	1.408	100.2
151	53.218	53.793	0.575	0.569	99.0
155	37.003	37.871	0.868	0.863	99.4
159	37.897	39.123	1.226	1.226	100.0
169	52.077	52.764	0.687	0.683	99.3
173	56.294	56.509	0.215	0.208	96.7
177	23.383	24.298	0.915	0.907	99.1
192	61.326	61.258	-0.068	-0.075	109.9
198	27.041	27.324	0.283	0.263	93.0
212	67.683	67.463	-0.220	-0.226	102.7
216	53.350	53.241	-0.109	-0.118	107.7
235	82.704	82.424	-0.280	-0.285	101.6
239	100.980	101.057	0.077	0.075	97.5
255	79.509	79.143	-0.366	-0.369	100.9
259	101.657	101.295	-0.362	-0.364	100.6
274	67.361	66.997	-0.364	-0.368	101.2
278	64.853	64.565	-0.288	-0.290	100.7
288	37.003	37.871	0.868	0.863	99.4
319	27.894	27.717	-0.177	-0.182	102.5
337	21.848	21.623	-0.225	-0.230	102.2
358	14.556	14.247	-0.309	-0.313	101.2
380	12.189	11.934	-0.255	-0.255	100.1
400	8.471	8.315	-0.156	-0.154	99.1

10. AN OPTIMIZATION PROBLEM

To demonstrate application of the design sensitivity analysis method for built-up structures in structural design optimization, the truss-beam-plate built-up structure of Section 9 is optimized using a sparse matrix symbolic factorization technique for iterative structural optimization [28] and Pshenichny's linearization method [29].

For numerical design sensitivity analysis and optimization, design variables are discretized. That is, each plate element has constant thickness and each beam element has constant width and height. Since the built-up structure is symmetric with respect to the center C, thickness t_i, i = 1,210, width d_i and height b_i, i = 1, 40, and the locations α_i, i = 1, 2 of transverse and longitudinal beams, measured from the center C, are taken as design parameters. Thus, the total number of design parameters is 292.

The optimal design problem of the built-up structure is to minimize weight of the structure, subject to the following constraints:

Displacement at C; $\psi_1 = z(C) \leq 0.105$ in.

Plate element von Mises stress; $\psi_i \leq 17500$ psi, i = 2,211

Beam element bending stress; $-70000 \text{ psi} \leq \psi_i \leq 70000$ psi, i = 212,251

Plate thickness; $0.05 \text{ in.} \leq t_i \leq 0.25$ in., i = 1,210

Beam width; $0.075 \text{ in.} \leq d_i \leq 0.30$ in., i = 1, 40

Beam height; $0.25 \text{ in.} \leq b_i \leq 1.00$ in., i = 1, 40

Beam position; $0 \leq \alpha_1 < \alpha_2 \leq \frac{L_1}{2}$

Thus, the total number of inequality constraints is 543.

Same numerical data as in Section 9 is used except weight density is 0.1 lb/in.^{3} and uniformly distributed load f = 17.5 lb/in.^{2} is applied to the plate components.

For numerical computation of the shape design sensitivity information, derivatives I_x, J_x, and h_x in Eq. 5.1 for beam component and ∇t in Eq. 5.8 for plate component must be computed. Since each plate element has constant thickness and each beam element has constant width and height, these derivatives are Dirac delta measures and computations of the shape design sensitivity information become complicated.

To avoid this difficulty, the design process is divided into two phases. In Phase I, each plate and beam components (not element) have constant thickness and constant width and height, respectively. Hence in Phase I, the design parameter set includes 6 plate thicknesses, 6 beam heights and widths, and 2 beam locations with total of 20 design parameters. To assign the same design parameter to elements in a component, design variable linking is used in Phase I. Once an optimum point is reached in Phase I, the design process is switched to Phase II where shape design parameters are fixed and each plate and beam elements are allowed to have different design parameters. Hence the total number of design parameter is 290 in Phase II.

For numerical computation, PRIME-750 and Cray-1S computers are used for design Phases I and II, respectively. The initial and final designs of Phase I is given in Table 10.1. The initial cost is 0.7875 lb and the final cost of Phase I is 0.5894 lb. There are 12 design iterations in

Table 10.1. Initial and Final Designs of Phase I

Design Parameter		Initial Design	Final Design
	t_1		0.0874
Plate	t_2		0.0500
Component	t_3	0.1	0.0518
Thickness	t_4		0.0501
	t_5		0.0815
	t_6		0.0910
	b_1		0.4234
Beam	b_2		0.6902
Component	b_3	0.5	0.4990
Height	b_4		0.3974
	b_5		0.4266
	b_6		0.4326
	d_1		0.1087
Beam	d_2		0.2214
Component	d_3	0.15	0.2499
Width	d_4		0.0754
	d_5		0.0838
	d_6		0.1286
Beam	α_1	1.50	1.4221
Position	α_2	4.50	4.5872

Phase I with average CPU time 15986 seconds per iteration on a PRIME-750 computer. It is observed in design Phase I that inner beam stiffners move inward (α_1 = 1.4211 in.) and outer beam stiffners move outward (α_2 = 4.5872 in.). Number of active stress and displacement constraints at the final design of Phase I is 31. Thus, it is necessary to calculate sensitivity information for 31 constraints out of 251.

There are 9 design iterations in Phase II with average CPU time 25.84 seconds per iteration on a Cray-1S computer. Cost function history of Phases I and II is shown in Fig. 10.1. The final cost of Phase II is 0.5388 lb. Number of active stress and displacement constraints at the final design of Phase II is 54. A profile of upper half of the final design is shown in Fig. 10.2

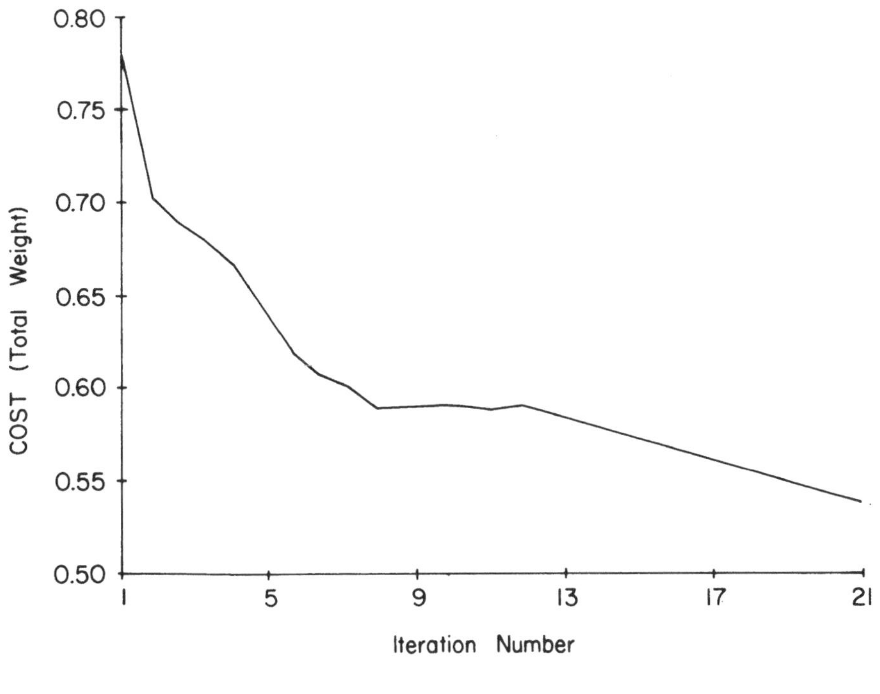

Figure 10.1 Cost Function History

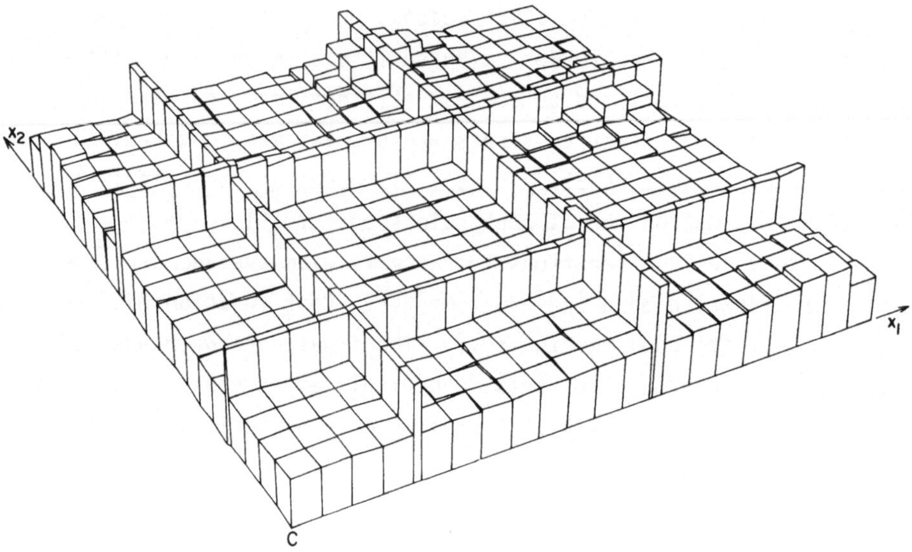

Figure 10.2 A Profile of the Final Design

REFERENCES

1. Haftka, R.T. and Kamat, M.P., Elements of Structural Optimization, Martinus Nijhoff Publishers, Boston, 1985.
2. Haug, E.J., Choi, K.K. and Komkov, V., Design Sensitivity Analysis of Structural Systems, Academic Press, New York, N.Y., March, 1986.
3. Haftka, R.T. and Grandhi, R.V., "Structural Shape Optimization-Survey," Computer Methods in Applied Mechanics and Engineering, to appear, 1986.
4. Botkin, M.E., "Shape Optimization of Plate and Shell Structures," AIAA J., Vol. 20, No. 2, 1982, pp. 268-273.
5. Francavilla, A., Ramakrishnan, C.V. and Zienkiewicz, O.C., "Optimization of Shape to Minimize Stress Concentration," J. of Strain Analysis, Vol. 10, No. 2, 1975, pp. 63-70.
6. Braibant, V. and Fleury, C., "Shape Optimal Design Using B-Splines," Computer Methods in Applied Mechanics and Engineering, Vol. 44, 1984, pp. 247-267.
7. Dems, K. and Mroz, Z., "Variational Approach by Means of Adjoint Systems to Structural Optimization and Sensitivity Analysis - II Structural Shape Variation," Int. J. Solids and Structures, Vol. 20, No. 6, 1984, pp. 527-552.
8. Cea, J. "Problems of Shape Optimal Design," Optimization of Distributed Parameter Structures, E.J. Haug and J. Cea (Eds.), Sijthoff & Noordhoff, Alpen ann den Rijn, The Netherlands, 1981, pp. 1005-1048.
9. Choi, K.K. and Haug, E.J., "Shape Design Sensitivity Analysis of Elastic Structures," J. of Struct. Mechanics, Vol. 11, No. 2, 1983, pp. 231-269.
10. Choi, K.K., "Shape Design Sensitivity Analysis of Displacement and Stress Constraints," J. of Struct. Mechanics, Vol. 13, No. 1, 1985, pp. 27-41.
11. Haug, E.J., Choi, K.K., Hou, J.W. and Yoo, Y.M., "A Variational Method for Shape Optimal Design of Elastic Structures," New Directions in Optimum Structural Design, E. Atrek, R.H. Gallagher, K.M. Ragsdell and O.C. Zienkiewicz (Eds.), John Wiley and Sons, Ltd., 1984, pp. 105-137.
12. Banichuk, N.V., "Optimization of Elastic Bars in Torsion," Int. J. of Solids and Structures, Vol. 12, 1976, pp. 275-286.
13. Na, M.S., Kikuchi, N. and Taylor, J.E., "Shape Optimization for Elastic Torsion Bars," Optimization Methods in Structural Design, H. Eschenauer and N. Olhoff (Eds.), Bibliographisches Institut, Zurich, Germany, 1983, pp. 216-223.
14. Babuska, I. and Aziz, A.K., "Survey Lectures on the Mathematical Foundations of the Finite Element Method," The Mathematical Foundations of the Finite Element Method with Applications to Partial Differential Equations, Academic Press, 1972, pp. 1-359.
15. Choi, K.K. and Seong, H.G., "A Domain Method for Shape Design Sensitivity Analysis of Built-Up Structures," Computer Methods in Applied Mechanics and Engineering, to appear, 1986.
16. Seong, H.G. and Choi, K.K., "Boundary Layer Approach to Shape Design Sensitivity Analysis," J. of Struct. Mechanics, to appear, 1986.
17. Choi, K.K. and Yao, T.M., "3-D Modeling and Automatic Regridding in Shape Design Sensitivity Analysis," NASA Symposium: Sensitivity Analysis in Engieering, NASA-Langley Research Center, Virginia, September 25-26, 1986.

18. Choi, K.K. and Seong, H.G., "Design Component Method for Sensitivity Analysis of Built-Up Structures," J. of Struct. Mechanics, to appear, 1986.
19. Choi, K.K., Santos, J.L.T., and Frederick, M.C., "Implementation of Design Sensitivity Analysis with Existing Finite Element Codes," ASME Journal of Mechanisms, Transmissions, and Automation in Design, 85-DET-77.
20. Zolesio, J-P., "The Material Derivative (or Speed) Method for Shape Optimization," Optimization of distributed Parameter Structures, E.J. Haug and J. Cea (Eds.) Sijthoff & Noordhoff, Alphen ann den Rijn, The Netherlands, 1981, pp. 1089-1151.
21. Zolesio, J-P., "Gradient des coute Governes par des Problems de Neumann
poses des Wuverts Anguleux en Optimization de Domain, CRMA-Report 116, University of Montreal, Canada, 1982.
22. Choi, K.K. and Seong, H.G., "A Numerical Method for Shape Design Sensitivity Analysis and Optimization of Built-Up Structures," The Optimum Shape: Automated Structural Design, (Eds. J.A. Bennett and M.E. Botkin), Plenum Press, New York, 1986.
23. Yang, R.J. and Choi, K.K., "Accuracy of Finite Element Based Design Sensitivity Analysis," J. of Struct. Mechanics, Vol. 13, No. 2, 1985, pp. 223-239.
24. Dym, C. and Shames, I.H., Solid Mechanics-A Variational Approach, McGraw-Hill, 1973.
25. Rogers, D.F. and Adams, J.A., Mathematical Elements for Computer Graphics, McGraw-Hill, 1976.
26. DeSalvo, G.J. and Swanson, J.A., ANSYS Engineering Analysis System, Users Manual, Swanson Analysis System, Inc., P.O. Box 65, Houston, PA, Vols. I and II, 1983.
27. Przemieniecki, J.S., Theory of Matrix Structural Analysis, McGraw-Hill, 1968.
28. Lam, H.L., Choi, K.K., and Haug, E.J., "A Sparse Matrix Finite Element Technique for Iterative Structure Optimization," Computers and Structures, Vol. 16, No. 1-4, 1983, pp. 289-295.
29. Choi, K.K., Haug, E.J., Hou, J.W., and Sohoni, V.N., "Pshenichny's Linearization Method for Mechanical System Optimization," ASME Journal of Mechanisms, Transmissions, and Automation in Design, Vol. 105, No. 1, 1983, pp. 97-103.

Adaptive Grid Design for Finite Element Analysis In Optimization: Part 1, Review of Finite Element Error Analysis*

Noboru Kikuchi

The University of Michigan, Ann Arbor, Michigan, 48109, USA

Abstract

Finite element gridding is regarded as an optimal design problem which yields the optimal grid adaptively by applying the optimality criteria method. Finite element approximation error analysis is critically reviewed to determine the effect of grid distortion which is a key factor of the irregular distribution of approximation error in flux and stress. Based on this study, appropriate error measures are defined for the optimal grid design problem as well as error indicators which estimate the total amount of approximation error.

---

*The present work is supported by NASA Lewis Research Center/ NAG 3-388 and ONR N001485K0799. The author expresses his sincere appreciation for these supports.

NATO ASI Series, Vol. F27
Computer Aided Optimal Design: Structural and Mechanical Systems. Edited by C. A. Mota Soares

1. Introduction

After thirty years of development, finite element methods have solidly established their existence in science and engineering. It is now almost impossible to talk about advanced engineering based on, e.g., stress, heat, and flow analyses without using finite element methods which are general enough so that many commercially available, general purpose codes can be developed, together with various supporting devices such as preprocessors containing automatic grid generation methods and post processors plotting computed results graphically. This development certainly extends the range of analysis capability for design and manufacturing in engineering as well as applied functional analysis in mathematics, which deals with initial boundary value problems governed by partial/ordinary differential equations. The science side of finite element methods, especially their mathematical theory, can be found in, e.g., Ciarlet [1] and Glowinski [2], while various engineering applications are discussed in Zienkiewicz [3]. See also, Finite Elements Handbook [4] which will soon be published. For educational purposes, a series of textbooks is published by Oden et. al. [5-7].

Our emphasis here is more toward preprocessing to finite element analysis. Especially, we shall study sensitivity of accuracy of finite element approximations to the direction of finite element gridding and to the shape and size of finite elements. Qualitatively, it has been established that finite element approximation error converges to zero as the grid size tends to zero, that is, the total number of finite elements goes to infinite. How can we then quantify the amount of error? How is it reduced? Or more precisely, what is the optimal gridding for a given number of the total degrees of freedom? How can the optimal refinement be achieved that involves the minimum amount of error with the least number of elements? Is it possible to predict the number of elements which yield a specified amount of error, say, 5% error? These have been answered mathematically by, e.g., Babuska, Rheinbolt, Sabo, [8-13] under the name of adaptive finite element methods. These have also been implemented by Shephard et. al. [14-15] and Diaz et. al. [16-17] as parts of pre- and post-processors to finite element analysis. Other references of adaptive methods will be given in Part 2.

It is noted that the study of adaptive methods was motivated while shape optimization problems were solved at The University of Michigan. As shown in Fig. 1, the optimal shape strongly depends on finite element gridding! There, the same analysis and shape optimization algorithms are applied to find the optimal shape of an initially triangular, thin plate which transmits the tension force to two sliding supports.

This paper consists of the following three parts:

Part 1: Review of Finite Element Error Analysis

Part 2: Optimal Grid Design Problem

Part 3: Application to Shape Optimization

In Part 1, we shall review finite element error analysis using one-dimensional problems in order to describe relations of the residual and interpolation errors to the finite element approximation error. This study enables us to define an appro-

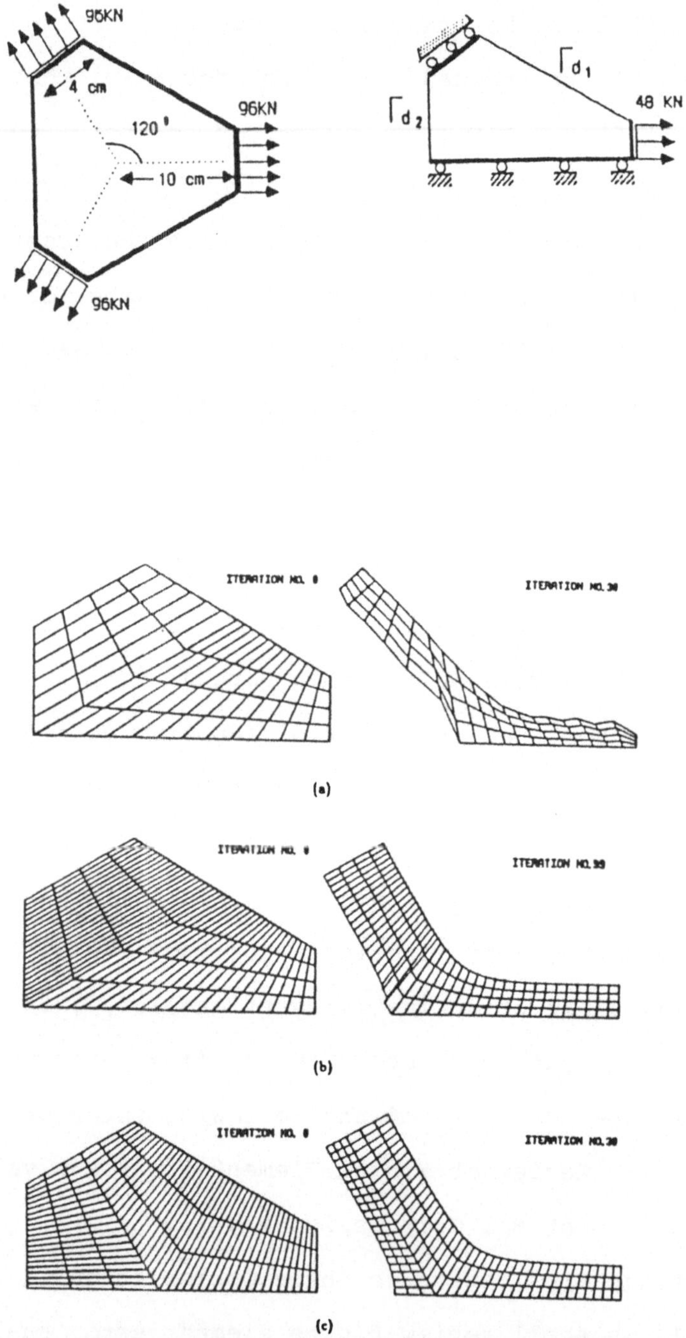

Figure 1. Optimal Shape by Different Finite Element Grids (by Dr. K.Y. Chung)

priate error measure to the grid optimization problem defined in the second part of this paper, Part 2.

2. As Usual, One-Dimensional Problems

Suppose that a boundary value problem is given as follows:

$$-d(kdT/dx)/dx + cdT/dx + hT = f \text{ in } (0,L) \tag{1}$$

$$T(0) = T_0, \quad \text{and} \quad T(L) = T_L$$

where k, c, and h are specified constants of f as a given function. Let us solve this using 2-node line elements, the length of which are h_e, $e = 1,\ldots,NE$, where NE is the total number of line elements. In this case, (1) is discretized as

$$\left((k/h_e)\begin{bmatrix} 1 & -1 \\ -1 & 1 \end{bmatrix} + (c/2)\begin{bmatrix} -1 & 1 \\ -1 & 1 \end{bmatrix} + (hh_e/6)\begin{bmatrix} 2 & 1 \\ 1 & 2 \end{bmatrix} \right) \begin{Bmatrix} T_1 \\ T_2 \end{Bmatrix} = (h_e/6)\begin{Bmatrix} 2f_1 + f_2 \\ f_1 + 2f_2 \end{Bmatrix} \tag{2}$$

in each finite element Ω_e, $e = 1, 2,\ldots,NE$, where T_1, T_2, f_1, and f_2 are the values of T and f at the 1-st and 2-nd nodes of each element, respectively.

<u>Residual r_e.</u> Residual r_e of finite element approximation of the differential equation (1) is defined in each finite element $\Omega_e = (x_e, x_{e+1})$ as

$$r_e: = f - hT_h - cdT_h/dx + d(kdT_h/dx)/dx \tag{3}$$

where T_h is the finite element approximation of T obtained by (2) for given gridding and elements. If 2-node linear line elements are applied, the second derivative of T_h vanishes in each element. Then the residual becomes

$$r_e = f - hT_h - cdT_h/dx + (dk/dx)(dT_h/dx) \tag{4}$$

If the heat conductivity k is constant in each element, i.e., $dk/dx = 0$, (4) can be simplified as

$$r_e = f - hT_h - cdT_h/dx \tag{5}$$

Since r_e is defined at every point x in Ω_e, let us introduce an error measure E_e of the finite element approximation error using the L^2-norm of r_e:

$$E_e: = h_e\{\int_{\Omega_e} |r_e|^2 d\Omega_e\}^{1/2} \tag{6}$$

where h_e is the size of the element Ω_e. The reason why h_e is multiplied to the L^2-norm of r_e is that E_e has the same dimension as the "energy" norm

$$||T||: = (\int_{\Omega}\{k(dT/dx)^2+hT^2\}d\Omega)^{1/2} \tag{7}$$

<u>Interpolation Error i_e</u>. Let T_I be the interpolation of T using 2-node line elements. <u>If the second derivative of T is assumed to be constant in each finite element, Ω_e, e</u> = 1,..., NE, the differences of $d(T_I-T)/dx$ and T_I-T are given by

$$d(T_I-T)/dx = ((x_e+x_{e+1})/2-x)d^2T/dx^2 \tag{8}$$

$$T_I - T = (1/2)(x-x_e)(x_{e+1}-x)d^2T/dx^2 \tag{9}$$

respectively. Indeed, Taylor's expansion yields

$$T_1 = T + (dT/dx)(x_e-x) + (1/2)(d^2T/dx^2)(x_e-x)^2 \tag{10}$$

$$T_2 = T + (dT/dx)(x_{e+1}-x) + (1/2)(d^2T/dx^2)(x_{e+1}-x)^2 \tag{11}$$

Subtracting (10) from (11) yields

$$(T_2-T_1)/(x_{e+1}-x_e) = dT/dx + ((x_e+x_{e+1})/2-x)d^2T/dx^2$$

that is,

$$d(T_I-T)/dx = ((x_e+x_{e+1})/2-x)d^2T/dx^2$$

On the other hand, addition of $(10)*(x_{e+1}-x)$ and $(11)*(x-x_e)$ yields

$$T_I(x_{e+1}-x_e) = T(x_{e+1}-x_e) + (1/2)(d^2T/dx^2)(x-x_e)$$
$$(x_{e+1}-x)(x_{e+1}-x_e)$$

that is,

$$T_I - T = (1/2)(x-x_e)(x_{e+1}-x)d^2T/dx^2$$

Using these interpolation errors and the energy norm, another error measure of the finite element approximation can be defined by

$$E_e: = ||i_e||_e: = (\int_{\Omega_e}\{k(d(T_I-T)/dx)^2 + h(T_I-T)^2\}d\Omega_e)^{1/2} \quad (12)$$

where $d(T_I-T)/dx$ and T_I-T are given by (8) and (9), respectively.

From the differential equation (1), the second derivative of T can be written as

$$d^2T/dx^2 = (f-hT-cdT/dx+(dk/dx)(dT/dx))/k \quad (13)$$

If the right hand side is approximated by the finite element approximation T_h, the second derivative d^2T/dx^2 is approximated by

$$d^2T/dx^2 \simeq (f-hT_h-cdT_h/dx+(dk/dx)(dT_h/dx))/k \quad (14)$$

Another approximation of the second derivative of T can be obtained by using the continuous piecewise linear gradient ε_h spanned by 2-node line elements computed by the least square method

$$\min_{\varepsilon_h}(1/2) \int_{\Omega} (\varepsilon_h-dT_h/dx)^2d\Omega \quad (15)$$

Since dT_h/dx is piecewise constant, d^2T_h/dx^2 is not well-defined in the whole domain Ω. Thus, we first compute a continuous approximation of dT_h/dx, and then its derivative is taken to approximate the second derivative, that is,

$$d^2T/dx^2 \simeq d\varepsilon_h/dx \quad (16)$$

In each finite element Ω_e, $e = 1,\ldots,NE$. The least square method (15) yields the system of linear equations

$$\sum_{e=1}^{NE} (h_e/6)\begin{bmatrix} 2 & 1 \\ 1 & 2 \end{bmatrix}\begin{Bmatrix} \varepsilon_1 \\ \varepsilon_2 \end{Bmatrix} = (dT_h/dx)_e(h_e/2)\begin{Bmatrix} 1 \\ 1 \end{Bmatrix} \tag{17}$$

where ε_1 and ε_2 are the values of ε_h at the 1-st and 2-nd node in Ω_e, respectively, and $(dT_h/dx)_e$ is the value of dT_h/dx in Ω_e.

To show the nature of the least square method (15) to obtain a continuous first derivative, let us consider a domain consisting of two line elements whose lengths are h_e and mh_e, respectively. If

$$p = (dT_h/dx)_1 \quad \text{and} \quad q = (dT_h/dx)_2$$

are assumed, (15), i.e., (17) implies

$$\varepsilon_1 = (2p+3mp-mq)/2(1+m), \quad \varepsilon_2 = (p+mq)/(1+m)$$

$$\varepsilon_3 = (3q-p+2mq)/2(1+m)$$

If $p = 1$ and $q = 0$ are assumed, these become

$$\varepsilon_1 = (2+3m)/2(1+m), \quad \varepsilon_2 = 1/(1+m)$$

$$\varepsilon_3 = -1/2(1+m)$$

For various m, ε_1, ε_2, and ε_3 are plotted in Fig. 2.

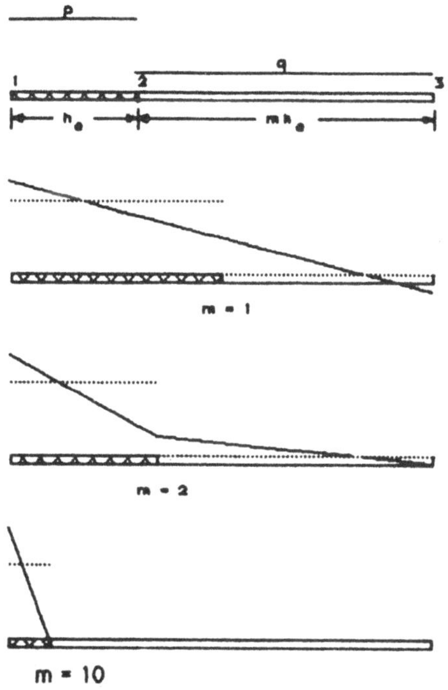

Figure 2. Least Square Method

It is clear that the least square method can provide reasonably good continuous function for various sizes of elements.

As an example, we shall compute distribution of error measures defined by the residual and interpolation error for a specific case of the boundary value problem (1). Suppose that $k = 1$, $c = 20$, $h = 1$, $f = 25$, $T_0 = 100$, $T_L = 0$, and $L = 20$.

For twenty elements, let us compute the finite element solution and the error measures defined above. It is clear that a large amount of error is accumulated at both ends of the domain (see Figs. 3, 4, and 5). As shown in Fig. 3, if c is large enough, oscillation appears in the finite element

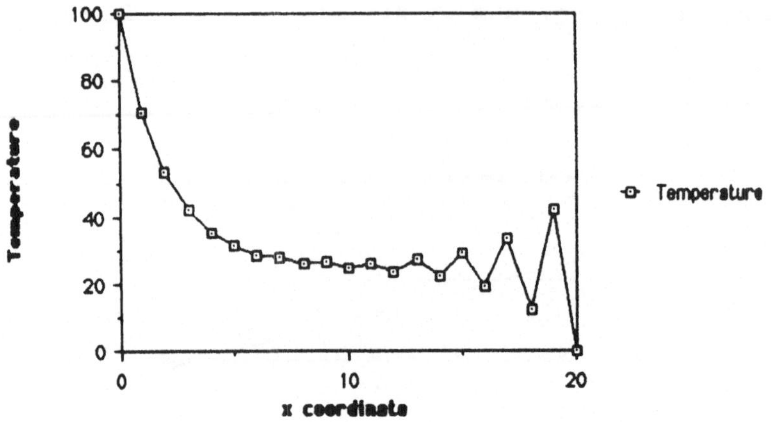

Figure 3. Temperature Distribution (k=1, c=20, h=10, L=20)

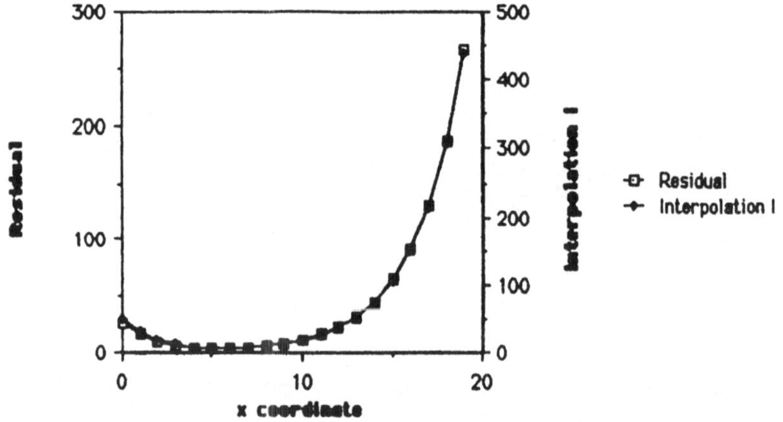

Figure 4. Error Measures by the Residual and Interpolation Error

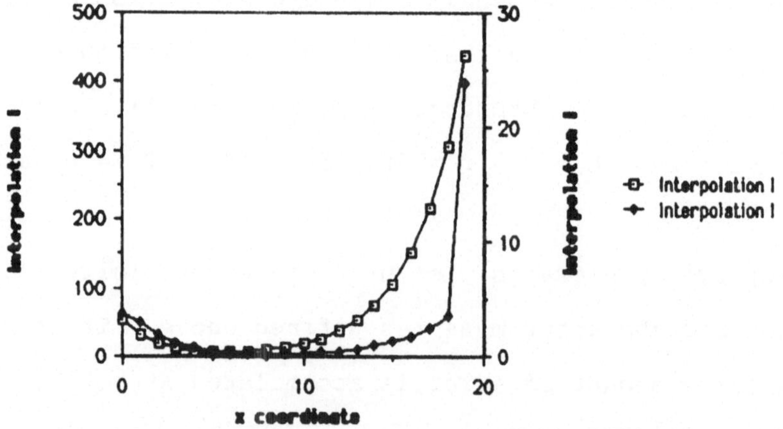

Figure 5. Error Measures by Two Different Interpolations

solution which is not acceptable at all in terms of its physics. Figs. 4 and 5 show the distribution of error measures due to the residual and the interpolation error. Both behave almost exactly the same, especially the relative distribution which is almost identical although their absolute values are different. It is clear that very refined grids should be allocated in the place where large error measure is computed.

3. How is the Residual r_e Related to the Approximation Error?

We shall briefly review the relation between the residual r_e and the finite element approximation error $e = T-T_h$, where T is the exact solution to (1) and T_h is its finite element approximation. For simplicity, the convection term is neglected, that is, $c = 0$ is assumed in this section. The following description is based on Babuska and Rheinboldt [9, 10].

Define

$$a(T,T) = \int_\Omega (kT'T'+hTT)d\Omega \tag{18}$$

and

$$||T|| = \sqrt{a(T,T)} \tag{19}$$

The weak form to the original problem (1) becomes

$$a(T,T) = \int_\Omega fT d\Omega \text{ for every } T \tag{20}$$

such that $T = 0$ at $x = 0$ and $x = L$. Its finite element approximation is then given by

$$a(T_h,T_h) = \int_\Omega fT_h d\Omega \text{ for every } T_h \tag{21}$$

such that $T_h = 0$ at $x = 0$ and $x = L$, and T_h is spanned by 2-node elements. It follows from (20) and (21) that

$$a(T-T_h,T_h) = 0 \text{ for every } T_h \tag{22}$$

such that $T_h = 0$ at $x = 0$ and $x = L$. This property is called the <u>orthogonality of the finite element approximation error</u> e to the finite element approximation space.

The key idea of Babuska and Rheinboldt is the application of the <u>elliptic projection</u> πe of the error e defined by

$$a(e-\pi e,w) = 0 \text{ for every } w \tag{23}$$

such that

$$w\varepsilon H: = \{v|v(x_e) = 0,\ e = 1,\ldots,NE + 1\} \tag{24}$$

in order to derive a <u>posteriori estimate</u>. It is noted that the difference between an arbitrary function w and its interpolation using 2-node line elements, belongs to H.

Moreover, (23) yields

$$\begin{aligned} a(\pi e,w) &= a(e,w) \\ &= a(T,w) - a(T_h,w) \\ &= \sum_{e=1}^{NE} \int_{\Omega_e} (f-hT_h+(kT_h')')wd\Omega_e \end{aligned}$$

for every $w\varepsilon H$. This means that

$$-(k(\pi e)')' + h(\pi e) = r_e \tag{25}$$

in each finite element Ω_e, $e = 1,\ldots,NE$ where r_e is the residual defined by (3) using finite element approximation T_h, i.e.,

$$r_e = f - hT_h + (kT_h')'$$

If the size of an element Ω_e is given by h_e, the minimum eigenvalue of the differential operator $T \rightarrow (kT')'$ is given by

$$\underline{k}_e\pi^2/h_e^2$$

where $\underline{k}_e$ is the minimum value of k in Ω_e. Using this, (25) yields

$$\{\int_{\Omega_e} (\pi e)^2 d\Omega_e\}^{1/2} \leq (\underline{k}_e\pi^2/h_e^2+\underline{h}_e)^{-1}\{\int_{\Omega_e} r_e^2 d\Omega_e\}^{1/2} \qquad (26)$$

where $\underline{h}_e$ is the minimum value of h in Ω_e. This means that the L^2-norm of the elliptic projection πe of the error is bounded by the L^2-norm of the residual in each finite element Ω_e. The upper bound is given by

$$C_e = h_e^2/(\underline{k}_e\pi^2+\underline{h}_e h_e^2) \qquad (27)$$

It also follows from (25) that

$$||\pi e||_e^2 = \int_{\Omega_e} r_e \pi e d\Omega_e \leq \{\int_{\Omega_e} r_e^2 d\Omega_e\}^{1/2}\{\int_{\Omega}(\pi e)^2 d\Omega_e\}^{1/2} \qquad (28)$$

where

$$||\pi e||_e: = (\int_{\Omega_e} \{k((\pi e)')^2 + h(\pi e)^2\}d\Omega_e)^{1/2} \qquad (29)$$

Applying inequality (26) into (28) yields

$$||\pi e||_e^2 \leq C_e \int_{\Omega_e} r_e^2 d\Omega_e$$

that is,

$$||\pi e||_e \leq C_e \{\int_{\Omega_e} r^2 d\Omega\}^{1/2} \qquad (30)$$

This means that the energy norm of the elliptic projection πe of the approximation error e is bounded by the L^2-norm of the residual in each finite element Ω_e. The upper bound constant is almost proportional to the size of finite element h_e.

The last step is to find relation between $||\pi e||$ and $||e||$. To do this we shall recall the orthogonality condition (22). Applying (22) implies

$$||e||^2 = a(e,e) = a(e,e-e_I)$$
$$= a(\pi e,e-e_I) + a(e-\pi e,e-e_I) \qquad (31)$$

for the interpolation e_I of e using 2-node line elements. Since $e-e^I$ belongs to H, the second term of the right hand side vanishes because of the definition of elliptic projection. Thus we have

$$||e||^2 \leq ||\pi e|| \; ||e-e_I|| \tag{32}$$

For 2-node line elements, the energy norm of the interpolation error of the error e is bounded by its energy norm, that is,

$$||e-e_I|| \leq ||e|| \tag{33}$$

Therefore, the finite element approximation error $||e||$ is bounded by the energy norm of the elliptic projection of the error e, i.e.,

$$||e|| \leq ||\pi e|| = \{\sum_{e=1}^{NE} ||\pi e||_e^2\}^{1/2} \tag{34}$$

Substitution of estimate (3) into (34) implies the upper bound of the finite element approximation error by the residual r_e:

$$||e|| \leq \{\sum_{e=1}^{NE} C_e \int_{\Omega_e} r_e^2 d\Omega\}^{1/2} \tag{35}$$

4. How is the Interpolation Error i_e Related to the Approximation Error?

Relation of the interpolation error i_e to the finite element approximation error is more straightforward than for the residual. Indeed, because of the orthogonality of the error, we have

$$a(e,e) = a(e,T-T_I) + a(e,T_I-T_h) = a(e,T-T_I) \tag{36}$$

where T_I is the interpolation of T. Then

$$||e|| \leq ||T-T_I|| = \{\sum_{e=1}^{NE} ||i_e||_e^2\}^{1/2} \tag{37}$$

That is, the finite element approximation error is bounded by the interpolation error in the energy norm.

The main difference of the upper bound by the interpolation error from that by the residual is the second derivative of T must be approximated using the finite element solution T_h, while the residual r_e does not require any approximation of the second derivative of T. In fact, T_h itself defines the residual. This indicates that application of the interpolation error requires one more step that computes the continuous first derivative ε_h from piecewise constant dT_h/dx. It is noted that if the heat conductivity k is discontinuous at a point, the first derivative of T must be discontinuous since the heat flux $\sigma = kdT/dx$ is expected to be continuous at that point. This implies that computing continuous ε_h is not a good idea. Thus, if k is discontinuous σ_h is computed by the least square method:

$$\min_{\sigma_h} \frac{1}{2} \sum_{e=1}^{NE} \int_{\Omega_e} (\sigma_h - k(dT_h/dx)_e)^2 d\Omega_e \tag{38}$$

Applying σ_h instead of ε_h, the second derivative of T is approximated by

$$d^2T/dx^2 \approx d(\varepsilon_h)/dx \text{ with } \varepsilon_h = \sigma_h/k \tag{39}$$

in each finite element Ω_e, e = 1,...,NE. If the heat conductivity k is piecewise constant, that is more precisely, if k is constant in each finite element,

$$d^2T/dx^2 \approx (1/k)d\sigma_h/dx$$

in Ω_e, e = 1,...,NE. This way is more promising in the engineering point of view, since continuous estimation of the heat flux σ is important, useful information in practice.

5. Extension to Multidimensional Problems

Extension of the upper bound by the interpolation error into multi-dimensional problems is very straightforward, while the bound by using the residual based on a posteriori estimate is not yet fully developed as in one-dimensional problems. In multi-dimensional problems the residual r_e consists of not only the difference of $f-L(T_h)$ but also the jumps of the normal derivative of T_h at the corners of each finite element, where the differential equation in Ω is represented by $L(T) = f$ in abstract form. Why must the jumps of the normal derivative be added into the residual? To explain this, let us consider a two-dimensional heat conduction problem

$$-\partial(k_x\partial T/\partial x+k_{xy}\partial T/\partial y)/\partial x-\partial(k_{xy}\partial T/\partial x+k_y\partial T/\partial y)/\partial y$$

$$= f \text{ in } \Omega \quad T = T_o \text{ on } \Gamma_1 \text{ and } q = q_o \text{ on } \Gamma_2 \quad (40)$$

where the boundary Γ of the domain Ω consists of two parts, Γ_1 and Γ_2, T_o is specified temperature, q_o is specified heat flux, and

$$q = n_x(k_x\partial T/\partial x+k_{xy}\partial T/\partial y) + n_y(k_{xy}\partial T/\partial x+k_y\partial T/\partial y)$$

The unit vector $\mathbf{n} = n_x\mathbf{e}_x + n_y\mathbf{e}_y$ is outward normal to the boundary Γ. The variational formulation to this boundary value problem is

$$a(T,\overline{T}) = f(\overline{T}) \text{ for every } \overline{T} \quad (41)$$

such that $\overline{T} = 0$ on Γ_1, where

$$a(T,\overline{T}) = \int\{(\partial\overline{T}/\partial x)(k_x\partial T/\partial x+k_{xy}\partial T/\partial y) + (\partial\overline{T}/\partial y)(k_{xy}\partial T/\partial x + k_y\partial T/\partial y)\}d\Omega$$

and

$$f(\overline{T}) = \int_\Omega \overline{T}fd\Omega + \int_{\Gamma_2} \overline{T}q_od\Gamma \quad (42)$$

For simple notation, let us define the heat flux vector **q** by

$$\mathbf{q} = q_x e_x + q_y e_y, \quad q_x = k_x \partial T/\partial x + k_{xy} \partial T/\partial y,$$

$$q_y = k_{xy} \partial T/\partial x + k_y \partial T/\partial y$$

For two-dimensional problems, the elliptic projection may be defined by

$$a(e-\pi e, w) = 0 \text{ for every } w \tag{44}$$

such that

$$w \varepsilon H: = \{v | v(x_I) = 0, \ I = 1,\ldots,N\} \tag{45}$$

where N is the total number of nodal points in a given finite element grid, and e is the approximation error defined by the difference of the exact solution T and its finite element approximation T_h. However, this elliptic projection would not yield a simple form of the local problem defined in each element Ω_e as (25). Indeed, the elliptic projection yields

$$\begin{aligned} a(\pi e, w) &= a(T,w) - a(T_h,w) \\ &= \sum_{e=1}^{NE} \int_{\Omega_e} (f + \partial q_{xh}/\partial x + \partial q_{yh}/\partial y) w d\Omega \\ &+ \sum_{e=1}^{NE} \int_{\Gamma_{2e}} (q_o - q_{oh}) w d\Gamma \\ &+ \sum_{e=1}^{NE} \int_{\Gamma_e} [|q_h|] w d\Gamma \end{aligned} \tag{46}$$

where q_{oh} is the interpolation of q_o using the shape functions defined on the boundary elements Γ_{2e}, e = 1,...,N2 on Γ_2, $[|q|]$ is the jump of the heat flux on the internal element boundaries Γ_e, e = 1,...,NE, excluding the portion common to Γ. In definition of the elliptic projection, w must be zero only on nodal points. It need not vanish along the element boundaries. Thus, the second and third terms are not zero in the right hand side of the above equation. This is the difference from the one-dimensional problem. Therefore, the residual r_e must be defined as a combination of

$$f + \partial q_{xh}/\partial x + \partial q_{yh}/\partial y \tag{47}$$

$$q_o - q_{oh} \tag{48}$$

and the jumps

$$[|q_h|] \tag{49}$$

The question is, how are these three to be combined to have a posteriori estimate? Applying relations in Sobolev norms and inverse theorem $||\pi e||_e$ may be bounded by

$$\begin{aligned} ||\pi e||_e^2 \le\ & C_1 A_e \{\int_{\Omega_e} (f+\partial q_{xh}/\partial x+\partial q_{yh}/\partial y)^2 d\Omega_e\}^{1/2} \\ & + C_2 \{\int_{\ell_{2e}\partial\Omega_e} h_e (q_o - q_{oh})^2 d\Gamma\}^{1/2} \\ & + C_3 \{\int_{\ell_e \partial\Omega_e} h_e [|q_h|]^2 d\Gamma\}^{1/2} \end{aligned} \tag{50}$$

where $\partial\Omega_e$ is the boundary Ω_e, A_e the area of Ω_e, and h_e is the length of the edges of Ω_e. Constants C_1, C_2 and C_3 are dependent on the shape of an element, trace theorem in Sobolev spaces, and inverse theorem in finite element approximation spaces. At this moment, there does not exist a clear way to estimate these three constants.

On the other hand, the upper bound using the interpolation error is straightforward. Indeed, the orthogonality of the finite element approximation to the finite element space is still valid, and then

$$\begin{aligned} a(e,e) &= a(e,T-T_I) + a(e,T_I-T_h) \\ &= a(e,T-T_I) \end{aligned}$$

that is,

$$||e|| = a(e,e) \leq ||E_I|| = a(E_I,E_I) \tag{51}$$

where $E_I = T-T_I$ is the interpolation error. Thus, finding E_I for each given finite element remains.

Under the assumption that the second derivatives of T are constant in each finite element, we have the following Taylor's expansion for 3-node triangular elements:

$$\begin{aligned} T_1 &= T + T_x(x_1-x) + T_y(y_1-y) + 1/2T_{xx}(x_1-x)^2 \\ &\quad + T_{xy}(x_1-x)(y_1-y) + 1/2T_{yy}(y_1-y)^2 \\ T_2 &= T + T_x(x_2-x) + T_y(y_2-y) + 1/2T_{xx}(x_2-x)^2 \\ &\quad + T_{xy}(x_2-x)(y_2-y) + 1/2T_{yy}(y_2-y)^2 \\ T_3 &= T + T_x(x_3-x) + T_y(y_3-y) + 1/2T_{xx}(x_3-x)^2 \\ &\quad + T_{xy}(x_3-x)(y_3-y) + 1/2T_{yy}(y_3-y)^2 \end{aligned} \tag{52}$$

where $T_x = \partial T/\partial x$, $T_{xx} = \partial^2 T/\partial x^2$, $T_{xy} = \partial^2 T/\partial x \partial y$, etc.

Defining

$$b_1 = (y_2-y_3)/J, \quad b_2 = (y_3-y_1)/J, \quad b_3 = (y_1-y_2)/J$$

$$c_1 = (x_3-x_2)/J, \quad c_2 = (x_1-x_3)/J, \quad c_3 = (x_2-x_1)/J$$

and

$$J = x_1(y_2-y_3) + x_2(y_3-y_1) + x_3(y_1-y_2), \tag{53}$$

we have

$$\begin{aligned} T_1b_1 + T_2b_2 + T_3b_3 &= T_x + (1/2)T_{xx}\{(x_1-x)^2b_1 + (x_2-x)^2b_2 \\ &\quad + (x_3-x)^2b_3\} + T_{xy}\{(x_1-x)(y_1-y)b_1 \\ &\quad + (x_2-x)(y_2-y)b_2 + (x_3-x)(y_3-y)b_3\} \\ &\quad + (1/2)T_{yy}\{(y_1-y)^2b_1 + (y_2-y)^2b_2 \\ &\quad + (y_3-y)b_3\} \end{aligned} \tag{54}$$

and

$$\begin{aligned} T_1c_1 + T_2c_2 + T_3c_3 &= T_y + (1/2)T_{xx}\{(x_1-x)^2c_1 + (x_2-x)^2c_2 \\ &\quad + (x_3-x)^2c_3\} + T_{xy}\{(x_1-x)(y_1-y)c_1 \\ &\quad + (x_2-x)(y_2-y)c_2 + (x_3-x)(y_3-y)c_3\} \end{aligned}$$

$$+ (1/2)T_{yy}\{(y_1-y)^2c_1 + (y_2-y)^2c_2 + (y_3-y)^2c_3\} \quad (55)$$

Since

$$\partial T_I/\partial x = T_1b_1 + T_2b_2 + T_3b_3, \text{ and}$$

$$\partial T_I/\partial y = T_1c_1 + T_2c_2 + T_3c_3, \quad (56)$$

the first derivatives of the difference of the finite element interpolation and the solution T are explicitly represented by the second derivatives of the solution T and geometric quantities b_i and c_i, i = 1, 2, 3, of an element. Therefore, if T_{xx}, T_{xy}, and T_{yy} are approximated by the finite element solution T_h, the interpolation error in each finite element Ω_e is given by

$$||E_I||_e: = \{\int_{\Omega_e} \nabla(T-T_I) \cdot \mathbf{k}\nabla(T-T_I)d\Omega e\}^{1/2} \quad (57)$$

where $\nabla T = (\partial T/\partial x)\mathbf{e}_x + (\partial T/\partial y)\mathbf{e}_y$, $\mathbf{e}_x$ and $\mathbf{e}_y$ are the unit vectors along the x and y axes, respectively.

A method to estimate the second derivatives of the solution T using the finite element solution T_h is again the least square method (38). Defining the heat flux $\mathbf{q} = q_{xh}\mathbf{e}_x + q_{yh}\mathbf{e}_y$ by

$$q_{xh} = k_x\partial T_h/\partial x + k_{xy}\partial T_h/\partial y \text{ and } q_{yh} = k_{xy}\partial T_h/\partial x + k_y\partial T_h/\partial y$$

a continuous q_{xh} and q_{yh} spanned by the same shape functions of the temperature are obtained by the minimizers of

$$\min \sum_{e=1}^{NE} \frac{1}{2} \int_{\Omega_e} (\sigma_{xh}-q_{xh})^2 d\Omega_e \quad (58)$$

and

$$\min \sum_{e=1}^{NE} \frac{1}{2} \int_{\Omega_e} (\sigma_{yh}-q_{yh})^2 d\Omega_e \quad (59)$$

respectively. Then, in each element Ω_e we solve the 2 x 2 matrix equation

$$\begin{bmatrix} \varepsilon_x \\ \varepsilon_y \end{bmatrix} = \begin{bmatrix} k_x & k_{xy} \\ k_{xy} & k_y \end{bmatrix}^{-1} \begin{bmatrix} q_{xh} \\ q_{yh} \end{bmatrix} \tag{60}$$

where ε_x and ε_y are the approximations of $\partial T/\partial x$ and $\partial T/\partial y$ which are piecewise linear polynominals, respectively. Using these, the second derivatives of T may be approximated as

$$\partial^2 T/\partial x^2 \approx \partial \varepsilon_x/\partial x, \quad \partial^2 T/\partial y^2 \approx \partial \varepsilon_y/\partial y,$$
$$\partial^2 T/\partial x \partial y \approx \partial \varepsilon_x/\partial y \quad \text{or} \quad \partial \varepsilon_y/\partial x \tag{61}$$

As shown above, extension of an upper bound by the interpolation error to multi-dimensional problems is straightforward. However, an upper bound by posteriori estimates using the residual is not very obvious in multi-dimensional problems. We have to speculate certain estimates of errors especially, due to the amount of jump of the flux across the internal finite element boundaries. Thus, in the following development of adaptive finite element methods, we shall apply the interpolation error as an error measure since it is a very tight upper bound of the finite element approximation error.

Figures 6, 7, and 8 show the finite element grid for a heat flux conduction problem, finite element approximated results, and the distribution of error measure, respectively. It is clear that a large error measure is computed in the place where large heat flux is expected.

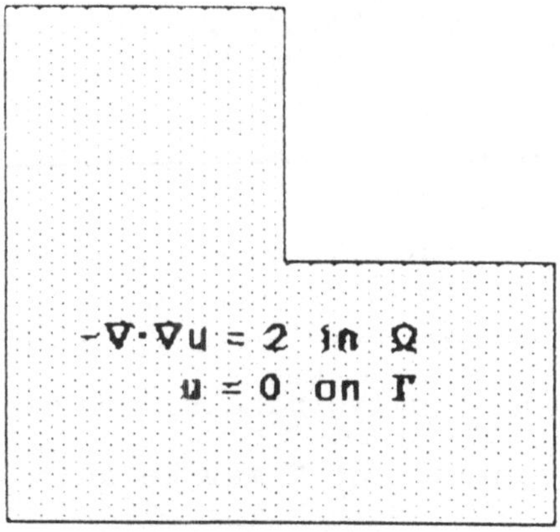

Figure 6. Finite Element Model of Heat Conduction L-Shape Domain (by Mr. T. Kishimoto)

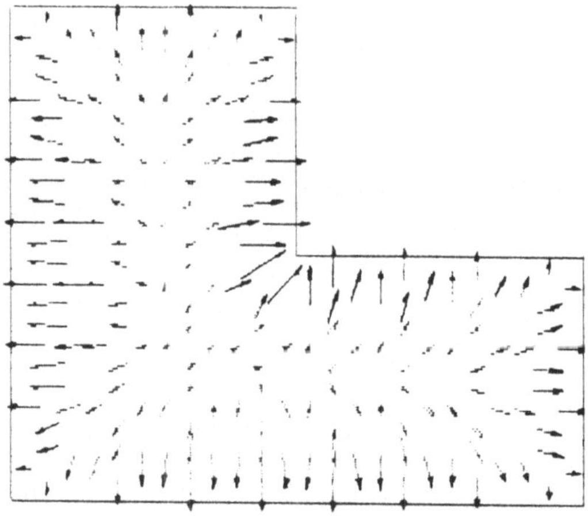

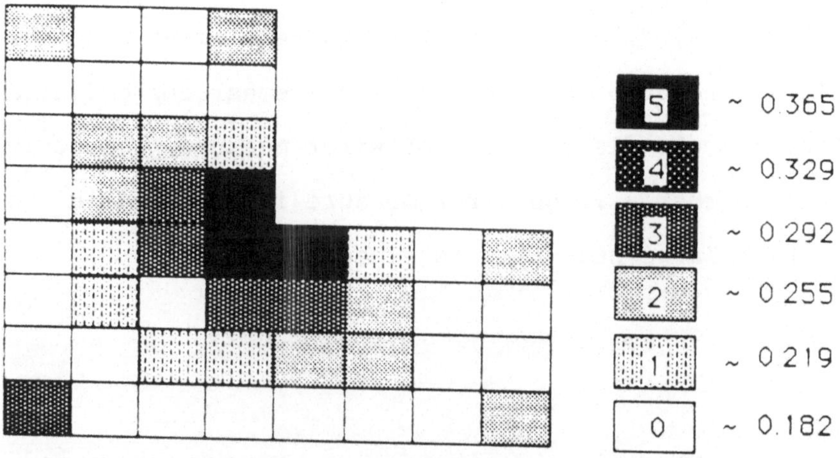

Figure 8. Distribution of Error Measure by a Uniform Grid

References

1. Ciarlet, P. G., The Finite Element Method for Elliptic Problems, North-Holland Publishing Company, Amsterdam, 1978.
2. Blowinski R., Numerical Method for Nonlinear Variational Problems, Spring-Verlag, New York, 1984.
3. Zienkiewicz, O. C., The Finite Element Method, 3rd Edition, McGraw-Hill, London, 1976.
4. Kardestunser, H., The Finite Element Handbook, McGraw-Hill, to be published in 1986.
5. Becker, E. B., Carey, G. F., and Oden, J. T., Finite Elements: An Introduction, Prentice-Hall, Englewood Cliffe, 1981.
6. Carey, G. F., and Oden, J. T., Finite Elements: A Second Course, Prentice-Hall, Englewood Cliffe, 1983.
7. Carey, G. F., and Oden, J. T., Finite Elements in Computational Aspects, Prentice-Hall, Englewood Cliffe, 1984.
8. Babuska, I., "The Self-Adaptive Approach in the Finite Element Method," in Mathematics of Finite Elements with Application: MATELAP 1975, Academic Press, New York, 1977.
9. Babuska, I., and Rheinboldt, W. C., "A-Posteriori Error Estimates for Finite Element Method," Int. J. Numer. Methd. Eng., 12, 1597-1615, 1978.
10. Babuska, I., and Rheinboldt, W. C., "Error Estimates for Adaptive Finite Element Computations, " SIAM J. Numer. Anal., 15, 736-754, 1978.

11. Babuska, I., and Rheinboldt, W. C., "Adaptive Approaches and Reliability Estimations in Finite Element Analysis," Comput. Methods Appl. Mech. Eng., 18, 519-540, 1979.

12. Babuska, I., and Szabo. B., "On the Rates of Convergence of the Finite Element Method," Int. J. Numer. Meth. Eng., 18, 323-341, 1982.

13. Szabo, B. A., and Mehta, A. K., "P-Convergent Finite Element Approximations in Fracture Mechanics," Int. J. Numer. Meth. Eng., 12, 551-560, 1978.

14. Shephard, M. S., Gallagher, R. H., and Abel, J. F., "The Synthesis of Near-Optimum Finite Element Meshes With Interactive Computer Graphics," Int. J. Numer. Meth. Eng., 15, 1021-1039, 1980.

15. Shephard, M. S., and Yerry, M. A., "An Approach to Automatic Finite Element Mesh Generation," Comput. Engr., 3, 21-28, 1982.

16. Diaz, A. R., Kikuchi, N., and Taylor, J. E., "A Method of Grid Optimization for Finite Element Methods," Comput. Meth. Appl. Mech. Eng., 37, 29-46, 1983.

17. Diaz, A. R., Kikuchi, N., and Taylor, J. E., "Optimal Design Formulations of the Finite Element Grid Adaptation," in V. Komkou ed., Sensitivity of Functionals with Applications To Engineering Science, Lecture Notes in Mathematics, 1086, 56-76, Springer-Verlag, 1984.

Adaptive Grid Design for Finite Element Analysis in Optimization: Part 2, Grid Optimization

Noboru Kikuchi

Department of Mechanical Engineering and Applied Mechanics

The University of Michigan, Ann Arbor, MI 48109, U.S.

Abstract

A finite element grid optimization problem is defined in the form of mini-max problem using an error measure which indicates the amount of finite element approximation error. A necessary condition to the optimization problem is given together with an iterative method based on the optimality criteria method. Various error measures are introduced for stress analysis of linearly elastic structures, and their mathematical characterization is described in order to provide information on choice of a particular error measure in a specific problem given. Three adaptive schemes; r- h-, and p-methods are introduced. Effectiveness of application of adaptive methods based on the grid optimization problem, is described using examples of stress analyses.

*The present work is supported by NASA Lewis Research Center/ NAG 3-388 and ONR N0001485K0799.

1. INTRODUCTION

In the previous part, finite element error analysis has been briefly reviewed using one dimensional boundary value problems. Finite element approximation error is related to either the residual or the interpolation error in a certain "energy" norm. Although the residual may not give an upper bound of the approximation error in multi-dimensional problems as clearly as in the one-dimension case, an upper bound by the interpolation error is the same in any dimensions. As shown in the previous paper, the residual in multi-dimensions may involve the amount of jump of the "flux" or "traction" which is not clearly bounded by certain estimates a posteriori obtained. On the other hand, the interpolation error, in general, requires computation of the second derivatives of the temperature/displacement. If lower order finite elements are applied in finite element approximation, the finite element solution would not be able to provide meaningful second derivatives by itself. Thus, the least square method is introduced to define a continuous "flux" or "traction" field using the finite element approximation which is discontinuous in Lagrange interpolation. The second derivatives are then computed using the "continuous" "flux" or "traction." It is noted that the second derivatives contains information of the amount of jump of "flux" and "traction." In this sense, the bound by the interpolation error is closely related to the upper bound by the residual. Since interpolation error can be evaluated independently of dimensionality of a given problem, we shall apply this to define an error measure in the finite element grid optimization problem.

Adaptive finite element methods are classified into three groups; r-, h-, and p-methods. The r-method involves relocation of nodes in a finite element grid to achieve the optimal location of nodes and shape of finite elements. The total number of nodes and elements are unchanged in the adaptive process. The h-method refines finite elements in which a large amount of finite element approximation error is accumulated. The same kind of finite elements is kept during the adaptation, while the total number of nodes and elements are increased. The p-method applies higher order polynomials in the shape functions of finite elements in which approximation error is large. Historically speaking, the r-method was first introduced. For example, Oliveira [1] obtained the optimal location of nodes in a finite element grid as well as Tang and Turcke [2], Melosh and Marcal [3], Turcke and McNeice [4], Shephard, Gallagher and Abel [5], Fellipa [6,7], Diaz et al. [8, 9, 10], and others. For the h- and p-methods, Babuska,Rheinboldt, and Szabo have worked, see references of Part 1. It is noted that Sewell [11] introduced the idea of h-method using three-node triangular elements in 1975. Although the combination of coarse and refined grids was applied even in the time of Southwell [12, 13], systematic concept of adaptivity based on finite element error analysis is introduced by Babuska and Rheinboldt. In finite difference methods, adaptive methods were studied together with grid generation which fits curved boundaries, by using truncation error which is similar to the interpolation error in finite elements methods. Adaptive methods were widely accepted in finite difference methods much earlier than in finite element methods, see Thompson and Warsi [14], and Ghia and Ghia [15]. The concept of adaptivity in finite difference is very close to the r-method which "deforms" grids so that refined elements (or grids) are assigned to the portion of high gradient of "flux" or "traction." Note that the truncation error of the 5 point difference scheme to the second order differential equations is bounded by the representative grid size and the second derivatives of the function, that is, the gradient of flux or traction. The p-method is, on the other hand, unique. Corresponding schemes have not been applied in finite difference methods. It uses the same finite element grid during the adaptation while the degree of polynomials of the shape functions are automatically increased for finite elements which have large amount of approximation error. To do this, hierarchial finite element [16, 17] are applied.

In Part 2, a finite element grid optimization problem is defined using an error measure related to the amount of approximation error in each finite element. Based on the necessary condition of the grid design problem, an iterative scheme is introduced using the optimality criteria method. We then restrict our attention to stress analysis of linearly elastic structures, and introduce various other possible error measures applicable to adaptive methods. The r- and h-methods are mainly discussed in this paper, while the p-method is just briefly mentioned. Examples are solved to demonstrate the capability of these adaptive methods.

2. OPTIMAL FINITE ELEMENT GRID DESIGN PROBLEM

Suppose that an appropriate error measure E_e is defined in each finite element that indicates the amount of approximation error. In general, this must be an upper bound of the approximation error in a given norm. Then the total amount of error may be bounded by

$$TE = \{\Sigma_e E_e{}^2\}^{1/2} \tag{1}$$

A natural optimization problem is defined by

$$\min_{\text{adaptive grid design}} TE \quad , \tag{2}$$

that is, minimizing the total amount of error in the whole domain. It is, however, very difficult to obtain the optimality condition because the minimization problem is subjected to finite element solutions which are strongly coupled among finite elements. Furthermore, the value TE itself is expected to be small if "reasonable" finite element approximations are considered. Thus performing minimization of TE may not be so effective in the sense that the minimum TE may not be so different from TE at the initial stage of adaptation. Although there is a singular point at which stress components are infinite, node relocation and refinement would not yield large difference of TE despite drastic change of the distribution of error measures E_e, $e = 1, \ldots, NE$ where NE is the total number of finite elements in a discrete model. To reflect this nature, the optimal grid design problem should be defined by the min-max concept:

$$\min_{\text{adaptive grid design}} \max_{e=1,\ldots,NE} E_e \tag{3}$$

It is clear that the maximum value of error measures defined in elements is very sensitive to nodal location, element shape, and element size. Thus, this quantity is appropriate to be applied to define an adaptive process. It is also clear that one of the necessary condition of the optimality to the optimal grid design problem (3) is given as

$$E_e = \text{constant for all } e,\ e = 1, \ldots, NE \tag{4}$$

This means that at the optimal grid error must be equally distributed among finite elements in the discrete model. Thus, adaptive grid design using the r-, h-, and p-methods can be defined as follows:

r-method Nodal points are relocated so that the necessary condition (4) is more likely. One of methods of node relocation is similar to the grid smoothing scheme introduced by Winslow [18]:

$$\mathbf{x} = \Sigma_e \mathbf{x}_e E_e A_e^{-m} / \Sigma_e A_e^{-m}, \tag{5}$$

where x is the coordinates of the nodal point P, x_e is the coordinate of the centroid of elements which share the node P, A_e is the area of such elements, m is a parameter for "smoothing," and the summation is taken over such elements. If m=1 for rectangular elements, constant E_e, e=1, ..., NE, yields no nodal relocation to the point P. In other words, if the necessary condition (4) is satisfied, there is no node relocation by the relocation scheme (5). It is, however, noted that we have not verified that infinite application of (5) yields the necessary condition (4) for two-dimensional problems, while for one dimensional problems (5) implies convergence to (4) after infinitely many times of application. If finite elements are not rectangular, (5) still relocates nodes for constant E_e although its movement is minor. This nature suggests that (5) may not imply the exact equality of error measures over the whole domain in spite of convergent location of nodes. It must, however, be noted that (5) is a sort of grid smoothing scheme which eliminates irregular finite elements and drastic change of size and shape of elements. Although it may still relocate for constant error measures, it does not destroy "good" nature of approximation.

Application of the r-method is recommended for a reasonably refined grid, since it is impossible to reduce to zero the total amount of approximation error while the maximum value of error measures is minimized unless the total number of degrees of freedom is increased. As far as practical application is concerned, it is not necessary to introduce a very refined finite element grid. Just a reasonably refined grid is sufficient to obtain accurate enough finite element approximation if the r-method is applied appropriately.

h-method Let the "energy" norm of the finite element approximation be denoted by $||u_h||$, and let QI be the ratio of TE with respect to $||u_h||$, i.e.,

$$QI = TE/||u_h|| \tag{6}$$

This indicates the relative amount of approximation error such that if the degree of the approximation is fine, the value of QI should be small. Roughly speaking, if QI is 0.05, then it might be said that 5% error is involved in the finite element approximation. But this is not the exact percentage of the error. For a given QI, define E_{exp} by

$$E_{exp} = QI*||u_h||/\sqrt{NE} \tag{7}$$

For two-dimensional problems, if error measure E_e of the eth finite element is more than or equal to the four times of E_{exp}, then a triangular or quadrilateral element is refined into four sub-elements. Using the adapted grid, new finite element approximation is obtained. Update $||u_h||$, E_{exp}, and E_e, e = 1, ..., NE, for new NE, and then find finite elements whose error measure is more than or equal to the four times of E_{exp}. Refine such elements if they exist, and repeat the process until there is no such element. It is clear that if this adaptation is applied for sufficiently small QI, all the values of E_e, e = 1, ..., NE, are located in a relatively narrow band so that the necessary conditions (4) is approximately satisfied. If more strict satisfaction of the necessary condition is desired, the r-method may be applied after the h-adaptation. Another useful quantity is the amount of deviation of E_e from the E_{exp}:

$$DV = \{\Sigma_e(E_e-E_{exp})^2\}^{1/2}/\{\Sigma_e E_e{}^2\}^{1/2} \tag{8}$$

If DV is large, E_e has a larger variation from the expected error measure E_{exp}.

p-method This adaptive method follows in a similar manner to the h-method. Instead of refinement, a degree is increased in polynomials of the shape functions of finite elements whose error measure E_e is larger than C/h_e where C is an appropriate constant and h_e is the representative element length of the e-th element. In practice, increasing degrees too much is not wise since development of a finite element analysis code for such p-adaptivity becomes too difficult. At most up to cubic elements are reasonable for plane elasticity problems. It is also noted that for plates and shells p-adaptive finite element methods may not be realistic because the degree of polynomials involved is too high. For a triangular element given in Fig. 1, linear, quadratic, and cubic shape functions are given by

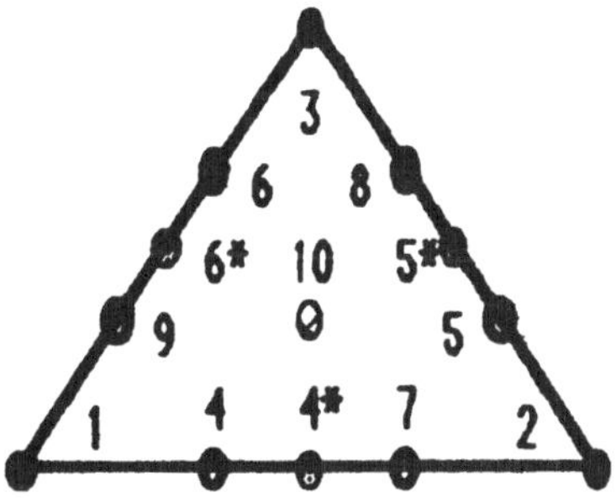

Figure 1. Linear, Quadratic, and Cubic Triangular Element

basis functions

$S_i = L_i$, $i = 1, 2$ and 3, $S_4^* = 4L_1L_2$, $S_5^* = 4L_2L_3$,

$S_6^* = 4L_3L_1$

$S_4 = -27/2\ L_1(L_2-2/3)L_2$, $S_5 = -27/2\ L_2(L_3-2/3)L_3$

$S_6 = -27/2\ L_3(L_1-2/3)L_1$, $S_7 = -27/2\ L_1(L_1-2/3)L_2$

$S_8 = -27/2\ L_2(L_2-2/3)L_3$, $S_9 = -27/2\ L_2(L_3-2/3)L_1$

$S_{10} = 27L_1L_2L_3$

shape functions

$N_{10} = \kappa_{10}S_{10}$

$N_i = N2_i + N3_i$, $i = 4, 5, 6, 7, 8$, and 9

$N_i = S_i - 1/3N3_j - 2/3N3_{j+3} - 2/3N3_k - 1/3N3_{k+3}$

$- 1/2N2_j - 1/2N2_k$, $i = 1, 2, 3$, and 4, $j = i + 3$,

$k = i + 2$, but $k = 6$ if $i = 1$,

where

$$N3_i = \kappa_i(S_i - 1/2N_{10}),\ i = 4, 5, 6, 7, 8, \text{ and } 9$$
$$N2_i = \kappa_i^* (S_i^* - 4/9N_{10}),\ i = 4, 5, \text{ and } 6$$
$$N2_i = 0,\ i = 7, 8, \text{ and } 9 \qquad (9)$$

using switch condition attached at each node such that

$$\kappa_i = 1 \text{ if } i \text{ node on } = 0 \text{ if } i \text{ node off},\ i = 1, \ldots, 10,$$
$$\kappa 4 = \kappa 7,\ \kappa 5 = \kappa 8,\ \kappa 6 = \kappa 9$$
$$\kappa^*_i = 1 \text{ if } \kappa_i = 0, \text{ and } \kappa^*_i = 0 \text{ if } \kappa_i = 1,\ i = 4, 5, \text{ and } 6 \qquad (10)$$

3. ERROR MEASURES FOR STRESS ANALYSIS

Stress analysis of linearly elastic structures is considered here. Let $\mathbf{u}$ be the displacement in equilibrium, and let σ be the corresponding stress in the contracted notation. For small deformation, the strain $\varepsilon(\mathbf{v})$ due to an arbitrary displacement $\mathbf{v}$ is defined by $\varepsilon(\mathbf{v}) = (\nabla\mathbf{v}+\nabla\mathbf{v}^T)/2$ where ∇ is the gradient operator in the global coordinate system, and $\mathbf{A}^T$ is the transpose of a matrix $\mathbf{A}$. Suppose that the constitutive relation of materials is represented by

$$\sigma = \mathbf{D}\varepsilon - \boldsymbol{\alpha}\Delta T \qquad (11)$$

where $\mathbf{D}$ is the inverse of the compliance matrix, $\boldsymbol{\alpha}$ is the "thermal expansion coefficient," and ΔT is the temperature difference from the reference. If a body force $\mathbf{b}$ is applied per unit mass, and if a traction $\mathbf{t}$ is applied on a part of the boundary Γ_t, the virtual displacement principle is written by

$$u \in V_g:\ a(\mathbf{v},\mathbf{u}) \ = \ f(\mathbf{v}) \text{ for every } \mathbf{v} \in V_0 \qquad (12)$$

Here $a(\mathbf{v},\mathbf{u})$ is the internal virtual work due to an arbitrary virtual displacement $\mathbf{v}$ at the equilibrium $\mathbf{u}$, $f(\mathbf{v})$ is the work done by the body force, applied traction, and thermal stress, V_g is the admissible set of displacements satisfying a given possibly nonhomogeneous essential boundary condition $\mathbf{v} = \mathbf{g}$ on $\Gamma_d = \Gamma - \Gamma_t$, and V_0 is its homogeneous case. More precisely,

$$a(\mathbf{v},\mathbf{u}) \ = \ \int_\Omega \varepsilon(\mathbf{v})\cdot D\varepsilon(\mathbf{u})\, d\Omega,\ f(\mathbf{v}) \ = \ \int_\Omega (\varepsilon(\mathbf{v})\cdot \boldsymbol{\alpha}\Delta T + \mathbf{v}\cdot \mathbf{b}\rho) d\Omega + \int_{\Gamma_t} \mathbf{v}\cdot \mathbf{t} d\Gamma$$

$$V_g \ = \ \{\mathbf{v} \in H^1(\Omega):\ \mathbf{v} = \mathbf{g} \text{ on } \Gamma_d\} \qquad (13)$$

where $H^1(\Omega)$ is a Sobolev space in which functions are square integrable and their generalized first derivatives are also square integrable.

Suppose that the subscript h means finite element aproximation, i.e., for example, $\mathbf{u}_h$ is a finite element approximation of the displacement $\mathbf{u}$ in equilibrium. The energy norm $||\mathbf{u}||$ is defined by

$$||\mathbf{u}|| \ = \ \sqrt{a(\mathbf{u},\mathbf{u})} \qquad (14)$$

The orthogonality of the approximation error $\mathbf{e} = \mathbf{u} - \mathbf{u}_h$ to the finite element approximation space is represented by

$$a(\mathbf{e},\mathbf{v}_h) \ = \ 0 \text{ for every } \mathbf{v}_h \in V_{0h}. \qquad (15)$$

Defining interpolation error $\mathbf{e}_I = \mathbf{u} - \mathbf{u}_I$ where $\mathbf{u}_I$ is the interpolation of $\mathbf{u}$ in V_{gh}, we have

$$a(\mathbf{e},\mathbf{e}) \ = \ a(\mathbf{e},\mathbf{e}_I) + a(\mathbf{e},\mathbf{u}_I - \mathbf{u}_h) \ = \ a(\mathbf{e},\mathbf{e}_I), \qquad (16)$$

and then Schwarz inequality of an inner product yields

$$||\mathbf{e}|| \leq ||\mathbf{e}_I||. \qquad (17)$$

This means that the finite element approximation error is bounded by the interpolation error as shown in Part 1. Since the interpolation error is written by

$$||e_I|| = \{\Sigma_e ||e_I||_e^2\}^{1/2} \text{ and}$$

$$||e_I||_e = \{\int_{\Omega_e} \varepsilon(e_I) \cdot D\varepsilon(e_I)\, d\Omega\}^{1/2}, \tag{18}$$

we can identify the energy norm of the interpolation error in an element as an error measure in adaptive methods. As shown in Part 1, the residual involves the jump of traction across the internal element boundaries, and since a posteriori estimates are not well established in stress analysis, only the interpolation error or its variations are considered here. That is, a natural choice is

$$E_e = ||e_I||_e. \tag{19}$$

In order to consider other possible error measures applicable to adaptive methods, 3-node triangular elements are introduced for simplicity. Suppose that σ_I is the piecewise constant interpolation of σ. Then if σ is assumed to be linear in a finite element Ω_e, the interpolation error is given by

$$\sigma - \sigma_I = (\mathbf{x}-\mathbf{x}_e) \cdot \nabla\sigma \tag{20}$$

in Ω_e, where $\mathbf{x}_e$ is the centroid of Ω_e. An error measure may be defined as the L^2 norm of the interpolation error of the stress, i.e.,

$$E_e = \{\int_{\Omega_\varepsilon} [(\mathbf{x}-\mathbf{x}_e) \cdot \nabla\sigma] \cdot [(\mathbf{x}-\mathbf{x}_e) \cdot \nabla\sigma] d\Omega\}^{1/2} \tag{21}$$

In this case QI must be redefined by

$$QI = TE/||\sigma_h|| \tag{22}$$

where $||\sigma_h||$ is the L^2 norm of the stress vector computed by the finite element approximation u_h. The expected value of error measure for a given QI for the h-method may be modified by

$$E_{exp} = QI*||\sigma_h||/\sqrt{NE} \qquad (23)$$

Instead of using σ, the strain ε is also applicable. Similarly, the equivalent stress σ_E or equivalent strain ε_E are also applicable to define error measures. It is, how, ever, noted that these error measures are not directly related to error analysis of the finite element approximation, whereas the natural choice using the energy norm is a clear upper bound of the approximation error. Computation of the second derivatives is, as shown in Part 1, based on the least square fitting of the stress σ_h using the same shape functions to the displacement $\mathbf{u}_h$, although it might not be the best.

In part 1, interpolation error is explicitly computed for 3-node triangular elements. For 4-node quadrilateral elements, similar evaluations are obtained for interpolation error, see Koh and Kikuchi [19] and Part 3 of this paper.

3. ADAPTIVE FINITE ELEMENT METHOD

In general finite element analysis consists of three processings: preprocessing, analysis processing and post processing. Adaptation would be inserted in between analysis and post processing for the r- and h-methods, and introduces an iterative process in the flow of macrocomputation.

For the p-method, adaptation is occasionally imbedded into analysis processing; analysis codes must be modified to manage hierarchical interpolation or introduction of nodeless variables while continuity of functions is ensured across internal element boundaries.

r-method As described in introduction, the r-method relocates nodes to satisfy the necessary condition of the optimality without increasing the total number of nodes and elements. Furthermore, element connectivities are unchanged during adaptation, and then continuity of interpolated functions is assured without special consideration while the h- and p-methods must introduce some methods to maintain continuity. In this sense the r-method is simplest among possible adaptive methods. A drawback is the possibility of element crash during node relocation process, while element distortion is not a serious disadvantage of the r-method. Indeed, element distortion itself does not provide approximation error. If large distortion is accompanied by high gradient of stresses, it is natural to generate large amounts of approximation error. But if distortion occurs in nearly constant stress fields, it would not generate significant error. Even in high gradient portion, distortion may not be so troublesome if it is "optimized." Note that the amount of error is given as (grid size)*(1 + distortion)*(second derivatives). Grid optimization must include all three factors. The relocation scheme given in (5) is not only reducing deviation of error measures in the finite element model but also smoothening given grids if error measures are almost constant.

Figure 2 shows a thin anisotropic linearly elastic plate with a circular hole subject to a tension force. The plate is a fiber reinforced lamina. Figures 3 and 4 describe the initial finite element grid and the circumferential stress along the hole, respectively. Figures 5 and 6 show the optimal grid and its result. It is clear that the adapted grid provides better results.

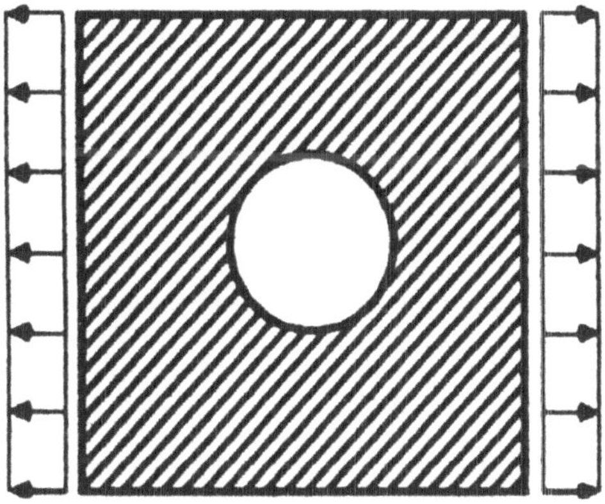

Figure 2 Fiber Reinforced Elastic Plate

The r-method described above would provide adapted grids similar to the ones by the

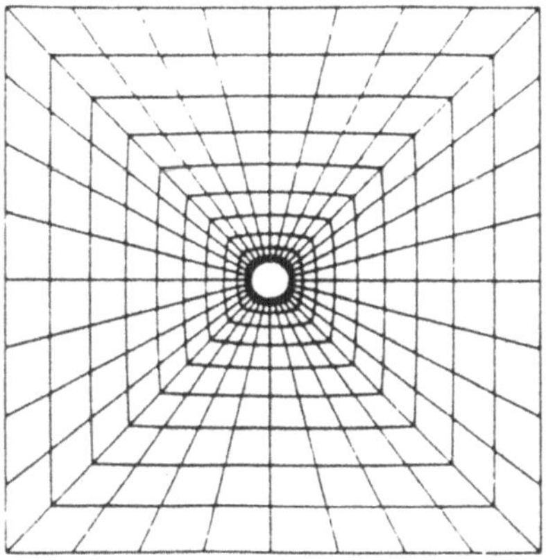

Figure 3 **Initial Finite Element Grid**

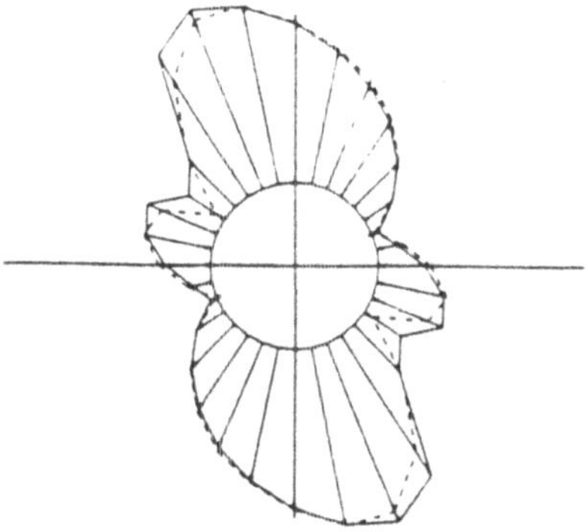

Figure 4 $\sigma_{\theta\theta}$ **around the Hole**

Real-F.E.M.

Dotted - Analytical

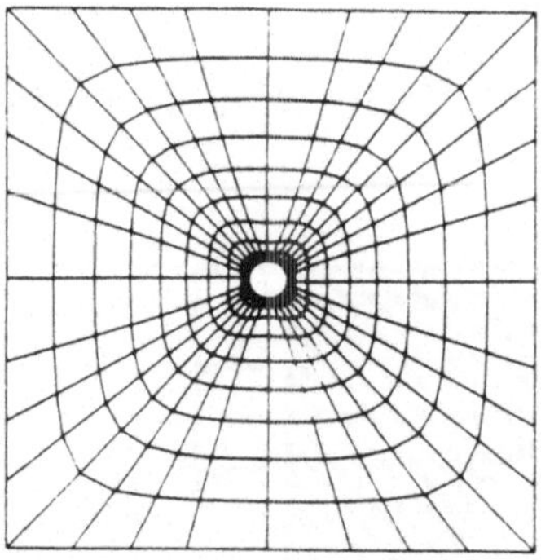

Figure 5. Adapted Grid by R-Method (by S.K. Lee)

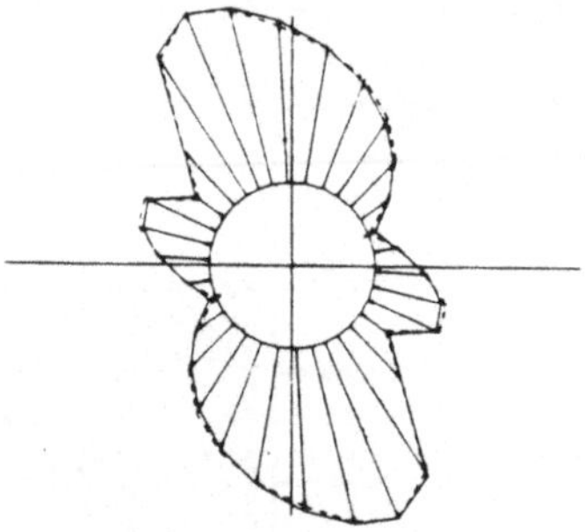

Figure 6 $\sigma_{\theta\theta}$ around the Hole 3% Error c.f.17% at the Initial

grid generation method using elliptic differential equations whose coefficient D and source terms P and Q are defined adaptively by error measures. More precisely, grid coordinates ξ and η defined on a rectangular or circular domain are transformed into the physical coordinates $\mathbf{x}$ to make a finite element grid using the elliptic differential equations

$$\nabla \cdot D \nabla \xi = P \quad \text{and} \quad \nabla \cdot D \nabla \eta = Q \tag{24}$$

Especially, if the initial grid for the r-adaptation is topologically the same to the grid in the mesh coordinates ξ and η, both end up with very similar results.

<u>h-method</u> There are, roughly speaking, two approaches in implementation of the h-method. The first one is to follow the process without regridding of a finite element model. That is, refinement will be performed based on the initial grid. If any grid smoothing schemes are not applied, the grid always contains that of the previous stage. For this approach, importance is to maintain continuity of interpolated functions. This leads to either the introduction of linear constraint equations or special refinement for transition from refined elements to coarse elements. Most of

the commercially available general purpose programs are capable to take account of linear constraint equations represented by some degree of freedom. For example, continuity restricts the degree of freedom at the midnode by

$$d_{2m-1} = (d_{2p-1}+d_{2q-1})/2, \quad d_{2m} = (d_{2p}+d_{2q})/2, \qquad (25)$$

where node m is located on the line consisting of nodes p and q. Another approach is to introduce special descretization to transmit a refined element to the remainder as shown in Fig. 7. New element connectivites must be reassigned in both cases. Application of the h-method to either plane or shell-like three dimensional domains are realistic, but for fully three dimensional solid domains, it may not be practical because of too rapid increasing in the total number of degrees of freedom. For example, if 20% of element of the 10x10x5 grid is refined, 800 elements are added in the original 500 elements. At the second adaptation 20% of elements are again refined, 3380 elements are generated. Then the grid adaptation is almost impossible.

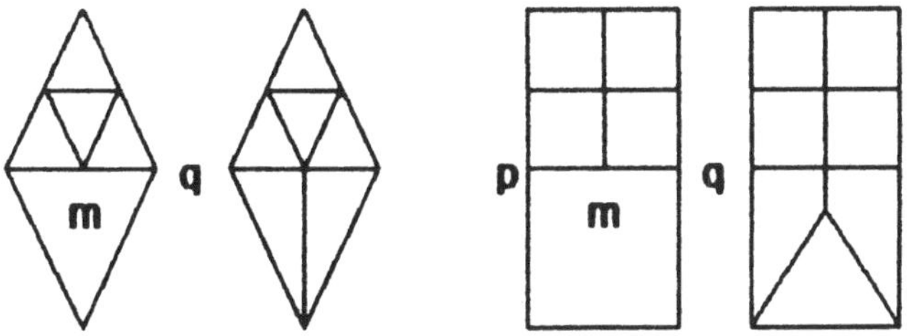

Figure 7 Typical h-Adaptive Grids

This refinement reduced the size of elements to h/2, h/4,...h/2m from the original size h, where m is the m-th adaptation. Since refinement is attributed to the amount of error in elements, "discrete" change of grid size is mathematically not the matter. However, as shown in Fig. 8 adapted grids are not smooth. Nonsmooth grids, in general, give a negative impression. The second approach is introduced by

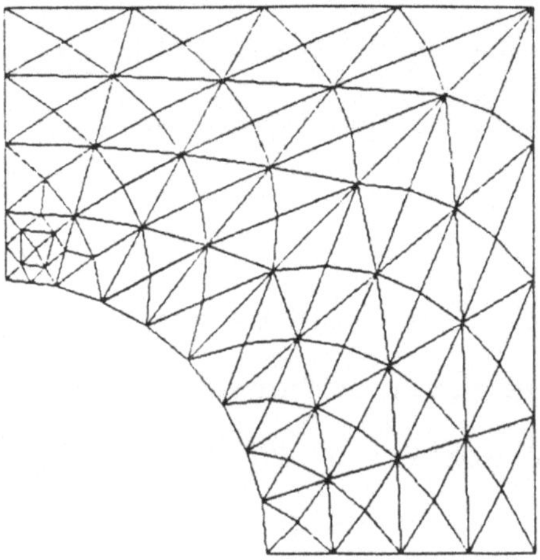

Figure 8. Application of the h-method

Thacher et al. [20], Joh [21], and Shuhara and Fukuhara [22], and provides smooth adapted grids by combining automatic grid generation and adaptive finite element methods. According to error measures computed, region 1, 2, ..., and m are assigned, where region k consists of element whose error measure is the k th largest. Node density is defined in each region, and then automatic triangulation algorithms are applied to develop adapted grids. Figure 9 shows a grid obtained by Joh . It is noted that the first approach can also provide smooth enough grids if some grid smoothening schemes are introduced, see Shephard. Figure 10 is a smoothed grid using the r-method by artificially setting all the error measures to constant in the adapted grid by the h-method.

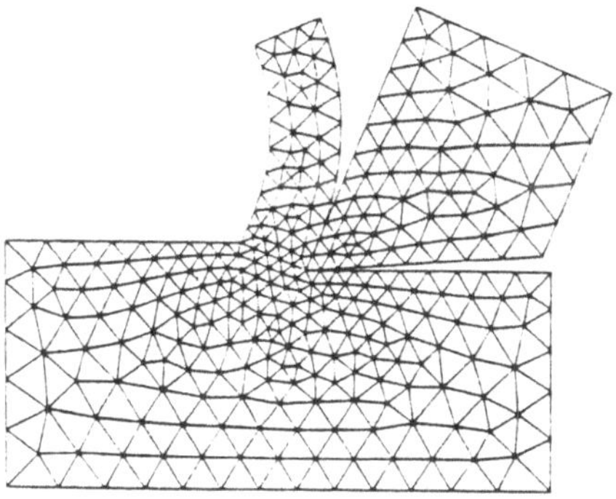

Figure 9 A Finite Element Grid by Joh and Akimoto

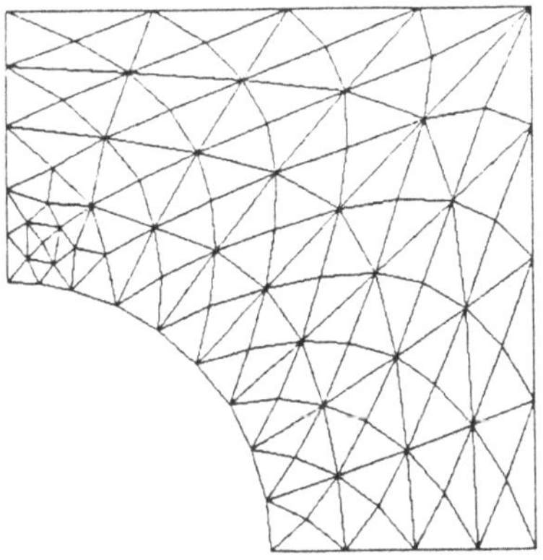

Figure 10. Application of the r-method with m = 2.5 for Smoothing

For time dependent problems, nonsmoothed h-adapted grids often provide oscillatory computed results in time, despite very smooth results in space if error measures are defined only by the "energy" norm in space. Error measures for time dependent problems may consist of "energy" both in space and in time. Appropriate application of the r-method, in general, provides smooth results both in time and space although it requires much more computing time.

It is also noted that if there are a reasonable number of elements, the r-method provides almost the same stress distribution as the h-method which involves considerably much more elements. In other words, although the r-method may not yield reduction of the total amount of approximation error, it reduces the maximum value of error measures significantly. Figure 11 shows comparison of the h-method to the r-method for stress analysis of a thin linearly elastic plate involving a crack.

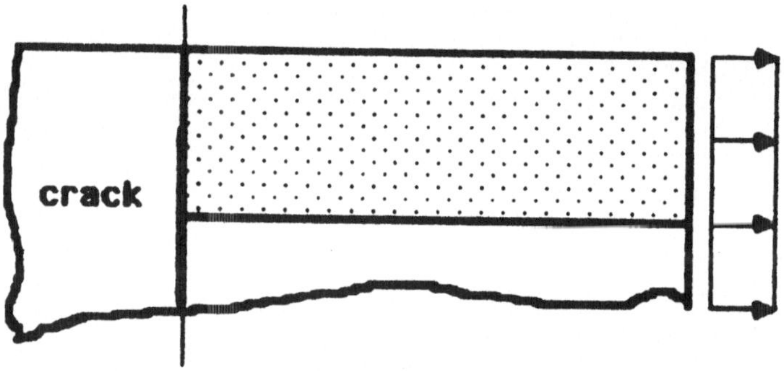

Linearly Elastic Thin Plate Subject to Tension

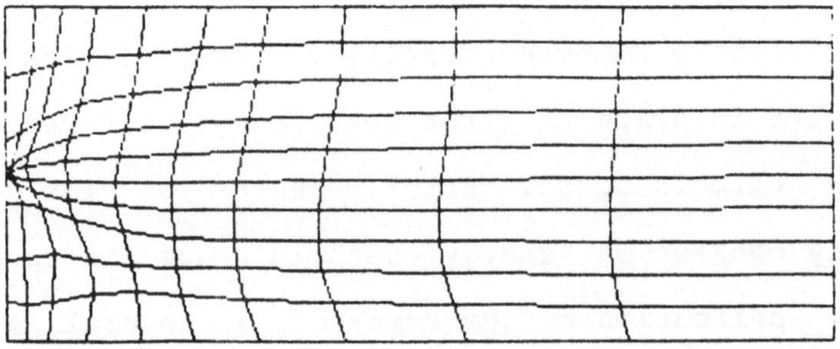

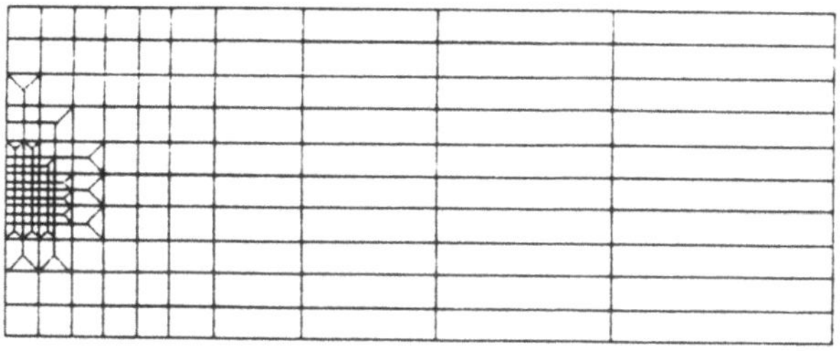

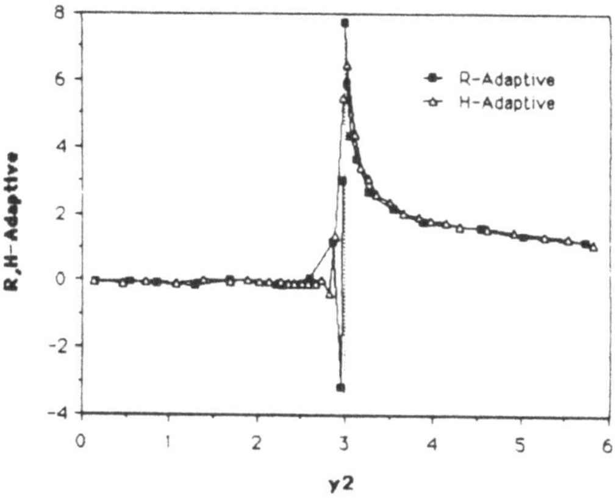

Figure 11. Application of the r- and h-adaptive Methods for Linear Fracture Mechanics (by Mr. T. Torigaki)

In shell structures, triangular elements which may be identified with special quadrilateral elements, are, in general, too stiff. For example, if QUAD4 elements in MSC/NASTRAN are applied, a shell structure becomes artificially very stiff in triangular elements which may be used to transmit refined elements to others. Thus, these elements should be applied in the area in which error measures are sufficiently small.

p-method Suppose that there are linear, quadratic, and cubic finite elements in the element library for plane stress analysis, and that linear constraint equations are taken in the linear equation solver. If hierarchical finite elements are not used which are based on the interpolation functions given in (9) together with switch indices at on/off nodes, then linear constraints equations must be assigned to the nodes along element boundaries of different kinds of elements: see Fig. 12,

(quadratic-linear)

$$d_{2m-1} = d_{2p-1} + d_{2q-1}, \quad d_{2m} = d_{2p} + d_{2q} \tag{26}$$

(cubic-quadratic)

$$d_{2m-1} = (2d_{2p-1}+8d_{2r-1}-d_{2q-1})/9, \quad d_{2m} = (2d_{2p}+8d_{2r} -d_{2q})/9$$

$$d_{2n-1} = (-d_{2p-1}+8d_{2r-1}+2d_{2q-1})/9, \quad d_{2n} = (-d_{2p}+8d_{2r} -2d_{2q})/9 \tag{27}$$

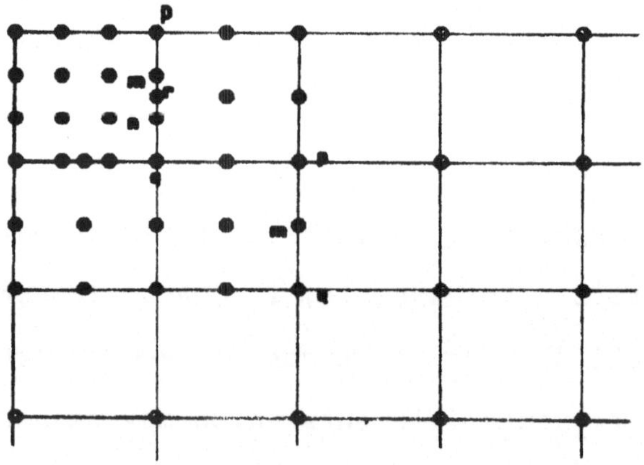

Figure 12 Typical p-Adapted Finite element Grid

The same node number is assigned for the corner nodes of the elements from which adjacent elements are of a different kind. If the hierarchical element using (9) is in the analysis program, the above constraints need not be introduced since transition elements are generated without any difficulty by setting up adequately on/off switches of nodes. Figure 13 shows an application of the p-method to a heat conduction problem.

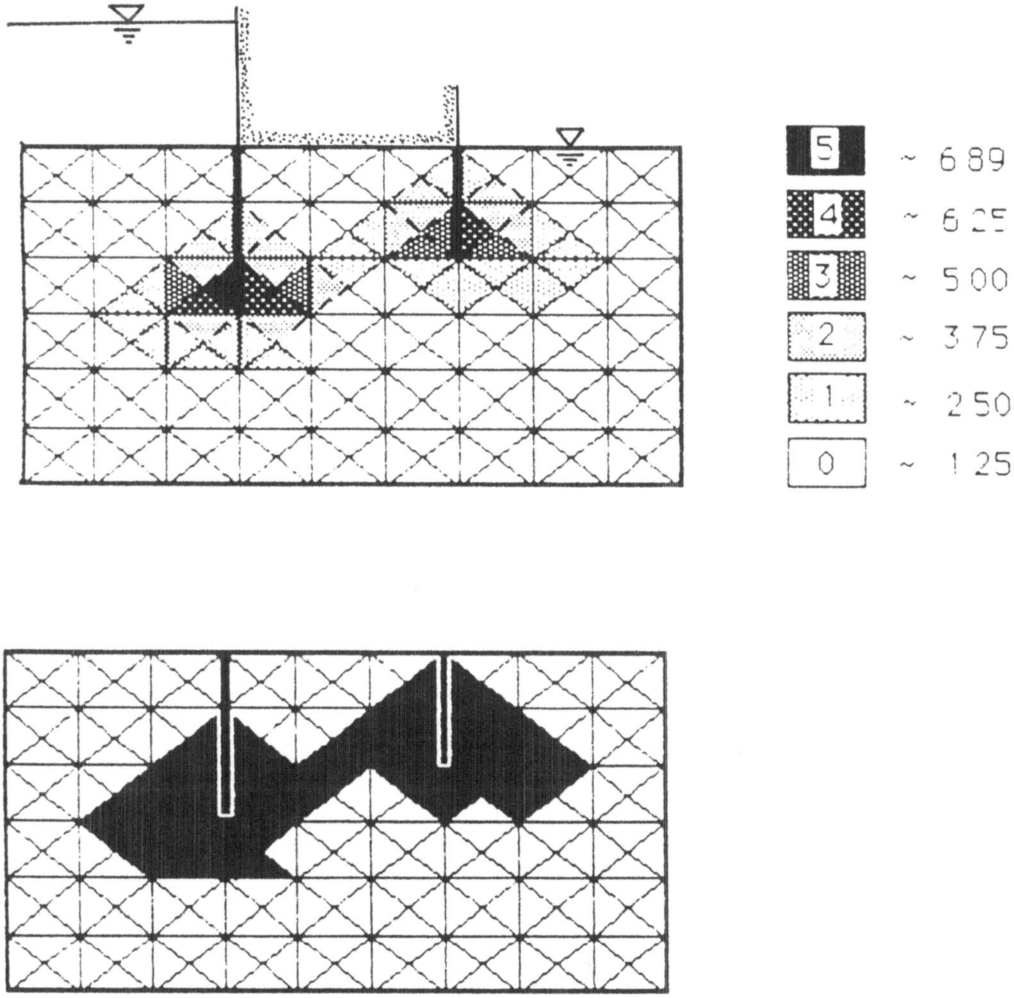

Figure 13. Application of the p-method of Flow through Porous Media (Flow Velocity, Distribution of Error Measure, p-adapted elements) (by Mr. T. Kishimoto)

In both the h- and p-methods, if quadrilateral elements

are used, rectangular elements are recommended in the initial grid in order to have high quality adaptation. If distorted finite elements are assigned near singular portions, refinement and sophistication possesses preferred directions. This, in general, destroys reproduction of the exact singular behavior of the solution, and they yields unnecessary local error, while the "energy" norm of the approximation error is slightly reduced by adaptation.

The above examples of adaptive methods are solved by Messrs. Toshikazu Torigaki, S. H. Lee, and Toshikazu Kishimoto using OPTIMESH which is a post processor of finite element analysis codes for adaptation developed in The University of Michigan. The author is grateful for their effort.

REFERENCES

1. Oliveira, E. R. de A., "Optimization of Finite Element Solution," Proc. 3rd Conf. Matrix Methods in Structural Mechanics, Wright-Patterson Air Force Base, Ohio, 1971.
2. Tang, J. W., and Turcke, D. J. "Characteristics of Optimal Grids," Comp. Meth Appl. Mech. Eng., 11, 1977, pp. 31-37.
3. Melosh, R. J., and Marcal, P. V., "An Energy Basis for Mesh Refinement of Structural Continua," Int. J. Num. Meth. Eng., 11, 1977, pp. 1083-1091.
4. Turcke, D. J., and McNeice, G. M., "Guidelines for Selecting Finite Element Grids Based on an Optimization Study," Computers & Structures, 4, 1974, pp. 499-519.

5. Shephard, M. S., Gallagher, R. H., and Abel, J. F., "The Synthesis of Near Optimum Finite Element Meshes with Interactive Computer Graphics," Int. J. Num Meth. Eng., 15, 1980, pp. 1021-1039.

6. Felippa, C. A., "Optimization of Finite Element Grids by Direct Energy Search," Appl. Math Modeling, 1, 1976, pp. 93-96.

7. Felippa, C. A., "Numerical Experiments in Finite Element Grid Optimization by Direct Energy Search," Appl. Math Modeling, 1, 1977, pp. 239-244.

8. Diaz, A. R., Optimization of finite element grids using interpolation error, Ph.D. Thesis in Aerospace Engineering, The University of Michigan, Ann Arbor, 1982.

9. Diaz, A. R., Kikuchi, N., Taylor, J. E., "A Method of Grid Optimization for Finite Element Methods," Comp. Method. Appl. Mech. Eng., 37, 1983, pp. 20-46.

10. Diaz, A. R., Kikuchi, N., and Taylor, J. E. "Optimal Design Formulations for Finite Element Grid Adaptation," in V. Komkov Ed. Sensitivity of Functionals with Applications to Engineering Science, Lecture Notes in Mathematics, #1086. Springer-Verlag, 1984, pp. 56-76.

11. Sewell, G., "An Adaptive Computer Program for the Solution of DIV(P(x,y)GRAD U) = F(x,y,U) on a Polygonal Region," in J. R. Whiteman ed., The Mathematics of Finite Elements and Applications, II, MAFELAP, 1975, Academic Press, New York.

12. Southwell, R. V., Relaxation Methods in Theoretical Physics, Vol. 1, Oxford University Press, 1943.

13. Southwell, R. V., Relaxation Methods in Theoretical Physics, Vol. 2, Oxford University Press, 1953.

14. Thompson, J. F. and Warsi, Z. U. A. Numerical Grid Generation Methods, North Holland, 1985, Ansterdam.

15. Ghia, K. N., and Ghia, U., Advances in Grid Generation, FED-Vol. 5, American Society of Mechanical Engineers, 1983, New York.

16. Rossow, M. P. and Katz, I. N., "Hierarchial Finite Elements and Precomputed Arrays," Int. J. Num. Meth. Eng., 12, 1978, pp. 977-999.

17. Peano, A., Pasini, A., Riccioni, R., and Sardella, L., "Adaptive Approximations in Finite Element Structural Analysis," Computer and Structures, 10, 1979, pp. 85-112.

18. Winslow, A., "Numerical Solution of the Quasilinear Poisson Equation," J. of Comp. Phy., 1, 1966, pp. 149-172.

19. Koh, B., and Kikuchi, N., Adaptive Methods for Directional Reduced Integration Scheme," in preparation to Comp. Meth. Appl. Mech. Eng., 1986.

20. Thacker, W. C., Gonzalez, A., and Putland, G. E., "A Method for Automating the Construction of Irregular Computational Grids for Storm Surge Forecast Models," J. Computational Physics, 37, 1980, pp. 371-387.

21. Joh , M., and Akimoto, Y., "Computer Program of Harmonized Mesh Generation for Finite Element Method," Pre-ession Mechanical Engineering, 46(3), 1980, pp. 8-13.

22. Shuhara, J. and Fukuda, J., "Automatic Mesg Generation for Finite Element Analysis," in Advances in Computational Methods in Structural Mechanics and Design, UAH Press, 1972, pp. 607-624.

Adaptive Grid Design for Finite Element Analysis in Optimization: Part 3, Shape Optimization*

Noboru Kikuchi
Department of Mechanical Engineering and Applied Mechanics
University of Michigan, Ann Arbor, MI 48109, U.S.

Abstract

Adaptive finite element methods based on the grid optimization are applied to solve optimal shape design problems of plane linearly elastic structures. Since most of shape optimization algorithms involve relocation of nodes on the design boundary, unnecessary element distortion is often introduced. For displacement finite element methods using lower order shape functions, accuracy of stresses is not great as for displacements. Especially, if finite elements are distorted, accuracy of stresses becomes very poor. We shall discuss details of element distortion effect using an explicit evaluation of the interpolation error for the four node isoparametric element. The r-and h-adaptive finite element methods are applied to reduce error due to element distortion during shape of optimization processes.

* The present work was supported by NASA Lewis Research Center under the grant NAG-3-388. The author expresses his appreciation to this support. This is a summary of the paper [18] by Kikuchi, Chung, Torigaki and Taylor.

1. Introduction

Optimal shape design of linearly elastic structures have been studied for this decade after extensive development of finite element and structural optimization methods. Especially, development of adaptive finite element and automatic grid generation methods enables us to solve some of the realistic shape design problems. In this paper the importance of adaptive finite element methods is discussed for optimal shape design problems to reduce finite element approximation error due to element distortion and stress "singularity".

Zienkiewicz and Cambell [1] applied finite element and penalty methods to solve shape optimization problems in 1973, and it was followed by Ramakrishnan and Francavilla [2], Tvergaard [3], Kristensen and Madsen [4], Quean and Trompette [5], Oda and Yamazaki [6], and others. Tvergaard applied finite difference methods in curvilinear coordinates for the shape optimization of a fillet to minimize the maximum elastic stress for a given load. Quean and Trompette used straight lines and circles to describe the design boundaries instead of applying (piecewise) polynomials to represent arbitrary shape of the boundaries.

Analysis and theory of shape optimization have been given by e.g. Banichuk [7], Dems [8], Dems and Mroz [9, 10, 11], Choi and Haug [12], Na et. al. [13], and others. Dems solved the problem of minimizing the cross-sectional area of torsion bar constraining the maximum torsional and bending rigidity. Dems and Mroz provided the first variations of an arbitrary stress/strain/ displacement functionals in terms of the stress and strain fields of primary and adjoint structures while the shape of the boundaries are varying. On the other hand, Choi and Haug introduced the idea of the material derivative in continuum mechanics to derive the design sensitivity. They express design change as design deformation velocity in the dynamic process of deformation of continuum. By employing the material derivative of variational equation and replacing the material derivative of the response by the virtual displacement they introduce an adjoint equation which leads to design sensitivity from the performance functional. Na et. al. applied the idea of structural remodeling introduced by Olhoff and Taylor [14] to solve shape optimization problems for a tension bar.

Necessity of application of adaptive finite element methods to shape optimization problems was recognized when the optimal shape of a fillet is obtained to minimize the volume of the thin elastic plate under the constraint that the maximum Mises stress on the boundary is less than or equal to a given upper bound. If the shape of the boundary of a fillet is defined by the location of nodes, and if they are searched using the optimality criteria method by specifying the moving direction of nodes on the design boundary, a physically unrealistic optimal shape is obtained as shown in Fig. 1. No matter how refined grids are applied, oscillation of the shape cannot be eliminated, and in fact, it tends to be larger as grid refinement is implemented. The reason for this could be explained very simply. Indeed, finite elements near to the left edge of the design boundary are severely distorted

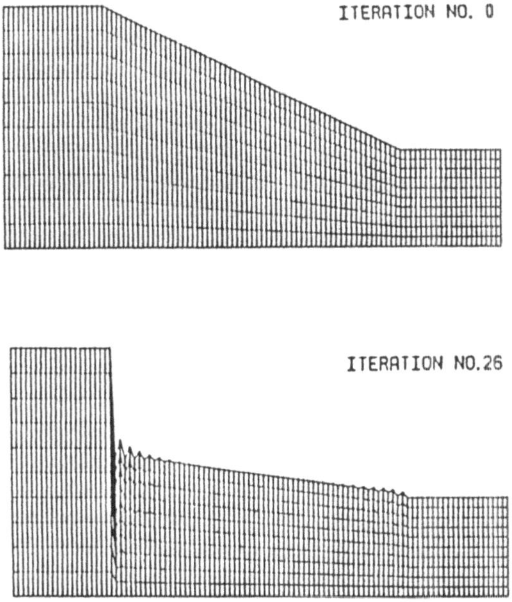

Figure 1. Shape Optimization using a Refined Finite Element Grid

no matter how refined grids are applied, since the optimal shape has the infinite slope at the left edge of the design boundary. With this severely distorted grid, stresses approximated by finite element methods must be poorly approximated. It is natural to consider that oscillation of the boundary shape would not be eliminated unless refined <u>regular shaped</u> finite elements are automatically assigned near to the portion where drastic change of the boundary shape is expected. How can we set up refined regular shaped finite elements in such a portion during an iteration process of optimization?

In this Part 3, we shall discuss difficulties in shape optimization and some of the remedies to them as well as a number of applications using the r-and h-adaptive finite element methods.

2. Shape Optimization Problems

Following Chung [19] let us consider a special case of shape optimization problems by minimizing the maximum value of a function of local measure. Examples of local measures are displaceements or von Mises equivalent stress, maximum shear stress etc.. In the formulation of the optimization problem, the design domain is, in general, restricted by an isoperimetric constraint of resource. A formulation of optimal design problem is given by

$$\underset{\Gamma_d}{\mathrm{Min}}(\mathrm{Area})$$
$$\text{subject to } |F(\underset{\sim}{u},\nabla \underset{\sim}{u})| - \beta \leq 0 \quad \text{on } \Gamma_d \tag{1}$$

or

$$\underset{\Gamma_d}{\mathrm{Min}}\{\underset{\underset{\sim}{x}\in\Gamma_d}{\mathrm{Max}}|F(\underset{\sim}{u},\nabla \underset{\sim}{u})|\}$$
$$\text{subject to } \int_\Omega d\Omega - A \leq 0 \tag{2}$$

The design variable is the shape of the boundary Γ_d. Here it is assumed that a boundary segment of Γ_d is a function either of the rectangular coordinate system (x,y) or of the polar coordinate system (r,Θ), that is, either y = f(x) or r = f(Θ). This choice already excludes design possibility shown in Fig.2. It may be appropriate

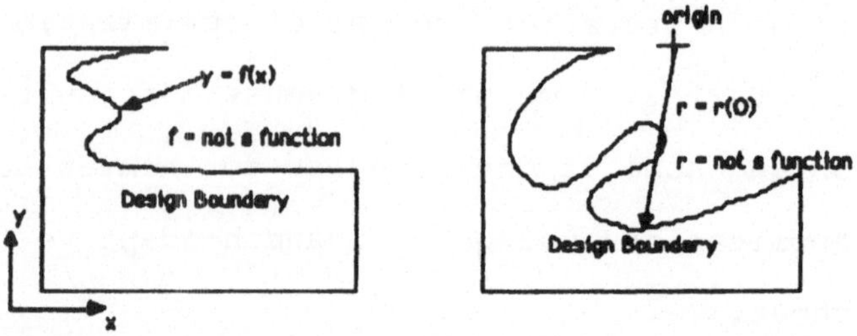

Figure 2. Excluding Shapes of the Design Boundary

to apply a parametric representation x = x(s) and y = y(s) using the coordinates along the boundary Γ_d in order to have wider possibility to represent a boundary segment. If the parametric representation is too flexible, low order spline approximation of a segment may be introduced to restrict possible shape design.

It must be specified whether the design boundary includes internal boundaries, more precisely, possibility to develop internal "holes" in a given structure. If so, we have to introduce some way to represent these internal boundaries. But, this task is extremely difficult.

A reason why the maximum is taken only on the design bounddary Γ_d is that if the maximum of the local criteria F is obtained at a point P inside of the domain Ω, it may not be possible to reduce the maximum value only by changing the shape of the boundary away from the point P. In other words, in despite of the optimization (1) and (2) may yield increase of the maximum value of F in Ω while it is reduced on the design boundary Γ_d. Thus (1) and (2) is not a perfect optimal design problem. Note that formulation (2) was used in the work by Francavilla et al. [2] as the approximate formulation of the minimization of stress concentration in domain Ω.

3. Computational Algorithm

In the present study we shall apply a finite element approximation based on the displacement method for plane problems using four-node quadrilateral isoparametric elements, in which each component of the displacement vector is approximated by a bilinear polynomial. In the displacement method, only the displaceement vector is the unknown to be solved, and then the stress tensor is computed a posteriori using a given constitutive equation.

Like may other free boundary problems, shape design problems of linearly elastic structures have been solved using geometric adaptive methods. The idea of geometric adaptive methods is stated in the form of a two step iteration algorithm for the purpose of satisfying the optimality conditions. The first step represents the calculation of some quantities under the assumption that the design boundaries are fixed, and the second step predicts the movement of nodes on the design boundary in the ratio of difference between the calculated quantity and the given or assumed constant at the nodes of a discrete model.

Optimality conditions of the design problem (1) yield the constant mutual energy and the saturation of the design functions F as $F - \beta = 0$ on the design boundary Γ_d. We shall develop a computational algorithm of the geometric adaptive method for shape optimization so that design boundaries are moved in the direction of satisfying these conditions (that is, constant mutual energy and the saturation of the design function F) until convergence to the final shape is reached within a certain tolerance.

The expression of the iterative scheme to find new coordinates of nodes on the design boundaries is

$$x_i^{k+1} = x_i^k + \Delta x_i^k \text{ , } i = 1,\ldots,N \text{ ,} \tag{3}$$

in k-th iteration, where N is the total number of nodes on the design boundary. Here, Δx_i^k is the movement of nodes in the specified direction and this value can be obtained from the movement in the normal direction:

$$\Delta x_i = \Delta x_{ni}/\cos\theta_i \tag{4}$$

in the k-th iteration. Θ_i is the angle between the unit normal vector n_i and unit direction vector α_i at the i-th node in the specified direction. For simplicity, superscript k is omitted in the following. The normal movement x_{mi} at the i-th node in the k-th iteration is obtained from

$$\Delta x_{ni} = \Delta A_i/S_i, \quad S_i = (L_{i-1}+L_i)/2 \tag{5}$$

where ΔA_i is the area allotted at the i-th node and L_i is the length of the i-th element on the boundary. The allotted area ΔA_i at the i-th node is obtained from the ratio between F (or mutual energy) calculated in the previous step of the k-th iteration and the prescribed constant β. Note that the value β in Min(Max F) can be assumed as the average of F value on the nodes along the design boundary.

$$\Delta A = \Delta Area*(F_i/\beta-1)S_i/\Sigma_{i=1,\ldots,N} |F_i/\beta-1|S_i \tag{6}$$

F_i is the value F at node i, which is extrapolated from the Gaussian points using least square methods, and ΔArea is the area between the design boundary Γ_d^{k+1} and Γ_d^k. The value of ΔArea must be large enough at the beginning and diminish as the iteration goes on in order to have convergence. For this purpose we define a percent deviation DTP from the optimum using L^2 norm. The value for ΔArea is given by

$$\Delta\text{Area} = (\text{Total Area of Domain}) \times C \times DTP. \quad (7)$$

From our experience, the value of C is between zero and 1.0, although the proper value must be adjusted according to the problem and the speed of convergence. That is, the bigger the C value is, the faster the converging speed is. However, faster convergence is sometimes accompanied with oscillations of the design boundaries, and possibly by oscillations of the percent deviation as well, as the iterations proceed. On the other hand, a small value of C may result in very slow convergence without oscillations.

The above algorithm possesses the property that if $F_i - \beta \leq 0$ is satisfied at a point $x \varepsilon \Gamma_d{}^k$, that is at node i, x moves in the the direction that the area A of the domain Ω is reduced. This may guarantee that although we have used only the necessary condition, the iteration algorithm yield the minimum automatically.

One disadvantage of this geometric adaptive methods is the possibility that the finite element grid set up can be distorted during adaptation process. Too much distortion of finite elements certainly yields unnecessary approximation error which may disturb results of the final shape of design boundaries.

More precisely, if four-node quadrilateral elements are applied for isotropic linearly elastic structures, the components of the strain tensor are approximated by

$$\varepsilon_\xi = A_1 + C_1\eta, \ \varepsilon_\eta = A_2 + B_1\xi, \ \gamma_{\xi\eta} = A_3 + B_2\xi + C_2\eta \quad (8)$$

in the normalized coordinate system (ξ,η). Approximation could be poor if element distribution is inconsistent to the equi-strain lines. To quantify error of finite element approximation, let us obtain the interpolation error of a function w approximated by four-node isoparametric elements. Suppose that the second derivatives of all the components of the true displacement w are constant in each finite element although their values need not be the same in different elements. Define

$$A = \partial^2 w/\partial x^2, \quad B = \partial^2 w/\partial x \partial y, \quad C = \partial^2 w/\partial y^2$$
$$\mathbf{x}^T = \{x_1, x_2, x_3, x_4\}, \quad \mathbf{y}^T = \{y_1, y_2, y_3, y_4\},$$
$$\mathbf{X}^T = \{X_1, X_2, X_3, X_4\}, \quad \mathbf{Y}^T = \{Y_1, Y_2, Y_3, Y_4\}$$
$$\mathbf{L_s}^T = (1/4)\{-1,1,1,-1\}, \quad \mathbf{L_t}^T = (1/4)\{-1,-1,1,1\}$$
$$\mathbf{h}^T = (1/4)\{1,-1,1,-1\}, \quad J_{11} = (\mathbf{L_s}\cdot\mathbf{x}) + (\mathbf{h}\cdot\mathbf{x})\eta,$$
$$J_{12} = (\mathbf{L_s}\cdot\mathbf{y}) + (\mathbf{h}\cdot\mathbf{y})\eta, \quad J_{21} = (\mathbf{L_t}\cdot\mathbf{x}) + (\mathbf{h}\cdot\mathbf{x})\xi,$$
$$J_{22} = (\mathbf{L_t}\cdot\mathbf{y}) + (\mathbf{h}\cdot\mathbf{y})\xi, \quad J = J_{11}J_{22} - J_{12}J_{21}, \tag{9}$$

where (x_α, y_α) are the nodal coordinates of the four corner nodes of an element in the physical coordinate system, and (X_α, Y_α) are four nodes inside of the element, say, the four nodes corresponding to the 2-2 Gaussian integration points. It is noted that if an element is a parallelogram the terms $\mathbf{h}\cdot\mathbf{x}$ and $\mathbf{h}\cdot\mathbf{y}$ become identically zero. Under the assumption stated in above, the first derivatives of the difference of the function w and its interpolation w_h can be written as

$$\begin{aligned}\partial(w_h - w)/\partial w = {} & (1/J)[-J_{22}(J_{11}{}^2A + 2J_{11}J_{12}B + J_{12}{}^2C)\xi + \\ & J_{12}(J_{21}{}^2A + 2J_{21}J_{22}B + J_{22}{}^2C)\eta + \{(1-\eta^2)\mathbf{L_t}\cdot \\ & \mathbf{y}(\mathbf{L_t} - \xi\mathbf{h}) + (1-\xi^2)\mathbf{L_s}\cdot\mathbf{y}(-\mathbf{L_s} + \eta\mathbf{h})\}\cdot\{\mathbf{h}\cdot\mathbf{x}(A\mathbf{X} + B\mathbf{Y}) + \\ & \mathbf{h}\cdot\mathbf{y}(B\mathbf{X} + C\mathbf{Y})\} + \{\xi(1-\eta^2)(\mathbf{L_t} - \xi\mathbf{h}) + \eta(1-\xi^2)(-\mathbf{L_s} + \\ & \eta\mathbf{h})\}\cdot \{\mathbf{h}\cdot\mathbf{x}\ \mathbf{h}\cdot\mathbf{y}(A\mathbf{X} + B\mathbf{Y}) + \mathbf{h}\cdot\mathbf{y}\ \mathbf{h}\cdot\mathbf{y}(B\mathbf{X} + C\mathbf{Y})\}]\end{aligned}$$

$$\partial(w_h-w)\partial y = (1/J)[-J_{21}(J_{11}{}^2A+2J_{11}J_{12}B+J_{12}{}^2C)\xi + J_{11}(J_{21}{}^2A+ 2J_{21}J_{22}B+J_{22}{}^2C)\eta + \{(1-\eta 2)L_t\cdot \mathbf{x}(-L_t-\xi \mathbf{h}) + (1-\xi^2)L_s\cdot \mathbf{x}(L_s+\eta \mathbf{h})\}\cdot\{\mathbf{h}\cdot \mathbf{x}(AX+BY) +\mathbf{h}\cdot\ y(BX+CY)\} + \{\xi(1-\eta^2)(-L_t-\xi \mathbf{h})+ \eta(1-\xi^2)(L_s+\eta \mathbf{h})\}\cdot\{\mathbf{h}\cdot \mathbf{x}\ \mathbf{h}\cdot \mathbf{x}(AX+BY)+\mathbf{h}\cdot \mathbf{x}\ \mathbf{h}\cdot \mathbf{y}(BX+CY)\}] \quad (10)$$

If an element is a parallelogram, then the interpolation error becomes very simple since $\mathbf{h}\cdot\mathbf{x} = 0$ and $\mathbf{h}\cdot\mathbf{y} = 0$, that is, the last half part are identically zero in each partial deriv-derivative. On the other hand, if an element is considerably distorted from a parallelogram, then the terms in the second and third lines in the interpolation error become large in the region where large strain is expected, since (AX+BY) and other similar terms are basically strain components in an element. This suggests that regular refined finite elements must be assigned in the neighborhood of singular points. Here regularity means that an element is close to a rectangle or parallelogram. Otherwise, error contribution would become large because of the terms in the second and third lines. This means grids generated by conformal mappings are appropriate. Similarly, grids by the elliptic differential equations method with the orthogonality condition and by the algebraic integer method, are suitable in the sense that contribution from the grid distortion and high strain (i.e., stress) can be restricted to be small.

4. Examples of Shape Optimization

The first example is a shape design problem of a highway

road pole that was solved by Oda and Yamazaki [16] where a hole was created at the place of minimum thickness to obtain a fully stressed shape although the fully stressed design has not been achieved in their results. Considering only a half portion of the highway road pole, let us introduce two design boundaries at the center symmetric line and the right hand side outside boundary. If the bottom line is allowed to move horizontally, the optimal shape is obtained as shown in Fig.3 starting from the initial grid specified in

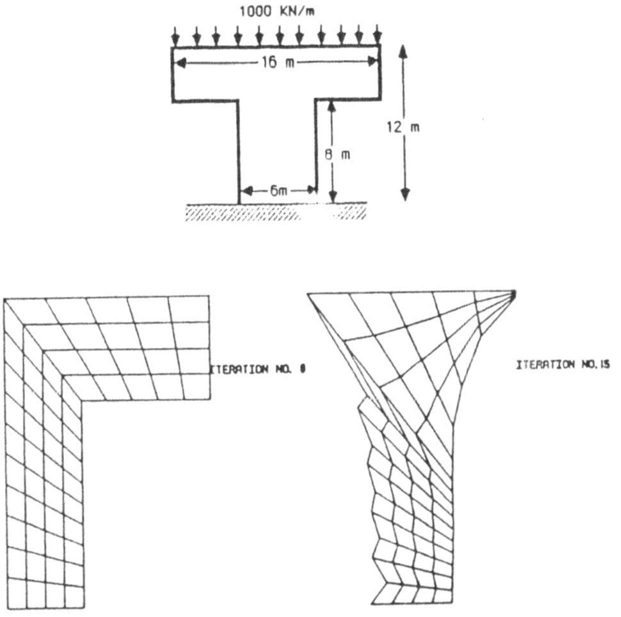

Figure 3. Road Pole Shape Optimization (Initial Grid, Optimal Shape) by Dr. K. Y. Chung

the same figure without applying a remeshing scheme during the geometric adaptive iteration for design change. Nodes on the design boundary are moved along the grid direction in the initial finite element grid. Thus, after certain number of iteration considerably large design change yields significant distortion of the grid. The final shape obtained is naturally unacceptable. Now let us continue the shape design process by applying the least square method to define a smooth design boundary in order to set up a finite element grid for restart of shape design. Further, suppose that we do not want to give up the grid in the final design in Fig.3. In other words, we do not give up the element connectivities defined at the initial grid. But the location of nodes will be modified by applying the r-adaptive method using the error measures computed at the final design stage. The grid shown in Fig.4 is

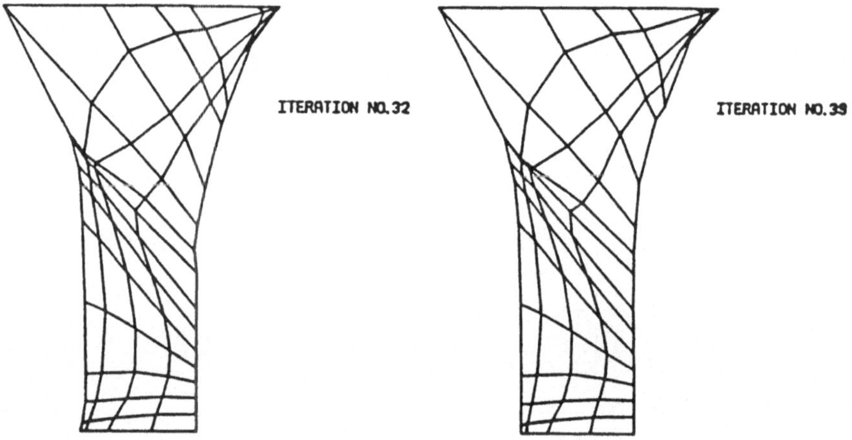

Figure 4. r-adapted Grid and the Optimal Shape

obtained that is assumed to be the second initial grid for shape optimization after "remeshing." If the geometric adaptive method is applied, the optimal shape is obtained as shown in the figure. If the bottom line is fixed, singular shape design is obtained at the end points of the bottom line. We cannot expect a hole inside of the road pole.

One of the most frequently used examples in shape optimization is the fillet problem because of its practical importance and its difficulty to be solved computationally. As shown in Fig.1, if remeshing is not performed during shape design, no matter how refined finite element grids are applied, the optimal shape computed is unacceptable. Thus, following the idea for the road pole problem, we shall remesh by applying the least square method to the design boundary to have a smooth boundary. For the initial grid in Fig.5, let us reconstruct finite element grids by applying the r-and h-adaptive methods as shown in Figs.6 and 7. If the geometric adaptive method is again applied to the second initial grids in Figs.6 and 7, the optimal shape of the fillet can be obtained without unreasonable physically nonsensical oscillation. In this case, the value of the maximum Mises stress in the whole domain is not the same to that on the design boundary, since the right side of the design boundary is also restricted. Thus, the maximum value of the Mises stress appears outside of the design boundary. If the right side of the design boundary is released from the design restriction, the maximum of the Mises stress is on the design boundary. In this case, the optimal shape is obtained as shown in Fig.8. Figure 9 is a comparison to the photoelastic result by Schnack [17]. It is clear that the optimal shape results in the same stress fringes as in photoelasticity case.

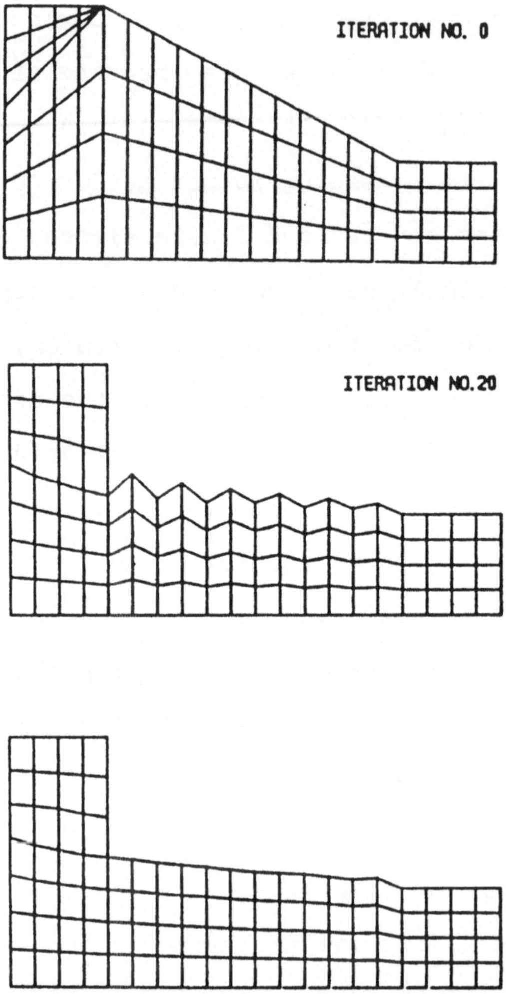

Figure 5. Shape Optimization of a Fillet (Initial Grid, Optimal Shape, Smoothed Shape) by Dr. K. Y. Chung

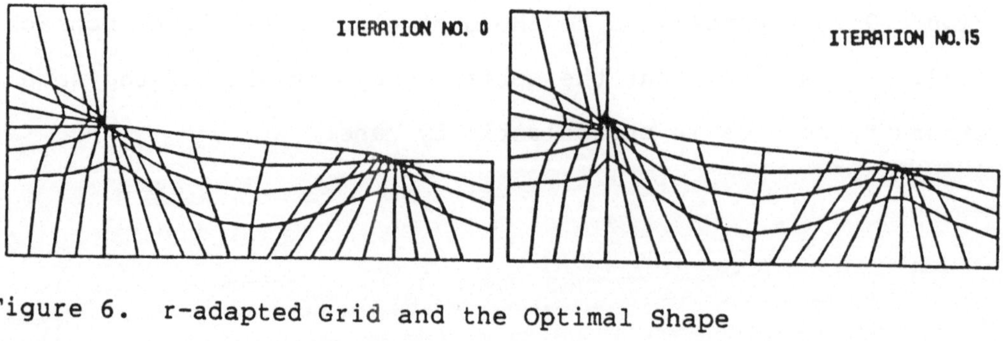

Figure 6. r-adapted Grid and the Optimal Shape

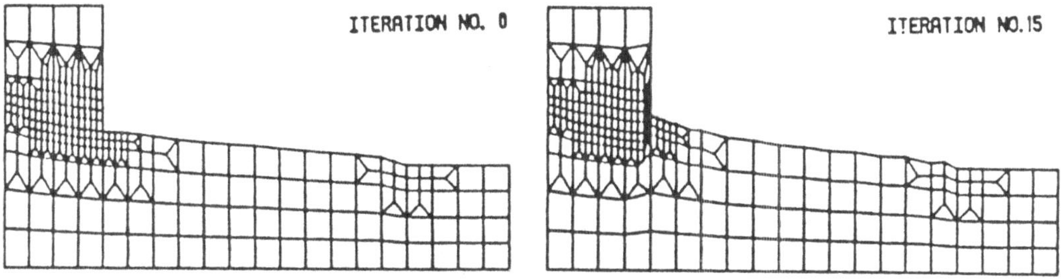

Figure 7. h-adapted Grid and the Optimal Shape

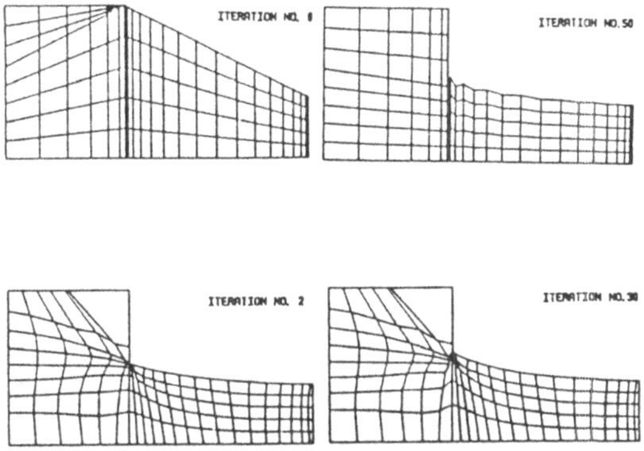

Figure 8. Optimal Design of a Fillet by Dr. K. Y. Chung

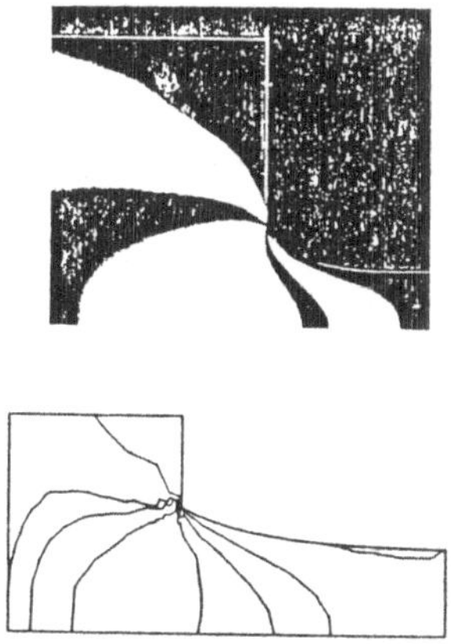

Figure 9. Comparison of the Finite Element Solution to Photoelasticity

References

1. Zienkiewicz, O. C. and Cambell, J. S., "Shape Optimization and Sequential Linear Programming," Chapter 7 in Optimal Structural Design, eds. Gallagher, R. H. and Zienkiewicz, O. C., Wiley, London.
2. Ramakrishnan, C. V. and Francavilla, A., "Structural Shape Optimization using Penalty Functions," J. Structural Mechanics, 3(4), pp. 403-422.
3. Tvergaard, V., "On the Optimal Shape of a Fillet in a Bar with Restrictions," Proc. IUTUM Symposium on Optimization in Structural Design, Sprig Verlag, 1973.
4. Kristensen, E. S. and Madsen, N. F., "On the Optimum Shape of Fillets in Plates Subjected to Multiple Inplane Loading Cases, " Int. J. Num. Meth. Engng., Vol.10, 1976, pp. 1006-1019.
5. Quean, J. P. and Trompette, P. H., "Two-Dimensional Shape Optimal Design by Finite Element Method," Int. J. Num. Meth. Engng., Vol. 15, 1980, pp. 1603-1612.
6. Oda, J. and Yamazaki, K., "A Procedure to Obtain a Fully Stressed Shape of Elastic Continuum," Int. J. Num. Meth. Engng., Vol.15, 1980, pp. 1095-1105.
7. Banichuk, N. V., Problems and Methods of Optimal Structural Design, Translated by Komkov, V. and Haug, E. J., Plenum Press, 1983, NY.
8. Dems, K., "Multiparameter Shape Optimization of Elastic Bars in Torsion," Int. J. Num. Meth. Engng., Vol.15, 1980, pp. 1517-1539.

9. Dems, K. and Mroz, Z., "Optimal Shape Design of Multi-composite Structures," J. Struct. Mech., Vol.8(3), 1980, pp. 309-329.

10. Dems, K. and Mroz, Z., "Variational Approach by Means of Adjoint Systems to Structural Optimization and Sensitivity Analysis-I," Int. J. Solids and Structures, Vol.19(8), 1983, pp. 677-692.

11. Dems, K., and Mroz, Z., "Variational Approach by Means of Adjoint Systems to Structural Optimization and Sensitivity Analysis-II, Int. J. Solids and Structures, Vol.20(6), 1984, pp. 527-552.

12. Choi, K. K. and Haug, E. J., "Shape Design Sensitivity Analysis of Elastic Structures," J. Struct. Mech., Vol. 11(2), 1983, pp. 231-269.

13. Na. M. S., Kikuchi, N. and Taylor, J. E., "Optimal Modification of Shape for Two Dimensional Elastic Bodies," J. Struct. Mech., Vol.11(1), 1983, pp. 111-135.

14. Olhoff, N. and Taylor, J. E., "Optimal Structural Remodeling," J. Opti. Theory and Applications, Vol.27(4), 1979, pp. 571-582.

15. Kikuchi, N., "Adaptive Grid Design Methods for Finite Element Methods," in Computer Methods in Applied Mechanics and Engineering, 55, 1986, pp. 129-59.

16. Oda, J. and Yamazaki, K., "On a technique to Obtain an Optimum Strength Shape of an Axisymmetric Body by the Finite Element Methods," Bulletin of the JSME, Vol. 20(150), pp. 1524-1532.

17. Schnack, E., "An Optimization Procedure for Stress Concentration by Finite Element Technique," Int. J. Num. Meth. Engng., Vol.14, 1979, pp. 115-124.

18. Kikuchi, N., Chung, K.Y., Torigaki, T., and Taylor, J.E., "Adaptive Finite Element Methods for Shape Optimization of Linearly Elastic Structures," Compt. Meth. Applied. Mech. Engin., 57, 1986, pp. 67-89.

19. Chung, K.Y., Shape Optimization and Free Boundary Problems with Grid Adaptation, Dissertation at The University of Michigan, Ann Arbor, MI, 1985.

SENSITIVITY ANALYSIS IN THERMOELASTICITY PROBLEMS

K. Dems
Łódź Technical University, 90924 Łódź, Poland.

1. Introduction

In many problems of structural mechanics there is a need to assess the effect of variation of one or several design functions or parameters /such as material properties, cross-sectional dimensions, shape or support conditions/ on the thermal and mechanical state fields within a structure. The methods of such sensitivity analysis have been explored in various fields of science and engineering. Numerous publications reveal the growing interest in the optimum design of structures subject to both mechanical and temperature constraints. When mathematical optimization techniques are used to design structures subjected to temperature, stress and/or displacement constraints, derivatives of these constraints with respect to design variables are usually required. The present paper extends the previous works [1 - 2] and is concerned with such class of problems for which the first and second variations of any functional that depends on displacements, strains, stresses and temperature can be explicitly expressed in terms of variations of design variables. We assume for our subsequent analysis that either the mechanical and thermal properties of material depend on a set of design parameters φ_k within fixed domain of a body, or the shape of a body is allowed to vary and it is described by a set of design variables φ_k, whereas the mechanical and thermal properties of a material are constant. Thus, we shall consider the problem of evaluating the sensitivities of an arbitrary functional G of the form:

$$G = \int_{V(\varphi_k)} \psi(\underline{\sigma}, \underline{\varepsilon}, \underline{u}, \theta, \varphi_k)\, dV + \int_{S(\varphi_k)} h(\underline{T}, \underline{u}, \theta, q)\, dS \qquad (1)$$

NATO ASI Series, Vol. F27
Computer Aided Optimal Design: Structural and Mechanical Systems. Edited by C. A. Mota Soares

depending on stress $\underline{\sigma}$, strain $\underline{\varepsilon}$, displacement $\underline{u}$ as well as temperature θ within structure domain V, and surface traction $\underline{T}$, displacement and temperature and heat flux q on its surface S, and on a set of design parameters φ_k describing the material properties or the shape of structure. The sensitivities of functional (1) can be obtained by using the adjoint or direct approaches, which can be equally applicable to analytical or numerical solution techniques applied to the problem considered.
In the foregoing Sections we shall briefly discuss the application of these both approaches to sensitivity analysis for any functional expressed in the general form (1).

2. Sensitivity analysis with respect to material parameters of structure

Let us consider first the case of a fixed domain of structure for which the stress and displacement conditions are specified on the portions S_T and S_u of its boundary, and the temperature and heat flux conditions are prescribed on the portions S_θ and S_q, so that $S=S_T \cup S_u = S_\theta \cup S_q$. Moreover, the structure is subjected to the imposed fields of body force $\underline{f}$ and heat source Q within its domain. We confine our analysis to small displacement and strain theory, and the non-linear stress-strain relation is assumed in a very general form:

$$\underline{\sigma} = \underline{S}(\underline{\varepsilon}, \theta, \varphi_k) \tag{2}$$

where φ_k is a set of design parameters. Furthermore, the thermal conductivity $\lambda = \lambda(\varphi_k)$ can also depend on design parameters φ_k. The first variation of functional (1) can be expressed as follows:

$$\delta G = \int (\psi,_{\underline{\sigma}} \cdot \delta \underline{\sigma} + \psi,_{\underline{\varepsilon}} \cdot \delta \underline{\varepsilon} + \psi,_{\underline{u}} \cdot \delta \underline{u} + \psi,_\theta \delta\theta + \psi,_{\varphi_k} \delta\varphi_k)\, dV + \\ + \int (h,_{\underline{T}} \cdot \delta \underline{T} + h,_{\underline{u}} \cdot \delta \underline{u} + h,_\theta \delta\theta + h,_q \delta q)\, dS \tag{3}$$

To derive the sensitivity of G with respect to design parameters φ_k, we have now to replace the terms involving variations of mechanical and thermal state fields with the terms depending on variations of design parameters. To do this we can apply the adjoint or direct approach.

2.1 Adjoint approach for sensitivity analysis

To eliminate in Eq.(3) the terms involving the variations of state fields, let us introduce two adjoint structures of the same shape as primary one, namely the mechanical adjoint structure and thermal adjoint structure. The mechanical adjoint structure is subjected to imposed fields of initial strain, stress and body force defined by:

$$\varepsilon^{ai} = \psi,_{\underline{\sigma}} \quad , \quad \underline{\sigma}^{ai} = \psi,_{\underline{\varepsilon}} \quad , \quad \underline{f}^{a} = \psi,_{\underline{u}} \quad \text{within V} \tag{4}$$

and satisfies the following set of boundary conditions:

$$\underline{T}^{ao} = h,_{\underline{u}} \quad \text{on } S_T \quad , \quad \underline{u}^{ao} = -h,_{\underline{T}} \quad \text{on } S_u \tag{5}$$

The linear stress-strain relation of this adjoint structure is assumed in the form:

$$\underline{\sigma}^{a} = \underline{S},^{T}_{\underline{\varepsilon}} \cdot (\underline{\varepsilon}^{a} - \underline{\varepsilon}^{ai}) - \underline{\sigma}^{ai} \tag{6}$$

For the purposes of our subsequent analysis, let us denote the stress, strain and displacement fields within this structure by $\underline{\sigma}^a$, $\underline{\varepsilon}^a$ and $\underline{u}^a$, respectively.
The thermal adjoint structure is subject to the following set of boundary conditions:

$$\theta^{ao} = h,_q \quad \text{on S} \quad , \quad q^{ao} = -h,_{\theta} \quad \text{on } S_q \tag{7}$$

with prescribed rate of heat generation within its domain, defined by:

$$Q^a = \psi,_\theta - (\underline{\varepsilon}^a - \psi,_{\underline{\sigma}})\cdot \underline{S},_\theta \qquad \text{within V} \tag{8}$$

The temperature field resulting from boundary conditions (7) and satisfying the conduction equation is denoted by θ^a.
In view of (4-5) and (7-8), Eq.(3) can be rewritten in the form:

$$\delta G = \int [\underline{\varepsilon}^{ai}\cdot\delta\underline{\sigma} + \underline{\sigma}^{ai}\cdot\delta\underline{\varepsilon} + \underline{f}^a\cdot\delta\underline{u} + Q^a\delta\theta + (\underline{\varepsilon}^a - \underline{\varepsilon}^{ai})\cdot\underline{S},_\theta + \\ + \psi,_{\varphi_k}\delta\varphi_k]\, dV - \int \underline{u}^{ao}\cdot\delta\underline{T}\, dS_u + \int \underline{T}^{ao}\cdot\delta\underline{u}\, dS_T + \int \theta^{ao}\delta q\, dS - \int q^{ao}\delta\theta\, dS_q \tag{9}$$

Using now Eq.(6) and conduction equation and noting that in view of (2) we have:

$$\delta\underline{\sigma} = \underline{S},_{\underline{\varepsilon}}\cdot\delta\underline{\varepsilon} + \underline{S},_\theta\delta\theta + \underline{S},_{\varphi_k}\delta\varphi_k \tag{10}$$

equation (9), after some transformations, yields the following expression for the first variation of G:

$$\delta G = \int [\underline{u}^a\cdot\delta\underline{f} - (\underline{\varepsilon}^a - \underline{\varepsilon}^{ai})\cdot\underline{S},_{\varphi_k}\delta\varphi_k - \theta^a,_i\lambda,_{\varphi_k}\theta,_i\delta\varphi_k + \theta^a\delta Q + \\ + \psi,_{\varphi_k}\delta\varphi_k]\, dV \tag{11}$$

Assuming now that the body forces and heat source within primary structure depend on design parameters, that is:

$$\delta\underline{f} = \underline{f},_{\varphi_k}\delta\varphi_k \quad , \quad \delta Q = Q,_{\varphi_k}\delta\varphi_k \tag{12}$$

we finally obtain the first variation of G in the form:

$$\delta G = \frac{dG}{d\varphi_k}\delta\varphi_k = \Big\{\int [\underline{u}^a\cdot\underline{f},_{\varphi_k} - (\underline{\varepsilon}^a - \psi,_{\underline{\sigma}})\cdot\underline{S},_{\varphi_k} - \theta^a,_i\lambda,_{\varphi_k}\theta,_i + \\ + \theta^a Q,_{\varphi_k} + \psi,_{\varphi_k}]\, dV\Big\}\delta\varphi_k \tag{13}$$

Thus, δG is explicitly expressed in terms of derivatives of the integrand of (1) and the strain and temperature fields within primary and both adjoint structures. It is then seen, that first variation of G and all its derivatives with respect to design parameters may be evaluated by using only two solutions of elasticity problems and two solutions of heat transfer problems, independently of the number of design parameters.

2.2 Direct approach for sensitivity analysis

When the direct approach is used for calculating the first variation of G, we assume that the variations of mechanical and thermal state fields occurying in Eq.(3) can be expressed in the form:

$$\delta \underline{\sigma} = \underline{\sigma},_{\varphi_k} \delta\varphi_k = \underline{\sigma}^k \delta\varphi_k \quad , \quad \delta\underline{\varepsilon} = \underline{\varepsilon},_{\varphi_k}\delta\varphi_k = \underline{\varepsilon}^k\delta\varphi_k \quad ,$$

$$\delta \underline{u} = \underline{u},_{\varphi_k}\delta\varphi_k = \underline{u}^k\delta\varphi_k \quad , \quad \delta\underline{T} = \underline{\sigma},_{\varphi_k}\cdot\underline{n}\,\delta\varphi_k = \underline{T}^k\delta\varphi_k \quad , \qquad (14)$$

$$\delta\theta = \theta,_{\varphi_k}\delta\varphi_k = \theta^k\delta\varphi_k \quad , \quad \delta q = q,_{\varphi_k}\delta\varphi_k = -(\lambda\theta,_i),_{\varphi_k} n_i \delta\varphi_k =$$

$$= (\tfrac{1}{\lambda}\lambda,_{\varphi_k} q + q^k)\delta\varphi_k$$

Using (14) in (3), the first variation of G takes the form:

$$\delta G = \frac{dG}{d\varphi_k}\delta\varphi_k = \Big\{ \int \big[\psi,_{\underline{\sigma}}\cdot\underline{\sigma}^k + \psi,_{\underline{\varepsilon}}\cdot\underline{\varepsilon}^k + \psi,_{\underline{u}}\cdot\underline{u}^k + \psi,_\theta\,\theta^k + \psi,_{\varphi_k}\big]\, dV +$$

$$+ \int h,_{\underline{T}}\cdot\underline{T}^k dS_u + \int h,_{\underline{u}}\cdot\underline{u}^k dS_T + \int h,_\theta\,\theta^k dS_q + \int h,_q(\tfrac{1}{\lambda}\lambda,_{\varphi_k} q + \qquad (15)$$

$$+ q^k) dS_\theta \Big\}\delta\varphi_k$$

The sensitivities $\underline{\sigma}^k$, $\underline{\varepsilon}^k$, $\underline{u}^k$, $\underline{T}^k$, θ^k and q^k of state fields can be obtained as the solutions of some auxiliary mechanical and thermal problems. Equations describing these problems are obtained by differentation, with respect to each design parameter φ_k, the governing sets of mechanical and thermal state equations together with proper sets of boundary conditions. The k-th thermal auxiliary problem is then described by the following set of equations:

$$\lambda\theta^k,_{ii} + Q^k = 0 \quad , \quad Q^k = Q,_{\varphi_k} - \tfrac{1}{\lambda}\lambda,_{\varphi_k} Q \ \text{ within } V$$

$$\theta^{ko} = 0 \quad \text{on } S \quad , \quad q^{ko} = -\frac{1}{\chi}\lambda,_{\varphi_k} q^{o} \quad \text{on } S_q \tag{16}$$

whereas the k-th mechanical auxiliary problem is described by the equations:

$$\begin{aligned} &\operatorname{div} \underline{\sigma}^k + \underline{f}^k = 0 \quad , \quad \underline{f}^k = \underline{f},_{\varphi_k} \quad \text{within } V \\ &\underline{T}^{ko} = 0 \quad \text{on } S_T \quad , \quad \underline{u}^{ko} = 0 \quad \text{on } S_u \end{aligned} \tag{17}$$

The stress-strain relation for the k-th mechanical auxiliary problem is obtained by differentiating Eq.(2) with respect to φ_k, and then has the form:

$$\underline{\sigma}^k = \underline{S},_{\underline{\varepsilon}} \cdot \underline{\varepsilon}^k - \underline{\sigma}^{ki} \quad , \quad \underline{\sigma}^{ki} = -(\underline{S},_{\theta} \theta^k + \underline{S},_{\varphi_k}) \tag{18}$$

The solutions of equations (16 - 18) have to be repeated for evaluating the sensitivity derivatives of state fields with respect to all φ_k. It is thus seen, that when there are n design parameters, the direct approach requires n+1 solutions of heat transfer problems and n+1 solutions of elasticity problems.
Comparing the adjoint and direct approaches, it is seen that the adjoint method, requiring only four solutions, is more economical than the direct method requiring 2 n+1 solutions. However, when there are m functionals and n design parameters, then the direct method would still require 2 n+1 solutions in order to generate the first variations of m functionals G, whereas the adjoint approach would need 2(m+1) solutions. Thus, the choice between the two methods can depend on ratio m to n as well as the relative difficulty of obtaining adjoint solutions versus sensitivity solutions.
It should be added that using simultaneously the direct and adjoint solutions we can easily calculate the second variation of any functional G. Such approach requires for m functionals and n design parameters only 2(n+m+1) solutions and is always more efficient than the adjoint approach that would need 2(n+1)(m+1) solutions, and than the direct approach requiring (n+1)(n+2) so-

lutions, unless $n \ll m$.

3. Sensitivity analysis with respect to shape parameters

Assume now that the shape of the body discussed in Section 2 is allowed to vary, whereas its material properties are specified in advance. Consider then the transformation process of body domain with the imposed transformation field $\underline{\varphi}(\underline{x})$, such that any point P initially placed at $\underline{x}$ is transformed to the position $\underline{x}^*$, according to the rule:

$$P \rightarrow P^*: \quad \underline{x}^* = \underline{x} + \underline{\varphi}(\underline{x}) \tag{19}$$

where the components φ_k /k=1,2,3/ of vector $\underline{\varphi}(\underline{x})$ can be regarded as the design variables describing the shape of the body. Similarly as previously, we shall derive the first variation of any functional G of the form (1) associated with the variation of structural shape. This variation can be obtained by using the adjoint or direct approaches.

3.1 Adjoint approach for shape sensitivity analysis

Since the variations of any state field can be now expressed as the sum of variation within unperturbated domain and the terms associated with the domain transformation, the first variation of any functional G can be expressed as follows:

$$\begin{aligned} \delta G = & \int (\psi,_{\underline{\sigma}} \cdot \delta\bar{\underline{\sigma}} + \psi,_{\underline{\varepsilon}} \cdot \delta\bar{\underline{\varepsilon}} + \psi,_{\underline{u}} \cdot \delta\bar{\underline{u}} + \psi,_{\theta} \delta\bar{\theta}) dV + \int (h,_{\underline{T}} \cdot \delta\bar{\underline{T}} + h,_{\underline{u}} \cdot \delta\bar{\underline{u}} + \\ & + h,_{\theta} \delta\bar{\theta} + h,_{q} \delta\bar{q}) dS + \int \Big\{ \psi n_k \delta\varphi_k + h(\delta_{kl} - n_k n_l) \delta\varphi_{k,l} + \\ & + h,_{T_i} [\sigma_{ij}(n_l n_j - \delta_{jl}) n_k \delta\varphi_{k,l} + \sigma_{ij,k} n_j \delta\varphi_k] + h,_{\underline{u}} \cdot \underline{u},_k \delta\varphi_k + \\ & + h,_{\theta} \theta,_k \delta\varphi_k + h,_q [\lambda\theta,_{ik} n_i \delta\varphi_k + \lambda\theta,_i (n_i n_l - \delta_{il}) n_k \delta\varphi_{k,l}] \Big\} dS \end{aligned} \tag{20}$$

where $\delta(\bar{\cdot})$ denotes the variation of proper state field within unperturbated domain of structure.

To eliminate in Eq.(20) the terms involving the variation of state fields, we shall use the solutions of the adjoint mechanical and thermal problems defined by Eqs.(4 - 8). Thus, after some transformations, the first variation of G with respect to variation of shape of structure takes the form:

$$
\begin{aligned}
\delta G = & \int \Big[\big(\psi - \underline{\underline{\sigma}} \cdot \underline{\underline{\varepsilon}}^a + \underline{f} \cdot \underline{u}^a - \lambda \theta_{,i} \theta^a_{,i} + Q\theta^a + (h + \underline{T} \cdot \underline{u}^a - q\theta^a)_{,n} + \\
& - 2(h + \underline{T} \cdot \underline{u}^a - q\theta^a) H \Big] \delta\varphi_n \, dS + \int \Big[(h^- + \underline{T}^- \cdot \underline{u}^a - q^- \theta^a) \delta\varphi_{\nu^-} + \\
& - (h^+ + \underline{T}^+ \cdot \underline{u}^a - q^+ \theta^a) \delta\varphi_{\nu^+} \Big] d\Gamma + \int (h_{,\underline{T}} + \underline{u}^a)(\delta \underline{T}^o - \underline{T}^o_{,k} \delta\varphi_k) \, dS_T + \\
& + \int (h_{,\underline{u}} - \underline{T}^a) \cdot (\delta \underline{u}^o - \underline{u}^o_{,k} \delta\varphi_k) \, dS_u + \int (h_{,q} - \theta^a)(\delta q^o - q^o_{,k} \delta\varphi_k) \, dS_q + \\
& + \int (h_{,\theta} + q^a)(\delta\theta^o - \theta^o_{,k} \delta\varphi_k) \, dS_\theta
\end{aligned}
\tag{21}
$$

where H denotes the mean curvature of external surface S, and $\delta\varphi_n$ is the normal component of boundary variation on S. Assuming that S is a piecewise regular surface and Γ denotes the intersection curve between two adjacent parts of S, $\delta\varphi_\nu$ is the component of boundary variation laying in a plane tangential to S and normal to intersection curve Γ. Note furthermore, that the variations $\delta\underline{T}^o$, $\delta\underline{u}^o$, δq^o and $\delta\theta^o$ on S_T, S_u, S_q and S_θ are known due to assumed form of boundary conditions on these boundary portions.

It follows from (21) that the first variation of G can be obtained as the result of solutions of heat transfer and elasticity problems for primary and adjoint structures, and then may be evaluated by using four solutions, independently on the number of shape parameters.

When the shape of structure depends on a set of shape parameters b_p, then the transformation vector field $\underline{\varphi}$ is a given function of space and is expressed by :

$$
\underline{\varphi} = \underline{\varphi}(b_p, \underline{x}) \tag{22}
$$

and the variation of this field appearing in Eq.(21) should be replaced by:

$$\delta\varphi_k = \frac{\partial\varphi_k}{\partial b_p}\delta b_p = v_k^p \delta b_p \tag{23}$$

where v_k^p denotes the k-th component of transformation velocity field associated with shape parameter b_p, treated as time-like parameter.

3.2 Direct approach for shape sensitivity analysis

Using the direct method to calculate the variation of G, we express the variation of state fields within unperturbated domain of structure in the form:

$$\delta(\bar{\cdot}) = \frac{\partial(\cdot)}{\partial b_p}\delta b_p = (\cdot)^p \delta b_p \tag{24}$$

where $(\cdot)$ denotes the stress, strain, displacement or temperature field, respectively. The sensitivities $(\cdot)^p$ of state fields can be obtained, similarly as in Section 2.2, as the solutions of some auxiliary thermal and mechanical problems. The governing equations of these problems are obtained by differentiation, with respect to design parameter b_p, the state equations of primary problem. Thus, the variation of G can be calculated directly from Eq.(20), in which variations of state fields are replaced with (24) and variation of transformation field $\delta\varphi_k$ is replaced with (23). It is then seen, that δG is obtained as result of n+1 solutions of heat transfer problems and n+1 solutions of elasticity problems.

Using simultaneously the direct and adjoint solutions it is also possible to calculate the second variation of any functional with respect to shape variation of structure.

4. Sensitivity of local quantities

The presented analysis was applicable for any functional of integral form. But it can be also extended to any local constraint of the form:

$$g(\underline{x}_o) = g[\underline{\sigma}(\underline{x}_o), \underline{\varepsilon}(\underline{x}_o), \underline{u}(\underline{x}_o), \theta(\underline{x}_o)] \tag{25}$$

Introducing the Dirac delta function $\delta(\underline{x} - \underline{x}_o)$, the local constraint (25) can be replaced by its integral form:

$$G = \int g(\underline{\sigma}, \underline{\varepsilon}, \underline{u}, \theta)\delta(\underline{x} - \underline{x}_o)\, dV \tag{26}$$

for which the analysis presented in Section 2 and 3 can be easily applied.

5. Concluding remarks

The present paper provides systematic adjoint and direct approaches to sensitivity analysis for thermoelasticity problems when the material properties or shape of structure is allowed to vary. We discussed here the steady-state case only, but the extension of the presented analysis to the time dependent problems is very simple and follows the similar steps.

Acknowledgement

This reseatch work was carryed out within Polish Academy of Science Grant No. CPBP 02.01.

References

1. K. Dems, Z. Mróz, Variational approach by means of adjoint systems to structural optimization and sensitivity analysis, Int. J. Solids Struct., 19, 677-692, 1983 and 20, 527-552, 1984
2. K. Dems, Sensitivity analysis in thermal problems, J. Therm. Stresses, 9, 301-322, 1986 and 9, 1986 /in press/

A NEW VARIATIONAL APPROACH TO STRUCTURAL SHAPE DESIGN SENSITIVITY ANALYSIS

Robert B. Haber
Associate Professor of Civil Engineering and
Theoretical and Applied Mechanics
University of Illinois at Urbana-Champaign
Urbana, IL 61801, USA

1. Introduction

This paper is concerned with a new variational approach to structural design sensitivity analysis, including the consideration of shape variation. Specifically, the mutual Hu-Washizu functional is introduced for obtaining explicit sensitivity expressions for response functionals; and the Eulerian-Lagrangian kinematic description is applied to the description of design geometry variations.

Sensitivity analysis is a key step in the process of structural optimization, in reliability analysis and in identification problems. Here we take sensitivity analysis to mean the determination of the rate of change with respect to variations of the design parameters of a differentiable, scalar functional involving general nonlinear functions of design and response field variables. Whether the functional represents cost or the value of a constraint, the problem of sensitivity analysis is the same; and the techniques presented below are intended to address all such cases. This development is restricted to linear elasticity, but extensions to large-deformation nonlinear problems are possible.

Two function spaces must be considered in design sensitivity analysis. The design function space includes field variables such as geometric coordinates $\mathbf{X}$, material coefficients, section properties, and designer-specified loads such as body force $\mathbf{b}$, surface tractions $\mathbf{t}$, initial stress $\boldsymbol{\tau}^{\mathbf{I}}$ or strain $\mathbf{e}^{\mathbf{I}}$ and prescribed boundary displacements. The design function space can be fully described by a set of M scalar functions $\boldsymbol{\theta}$: $\theta_\alpha(X)$: $\alpha = 1,M$. Each design function is either prescribed by the design problem statement or can be independently specified by the designer. The response function space includes the elasticity solution field variables: displacement $\mathbf{u}$, strain $\mathbf{e}$, elastic stress $\boldsymbol{\tau}^{\mathbf{E}}$ and reaction surface tractions. The components of these tensor quantities can be represented by a set of N scalar response functions $\mathbf{U}$: $U_\alpha(\mathbf{X})$; $\alpha = 1,N$.

NATO ASI Series, Vol. F27
Computer Aided Optimal Design: Structural and Mechanical Systems. Edited by C. A. Mota Soares

It is important to note that the function space **U** is determined implicitly by the function space **θ** and the governing equations and boundary conditions of linear elasticity. The response functions must satisfy the governing equations of strain-displacement compatibility, the material stress-strain relation and equilibrium over the volume V of the structure.

$$e_{ij} = \frac{1}{2}(u_{i,j} + u_{j,i}) \tag{1}$$

$$\tau_{ij} = C_{ijk\ell}\,(e_{k\ell} - e^{I}_{k\ell}) + \tau^{I}_{ij} \tag{2}$$

$$\tau_{ij,j} + b_i = 0 \tag{3}$$

C is the fourth-order elasticity tensor. The stresses and surface tractions must satisfy Cauchy's relation at the boundary A of the structure;

$$\tau_{ij}\, n_j - t_i = 0 \tag{4}$$

where **n** is the unit normal vector to the surface A. Prescribed displacement and surface traction boundary conditions must be satisfied on the surface regions A_u and A_t respectively.

$$t_i - t^{P}_i = 0 \qquad \text{on } A_t \tag{5}$$

$$u_i - u^{P}_i = 0 \qquad \text{on } A_u \tag{6}$$

Once the design function space **θ** is specified, equations (1)-(6) uniquely determine the solution in the response function space **U**.

Suppose we are seeking the sensitivity of a scalar functional Γ to design variations $\delta\boldsymbol{\theta}$, where Γ is written as

$$\Gamma = \Gamma\,(\boldsymbol{\theta},\mathbf{U})$$

$$= \int_V f\,(\mathbf{e}, \boldsymbol{\tau}^{E}, \mathbf{u}, \boldsymbol{\theta})\,dV + \int_{A_t} g(\mathbf{u}, \boldsymbol{\theta})\,dA + \int_{A_u} h(\mathbf{t}, \boldsymbol{\theta})\,dA \tag{7}$$

in which f, g and h are scalar, nonlinear functions and $\boldsymbol{\tau}^{E}$ is the elastic part of the stress tensor with components $\tau^{E}_{ij} = C_{ijk\ell}e_{k\ell}$. Direct application of the calculus of variations leads to

$$\delta\Gamma = \delta\Gamma\,(\mathbf{U},\,\delta\mathbf{U},\,\boldsymbol{\theta},\,\delta\boldsymbol{\theta})$$

$$= \int_V \frac{\partial f}{\partial \theta_\alpha}\,\delta\theta_\alpha\, dV + \int_{A_t} \frac{\partial g}{\partial \theta_\alpha}\,\delta\theta_\alpha\, dA + \int_{A_u} \frac{\partial h}{\partial \theta_\alpha}\,\delta\theta_\alpha\, dA + \delta R; \quad \alpha = 1,M \tag{8}$$

where

$$\delta R = \int_V \left[\frac{\partial f}{\partial e_{ij}}\,\delta e_{ij} + \frac{\partial f}{\partial \tau_{ij}^E}\,\delta\tau_{ij}^E + \frac{\partial f}{\partial u_i}\,\delta u_i\right] dV$$

$$+ \int_{A_t} \frac{\partial g}{\partial u_i}\,\delta u_i\, dA + \int_{A_u} \frac{\partial h}{\partial t_i}\,\delta t_i\, dA \tag{9}$$

In practice, the most difficult and costly step in the evaluation of $\delta\Gamma$ is determining the response function variations appearing in δR (δe_{ij}, $\delta\tau_{ij}^E$, δu_i and δt_i) as implicitly defined by the design variations $\delta\boldsymbol{\theta}$ and equations (1)-(6).

A more attractive approach is to seek an explicit functional variation $\delta\Gamma^*$ such that

$$\delta\Gamma = \delta\Gamma^*(\mathbf{U},\boldsymbol{\theta},\delta\boldsymbol{\theta}) \tag{10}$$

when equations (1)-(6) are implicitly satisfied. If the explicit form $\delta\Gamma^*$ is available; then the design sensitivities can be evaluated directly, without the need to determine the response function variations $\delta\mathbf{U}$.

This approach has been exploited successfully by several researchers in related variational methods employing adjoint function spaces [Taylor and Bendsoe, 1984], [Haug, Choi and Komkov, 1986] and [Dems and Mröz, 1985]. The most general result was obtained by Dems and Mröz, who identified the correct physical interpretation of the adjoint field variables. In the following section it is shown that a result identical to that of Dems and Mröz can be obtained directly from a mutual form of the Hu-Washizu principle.

The design sensitivity analysis problem becomes more complex when variation of structural form is considered. Shape design sensitivity analysis corresponds to a calculus of variations problem defined on a varying domain. Haug and Choi and Dems and Mröz employed material derivative approaches to address this problem. In section 3 the Eulerian-Lagrangian kinematic description is used to describe continuous variations in structure geometry. This leads to a simple extension of the mutual Hu-Washizu energy method for explicit shape design sensitivity analysis.

2. The Mutual Hu-Washizu Principle in Sensitivity Analysis

Mutual energy functionals were first introduced as a tool in sensitivity analysis by Shield and Prager [1970]. The mutual complementary energy [Huang, 1971] provides explicit sensitivities for the full set of force response variables, while the mutual potential energy [Hegemeir and Tang, 1973] yields explicit sensitivities for the full set of kinematic response variables. This section introduces the mixed-form mutual Hu-Washizu principle to obtain explicit sensitivity expressions for the combined set of force and kinematic response variables.

The Hu-Washizu energy functional [Washizu, 1975] for a linearly elastic structure with initial stress and strain is given by

$$\Pi = \int_V [\tfrac{1}{2} e_{ij} C_{ijk\ell} e_{k\ell} + e_{ij} \tau^I_{ij} - e^I_{ij} C_{ijk\ell} e_{k\ell}]dV$$
$$- \int_V [\tau_{ij} (e_{ij} - \tfrac{1}{2} (u_{i,j} + u_{j,i}))]dV$$
$$- \int_V b_i u_i dV - \int_{A_t} t^P_i u_i dA - \int_{A_u} t_i (u_i - u^P_i)dA \tag{11}$$

where **e**, **τ**, **u** and **t** are independent tensor fields. Satisfaction of equations (1) - (6) leads directly to the stationary condition $\delta\Pi = 0$ for arbitrary, admissible variations δ**e**, δ**τ**, δ**u** and δ**t**.

Now consider a structure subjected to two independent sets of loads: the real loads acting on the structure and an imaginary adjoint load set. The mutual Hu-Washizu energy Π^M is the difference between the Hu-Washizu energy of the superposed load sets and the sum of the energies of the real and adjoint load sets acting alone. $\Pi(\text{Real + Adjoint}) = \Pi(\text{Real}) + \Pi(\text{Adjoint}) + \Pi^M$.

$$\Pi^M = \int_V [e_{ij}C_{ijk\ell}\bar{e}_{k\ell} + e_{ij}\bar{\tau}^I_{ij} + \bar{e}_{ij}\tau^I_{ij} - \bar{e}^I_{ij}C_{ijk\ell}e_{k\ell} - e^I_{ij}C_{ijk\ell}\,\bar{e}_{k\ell}$$
$$- b_i\bar{u}_i - \bar{b}_i u_i - \tau_{ij}(\bar{e}_{ij} - \tfrac{1}{2} (\bar{u}_{i,j}+\bar{u}_{j,i})) - \bar{\tau}_{ij}(e_{ij} - \tfrac{1}{2} (u_{i,j}+u_{j,i}))]dV$$

$$- \int_{A_t} (t_i^P \bar{u}_i + \bar{t}_i^P u_i) dA - \int_{A_u} [t_i (\bar{u}_i - \bar{u}_i^P) + \bar{t}_i (u_i - u_i^P)] dA \qquad (12)$$

Superposed bars indicate field variables associated with the adjoint load set.

When both the real and adjoint field solutions satisfy equations (1)-(6), two important results are obtained. First, the mutual Hu-Washizu energy is stationary ($\delta\Pi^M = 0$) for arbitrary, admissible variations of both the real response variables ($\delta\mathbf{e}$, $\delta\boldsymbol{\tau}$, $\delta\mathbf{u}$ and $\delta\mathbf{t}$) and the adjoint response variables ($\delta\bar{\mathbf{e}}$, $\delta\bar{\boldsymbol{\tau}}$, $\delta\bar{\mathbf{u}}$ and $\delta\bar{\mathbf{t}}$). Second, Π^M can be written in a simplified form Π^{M1}.

$$\Pi^{M1} = \int_V [\bar{\tau}_{ij}^I e_{ij} - \bar{e}_{ij}^I \tau_{ij}^E - \bar{b}_i u_i] dV - \int_{A_t} \bar{t}_i^P u_i \, dA + \int_{A_u} \bar{u}_i^P t_i dA \qquad (13)$$

Note that these results only require <u>weak</u> satisfaction of equations (1)-(6). If equations (1)-(6) are implicitly satisfied for both the current design and all designs in the neighborhood of the current design, then $\delta\Pi^M = \delta\Pi^{M1}$.

Next, prescribe the adjoint load set as

$$\bar{\tau}_{ij}^I = \frac{\partial f}{\partial e_{ij}} ; \ \bar{e}_{ij}^I = \frac{-\partial f}{\partial \tau_{ij}^E} ; \ \bar{b}_i = \frac{-\partial f}{\partial u_i} ; \ \bar{t}_i^P = \frac{-\partial g}{\partial u_i} \text{ on } A_t \text{ and } \bar{u}_i^P = \frac{\partial h}{\partial t_i} \text{ on } A_u .$$

Take the variation of (13), combine with (9), replace $\delta\Pi^{M1}$ with $\delta\Pi^M$ and use the stationary conditions on Π^M to obtain

$$\delta R = \delta\Pi^{M1} - \int_V [\delta\bar{\tau}_{ij}^I e_{ij} - \delta\bar{e}_{ij}^I \tau_{ij}^E - \delta\bar{b}_i^I u_i] dV + \int_{A_t} \delta\bar{t}_i^P u_i dA - \int_{A_u} \delta\bar{u}_i^P t_i dA$$

$$= \int_V \{\bar{e}_{ij} [\delta C_{ijk\ell} (e_{k\ell} - e_{k\ell}^I) + \delta\tau_{ij}^I - C_{ijk\ell} \delta e_{k\ell}^I] - \bar{e}_{ij}^I \delta C_{ijk\ell} e_{k\ell} - \delta b_i \bar{u}_i\} dV$$

$$- \int_{A_t} \delta t_i^P \bar{u}_i dA + \int_{A_u} \delta u_i^P \bar{t}_i dA \qquad (14)$$

Combination of (14) with (8) produces the explicit design sensitivity expression for $\delta\Gamma$. The tensor components $\delta C_{ijk\ell}$, $\delta\tau_{ij}^I$, δe_{ij}^I, δb_i, δt_i^P and δu_i^P are the elements of the set of design function variations $\delta\boldsymbol{\theta}$. Furthermore, (14) is identical to the results of Dems and Mróz, except that variations of body force, surface traction and prescribed displacements are included.

To evaluate $\delta\Gamma$ the analyst must obtain a solution to the real load set, form the adjoint load set, solve the adjoint system, and evaluate the scalar integral expressions in (14). This requires only one assembly and reduction of the system stiffness matrix. Most often, $\delta\Gamma$ will not depend on the full sets of design and response field variables; and many of the terms in (8) and (14) will vanish.

3. Design Shape Variation with the Eulerian-Lagrangian Kinematic Description

The preceding section developed a method for explicit design sensitivity analysis, including general response functionals. This section extends the method to include shape as a design variable. In other words, the variation of the coordinates δX is included within the set of design function variations $\delta\boldsymbol{\theta}$. The Eulerian-Lagrangian kinematic description (ELD) [Haber, 1984] [Haber and Koh, 1985] provides a convenient method to describe shape variations δX in terms of a mapping from an independent reference geometry. Here the ELD is developed as an alternative to material derivative methods for shape sensitivity analysis.

The ELD kinematic model is illustrated in Figure 1. A fixed Cartesian coordinate system is used to describe the material and current configurations, denoted by volumes V and v. Superscripts ^{1} and ^{2} denote two distinct designs. An invariant, independent spatial reference configuration v^r is selected. A position vector $\mathbf{r}$ in a separate reference coordinate system is associated with each location in the reference configuration. For each design a unique mapping $\mathbf{X}$:$\mathbf{X}(\mathbf{r})$ is established that maps each location $\mathbf{r}$ in v^r onto a material particle with coordinate $\mathbf{X}$ in V. This mapping changes with the shape design, so that the material volume associated with v^r changes. Changes in the material particle associated with a fixed coordinate $\mathbf{r}$ are the Eulerian part of the model; the displacement of a particle $\mathbf{u}$ is the Lagrangian part. The reference coordinates $\mathbf{r}$ are the only independent spatial variables; and the displacement, strain and stress fields and the mapping to the material configuration are expressed as functions of $\mathbf{r}$.

$$\mathbf{u} = \mathbf{u}(\mathbf{r}) \tag{15}$$

$$\mathbf{e} = \mathbf{e}(\mathbf{r}) \tag{16}$$

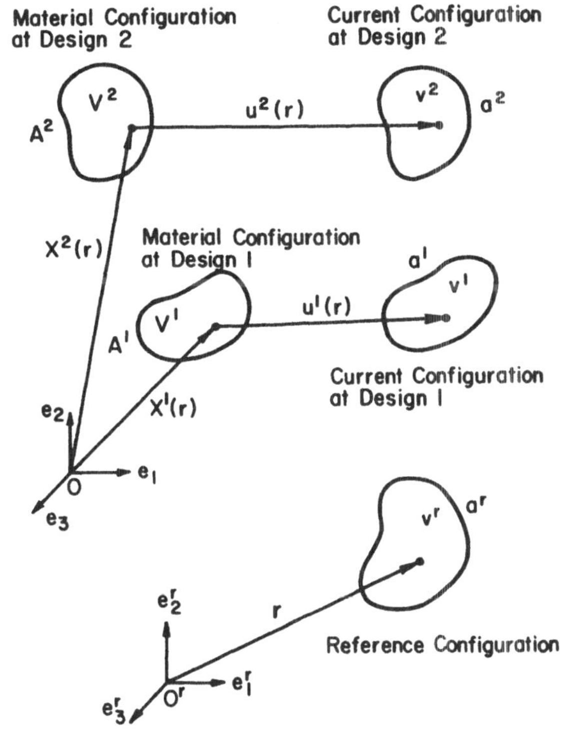

Figure 1. The Eulerian-Lagrangian description for shape variations.

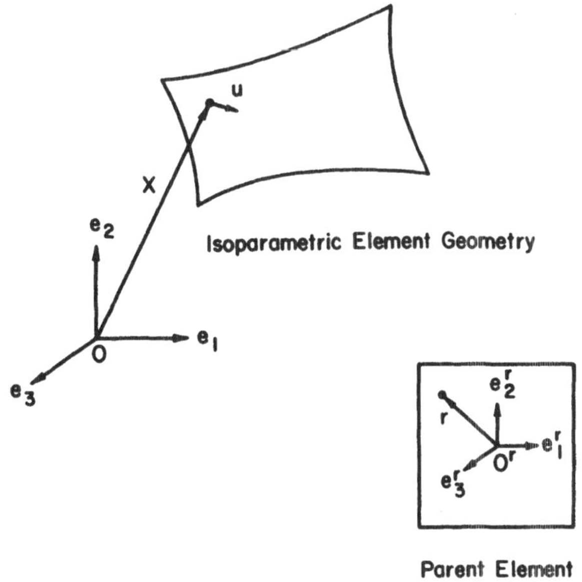

Figure 2. Isoparametric finite element version of the ELD.

$$\tau = \tau(\mathbf{r}) \tag{17}$$

$$X = X(\mathbf{r}) \tag{18}$$

Similar representations are used for the other design and response variables. The Jacobian of the mapping from v^r to V and its inverse are defined as follows.

$$J_{ij} = \frac{\partial X_i}{\partial r_j} = X_{i.j} \tag{19}$$

$$\bar{J}_{ij} = \frac{\partial r_i}{\partial X_j} \tag{20}$$

In this and following sections a period between subscripts indicates differentiation with respect to the reference coordinates **r**. Differential volumes and areas are expressed in the reference configuration as

$$dV = J\, dv^r \tag{21}$$

$$dA = \kappa\, da^r \tag{22}$$

in which J is the determinant of J_{ij} and κ is an area metric on the surface. It is assumed that both metrics are positive at all locations. While finite changes in the volume V are permitted, we assume that the deformation from V to v is adequately described by the engineering strain tensor.

The strain tensor components, written in the reference configuration are

$$e_{ij} = \frac{1}{2}(u_{i.k}\, \bar{J}_{kj} + u_{j.k}\, \bar{J}_{ki}) \tag{23}$$

The variation of the strain tensor splits naturally into two parts:

$$\delta e_{ij} = \frac{1}{2}[(\delta u_{i.k}\, \bar{J}_{kj} + \delta u_{j.k}\bar{J}_{ki}) + (u_{i.k}\delta\bar{J}_{kj} + u_{j.k}\delta\bar{J}_{ki})] \tag{24}$$

where [Haber]

$$\delta\bar{J}_{ij} = -\bar{J}_{ik}\, \delta X_{k.\ell}\bar{J}_{\ell j} \tag{25}$$

The first term in (24) is the strain variation due to changes $\delta u_{i.j}(\mathbf{r})$ with the geometry fixed; and the second term is the variation due to geometry changes $\delta\mathbf{X}(\mathbf{r})$ with $u_{i.j}(\mathbf{r})$ fixed. Explicit evaluation of the first term is possible (as explained below) using the mutual Hu-Washizu energy method of section 2; and the second term can be evaluated directly for given design variations $\delta\boldsymbol{\theta}$.

Equations (7) and (8) are rewritten using the ELD.

$$\Gamma = \int_{v^r} f^r J\, dv^r + \int_{a_t^r} g^r \kappa\, da^r + \int_{a_u^r} h^r \kappa\, da^r \tag{26}$$

$$\delta\Gamma = \int_{v^r} (\frac{\partial f^r}{\partial \theta_\alpha} \delta\theta_\alpha J + f^r \delta J) dv^r + \int_{a_t^r} (\frac{\partial g^r}{\partial \theta_\alpha} \delta\theta_\alpha \kappa + g^r \delta\kappa) da^r$$

$$\int_{a_u^r} (\frac{\partial h^r}{\partial \theta_\alpha} \delta\theta_\alpha \kappa + h^r \delta\kappa) da^r + \delta R \tag{27}$$

where f^r, g^r, and h^r are the functions f, g and h rewritten in the reference configuration; a_t^r and a_u^r are the regions of the surface of v^r that map onto A_t and A_u, respectively; and $\delta J = \frac{\partial J}{\partial \theta_\alpha} \delta\theta_\alpha$ and $\delta\kappa = \frac{\partial \kappa}{\partial \theta_\alpha} \delta\theta_\alpha$ only involve the geometry variations $\delta\mathbf{X}$, which are now non-zero elements within the complete set of design variations $\delta\boldsymbol{\theta}$. Equations (21)-(23) are used to rewrite the mutual Hu-Washizu functional and its simplified form with the adjoint loads defined as in section 2 (but with f^r, g^r, and h^r replacing f, g and h).

$$\Pi^M = \int_{v^r} [e_{ij} C_{ijk\ell} \bar{e}_{k\ell} + e_{ij} \bar{\tau}^I_{ij} + \bar{e}_{ij} \tau^I_{ij} - \bar{e}^I_{ij} C_{ijk\ell} e_{k\ell} - e^I_{ij} C_{ijk\ell} \bar{e}_{k\ell} - b_i \bar{u}_i - \bar{b}_i u_i$$

$$- \tau_{ij} (\bar{e}_{ij} - \frac{1}{2}(\bar{u}_{i.k} \bar{J}_{kj} + \bar{u}_{j.k} \bar{J}_{ki})) - \bar{\tau}_{ij} (e_{ij} - \frac{1}{2}(u_{i.k} \bar{J}_{kj} + u_{j.k} \bar{J}_{ki}))] J dv^r$$

$$- \int_{a_t^r} (t^P_i \bar{u}_i + \bar{t}^P_i u_i) \kappa da^r - \int_{a_u^r} [t_i (\bar{u}_i - \bar{u}^P_i) + \bar{t}_i (u_i - u^P_i)] \kappa da^r \tag{28}$$

$$\Pi^{M1} = \int_{v^r} (\bar{\tau}^I_{ij} e_{ij} - \bar{e}^I_{ij} C_{ijk\ell} e_{k\ell} - \bar{b}_i u_i) J dv^r - \int_{a_t^r} \bar{t}^P_i u_i \kappa da^r + \int_{a_u^r} \bar{u}^P_i t_i \kappa da^r \tag{29}$$

Again, $\Pi^M = \Pi^{M1}$ and $\delta\Pi^M = \delta\Pi^{M1}$ when equations (1)-(6) are satisfied in both the current design and all neighboring designs. The variation of Π^{M1} now includes variations of the metrics J and κ, so δR is written as

$$\delta R = \delta\Pi^{M1} - \int_{v^r} [(\delta\bar{\tau}^I_{ij} e_{ij} - \delta\bar{e}^I_{ij}\tau^E_{ij} - \delta\bar{b}_i u_i)J + (\bar{\tau}_{ij}e_{ij} - \bar{e}^I_{ij}\tau^E_{ij} - \bar{b}_i u_i)\delta J]dv^r$$

$$+ \int_{a^r_t} (\delta\bar{t}^P_i u_i\kappa + \bar{t}^P_i u_i \delta\kappa)da^r - \int_{a^r_u} (\delta\bar{u}^P_i t_i\kappa + \bar{u}^P_i t_i\delta\kappa)da^r \tag{30}$$

Next, replace $\delta\Pi^{M1}$ with the variation of (28); and utilize the stationary conditions on Π^M and symmetry of the stress tensor to obtain

$$\delta R = \int_{v^r} \{\bar{e}_{ij}[\delta C_{ijk\ell}(e_{k\ell} - e^I_{k\ell}) + \delta\tau^I_{ij} - C_{ijk\ell}\delta e^I_{k\ell}] - \bar{e}^I_{ij}\,\delta C_{ijk\ell}e_{k\ell}$$

$$+ (\tau_{ij}\bar{u}_{i.k} + \bar{\tau}_{ij}u_{i.k})\delta\bar{J}_{kj} - \delta b_i\bar{u}_i\}Jdv^r$$

$$+ \int_{v^r} [e_{ij}C_{ijk\ell}\bar{e}_{k\ell} + \bar{e}_{ij}\tau^I_{ij} - e^I_{ij}C_{ijk\ell}\bar{e}_{k\ell} - \tau_{ij}(\bar{e}_{ij} - \bar{u}_{i.k}\bar{J}_{kj})$$

$$- \bar{\tau}_{ij}(e_{ij} - u_{i.k}\bar{J}_{kj}) - b_i\bar{u}_i]\delta Jdv^r - \int_{a^r_t} (\delta t^P_i\bar{u}_i\kappa + t^P_i\bar{u}_i\delta\kappa)da^r$$

$$+ \int_{a^r_u} [\delta u^P_i\bar{t}_i\kappa + (t_i\bar{u}_i + \bar{t}_i(u_i - u^P_i))\delta\kappa]da^r \tag{31}$$

Equations (27) and (31) provide explicit sensitivity expressions for $\delta\Gamma$; and can be used with any pair of real and adjoint weak solutions of equations (1)-(6), including mixed solutions. When displacement-based stiffness solutions are used, (1) and (6) are satisfied identically; so that (31) simplifies.

$$\delta R = \int_{v^r} \{\bar{e}_{ij}[\delta C_{ijk\ell}(e_{k\ell} - e^I_{k\ell}) + \delta\tau^I_{ij} - C_{ijk\ell}\delta e^I_{k\ell}] - \bar{e}^I_{ij}\delta C_{ijk\ell}e_{k\ell}$$

$$+ (\tau_{ij}\bar{u}_{i.k} + \bar{\tau}_{ij}u_{i.k})\delta\bar{J}_{kj} - \delta b_i\bar{u}_i\}Jdv^r$$

$$+ \int_{v^r} (e_{ij}C_{ijk\ell}\bar{e}_{k\ell} + \bar{e}_{ij}\tau^I_{ij} - e^I_{ij}C_{ijk\ell}\bar{e}_{k\ell} - b_i\bar{u}_i)\delta Jdv^r$$

$$- \int_{a_t^r} (\delta t_i^P \bar{u}_i \kappa + t_i^P \bar{u}_i \delta\kappa) da^r + \int_{a_u^r} (\delta u_i^P \bar{t}_i \kappa + t_i \bar{u}_i^P \delta\kappa) da^r \tag{32}$$

Equations (27) and (32) are essentially the same as the domain version sensitivity expressions obtained with the material derivative method by Haug, Choi and Komkov; but allow for initial stress and strain effects. These expressions should only be used with stiffness solutions or exact solutions of the real and adjoint systems. When a mixed solution is used, equations (1) and (6) are approximated; and equations (27) and (31) must be used for the explicit sensitivity expression.

4. Finite Element Sensitivity Expressions

This section specializes the sensitivity analysis method for use with finite element models. The finite element version of the ELD, depicted in Figure 2, is a natural extension of conventional isoparametric element formulations. The mapping between the reference and the material configurations is established locally within each element. The parent element geometry is selected as the reference domain; and the element natural coordinates serve as the reference coordinates. Node coordinates in the material configuration and the isoparametric shape functions define the geometry mapping.

Geometry variations are represented by variations of the node coordinates, with the shape functions held fixed in the reference system. These geometry variations change the material volume associated with each element and change the material particle associated with each node. Motion of nodes normal to a structure surface describe true shape changes, with material either added or deleted at the surface. Node motion tangent to a surface and node motion on the interior of a structure do not imply shape change, but still can affect the approximate solution because they alter the finite element discretization. Care must be taken to avoid situations where changes in response due to alterations of the discretization mask the response changes that are associated with true shape variation.

Design field variables can be interpolated within each element using the isoparametric shape functions. For a generic design field variable θ and the specific design variables appearing in (31) and (32) we have

$$\theta(\mathbf{r}) = h_\alpha(\mathbf{r})\theta_\alpha; \; C_{ijk\ell} = h_\alpha C_{ijk\ell\alpha}; \; \tau_{ij}^I = h_\alpha \tau_{ij\alpha}^I; \; e_{ij}^I = h_\alpha e_{ij\alpha}^I; \; b_i = h_\alpha b_{i\alpha};$$

$$u_i^P = h_\alpha u_{i\alpha}^P \text{ on } a_u^r; \quad t_i^P = h_\alpha t_{i\alpha}^P \text{ on } a_t^r; \quad X_i = h_\alpha X_{i\alpha} \tag{33}$$

in which h_α is the shape function associated with node α; a design variable with a subscript α indicates the value of that variable at node α; and α ranges from one to the number of nodes in an element. Variations of the design variables are written as $\delta\theta = h_\alpha \delta\theta_\alpha$; $\delta X_i = h_\alpha \delta X_{i\alpha}$; etc. The response field variables are interpolated in the usual way; e.g. $u_i = h_\alpha u_{i\alpha}$ and $\delta u_i = h_\alpha \delta u_{i\alpha}$; where $u_{i\alpha}$ is the unknown displacement at node α. It is often desirable to constrain the design space, either to reduce the number of independent design parameters or to reflect real design constraints. In these cases the nodal design variables can be expressed as functions of a reduced set of master design parameters.

Substitution of (33) into (8), (14), (27), (31) and (32) and the use of finite element approximate solutions for the real and adjoint systems leads to discrete sensitivity expressions for Γ with respect to the independent design parameters. It must be emphasized that these sensitivities represent the "exact" sensitivities of the approximate finite element model--not the sensitivities of the actual continuum problem. Finite difference methods only approximate the sensitivity of the finite element model.

5. Numerical Example

Figure 3 shows a fillet problem taken from [Haug, Choi and Komkov, 1986]. The function Γ is the normalized excess stress $(\frac{\sigma_E - \sigma_A}{\sigma_A})$ averaged over each of a series of test regions; where σ_E is the Von Mises effective stress and σ_A is an allowable stress. The profile of the fillet is varied by adding a multiple of the vector **B** = [0 5.55 5.10 4.65 4.20 3.75 3.30 2.85 2.40 1.95 0] to the vertical coordinates of the nodes between A and B as shown in the figure. Sensitivities for selected regions using both 3-node (CST) and 6-node (LST) triangular elements are shown in Table 1. Finite difference approximations to the sensitivities using an increment of **B** x 10^{-7} are given for comparison.

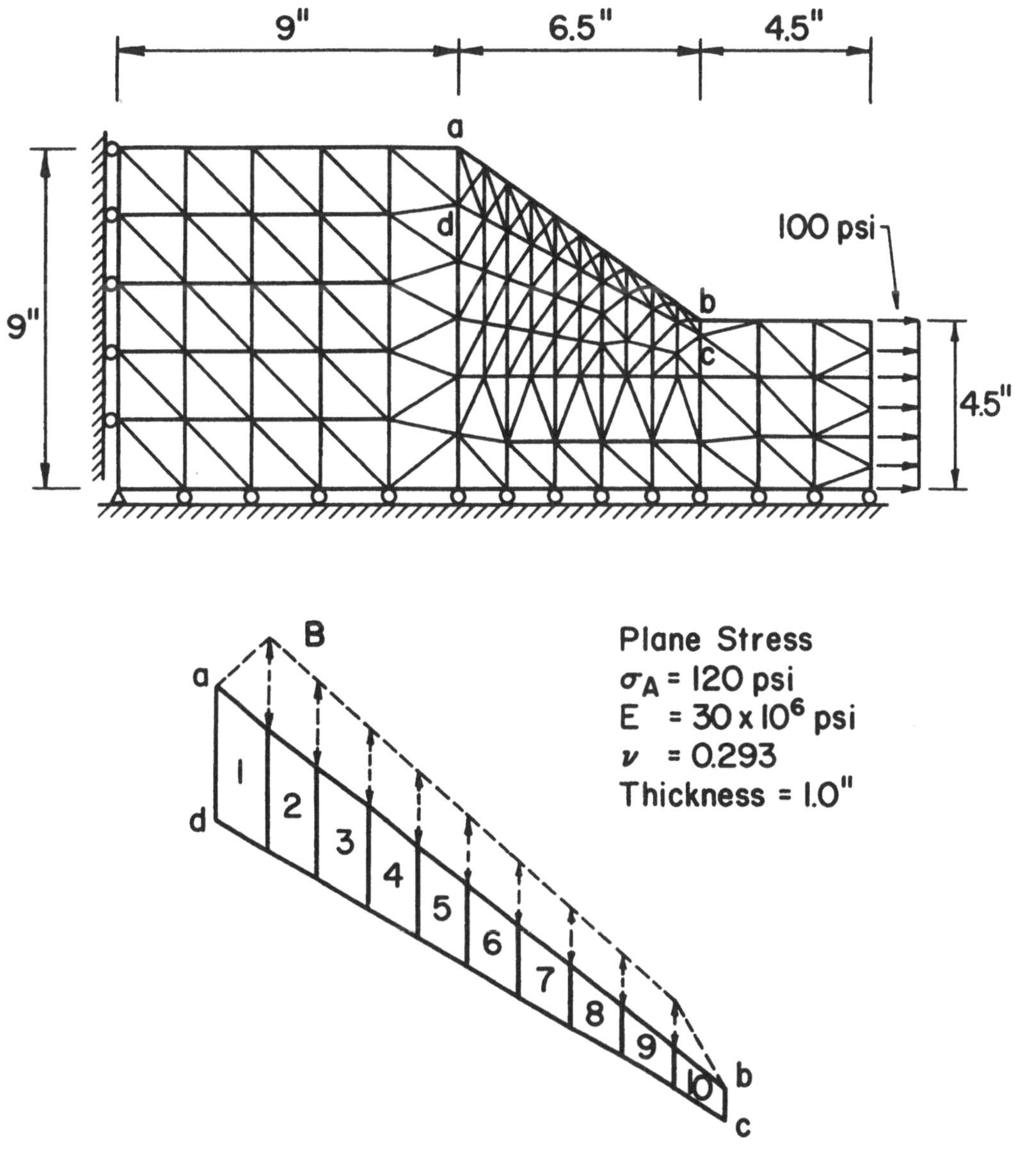

Figure 3. Fillet design shape sensitivity analysis problem.

Table 1. Sensitivites for Fillet Problem

Region/ El. Type		Average Excess Stress Functional	Sensitivity	
			Finite Difference	Explicit
1	CST	-0.792149	1.39058×10^{-5}	1.39058×10^{-5}
	LST	-0.796613	-8.22999×10^{-5}	-8.23000×10^{-5}
3	CST	-0.700975	-5.65964×10^{-4}	-5.65964×10^{-4}
	LST	-0.701107	-5.93831×10^{-4}	-5.93831×10^{-4}
5	CST	-0.591659	-6.17673×10^{-4}	-6.17673×10^{-4}
	LST	-0.592886	-6.17699×10^{-4}	-6.17699×10^{-4}
10	CST	0.156276	-4.27072×10^{-4}	-4.27072×10^{-4}
	LST	0.126887	-5.23812×10^{-4}	-5.23811×10^{-4}

In all cases, excellent agreement is obtained between the finite difference and explicit sensitivities. Great care was required in the selection of the finite difference step size to obtain good approximations. At the center of the fillet (test region 5) there is reasonably good agreement between the CST and LST solutions. However, toward the corners of the fillet (test regions 1 and 10) there are significant differences between the CST and LST predictions for the average excess stress functional itself and for its sensitivity. Still, the finite difference and explicit sensitivities are in close agreement. This demonstrates the fact that the computed values in each case are the sensitivities of a specific finite element model, which might or might not represent accurately the desired continuum problem. Discrepancies of this sort can be expected to be most severe near reentrant corners (region 10).

6. Conclusions and Acknowledgements

The combination of the mutual Hu-Washizu principle and the Eulerian-Lagrangian kinematic description yields a robust approach for sensitivity analysis of linearly elastic structures. Extensions for large-deflection problems and for second-order sensitivity calculations are possible. The explicit sensitivity expressions derived in this paper are quite general--in most practical applications many of the terms will vanish. The sensitivity expressions are more reliable and far less expensive to evaluate than finite difference estimates of the response sensitivity. In finite element applications the required computations are: 1) stiffness analysis of the structure; 2) assembly of the adjoint load vector; 3) solution of the adjoint problem by backsubstitution using the reduced matrix from step 1; and 4) integration of the explicit sensitivity expressions. Note that steps 2-4 are inexpensive relative to the original analysis.

Response sensitivities for finite element models can also be obtained by direct differentiation of the system stiffness matrix. The results will be identical to those obtained with the corresponding explicit sensitivity expressions presented above. Still, the explicit expressions require less computer storage and execution time, and are easier to formulate for complex elements. It is also possible to combine analytic solutions for the adjoint system with discrete solutions for the real system as in [Haber and Koh, 1985]. Because the sensitivities of a given finite element model might not accurately represent the sensitivity of the intended continuum model, adaptive finite element modeling is a desirable addition to the sensitivity analysis method.

The author wishes to acknowledge the aid of David Phelan, Yen Ling Chung and Creto Vidal in developing and implementing the techniques described in this paper.

7. References

Dems, K. and Mröz, Z., 1985, "Variational Approach to First- and Second-Order Sensitivity Analysis," Int. J. Num. Methods Engng., Vol 21, pp. 637-661.

Haber, R.B., 1984, "A Mixed Eulerian-Lagrangian Displacement Model for Large-Deformation Analysis in Solid Mechanics," Int. J. Num. Methods Engng., Vol. 43, pp. 277-292.

Haber, R.B. and Koh, H.M., 1985, "Explicit Expressions for Energy Release Rates Using Virtual Crack Extensions," Int. J. Num. Methods Engng., Vol. 21, pp. 301-315.

Haug, E.J., Choi, K.K. and Komkov, V., 1986, Design Sensitivity Analysis of Structural Systems, Academic Press, Inc., Orlando, Florida.

Hegemeir, G.A. and Tang, H.T., 1973, "A Variational Principle, the Finite Element Method, and Optimal Structural Design for Given Deflection," in Optimization in Structural Design, IUTAM Symp., Warsaw, pp. 464-483.

Huang, N.-C., 1971, "On Principle of Stationary Mutual Complementary Energy and Its Application to Optimal Structural Design," ZAMP, Vol. 22, pp. 608-620.

Shield, R.T. and Prager, W., 1970, "Optimal Structural Design for Given Deflection," ZAMP, Vol 21, pp. 513-523.

Taylor, J.E. and Bendsoe, M.P., 1984, "An Interpretation for Min-Max Structural Design Problems Including a Method for Relaxing Constraints," Int. J. Solids Structs., Vol. 20, pp. 301-314.

Washizu, K., 1975, Variational Methods in Elasticity and Plasticity, 2nd ed., Pergamon Press, Oxford.

DESIGN SENSITIVITY ANALYSIS AND OPTIMIZATION OF NONLINEAR STRUCTURES

J.S. Arora and C.C. Wu
OPTIMAL DESIGN LABORATORY
College of Engineering
The University of Iowa
Iowa City, Iowa 52242 USA

1. INTRODUCTION

Methods of design sensitivity analysis with linear response under static and dynamic loads have been developed and documented over the last fifteen years [1-5]. However, they are just beginning to be developed with nonlinear response [6-8]. Purpose of this paper is to describe a method of design sensitivity analysis of static nonlinear response using incremental finite element procedures. To accomplish this objective, nonlinear analysis of the structure must be performed which is usually quite tedious. The most effective procedure is to use load incrementation coupled with an iteration. With such a procedure, geometric as well as material nonlinearities (different material models) can be consistently and uniformly treated. However, the structure can collapse before the full load level is reached. In that case, design must be improved to have a stable structure. Thus procedures for calculation of collapse load and its sensitivity to design changes must be included in structural optimization problem with nonlinear response.

2. DESIGN SENSITIVITY ANALYSIS: GENERAL FORMULATION

The problem is to optimize a nonlinear structure with constraints on stresses, displacements, strains and the buckling load. These constraints can be described by the functional inequality, $\psi(\mathbf{b}, {}^t\mathbf{U}) \leq 0$, where $\mathbf{b} \varepsilon R^k$ is a design variable vector, and ${}^t\mathbf{U} \varepsilon R^n$ is the displacement vector with the left superscript indicating the load level. The constraint functionals usually depend explicitly on $\mathbf{b}$ and ${}^t\mathbf{U}$ as well as implicitly on $\mathbf{b}$ (because ${}^t\mathbf{U}$ depends implicitly on $\mathbf{b}$). Various constraints can be expressed as: stress, $|\sigma_i| - \sigma_a \leq 0$; displacement, $|{}^tU_i| - {}^tU_{i0} \leq 0$; strain, $|\varepsilon_i| - \varepsilon_{ia} \leq 0$; and buckling load, $1-p^{cr} \leq 0$; where σ_i is an effective stress at point i and σ_a is its limiting value, ${}^tU_i$ is the

NATO ASI Series, Vol. F27
Computer Aided Optimal Design: Structural and Mechanical Systems. Edited by C. A. Mota Soares

displacement at the ith point and ${}^tU_{i0}$ is its limiting value, ε_i is effective strain and ε_{ia} is its limiting value, and p^{cr} is the buckling load factor. Detailed numerical procedures for calculating these quantities are discussed in Refs. 9-11.

The problem of design sensitivity analysis is to calculate gradient of ψ at given design $\mathbf{b}^*$ and calculated displacement ${}^t\mathbf{U}^*$ as (the arguments $\mathbf{b}^*$, ${}^t\mathbf{U}^*$ will be omitted for brevity):

$$d\psi/d\mathbf{b} = \partial\psi/\partial\mathbf{b} + \left(\partial\psi/\partial^t\mathbf{U}\right) d^t\mathbf{U}/d\mathbf{b} \tag{1}$$

The partial derivatives $\partial\psi/\partial\mathbf{b}$ and $\partial\psi/\partial^t\mathbf{U}$ are relatively easy to calculate since explicit dependence of ψ on $\mathbf{b}$ and ${}^t\mathbf{U}$ is usually known. It will be also seen later that $\partial\psi/\partial^t\mathbf{U}$ is easy to calculate numerically when incremental finite element analysis is used. Calculation of the matrix $d^t\mathbf{U}/d\mathbf{b}$ is the major computational burden needing more closer scrutiny. Two numerical procedures [2-6] have been used for linear structures: direct differentation and adjoint variable methods. These methods can also be used for design sensitivity analysis of nonlinear structures. The criteria for choosing one over the other are the same as for the linear case [2,6], and it is assumed that the efficient procedure is used in numerical implementations.

In order to calculate $d^t\mathbf{U}/d\mathbf{b}$, the equilibrium equation for the structural system should be considered:

$${}^t\mathbf{Q}(\mathbf{b},{}^t\mathbf{U}) \equiv {}^t\mathbf{R} - {}^t\mathbf{F} = \mathbf{0} \tag{2}$$

where ${}^t\mathbf{R}$ is the externally applied equivalent nodal load vector which may be an explicit function of design variables, (it is assumed independent of displacements, although displacement dependent loads can be treated [9]), and ${}^t\mathbf{F}$ is the internal nodal force vector obtained from the calculated stress distribution. For linear problems, explicit form of the function ${}^t\mathbf{Q}$ in terms of $\mathbf{b}$ and ${}^t\mathbf{U}$ is known. For nonlinear problems, however, the explicit form is usually not known. It is not needed in the incremental procedures.

Taking total derivative of Eq. (2) with respect to design, we obtain

$$\frac{\partial^t\mathbf{Q}}{\partial\mathbf{b}} + \frac{\partial^t\mathbf{Q}}{\partial^t\mathbf{U}}\frac{d^t\mathbf{U}}{d\mathbf{b}} = 0; \quad \frac{\partial^t\mathbf{Q}}{\partial^t\mathbf{U}} = \frac{\partial^t\mathbf{R}}{\partial^t\mathbf{U}} - \frac{\partial^t\mathbf{F}}{\partial^t\mathbf{U}}; \quad \frac{\partial^t\mathbf{Q}}{\partial\mathbf{b}} = \frac{\partial^t\mathbf{R}}{\partial\mathbf{b}} - \frac{\partial^t\mathbf{F}}{\partial\mathbf{b}} \tag{3}$$

It can be shown that $-\partial^{t}\mathbf{Q}/\partial^{t}\mathbf{U}$ is the tangential stiffness matrix $^{t}\mathbf{K}$. The term $\partial^{t}\mathbf{R}/\partial^{t}\mathbf{U}$ is actually the load correction matrix (when the load depends on displacements), and is a nonsymmetric contribution to the tangential stiffness matrix. If we assume that the external load is deformation independent and the associate flow rule is applicable for the material of the structure, then the tangential stiffness matrix is symmetric. Therefore, derivative of the displacement vector can be computed if we can calculate $\partial^{t}\mathbf{Q}/\partial\mathbf{b}$ in Eq. (3). Calculation of $\partial^{t}\mathbf{R}/\partial\mathbf{b}$ in Eq. (3) is quite straight forward, since explicit dependence of $^{t}\mathbf{R}$ on design is known. Partial derivatives of internal forces $^{t}\mathbf{F}$ with respect to the design variables need to be calculated.

3. DESIGN SENSITIVITY ANALYSIS: NUMERICAL PROCEDURES

Nonlinear Analysis. A difference between linear and nonlinear analyses is that the virtual work principle must be written in a deformed configuration for the latter case which is not known. In the incremental procedure, it is assumed that the equilibrium configuration at the load level t is known and it is desired at the level t+Δt. To accomplish this, an incremental form of the principle of virtual displacements is obtained by introducing incremental decompositions for stresses, strains and displacements. Then the usual finite element approximation relating incremental displacement field to incremental nodal displacements is introduced. Thus, a matrix equation for incremental displacements is obtained. Since linear approximations are also used in the incremental virtual work principle, the equilibrium at t+Δt will not be exactly satisfied. Therefore, an iteration within the load increment is necessary to satisfy the equilibrium at t+Δt.

Two slightly different numerical approaches for nonlinear analysis have been pursued in the recent literature. The first one relates all the static and kinematic variables to the initial configuration, and is generally called the Total Lagrangian (TL) formulation. The second formulation called the Updated Lagrangian (UL) refers all the variables to an updated configuration. If consistent derivations are used, then both formulations give same final response. The TL formulation uses 2nd Piola-Kirchhoff stress and Green-Lagrange strain tensors and the UL formulation uses Cauchy stress and infintesimal strain tensors. One advantage of the

TL formulation is that the 2nd Piola-Kirchhoff stress tensor is invariant under a rigid body rotation where as Cauchy stress tensor is not. If large strains are present in an inelastic analysis, the UL formulation has to use a spin-invariant stress measure, such as the Jaumman stress rate. It is possible to transform one stress measure to the other one.

The design sensitivity analysis can be performed using TL or UL formulation [6,11]. However, only UL formulation will be summarized here. The TL formulation follows exactly similar steps. In the following, the equations are given for a finite element, and usual assembly procedures are used to obtain the equilibrium equation for the entire finite element model.

The equation with U as the unknown incremental displacement is given as

$$({}^{t}_{t}K_{L} + {}^{t}_{t}K_{NL})\mathbf{U} = {}^{t+\Delta t}_{t}\mathbf{R} - {}^{t}_{t}F \tag{4}$$

$${}^{t}_{t}\mathbf{K}_{L} = \int_{{}^{t}V} {}^{t}_{t}\mathbf{B}_{L}^{T}\ {}^{t}_{t}\mathbf{D}\ {}^{t}_{t}\mathbf{B}_{L}\ {}^{t}dV; \quad {}^{t}_{t}\mathbf{K}_{NL} = \int_{{}^{t}V} {}^{t}_{t}\mathbf{B}_{NL}^{T}\ {}^{t}_{t}\mathbf{D}\ {}^{t}_{t}\mathbf{B}_{NL}\ {}^{t}dV \tag{5}$$

${}^{t}_{t}K_{L}$ is the linear strain incremental stiffness matrix, ${}^{t}_{t}\mathbf{D}$ is the tangential material property matrix measured at load level t, ${}^{t}_{t}K_{NL}$ is the initial stress, geometric stiffness, or the nonlinear strain incremental stiffness matrix, V is the volume, ${}^{t}_{t}\mathbf{B}_{L}$ and ${}^{t}_{t}\mathbf{B}_{NL}$ are the linear and nonlinear strain-displacement matrices at the load level t, the left superscript indicates the configuration in which the quantity is calculated, and left subscript indicates the reference configuration. ${}^{t+\Delta t}_{t}\mathbf{R}$ is the equivalent nodal force. The second term on the R.H.S. of Eq. (4) is the internal force vector given in terms of the Cauchy stress vector ${}^{t}\boldsymbol{\tau}$ as

$${}^{t}_{t}\mathbf{F} = \int_{{}^{t}V} {}^{t}_{t}\mathbf{B}_{L}^{T}\ {}^{t}\boldsymbol{\tau}\ {}^{t}dV \tag{6}$$

Buckling Load Calculation. For some problems where the pre-collapse displacements are negligible, the critical load can be estimated by solving a linearized buckling eigenvalue problem as $\det({}^{\tau}\mathbf{K}) = 0$, where ${}^{\tau}\mathbf{K} = {}^{\tau}_{t}K_{L} + {}^{\tau}_{t}K_{NL}$ is the tangential stiffness matrix at load level τ. It is calculated as ${}^{\tau}\mathbf{K} = {}^{t-\Delta t}\mathbf{K} + \lambda({}^{t}\mathbf{K} - {}^{t-\Delta t}\mathbf{K})$, where λ is a scalar quantity [9,10]. Accordingly, an approximation to the buckling load $\mathbf{R}_{b}$ is given as $\mathbf{R}_b = {}^{t-\Delta t}\mathbf{R} + \lambda({}^{t}\mathbf{R} - {}^{t-\Delta t}\mathbf{R})$. These equations can be applied at any load level t. However, it is assumed that the tangential stiffness matrices change proportionally with additional load increments from load level ${}^{t}\mathbf{R}$. This assumption is severe leading to over estimation of the buckling load. This gives difficulties in numerical design optimization. The best procedure for design optimization is to use load incrementation to precisely calculate the critical load. The eigenvector $\mathbf{y}$ is calculated from ${}^{t}\mathbf{K}\mathbf{y} = \mathbf{0}$ using inverse iteration procedure. This approach is used in the present work to impose a collapse load constraint for the structure.

Gradient Calculations. To complete the calculation of gradients in Eqs. (1) and (3), we need to calculate $\partial\psi/\partial\mathbf{b}$, $\partial\psi/\partial\mathbf{U}$, and $\partial{}^{t}\mathbf{Q}/\partial\mathbf{b}$. The main quantities needed to accomplish these calculation are the internal force partial derivatives w.r.t. state and design variables. The partial derivative w.r.t. the state variable is quite easy to calculate using the incremental equation:

$$\delta{}^{t}\mathbf{F} = {}^{t}\mathbf{K}(\mathbf{b},{}^{t}\mathbf{U})\,\delta{}^{t}\mathbf{U} \tag{7}$$

where $\delta{}^{t}\mathbf{F}$ is the increment in the internal force ${}^{t}\mathbf{F}$ corresponding to an increment $\delta{}^{t}\mathbf{U}$ in the vector ${}^{t}\mathbf{U}$, ${}^{t}\mathbf{K}$ is the tangential stiffness matrix, $\mathbf{b}$ is a given design which is fixed during the analysis phase. Therefore, $\partial{}^{t}\mathbf{F}/\partial{}^{t}\mathbf{U} = {}^{t}\mathbf{K}$. There can be a couple of procedures to calculate partial derivatives w.r.t. the design variables. One way is to first calculate the total internal force as

$$ {}^{t}\mathbf{F} = \int_0^{{}^{t}\mathbf{U}} {}^{\tau}\mathbf{K}(\mathbf{b},{}^{t}\mathbf{U})\,d{}^{\tau}\mathbf{U} = {}^{t}\mathbf{K}_s(\mathbf{b},{}^{t}\mathbf{U})\,{}^{t}\mathbf{U} \tag{8}$$

where ${}^{t}\mathbf{K}_s$ is the secant stiffness matrix. The secant stiffness matrix relates total displacements to total forces. A relationship between tangential matrix ${}^{t}\mathbf{K}$ and the secant matrix ${}^{t}\mathbf{K}_s$, can be obtained by taking the first order variation of Eq. (8) w.r.t. ${}^{t}\mathbf{U}$ and identifying

$$ {}^{t}\mathbf{K} = \partial({}^{t}\mathbf{K}_s\,{}^{t}\mathbf{U})/\partial{}^{t}\mathbf{U} + {}^{t}\mathbf{K}_s \tag{9}$$

where the given design $\mathbf{b}$ is held fixed. Also, from Eq. (8) $\partial{}^{t}\mathbf{F}/\partial\mathbf{b}$ is given as

$$\frac{\partial {}^tF}{\partial b} = \frac{\partial}{\partial b} \int_0^{{}^tU} {}^\tau K(b, {}^\tau U) d{}^\tau U = \frac{\partial}{\partial b} ({}^t K_s (b, {}^tU) {}^tU) \tag{10}$$

Thus, calculation of the total internal force in Eq. (8) or its design derivative in Eq. (10) needs integration over the entire displacement history.

The design derivative of internal forces is also obtained by taking partial derivative of Eq. (6) with respect to the design variables as

$$\frac{\partial}{\partial b} {}^t_tF = \frac{\partial}{\partial b} \int_{{}^tV} {}^t_tB_L^T \, {}^t\tau \, {}^tdV = \frac{\partial}{\partial b} \int_{{}^0V} {}^t_tB_L^T \, {}^t\tau \, {}^t_0X \, {}^0dV \tag{11}$$

where ${}^t_0X$ is determinant of the element deformation gradient which is given as ${}^0\rho/{}^t\rho$ and ρ is the mass density [9]. It is convenient to perform calculations in Eq. (11) with the isoparametric element formulation. In such a formulation, the integrand is expressed in terms of the element intrinsic coordinates r, s, and t such that ${}^0dV$ for various elements is given as ${}^0dAJdr$ (beam), ${}^0AJdr$ (truss and cable), ${}^0dzJdrds$ (plate), hJdrds (membrane and shear panel), and Jdrdsdt (brick) where J is the determinant of Jacobian of transformation equations, A is the cross-sectional area and h is the thickness measured along the z axis. All the preceeding expressions can be concisely written as ${}^0dV = Jd\Gamma$. Thus, Eq. (11) can be expanded as

$$\frac{\partial}{\partial b} {}^t_tF = \int_\Gamma \frac{\partial}{\partial b}({}^t_tB_L^T \, {}^t\tau) {}^t_0XJd\Gamma + \int_\Gamma {}^t_tB_L^T \, {}^t\tau \, \frac{\partial}{\partial b}({}^t_0XJ) d\Gamma$$
$$+ \int_\Gamma {}^t_tB_L^T \, {}^t\tau ({}^t_0XJ) \frac{\partial}{\partial b}(d\Gamma) \tag{12}$$

For design with fixed domain ${}^t_0X$ does not explicitly depend on design variables. For trusses, membranes and shear panels, the strain-displacement matrix ${}^t_tB_L$ and stress ${}^t\tau$ do not explicitly depend on cross-sectional properties and, therefore, design variables. For bending type elements (e.g., beam and plate), the strain-displacement matrix and the stress depend explicitly on the element cross-sectional geometry which must be accounted for in derivative calculations. For the brick element the parent cube drdsdt is not a function of design variables. For other elements, the term $d\Gamma$ depends on the cross-sectional geometry. Note that calculations in Eqs. (10) and (12) are essentially the same; in Eq. (10)

integration over the volume is carried out first to calculate stiffness matrices, where as in Eq. (12) integration over the displacement history is carried out first to calculate the total stress.

Stress and Strain Derivatives. The stress or strain constraint functions are generally expressed as explicit functions of stress or strain components. Let $\boldsymbol{\sigma}$ and $\mathbf{e}$ be the stress and strain vectors ${}^{t}\boldsymbol{\tau}$ and ${}^{t}_{t}\boldsymbol{\varepsilon}$, respectively. The design sensitivities of the stress and strain constraints are then given as

$$\frac{d\psi^{\sigma}}{d\mathbf{b}} = \frac{d\psi^{\sigma}}{d\boldsymbol{\sigma}} \frac{d\boldsymbol{\sigma}}{d\mathbf{e}}\left(\frac{\partial \mathbf{e}}{\partial \mathbf{b}} + \frac{\partial \mathbf{e}}{\partial {}^{t}\mathbf{U}} \frac{d {}^{t}\mathbf{U}}{d\mathbf{b}}\right); \quad \frac{d\psi^{\varepsilon}}{d\mathbf{b}} = \frac{d\psi^{\varepsilon}}{d\mathbf{e}}\left(\frac{\partial \mathbf{e}}{\partial \mathbf{b}} + \frac{\partial \mathbf{e}}{\partial {}^{t}\mathbf{U}} \frac{d {}^{t}\mathbf{U}}{d\mathbf{b}}\right) \tag{13}$$

where ψ^{σ} and ψ^{ε} denote the stress and strain constraint functions, respectively, and the stress is assumed related to the element strain only. In Eq. (13), calculations for the three quantities $\partial\mathbf{e}/\partial\mathbf{b}$, $\partial\mathbf{e}/\partial{}^{t}\mathbf{U}$ and $\partial\boldsymbol{\sigma}/\partial\mathbf{e}$ depend on the finite element formulation used.

The design derivative of element strains cannot be obtained directly from the incremental form of strain-displacement relations. Accordingly, the basic equation of Green-Lagrange strain or Almansi strain should be used for the derivation. The Almansi strain vector by a secant relationship is given as [9]

$${}^{t}_{t}\boldsymbol{\varepsilon} = ({}^{t}_{t}\mathbf{B}_{L0} - {}^{t}_{t}\mathbf{B}_{L1})\,{}^{t}\mathbf{U} \tag{14}$$

where components of the strain-displacement matrices ${}^{t}_{t}\mathbf{B}_{L0}$ and ${}^{t}_{t}\mathbf{B}_{L1}$ are obtained by proper differentiation of the finite element idealization of the displacement field. Then, partial derivative of Almansi strain vector w.r.t. design variables is easily obtained from Eq. (14). The derivatives of strains with respect to the displacements $\partial\mathbf{e}/\partial{}^{t}\mathbf{U}$ can be obtained from the incremental strain-displacement relations. Since the strain derivative is computed at the final response ${}^{t}\mathbf{U}^{*}$, the displacement increment $\mathbf{U}^{*}$ is zero; thus the nonlinear incremental strains vanish. Then strain derivative w.r.t. state variables is given as $\partial{}^{t}_{t}\varepsilon/\partial{}^{t}\mathbf{U} = {}^{t}_{t}\mathbf{B}_{L}$, where ${}^{t}_{t}\mathbf{B}_{L}$ is the linear incremental tangential strain-displacement matrix. The total stress derivative $d\boldsymbol{\sigma}/d\mathbf{e}$ can be obtained from the incremental stress-strain relations as $d{}^{t}\boldsymbol{\tau}/d{}^{t}\mathbf{e} = {}^{t}_{t}\mathbf{D} \equiv [{}^{t}_{t}C_{ijk\ell}]$, where ${}^{t}_{t}\mathbf{D}$ is the incremental constitutive matrix.

Sensitivity of Buckling Constraint. The sensitivity of buckling constraint ($\psi^b \equiv 1-p^{cr} \leq 0$) is $-dp^{cr}/d\mathbf{b}$ at given $\mathbf{b}$ and calculated $\mathbf{U}^{cr}$. The critical displacement $\mathbf{U}^{cr}$ corresponds to the critical load level $\mathbf{R}_b = p^{cr}\mathbf{R}$. Several ways to calculate sensitivity vector for the critical load are described in Ref. 11. Only the formulation that is simple and numerically most effective is described here. We define the unbalanced load ${}^t\mathbf{Q}^r$ as

$$ {}^t\mathbf{Q}^r = \mathbf{R} - \mathbf{R}_b = \mathbf{R}(1-p^{cr}) \tag{15} $$

and the equilibrium equation at critical load level as $\mathbf{R}_b - {}^t\mathbf{F}^b = 0$, where ${}^t\mathbf{F}^b$ is computed from the assemblage of element internal force vectors given in Eq. (6). Therefore, since $\mathbf{R}_b = {}^t\mathbf{F}^b$, the residual force ${}^t\mathbf{Q}^r$ in Eq. (15) is expressed as

$$ {}^t\mathbf{Q}^r = \mathbf{R} - {}^t\mathbf{F}^b \tag{16} $$

Taking total design derivative of Eq. (15) at the critical displacement $\mathbf{U}^{cr}$ and current design $\mathbf{b}^*$, one obtains

$$ \frac{\partial\, {}^t\mathbf{Q}^r}{\partial \mathbf{b}} + \frac{\partial\, {}^t\mathbf{Q}^r}{\partial \mathbf{U}^{cr}} \frac{d\mathbf{U}^{cr}}{d\mathbf{b}} = - \mathbf{R}\frac{dp^{cr}}{d\mathbf{b}} + \left(1-p^{cr}\right)\frac{d\mathbf{R}}{d\mathbf{b}} \tag{17} $$

Assuming that the external load $\mathbf{R}$ is not dependent on the deformation, the partial derivative of unbalanced load in Eq. (16) with respect to state variables is given as $\partial\, {}^tQ^r/\partial\, {}^t\mathbf{U}^{cr} = - \partial\, {}^t\mathbf{F}^b/\partial \mathbf{U}^{cr}$. It can be shown that $\partial\, {}^t\mathbf{F}^b/\partial \mathbf{U}^{cr}$ is the tangential stiffness matrix ${}^t\mathbf{K}$ computed at the critical displacement $\mathbf{U}^{cr}$. Thus it is a singular matrix. Now, premultiplying both sides of Eq. (17) by the transpose of an eigenvector $\mathbf{y}$, using the condition that $\mathbf{y}$ is calculated from ${}^t\mathbf{K}\mathbf{y} = \mathbf{0}$, and simplifying one obtains

$$ \frac{dp^{cr}}{d\mathbf{b}} = \mathbf{y}^T \left[\frac{\partial\, {}^t\mathbf{F}^b}{\partial \mathbf{b}} - p^{cr}\frac{d\mathbf{R}}{d\mathbf{b}}\right]/(\mathbf{y}^T\mathbf{R}) \tag{18} $$

This is a simple expression for design sensitivity coefficients of the critical load factor. Note that the derivatives of the element internal forces $\partial\, {}^t\mathbf{F}^b/\partial \mathbf{b}$ and $d\mathbf{R}/d\mathbf{b}$ are also needed in the sensitivity analysis of other constraints; see Eq. (3). Therefore, calculation of gradient of the critical load factor p^{cr} in Eq. (18) is quite straight forward and does not need much additional computational effort.

4. EXAMPLE PROBLEMS

Using the design sensitivity expressions a computer program is developed and several numerical examples are solved [11]. Geometric as well as material nonlinearities are treated. Constraints on stresses, displacements, strains and the buckling load are imposed simultaneously. Several problems have been solved with the buckling load constraint only to compare results with those available in the literature [12]. Only three example problems are discussed here. All results are obtained by routinely using the program IDESIGN3 [13] which uses a sequential quadratic programming algorithm with potential constraint strategy [14].

Two Bar Truss. This simple example has been used in the literature to treat buckling load problem [12]. The geometry, dimension and loading for the structure are shown in Fig. 1. The cross-sectional area of each member is treated as a design variable. The problem is to minimize volume of the structure such that it does not snap through. Only geometric nonlinearity is considered. The elastic modulus for the material is 10,000 ksi. For calculation of sensitivity coefficients cross-sectional areas are taken as 0.5 in^2. The following solution is obtained: p^{cr} = 0.62582943, $\mathbf{U}^{cr}$ = (-1.05644845, -0.03121569) in. and $\mathbf{y}$ = (1.0, 0.02165292). The critical load factor is obtained by using a very simple bisection approach and the eigenvector is computed by inverse power method [11]. The design sensitivity vector using Eq. (18) is obtained as (-1.002, -0.2498). Using the analytical expressions given in Ref. 12, it is obtained as (-1.003, -0.2507). The two vectors are quite close.

The optimal design for the problem is obtained by starting from (0.952381, 0.190476)in^2 with 0.1 and 10.0in^2 as the lower and upper limits. The program IDESIGN3 [13] finds the optimum in five iterations as (0.7993, 0.7995)in^2 with volume as 199.8 in^3. This is essentially the solution given in Ref. 12; (0.7997502, 0.7997502)in^2 with volume as 200 in^3. Thus the design sensitivity formula given in the paper is reasonable and works well. Several other examples with buckling load constraint are given in Ref. 11.

Optimal results for the above two cases, and the linear cases are given in Table 1. For the case of stiffness hardening (Case 1) the optimum volume is 509.866 in^3. For the case of stiffness softening (Case 2), the optimum volume is 981.419 in^3. For the linear case, the optimum volumes are 714.385 in^3 for Case 1 and 837.640 in^3 for Case 2. The results show that the optimum volume of the structure with stiffness hardening (inclusion of geometric nonlinearities) is less than the optimum volume with the linear case, whereas with stiffness softening it is higher than the one for the linear case. Therefore, it is not always true that lighter designs can be obtained by considering nonlinearity, as stated in Ref. 5. Also, with larger tolerance for displacement constraints, the optimum designs with nonlinear behavior are more realistic. It is important to note that optimum designs considering linear response will fail catastrophically when the structure tends to soften.

Nine Bar Plane Truss. The statically determinate truss shown in Fig. 3 is to be optimized with the constraints on displacement at each node and the strain in each element. The design variable numbers (five) and member

Table 1. Optimum Results for Six Bar Truss (Areas in in^2)

Element	Linear		Nonlinear Geometric		
	Case 1	Case 2	Case 1	Case 2	Case 2*
1	2.1922	2.7086	1.6232	3.1188	3.14396
2	1.2144	1.3162	1.0892	1.5422	1.53072
3	1.1781	1.2769	0.077224	1.5327	1.521454
4	1.0000	1.0000	0.500000	0.70566	0.70566
5	1.0200	1.4276	0.57271	1.6511	1.66439
6	1.2144	1.3162	0.85646	1.6934	1.6823316
Vol, in^3	714.385	837.640	509.866	981.4192	981.8200
NAC	1	1	1	1	1
MCV	0.45E-04	0.0	2.4774E-09	0.0	0.0
NIT	22	16	17	24	21

Notes: Case 2* is obtained using the program ADINA which is a general nonlinear analysis computer program; optimum solution is also obtained using this analyzer [7]. NAC = number of active constraints, MCV = maximum constraint violation at optimum, and NIT = number of iterations to satisfy a strict convergence criterion of 0.0001.

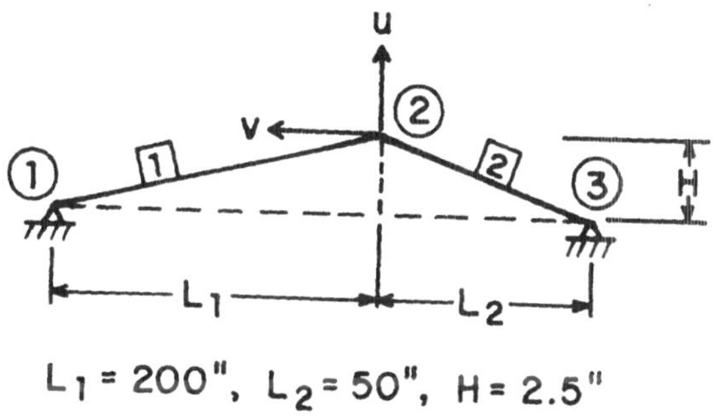

Figure 1. Two Bar Truss (Unsymmetric)

Six Bar Truss. The six-member cantilever truss shown in Fig. 2 is taken as a numerical example to compare optimal designs for linear and geometrically nonlinear response. It is designed with two cases: Case 1 - with tension force in the horizontal direction, and Case 2 - with the compression force at node 5. In addition a downward force at node 5 is also applied in both the cases. The cross-sectional area of each member is chosen as a design variable and volume of the truss is to be minimized. Starting design, and lower and upper bounds for all the variables are 2.0, 0.5 and 10.0 in^2, respectively. For both the cases, the allowable displacement limits in x_1 and x_2 directions at all nodes are 20 and 40 inches respectively. The elastic modulus is 30,000 ksi for all members. For Case 1, the tensile force induced in each member tends to increase the tangential stiffness. Thus, the structure hardens as the deflection increases. Whereas in Case 2 the compression force induced in each member tends to decrease the tangential stiffness. Thus, the structure softens as the deflection increases.

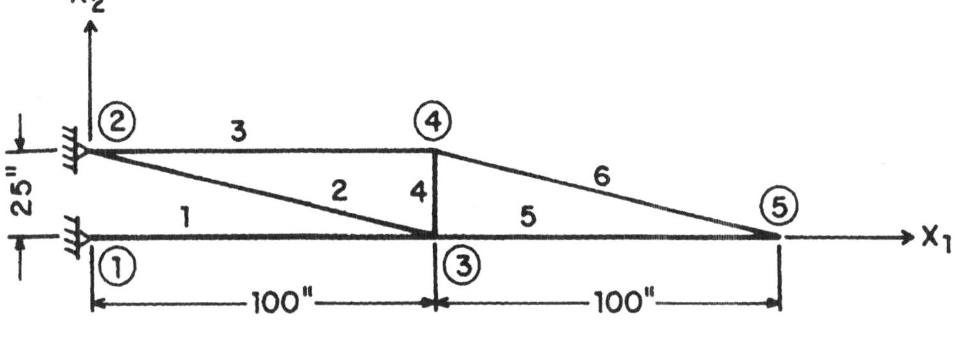

Figure 2. Six Bar Truss

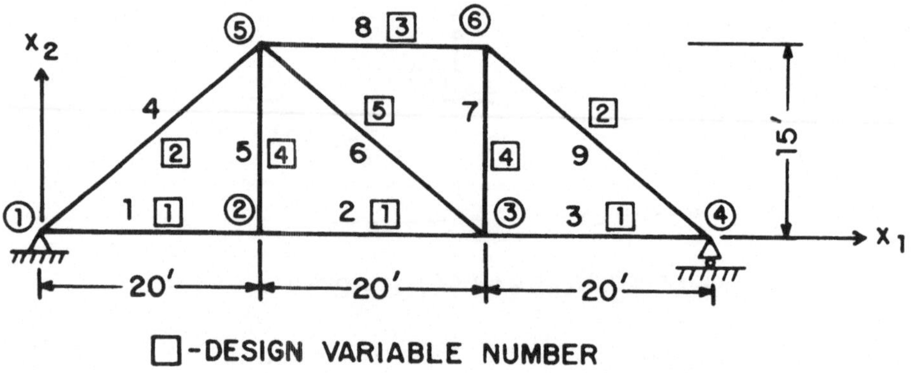

Figure 3. Nine Bar Plane Truss

numbers are shown in the figure. The constraint information is as follows: strain limits for members 1-3; 0.005; members 4, 8, 9; 0.01, members 5-7; 0.0075; displacement limits as 5 and 10 inches in the x_1 and y_2 directions, respectively, at all nodes. The structure is analyzed by considering geometric as well as material nonlinearities. The material is considered to be elastic-plastic with elastic modulus, E = 200,000 ksi, tangent modulus, E_T = 50,000 ksi and initial yielding stress, σ_y = 100 ksi. To investigate the role of the nonlinear buckling constraint in the optimal design process two cases are considered. In Case 1 the starting design of smaller cross sectional areas of 0.2 in^2 for each member is used. In this case the structure collapses before reaching the final load level at the first iteration. On the other hand, in Case 2 starting design of larger cross sectional areas of 1.0 in^2 for each member is chosen. The upper and lower bounds on the design variables are 0.05 and 10 in^2, respectively.

The optimum results for both the cases are the same: Design Variables = (0.18623, 3.3501, 2.656, 0.095745, 0.2256), Vol = 240.3086 in^3, MCV = 4.95E-7 (Case 1) and 2.63E-7 (Case 2), NIT = 26 (Case 1) and 22 (Case 2). The detailed design history shows that the structure collapses at the 1st, 3rd and 4th iterations for Case 1. For Case 2 the starting design consists of larger member sizes and the structure does not collapse during the entire iterative process. The strain constraint for members 4 and 8 is active at the optimum.

5. DISCUSSION AND CONCLUSIONS

Design sensitivity analysis of nonlinear structures using incremental/iterative solution procedure is derived. In such a procedure, geometric as well as material nonlinearities are treated quite consistently and routinely. Whereas nonlinear analysis of structures requires solution of nonlinear equations using load incrementation and iterations with in it, the design sensitivity analysis requires only the solution of linear equations. In addition, most of the quantities needed in sensitivity analysis are already calculated in the incremental analysis phase. Specifically, derivatives of internal forces, stresses and strains with respect to the state variables are obtained numerically without any additional calculations. Thus design sensitivity analysis of nonlinear structures requires only a fraction of the computational effort needed to calculate their response. In fact, recent experience with optimal design of nonlinear structures using ADINA [7] shows that 90 percent of the total computational effort is spent in the analysis phase only. Thus, with a little additional computational effort, nonlinear structures can be optimized.

The preceeding discussion also points out a need to make the entire analysis/optimization process more efficient for nonlinear structures. In this regard it is desirable to use sensitivity information to predict the displacements for the next optimization iteration. Such an approach which can be highly efficient has been evaluated on small scale problems [11]. It shows considerable promise for large scale applications.

A very simple and effective procedure for design sensitivity analysis of the buckling load has been discovered and evaluated. It needs derivatives of internal forces which are also needed for other constraints.

Conclusions based on the present investigation are:

1. Design sensitivity analysis of nonlinear response including the buckling load is quite efficient using the incremental structural analysis procedure.
2. Optimization of nonlinear structures cannot be performed without the buckling load constraint.

3. Optimal design of structures with the assumption of linearity can be dangerous if the structure is likely to soften. The optimized structure can fail catestrophically. If the structure is likely to harden, then lighter optimal designs can be obtained with the inclusion of geometric nonlinearities.

REFERENCES

1. Adelman, H.M. and Haftka, R.T., "Sensitivity Analysis for Discrete Structural Systems," AIAA Journal, Vol. 24, No. 5, pp. 823-832, 1986.

2. Arora, J.S. and Haug, E.J., "Method of Design Sensitivity Analysis in Structural Optimization," AIAA Journal, Vol. 17, No. 9, pp. 970-974, 1979.

3. Haug, E.J. and Arora J.S., Applied Optimal Design, John Wiley and Son, New York, N.Y., 1979.

4. Hsieh, C.C. and Arora, J.S., "Design Sensitivity Analysis and Optimization of Dynamic Response," Computer Methods in Applied Mechanics and Engineering, Vol. 43, pp. 195-219, pp. 195-219, 1984.

5. Hsieh, C.C. and Arora, J.S., "Structural Design Sensitivity Analysis with General Boundary Conditions: Static Probelm," Int. J. for Num. Methods in Engrg., Vol. 20, No. 9, pp. 1661-1670, 1984; Dynamic Problem, Vol. 21, No. 2, pp. 267-283, 1985.

6. Ryu, Y.S., Haririan, M., Wu, C.C. and Arora, J.S., "Structural Design Sensitivity Analysis of Nonlinear Response," Computers and Structures, Vol. 21, No. 1/2, pp. 245-255, 1985.

7. Haririan, M., Wu, C.C. and Arora, J.S., "Structural Optimization of Nonlinear Response with ADINA," Proceedings of the 9th Conference on Electronic Computations, American Society of Civil Engineers, held at University of Alabama, Birmingham, AL, Feb. 23-26, pp. 495-506, 1986.

8. Mroz, A., "Sensitivity Analysis and Optimal Design with Account for Varying Shape and Support Conditions," in these proceedings, 1986.

9. Bathe, K.J., Finite Element Procedures in Engineering Analysis, Prentice-Hall, Englewood Cliffs, N. J., 1982.

10. Bathe, K.J. and Dvorkin, E.N., "On the Automatic Solution of Nonlinear Finite Element Equations," Computer and Structures, Vol. 17, No. 5-6, pp. 871-879, 1983.

11. Wu, C.C. and Arora, J.S., "Design Optimization of Geometric and Material Nonlinear Problems," Technical Report No. ODL-86.8, Optimal Design Laboratory, College of Engineering, The University of Iowa, Iowa City, Iowa, 1986.

12. Kamat, M.P., Khot, N.S. and Venkayya, V.B., "Optimization of Shallow Truss Against Limit Point Instability," AIAA Journal, Vol. 22, No. 3, pp. 403-408, 1984.

13. Arora, J.S. and Tseng, C.H., User's Manual for Program IDESIGN, Report No. ODL-85.10, Optimal Design Laboratory, College of Engineering, The University of Iowa, Iowa City, Iowa, 1986.

14. Lim, O.K. and Arora, J.S., "An Active Set RQP Algorithm for Optimal Deisgn," Computer Methods in Applied Mechanics and Engineering, 1986.

BOUNDARY ELEMENTS IN SHAPE OPTIMAL DESIGN OF STRUCTURAL COMPONENTS

C. A. MOTA SOARES and R.P. LEAL

Centro de Mecânica e Materiais da Universidade
Técnica de Lisboa (CEMUL), Instituto Superior
Técnico, Av. Rovisco Pais, 1096 LISBOA CODEX
PORTUGAL

K. K. CHOI

Center for Computer Aided Design, College of
Engineering, The University of Iowa, Iowa City,
IA 52242
U.S.A

ABSTRACT

The shape optimal design of shafts and two-dimensional elastic structural components is formulated using boundary elements. The design objective is to maximize torsional rigidity of the shaft or to minimize compliance of the structure, subject to an area constrain. Also a model based on minimum area and stress constraints is developed, where the real and adjoint structures are identical, but with different loading conditions. All degrees of freedom of the models are at the boundary and there is no need for calculating displacements and stresses in the domain. Formulations based on constant, linear and quadratic boundary elements are developed. A method for calculating accurately the stresses at the boundary is presented, which improves considerably the design sensitivity information. It is developed a technique for an automatic mesh refinement of boundary element models. The corresponding nonlinear programming problems are solved by Pshenichny's linearization method. The models are applied to shape optimal design of several shafts and elastic structural components. The advantages and disadvantages of the boundary element method over the finite element technique for shape optimal design of structures are discussed with reference to applications. A literature survey of the development of the boundary element method for shape optimal design is presented.

NATO ASI Series, Vol. F27
Computer Aided Optimal Design: Structural and
Mechanical Systems. Edited by C. A. Mota Soares

INTRODUCTION

The finite element method has been extensively used in structural optimization during the last decade, including successful application to shape optimal design of shafts and elastic structural components. In contrast, the boundary element method has only recently been applied for shape optimal design of structures.

Application of the finite element method for shape optimal design of structural components has been successfully demonstrated, but with some disadvantages. It is often required to redefine new finite element meshes as the geometry of the structure changes. An inaccurate evaluation of stresses at the boundary can be responsible for the calculations of very inaccurated design sensitivity analysis, thus leading to a large number of optimization iterations or even unrealistic designs.

These difficulties with the finite element formulation can be partially overcome by using the boundary element method to discretize the structure. Results of the boundary element analysis of elasticity problems are more accurate than the corresponding solutions of the finite element models and it is expected to yield improved design sensitivity information. Consequently, a smaller number of iterations are needed to find the optimum shape.

In the last years about thirty papers have been published in the development of the boundary element method for shape optimal design of engineering systems. A literature survey and a review of the state of art is presented.

The boundary element method is less versatible for structural analysis than the finite element technique. Its applicability to shape optimal design of structures is at present limited to elasticity problems, subject to static constraints. However, with continuing development of the boundary element method, the range of shape optimal design problems that can be efficiently solved is expected to increase in the near future.

In this chapter the shape optimal design of shafts and two dimensional elastic structural components is formulated using boundary elements. The optimal design objective is to maximize torsional rigidity of the shaft or to minimize compliance of the structural component, subject to a fixed amount of material. Also, a model based on minimum area and stress constraints is developed, where the real and adjoint structures are identical, but with diferent loading conditions. All the degrees of freedom of the boundary element models are at the boundary of structural systems and there is no need of internal cells. Displacements and stresses are only calculated at the boundary. The boundary element models are based on constant, linear and quadratic elements.

A method of calculating accurately the stresses at the boundary is presented, which improves considerably the design sensitivity information. It is developed a technique for an automatic refinement of the boundary discretization based on global equilibrium and on the continuity of the tangencial boundary stresses for unloaded smooth surfaces.

The shape optimization nonlinear programming problem is solved by Pshenichny's linearization method. The models are applied to the shape optimal design of several shafts and elastic structural components. The advantages and disadvantages of the boundary element technique for shape optimal design are discussed with references to applications.

LITERATURE SURVEY

Extensive literature has been published on numerical methods for optimization of structures whose shapes are defined by cross section and thickness variables. Only limited literature has appeared in the area of shape optimal design. Recently Pironneau [1] and Haug, Choi, and Komkov [2] have published books dedicated to this subject.

The finite element method has been applied extensively to shape optimal design of structures since 1973 [3], while it is only in the last years that the boundary element method has been used in this field.

Mota Soares, Rodrigues, Oliveira Faria and Haug [4-8] developed models for shape optimal design of solid and hollow shafts, based on constant, linear and quadratic boundary elements and nonlinear programming techniques. The design objective is to choose a shaft with a given area, which has a maximal torsional stiffness. These models are much more efficient and robust than the corresponding finite element discretizations, since the sensitivity information is more accurate and there is no need of calculatinmg the state variable in the domain.

A similar model for the shape optimal design of shafts. based on the boundary element method, has also been developed by Burczynski and Adamczyk [9]. Optimality conditions are generated and the Newton - Raphson method is used to solve a set of nonlinear algebric equations. The examples show that the number of analysis required is smaller than a corresponding finite element discretization.

Models for the shape optimal design of bidimensional elasticity problems based on the boundary element method and linear programming technique has been developed by Zochwski and Mizukami [10]. The design objective is to minimize the area, subject to displacement and geometrical constraints. The adjoint structure generated is not identical to the real structure. The boundary element model is compared with equivalent finite element models, and it concluded that the boundary element technique is more accurated but less efficient in computational time than the finite element method.

Mota Soares, Rodrigues and Choi [11-12] have developed models for the shape optimal design of bidimensional elastic structural components based on linear and quadratic boundary elements and nonlinear programming techniques. The design objective is to minimize compliance, subject to a constant area. The adjoint and real structures are identical and subjected to the same loading conditions. Applications show that the boundary element model is more accurate and efficient than the corresponding finite element model. For general shapes, the technique used to calculated the stresses at the boundary was not very accurate. This problem has been overcome by Leal [13]. Also in Ref. 13 an automatic technique for mesh refinement has been developed. This adaptive scheme improves the discretization and the accuracy of the boundary stresses. This technique is based on the continuity of the boundary stresses for smooth unloaded surfaces.

Also Mota Soares and Choi [14] have recently reviewed the progress of boundary element methods in shape optimal design of structures A model for bidimensional elasticity is developed based on minimum area and Von Mises stress constraint at the boundary. All the necessary information is only calculated at the boundary. This chapter is an update extension of this paper.

Shape optimal design models for two and three dimensional elasticity problems, based on the boundary element method, has recently been developed by Burczynski and Adamczyk [15-18]. The design objective is to maximize stiffness, subjected to constant volume. The optimality conditions are derived for an optimal boundary. An interative process is used, based on finite differences and Newton-Raphson method to solve a set of nonlinear algebraic equations, enabling the determination of the unknown optimal shape. These authors [19] have recently extended the boundary element method formulation to the shape optimal design of elastic components subjected to dynamic constraints.

Eizadian and Trompette [20-21] have also developed a model for shape optimal design of two dimensional structures, based on the boundary element method and nonlinear programming techniques. The design objective is to minimize the tangencial stress subjected to geometrical constraints. The geometry is defined by linear and circular elements. Substructures are used to represent the fixed and moving boundaries. The multiplier method is used to solve the nonlinear programming problem . Several applications are presented, including the shape design of a connecting rod and a rotor. Numerical instabilities are reported

Kane [22-23] is a recent Ph.D. Thesis has developed the sensitivity analysis of a discrete boundary element model for bidimensional elasticity problems. Isoparametric quadratic elements and substructural analysis are used. Several shape optimal design problems are solved and it is demosntrated the effectiveness of the method used. It is shown that the boundary element model is competitive, if not superior, to the finite element formulation for shape optimization of elastic structural components.

The boundary element method has also been applied to the shape optimal design of heat transfer problems. Futagami [24-26] presented a model for steady state and transient optimal heat conduction control based on linear and dynamic programming. A coupled boundary element and finite element model is also developed. The applications show that the boundary elements is a powerful technique for these type of problems.

Barone and Caulk [27] optimized the position, size and surface temperature of circular holes inside a two dimensional heat conductor to produce a minimum variation in surface temperature over a portion of the outer boundary. In this problem, which arises in thermal design of moulds and dies, the internal geometry of the heat conductor depends on the design variables. Since the objective function depend only on the boundary temperatures, there is no need of determining temperature in the interior. Also, it is not required to regenerate a boundary mesh everytime the boundary is changed. The model is applied to the termal design of compression moulds.

Boundary elements have been used by Meric [28-30] for the optimal heating of solids. The design objective is to achieve a desired temperature profile along a segment of a solid boundary with a minimum amount of boundary heat flux. Adjoint equations and the necessary optimality conditions are derived. The conjugate gradient method is used to solve the mathematical programming problem. Numerical results show the efficiency and the accuracy of the boundary element model.

Meric [31] has also developed in this book, the shape design sensitivity analysis of thermoelastic solids using material derivative and adjoint variable method. Coupled thermal and elastic fields are taken into account. The boundary element method has been proposed for spacial discretization of relevant equations. Domain integral terms will exist in the integral equation associated with the adjoint temperature equation due to thermoelastic coupling effects. The author argues that the boundary element method has advantages over the domain type methods, especially with its inherent higher accuracy in the evaluation of boundary stresses, which is crucial for shape design sensitivity analysis. Meric [32] has recently extended this work to shape design sensitivity analysis for nonlinear anisotropic heat conducting solid body.

Kwak and Choi [33-34] have developed a general method for shape design sensitivity analysis of ellipitic boundary value problems using a direct boundary integral equation formulation. The material derivative concept and adjoint variable method are employed to obtain an explicit expression for the variations of the performance functional in terms of the boundary variation. The adjoint problem, although defined in the indirect boundary integral equation form, can also be solved using the same direct boundary integral equation of primal problem. The formulation is obtained for potencial, plate and elasticity problems. The accuracy of the sensitivity formula is studied with reference to a seepage problem. Good accuracy is obtained and the numerical results are compared with those of finite differences.

The boundary element method has also being applied to the shape optimization of airfoils and wings by Pironneau [1]. The author argues that the boundary element method is advantageous over the finite element or finite difference techniques, when the solution of the partial difference equations is needed only at the boundary, and consequently its range of applicability is limited.

BOUNDARY ELEMENT METHOD IN THE TORSION OF SHAFTS

The boundary element method for the torsion of shafts is based on Green's formula, which allows the formulation of the certain boundary value problems as integral equations, involving the solution of the state variable and its normal derivative only on the boundary Consider a shaft defined in Figure 1.

Γ = Piecewise smooth boundary
Ω = Domain
$\bar{\Omega} = \Omega \cup \Gamma$
$Q \in \Gamma$
$q \in \Omega$
$P \in \bar{\Omega}$
n = Outwards normal to Γ
r(P,Q) = Distance between P and Q
x,y = Coordinate systems

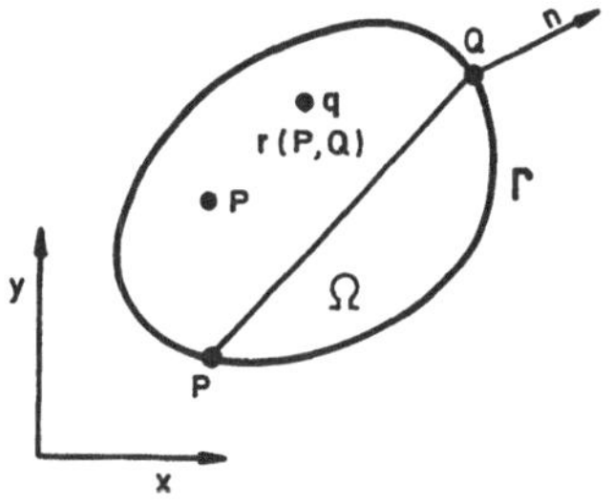

Figure 1 . Definitions.

Let z be the stress function of the torsion problem. This stress function must satisfy the Poisson state equation:

$$\frac{\partial^2 z}{\partial x^2} + \frac{\partial^2 z}{\partial y^2} = \nabla^2 z = -2 \quad in \quad \Omega \qquad z = 0 \quad on \quad \Gamma \tag{.1}$$

This equation can be transformed by considering a new variable

$$u = z + \frac{1}{2}x^2 + \frac{1}{2}y^2 \tag{.2}$$

into a Laplace state equation:

$$\nabla^2 u = 0 \quad in \quad \Omega \qquad u = \frac{1}{2}x^2 + \frac{1}{2}y^2 \quad on \quad \Gamma \tag{.3}$$

Using Green's identity,

$$\int_\Omega (v\nabla^2 u - u\nabla^2 v)d\Omega = \int_\Gamma (v\frac{\partial u}{\partial n} - u\frac{\partial v}{\partial n})d\Gamma \tag{.4}$$

where u and v are solutions of the Laplace equation (3), and assuming that

$$v = \ln r \tag{.5}$$

equation (4) becomes

$$\int_\Omega u\nabla^2 \ln r d\Omega = \int_\Omega \ln r \nabla^2 u d\Omega - \int_\Gamma \ln r \frac{\partial u}{\partial n} d\Gamma + \int_\Gamma u \frac{\partial}{\partial n} \ln r d\Gamma. \tag{.6}$$

For any sufficiently smooth function u(x,y) defined in $\bar{\Omega}$, equation (6) becomes

$$c(P)u(P) = \int_\Omega \nabla^2 u(q) \ln r(P,q) d\Omega + \int_\Gamma u(Q)\frac{\partial}{\partial n} \ln r(P,Q) d\Gamma - \int_\Gamma \ln r(P,Q) \frac{\partial u}{\partial n}(Q) d\Gamma \tag{.7}$$

where

$$c(P) = \int_\Gamma \frac{\partial}{\partial n} \ln r(P,Q) d\Gamma \tag{.8}$$

Since

$$\frac{\partial}{\partial n} \ln r(P,Q) = \frac{d\theta}{d\Gamma}(P,Q) \tag{.9}$$

where

$$\theta(P,Q) = tan^{-1} \frac{y(Q) - y(P)}{x(Q) - x(P)} \tag{.10}$$

it follows that

$$c(P) = \begin{matrix} 2\pi & for\ P & inside\ \ \Omega \\ 0 & for\ P & outside\ \ \overline{\Omega} \end{matrix} \tag{.11}$$

and if Γ has an unique tangent at P, then

$$c(P) = \pi \ \ in \ \ \Gamma \tag{.12}$$

For the Laplace equation, $\nabla^2 u(P) = 0$, and consequently equation (7) becomes

$$c(P)u(P) - \int_\Gamma u(Q) \frac{\partial}{\partial n} \ln r(P,Q) d\Gamma = - \int_\Gamma \ln r(P,Q) \frac{\partial u}{\partial n}(Q) d\Gamma \tag{.13}$$

This equation can be approximated numerically by dividing the boundary into N segments Γ_j and on each segment u and $\partial u/\partial n$ are assumed to be constant. Writing equation (13) at the middle point of each segment, N equations can be obtained of the form

$$cu_i - \sum_{j=1}^{N} u_j \int_{\Gamma_j} \frac{\partial}{\partial n} \ln r d\Gamma = - \sum_{j=1}^{N} \frac{\partial u_j}{\partial n} \int_{\Gamma_j} \ln r \ d\Gamma \qquad i = 1, 2, ..., N \tag{.14}$$

where u_i and $\partial u_i/\partial n$ are nodal values. This equation can be written as

$$\mathbf{H}\ \mathbf{u} \ = \ \mathbf{G}\ \mathbf{p} \tag{.15}$$

where

$$\mathbf{u} = \left\{ \begin{matrix} u_1 \\ \cdot \\ \cdot \\ \cdot \\ u_N \end{matrix} \right\} \mathbf{p} = \left\{ \begin{matrix} \partial u_1/\partial n \\ \cdot \\ \cdot \\ \cdot \\ \partial u_N/\partial n \end{matrix} \right\} \tag{.16}$$

and **H** and **G** are full unsymmetric matrices.

Imposing the boundary conditions, equation (15) becomes

$$\mathbf{G}\ \mathbf{p} \ = \ \mathbf{f} \tag{.17}$$

where

$$\mathbf{f} \ \ = \mathbf{H}\ \mathbf{u} \tag{.18}$$

The solution of equation (17) gives the values of $\partial u/\partial n$ at the boundary nodes. It should be noticed that for the solution of the Laplace equation it is only necessary to discretize the boundary.

Although the boundary element equations are based on full and unsymmetric matrices, the number of degrees of freedom are small when compared with finite element models.

For the constant element, the integrals of equation (14) can be evaluated in closed form, transforming it to a system (ξ, η) centered at the point P and with ξ parallel to the segment in which the integration is performed, Figure 2.

Γ_j = Boundary of segment
ξ, η = Coordinate system
a = Perpendicular distance from P to the segment in which the integration is performed

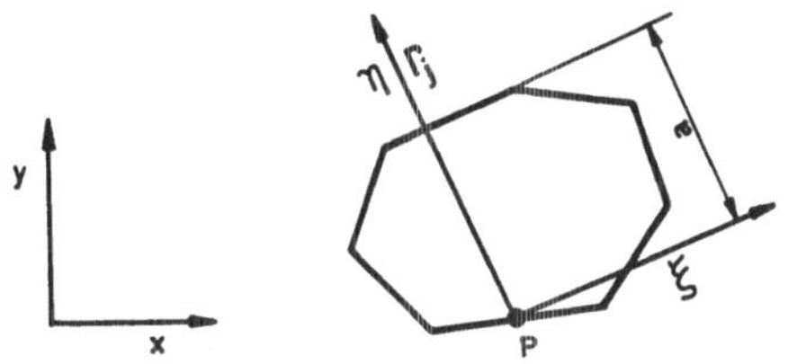

Figure 2. Transformation of the coordinate system

The integrals of equation (14) are of the type:

$$\int_{\Gamma_j} \ln(\xi^2 + a^2) d\xi \qquad \int_{\Gamma_j} \frac{a}{\xi^2 + a^2} d\xi \tag{.19}$$

For a linear boundary element, it is assumed a linear variation of u and $\partial u/\partial n$ in each segment:

$$u = N_1 u_1 + N_2 u_2$$
$$\frac{\partial u}{\partial n} = N_1 \frac{\partial u_1}{\partial n} + N_2 \frac{\partial u_2}{\partial n} \tag{.20}$$

where Ni are the unidimensional linear shape functions, and u_i and $\partial u_i/\partial n$ are the nodal values of u and $\partial u/\partial n$ at the extremes of the element. Thus, equation (13) becomes

$$c_i u_i - \sum_{j=1}^{N} \int_{\Gamma_j} u \frac{\partial}{\partial n} \ln r d\Gamma = - \sum_{j=1}^{N} \int_{\Gamma_j} \ln r \frac{\partial u}{\partial n} d\Gamma \qquad i = 1, 2, ...N \tag{.21}$$

and

$$c_i u_i - \sum_{j=1}^{N} \int_{\Gamma_j} [N_1 N_2] \frac{\partial}{\partial n} \ln r d\Gamma \left\{ \begin{array}{c} u_1 \\ u_2 \end{array} \right\} = - \sum_{j=1}^{N} \int_{\Gamma_j} [N_1 N_2] \ln r d\Gamma \left\{ \begin{array}{c} \frac{\partial u_1}{\partial n} \\ \frac{\partial u_2}{\partial n} \end{array} \right\} \tag{.22}$$

$$i = 1, 2, \ldots, N$$

where c_i is the constant c for node i.

The integrals of this equation can be evaluated using numerical and analytical integration. Equation (22) can be written in the same form as equation (15).

When the boundary is not smooth, equation (12) is not valid and the diagonal values of matrix **H** are calculated analytically or from rigid body considerations [35]:

$$\mathbf{H} \ \mathbf{I} = \mathbf{O} \qquad \mathbf{I} = \left\{ \begin{array}{c} 1 \\ \cdot \\ \cdot \\ 1 \end{array} \right\} \qquad \mathbf{0} = \left\{ \begin{array}{c} 0 \\ \cdot \\ \cdot \\ 0 \end{array} \right\} \tag{.23}$$

The boundary element method can also be applied to the solution of the Poisson equation (1). Following Fairweather, Rizzo, Shippy and Wu [35] the domain integral of equation (7) can be transformed to a boundary integral. Thus, the boundary element model for the solution of the Poisson equation does not need internal cells and all the calculations are at the boundary. The boundary integral equation for the torsion of shafts, in terms of the stress function, is:

$$c(P)z(P) - \int_{\Gamma} z \frac{\partial}{\partial n} \ln r d\Gamma = - \int_{\Gamma} \ln r \frac{\partial z}{\partial n} d\Gamma - \frac{1}{2} \int_{\Gamma} \frac{\partial}{\partial n} (r^2 (\ln r - 1)) d\Gamma \tag{.24}$$

This equation can be used for the development of boundary elements.

Full details of the boundary element method are presented in the books of Banerjee and Butterfield [35] and Brebbia, Telles and Wrobel [37].

The boundary element method for elasticity is based on Somigliana's identity [35]. This boundary integral equation is, with reference to the nomenclature of Figure 3, given by

$$C_{ij}(P)u_i(P) + \int_\Gamma u_j(Q)T_{ij}(P,Q)d\Gamma = \int_\Gamma t_j(Q)U_{ij}(P,Q)d\Gamma + \int_\Omega b_j(q)U_{ij}(P,q)d\Omega \qquad (.25)$$

where U_{ij} and T_{ij} are the fundamental Kelvin solutions for displacements and tractions, due to a unit concentrated force in an elastic infinite space: C_{ij} is the coefficient that depends on the geometry of the boundary at point **P**.

When body forces are not presented, equation (25) is only dependent on the boundary displacements and tractions. In this case, there is no need for internal cells in the domain. The boundary can be divided into N segments, or elements, with surfaces Γ_k, k = 1,..., N. Within each element, geometry, displacement, and traction fields can be assumed to be linear or quadratic, as shown in Figure 4.

Γ = Piecewise smooth boundary
Ω = Domain
$Q \in \Gamma$
$P \in \Gamma$
$q \in \Omega$
n = Outwards normal to Γ at Q
s = Tangencial direction
x_i = Coordinates of System
u_i = Displacements in x_i direction
t_i = Tractions on the Γ surface
b_i = Body forces

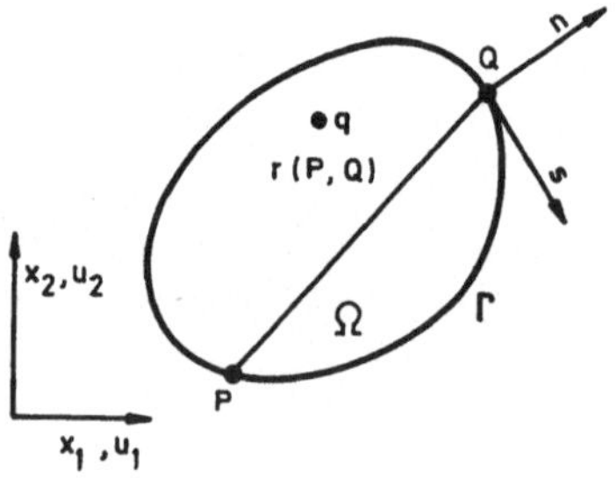

Figure 3. Nomenclature.

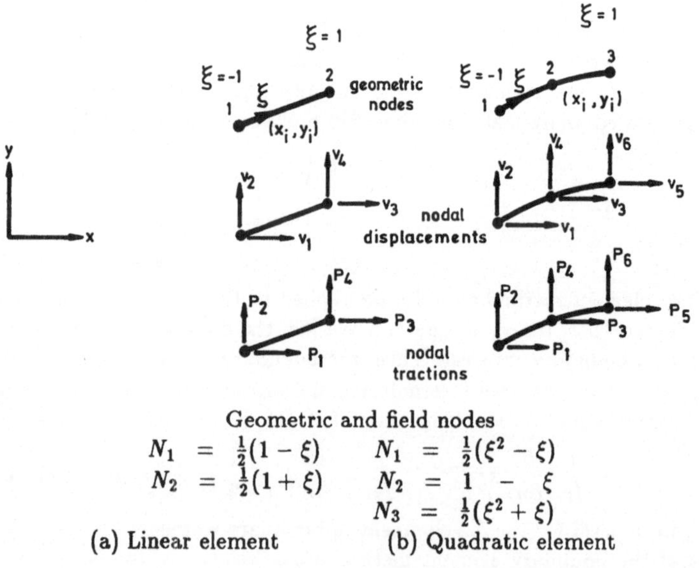

Geometric and field nodes

$$N_1 = \tfrac{1}{2}(1-\xi) \qquad N_1 = \tfrac{1}{2}(\xi^2-\xi)$$
$$N_2 = \tfrac{1}{2}(1+\xi) \qquad N_2 = 1 - \xi$$
$$N_3 = \tfrac{1}{2}(\xi^2+\xi)$$

(a) Linear element (b) Quadratic element

Figure 4. Linear and quadratic boundary elements for two-dimensional elasticity

Any variable within an element is assumed to be given by

$$\Theta(\xi) = \sum_{m=1}^{l} N_m(\xi)\Theta^m \tag{.26}$$

where $N_m(\xi)$ are the shape functions in local nondimensional coordinates $(-1 \leq \xi \leq 1)$. For the linear ($l = 2$) or quadratic element ($l = 3$) and Θ^m are the nodal values of the variable.

For problems without body forces, equation (25) becomes

$$C_{ij}(P_n)u_i(P_n) + \sum_{k=1}^{N} \int_{\Gamma_k} N_m(\xi)T_{ij}(P_n,Q)J(\xi)d\xi u_j^m = \sum_{k=1}^{N} \int_{\Gamma_k} N_m(\xi)U_{ij}(P_n,Q)J(\xi)d\xi t_j^m \tag{.27}$$

where u_j^m is the value of u_j at local node m, t_j^m is the value of t_j at local node m, and J is the Jacobian for the transformation of coordinates.

Nothing that P_n refers to a particular node, for all nodes equation (27) can be expressed in matrix form as

$$\mathbf{C}\ \mathbf{u} + \hat{\mathbf{H}}\ \mathbf{u} = \mathbf{G}\ \mathbf{t} \tag{.28}$$

$$\mathbf{H}\ \mathbf{u} = \mathbf{G}\ \mathbf{t} \tag{.29}$$

where $\mathbf{u}$ and $\mathbf{t}$ are the boundary nodal displacements and tractions. The elements of matrices $\hat{\mathbf{H}}$ and $\mathbf{G}$ can be obtained from the integrals

$$\hat{H}_{ij}^{nm} = \int_{-1}^{1} N_m(\xi)T_{ij}(P_n,Q(\xi))J(\xi)d\xi \tag{.30}$$

$$G_{ij}^{nm} = \int_{-1}^{1} N_m(\xi)U_{ij}(P_n,Q(\xi))J(\xi)d\xi \tag{.31}$$

The strong singular integral of equation (30) and the corresponding coefficients C_{ij} can be evaluated by rigid body considerations [35].

The weak singular integrals of equation (31) lead to integrals of the type

$$\int_{-1}^{1} \ln\frac{1}{r} f(\xi)d\xi \tag{.32}$$

which can be transformed to

$$\int_{-1}^{1} \ln\frac{1}{r} f(\xi)d\xi = \int_{-1}^{1} \ln(\frac{1+\xi}{2r}) f(\xi)d\xi + 2\int_{0}^{1} \ln\frac{1}{\varsigma} f(\varsigma)d\varsigma \tag{.33}$$

where

$$\varsigma \doteq \frac{1}{2}(1+\xi) \tag{.34}$$

The first integral on the right side of equation (33) is evaluated using standard Gaussian quadrature with four integration points, while this second integral is calculated numerically by formulas given by Banerjee and Butterfield [35].

OPTIMIZATION OF THE GEOMETRY OF SOLID SHAFTS

The design objective is to choose the shape of a solid shaft, with a given cross-sectional area and subjected to constraints on the design variables, which has maximal torsional stiffness. Full details of the theory is presented by Haug, Choi and Komkov [2].

With reference to Figure 5, let z be the stress function of the torsion problem.

Γ = Boundary
Ω = Domain
x,y = Coordinate system
n = Normal to the boundary
b_i = Design variables
N = Number of design variables
θ = $2\pi/N$
A = Given area of shaft

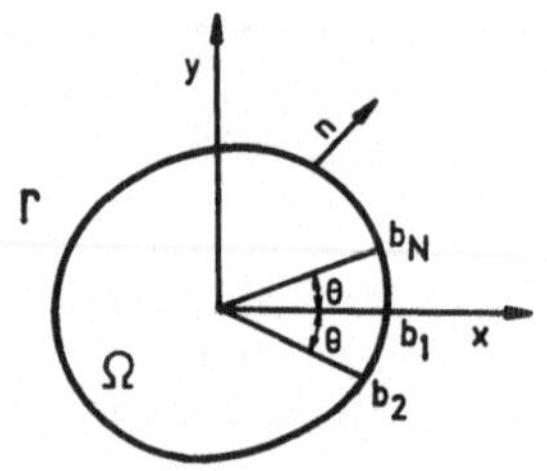

Figure 5. Design variables of solid shaft.

The torsional rigidity is given by the negative of

$$\psi_0 = -\int_\Omega 2z d\Omega \tag{.35}$$

which using Green's identity can be transformed to:

$$\psi_0 = -\int_\Omega (x^2 + y^2) d\Omega + \int_\Gamma u \frac{\partial u}{\partial n} d\Gamma \tag{.36}$$

For all simply connected domains, the problem is expressed as finding Ω to minimize ψ_0 subject to the constraint

$$\psi_1 = \int_\Omega d\Omega - A = 0 \tag{.37}$$

and to prescribed constraints on the design variables.

It has been shown by Haug, Choi and Komkov [2], by using total derivatives and variational calculus, that the variations of the objective and constraint functionals in terms of the design variations are given by

$$\delta\psi_0 = -\int_\Gamma \left(\frac{\partial z}{\partial n}\right)^2 v_n d\Gamma = -\int_\Gamma \left(\frac{\partial u}{\partial n} - x n_x - y n_y\right)^2 v_n d\Gamma \tag{.38}$$

$$\delta\psi_1 = \int_\Gamma v_n d\Gamma \tag{.39}$$

where n_x and n_y are the direction cosines of the boundary and v_n is the "normal perturbation" of the boundary.

With reference to Figure 6 and assuming that the boundary is divided into constant or linear boundary elements, the first order approximation of the sectorial area change due to a perturbation in the design variables is given by

$$\int_{\Gamma_i} v_n d\Gamma = \frac{1}{2} sin\theta_i (b_i \delta b_j + b_j \delta b_i) \tag{.40}$$

b_i = Design variable
Γ_i = Boundary of element
δb_i = Variation of design variable

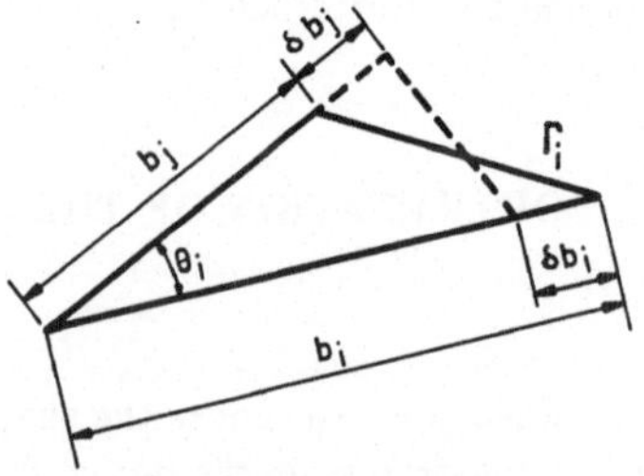

Figure 6. Perturbation of the boundary of an element.

It should be observed that the equations (36-39) are boundary integrals of the state variable and domain integrals of the geometry of the shaft. These are the necessary equations for the solution of the nonlinear programming problem by the Pshenichny's linearization method [38]. Also, for the solutions of the Laplace and Poisson torsion equations by the boundary element method, it is only necessary to evaluate the state variable at the boundary. Thus for the shape optimal design of solid shafts all the calculation of the state variable are at the boundary.

OPTIMIZATION OF THE GEOMETRY OF HOLLOW SHAFTS

The design objective is to maximize the rigidity of an hollow shaft with a known hole and a given cross-sectional area and subjected to geometrical constraints.

With reference to Figure 7, the state equation in terms of the stress function is:

$$\begin{aligned} \nabla^2 z &= -2 \quad in \quad \Omega \\ z &= 0 \quad on \quad \Gamma \\ z &= z_0 \quad on \quad \Gamma_0 \end{aligned} \tag{.41}$$

$$\int_{\Gamma_0} \frac{\partial z}{\partial n} d\Gamma = 2\Omega_0$$

where z_0 is a constant to be determined. This Poisson equation can be transformed into a Laplace equation using equation (2).

Γ = Outer boundary
Ω = Domain
n = Normal to outer boundary
Γ_0 = Inner boundary
Ω_0 = Area of hole
n_0 = Normal to inner boundary
x,y = Coordinate system
b_i = Design variables
N = Number of design variables
A = Area of shaft
$\theta = 2\pi/N$

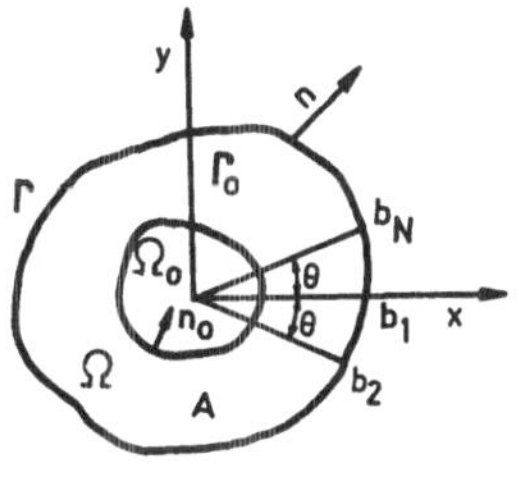

Figure 7. Design variables of hollow shafts.

The torsional rigidity is given by the negative of

$$\psi_0 = -\int_{\Omega} 2z d\Omega - 2z_0\Omega_0 \tag{.42}$$

which, using Green's theorem, can be transformed to

$$\psi_0 = -\int_{\Omega} (x^2 + y^2) d\Omega + \int_{\Gamma} u \frac{\partial u}{\partial n} d\Gamma + \int_{\Gamma_0} u \frac{\partial u}{\partial n} d\Gamma \tag{.43}$$

The area constraint for the hollow shaft is identical to equation (37). Following Gelfand and Fomin [39], it can be shown that the first variation of the objective function of the hollow shaft is identical to the first variation for the solid shaft (38).

It should notice that equations (43) is a boundary integral of the state variable and a domain integral of the geometry of the shaft. Thus for the shape optimal design of hollow shafts it is only necessary to evaluate the state variable at the boundaries.

SHAPE OPTIMAL DESIGN OF STRUCTURES BASED ON MINIMUM COMPLIANCE

The design objective is to find the shape of an unloaded boundary of a specified structure, with a given area and subjected to constraints in the design variables, which has minimum compliance.

Consider an elastic body that is rigidly supported on a boundary Γ_0 and loaded by tractions on boundary Γ_1(see Figure 8). Also let the design boundary Γ_2 be free from loading. It is assumed that there are no body forces. The nomenclature used is shown in Figure 8.

Γ_0= Boundary where displacement are zero
Γ_1 = Boundary where tractions are at t_i^0
Γ_2 = Design boundary (unloaded surface)
Ω = Domain; Γ_0 U Γ_1 U Γ_2
u_1 = Displacements
t_i = Tractions
x_i = Coordinates

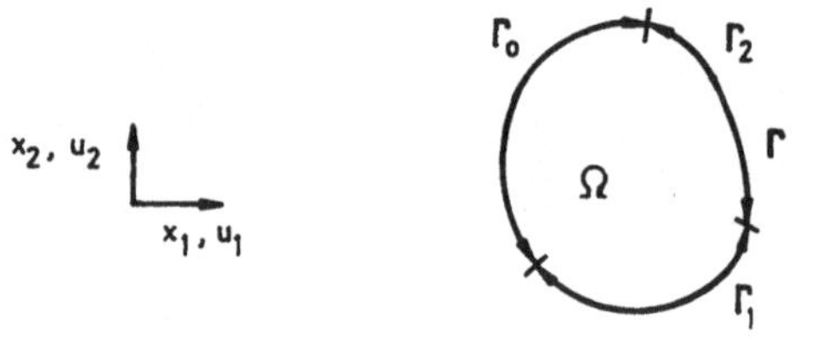

Figure 8. Definition of domain.

The objective function is the compliance of the structure, which is given by

$$\psi_0 = \frac{1}{2}\int_\Gamma t_i u_i d\Gamma \tag{.44}$$

The optimization problem is expressed as finding Γ_2 to minimize ψ_0 subjected to the area constraint

$$\psi_1 = \int_\Omega d\Omega - A \leq 0 \tag{.45}$$

where **A** is the given area of the domain.

To solve this nonlinear programming problem numerically, it is necessary to evaluate the first variation of the objective and constraint functionals. A general formulation for sensitivity analysis of volume and boundary functionals is presented by Haug, Choi and Komkov [2]. For unloaded design boundaries of linear materials the first variation of the objective functional of equation (44) becomes

$$\delta\psi_0 = -\int_{\Gamma_2} U v_n d\Gamma \tag{.46}$$

where U is the strain energy density and v_n is the "normal perturbation" of the boundary. The variation of the constraint functional of equation (45) is

$$\delta\psi_1 = \int_{\Gamma_2} v_n d\Gamma \tag{.47}$$

The boundary can be divided into N linear and/or quadratic boundary elements. It is assumed that the design boundary is represented by M geometrical linear elements. With reference to Figure 6, the first order approximation of area change due to a small perturbation in the design variables is given by

$$\int_{\Gamma_i} v_n d\Gamma = \frac{1}{2} sin\theta_i (b_i \delta b_j + b_j \delta bi) \quad (.48)$$

The variation in compliance given by equation (46) can be accurately and efficiently calculated by the boundary element method. For linear elements, the strain energy density is constant on each element. Thus, equation (46) becomes

$$\delta\psi_0 = -\sum_{m=1}^{M} U_m \int_{\Gamma_m} v_n d\Gamma \quad (.49)$$

where U_m is the strain energy density of a boundary element. This energy can be evaluated from the boundary tangential stress. For an unloaded design boundary, the only non-zero stress component is the tangential stress.

For shape optimal design, all equations that are necessary (equations (44)- (47)) to implement the Pshenichny linearization method [38] of nonlinear programming are boundary functionals of the displacement, tractions and stresses, and domain functionals of the geometry. Consequently, shape optimal design of structures, based on minimum compliance, can be efficiently and accurately solved using the boundary element method to discretize the structure. Also, there is no need of calculating displacement and stress in the domain.

SHAPE OPTIMAL DESIGN OF STRUCTURES BASED ON STRESS CONSTRAINTS

The design objective is to find the shape on an unloaded boundary for minimum area, subjected to constraints on the stresses and design variables. The nomenclature of Figure 8 is used.

The objective function

$$\psi_0 = \int_\Omega d\Omega \quad (.50)$$

should be a minimum. The variation of this area functional is given by equation (47).

The stress should be less than the allowable stress σ_a in the domain or boundary. For plane stress or strain problems with smooth boundaries and without body forces, it can be proved [40] that the maximum Von Mises stress is always at the boundary. The Von Mises yield stress constraint functional, average over a small region Ω_k, defined in Figure 9, can be represented by

$$\psi_k = \int_\Omega \phi m_k d\Omega = \frac{\int_{\Omega_k} \phi d\Omega}{\int_{\Omega_k} d\Omega} \quad (.51)$$

where

$$\phi = \frac{\sigma_y - \sigma_a}{\sigma_a} \quad (.52)$$

$$\sigma_y = \sqrt{\sigma_{11}^2 + \sigma_{22}^2 + 3\sigma_{12}^2 + \sigma_{11}\sigma_{22}} \quad (.53)$$

and σ_{ij} are the components of the stress tensor. In equation (51) m_k is a characteristic function defined as

$$m_k = \overline{m}_k = \frac{1}{\int_{\Omega_k} d\Omega} \; in \; \Omega_k \qquad m_k = 0 \; in \; \Omega \quad \Omega_k \quad (.54)$$

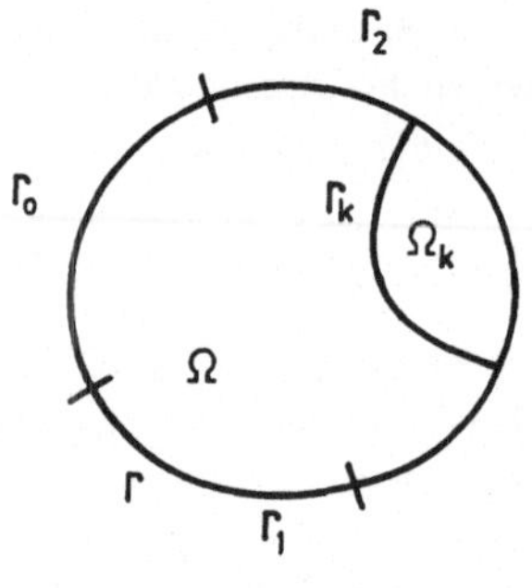

Figure 9. Domain of the stress constraint.

Following Haug, Choi and Komkov [2], the variation of equation (51) is given by

$$\delta\psi_k = -\int_{\Gamma_2} \sigma_{ij}\epsilon_{ij}(\lambda)v_n d\Gamma + \overline{m}_k \int_{\Gamma_k} (\phi - \psi_k)v_n d\Gamma \tag{.55}$$

where $\epsilon_{ij}(\lambda)$ are the components of the strain tensor of the adjoint structure. The adjoint structure is identical to the real structure, but with different loading conditions. For each constraint, the adjoint problem is defined by the equilibrium equation:

$$\sigma_{ij,j}(\lambda) + F_i^* = 0 \tag{.56}$$

where λ is the adjoint variable, and F_i^* are the adjoint body forces

$$F_i^* = -\sum_{j=1}^{2}(\sum_{k,l=1}^{2} \frac{\partial\phi}{\partial\sigma_{kl}} D_{klij} m_P), j \tag{.57}$$

where D_{klij} are the stress/strain relations. The boundary conditions of the adjoint structures are:

$$\begin{array}{l} \lambda_i{=}0 \;\; on \;\; \Gamma_0 \\ \sigma_{ij}n_j{=}0 \;\; on \;\; \Gamma_1 \;\; U \;\; \Gamma_2 \setminus \Gamma_k \\ \sigma_{ij}n_j{=}t_i^* \;\; on \;\; \Gamma_k \end{array} \tag{.58}$$

where t_i^* are the adjoint tractions

$$t_i^* = \sum_{j=1}^{2}(\sum_{k,l=1}^{2} \frac{\partial\phi}{\partial\sigma_{kl}} D_{klij} m_P)n_j \tag{.59}$$

and n_j are the normal components to the boundary.

It should be noticed that the sensitivity equation (55) depends on stress at the boundary of the real structure and on strains at the boundary of the adjoint structures. Also, there is no need of calculating displacements and stresses of the real and adjoint structures in the domain

Thus, the boundary element method should be efficient and accurate in the shape optimal design of structures based on minimum area and stress constraints.

It is assumed that the design boundary is represented by quadratic elements with straight geometries. For the boundary element model, the small area Ω_k is defined in Figure 10.

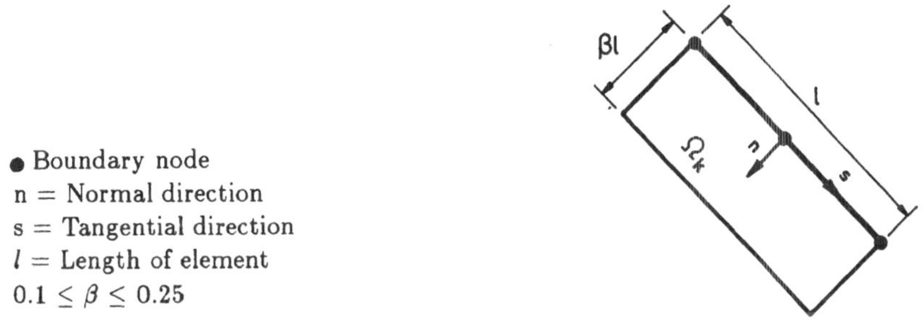

Figure 10. Domain of the stress constraint for the boundary element model

Also within Ω_k the tangential stress σ_{ss} and strain, and the shear σ_{ns} and normal σ_{nn} stresses are linear in the s direction, but constant in the n direction. The shear and normal stresses are only dependent of the boundary tractions. These assumptions are only accurate, if the parameter β in the boundary element model is small, preferably much less than 0.25.

With these approximations, the distributed adjoint body forces are zero. Also, the concentrated adjoint body forces are identical to the adjoint boundary tractions. In Figure 11 it is represented the loading conditions of the adjoint structure.

The application of the boundary element method to calculate the stresses and strains of the adjoint structure is almost standard. However, it should consider the integration of the pseudo tractions in surface 2, 3 and 4 of Figure 11. Because of the almost singularity of the integrals when i ϵ Ω_k it is necessary to integrate the pseudo adjoint tractions with nine Gaussian points. For this reason the parameter β is the boundary element model should be larger than 0.1.

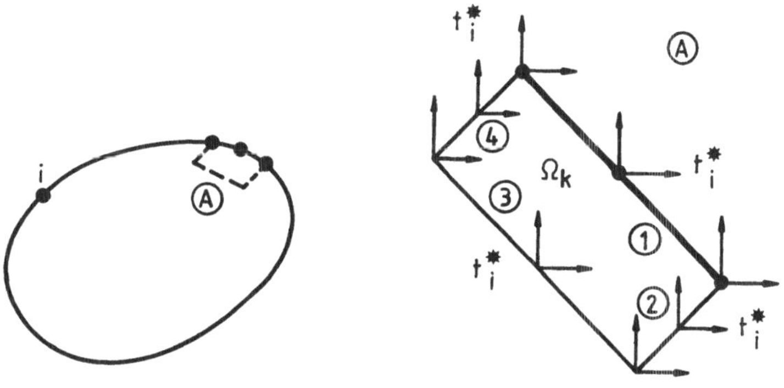

Figure 11. Adjoint loading for stress constraint.

BOUNDARY STRESSES

The accuracy of the boundary stresses ans strains is crucial for the shape optimal design of elastic structures. It is well known that the stresses at the boundary are more accuratelly calculated by the boundary element method than the finite element technique.

Contrary to the stresses at interior points, the stresses at the boundary are not calculated directly by the boundary element method. In reference [35] a technique for calculating boundary stresses is presented. However, this technique has been found not to be accurate as design changes [13]. Another technique has been developed by Hartmann [41], which calculates accurate boundary stresses [13].

The boundary element method calculates accuratelly the displacement u_i and the tractions t_i at the boundary. Also the derivative of the displacements with respect to tangential coordinate s $(\partial u_i/\partial s)$ can be calculated accuratelly.

The stresses at the boundary must also obey the Hooke and Cauchi Laws. Thus the following equations can be derived:

$$\sigma_{ij} = \frac{2\mu\nu}{1-2\nu}\delta_{ij}\epsilon_{kk} + 2\mu\epsilon_{ij} \tag{.60}$$

$$t_i = \sigma_{ij} n_j \tag{.61}$$

$$\frac{\partial u_i}{\partial s} = \frac{\partial u_i}{\partial x_j}\frac{\partial x_j}{\partial s} \tag{.62}$$

where μ is the rigidity modulus, ν is the Poisson ratio, δ_{ij} is the Kronecker delta, and n_j are the normal components of the boundary.

For two dimensional elasticity these equations form a system of 7 algebric equations:

$$\begin{bmatrix} 1 & 0 & 0 & a & 0 & 0 & -\lambda \\ n_1 & 0 & n_2 & 0 & 0 & 0 & 0 \\ 0 & n_2 & n_1 & 0 & 0 & 0 & 0 \\ 0 & 1 & 0 & -\lambda & 0 & 0 & a \\ 0 & 0 & 0 & -n_2 & 0 & n_1 & 0 \\ 0 & 0 & 1 & 0 & -\mu & -\mu & 0 \\ 0 & 0 & 0 & 0 & -n_2 & 0 & n_1 \end{bmatrix} \begin{Bmatrix} \sigma_{11} \\ \sigma_{22} \\ \sigma_{12} \\ u_{1,1} \\ u_{2,1} \\ u_{1,2} \\ u_{2,2} \end{Bmatrix} = \begin{Bmatrix} 0 \\ t_1 \\ t_2 \\ 0 \\ u_{1,s} \\ 0 \\ u_{2,s} \end{Bmatrix} \tag{.63}$$

where

$$a = -\lambda - 2\mu \qquad \lambda = \frac{2\mu\, nu}{1-2\nu} \tag{.64}$$

The solution of equation (63) at any boundary point gives the boundary stresses. Since the right hand side of this equation is calculated accuratelly by the boundary element method, the boundary stresses are accurate. It should be noticed that equation (63) is only valid for smooth boundaries.

AUTOMATIC MESH REFINEMENT

Although the boundary element discretization is able to adapt itself to a new configuration without major distorsion of the boundary elements, it is convenient to have an automatic generator of boundary element meshes. Contrary to the adaptive finite element method which had been developed for more than a decade, the adaptive boundary element technique has just been iniciated. Rank [42-43] have developed an h-adaptive boundary element method using a local indicator based on a-posteriori estimated error derived from the Galerkin boundary element method presented by Hsiao and Wendland [44]. The author proved that uniform refinement presents a lower convergence rate than an adaptive refinement. The main drawback of the technique developed is that it can not be applied to the commonly used boundary element method.

Rencis and Mullen [45-47] have presented another h-adaptive method based on assymptotic error estimate in terms of the characteristic dimension of the uniform mesh This technique applied at local level allows the calculation of the number of elements necessary for a given accuracy. The authors have applied this technique to potencial and to elasticity problems.

A p-adaptive boundary element method has been developed by Alarcon and colaborators [48-51] using a modified criterion defined for finite element analysis by Peano et al. [52]. The hierarchical process terminates when global equilibrium is achieved. The authors concluded that the use of hierachical functions are effficient since the introduction of new degrees of freedom do not destroy the previous system matrices. The technique has been applied to potential problems with some advantages.

In this paper a technique for automatic boundary element mesh refinement is introduced based on a global and local equilibrium criterion.

In a good discretization global equilibrium is achieved. For systems without body forces, the integrals

$$\int_{\Gamma} t_i \; d\,\Gamma \qquad i = 1,2 \tag{.65}$$

are almost null. If the integrals of equation (65) are more than a prescribed value means that equilibrium has not been obeyed and consequently discretization is not acceptable. In this case a global refinement is performed, being each element divided into two elements. This procedure is used until global equilibrium is obtained.

After global equilibrium is achieved a local criterion is applied based on the continuity of the tangencial stresses for unloaded boundaries. The tangencial stress calculated for the same node in two adjacent elements should be almost the same; if the difference is greater than an acceptable value, the adjacent elements are divided into two elements. The number of divisions is limited to avoid numerical integration instabilities.

The application of this technique to the gear tooth presented by Lachat [51] shows the effectiveness of the process for problems with large stress gradient. An acceptable discretization was achieved with one global and three local divisions (Figure 12).

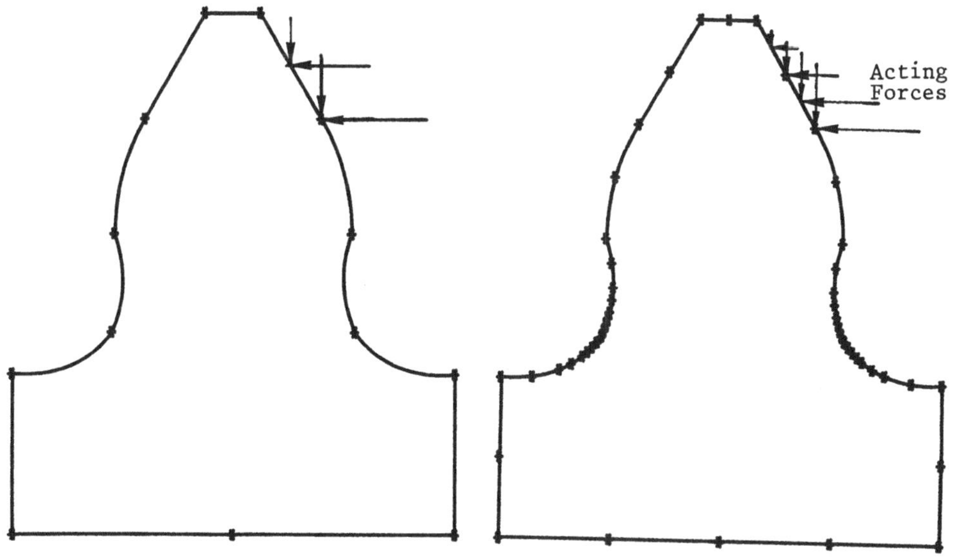

Figure 12. Adaptive process of boundary element mesh for gear tooth.

APPLICATIONS

The models developed are applied to the shape optimal design of several shafts. In all applications, the boundary element model is based on the boundary integral formulation (13) of the Laplace equation (3).

The constant element model is applied to the shape optimal design of a shaft presented by Choi, Haug, Hou and Sohoni [54]. The design objective is to choose the shape of a shaft with a cross-section of 14.0, that must fit a square housing of 16.0 and has a maximal torsional stiffness. This problem has been solved using finite elements and the model is shown in Figure 13.a. The discretization used for the boundary element formulation is shown in Figure 13.b.

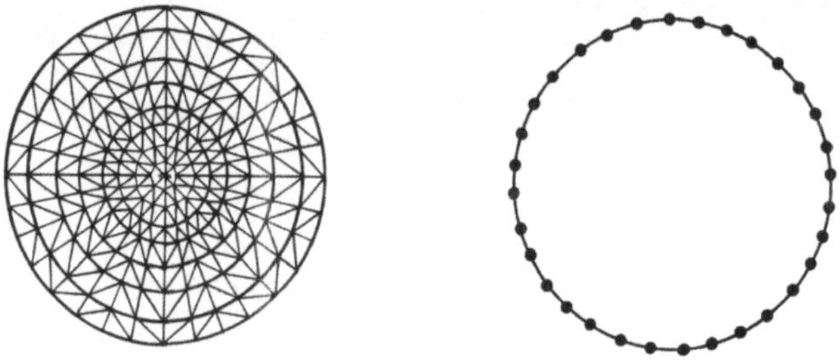

384 linear finite elements | 32 constant boundary elements
209 degrees of freedom [54] | 32 degrees of freedom
(a) Finite element model | (b) Boundary element model

Figure 13. Finite element and boundary element models

Table 1 shows the results obtained and compare them with a finite element solution. It can be concluded that the boundary element results converges faster than the finite element solution, because the nodal values of the normal derivative of the state variable at the boundary are more accurate.

Iteration number	b_1	b_2	b_3	b_4	b_5	Torsional Rigidity	Area
1	2.0000	2.0392	2.1648	2.1865	2.2109	31.0226	14.0906
2	2.0000	2.0392	2.1648	2.1669	2.2218	30.8578	14.0451
3	2.0000	2.0392	2.1648	2.1639	2.2212	30.8162	14.0338
4	2.0000	2.0392	2.1648	2.1593	2.2206	30.7538	14.0169
5	2.0000	2.0392	2.1648	2.1581	2.2206	30.7382	14.0126
6	2.0000	2.0392	2.1648	2.1544	2.2203	30.6912	14.0000
Finite Elements (30 iterations) [54]	2.0000	2.0392	2.1648	2.1675	2.1953	30.4541	14.0021

Initial Design: Circle with 2.2 Radius; b_i is the design variable at 11.25(i-1) degrees

TABLE 1 - Numerical results for optimal design of shaft

The boundary element constant model is also applied to the same problem but with different initial shapes. In all the applications the model converges to the correct shape, without redefining the boundary element mesh, in a few iterations, as shown in Figure 14 and 15.

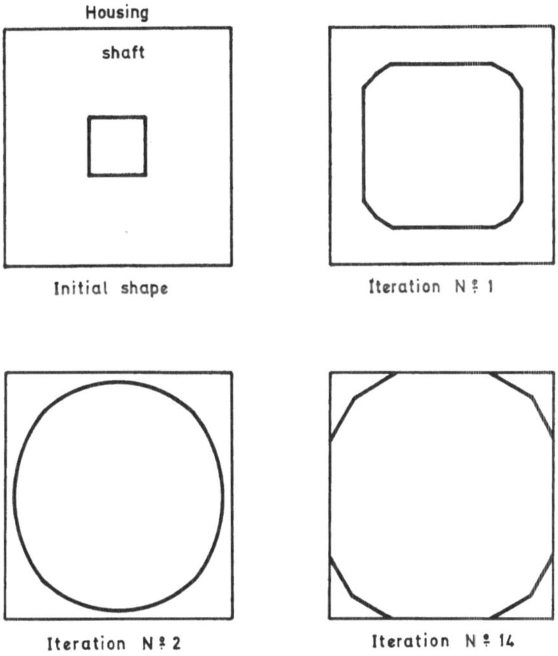

Figure 14. Modification of the geometry of shaft with iteration process: square initial shape.

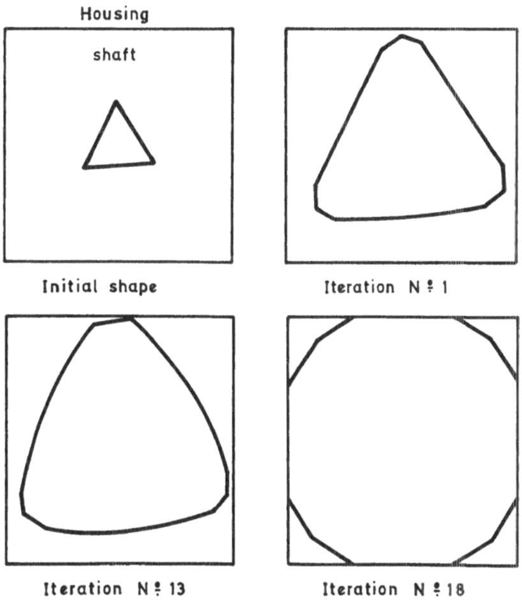

Figure 15. Modification of the geometry of shaft with iteration process: triangular initial shape.

The constant element model for hollow shafts is applied to the shape optimal design of several shafts. The discretization used is presented in Figure 16. All the shafts have an elliptical, square or round hole and an initial circular or elliptical outside boundary. Several cross-sectional areas of the shaft are considered. In all the applications, the final shape is found in a few iterations and without redefining the mesh. Some results are presented in Figure 17 and these values are in accordance with the exact solutions of Banichuk [40] and the finite element results of Hou, Haug and Benedict [55]. The iteration process for two particular shafts are shown in Figure 18 and 19.

32 design variables
64 constant boundary elements
64 degrees of freedom
● Boundary nodes

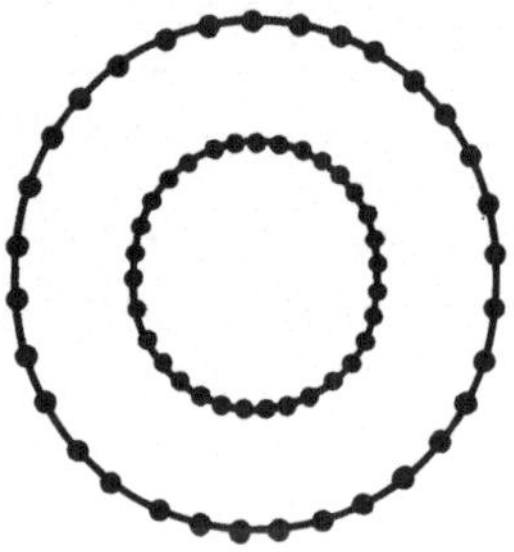

Figure 16. Boundary element model for hollow shafts.

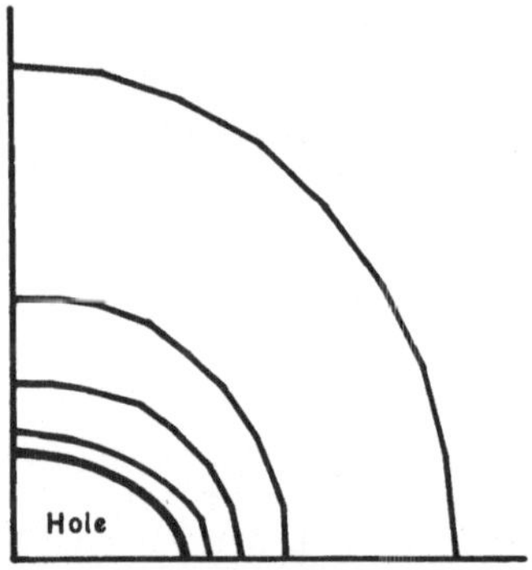

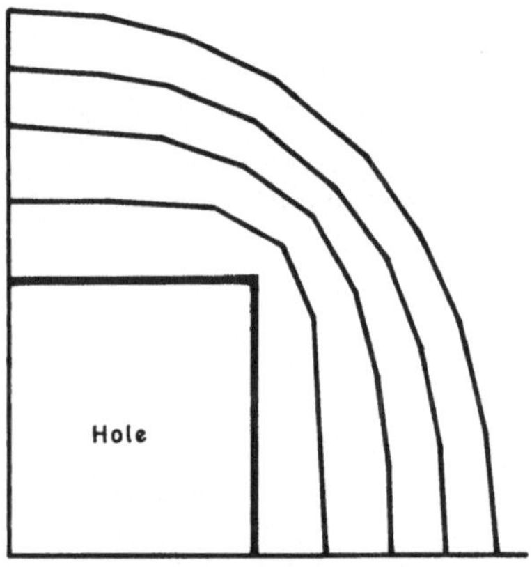

Figure 17. Hollow shafts results for several cross-section areas.

The linear boundary element model for hollow shafts is applied to the shape optimal design of shafts with an elliptical hole and subjected to constraints on the cross section area. The boundary element model has 48 degrees of freedom and 24 design variables. The results are in accordance with Figure 17.a.

The boundary element model for structures based on minimum compliance is applied for the shape optimal design of several elastic structural components. The square plate of Figure 20 is subject to uniform in-plane tensile loads along its four edges. The design objective is to fing the shape of an initially square hole, of area 1.0 percent of the plate, that minimizes compliance.

The boundary element model is represented in Figure 21. The model has 12 quadratic boundary elements, 12 linear design boundary elements, 13 design variables and 72 degrees of freedom. The final design is achieved after 7 iterations, 14 structural analysis, and 5 CPU minutes on a PRIME 750 super mini-computer. The evolution of the design of the hole is represented in Figure 22. It should be noticed that after only two iterations, the design is almost optimal.

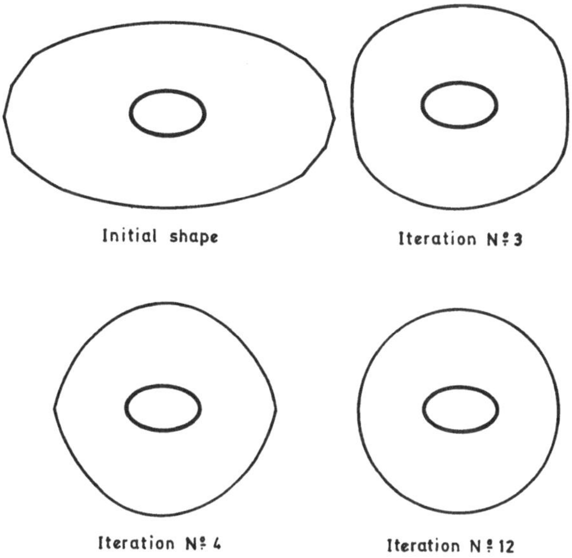

Figure 18. Modification of the shape of shaft with iteration process: elliptical hole.

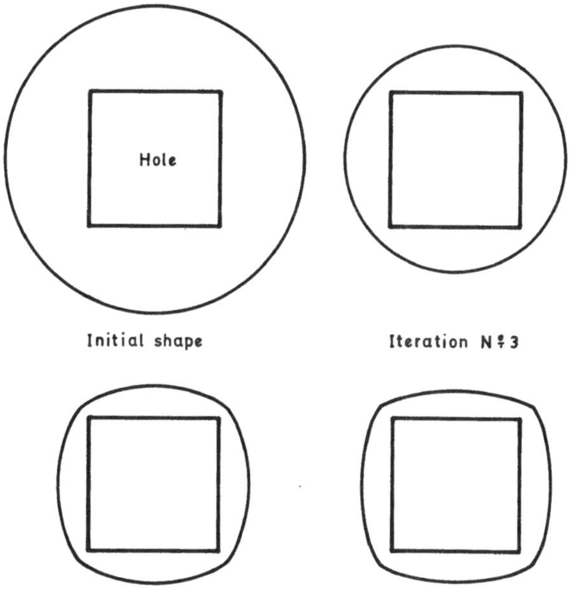

Figure 19. Modification of the shape of shaft with iteration process: square hole.

f_x, f_y = Biaxial of applied loads
Area of hole = 1.0
Plane stress
Poisson ratio = 0.3

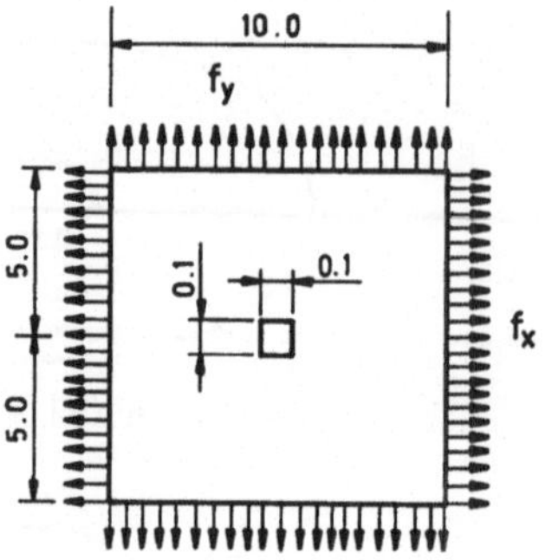

Figure 20. Square plate subjected to biaxial applied loads.

12 quadratic elements
72 degrees of freedom
13 design variables

design variables

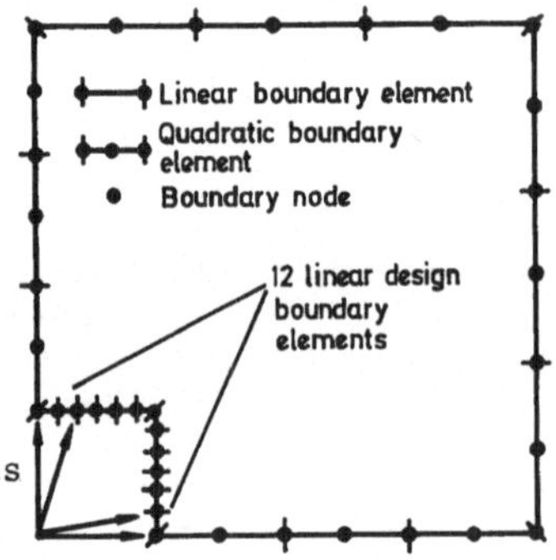

Figure 21. Boundary element model of one quarter of plate.

Convergence obtained
7 iterations
14 boundary element analyses
5 CPU minutes on PRIME 750

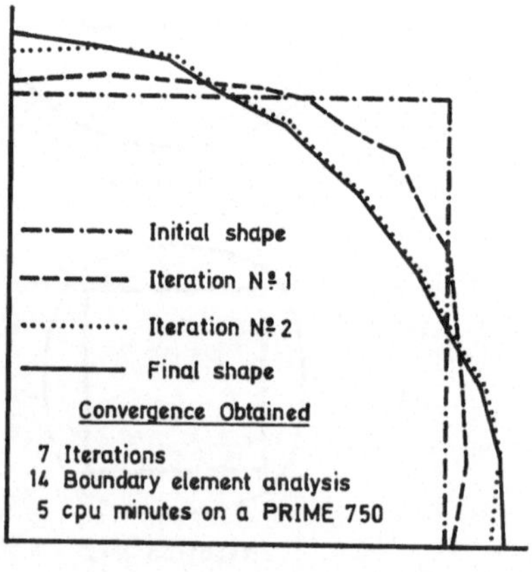

Figure 22. Evolution of design of hole with iteration process.

The final design of the hole is almost a circle, whose radius is constant to within 1.8 percent. Also, the tangential stress at the boundary of the hole is constant to within 1.0 percent. The stress concentration factor of the hole is 2.02, which an error of only 1.0 percent. The final design is almost identical to the analytical solution given by Banichuk [40].

The same problem is also solved using a simpler model with 8 quadratic elements, 8 linear design elements, 9 design variables, and 48 degrees of freedom. A practically identical design is achieved.

The formulation developed is applied to shape optimal design of the fillet shown in Figure 23. The boundary element model is shown in Figure 24. Starting from an initially straight line design and with a constraint on the area, the design process takes six iterations and eleven boundary element analysis to achieve the final design. The evolution of shape of the fillet is shown in Figure 25. The iterative process takes 1.5 CPU minutes on a PRIME 750 super mini-computer. The stress concentration factor of the final design is 1.37.

f_a, f_b= Applied loads
Γ_2 = Design boundary: unknown
Modulus of elasticity = 2.26 x $10^{11} N/m^2$
Poisson Ratio = 0.3

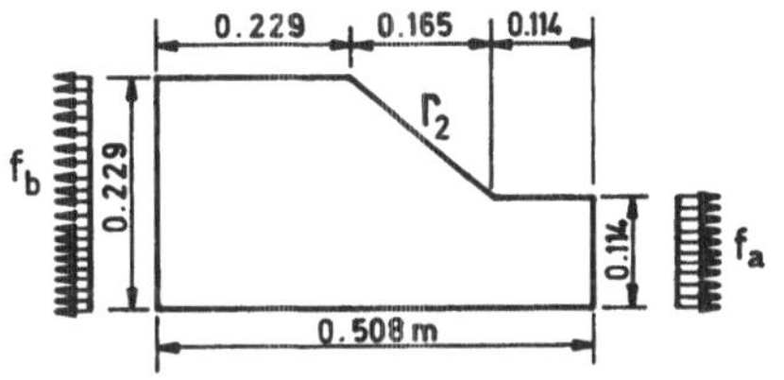

Figure 23. The fillet problem.

9 design variables
14 quadratic eelements
10 linear elements
76 degrees of freedom

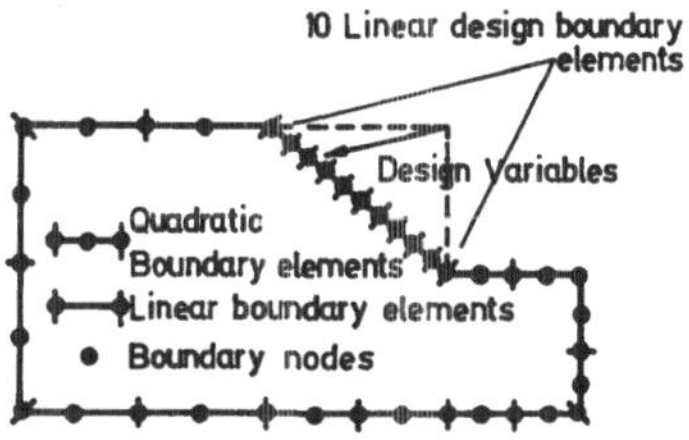

Figure 24. Boundary element model for fillet.

Number of iterations = 6
Number of analyses = 11
Time of computation = 1.5 CPU minutes
Stress concentration factor = 1.37

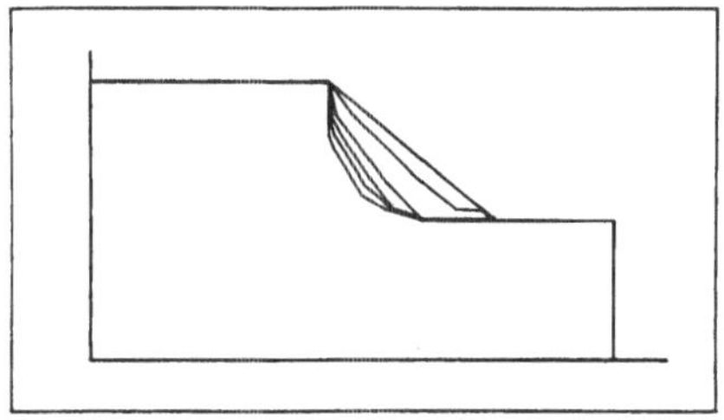

Figure 25. Evolution of design of fillet with iteration process.

The boundary element solution are similar to finite element results presented in references [56-57]. The number of degrees of freedom of the boundary element model, however, is about five times smaller than the equivalent finite element model. The computer time required by the boundary element method is a factor of approximately eight less than that required by the finite element method.

In each application with the compliance model, the final design is essentially achieved in a few iterations. However, due to the lack in sensitivity of compliance to a small perturbation in a remote boundary, the final iteration process converges slowly. This fact is a limitation of compliance as an objective function, especially for very stiff structures, for which the applied forces are very far from the design boundaries.

ACKNOWLEDGEMENT

This research was suportted by the Advisory Group for Aerospace Research & Development Project AGARD-SMP-P31 and National Science Foundation Project No. CEE 83-19871.

REFERENCES

1. O. Pironnneau, *Optimal Shape Design for Elliptical Systems*, Springer-Verlag, (1984).

2. E.J. Haug, K.K. Choi and V. Komkov, *Design Sensitivity Analysis of Structural Systems*, Academic Press, (1986).

3. O.C. Zienkiewicz and J.S. Campbell, Shape Optimization and Sequential Linear Programming, *Optimum Structural Design* (Ed. R.H. Gallagher and O.C. Zienziewicz), Wiley, 109-126 (1973).

4. C.A. Mota Soares, H.C. Rodrigues, L.M. Oliveira Faria and E.J. Haug, Optimization Of the Geometry of Shafts Using Boundary Elements. *ASME Journal of Mechanisms, Transmissions and Automation in Design*, 106, 199-203 (1984).

5. C.A. Mota Soares, H.C. Rodrigues, L.M. Oliveira Faria and E.J. Haug, Optimization of the Shape of Solid and Hollow Shafts Using Boundary Elements, *Boundary Elements* (Ed. C.A. Brebbia), Springer- Verlag, 883-889 (1983).

6. C.A. Mota Soares, H.C. Rodrigues, L.M. Oliveira Faria and E.J. Haug, Boundary Elements in Shape Optimal Design of Shafts, *Optimization in Computer Aided Design* (Ed. J.S. Gero), North- Holland, 155-175 (1985).

7. H.C. Rodrigues and C.A. Mota Soares, Shape Optimization of Shafts, *Third National Congress of Theoretical and Applied Mechanics*, Lisbon, (1983) (in Portuguese).

8. H.C. Rodrigues, *Shape Optimization of Shafts Using Boundary Elements*, M.Sc. Thesis, Technical University of Lisbon, (1984) (in Portuguese).

9. T. Burczynski and T. Adamczyk, Multiparameter Shape Optimization of a Bar in Torsion by the Boundary Element Method, *Proc. 23rd Symposium on Modelling in Mechanics*, The Polish Society of Theoretical and Applied Mechanics, Gliwice, (1984) (in Polish).

10. A. Zochowski and K. Mizukami, A Comparison of BEM and FEM in Minimum Weight Design, *Boundary Elements* (Ed. C.A. Brebbia), Springer- Verlag, 901-911 (1983).

11. C.A. Mota Soares, H.C. Rodrigues and K.K. Choi, Shape Optimal Design of Elastic Structural Components Using Boundary Elements, *Tenth International Congress on the Applications of Mathematics in Engineering Science*, Weimar, 5, 80-82 (1984).

12. C.A. Mota Soares, H.C. Rodrigues and K.K. Choi, Shape Optimal Structural Design Using Boundary Elements and Minimum Compliance Techniques, *ASME Journal of Mechanisms, Transmissions and Automation in Design*, 106, 518-523 (1984).

13. R.P. Leal,*Boundary Elements in Bidimensional Elasticity*, M.Sc. Thesis, Technical University of Lisbon, (1985) (in Portuguese).

14. C.A. Mota Soares and K.K. Choi. Boundary Elements in Shape Optimal Design of Structures, *The Optimum Shape-Automated Structural Design* (Editor J.A. Bennett and M.E. Botkin), Plenum Press, 199 - 236 (1986).

15. T. Burczynski and T. AdamczyK, The Application of the Boundary Element Method to Optimal Design of Shape of Structures, *Proc. 4th Conference on Methods and Instrumentations of Computer Aided Design*, Warsaw, (1983) (in Polish).

16. T. Burczynski and T. Adamczyk, The Boundary Element Formulation for Multiparameter Structure Shape Optimization, *Applied Mathematical Modelling*, 9, 195-200 (1985).

17. T. Burczynski and T. Adamczyck, The Boundary Element Method for Shape Design Synthesis of Elastic Structures, *Seventh International Conference on Boundary Element Methods* (Ed. C.A. Brebbia), Springer- Verlag, (1985).

18. T. Burczynski and T. Adamczyck, *Generation of Optimal Shape of Structures by Means of the Boundary Element Method*, Zeszyty Naukowe Politechniki Slaskiej, Mechanika Z.82, Slaska, (1985) (in Polish).

19. T. Burczynski and T. Adamczyck, The Boundary Element Procedure for Dependence of Eigenvalues with Respect to Stochastic Shape of Elastic Systems, *Proc. 25th Symposium in Mechanics*, Gliwice, (1986).

20. D. Eizadian, *Optimization of The Shape of Bidimensional Structures by the Boundary Integral Equation Method*, Ph.D. Thesis, National Institute of Applied Science of Lyon, (1984) (in French).

21. D. Eizadian and M. Trompette, Shape of Bidimensional Structures by the Boundary Element Method, *Conference on CAD/CAM, Robotics and Automation in Design*, Tucson, Arizona, (1985).

22. J.H. Kane, *Optimization of Continuum Structures Using a Boundary Element Formulation*, Ph.D. Thesis, University of Connecticut, (1986).

23. J.H. Kane, Shape Optimization Utilizing a Boundary Element Formulation, to appear.

24. T. Futagami, Boundary Element and Linear Programming Method in Optimization of Partial Differential Systems, *Boundary Element Methods* (Ed. C.A. Brebbia), Springer-Verlag, 457-471 (1981).

25. T. Futagami, Boundary Element and Dynamic Programming Method in Optimization of Transient Partial Differential Systems, *Boundary Elements Methods in Enginnering* (Ed. C .A. Brebbia), Springer-Verlag, 58-71 (1982).

26. T. Futagami, Boundary Element Method - Finite Element Method Coupled with Linear Programming for Optimal Control of Distributed Parameter Systems, *Boundary Elements* (Ed. C.A. Brebbia), Springer-Verlag, 891-900 (1983).

27. M.R. Barone and D.A. Caulk, Optimal Arrangement of Holes in a Two- Dimensional Heat Conductor by a Special Boundary Integral Method, *International Journal of Numerical Methods in Engineering*, 18, 675-685 (1982).

28. R.A. Meric, Boundary Integral Equation and Conjugate Gradient Methods for Optimal Boundary Heating of Solids, *International Journal of Heat and Mass Transfer*, 26, 261-264 (1983).

29. R.A. Meric, Boundary Element Methods for Static Optimal Heating of Solids, *ASME Journal of Heat Transfer*, 106, 876-880 (1984).

30. R.A. Meric, Boundary Element Methods for Optimization of Distributed Parameter Systems, *International Journal of Numerical Methods in Engineering*, 20, 1291-1306 (1984).

31. R.A. Meric, Boundary Elements in Shape Design Sensitivity Analysis of Thermoelastic Solids, *Computer Aided Optimal Design: Structural and Mechanical Systems* (Ed. C.A. Mota Soares), Springer-Verlag, (1987).

32. R.A. Meric, Shape Design Sensitivity Analysis for Nonlinear Anisotropic Heat Conducting Solids and Shape Optimization by BEM, *International Journal of Numerical Methods in Engineering*, to appear.

33. B.M. Kwak and J.H. Choi, Shape Design Sensitivity Analysis Using Boundary Integral Equation for Potential Problems,*Computer Aided Optimal Design: Structural and Mechanical Systems*, (Ed. C.A. Mota Soares), Springer-Verlag, (1987).

34. J.H. Choi and B.M. Kwak, Boundary Integral Equation Method for Shape Design Sensitivity Analysis, *Journal of Structural Mechanics*, to appear.

35. P.K. Banerjee and R. Butterfield, *Boundary Element Methods in Engineering Science*, Mcgraw-Hill, (1981).

36. G. Fairweather, F.J. Rizzo, D.J. Shippy and Y.S. Wu, On the Numerical Solution of Two-Dimensional Potential Problems by an Improved Boundary Integral Equation Method, *Journal of Computational Physics*, 31, 96-112 (1979).

37. C.A. Brebbia, J. Telles and L. Wrobel, *Boundary Element Techniques: Theory and Applications in Engineering*, Springer-Verlag, (1984).

38. B.N. Pshenichny and J.M. Danilin, *Numerical Methods in Extremal Problems*, MIR, (1978).

39. I.M. Gelfand and S.W. Fomin, *Calculus of Variations*, Prentice- Hall, (1961).

40. N.W. Banichuk, *Problems and Methods of Optimal Structural Design*, Plenum Press, (1983).

41. F. Hartmann, Elastostatics, *Progress in Boundary Element Methods Volume I* (Ed. C.A. Brebbia), Wiley, 84-167 (1981).

42. E. Rank, A-Posteriori Error Estimates and Adaptive Refinement for Some Boundary Integral Element Methods, *Proceeddings of the International Conference on Accuracy Estimates and Adaptive Refinement in Finite Element Computations*, Lisboa, (1984).

43. E. Rank, Adaptivity and Accuracy Estimation for Finite Element and Boundary Integral Element Methods, *Accuracy Estimates and Adaptivity for Finite Element*, (Ed. J. Babuska, O.C. Zienckiencz, J.P. Gago and E. Arantes e Oliveira), Wiley, (1986).

44. G.C. Hsiao, P. Kopp and W.C. Wendland, A Galerkin Colocation Method for Some Integral Equations of the First Kind, *Computing*, 25, 88 - 130 (1980).

45. J. Rencis and R.L. Mullen, A Self Adaptive Mesh Refinement Technique for Boundary Element Solution of the Laplace Equation, *21st Annual Meeting of the Society of Engineering Science*, Blacksburg, Virginia, (1984).

46. J.J. Rencis and R.L. Muellen, A Self-Adaptive Mesh Refinement Technique for Boundary Element Solution of the Laplace Equation, *Computer Methods in Applied Mechanics and Engineering*, (1986).

47. J.J. Rencis e R. L. Muellen, Solutation of Elasticity Problems by a Self-Adaptive Mesh Refinement Technique for Boundary Element Computation, *International Journal for Numerical Methods in Engineering*, 23, 1509-1527 (1986).

48. E.Alarcon, L. Avia and A. Revester, On the Possibility of Adaptive Boundary Element, *International Conference on Accuracy Estimates and Adaptive Refinements in Finite Element Computations*, Lisbon, 25- 34 (1984).

49. E. Alarcon, A. Revester e J. Molina, Hierarchical Boundary Elements, *Computer and Structures*, 20, 151-156 (1985).

50. E. Alarcon and A. Revester, p-Adaptive Boundary Elements, *International Journal for Numerical Methods in Engineering*, 23, 801- 829, (1986).

51. A. Gonzalez and E. Alarcon. Adaptive Refinements in Boundary Elements, *Proceedings of the First World Congress on Computational Mechanics*, Austin, Texas, 22-26 September, (1986).

52. A. Peano, A. Pasini, R. Riccioni and L. Sardella, Adaptive Approximation in Finite Element Structural Analysis, *Computer and Structures*, 10, 333-342 (1979).

53. J.C. Lachat, *A Further Developement of the Boundary Integral Technique for Elastostatics*, Ph.D. Thesis, Univ. of Southampton, (1973).

54. K.K. Choi, E.J. Haug, J.W. Hou and V.M. Sohoni, Pshenichny's Linearizations Method for Mechanical Systems Optimization, *ASME Journal of Mechanisms, Transmissions and Automation in Design*, 105, 97-104 (1983).

55. J.W. Hou, E.J. Haug and R.L. Benedict, Shape Optimization of Elastic Bars in Torsion, *Sensitivity of Functionals with Applications to Engineering Problems* (Ed. V. Komkov), Springer-Verlag, 31-55 (1984).

56. E.J. Haug, K.K. Choi, J.W. Hou and Y.M. Yoo, A Variational Method for Shape Optimal Design of Elastic Structures,*New Directions in Optimum Structural Design* (Ed. E. Atrek, R.H. Gallagher, K.M. Ragsdell and O.C. Zienziewicz), Wiley, 105-137 (1984).

57. R.J. Yang, K.K. Choi and E.J. Haug, Numerical Considerations in Structural Component Shape Optimization, *ASME Journal of Mechanisms, Transmissions and Automation in Design*, to appear.

SHAPE DESIGN SENSITIVITY ANALYSIS USING BOUNDARY INTEGRAL EQUATION FOR POTENTIAL PROBLEMS

B. M. Kwak and J. H. Choi

Department of Mechanical Engineering
Korea Advanced Institute of Science and Technology
Seoul, Korea

ABSTRACT

A general method for shape design sensitivity analysis as applied to potential problems is developed with the standard direct boundary integral equation (BIE) formulation. The material derivative concept and adjoint variable method are employed to obtain an explicit expression for the variation of the performance functional in terms of the boundary shape variation. The adjoint problem defined in the present method takes a form of the indirect BIE. This adjoint problem can be solved using the same direct BIE of the original problem with a different set of boundary values, which brings about computational simplicity. The accuracy of the sensitivity formula is studied with a seepage problem. The detailed derivation of the formulas for general elliptic problems and a more elaborate numerical scheme will be described elsewhere.

NATO ASI Series, Vol. F27
Computer Aided Optimal Design: Structural and Mechanical Systems. Edited by C. A. Mota Soares

1. Statement of the Potential Problem

Consider the following mixed boundary value problem for a potential u defined on an arbitrary domain Ω with a sufficiently smooth boundary $\partial\Omega$

$$\nabla^2 u = 0 \qquad \text{in } \Omega \tag{1}$$

$$\left.\begin{aligned} u(x) &= b(x) && \text{on } \partial\Omega_0 \\ \frac{\partial}{\partial n} u(x) &= c(x) && \text{on } \partial\Omega_1 \end{aligned}\right\} \tag{2}$$

where b(x) and c(x) are smooth functions prescribed on the boundary, and $\partial\Omega = \partial\Omega_0 \cup \partial\Omega_1$.

By introducing the fundamental solution G, and applying the Green's formula for u and G, the following direct BIE is obtained:

$$\alpha u(x_0) + \int_{\partial\Omega} \{u(x) \frac{\partial}{\partial n} G(x,x_0) - \frac{\partial}{\partial n} u(x) G(x,x_0)\} \, ds = 0$$

$$\alpha = \begin{cases} 1, & x_0 \in \Omega \\ \tfrac{1}{2}, & x_0 \in \partial\Omega \end{cases} \tag{3}$$

where ds means that the integration is done with respect to x along $\partial\Omega$.

For $x_0 \in \partial\Omega$, the direct BIE defined by Eq (3) can be transformed to the following boundary integral identity by multiplying an arbitrary function $\mu(x_0)$ defined on $\partial\Omega$ and integrating over $\partial\Omega$ with respect to x_0

$$\int_{\partial\Omega} \Big[u(x) \{ \tfrac{1}{2}\mu(x) + \int_{\partial\Omega} \mu(x_0) \frac{\partial}{\partial n} G(x,x_0) \, ds_0 \} - \frac{\partial}{\partial n} u(x) \int_{\partial\Omega} \mu(x_0) G(x,x_0) \, ds_0 \Big] \, ds = 0 \tag{4}$$

where ds_0 means the integration with respect to x_0 along $\partial\Omega$.
It is noted from the indirect BIE formulation [1] that an arbitrary potential w(x) can be expressed in terms of $\mu(x_0)$ as follows

$$w(x) = \int_{\partial\Omega} \mu(x_0)\, G(x,x_0)\, ds_0 \tag{5}$$

and the normal derivative on the boundary is expressed as

$$\frac{\partial}{\partial n} w(x) = \int_{\partial\Omega} \mu(x_0) \frac{\partial}{\partial n} G(x,x_0)\, ds_0 + \tfrac{1}{2}\mu(x) \tag{6}$$

Substitution of Eqs (5) and (6) into Eq (4) yields the following simple expression

$$\int_{\partial\Omega} \{ u(x) \frac{\partial}{\partial n} w(x) - \frac{\partial}{\partial n} u(x)\, w(x) \}\, ds = 0 \tag{7}$$

2. Method of Shape Design Sensitivity Analysis

Consider now a general functional defined as an integral over the boundary in the following form

$$\psi = \int_{\partial\Omega} f(u, \frac{\partial u}{\partial n})\, ds \tag{8}$$

where f is a function of u and $\partial u/\partial n$ defined on $\partial\Omega$. The case of volume integrals can also be considered in this form since they can be transformed to boundary integrals with the application of Eq (3).

Now the objective is to get an explicit expression for the variation of ψ in terms of the boundary shape variations with the aid of the direct BIE formulation. Following the material derivative idea [2,3], a velocity field V(x) at t=0 is considered and the varied domain is described by the following transformation:

$$\begin{aligned} \Omega_t &= T(\Omega, t) \\ x_t &= T(x,t) \cong x + t\,V(x) \end{aligned} \tag{9}$$

Then the material derivatives at time t=0:

$$\dot{u} = u' + V \nabla u \tag{10}$$

$$\dot{\psi} = \int_{\partial\Omega} \{ P \dot{u} + Q \dot{\widehat{\frac{\partial u}{\partial n}}} + f H V_n \} ds \tag{11}$$

where $P = \partial f/\partial u$, $Q = \partial f/(\partial u/\partial n)$, H is the mean curvature of the boundary and V_n is the normal component of V on the boundary, which will be termed as the boundary variation. Since a changed boundary can be described by specifying only V_n, the tangential component V_s is unimportant in general, as long as design of a shape is considered.

Decomposing the right hand side of Eq (11) as a sum of integration over $\partial\Omega_0$ and $\partial\Omega_1$, and taking into consideration of the prescribed boundary conditions,

$$\dot{\psi} = \int_{\partial\Omega_1} P \dot{u} \, ds + \int_{\partial\Omega_0} Q \dot{\widehat{\frac{\partial u}{\partial n}}} ds$$
$$+ \int_{\partial\Omega_0} P \frac{\partial b}{\partial n} V_n \, ds + \int_{\partial\Omega_1} Q \frac{\partial c}{\partial n} V_n \, ds + \int_{\partial\Omega} f H V_n \, ds \tag{12}$$

Now $\dot{u}$ and $\dot{\widehat{\partial u/\partial n}}$ are related implicitly to V_n through the governing BIE (3). To represent these terms in terms of V_n, the identity (7) is utilized. After a lengthy manipulation with several simplifying relations among various derivatives, the following relation is obtained:

$$\int_{\partial\Omega_1} \dot{u} \frac{\partial w}{\partial n} ds - \int_{\partial\Omega_0} \dot{\widehat{\frac{\partial u}{\partial n}}} w \, ds = \int_{\partial\Omega_1} w \frac{\partial c}{\partial n} V_n \, ds - \int_{\partial\Omega_0} \frac{\partial w}{\partial n} \frac{\partial b}{\partial n} V_n \, ds$$
$$- \int_{\partial\Omega} \{ \nabla_s u \cdot \nabla_s w - \frac{\partial u}{\partial n} (\frac{\partial w}{\partial n} + w H) \} V_n \, ds \tag{13}$$

In order to eliminate $\dot{u}$ and $\dot{\widehat{\partial u/\partial n}}$ in Eq (12), an adjoint system is introduced, which has the same form as Eqs (5) and (6) and satisfies the following condition

$$\left.\begin{aligned} u^* &= -Q \quad \text{on} \quad \partial\Omega_0 \\ \frac{\partial u^*}{\partial n} &= P \quad \text{on} \quad \partial\Omega_1 \end{aligned}\right\} \tag{14}$$

Although this adjoint system is expressed in a conventional indirect BIE, it can be shown that a direct BIE of the form as the original BIE, Eq (3), is equivalent to this. This direct BIE formulation is employed for numerical solution of u^*. Substituting u^* in place of the arbitrary potential w in Eq (13) and utilizing the condition (14), Eq (12) takes the form sought for:

$$\dot{\psi} = \int_{\partial\Omega} \{ -\nabla_s u \cdot \nabla_s u^* + \frac{\partial u}{\partial n} (\frac{\partial u^*}{\partial n} + u^* H) + f H \} V_n \, ds$$
$$+ \int_{\partial\Omega_0} (P - \frac{\partial u^*}{\partial n}) \frac{\partial b}{\partial n} V_n \, ds + \int_{\partial\Omega_1} (Q + u^*) \frac{\partial c}{\partial n} V_n \, ds \qquad (15)$$

The evaluation of this sensitivity formula requires solutions for the potential u and the adjoint variable u^*. The variable u^* is obtained by means of a direct BIE instead of indirect BIE, as noted above. This facilitates an efficient calculation, since the same BIE is used with a different set of boundary values.

3. Numerical Example for a Seepage Problem

To illustrate a numerical implementation of the shape design sensitivity formula derived above, a seepage problem is considered.

Consider a two dimensional flow through a porous medium as shown in Fig. 1. With some simplifying assumptions such as homogeneity and isotropy of the medium, the governing equation based on Darcy's law and the boundary conditions are given as follows

$$\nabla^2 u = 0 \qquad \text{in} \quad \Omega \qquad (16)$$

$$u = y, \quad \frac{\partial u}{\partial n} = 0 \quad \text{on} \quad \Gamma_1$$
$$u = 1 \qquad \text{on} \quad \Gamma_2$$

$$\left.\begin{aligned} \frac{\partial u}{\partial n} &= 0 && \text{on} \quad \Gamma_3 \\ u &= 0.3 && \text{on} \quad \Gamma_4 \\ u &= y && \text{on} \quad \Gamma_5 \end{aligned}\right\} \tag{17}$$

where u is the total head. The over-determined conditions on Γ_1 make Γ_1 a free boundary. This problem can be transformed to an equivalent shape optimization problem by defining a proper objective functional associated with one of the conditions given on the free boundary. The following functional is taken for this purpose:

$$\psi = \int_{\Gamma_1} (u - y)^2 \, ds \tag{18}$$

For determination of the free boundary, the height is chosen as the design variable, and the velocity field is assumed unidirectional so that Γ_1 stays within the dam width. Then, the normal and tangential movement of the boundary can be written as

$$V_n = \delta y \, n_y , \qquad V_s = \delta y \, n_x \tag{19}$$

where n_x and n_y represent the x and y component of the normal vector on the boundary. Since the velocity field is taken to be vertical in the present problem, the sensitivity formula derived above is modified to include the tangential boundary variation V_s. Following a similar procedure as used above, the following result can be derived

$$\dot{\psi} = \int_{\Gamma_1} \left[\{ -\frac{\partial u}{\partial s}\frac{\partial u^*}{\partial s} + (u-y)^2 H \} V_n + 2(u-y)\frac{\partial y}{\partial s} V_s - 2(u-y)\,\delta y \right] ds \tag{20}$$

where δy is the boundary variation in vertical direction as given in Fig. 2, and $\partial y/\partial s = n_x$.

If the formula (15) is used, one simply obtains

$$\dot{\psi} = \int_{\Gamma_1} \left[\{ -\frac{\partial u}{\partial s}\frac{\partial u^*}{\partial s} + (u-y)^2 H \} V_n - 2(u-y)\,\delta y \right] ds \tag{21}$$

Table 3. Comparison of ψ and u-y values for initial and final design

	ψ value	u-y value at each node 2	3	4	5	6	7	8	9	10	11	12
Initial	2.497E-03	-1.677E-02	-2.784E-02	-3.848E-02	-5.624E-02	-6.747E-02	-6.478E-02	-5.600E-02	-4.537E-02	-3.589E-02	-2.380E-02	-1.118E-02
Final	2.596E-06	5.061E-05	6.693E-04	6.848E-04	5.981E-04	6.829E-05	2.228E-04	2.393E-04	3.396E-04	1.298E-04	2.268E-05	1.109E-04

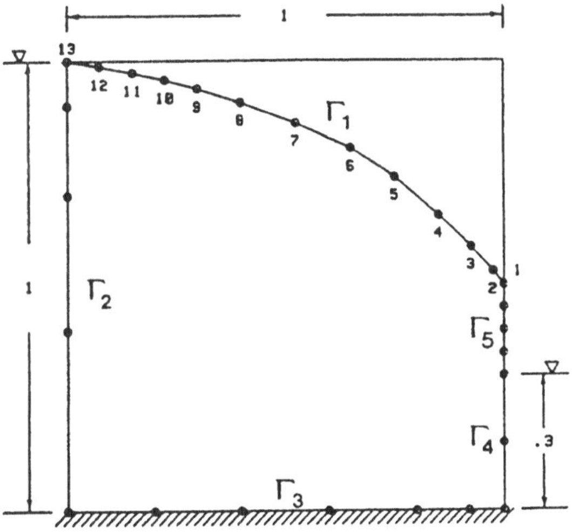

Fig. 1 Initial geometry for seepage problem

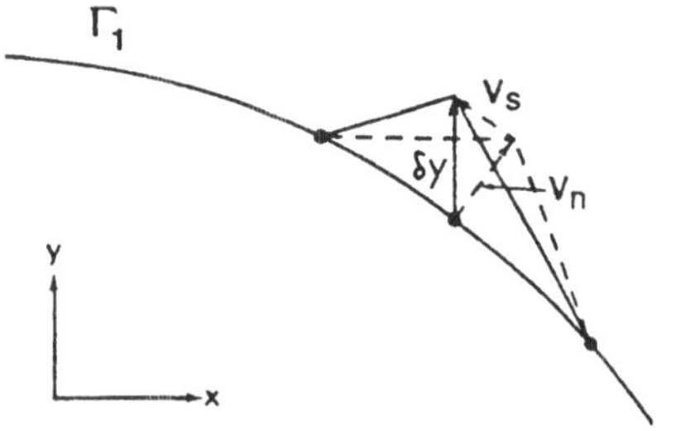

Fig. 2 Shape variation corresponding to vertical nodal change δy and normal nodal change V_n

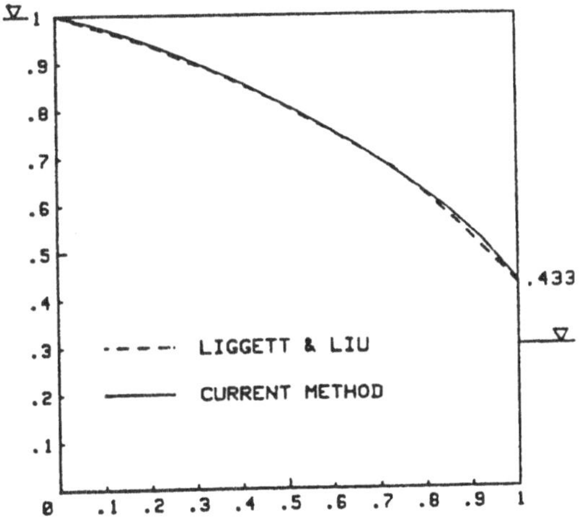

Fig. 4 Optimal solution of seepage problem

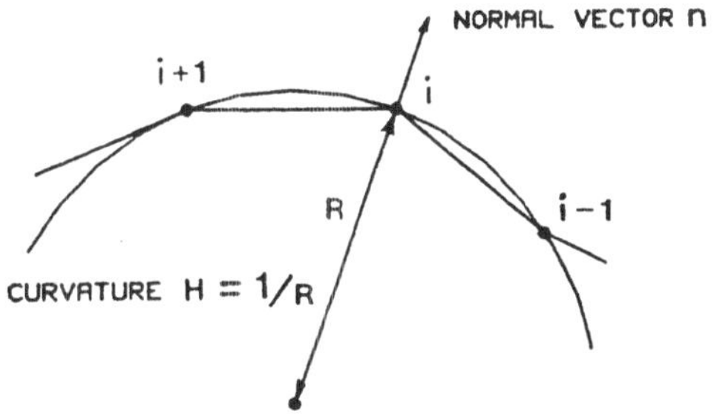

Fig. 3 Approximation of the normal vector n and curvature H at i-th node

where δy is given by $\frac{\partial y}{\partial n} V_n = n_y V_n$. In these equations, the last term in the integrand comes from the variation of y appearing in the objective functional, and the adjoint variable u* is determined by the condition:

$$\left.\begin{aligned} \frac{\partial u^*}{\partial n} &= 2(u-y) && \text{on } \Gamma_1 \\ &= 0 && \text{on } \Gamma_3 \\ u^* &= 0 && \text{on } \Gamma_2, \Gamma_4, \Gamma_5 \end{aligned}\right\} \tag{22}$$

It is noted that the variations from (20) and (21) represent different situations; the sensitivity from (20) refers to the change of shape depicted in solid line in Fig. 2, while that from (21) corresponds to the change in dashed line.

For a numerical experimentation for the sensitivity analysis, a linear boundary element model is used with 28 nodes and 11 design variables which are the y coordinates of node 2 through 12 as shown in Fig. 1. The y value of node 1 has been determined by extrapolation of those of node 2 and 3 because of poor accuracy of solution at the corner. A simple numerical scheme is used in the following, and a more elaborate scheme will be reported later, with a proper treatment of corners and discontinuities. The nodal values of normal vector and curvature appearing in Eqs (20) and (21) are calculated by approximating the two line segments with 3 nodes as a circle passing through those 3 points as shown in Fig. 3. After all these calculations are performed, one obtains the desired sensitivity coefficients for each node on the free boundary.

In order to compare the accuracy of the results by present method with those of finite differences, define ψ_0 and ψ_k as the functional values for initial and modified designs for k-th node, respectively, and let $\Delta\psi_k$ be $\psi_k - \psi_0$, and let $\dot{\psi}$ be the predicted difference calculated by either (20) or (21), depending on the change taken. The ratio $\dot{\psi}/\Delta\psi_k$ times 100 is used as a measure of accuracy. Numerical results for the change by solid line in Fig. 2 are shown in Table 1. Here, 0.1% change is used such that $\delta y = 0.001y$. In Table 2, the sensitivities for design changes in the normal direction, corresponding to the dashed line in Fig. 2, are shown for comparison. Even with a simple numerical scheme, these results show good accuracy, except at node 11, where the absolute value of $\Delta\psi$ is an order smaller than the others.

Utilizing the sensitivity results given by Eq (20), optimum free boundary is now obtained. A conjugate gradient method is used, and eight iterations are made from the initial shape given by Fig. 1. The result in [4] is compared in Fig. 4, and the two match very well. The objective functional and differences u-y at each node are compared in Table 3 for the initial and final design.

4. Discussions and Conclusions

The present work provides a procedure and formulas for the shape design sensitivity analysis using the BIE formulation. The method is illustrated by deriving explicit formulas for potential problems. Despite that the formulated BIE is not in symmetric form, the adjoint variable method has been successfully applied by taking advantage of the relations between the direct and indirect BIE's. Design sensitivity analysis is carried out for a seepage problem, and the numerical results are compared with those of finite differences. Good accuracy has been obtained even with a simple numerical scheme. A detailed method extended to general elliptic type problems and a more elaborate numerical scheme will be reported elsewhere.

REFERENCES

1. P. K. Banerjee and R. Butterfield, Boundary Element Methods in Engineering Science, McGraw-Hill, 1981

2. J. P. Zolesio, The Material Derivative Method for Shape Optimization, in Optimization of Distributed Parameter Structures (E.J. Haug & J. Cea, eds.) Sijthoff & Noordhoff, Alphen aan den Rijn, Netherlands, 1981

3. K. K. Choi and E. J. Haug, Shape Design Sensitivity Analysis of Elastic Structures, J. Struct. Mech. 11:231-269 (1983)

4. J. A. Liggett and P. L-F. Liu, The Boundary Integral Equation Method for Porous Media Flow, George Allen & Unwin, 1983

Table 1. Sensitivity result for $\dot{\psi}$ calculated by Eq (20)

Node	ψ_k	$\Delta\psi_k$	$\dot{\psi}$	$\dot{\psi}/\Delta\psi_k(\%)$
2	2.499E-03	2.423E-06	2.419E-06	99.8
3	2.501E-03	4.340E-06	4.378E-06	100.9
4	2.504E-03	6.481E-06	6.754E-06	104.2
5	2.507E-03	9.653E-06	9.355E-06	96.9
6	2.508E-03	1.141E-05	1.141E-05	99.9
7	2.509E-03	1.161E-05	1.154E-05	99.4
8	2.506E-03	8.539E-06	8.273E-06	96.9
9	2.502E-03	4.537E-06	4.414E-06	97.3
10	2.499E-03	1.990E-06	2.054E-06	103.3
11	2.497E-03	2.714E-07	3.096E-07	114.1
12	2.496E-03	-1.265E-06	-1.239E-06	97.9

Table 2. Sensitivity result for $\dot{\psi}$ calculated by Eq (21)

Node	ψ_k	$\Delta\psi_k$	$\dot{\psi}$	$\dot{\psi}/\Delta\psi_k(\%)$
2	2.504E-03	6.911E-06	6.900E-06	99.8
3	2.508E-03	1.056E-05	1.068E-05	101.2
4	2.510E-03	1.340E-05	1.374E-05	102.5
5	2.513E-03	1.617E-05	1.564E-05	96.8
6	2.513E-03	1.606E-05	1.621E-05	101.0
7	2.512E-03	1.466E-05	1.459E-05	99.5
8	2.507E-03	9.972E-06	9.652E-06	96.8
9	2.502E-03	5.043E-06	4.899E-06	97.1
10	2.499E-03	2.141E-06	2.207E-06	103.1
11	2.497E-03	2.891E-07	3.260E-07	112.8
12	2.496E-03	-1.299E-06	-1.271E-06	97.8

ψ_0= 2.497E-03

BOUNDARY ELEMENTS IN SHAPE DESIGN SENSITIVITY ANALYSIS OF THERMOELASTIC SOLIDS

R.Alsan Meriç
Department of Applied Mathematics
Research Institute for Basic Sciences, TÜGAM
Gebze, Kocaeli, Turkey

Abstract. Shape design sensitivity analysis of thermoelastic solids is done by using the material derivative idea and the adjoint variable method. Boundary element methods are proposed for spatial discretization of system of equations and relevant functionals.

1. INTRODUCTION

Inverse (or synthesis) problems of engineering science, as against direct (or analysis) problems, find important applications in practice. Optimization or optimal control problems, free and/or moving boundary problems, material/load/shape/eigenvalue design or identification problems may be cited as examples.

Shape design sensitivity analysis (SDSA) of elastic structures was developed recently by several authors [1,2]. In reference [1], the material derivative idea of continuum mechanics and the adjoint variable method of design sensitivity analysis were utilized effectively to obtain computable expressions for the effect of shape variation on the objective functions.

For spatial discretizations, the finite element method (FEM) has been extensively used for a variety of problems in structures. The boundary element method (BEM), however , is attracting more and more researchers as an alternative technique in the field of shape sensitivity analysis [3,4]. The main reason for this is that more accurate stress values may be obtained by this method, thus requiring a smaller number of iterations needed to find optimal shapes.

The BEM is especially suitable for the "boundary method" of SDSA, as against the "domain method" [5]. In the boundary method of SDSA the required sensitivity information is evaluated in terms of integrals defined on the boundary, where accurate data on functions is already available by the BEM.

In this paper, the SDSA of thermoelastic solids is performed by the same scheme as given in [1] for elastic structures. The BEM is outlined in terms of an example problem for the spatial discretizations.

NATO ASI Series, Vol. F27
Computer Aided Optimal Design: Structural and Mechanical Systems. Edited by C. A. Mota Soares

2. PROBLEM STATEMENT

Consider a thermoelastic, isotropic and homogeneous solid body in 3-dimensions. The equilibrium equations may be written as

$$\text{in } D \ : -q_{i,i} + Q = 0 \quad \text{and} \quad \sigma_{ij,j} + b_i = 0 \tag{1}$$

where D is the domain of interest which is to be varied; q_i and σ_{ij} are the heat flux vector and stress tensor; Q and b_i are the distributed heat source and body force vector, respectively. The constitutive equations may be written as

$$q_i = -k\, T_{,i} \quad \text{and} \quad \sigma_{ij} = \lambda\, u_{k,k}\, \delta_{ij} + \mu(u_{i,j} + u_{j,i}) - \gamma T \delta_{ij} \tag{2}$$

where k is the thermal conductivity; λ and μ are Lamé's constants ; T and u_i are the temperature and displacements, respectively, and $\gamma=(3\lambda+2\mu)\,\alpha$, α being the coefficient of linear thermal expansion.

Mixed boundary conditions are applied on the surface $S = S_1 + S_2$ of D, namely

$$\text{on } S_1 \ : \ T = T_o \quad \text{and} \quad u_i = u_{io} \tag{3}$$

$$\text{on } S_2 \ : \ q_n = q_i\, n_i = q \quad \text{and} \quad t_i = \sigma_{ij}\, n_j = p_i \tag{4}$$

where q_n is the normal heat flux; n_i is the unit vector normal to the boundary, and t_i are the boundary tractions. Conservative loading that depends on the position, but not the shape of boundary is considered on S_2.

A general integral performance functional, representing either an objective functional to be minimized or a constraint to be satisfied, is adopted for the present SDSA of thermoelastic solids:

$$I = \int_D f(x_i,T,q_i,\sigma_{ij})\, dD + \int_{S_1} g(x_i,q_n,t_i)\, dS + \int_{S_2} h(x_i,T,u_i)\, dS \tag{5}$$

It is now desired to find the effects of shape variation on the functional I. Once design derivative of the performance functional is calculated, iterative direct methods can be utilized for any structural optimization or shape identification problem in thermoelasticity.

3. SHAPE DESIGN SENSITIVITY ANALYSIS

Following reference [1], with a slight modification in the procedure, the adjoint variable method and the material derivative idea will be utilized for the present sensitivity analysis in thermoelasticity.

3.1. Augmented Functional

By adopting an optimal control methodology, an augmented functional $\bar{I}$ is defined as

$$\bar{I} = I + \int_D [T^*(-q_{i,i} + Q) + u_i^* (\sigma_{ij,j} + b_i)] \, dD \tag{6}$$

where T^* and u_i^* are the adjoint temperature and displacements, respectively.

By inserting eqn. (5) into (6) and using integration by parts twice, $\bar{I}$ may take the form of $\bar{I} = \bar{I}_o + \bar{I}_1 + \bar{I}_2$ where

$$\bar{I}_o = \int_D (f + q_i T^*_{,i} + QT^* - \sigma_{ij} u^*_{i,j} + b_i u_i^*) \, dD \tag{7}$$

$$\bar{I}_1 = \int_{S_1} (g + t_i u_i^* - q_n T^*) \, dS \tag{8}$$

$$\bar{I}_2 = \int_{S_2} (h + t_i u_i^* - q_n T^*) \, dS \tag{9}$$

3.2. Material Derivative of Integrals

Since the shape of domain D is treated as the design variable in the shape sensitivity analysis, it is convenient to think of D as a continuous medium and utilize the material derivative idea from continuum mechanics. Following reference [1], the material derivatives of domain and surface integrals are given as follows :

$$\phi_1 = \int_D w_1 \, dD : \quad \dot{\phi}_1 = \int_D w_1' \, dD + \int_S w_1 V_n \, dS \tag{10}$$

$$\phi_2 = \int_S w_2 \, dS : \quad \dot{\phi}_2 = \int_S [w_2' + (w_{2,n} + Hw_2) V_n] \, dS \tag{11}$$

where w_1 and w_2 are general regular functions defined in D and on S; $(\dot{\ })$, $(\ )'$ and $(\ )_{,n}$ denote the material, partial and normal derivatives of $(\ )$, respectively; H is the curvature of the boundary S in R^2 and twice the mean curvature of S in R^3; and V_n is the normal component of the design perturbation velocity V_i.

By using eqns. (10) and (11), and noting the functional forms of f, g and h as given in eqn. (5), the material derivatives of eqns. (7)-(9) may be obtained in the following form:

$$\dot{\bar{I}}_o = \int_D \Big(\frac{\partial f}{\partial T} T' + \frac{\partial f}{\partial q_i} q_i' + \frac{\partial f}{\partial u_i} u_i' + \frac{\partial f}{\partial \sigma_{ij}} \sigma_{ij}' + q_i' T^*_{,i} + q_i T^{*'}_{,i}$$
$$+ QT^{*'} - \sigma_{ij}' u^*_{i,j} - \sigma_{ij}' u^{*'}_{i,j} + b_i u_i^{*'} \Big) \, dD$$

$$+ \int_S (f + q_i T^*_{,i} + QT^* - \sigma_{ij} u^*_{i,j} + b_i u^*_i) V_n \, dS \tag{12}$$

$$\dot{I}_1 = \int_{S_1} \{ \frac{\partial g}{\partial q_n} q'_n + \frac{\partial g}{\partial t_i} t'_i + t'_i u^*_i + t_i u^{*'}_i - q'_n T^* - q_n T^{*'}$$

$$+ [\frac{\partial g}{\partial q_n} q_{n,n} + \frac{\partial g}{\partial t_i} t_{i,n} + (t_i u^*_i)_{,n} - (q_n T^*)_{,n}$$

$$+ H(g + t_i u^*_i - q_n T^*)] V_n \} \, dS \tag{13}$$

and

$$\dot{I}_2 = \int_{S_2} \{ \frac{\partial h}{\partial T} T' + \frac{\partial h}{\partial u_i} u'_i + t'_i u^*_i + t_i u^{*'}_i - q'_n T^* - q_n T^{*'}$$

$$+ [\frac{\partial h}{\partial T} T_{,n} + \frac{\partial h}{\partial u_i} u_{i,n} + (t_i u^*_i)_{,n} - (q_n T^*)_{,n}$$

$$+ H (h + t_i u^*_i - q_n T^*)] V_n \} \, dS \tag{14}$$

It is noted that the curve ∂S_2 that bounds the loaded boundary surface S_2 is taken as fixed, so that the velocity field V_i at ∂S_2 is zero.

At this point, partial derivative form of the constitutive eqn. (2) may be introduced into eqns. (12)-(14), and integration by parts may be used repeteadly. The boundary conditions (3)-(4) are also substituted into the resulting equations in their partial derivative forms.

3.3. Adjoint Problem

It is now convenient to define adjoint heat flux vector q^*_i and adjoint stress tensor σ^*_{ij} as

$$q^*_i = -k(T^*_{,i} + \frac{\partial f}{\partial q_i}) \tag{15}$$

$$\sigma^*_{ij} = \lambda(u^*_{k,k} - \frac{\partial f}{\partial \sigma_{kk}}) \delta_{ij} + \mu[(u^*_{i,j} - \frac{\partial f}{\partial \sigma_{ij}}) + (u^*_{j,i} - \frac{\partial f}{\partial \sigma_{ji}})] \tag{16}$$

The adjoint problem then satisfies the following equations :

$$\text{in D} : -q^*_{i,i} + \alpha\sigma^*_{ii} + \frac{\partial f}{\partial T} = 0 \quad \text{and} \quad \sigma^*_{ij,j} + \frac{\partial f}{\partial u_i} = 0 \tag{17}$$

$$\text{on } S_1 : T^* = \frac{\partial g}{\partial q_n} \quad \text{and} \quad u^*_i = - \frac{\partial g}{\partial t_i} \tag{18}$$

$$\text{on } S_2 : q^*_n = - \frac{\partial h}{\partial T} \quad \text{and} \quad t^*_i = \frac{\partial h}{\partial u_i} \tag{19}$$

3.4. Material Derivative of the Performance Functional

If the primary and adjoint problems are satisfied, the material derivative of the performance functional I can be expressed in terms of integrals defined only on the moving boundary portions S_1 and S_2, i.e.,

$$\begin{aligned}\dot{I} = \int_{S_1} \{ & [f + q_i T^*_{,i} + QT^* - \sigma_{ij} u^*_{i,j} + b_i u^*_i + t_i u^*_{i,n} - q_n T^*_{,n} \\ & + H (g + t_i u^*_i - q_n T^*)] V_n + (t^*_i u_{i,k} - q^*_n T_{,k}) V_k \} dS \\ & + \int_{S_2} \{ [f + q_i T^*_{,i} + QT^* - \sigma_{ij} u^*_{i,j} + b_i u^*_i + \frac{\partial h}{\partial T} T_{,n} + \frac{\partial h}{\partial u_i} u_{i,n} \\ & + (t_i u^*_i)_{,n} - (q_n T^*)_{,n} + H (h + t_i u^*_i - q_n T^*)] V_n \\ & - (u^*_i t_{i,k} - T^* q_{n,k}) V_k \} dS \end{aligned} \tag{20}$$

It is noted that if the boundary inputs of T and u_i on S_1, and of q_n and t_i on S_2 do not vary along the corresponding boundary portions, or if the tangential component of V_k is taken as negligible, then it is sufficient to consider only the normal component V_n of the design velocity field on the boundary for the derivative calculations.

4. EXAMPLE PROBLEM

A two-dimensional example problem is now given in order to explain specifically some of the details of the BEM of discretization. The quantities Q, b_i and u_{io} are taken as zero; S_1 is held fixed (i.e., $V_n=0$ on S_1), while S_2 is varied. It is also assumed that T_o, q and p_i are constant along the corresponding boundary portions.

A relevant performance functional is expressed as follows :

$$\begin{aligned} I = & \frac{1}{2} \int_D (2\beta_1 T + \beta_2 q_i q_i + \beta_3 \sigma_{ij} \varepsilon_{ij}) dD \\ & + \frac{1}{2} \int_{S_1} (\beta_4 q_n^2 + \beta_5 t_i t_i) dS + \frac{1}{2} \int_{S_2} (\beta_6 T^2 + \beta_7 u_i u_i) dS \end{aligned} \tag{21}$$

where β_i (i=1,2,..7) are given weighting parameters. The functional forms of the integrands in the above equation have been chosen such that the BEM especially proves to be efficient for numerical calculations.

The primary and adjoint problems for the example problem can be given in terms of the primitive quantities, i.e., (primary and adjoint) temperature and displacements, as follows :

$$\text{in } D \ : \ kT_{,ii} = 0 \tag{22}$$

$$\lambda u_{j,ji} + \mu(u_{i,j} + u_{j,i})_{,j} - \gamma T_{,i} = 0 \tag{23}$$

$$\lambda u^*_{j,ji} + \mu(u^*_{i,j} + u^*_{j,i})_{,j} = 0 \tag{24}$$

$$k T^*_{,ii} + \gamma(u^*_i - \beta_4 u_i)_{,i} + \beta_1 = 0 \tag{25}$$

$$\text{on } S_1 \ : \ T = T_o \quad \text{and} \quad u_i = 0 \tag{26}$$

$$u^*_i = -\beta_5 t_i \quad \text{and} \quad T^* = \beta_4 q_n \tag{27}$$

$$\text{on } S_2 \ : \ -k T_{,i} n_i = q \tag{28}$$

$$\lambda u_{j,j} n_i + \mu(u_{i,j} + u_{j,i}) n_j = P_i + \gamma T n_i \tag{29}$$

$$\lambda u^*_{j,j} n_i + \mu(u^*_{i,j} + u^*_{j,i}) n_j = \beta_3 t_i + \beta_7 u_i \tag{30}$$

$$-k T^*_{,i} n_i = k \beta_2 q_n - \beta_6 T \tag{31}$$

The field equations (22)-(31) are written in an order, suggesting that they must be solved for T, u_i, u^*_i and T^* consecutively following the same order at each iteration level in an iterative procedure. The material derivative of I is thus given by

$$\dot{I} = \int_{S_2} [\beta_1 T + \frac{\beta_2}{2} q_i q_i + \frac{\beta_3}{2} \sigma_{ij}\varepsilon_{ij} + q_i T^*_{,i} - \sigma_{ij} u^*_{i,j} + \beta_6 TT_{,n} + \beta_7 u_i u_{i,n} + t_i u^*_{i,n} - q_n T^*_{,n} + H(\frac{\beta_6}{2} T^2 + \frac{\beta_7}{2} u_i u_i + t_i u^*_i - q_n T^*)] V_n \, dS \tag{32}$$

5. BOUNDARY ELEMENT METHODS

Accurate evaluation of boundary stresses and strains are essential for design sensitivity information. For the last several years there have been some attempts to use the BEM for shape optimal design of engineering systems [3,4]. The BEM can be viewed as consisting fundamentally of two major steps: (i) transformation of the differential equations in the domain D into integral equations on the boundary S; and (ii) discretization of these integral equations via finite element concepts and subsequent solution by the collocation method.

The spatial discretization of the primary and adjoint problems can be achieved by means of the BEM. In the following subsections, some remarks on the utilization of the method for the specific problem at hand will be given.

5.1. BEM for T and T^* Problems

The primary and adjoint temperature problems, i.e., the so-called T and T^* problems, are described by the Poisson's type of scalar elliptic PDE's. The T-problem is defined by eqns. (22), (26) and (28), while the T^*-problem is given by eqns. (25), (27) and (31). It is reminded here that the T^*-problem is solved after the solution of T, u_i and u_i^* problems. Hence, the domain and boundary "inputs" in the relevant equations are assumed to be known for the current design estimate.

In the case of Poisson's equation, the transformation of the differential equation into an integral equation is achieved through Green's second identity as follows :

$$c(\xi)\ T(\xi) = \int_D Q\hat{T}\ dD + \int_S (\hat{q}_n\ T - q_n\ \hat{T})\ dS \qquad (33)$$

where T and q_n may denote either the primary or the adjoint temperature and normal heat flux, respectively; $\hat{T}$ is the "two-point" fundamental solution of Laplace's equation; $\hat{q}_n$ is the normal heat flux corresponding to $\hat{T}$; ξ is any boundary point on S; $c(\xi)$ is the internal solid angle for any ξ.

For the solution of T and T^* problems corresponding expressions for Q in D, T on S_1 and q_n on S_2 should be inserted in eqn. (33). It is noted that the domain integral in the above equation vanishes for the solution of the T-problem, since Q is prescribed as zero in the example problem. Hence, eqn. (33) represents a boundary integral equation for the T-problem.

In the case of the T^*-problem, the domain integral term in eqn. (33) may be put in the following form by noting eqn. (25) :

$$\int_D [\gamma(u_i^* - \beta_4 u_i)_{,i} + \beta_1]\ \hat{T}\ dD$$

$$= \int_D [\beta_1 \hat{T} - \gamma(u_i^* - \beta_4\ u_i)\ \hat{T}_{,i}]\ dD + \int_S \gamma(u_i^* - \beta_4 u_i)\ \hat{T}\ n_i\ dS \qquad (34)$$

It is well known that the main advantage of the BEM is obtained when there exist no domain integrals in the relevant integral formulations [6] . For the solution of the T^*-problem, this advantage is lost due to the presence of the domain integral term in the above equation. It has been observed that for the SDSA of thermoelastic solids, this type of domain integrals will always exist, coupling the primary and adjoint fields in the domain.

For the implementation of the BEM, eqn. (33) represents the starting point. Standard BEM procedure [6] is then followed which is summarized as follows :

i) Divide the boundary S into boundary elements
ii) Approximate T and q_n on each element by using finite element concepts
iii) Divide the domain D into internal cells (needed only for the T^*-problem)
iv) Use point collocation method (using the element nodes as the collocation points)
v) Introduce the known boundary values for T on S_1 and q_n on S_2 .
vi) Solve the resulting set of algebraic equations for the unknown values of q_n on S_1 and T on S_2.

5.2. BEM for u_i and u_i^* Problems

The primary and adjoint displacement problems, i.e., the so-called u_i and u_i^* problems, are described by vector elliptic PDE's, which are given by eqns. (23), (26), (29) and (24), (27), (30), respectively. The BEM for elasticity problems is based on Somigliana's identity [6]. The relevant boundary integral is given by

$$c_{ij}(\xi)\ u_i(\xi) = \int_D b_i\ \hat{u}_{ij}\ dD + \int_S (t_i\ \hat{u}_{ij} - u_i\ \hat{t}_{ij})\ dS \qquad (35)$$

where u_i and t_i may represent either the primary or adjoint displacements and tractions, respectively; $\hat{u}_{ij}$ and $\hat{t}_{ij}$ are the fundamental solution for displacements and tractions; c_{ij} is the coefficient that depends on the geometry of the boundary at point ξ.

It is noted that the thermal effects in the present thermoelastic problem are taken into account in the u_i-problem by considering them as body forces in D (i.e., $-\gamma T_{,i}$) and surface tractions on S (i.e., $\gamma T n_i$). The u_i^* - problem has no body forces involved as can be seen from eqn (24). Hence, the domain integral term drops out for this case in eqn. (35).

Since the body forces in eqn. (23) are generated by the steady-state temperature field satisfying eqn. (22), it is possible to use a technique introduced by Rizzo and Shippy [7] to convert the domain integral into a surface integral. For the present 2-dim. problem, the domain integral term is thus given by the following [6] :

$$-\int_D \gamma T_{,i}\, \hat{u}_{ij}\, dD = -\int_S \gamma T\, n_i\, \hat{u}_{ij}\, dS + \frac{\gamma\mu}{2(\lambda+3\mu)} \int_S [T_{,n} \frac{\partial}{\partial \xi} (r^2\, \ell nr)$$

$$- T \frac{\partial}{\partial n} \frac{\partial}{\partial \xi_j} (r^2 \ell nr)\; dS \qquad (36)$$

where r is the euclidean distance between the field and source points. By introducing the above expression into eqn. (35) a boundary integral equation defined solely on S may be obtained.

Spatial discretization of eqn. (35) by the BEM, either for u_i or u_i^* -problem, follows a similar procedure as is given in the preceding subsection, except that no internal cells are now required.

5.3. Evaluation of I and $\dot{I}$

Domain integral terms in eqn. (21) may be converted into surface integrals by using integration by parts and the homogeneous thermoelastic equilibrium equations resulting in the following expression for I :

$$I = \frac{1}{2} \int_S [\beta_1 (x_i n_i\, T - \frac{1}{2}\, x_i x_i\, T_{,n}) - \beta_2 kTq_n + \beta_3 t_i u_i\,]\; dS$$

$$+ \frac{1}{2} \int_{S_1} (\beta_4\, q_n^2 + \beta_5 t_i t_i)\; dS + \frac{1}{2} \int_{S_2} (\beta_6 T^2 + \beta_7 u_i u_i)\; dS \qquad (37)$$

The derivative $\dot{I}$ is, on the otherhand, given by eqn. (32), where the curvature H of the linear boundary elements can be taken as zero.

After the solution of T, u_i, u_i^* and T^* problems by the BEM for the current design estimate of domain D, the evaluation of I,eqn. (37), is straightforward. To find $\dot{I}$, however, it is necessary to calculate boundary stresses and the derivatives of displacements. This can be done through the use of thermoelastic constitutive equations and Cauchy law on the boundary. Further details on these calculations and also the treatment of V_n on the boundary as an expression in terms of design variable perturbations can be found in references [3,4].

6. CONCLUSIONS

The SDSA for thermoelastic solids has been performed by using the material derivative idea and the adjoint variable method. Coupled thermal and elastic fields are taken into account in obtaining the material derivative of a general performance functional.

The BEM has been proposed for spatial discretization of relevant equations.

It has been found that domain integral terms will always exist in the integral equation associated with the adjoint temperature equation due to the thermoelastic coupling effects. It is maintained, however, that the BEM still has advantages over the other domain type of methods, especially with its inherent higher accuracy in the evaluation of boundary stresses, which is crucial for any shape design sensitivity analysis.

REFERENCES

1. Choi, K.K. and Haug, E.J., "Shape Design Sensitivity Analysis of Elastic Structures", J.Struct. Mech. 11, 231-269 (1983)
2. Dems, K. and Mroz, Z., "Variational Approach by Means of Adjoint Systems to Structural Optimization and Sensitivity Analysis - II. Structure Shape Variation, Int. J.Solids Structures 20, 527-552 (1984)
3. Mota Soares, C.A. and Choi K.K., "Boundary Elements in Shape Optimal Design of Structures", The NATO ASI on Computer Aided Optimal Design: Structural and Mechanical Systems, Troia, Portugal (1986)
4. Burczynski, T. and Adamczyk, T., "The Boundary Element Formulation for Multiparameter Structural Shape Optimization", Appl.Math.Modelling 9, 195-200 (1985).
5. Choi, K.K., "Shape Design Sensitivity Analysis and Optimal Design of Structural Systems", The NATO ASI on Computer Aided Optimal Design: Structural and Mechanical Systems, Troia, Portugal (1986).
6. Banerjee, P.K. and Butterfield, R., "Boundary Element Methods in Engineering Science", McCraw-Hill (1981)
7. Rizzo, F.J. and Shippy, D.J., "An Advanced Boundary Integral Equation Method for Three-Dimensional Thermo-Elasticity", Int.J.Num.Meth. in Eng'g, 11, 1753 (1977).

PART IV

MULTILEVEL AND INTERDISCIPLINARY OPTIMAL DESIGN

INTERDISCIPLINARY AND MULTILEVEL OPTIMUM DESIGN

Jaroslaw Sobieszczanski-Sobieski
NASA Langley Research Center
Hampton, Virginia 23665 USA

and

Raphael T. Haftka
Virginia Polytechnic Institute and State University
Blacksburg, Virginia 24061 USA

INTRODUCTION

Engineering system design used to be compartmentalized by discipline. Material specialists would design better materials, fluid mechanics specialists would design optimum shapes, structural analysts would produce optimum structural designs based on materials and loads obtained by material and fluid mechanics specialists, and so on. Occasionally, interdisciplinary effects forced cooperation between disciplines. Aeroelastic phenomena such as flutter or loss of control-surface effectiveness forced aerodynamic and structural analysts to cooperate in the creation of the new discipline of aeroelasticity. However, when such interdisciplinary phenomena did not force cooperation, very little existed, beyond the conceptual design level.

While integrated design is more truly optimal than compartmentalized design (e.g., ref. 1), the difference in performance was not enough of an incentive to overcome the difficulties associated with design integration until two modern developments provided an additional incentive to do so.

The first development is the advent of tailored materials such as graphite-epoxy composites which permit the designer to tailor material properties to suit the specific requirements of the system being designed. The second development is the introduction of active control systems which permit a designer to improve performance through the use of a control system rather than by improving structural, aerodynamic, acoustic or other system characteristics.

The increasing interdisciplinary nature of the design process is most noticeable in the aerospace industry. A case

NATO ASI Series, Vol. F27
Computer Aided Optimal Design: Structural and
Mechanical Systems. Edited by C. A. Mota Soares

in point is the Grumman X-29A forward-swept-wing fighter for which composite materials were tailored to produce favorable aerodynamic-structure interaction. Because a metal swept-forward wing has an inherently destabilizing interaction between bending and twisting, it is not practical to build this type of wing with metal. However, a composite material was developed to reverse this destabilizing interaction and make the X-29A design feasible.

Integration in the design of complex engineering systems can be achieved at the governing equations level, by decomposition into self-contained but coupled tasks, or by judicious use of both approaches.

Equation-level integration in analysis typically begins with a realization that a number of disciplines contribute terms to equations that describe a particular physical phenomenon. Then, it is logical to form a unified set of equations from the terms contributed by the participating disciplines, research the best ways of solving these equations, and build operational experience by verification tests and applications.

In most cases equation-level integration is not required to describe physical phenomena, and integration is useful only for obtaining superior designs. Then integration by decomposition is in order. Each discipline remains a self-contained task. The integration is achieved by defining the interdisciplinary information channels, and finding the best ways of sequencing (iterating) the disciplinary computations. The sequencing can be strictly serial, figure 1, or it can exploit parallelism that leads to a hierarchial arrangement shown in figure 2. Frequently, a mix of the serial and parallel schemes is appropriate.

The same two approaches may be distinguished with regard to synthesis. Equation-level synthesis relies on equation-level integrated analysis in the same manner as single discipline synthesis - as a source of data describing the behavior and sensitivity of the object being optimized. The optimization procedure is shown in figure 3.

The same scheme can accommodate synthesis based on analysis decomposed serially by placing the content of figure 1 in the "ANALYSIS" box in figure 3. Examples of this type of multidisciplinary optimization applied to a large space-based antenna structure and a glider configuration are provided later.

However, if the analysis is decomposable in a way shown in figure 2, then the optimization can also be decomposed as shown in references 2, 3, giving rise to a multilevel optimization scheme illustrated in figure 4. According to reference 4, each box in the scheme can represent a physically separable subsystem (object decomposition), or a discipline analyzing one of many aspects of the same object (aspect decomposition).

The scheme relies on the separate optimization subtasks self-contained within subsystems or disciplines, and on sensitivity derivatives of the optimum to the inputs coming from the next higher level in the decomposition hierarchy. An example of one such sensitivity derivative is a derivative of minimum structural weight with respect to the wing aspect ratio. An algorithm for computing optimum sensitivity derivatives without engaging in a costly finite difference procedure is given in reference 5. Optimization of the entire system uses these derivatives for approximate assessment of the effects of system-level design decisions on the subsystems and contributing disciplines.

The objective of the present paper is to survey multidisciplinary optimization applications and focus on multilevel optimization as a means for integrating the design process. The paper begins with a survey of multidisciplinary optimization problems, continues by reviewing one practical multilevel optimization technique applied to a generic system and concludes with an example of a multilevel multidisciplinary optimization.

SURVEY OF INTERDISCIPLINARY OPTIMIZATION PROBLEMS

In some distant future we may expect that engineering system design will be fully integrated. At the present, design integration is typically proceeding by combining the design process of two or three disciplines. The following survey discusses briefly several areas of multidisciplinary optimization and then describes in more detail two multidisciplinary design studies.

Controls and Structures

Active control systems are intended to reduce the demands on the structure by load alleviation or by active damping. Load alleviation systems anticipate naturally occurring loads and add loads that tend to cancel some of these original loads. A typical example is the load alleviation system designed for the B-52 Bomber which senses gusts ahead, and deflects control surfaces to alleviate them. Similar systems (called active suspension) are envisioned for future cars which will sense the bumpiness of the road and apply compensating loads to improve ride quality.

Active suspension systems also include an active damping component in that they sense vibrations and apply forces to damp them out. This active vibration damping is particularly important in applications to large and flexible space structures (e.g,. ref. 6).

At present the control and structure design are compartmentalized with the control system designer assuming that the structural design is given. There is, however, a growing interest in simultaneous control/structure design (see ref. 6 for additional references). As shown in reference 6, the compartmentalized design approach can result in very large weight penalties if it leads to a structure which is too flexible.

Material and Structure

Tailored materials such as graphite-epoxy composites permit the designer to tailor material behavior to suit the structural application. For example, by proper selection of ply orientations it is possible to produce a composite laminate with a miniscule coefficient of thermal expansion, suitable for minimizing thermal deformations in large space antennas. Similarly, it is possible to take advantage of the different failure characteristics of various ply combinations. For example, reference 7 shows that the failure load of compression-loaded plates with holes can be increased by removing the zero plies from a strip containing the hole.

At the present, while there is substantial interaction between composite material designers and structural designers, the design process is still disjoint. This is not acceptable because in composite materials, both material response and failure characteristics depend on the structural applications. That is, the same composite material can display vastly different characteristics when used in different structures. Therefore, optimum structural design must be coupled with the material design process to produce a true optimum, and to prevent unexpected failure modes.

Structures and Aerodynamics

The interaction between structures and aerodynamics is strong enough so that even in the traditional design process it was considered a separate superdiscipline - aeroelastic design. However, the aeroelastician worried only about aeroelastic interactions where the deformations of the structure affect the aerodynamic loads. Thus aeroelastic design considered phenomena such as divergence, flutter, and control surface effectiveness. Aeroelasticians did not consider the possibility of using the structural deformation to improve aerodynamic performance (because the effect is too small in metal aircraft), or changing the aerodynamic design to reduce structural weight. This latter trade-off between aerodynamic and structural performance was considered in an approximate way in the conceptual design stage.

With the growing use of composite materials designers are beginning to consider using structural deformations to improve aerodynamic performance. The next section of this paper contains an example of combined aerodynamic/structural design of a glider. It was performed by a cooperative effort between aerodynamicists and structural analysts, rather than by a person or a group who mastered both disciplines.

Structure and Heat Conduction

Currently there is little interaction between the thermal design and structural design of systems subjected to high thermal loads. A typical example is a reentry vehicle which requires thermal protection systems such as insulation, ablation shields, or passive and active cooling systems. In the current design process the thermal protection system is designed to keep the temperature of the structure below a specified limit. The structure is then designed to carry the thermal and mechanical loads at that temperature. The current interaction between the two design processes is due to the fact that the structural material distribution affects the temperature distribution.

Reference 8 showed that the sequential thermal/structural design process is not always optimal. The combined design process can reduce the total weight of the system by overdesigning the thermal protection system to produce a structure operating at lower temperatures where its strength is higher. The combined design approach is facilitated by finite element software packages which permit the analyst to perform both the structural and thermal analysis simultaneously.

MULTIDISCIPLINARY DESIGN EXAMPLES

To complete this survey of interdisciplinary optimization problem, we consider in more detail two multidisciplinary design studies. The two have in common a serial decomposition approach (see fig.3).

Aerodynamic/Structural Optimization of Glider Wing

The integrated aerodynamic/structural design of the glider wing (ref. 9) is an example of combined optimization where the disciplinary analyses are performed separately and integrated through the optimizer. This case also provides an example of the pay-offs of integrated design.

The glider mission is to fly over a distance by gaining altitude circling in a thermal, and then glide to the next thermal, losing altitude in the process (see fig. 5). One measure of performance used in the study was the cross-country speed of the glider which is the average speed considering both phases of the flight. A second measure of performance was the weight of the glider for a given cross-country speed. The design variables and constraints are summarized in tables 1, 2 and figures 6, 7. Two optimization procedures used to demonstrate the advantage of the integrated approach are shown schematically in figure 8. The first is a sequential approach, typical of the traditional compartmentalized approach. The aerodynamic design is first obtained for an initial estimate of the weight by varying the aerodynamic variables to maximize the cross country speed. The loads based on this aerodynamic design are then used to optimize the structure for minimum weight, and the new weight is used to restart the aerodynamic design. The process was considered converged when the change in weight from one iteration to the next was less than 0.2 percent. The combined optimization varies simultaneously both the aerodynamic and structural parameters to obtain the optimum design. The combined approach is able to take advantage of two interactions that the sequential approach cannot. The first interaction is the reduction in structural weight that can be achieved by modifications in aerodynamic shape, and the second is the improvement in aerodynamic performance which can be achieved by tailoring structural deformations.

The two design procedures were applied to a simple model of the wing. The aerodynamic analysis was based on lifting line theory, the aerodynamic design variables controlled the planform and twist distribution, and constraints were placed on maximum angle of attack and bank angle. The structural analysis was based on a beam model, the design variables were skin and web thickness and spar-cap areas, and constraints were placed on stresses and the divergence speed. The results of the two optimization procedures are compared in table 3. The iterated sequential design performance was only one percent inferior in the performance to the integrated design. However, this one percent in performance was parlayed into an 11 percent weight gain, when the combined design was optimized for minimum weight. The reason for the disparity is that the structural design was limited to orthotropic skin (no changes in ply orientations or percentages of various plies allowed) so that no anisotropic aeroelastic tailoring for improving aerodynamic performance was available. On the other hand, there was complete freedom to tailor the aerodynamics to help the structure.

Optimization of Antenna Parabolic Dish Structure for Minimum Weight and Prescribed Emitted Signal Gain

This particular optimization application has been described in reference 10. The object of optimization is the minimum weight design of the support structure of a large (55 m diameter) parabolic dish antenna shown in figure 9. The support structure is made up of two surface lattices held apart by connecting struts forming a tetrahedral-cell truss. The concave side lattice is overlaid with a fine wire mesh that forms a parabolic reflector converting the electromagnetic radiation emitted from the feed placed in the focus into a coherent beam.

In orbit, the antenna moves through the Earth shadow and changes its orientation relative to the Sun. The resulting heating which varies over the structure and also in time distorts the support structure and the parabolic reflector surface causing a loss of emitted signal strength. The optimization calls for finding the cross-sectional areas of the support trusses such that the structural weight is minimum while not permitting the surface distortion to rise above the level that would weaken the electromagnetic radiation below a prescribed limit.

Two design variables were chosen to control three cross-sectional areas: one for all members in both of the two surface lattices, and one for all the connecting truss members. The analysis begins with thermal analysis to determine the member temperatures at a particular location on orbit. The temperatures are functions of the member cross-sections and generate stresses and deformations which are calculated next. The deformations are passed to the electromagnetic radiation analysis program to obtain the resulting weakening of the emitted signal.

Thus, the analyses are arranged serially as shown in figure 10 (the thermal and thermal-structural analyses were executed in this implementation as processors of the same finite element programming system, reference 11). Derivatives required for optimization, performed by a useable-feasible directions algorithm (ref. 12), were obtained by a finite difference procedure. In keeping with the programming system approach (ref. 13), the optimizer was coupled directly not with the full analysis (boxes TA, SA, EMRA in fig. 10), but with an approximate analysis (box AA) to conserve the computer resources and to leave their control in the hands of the user. The approximate analysis is a linear, derivative-based extrapolation with an automatic switching to reciprocal variables as proposed in reference 14.

The results shown in figure 9 indicate a weight reduction by more than 1/3 from the initial value representing the best design achieved without systematic optimization. The surface precision as measured by the deflection RMS value has also improved, and the emitted signal strength measured by the gain value was kept above the required minimum of 19000. Judging by the results the optimizer reduced weight and distortion together. This occurred because in a thermally loaded structure, the internal forces may be reduced by reducing the structural sizes. The optimizer took advantage of this and achieved increased performance and reduced weight.

MULTILEVEL OPTIMIZATION

The last two examples, the glider and the antenna, demonstrated benefits attainable when optimization of engineering systems is carried out by means of stringing out disciplinary analyses in a sequence coupled to an optimizer set to improve a measure of system performance under all the appropriate constraints. The set of analyses included in the sequence responds to the optimizer requests for information as if it was a single analysis; therefore, one may call this arrangement optimization with integrated analysis.

Although this arrangement is demonstrably effective, it may not be practical for very large design tasks involving numerous engineering staff. Engineers tend to cluster into specialty groups operating concurrently. This time-honored mode of operation, which results in a broad work front reducing the design elapsed time, requires decomposition of the system optimization into several smaller sub-optimizations each assigned to an engineering group. The remainder of the paper is devoted to an algorithm that supports such decomposition while preserving the internal couplings of the system optimization problem.

This section introduces multilevel optimization by decomposition in a particular formulation that applies to structures. As shown in reference 15, it is natural to base that formulation on the well-known analysis by substructuring which is a form of the object decomposition.

<u>Optimization Terminology</u>

An optimization formulation without decomposition serves as a reference from which the multilevel optimization algorithm is derived. The optimization is defined in terms of: the design variables, Z_b (see Table 4 for notation), which are the cross-sectional dimensions of the structural components; the objective function F(Z) that can be any computable function of these variables (structural mass is the frequent choice);and the constraints, $g_w(Z)$ imposed on the behavior variables to account for the potential failure modes. Writing constraint functions as

$$g = d/c - 1 \leq 0 \qquad (1)$$

the optimization problem in a standard formulation is

$$\min F(Z);\ \text{such that}\ g_w(Z) \leq 0 \qquad (2)$$

and requires a search of the design space considering all the design variables and constraints concurrently. In contrast, the algorithm presented in the next section breaks the problem into a number of search and analysis operations, each concerned with a smaller number of design variables and constraints.

<u>Preliminary Definitions</u>

The diagram in figure 11 shows a structure decomposed into several levels of substructures. The term "substructure" will refer to any entity in this decomposition scheme including the extremes of the full, assembled structure represented by the box on the top of the pyramid and single structural components representing the ultimate geometrical details appropriate to the problem at hand. The substructure levels are numbered from 1 on the top to i_{max} at the bottom.

The hierarchical nature of the scheme instigates the use of the term "parent" to the structure at level i which, in turn, is decomposed into a number of "daughter" substructures at level i+1. A daughter may have only one parent and that parent must be at the level immediately above. Thus, it will be convenient to label each substructure SSijk, where i denotes the level, j defines the position at the level i counting from the left, and k identifies the parent's position at the level i-1. The substructure occupying the lowest position in a particular parent-daughter succession represents the ultimate level of detail at which the decomposition stops. There is no requirement that all such substructures must be at the same bottom level i_{max}. In discussions involving more than one substructure, the triplets nlp, mkl, ijk, are used to distinguish among the substructures forming the the hierarchy shown in figure 12.

Substructuring analysis (e.g., refs. 16, 17, 18) establishes the following functional relations (note that the subscript for the parent substructure is omitted):

$$Q^{ij} = f(K^{bij}, P^{bij}) \tag{3}$$

$$K^{bij} = f(K^{ij}) \tag{4}$$

$$K^{ij} = f(K^{bi+1,j}) = S_K(K^{bi+1,j}) \tag{5}$$

$$M^{ij} = f(M^{i+1,j}) = \sum_j M^{i+1,j} \tag{6}$$

$$P^{bij} = f(P^{ij}) \tag{7}$$

$$P^{ij} = S_P(P^{bi+1,j}) \tag{8}$$

The symbol f appearing in equations 3-8 denotes a general functional relationship which is different for each equation, and is computable in a manner prescribed by the particular substructuring algorithm chosen. For example, equations 4 and 7 take the form of matrix equations given in reference 17, Ch.9, Sec.1, as equations 9.13 and 9.14, respectively.

For SSijk at the ultimate level of detail, the distinctions between K^{bij}, P^{bij}, and K^{ij}, P^{ij} vanish, and K^{ij}, M^{ij}, derive directly from Z^{ij}. Consequently:

$$K^{bij} = K^{ij} \tag{9}$$

$$P^{bij} = P^{ij} \tag{10}$$

$$K^{ij} = f(Z^{ij}) \tag{11}$$

$$M^{ij} = f(Z^{ij}) \tag{12}$$

The local constraints that arise in SSijk at the ultimate level of detail involve calculation of stresses, strains, local buckling, etc., from Q^{ij} and Z^{ij}. In addition, constraints may be imposed on the internal forces, critical forces, and displacements of SSijk to account fully for all the constraints that would have been included in the one-level optimization problem represented by equation 2.

Although the foregoing definition of substructuring analysis is based on the finite element stiffness method, the use of a finite element analysis is not mandatory for the multilevel optimization algorithm presented here. As far as that algorithm is concerned, the analysis is a "black box" where only the inputs and outputs are important.

Multilevel Optimization Algorithm

With the substructuring scheme and analysis established in the foregoing, this section describes the optimization algorithm itself. The essentials of the computer implementation are also given.

Basic Concept.- The basic idea for the proposed multilevel optimization by substructuring stems from the elementary observation, based on equations 3 through 8, that the effect of a daughter SSijk on its parent SSi-1,kl is felt only through K^{bij}, M^{ij}, and P^{bij} which depend on $K^{b,i+1,j}$, $M^{i+1,j}$, and $P^{b,i+1,j}$, respectively. Consequently, the entries of $K^{b,i+1,j}$, $M^{i+1,j}$, and $P^{b,i+1,j}$ may be manipulated as generalized design variables without disturbing the results of the SSi-1,kl analysis as long as the entries of K^{bij}, M^{ij}, and

P^{bij} are held constant. If these entries are held constant, then the boundary forces Q^{ij} acting on every SSijk in SSi-1,kl remain constant and the effect of manipulating the generalized design variables in a particular SSijk is limited to that SSijk itself and its daughters. As explained later, the purpose of the above manipulation of the matrix entries is not to minimize the substructure mass M^{ij} which, as stated above, remains constant. Instead, the purpose is to improve satisfaction of the constraints in the SSijk and its daughters, while performing the task of the total mass optimization at the assembled structure level.

Invariance of the entries K^{bij}, M^{ij}, and P^{bij} can be enforced by rewriting equations 4 through 8 as equality constraints.

$$h_K^{ij} = K^{bij} - f(K^{bi+1,j}) = 0 \tag{13}$$

$$h_M^{ij} = M^{ij} - f(M^{i+1,j}) = 0 \tag{14}$$

$$h_P^{ij} = P^{bij} - f(P^{bi+1}) = 0 \tag{15}$$

Equations 13, 14, and 15 establish the entries of K^{bij}, M^{ij}, and P^{bij} as given parameters in optimization of SSijk. Simple replacement of indices renders these equations valid for SSi-1,kl and redefines the optimization parameters of the daughter SSijk as generalized design variables in the optimization of its parent SSi-1,kl, so that

$$\{X^{i-1,k}\} = \{Y^{ij}\} \tag{16}$$

$$\{Y^{ij}\}^t = \{K^{bij} | M^{ij} | P^{bij}\}^t \tag{17}$$

These equations define a recursive relation of the variables and parameters that extends from the top of the substructuring scheme to the bottom.

Of course, the number of design variables T^{ij} must exceed the number of constraints V^{ij} (which is equal to the number of individual equations in the vector equations 13, 14, and 15).

$$T^{ij} > V^{ij} \tag{18}$$

for a design freedom to exist, allowing for the symmetry of the stiffness matrices. Otherwise, if

$$T^{ij} \leq V^{ij} \tag{19}$$

then the equality constraints of equations 13 through 15 either define the SSijk design variables uniquely or overdetermine them.

The basic concept outlined above translates into an algorithm to be introduced now in detail.

Optimization at the most detailed level.- Introduction of the optimization algorithm begins at the level of the most detailed substructures. Consequently, equations 9 through 12 apply and the design variables are the cross-sectional dimensions so that

$$X^{ij} = Z^{ij} \qquad (20)$$

and the parameters (held constant during optimization at this level) are

$$\{Y^{ij}\}^t = \{K^{ij}|M^{ij}|P^{ij}\}^t \qquad (21)$$

It is assumed that a complete, top-down, substructuring analysis for an initial structure has been carried out so that for an SSijk one has computed its Q^{ij}, while its M^{ij}, Z^{ij}, K^{ij}, and P^{ij} are given.

Optimization for improvement of inequality constraint satisfaction is achieved by minimizing a single measure representing all the constraints and called the cumulative constraint, a concept similar to the use of a penalty function. A differentiable cumulative constraint function can be obtained (as it was in ref. 19) by means of the Kreisselmeier-Steinhauser function (KS) defined in reference 20.

$$C^{ij} = KS(g_w^{ij}) = 1/\rho \ln (\sum_w \exp (\rho g_w^{ij})) \qquad (22)$$

that has the property of approximating the maximum constraint so that

$$\underset{w}{MAX}(g_w^{ij}) < KS(g_w^{ij}) < \underset{w}{MAX}(g_w^{ij}) + 1/\rho \ln (W^{ij}) \qquad (23)$$

with the factor ρ controlled by the user. Thus, the KS function serves as a convenient single measure of the degree of constraint violation (or satisfaction).

Analysis of SSijk yields the local inequality constraints as

$$g_w^{ij} = f(X^{ij}, Y^{ij}, Q^{ij}) \qquad (24)$$

Based on the above definitions, the optimization problem is formulated.

$$\min_{X^{ij}} C^{ij}(X^{ij}, Y^{ij}, Q^{ij}) \text{ such that} \qquad a) \qquad (25)$$

$$h_K^{ij} = 0 \qquad b1)$$

$$h_M^{ij} = 0 \qquad b2)$$

$$h_P^{ij} = 0 \qquad b3)$$

$$L^{ij} \le X^{ij} \le U^{ij} \qquad c)$$

Solution of this optimization problem (by any technique available) yields a constrained optimum described by a vector π^{ij} composed of the minimum value of the cumulative constraint, $\bar{C}^{ij}$, and the optimal vector of the design-variables, $\bar{X}^{ij}$

$$\pi^{ijt} = \{\bar{C}^{ij} | \bar{X}^{ij}\}^t \qquad (26)$$

This solution depends on Y^{ij} and Q^{ij}, and the derivatives $d\pi^{ij}/dY_z^{ij}$ may be expressed by a chain differentiation to account for equations 3 and 21 that tie Q^{ij} to Y^{ij}

$$d\bar{C}^{ij}/dY_z^{ij} = \partial\bar{C}^{ij}/\partial Y_z^{ij} + \sum_r (\partial\bar{C}^{ij}/\partial Q_r^{ij})(\partial Q_r^{ij}/\partial Y_z^{ij}) \qquad (27)$$

$$d\bar{X}^{ij}/dY_z^{ij} = \partial\bar{X}^{ij}/\partial Y_z^{ij} + \sum_r (\partial\bar{X}^{ij}/\partial Q_r^{ij})(\partial Q_r^{ij}/\partial Y_z^{ij}) \qquad (28)$$

In equations 27 and 28, the partials of $\bar{C}^{ij}$ with respect to Y_z^{ij} and with respect to Q_r^{ij} are obtained from the algorithm described in reference 5, and the partial Q_r^{ij} with respect to Y_z^{ij} by conventional structural sensitivity analysis. Parenthetically, one may add that the algorithm of reference 5 uses second derivatives of constraints that may be expensive to calculate. However, a modified version of the algorithm is available in reference 21 that avoids the cost of second derivatives and calculates the sensitivity derivatives for $\bar{C}^{ij}$, but not for $\bar{X}^{ij}$.

Optimization of the lowest parent substructure.- The design variables for all parent substructures control the stiffness and mass distribution in that substructure. They could be elements of the substructure boundary or mass matrices, or quantities which control these entries. Because these substructure design variables are not necessarily tangible quantities, they are referred to in the following as "generalized" design variables. As shown in figure 12, the parent substructure SSmkl, m=i-1, receives from its daughters, SSijk, the minimized values of their cumulative constraints, $\bar{C}^{ij}$, optimal values of their design variables, $\bar{X}^{ij}$, and the optimum sensitivity derivatives of these quantities with respect to parameters, Q^{ij} and Y^{ij}, calculated from equations 27 and 28.

Preparing for the formulation of the optimization problem for the parent substructure, we consider the recursive relation between the design variables and parameters according to equations 16 and 17, and recognize that equations 9 through 12 do not apply. When optimizing the parent substructure, we want to improve satisfaction of the assembled substructure inequality constraints, such as limits on its elastic deformations and stability that depend on the substructure stiffness, mass, and boundary forces:

$$g^{mk} = g^{i-1,k} = f(X^{mk}, Y^{mk}, Q^{mk}) \tag{29}$$

At the same time, we want to improve constraint satisfaction in all the substructure daughters. These can be approximated (as in ref. 19) by linear extrapolation of their cumulative constraints using the derivatives from equation 27 and replacing Y^{ij} with X^{mk} according to equation 16.

$$\bar{C}_e^{ij} = \bar{C}_o^{ij} + \sum_t (d\bar{C}^{ij}/dX_t^{mk})\Delta X_t^{mk} \tag{30}$$

This extrapolation plays a key role in the algorithm because it approximates the daughter-parent coupling without incurring the expense of reoptimizing the daughters (repeating eq. 25) for every change of the parent design variables.
Including the $\bar{C}_e^{ij}$ values together with g^{mk} in a cumulative constraint formed by the KS function we have

$$C^{mk} = 1/\rho \ln \left(\sum_{w} \exp(\rho g_w^{mk}) + \sum_{j} \exp(\rho \bar{C}_e^{ij})\right) \tag{31}$$

and the optimization problem to be solved for the parent SSi-1,kl is

$$\min_{X^{mk}} C^{mk}(X^{mk}, Y^{mk}, Q^{mk}) \text{ STOC} \qquad a) \tag{32}$$

$$h^{mkt} = \{h_K^{mk} | h_M^{mk} | h_P^{mk}\}^t = 0 \qquad b)$$

$$L^{mk} \leq X^{mk} \leq U^{mk} \qquad c)$$

$$L^{ij} \leq X_e^{ij} \leq U^{ij} \qquad d)$$

where

$$X_e^{ij} = X_o^{ij} + \sum_{t} (d\bar{X}^{ij}/dX_t^{mk}) \, \Delta X_t^{mk} \tag{33}$$

The increment ΔX^{mk} is defined as

$$\Delta X^{mk} = X^{mk} - X_o^m \tag{34}$$

The constraints of equation 32b are analogous to equations 25b1, b2, b3 written in a compact format. The constraints of equation 32c incorporate the side constraints to prevent the design variables from attaining physically impossible values (e.g., negative diagonal entries in a stiffness matrix) and include the move limits to control the extrapolation errors introduced by equation 30. The constraints of equation 32d are introduced to keep the design variables in the daughters from exceeding their side constraints. These constraints are not essential because their function may be performed directly by the daughter side constraints. In fact, omitting the constraints of equation 32d eliminates the need for the derivatives of $\bar{X}^{ij}$ and allows replacing the algorithm of reference 5 by the much less costly algorithm of reference 21. However, these constraints are included in this description for completeness.

Solution of the problem of equation 32 generates the result vector and its derivatives that are analogous to those of equations 26, 27, and 28 with the indices ij replaced by m=i-1, and k.

Optimization of the next parent structure.- Moving on to the substructure SSnlp, everything stated in the preceding subsection on optimization of SSmkl applies to SSnlp directly, provided that: the indexes n, l, and p are replaced by another triplet, say, α, β, γ, that identifies the parent of SSnlp at the level α = n-1; and the indexes m, k, l are replaced by n, l, p. For consistency, equation 32d, if used, should be replicated to encompass fully each line of succession emanating downward from SSnlp. Beyond these changes, no new conceptual elements are introduced, and no additional definitions or discussion are needed at the junctions between the levels until one arrives at the top level. Hence, any number of intermediate levels of substructuring can be inserted, if physically justified, into a line of succession extending downward from the assembled structure on the top; i.e., the algorithm is recursive.

Optimization of the assembled structure.- The assembled structure is designated SS110. Its optimization problem is similar to the one described for a parent substructure SSmkl with the following differences:

1. No parameters are defined solely for the decomposition purposes; therefore, there is no need for the equality constraints to enforce constancy of the mass and the boundary stiffnesses.
2. The objective function is the mass of the assembled structure.
3. There is no need for a single cumulative constraint (unless one needs it to reduce the number of constraints to be processed at that level).
4. The boundary forces are the external loads on the assembled structure.

Accounting for these differences, the optimization problem for the top level is

$$\min_{X^{11}} M^{11}(X^{11}) \text{ such that} \qquad \text{a)(35)}$$

$$g^{11} \leq 0 \qquad \text{b)}$$

$$\bar{C}_e^{2j} \leq 0 \qquad \text{c)}$$

$$L^{11} \leq X^{11} \leq U^{11} \qquad \text{d)}$$

$$L^{2j} \leq \bar{X}_e^{2\,j} \leq U^{2j} \qquad e)$$

where equation 35e is analogous to equation 32d with the limits L^{2j}, U^{2j} applied in conjunction with extrapolations of the type expressed by equation 33, extended recursively to encompass all the levels below as mentioned in the subsection on SSnlp. Unlike in the daughters SSijk, the optimization of SS110 does not have to be analysed for the optimum sensitivity. Information transmitted to the top level optimization problem is indicated in figure 12.

Iterative procedure.- When the SS110 optimization is completed, the entire structure has acquired a new distribution of stiffness and mass within the move limits. Hence, the analysis must be repeated and followed by a new round of substructure optimizations in an iterative manner until convergence. Accordingly, the procedure follows these steps:

1. Initialize all cross-sectional dimensions.
2. Perform a substructuring analysis, including for each substructure at each level the transformation of the stiffness matrix into the boundary stiffness matrix and the transformation of the forces applied to the interior degrees of freedom to the forces coinciding with the boundary degrees of freedom. Calculations of the behavior derivatives needed for the ensuing optimizations and for the optimum sensitivity analyses are included in the substructuring analysis.
3. Perform the operations of optimization and optimum sensitivity analysis as defined by equations 25 through 34.
4. Optimize the assembled structure as defined by equation 35.
5. Repeat from step 2 and terminate only when: all constraints g^{ij} are satisfied at all levels and M^{11} has entered a phase of diminishing returns.

This procedure is illustrated in figure 13 by a flow chart in the Chapin's chart format (ref. 22).

Salient features of the algorithm.- In perspective, the multilevel algorithm differs from a single-level one in a number of the following salient features.

A multitude of smaller problems, that may be processed concurrently, replace a single large problem. Although the subproblems are isolated, their coupling is preserved because the influence of the changes in the parent on the daughters is represented by linear extrapolation based on the optimum sensitivity and behavior sensitivity derivatives. With the exception of the most detailed level, the stiffness and mass distributions are controlled directly by generalized design variables. Mass is the objective at the top level, while the constraint satisfaction improvement is the objective at all levels below.

Selection of the generalized design variables is a matter of judgment. In the extreme case, one may choose to control as design variables all entries of the boundary stiffness matrix, boundary forces vector, and mass of each daughter; although, intuitively, this would seem impractical. Experience will probably show that a limited control, e.g., over the diagonal entries of the stiffness matrix only, will suffice in most cases.

The overall procedure building blocks; i.e., the operations of substructure analysis, constraint calculations, optimization, and the behavior and optimum sensitivity analyses are "black boxes" whose algorithmic contents may be freely replaced provided that the input/output definitions remain unchanged. For example, different types of structural analysis may be used at each level and even for each substructure, as it will be shown in the numerical example.

PORTAL FRAME EXAMPLE

Problem Description.- The subject algorithm was tested by optimizing, with and without decomposition, a framework structure similar to the one used in references 19, 23, and 24. As shown in figures 14 and 15, the framework assembled at level 1 decomposes into three box beams, each beam being a substructure at level 2. Finally, each beam decomposes into three walls (the fourth wall is symmetric), each wall being the most detailed substructure at level 3. The external loads were applied at one corner of the framework as shown in figure 14. There were no interior loads on the substructures.

The objective was to minimize the structural material volume subject to constraints on the displacements of the loaded point, the in- and out-of-plane elastic stability of each beam treated as a column, and the stresses and local buckling of the wall panels treated as stringer-reinforced plates. There were also minimum gage constraints and the physical realizability constraints on the cross-sectional dimensions.

The objective functions, design variables, parameters, and constraints are defined for the multilevel optimization in table 5. A comprehensive description of all the physical and computational details of the test problem is given in reference 15.

Tools for Analysis and Design Space Search.- A finite element analysis was used to calculate the framework's displacements and the beam end-forces. Stresses in the beams loaded with the end-forces were computed by engineering beam theory. The beams were treated as columns for stability analysis, and local buckling of the walls was based on closed form "designer handbook" formulas provided in references 25, 26, and implemented as described in reference 27.

At each level, the optimization was conducted by the same general-purpose nonlinear mathematical programming code CONMIN, based on the useable-feasible directions technique and documented in reference 28.

Three-Level Optimization.- The framework was first optimized without decomposition to establish reference results. Then, the multilevel optimization algorithm was applied to the structure decomposed as shown in figures 14 and 15. In the decomposition, the stiffened panels are daughters clustered in triplets under a parent box beam. The beams, in turn, are daughters of the assembled structure.

As shown in table 5, the top level optimization manipulates the beam extensional and bending stiffnesses through the cross-sectional areas and bending moments of inertia. The cross-sectional area also controls the beam volume which contributes directly to the objective function.

At the middle level, the stiffnesses expressed by the area and moment of inertia become fixed parameters and the variables are the wall membrane stiffnesses controlled by the geometrical dimension variables. These variables, and consequently the membrane stiffnesses become fixed parameters at the bottom level at which the ultimate detail dimensions are engaged as variables. The equality constraints arise between the parameters and variables. Owing to relative simplicity of the expressions involved, (see Appendix, ref. 15), these constraints were solved explicitly.

Examination of table 5 in conjunction with the previous description of the analysis tools illustrates the point that dissimilar analyses may be used as needed at different places in a decomposition scheme.

The sensitivity analysis of behavior was carried out by a single step forward finite difference technique. The optimum sensitivity analysis was based on the algorithm given in reference 5.

Results and Remarks on the Method Performance.- Figure 16 shows a sample of results obtained with and without decomposition. The starting points for both methods are the same. The normalized plots illustrate the objective function, a selected individual constraint, and a cumulative constraint containing the above individual constraint as they varied over the iterations. An iteration is defined in the optimization without decomposition as the following set of operations: one analysis including gradients, computation of a useable-feasible search direction, and finding a constrained minimum in that direction. In the three-level optimization, it is defined as one execution of the series of steps listed in the procedure definition in the previous section.

The results verified that the multilevel algorithm was capable of finding a feasible design having an objective function close to and, in some cases lower than, the reference optimization without decomposition. As in reference 19, differences up to 72.1% were observed among the detailed design variables obtained by the two methods. However, these differences were no larger than those observed by comparing the designs obtained without decomposition starting from different initial design points. Therefore, these differences can be attributed to the problem non-convexity. The jagged appearance of the graphs in figure 16 is a characteristic of the usable-feasible directions search algorithm, amplified in the multilevel optimization by the extrapolation errors. A detailed comparison of the results from both methods is given in reference 15.

Regarding computational efficiency, the main intrinsic advantage of the multilevel algorithm is in its capability to process the subproblems concurrently. Demonstration of this advantage would require a large application, distributed computing, and division of work among many people. Consequently, computational efficiency was not one of the goals in execution of the relatively small numerical example on a conventional serial computer. However, the example showed that the amount of computational labor per iteration was less in the multilevel algorithm than in the single-level, conventional one, and that both algorithms required about the same number of iterations for convergence. The example also showed that for the multilevel algorithm programming of the operations of data moving and bookkeeping was the dominant effort.

DECOMPOSITION APPROACH IN OPTIMIZATION OF A GENERIC ENGINEERING SYSTEM

In the preceding discussion, the multilevel optimization by decomposition was introduced using a structure that was partitioned into components - an example of an object decomposition. This section describes that approach as it was extended in reference 3 to a case of a generic engineering system decomposable in both the object and aspect sense.

Decomposition of Two-Level System.- The key to the proposed approach is a formalized decomposition of the large design problem into a set of smaller manageable subproblems coupled by means of the densitivity data that measure the change of the subsystem design due to a change in the system design.

Let ES be an engineering system composed of the subsystems $SS_1, \ldots SS_2, \ldots SS_i$, SS_n as shown in figure 17 (the abbreviations are defined in table 6, and table 7 gives examples for the generic quantities in the context of aircraft design). The design variables are grouped in a vector SV for ES and the vectors DV_i for SS_i. The ES has a performance index PS that should be maximized within the system constraints collected in a vector GS. The ES imposes demands on each SS_i. These demands are quantified by entries of a vector DS_i which depends on SV through analysis of ES. Each SS_i is designed by manipulating DV_i so that it meets its DS_i, regarded as constants, while maximizing its safety margin SM_i representing (e.g., by using the KS function, ref. 20) a set of subsystem constraints GSS_i. These tasks, separate for each SS_i, can be carried out concurrently by whatever means the SS_i designers choose, including the appropriate analysis, optimization, and also judgment and experimentation.

A new element required under the proposed approach is evaluation of the sensitivity of the maximum (optimum) SM_i to changes in DS_i in the form of optimum sensitivity derivatives $\partial SM_i/\partial DS_i$. At the ES level, these derivatives combined with the derivatives $\partial DS_i/\partial SV$ in chain differentiation yield the sensitivity of SM_i to changes in SV in the form of derivatives $\partial SM_i/\partial SV$. The maximum SM_i and its derivatives show the ES designer, with a linear extrapolation accuracy, how the change of SV that he controls will affect the SM_i for each SS_i. Guided by this information and by the ES analysis, the ES designer can decide which variables in SV to change and by how much in order to move toward the goal of satisfying all the constraints GS and GSS_i while maximizing the PS. The SV change will alter the DS_i. Responding to that, the SS_i designers modify their designs and pass updated information to the ES designer who, then, changes the SV again, and so on. In this manner, the ES and the SS_i designers carry on a systematic iteration toward an improved system design, trading data in the form of DS_i, SM_i, and their derivatives. Each designer works on a separate assignment with the control of PS vested in the ES designer, while the SS_i designers focus on their SS_i feasibility. The whole problem is decomposed, yet remains coupled by the ES-SS_i data exchange shown in figure 17.

Overall Procedure. - Based on the above qualitative description, one may now formulate a step-by-step procedure to implement the decomposition approach.

Step 1. Initialize the system.

Step 2. Analyze the system. Calculate PS, GS, DS_i, and $\partial DS_i/\partial SV$.

Step 3. Design subsystems SS_i. The DV_i are manipulated within upper and lower bounds, L_i and U_i, so as to maximize SM_i for given DS_i. The latter requires vector of equality constraints GE_i for those DS_i that are also functions of DV_i. These constraints enforce equality of the DS_i values prescribed at the system level and computed as a function of DV_i so that
$GE_i = DS_i(SV) - DS_i(DV_i) = 0$. Formally, the task may be formulated as an optimization problem

$$\max_{DV_i} SM_i\ (DV_i,\ DS_i) \quad \text{subject to constraints} \tag{36}$$

$$GE_i(DV_i, DS_i) = 0$$

$$L_i \leq DV_i \leq U_i$$

The output of the operation is: $\overline{SM}_i = (SM_i)_{max}$, and the optimal subsystem design variables, $\overline{DV}_i$.

Step 4. Analyze each SS_i design for sensitivity to the inputs received from the system to obtain the $\partial SM_i/\partial DS_i$.

Step 5. Modify the SV to improve the system design. In this operation, one uses the $\partial DS_i/\partial SV$, $\overline{SM}_i$, and $\partial \overline{SM}_i/\partial DS_i$ obtained in Steps 2, 3, and 4, to extrapolate each SM_i as a function of the increment ΔSV

$$SM_i(\Delta SV) = \overline{SM}_i + \frac{\partial \overline{SM}_i}{\partial DS_i}\frac{\partial DS_i}{\partial SV}\Delta SV \tag{37}$$

Improvement of the system design may be formalized as an optimization:

a) $\max_{SV} PS(SV)$ subject to constraints (38)

b) $GS(SV) \leq 0$, c) $\overline{SM}_i(SV) \geq 0$ (for all i)

d) $L \leq SV \leq U$

in which the system level analysis provides the PS and GS, and the $\overline{SM}_i$ in equation 38c is approximated by equation 37. The bounds in equation 38d include "move limits" protecting the accuracy of the extrapolation in equation 38c. The above optimization problem may have no feasible solution within the move limits in equation 38d if it begins with significant constraint violations in equations 38b and c. If a feasible solution cannot be found, an acceptable outcome of equation 38 is a new design point moved as close to the constraint boundary as possible. The result of this step is a new SV defining a modified design of the system.

Step 6. Repeat from Step 2 until all the constraints GS are satisfied, all safety margins SM_i are non-negative, and the performance index PS has converged.

In the above procedure, also shown in figure 18, the analyses in Step 1 and 2 are problem-dependent. The behavior sensitivity analysis required to obtain the $\partial DS_i/\partial SV$ can be obtained by either a finite difference technique or, preferably, by a quasi-analytical method (e.g., ref. 29). The optimization defined by equations 36 and 38 can be carried out by any suitable algorithm. The extension of the above two-level algorithm to multilevel systems is given in ref. 2, and its application to aerospace systems is discussed in reference 30.

MULTILEVEL OPTIMIZATION STUDY OF A TRANSPORT AIRCRAFT

The general algorithm introduced in the preceding section has been tested in a design optimization study of a transport aircraft reported in reference 31. The procedure was applied to an existing transport aircraft, and the fuel for a particular mission was selected as the objective function. Everything in the aircraft system was fixed as in the existing design, except for the airfoil depth-to-chord ratio, h, and the cross-sectional dimensions of the stringer-stiffened wing cover panels. Constraints included those typical for the aircraft performance requirements; e.g., runway length, climb rate, cruising speed, etc., and the strength and local buckling limits on stresses in the wing box covers.

The optimization was predicated on the trade-off between the structural wing weight and the drag, both being functions of "h." In order to intensify that trade-off to obtain conclusive study results, the cruise Mach number was set at .90, significantly greater than in the subject aircraft That artifically high Mach number made the wave drag a larger fraction of the total drag. The problem was a natural candidate for decomposition approach because it contained a very large number of detailed design variables (6 per each of 216 panels for a total of 1296 variables) which were distinct from the system-level configuration variable "h." The analyses involved also differed in their nature, and ranged from a semi-empirical performance aerodynamics for entire aircraft, through a highly detailed finite element analysis of the wing box, to a handbook level stress and buckling analysis of each stiffened panel.

Following the approach described in the previous section, the problem was decomposed as shown in figure 19, and an iterative procedure was implemented, with each iteration consisting of top-down analyses and bottom-up optimizations. The existing aircraft data initialized the procedure.

The top, system-level analysis was carried out by a performance analysis program, reference 32, that included a semi-empirical aerodynamic analysis. The middle-level subsystem - the wing box - was analyzed by a finite element program, reference 11, and the resulting edge forces were applied to individual panels at the bottom level.

Optimizations began at the bottom level, separately for each panel. The objective function was the panel cumulative constraint representing all the stress and buckling constraints by means of the KS function, reference 20. The constraints included side constraints and equality constraints on panel skin thickness and equivalent, smeared stringer thickness which preserved the thicknesses set at the wing box level. These equality constraints assured that the panel membrane stiffnesses stayed constant; hence, the edge forces remained constant, and the panel was isolated from its neighbors for the duration of its optimization. Sensitivity analysis was performed on each optimized panel using algorithms described in reference 5 to obtain derivatives of the minimized cumulative constraint with respect to the thicknesses and edge forces that were defined above as the optimization parameters.

The middle level optimization designed the skin thickness and the equivalent, smeared stringer thickness. Spanwise distributions of these thicknesses were described by polynomial functions whose coefficients were the design variables. The objective function was a cumulative constraint formed from the cumulative constraints that were minimized for each panel. At the middle-level, these constraints were extrapolated linearly using the optimum sensitivity derivatives with respect to the wing box thickness variables. The equality constraint on the wing box weight kept it constant at the value set at the top, system level. Optimum sensitivity derivatives were computed for the objective function with respect to "h" and the wing-box weight.

Finally, the optimization at the highest level used only two design variables: the depth-to-chord ratio, "h," and the wing box structural weight. Its objective function was the mission block fuel, and the inequality constraints included, in addition to the performance constraints, the wing box cumulative constraint that was minimized at the middle level. The latter constraint was extrapolated with respect to the design variables using the optimum sensitivity derivatives calculated at the middle level.

Nonlinear mathematical programming was used for optimization at all levels. The bottom and middle levels employed the usable-feasible directions algorithm. It was coupled directly to the analysis program at the bottom level, but at the middle level, it was coupled to an approximate analysis (derivative-based extrapolation). A SUMT procedure incorporating the Davidon-Fletcher-Powell algorithm was implemented at the top level.

The study demonstrated that the procedure converged well, in 4 to 5 cycles, to the same end result when started from different initial design points (including the existing design). As seen in a sample of the optimization history, shown in figure 20, the convergence was reasonably smooth. Some improvements of both the fuel consumption and the wing-box structural weight were achieved relative to the existing design. The improvement of the fuel consumption was small, as expected when starting the optimization with an already refined design. Also, it has to be emphasized that the improvement should not be interpreted as an indication of the actual potential still remaining in the subject aircraft because the analysis was not as complete as the one that was used in support of the actual design (e.g., the gust loads were not considered, and manufacturing constraints were excluded). However, the study demonstrated a multilevel, multidisciplinary optimization system in operation.

CONCLUDING REMARKS

Modern developments such as the increasing use of composite materials tend to increase the interactions between various disciplines in the design of engineering systems. Interdisciplinary design approach will yield, in general, a better design, but requires a systematic algorithm to account for the interactions and to ensure convergence and efficiency. The paper presents a survey of some of the more important interdisciplinary interactions and examples of the benefits of interdisciplinary design. It then reviews multilevel optimization as a tool of breaking down the multidisciplinary design problem to a set of manageable tasks. A specific multilevel algorithm is first presented in the context of structural optimization and then generalized to engineering system design.

REFERENCES

1. Sobieszczanski-Sobieski, J.; Barthelemy, J. F.; and Giles, G. L.: Aerospace Engineering Design by Systematic Decomposition and Multilevel Optimization; International Council of Aeronautical Science, 14th Congress, Paper No. ICAS-84-4.7.3, 1984.

2. Sobieszczanski-Sobieski, J.: A Linear Decomposition Method for Large Optimization Problems. NASA TM-83248, 1982.

3. Sobieszczanski-Sobieski, J.; and Barthelemy, J-F.: Improving Engineering System Design by Formal Decomposition, Sensitivity Analysis and Optimization. Proceedings of International Conference of Engineering Design, Hamburg, West Germany, 1985. (R. Hubka, Editor), Vol. 1, pp. 314-321. Publisher: Heurista, Zurich.

4. Archer, B.: The Implication for the Study of the Design Methods of Recent Developments in Neighboring Disciplines. Proceedings of International Conference of Engineering Design, ibid, 161, pp. 833-840.

5. Sobieszczanski-Sobieski, J.; Barthelemy, J-F.; and Riley, K. M.: Sensitivity of Optimum Solutions to Problem Parameters. AIAA Journal, Vol. 20, 1982, pp. 1291-1299.

6. Haftka, R. T.: Optimum Control of Structures. Previous lecture, ASI.

7. Haftka, R. T.; and Starnes, J. H., Jr.: Use of Optimum Stiffness Tailoring to Improve the Compressive Strength of Composite Plates with Holes. AIAA/ASME/ASCE/AHS 26th Structures, Structural Dynamics and Materials Conference, Orlando, FL, April 1985.

8. Adelman, H. M.: Preliminary Design Procedure for Insulated Structures Subject to Transient Heating. NASA TP-1534, 1979.

9. Grossman, B.; Strauch, G.; Epperd, W. M.; Gurdal, Z.; and Haftka, R. T.: Integrated Aerodynamic/Structural Design of a Sailplane Wing. AIAA Paper No. 86-2623. AIAA Aircrft Systems Design and Technology Meeting, Dayton, OH, October 1986.

10. Adelman, H. M.; and Padula, S. L.: Integrated Thermal Structural Electromagnetic Design Optimization of Large Space Antenna Reflectors. NASA TM-87713, 1986.

11. Whetstone, W. D.: EISI-EAL: Engineering Analysis Language. Proceedings of the Second Conference on Computing in Civil Engineering, ASCE, 1980, pp. 276-285.

12. Vanderplaats, G. N.: The Computer Design and Optimization. Computing in Applied Mechanics (R. F. Hartung, Editor), AMD, Vol. 18, American Society of Mechanical Engineering, 1976, pp. 25-48.

13. Sobieszczanski-Sobieski, J.: From a Black Box to a Programming System. Chapter 11, Foundations of Structural Optimization: A Unified Approach. Edited by A. J. Morris, J. Wiley & Sons, New York, 1982.

14. Starnes, J. H., Jr.; and Haftka, R. T.: Preliminary Design of Composite Wings for Buckling, Stress, and Displacement Constraints. Journal of Aircraft, Vol. 16, 1979, pp. 564-570.

15. Sobieszczanski-Sobieski, J.; James, B. B.; and Riley, M. F.: Structural Optimization by Generalized Multilevel Optimization. AIAA Paper No. 85-0698. Proceedings of AIAA/ASME/ASCE/AHS 26th Structures, Structural Dynamics and Materials Conference, Orlando, FL, April 15-17, 1985. Also published as NASA TM-87605, October 1985.

16. Noor, Ahmed K.; Kamel, Hussein, A.; and Fulton, Robert E.: Substructuring Techniques--Status and Projections. Computers & Structures, Vol. 8, No. 5, May 1978, pp. 621-632.

17. Przemieniecki, J. S.: Theory of Matrix Structural Analysis. Chapter 9, McGraw-Hill Book Co., 1968.

18. Aaraldsen, P. O.: The Application of the Superelement Method in Analysis and Design of Ship Structures and Machinery Components. Presented at the National Symposium on Computerized Structural Analysis and Design, George Washingon University, Washington, DC, March 27-29, 1972.

19. Sobieszczanski-Sobieski, J.; James, B.; and Dovi, A.: Structural Optimization by Multilevel Decomposition. AIAA Journal, Vol. 23, No. 11, 1985, pp. 1775-1782.

20. Kreisselmeier, G.; and Steinhauser, G.: Systematic Control Design by Optimizing a Vector Performance Index. Proceedings of IEAC Symposium on Computer Aided Design of Control Systems, Zurich, Switzerland, 1971.

21. Barthelemy, J.-F. M.; and Sobieszczanski-Sobieski, J.: Optimum Sensitivity Derivatives of Objective Function in Nonlinear Programming. AIAA Journal, Vol. 21, No. 6, 1983, pp. 913-915.

22. Chapin, Ned: New Format for Flowcharts. Software - Practice and Experience, Vol. 4, 1974, pp. 341-357.

23. Lust, R. V.; and Schmit, L. A.: Alternative Approximation Concepts for Space Frame Synthesis. AIAA Paper No. 85-0696-CP. AIAA/ASME/ASCE/AHS 26th Structures, Structural

Dynamics and Materials Conference, Orlando, FL, April 15-17, 1985.

24. Haftka, R. T.: An Improved Computationl Approach for Mutilevel Optimum Design. Journal of Structural Mechanics, 12(2), 1984, pp. 245-261.

25. Angermayer, K.: Structural Aluminum Design. Reynolds Metals Company, Richmond, VA, 1965.

26. Timoshenko, S. P.; and Gere, J. M.: Theory of Elastic Stability. McGraw-Hill, New York, 1961.

27. Sobieszczanski-Sobieski, J.: An Integrated Computer Procedure for Sizing Composite Air Frame Structures. NASA TP-1300, Hampton, VA, Feb. 1979.

28. Vanderplaats, G. N.: CONMIN--A FORTRAN Program for Constrained Function Minimization: User's Manual. NASA TM X-62282, August 1973.

29. Adelman, H. M.; and Haftka, R. T.: Sensitivity Analysis of Discrete Structural Systems. AIAA Journal, Vol. 24, No. 5, May 1986, pp. 823- 832.

30. Sobieszczanski-Sobieski, J.; Barthelemy, J. F.; and Giles, G. L.: Aerospace Engineering Design by Systematic Decomposition and Multilevel Optimization. International Council of Aeronautical Science, 14th Congress,. Paper No. ICAS-84-4.7.3, September 1984.

31. Wrenn, G. A.; and Dovi, A. R.: Multilevel/Multidisciplinary Optimization Scheme for Aircraft Wing. NASA CR-178077, 1986.

32. McCullers, L. A.: Aircraft Configuration Optimization Including Optimized Flight Profiles. Proceedings of Symposium on Recent Experiences in Multidisciplinary Analysis and Optimization. NASA CP- 2327, Part 1, 1984, pp. 395-412.

Table 1: Design Variables for Glider Design

3 Performance Design Variables	1.	Angle of attack at the root during the turn.
	2.	Angle of attack at the root during cruise.
	3.	Radius of the turn.
6 Geometric Design Variables	4.	Angle of twist at the break relative to the root.
	5.	Angle of twist at the tip relative to the root.
	6.	Chord length at the root.
	7.	Chord length at the break.
	8.	Chord length at the tip.
	9.	Distance to the break.
24 Structural Design Variables	10-17.	Spar cap thickness for each wing section.
	18-25.	Spar web thickness for each wing section
	26-33.	Skin thickness for each wing section.

Table 2: Design Constraints for Glider Wing

3	Stall Constraints during turning maneuver.	1. No stall at the root. 2. No stall at the break. 3. No stall at the tip.
3	Performance Constraints	4. Bank angle less than 50%. 5. Climb speed greater than zero. 6. Minimum divergence speed.
24	Structural Constraints (at 43 m/sec, 5.9 g)	7-14. Maximum spar cap strain for each wing section, .3%. 15-22. Maximum shear stress for each wing section, web shear ≤ 6000 N/mm^2. 23-30. Wing skin must satisfy Tsai-Hill strength constraint for each wing section.

Minimum average cross-country speed was also used as a constraint for weight minimized designs.

Table 3: Optimal Glider Designs

	Iterated Sequential	Integrated Design	
		Maximum Cross-country Speed	Minimum Mass
Cross-country speed (m/s)	3.44	3.48	3.44
Mass of one wing (kg)	13.0	12.5	11.6

Table 4: Nomenclature for Multilevel Structural Optimization Algorithm

Quantities

A	Cross-sectional area.
C	Cumulative constraint (equation 22).
c	Capacity: limitation on the ability to meet a particular demand d (e.g., allowable stress).
d	Demand: a physical quantity the structure is required to have, to support, or to be subjected to in order to perform its function (e.g., stress).
F	Objective function.
f( )	Functional relation.
g^{ij}	Vector of constraint functions, g_w^{ij} ; $w = 1 \rightarrow W^{ij}$.
h^{ij}	Vector of partitions h_K^{ij} , h_M^{ij}, h_P^{ij} (eq. 25b).
h_K^{ij} , h_M^{ij} , h_P^{ij}	Vectors of the equality constraints defined by equations 13, 14, and 15, respectively. The vector elements are, respectively: h_{KS1}^{ij} , h_{MS2}^{ij} , h_{PS3}^{ij} , where $S1 = 1 \rightarrow S_1^{ij}$, $S2 = 1 \rightarrow S_2^{ij}$, $S3 = 1 \rightarrow S_3^{ij}$.
I	Cross-sectional moment of inertia.
K^{ij}	Stiffness matrix of SSijk.
K^{bij}	Boundary stiffness matrix for SSijk.
L^{ij}	Lower bound on X^{ij} including move limits.
M^{ij}	Mass of SSijk (a scalar).
P^{ij}	Vector of the External loads applied to interior and/or boundary of SSijk.
Q^{ik}	Boundary forces, Q_r^{ij}, of SSijk, $r = 1 \rightarrow R^{ij}$.
P^{bij}	Vector of P^{ij} transferred to the boundary of SSijk.
Q^{ij}	Vector of the forces, Q_r^{ij}, $r = 1 \rightarrow R^{ij}$, acting on the boundary of SSijk.
Q^{ik}	Boundary forces, Q_r^{ij}, of SSijk, $r = 1 \rightarrow R^{ij}$.

S_K	Summation of stiffnesses contributed by substructures SSijk assembled in a parent substructure SSmkl.
S_P	Summation of the boundary loads contributed by substructures, SSijk, assembled in a parent substructure SSmkl.
SSijk	A substructure (including the extremes of the assembled structure and a single structural element).
SSmkl	A substructure-parent of SSijk, m = i-1, see fig. 12.
SSSnlp	A substructure-parent of SSmkl, n = m-1, see fig. 12.
STOC	Acronym: subject to constraints.
T^{ij}	Total number of design variables for SSijk.
W^{ij}	Number of inequality constraints.
U^{ij}	Upper bound on X^{ik} including move limits.
V^{ij}	Number of constraints defined by eqs. (13)-(15).
X^{ij}	Vector of design variables, X_t, in SSijk, $t = 1 \to T^{ij}$.
Y^{ij}	Vector of the entries in K^{bij}, M^{ij}, and the entries in P^{bij} that are held constant as parameters in optimization of SSijk. The vector Y^{ij} contains V^{ij} elements Y_v^{ij}.
Z^{ij}	Vector of cross-sectional dimensions, Z_b^{ij}, $b = 1 \to B^{ij}$, used as design variables in SSijk that corresponds to a single structural element.
π	A vector defined by equation 26.
ρ	A user-controlled constant in the KS function (eq. 22).
Δ	Increment of a variable (see definition of subscript o)

Indices, Subscripts, and Superscripts Not Included in the Definitions Above

Overbar	Denotes an optimal quantity

b	Superscript to denote an association with the SS boundary
e	Subscript to identify an extrapolated value
o	Subscript to identify an original (reference) value from which an increment is measured.

Table 5: Quantities Defined for the Multilevel Test Case Optimization

	TOP LEVEL
OBJECTIVE:	The framework material volume.
DESIGN VARIABLES:	A and I of the beams.
CONSTAINTS:	Displacements of the loaded corner and C_e for the beams.
	MIDDLE LEVEL
OBJECTIVE:	Cumulative constraint C representing the column buckling and C_e for the walls.
DESIGN VARIABLES:	Wall membrane stiffness contributing to the beam axial and bending stiffnesses controlled through the dimensions shown in Fig. 14, Section A-A.
CONSTRAINTS:	Equality - beam cross-sectional area and moment of inertia.
	BOTTOM LEVEL
OBJECTIVE:	Cumulative constraint C representing a set of stress and local buckling constraints of the wall.
DESIGN VARIABLES:	Cross-sectional dimensions shown in Fig. 14, DETAIL B.
CONSTRAINTS:	Inequality - minimum gages, geometrical proportions, and geometrical realizability.
	Equality - membrane stiffnesses for tension-compression and bending of the wall in its own plane.

Table 6: Notation for Multilevel Generic Optimization

DS_i	vector of demand quantities imposed by the system on subsystem i.
DV_i	vector of design variables for subsystem i.
ES	(engineering) system.
GE_i	vector of equality constraints for subsystem i.
GS	vector of system inequality constraints; an inequality constraint is defined as $g = k(DEMAND/CAPACITY)-1$, satisfied when $g \leq 0$.
GSS_i	vector of inequality constraints for subsystem i.
L, L_i	vector of lower limits on SV, and DV_i, respectively (move limits included).
PS	performance index for ES (a scalar).
SM_i	safety margin for SS_i (a scalar), defined as $SM_i = \max\ (CAPACITY/DEMAND)-1$.
SS_i	subsystem i.
SV	vector of system design variables.
U, U_i	vector of upper limits on SV, and DV, respectively (move limits included).

Table 7: Examples of the Equivalents of the Generic Terms Typical for an Aircraft Application

DS_i	at the middle level: lift required of the wing; at the bottom level: edge loads N_x, N_Y, N_{xy} on a wing cover panel.
DV_i	at the middle level: wing bending stiffness distribution; at the bottom level: detailed wing panel dimensions.
ES	aircraft, top (system)level.
GE_i	at the middle level: wing structure weight prescribed at the top level; at the bottom level: panel spanwise membrane stiffness prescribed at the middle level.
GS	runway length.
GSS_i	at the middle level: wing tip deflection; at the bottom level: panel local buckling.
PS	fuel economy for a given mission.
SS_i	the wing box, middle level; the wing cover stiffened panels, third (bottom) level.
SV	wing structural weight, and airfoil thickness to chord ratio.
$\partial DS_i/\partial SV$	derivative of wing lift with respect to structural weight.
$\partial SM_i/\partial DS_i$	derivative of wing panel safety margin with respect to edge loads.

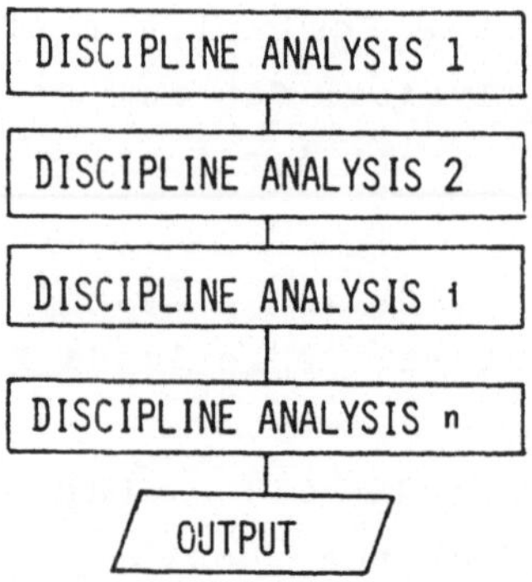

Figure 1. Many disciplinary analyses performed in series.

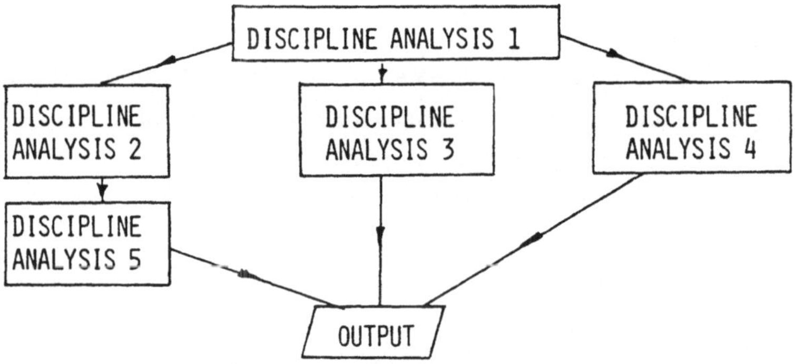

Figure 2. Disciplinary analyses in a hierarchical framework.

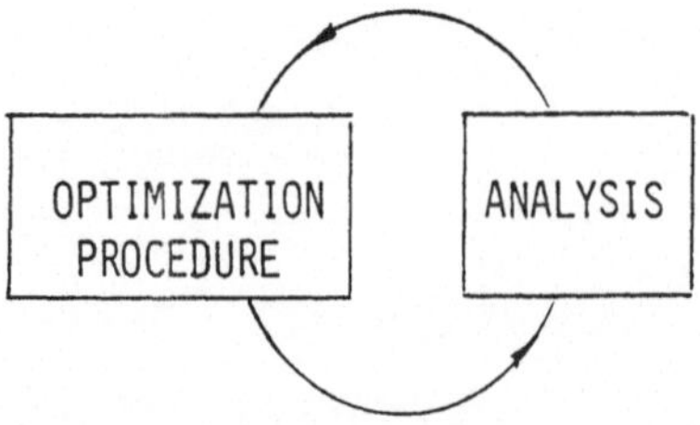

Figure 3. Multidisciplinary analysis coupled to an optimization procedure.

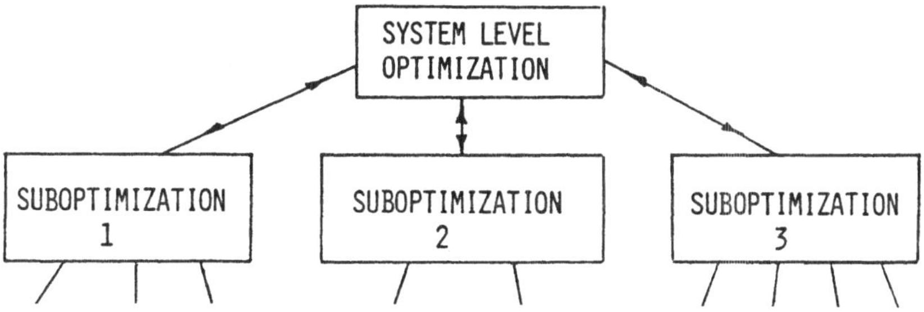

Figure 4. Multidisciplinary optimization as a hierarchy of subtasks.

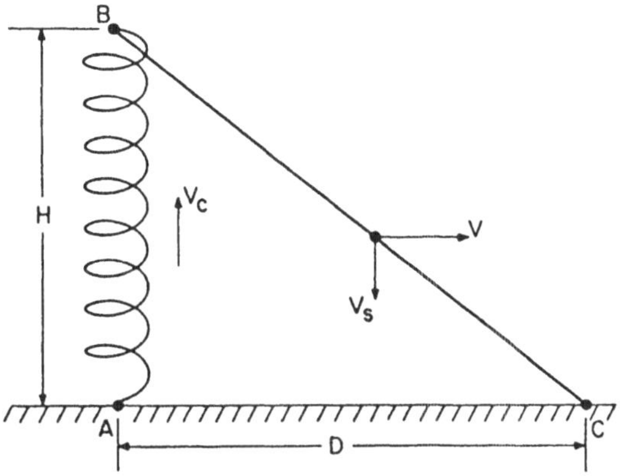

Figure 5. Glider mission profile.

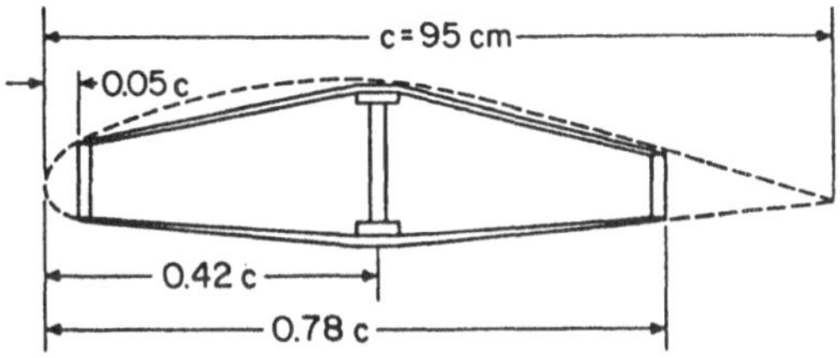

Figure 6. Typical cross-section of glider wing element.

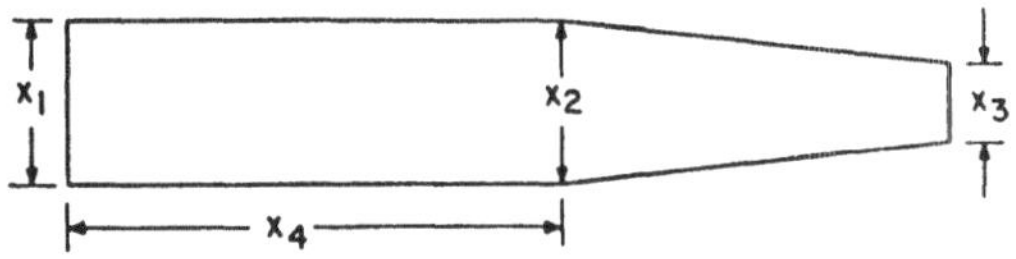

Figure 7. Planform geometry variables.

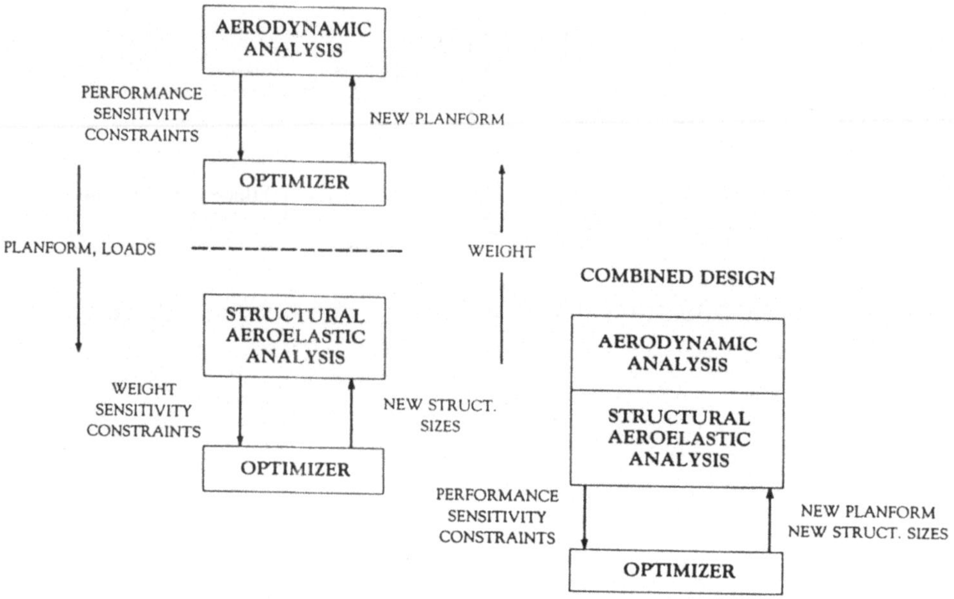

Figure 8. Schematic of sequential and integrated (combined) optimization procedures for glider wing.

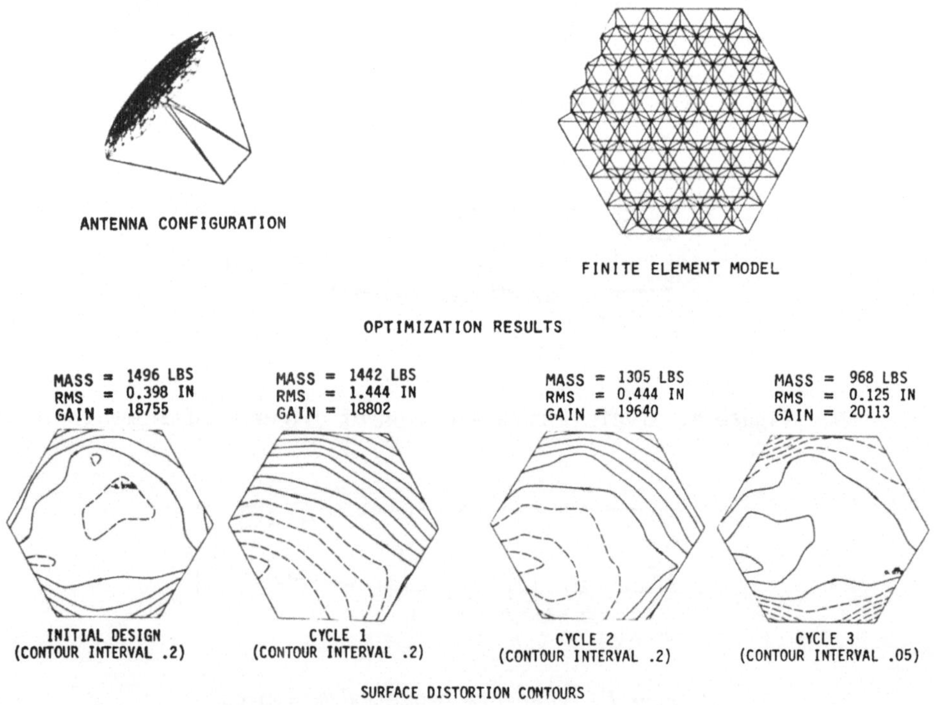

Figure 9. Antenna structure and optimization results.

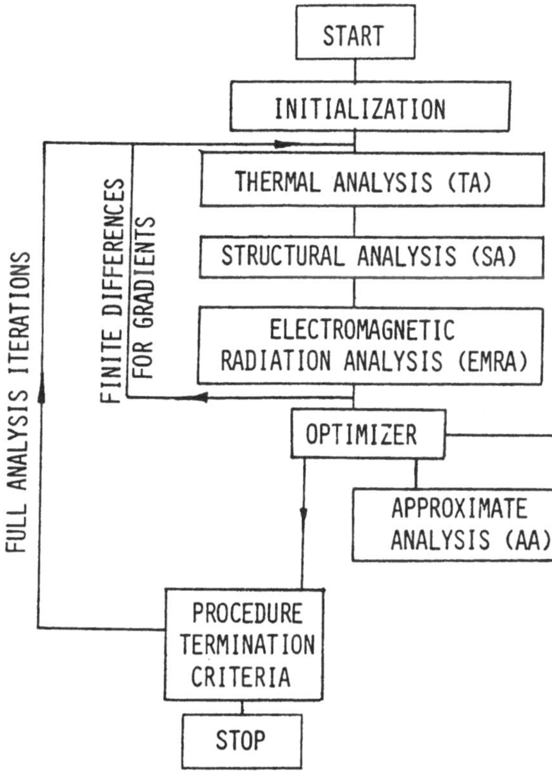

Figure 10. Antenna optimization procedure.

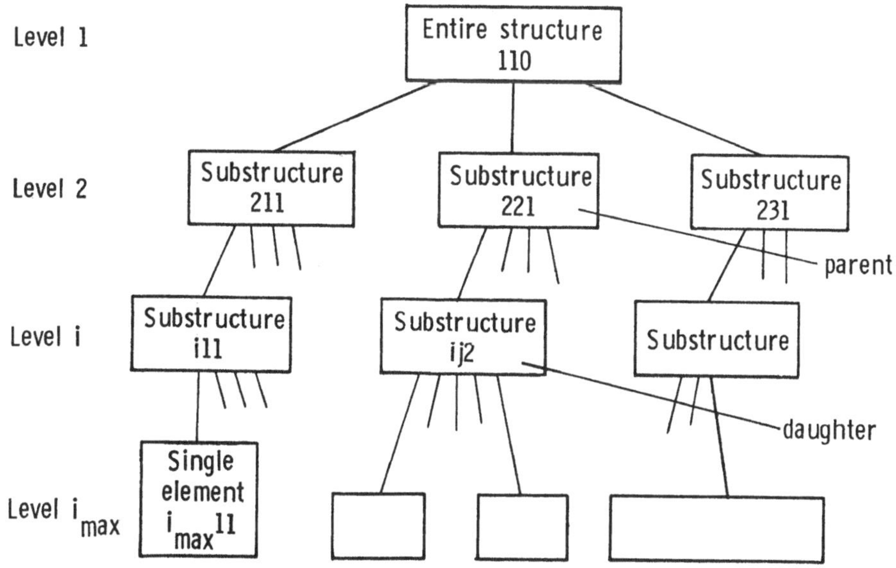

Figure 11. Multilevel substructuring.

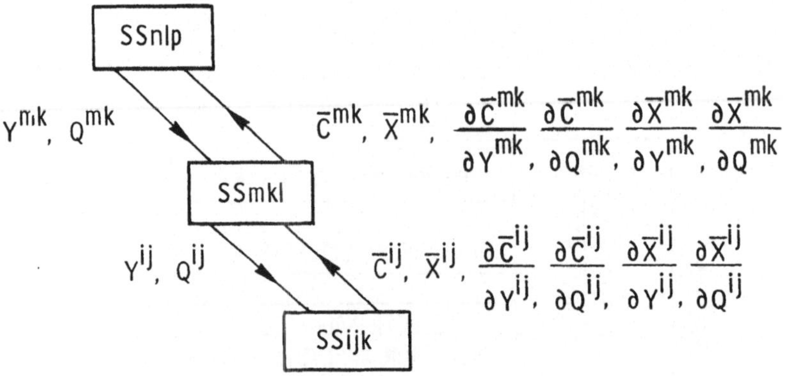

Figure 12. Flow of information.

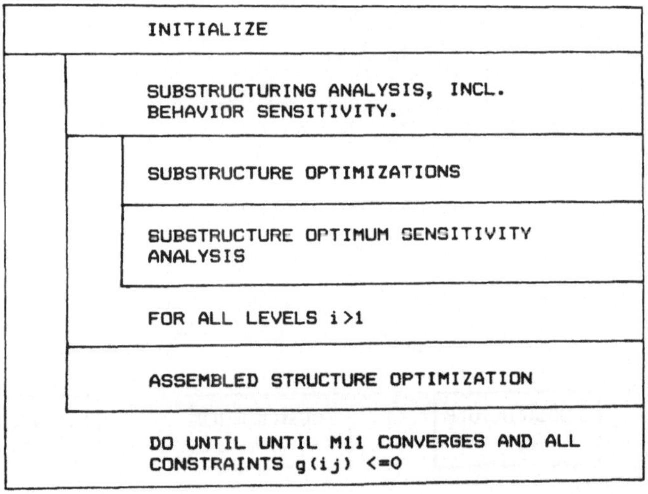

Figure 13. Multilevel optimization procedure flowchart.

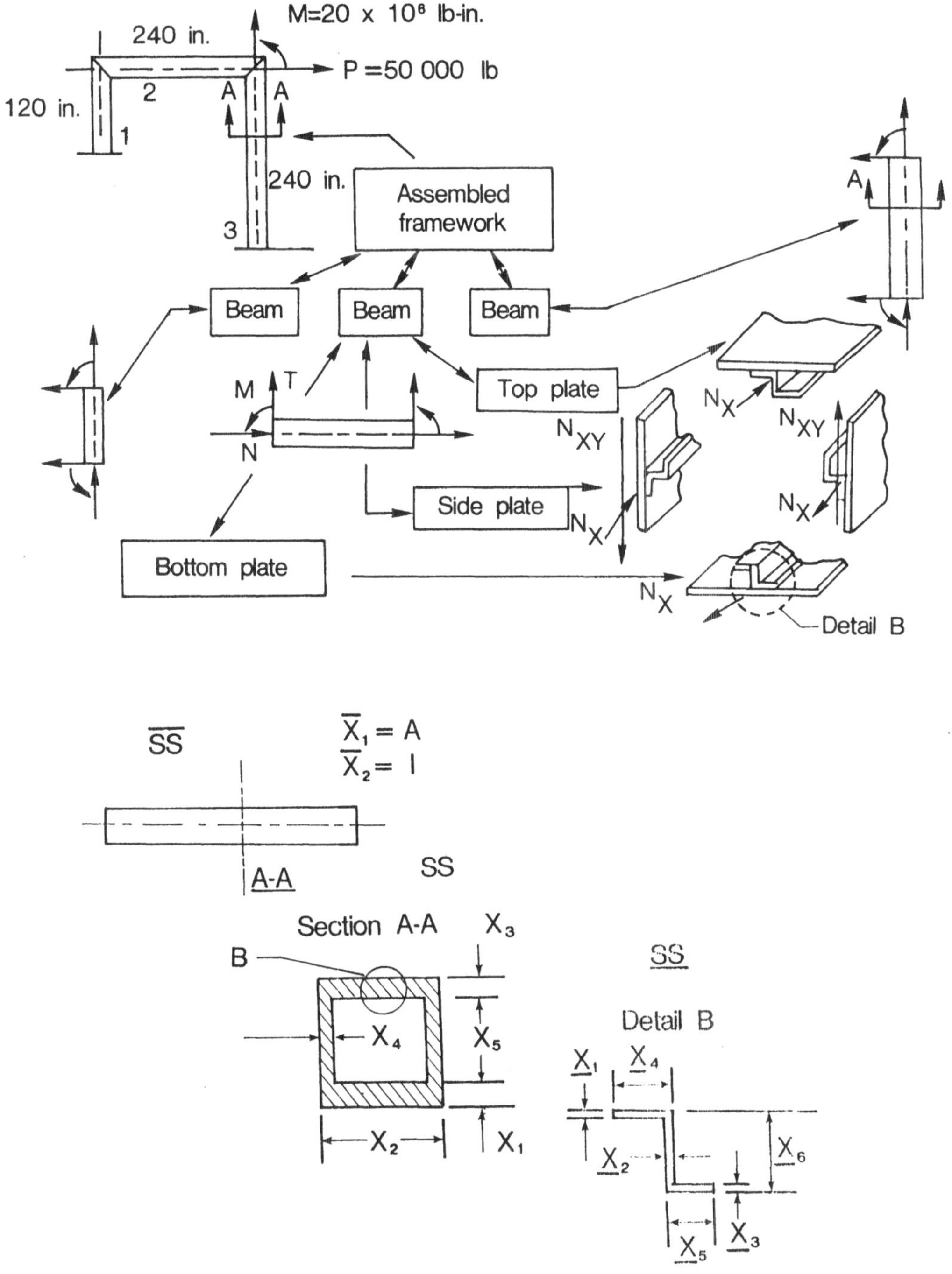

Figure 14. A portal framework.

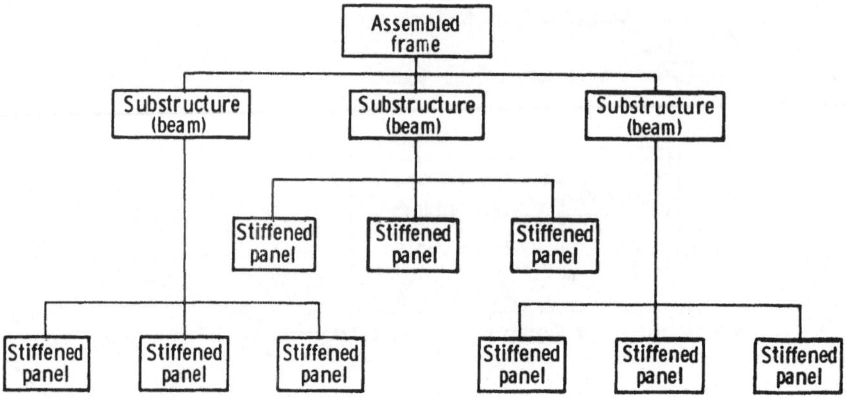

Figure 15. Hierarchical decomposition of the framework structure.

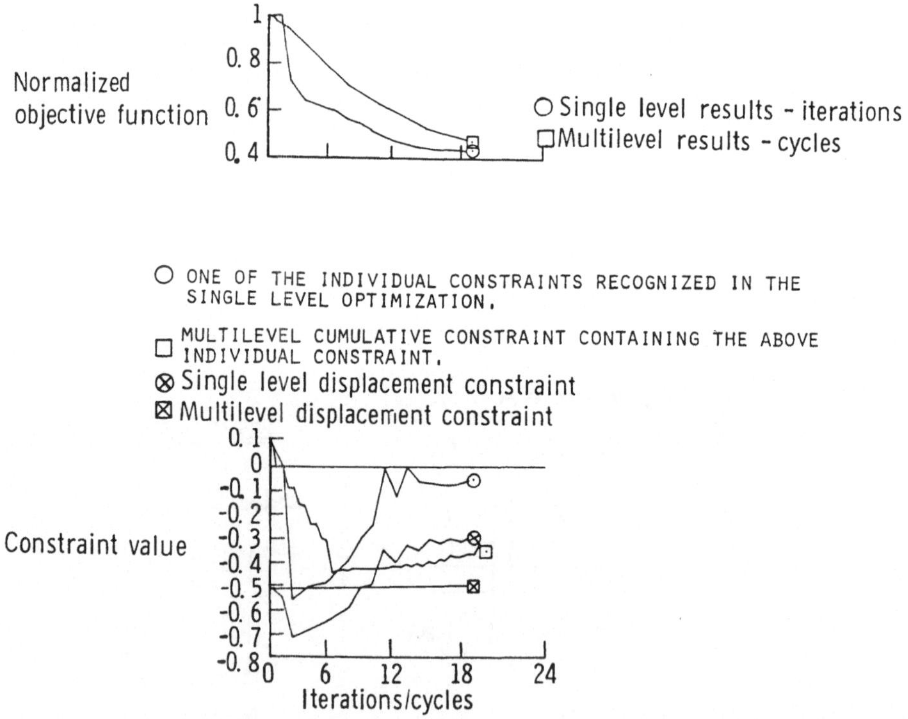

Figure 16. Representative results.

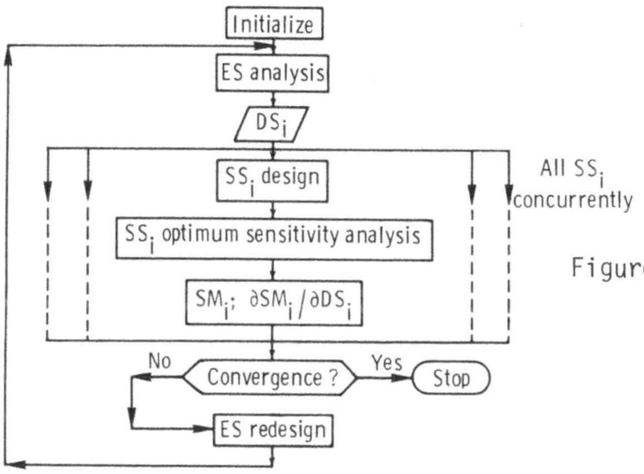

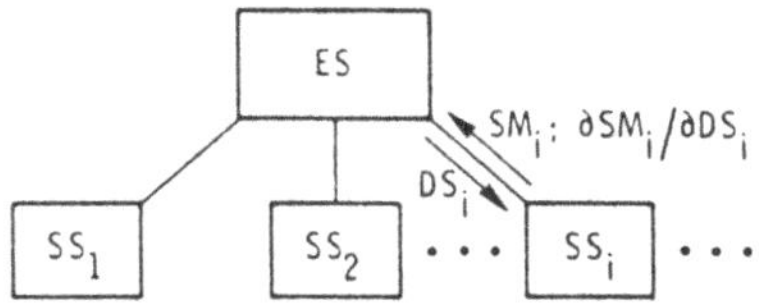

Figure 17. Typical two-level system.

Figure 18. Two-level system optimization procedure.

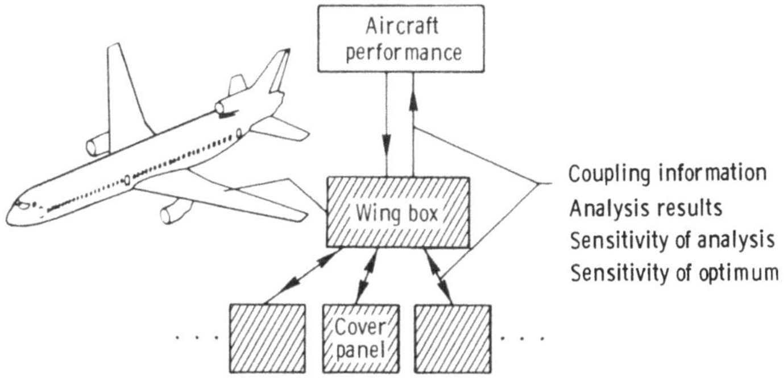

Figure 19. Aircraft and a schematic of its multilevel optimization.

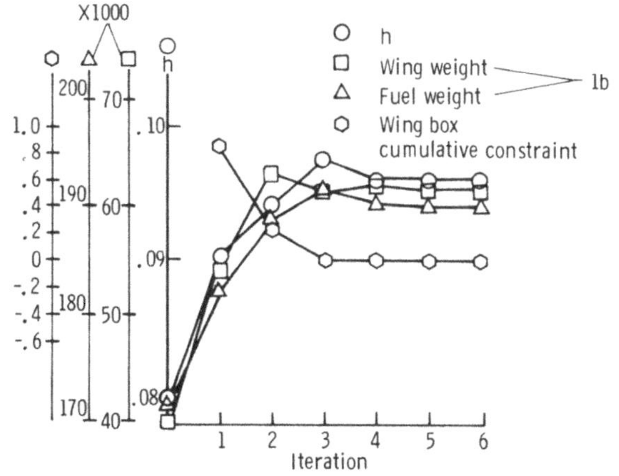

Figure 20. Histogram of an aircraft multilevel optimization.

PART V

OPTIMAL DESIGN OF MECHANICAL SYSTEMS

DESIGN SENSITIVITY ANALYSIS OF DYNAMIC SYSTEMS

Edward J. Haug
Center for Computer Aided Design
and Department of Mechanical Engineering
University of Iowa
Iowa City, Iowa 52242

Abstract. Methods for calculation of first and second design derivatives of performance measures for nonlinear dynamic systems are presented. Design sensitivity analysis formulations for dynamic systems are presented in two alternate forms; (1) equations of motion are written in terms of independent generalized coordinates and are reduced to first order form and (2) equations of motion of constrained systems are written in terms of a mixed system of second order differential and algebraic equations. Both first and second order design sensitivity analysis methods are developed, using a theoretically simple direct differentiation approach and a somewhat more subtle, but numerically efficient, adjoint variable method. Detailed derivations are presented and computational algorithms are discussed. Examples of first and second order design sensitivity analysis of mechanisms and machines are presented and analyzed.

1. INTRODUCTION

The theory of sensitivity analysis for dynamic systems has been well developed in the theory of differential equations and control systems [1,2]. Design sensitivity analysis theory that was originally developed in optimal control theory [2] has been extended to a unified theory for structural and dynamic systems in Refs. 3 and 4. Second order differential-algebraic equation formulations have been employed to develop a more practical and directly useable formulation for system dynamic design sensitivity analysis [5,6]. In a related

NATO ASI Series, Vol. F27
Computer Aided Optimal Design: Structural and Mechanical Systems. Edited by C. A. Mota Soares

development, second order dynamic design sensitivity analysis has been treated in Ref. 7. While general formulas for second derivative calculation are quite complex, the advent of symbolic computation techniques makes calculation of second design derivatives practical.

Alternative formulations of initial-value problems that describe system dynamics are employed in this paper, to yield algorithms for computation of design sensitivities. In Section 2, direct differentiation and adjoint variable methods are presented for design sensitivity analysis of mechanical systems that are described by nonlinear systems of first order differential equations. This theory is extended in Section 3 to present a new and remarkably efficient approach for second order design sensitivity analysis. This new method is based on combining the direct differentiation and adjoint variable methods, an idea that was introduced by Haftka for structural static design sensitivity analysis [8].

Elementary examples are studied in Sections 2 and 3 to illustrate use of the equations derived. A more realistic automatic cannon second order design sensitivity analysis example is treated in Section 4. Design sensitivity calculations are presented and the validity of first and second order approximations, using design sensitivity results, are analyzed.

An extension of the methods of Section 2 for first order design sensitivity analysis of large scale systems that are described by mixed differential-algebraic equations is presented in Section 5. As in the preceding, both the direct differentiation and adjoint variable approaches are presented. It is shown that both formulations lead to linear adjoint systems of mixed differential-algebraic equations that have similar structure to the original system. Examples using this second order formulation, with a computer code that automatically generates and solves the system equations of motion, are presented to illustrate feasibility of automated design sensitivity analysis. A related recently developed singular value decomposition numerical integration method has been presented in Ref. 9 for accurate solution of both the dynamic and adjoint equations.

2. FIRST ORDER DESIGN SENSITIVITY ANALYSIS FOR SYSTEMS DESCRIBED BY FIRST ORDER DIFFERENTIAL EQUATIONS

Design sensitivity analysis of mechanical system dynamics for formulations in which the equations of motion have been put in first order form has progressed to the point that first derivatives of dynamic response measures with respect to design parameters can be calculated [1-3]. Direct differentiation of equations of motion was initially used to obtain state sensitivity functions [1]. More recently, an adjoint variable method that was introduced in optimal control theory [2] and developed for mechanical design [3] has been successfully employed for design sensitivity analysis of large scale planar dynamic systems [5] and smaller scale dynamic systems with intermittent motion [6].

First Order Formulation of the Problem

Dynamic systems treated here are described by a design variable vector $\mathbf{b} = [b_1,\ldots,b_k]^T$ and a state variable vector $\mathbf{z}(t) = [z_1(t),\ldots,z_n(t)]^T$, which is the solution of an initial-value problem of the form

$$\begin{aligned} \dot{\mathbf{z}} &= \mathbf{f}(\mathbf{z},\mathbf{b}) \quad , \qquad t^1 \leq t \leq t^2 \\ \mathbf{z}(t^1) &= \mathbf{h}(\mathbf{b}) \end{aligned} \tag{2.1}$$

where $\dot{\mathbf{z}} = \frac{dz}{dt}$, t^1 is given, and t^2 is determined by the condition

$$\Omega\left(t^2,\mathbf{z}(t^2),\mathbf{b}\right) = 0 \tag{2.2}$$

The function $\mathbf{f}$ that appears on the right side of the differential equation of Eq. 2.1, the function $\mathbf{h}$ in the initial condition of Eq. 2.1, and the function Ω in Eq. 2.2 are assumed to be twice continuously differentiable in their arguments. Classical results of the theory of ordinary differential equations [10] assure that the solution of Eq. 2.1, denoted $\mathbf{z}(t;\mathbf{b})$, exists and is twice continuously differentiable with respect to both t and $\mathbf{b}$.

Consider a typical functional that may arise in a design formulation,

$$\psi = g\left(t^2, \mathbf{z}(t^2), \mathbf{b}\right) + \int_{t^1}^{t^2} F(t, \mathbf{z}, \mathbf{b})\,dt \qquad (2.3)$$

where the first term involves only behavior of the state of the system at the terminal time and design. The second term involves an integral measure of state and design over some period of motion. This form of functional is adequate for treating a large class of dynamic system optimal design problems [2,3].

Note that dependence on the design variable $\mathbf{b}$ in Eq. 2.3 arises both explicitly and through the state variable, which is written in the form $\mathbf{z}(t;\mathbf{b})$ to emphasize that it is a function of time that depends on design. In order to obtain the derivative of ψ with respect to $\mathbf{b}$, Leibniz rule of differentiation [11] may be applied to obtain

$$\frac{d\psi}{d\mathbf{b}} = g_{t^2} t^2_{\mathbf{b}} + g_{\mathbf{z}}\left[\mathbf{z}_{\mathbf{b}}(t^2) + \dot{\mathbf{z}}(t^2) t^2_{\mathbf{b}}\right] + g_{\mathbf{b}} + \int_{t^1}^{t^2} \left[F_{\mathbf{z}}\mathbf{z}_{\mathbf{b}} + F_{\mathbf{b}}\right] dt + F(t^2) t^2_{\mathbf{b}} \qquad (2.4)$$

where a subscript denotes derivative with respect to the subscripted variable. For a summary of matrix differentiation notation employed in this paper, the reader is referred to the Appendix. It is important to note that $\mathbf{z}_{\mathbf{b}}$ is the derivative of a vector function with respect to a vector variable. It is thus a matrix, and the order of terms in matrix products is important.

Equation 2.4 may be reduced to a form that depends only on $\mathbf{z_b}$ and not on $t^2_{\mathbf{b}}$. Differentiating Eq. 2.2 with respect to $\mathbf{b}$ yields

$$[\Omega_{t^2} + \Omega_{\mathbf{z}}\dot{\mathbf{z}}(t^2)]t^2_{\mathbf{b}} + \Omega_{\mathbf{z}}\mathbf{z_b}(t^2) + \Omega_{\mathbf{b}} \equiv \dot{\Omega}(t^2)t^2_{\mathbf{b}}$$

$$+ \Omega_{\mathbf{z}}\mathbf{z_b}(t^2) + \Omega_{\mathbf{b}} = \mathbf{0}$$

Since Eq. 2.2 must determine t^2, the coefficient $\dot{\Omega}(t^2)$ of $t^2_{\mathbf{b}}$ cannot be zero. Using Eq. 2.1 for $\dot{\mathbf{z}}$,

$$\dot{\Omega}(t^2) = \Omega_{t^2} + \Omega_{\mathbf{z}}\mathbf{f}(\mathbf{z}(t^2),\mathbf{b})$$

and

$$t^2_{\mathbf{b}} = -\left[\frac{\Omega_{\mathbf{z}}}{\dot{\Omega}(t^2)}\right]\mathbf{z_b}(t^2) - \frac{\Omega_{\mathbf{b}}}{\dot{\Omega}(t^2)} \tag{2.5}$$

Substituting this result into Eq. 2.4 yields

$$\frac{d\psi}{d\mathbf{b}} = \mathbf{G}^T(t^2,\mathbf{z}(t^2),\mathbf{b})\mathbf{z_b}(t^2) + g_{\mathbf{b}}$$

$$- \left[\frac{g_{t^2} + g_z\mathbf{f}(t^2) + F(t^2)}{\dot{\Omega}(t^2)}\right]\Omega_{\mathbf{b}}$$

$$+ \int_{t^1}^{t^2} [F_{\mathbf{z}}\mathbf{z_b} + F_{\mathbf{b}}]dt \tag{2.6}$$

where

$$\mathbf{G}(t^2,\mathbf{z}(t^2),\mathbf{b}) = g^T_{\mathbf{z}}(t^2) - \left[\frac{g_{t^2} + g_{\mathbf{z}}\mathbf{f}(t^2) + F(t^2)}{\dot{\Omega}(t^2)}\right]\Omega^T_{\mathbf{z}} \tag{2.7}$$

In order to make use of Eq. 2.6 practical, terms that involve $\mathbf{z_b}$ must be calculated or replaced by terms that are written explicitly as functions of computable quantities. Two very different methods of achieving this objective are now presented. The first and most classical method [1] uses direct differentiation of Eq. 2.1 to obtain an initial-value problem for $\mathbf{z_b}$, which is solved and the result is substituted in Eq. 2.6. The second method, which has attractive computational features, introduces an adjoint variable [2,3] that is used to write an alternate form of Eq. 2.6, which requires less computation than the direct differentiation method.

Direct Differentiation Method

Direct differentiation of Eq. 2.1 yields

$$\begin{aligned} \dot{\mathbf{z}}_\mathbf{b} - \mathbf{f_z z_b} &= \mathbf{f_b}\,, \qquad t^1 < t < t^2 \\ \mathbf{z_b}(t^1) &= \mathbf{h_b} \end{aligned} \tag{2.8}$$

Note that since $\mathbf{z_b}$ is an n×k matrix, Eq. 2.8 is in fact a system of k first order linear differential equations for the k columns of the matrix $\mathbf{z_b}$.

The initial-value problem of Eq. 2.8 can be solved numerically by forward numerical integration, to obtain the solution $\mathbf{z_b}(t)$ on the interval $t^1 < t < t^2$. The result may be substituted directly into Eq. 2.6 to obtain the first derivative of the functional ψ of Eq. 2.3 with respect to design. While this computational algorithm is conceptually very simple and can be implemented with a minimum of programming difficulty, if k is large it requires a substantial amount of numerical computation and data storage.

Adjoint Variable Method

It is desirable to find an alternate method of first order design sensitivity analysis that reduces the amount of computation associated with the direct differentiation method of the preceding subsection. To meet this objective, an adjoint variable $\boldsymbol{\lambda}$ is introduced by multiplying both sides of Eq. 2.1 by $-\boldsymbol{\lambda}^T$ and integrating from t^1 to t^2 to obtain the identity

$$-\int_{t^1}^{t^2} \lambda^T[\dot{\mathbf{z}}_{\mathbf{b}} - \mathbf{f}_{\mathbf{z}}\mathbf{z}_{\mathbf{b}} - \mathbf{f}_{\mathbf{b}}]dt = 0 \tag{2.9}$$

Integrating the first term by parts gives

$$\int_{t^1}^{t^2} [(\dot{\lambda}^T + \boldsymbol{\lambda}^T\mathbf{f}_{\mathbf{z}})\mathbf{z}_{\mathbf{b}} + \lambda^T\mathbf{f}_{\mathbf{b}}]dt - \boldsymbol{\lambda}^T(t^2)\mathbf{z}_{\mathbf{b}}(t^2)$$

$$+ \boldsymbol{\lambda}^T(t^1)\mathbf{h}_{\mathbf{b}} = 0 \tag{2.10}$$

where $\mathbf{z}_{\mathbf{b}}(t^1) = \mathbf{h}_{\mathbf{b}}$ has been substituted from the initial condition of Eq. 2.1.

Recall that Eq. 2.10 holds for any function $\boldsymbol{\lambda} = \boldsymbol{\lambda}(t)$. In order to obtain a useful identity, the function $\boldsymbol{\lambda}$ is chosen so that the coefficient of $\mathbf{z}_{\mathbf{b}}$ in the integrands of Eqs. 2.6 and 2.10 are equal; i.e.,

$$\dot{\lambda} + \mathbf{f}_{\mathbf{z}}^T\boldsymbol{\lambda} = F_{\mathbf{z}}^T \tag{2.11}$$

Further, equating coefficients of $\mathbf{z}_{\mathbf{b}}(t^2)$ in Eqs. 2.6 and 2.10,

$$\boldsymbol{\lambda}(t^2) = -\mathbf{G}(t^2,\mathbf{z}(t^2),\mathbf{b}) \equiv -g_{\mathbf{z}}^T$$

$$+ \left[\frac{F(t^2) + g_{t^2} + g_{\mathbf{z}}\mathbf{f}(t^2)}{\dot{\Omega}(t^2)}\right] \Omega_{\mathbf{z}}^T \tag{2.12}$$

The sum of terms in Eqs. 2.6 and 2.10 involving $\mathbf{z}_{\mathbf{b}}$ are now equal. The terms in Eq. 2.6 that involve $\mathbf{z}_{\mathbf{b}}$ may now be replaced by terms in Eq. 2.10 that involve only explicit derivatives with respect to design and the adjoint variable $\boldsymbol{\lambda}$. Thus, Eq. 2.6 becomes

$$\frac{d\psi}{d\mathbf{b}} = g_{\mathbf{b}} - \frac{[g_{t^2} + g_{\mathbf{z}}\mathbf{f}(t^2) + F(t^2)]\Omega_{\mathbf{b}}}{\dot{\Omega}(t^2)}$$

$$- \boldsymbol{\lambda}^T(t^1)\mathbf{h}_{\mathbf{b}} + \int_{t^1}^{t^2} [F_{\mathbf{b}} - \boldsymbol{\lambda}^T \mathbf{f}_{\mathbf{b}}]dt \qquad (2.13)$$

Thus, the first derivative of ψ with respect to design can be evaluated. The computational cost associated with evaluating this derivative vector includes backward numerical integration of the terminal-value problem of Eqs. 2.11 and 2.12, to obtain $\boldsymbol{\lambda}(t)$. Numerical integration in Eq. 2.13 then yields the desired design derivative. This method is called the adjoint variable method of design sensitivity analysis. It has been used extensively in optimal control and mechanical design [2,3].

Note that the derivative of ψ with respect to design is obtained in Eq. 2.13, with only backward numerical integration of a single terminal-value problem of Eqs. 2.11 and 2.12. If several design variables are involved; i.e., $k \gg 1$, then the adjoint variable method is substantially more efficient than the direct differentiation method. If, on the other hand, the design variable is a scalar, then the same number of differential equations must be integrated. For large numbers of design variables, the adjoint variable is computationally more efficient.

There is one practical complication associated with the adjoint variable method. The state variable $\mathbf{z}(t)$ must be stored at discrete times and interpolated for use in subsequent backward integration.

First Order Design Sensitivity Analysis of a Simple Oscillator

Consider the simple oscillator of Fig. 2.1 as an example. The objective is to derive first order design derivatives of position at a given terminal time $t_2 = \pi/2$.

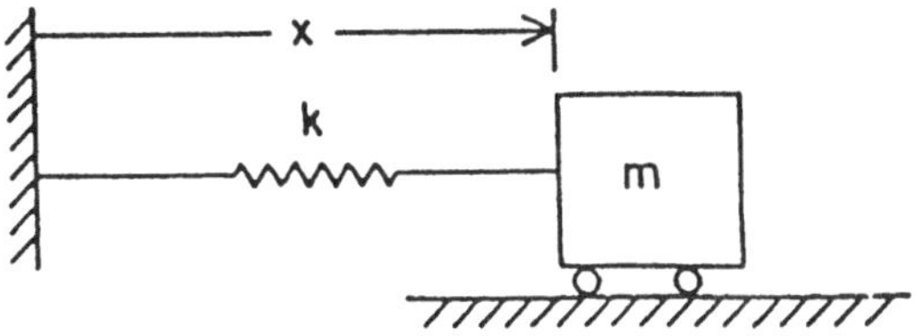

Figure 2.1 Simple Oscillator

The second order equation of motion and initial conditions are

$$m\ddot{x} + kx = 0, \qquad 0 \leq t \leq \frac{\pi}{2}$$

$$x(0) = 0 \tag{2.14}$$

$$\dot{x}(0) = v$$

where v is the initial velocity. Equation 2.14 can be written in the first order form of Eq. 2.1 as

$$\dot{\mathbf{z}} \equiv \begin{bmatrix} \dot{z}_1 \\ \dot{z}_2 \end{bmatrix} = \begin{bmatrix} z_2 \\ -\dfrac{b_1}{m} z_1 \end{bmatrix} \equiv \mathbf{f}(\mathbf{z},\mathbf{b}) \;, \; 0 \leq t \leq \frac{\pi}{2} \tag{2.15}$$

$$\mathbf{z}(0) \equiv \begin{bmatrix} z_1(0) \\ z_2(0) \end{bmatrix} = \begin{bmatrix} 0 \\ b_2 \end{bmatrix} \equiv \mathbf{h}(\mathbf{b})$$

where $z_1 \equiv x$ and $z_2 \equiv \dot{x} = \dot{z}_1$. The design variable vector is taken as $\mathbf{b} = [k,v]^T$.

To simplify the problem, let m = 1. Then the closed form solution of Eq. 2.15 is

$$\begin{bmatrix} z_1(t) \\ z_2(t) \end{bmatrix} = \begin{bmatrix} \dfrac{b_2}{\sqrt{b_1}} \sin\sqrt{b_1}\, t \\ b_2 \cos\sqrt{b_1}\, t \end{bmatrix} \tag{2.16}$$

The functional treated in this simple example is the position of the mass at terminal time $t^2 = \pi/2$; i.e.,

$$\psi = z_1(\tfrac{\pi}{2}) \equiv g(\mathbf{z}(\tfrac{\pi}{2}),\mathbf{b}) = \frac{b_2}{\sqrt{b_1}} \sin \frac{\pi\sqrt{b_1}}{2} \tag{2.17}$$

From Eq. 2.2, since t^2 has a given fixed value,

$$\Omega = t^2 - \pi/2$$

and

$$\Omega_{\mathbf{z}} = \mathbf{0}$$

$$\Omega_{t^2} = 1 \tag{2.18}$$

From Eq. 2.7,

$$\mathbf{G}(t^2,\mathbf{z}(t^2),\mathbf{b}) = \begin{bmatrix} 1 \\ 0 \end{bmatrix} \tag{2.19}$$

Direct Differentiation Method: To illustrate use of the direct differentiation method, the initial-value problem of Eq. 2.15 may be differentiated with respect to design, using the notation $y_i^j = (z_i)_{b_j}$, to obtain the matrix differential equations and initial conditions

$$\begin{bmatrix} \dot{y}_1^1 & \dot{y}_1^2 \\ \dot{y}_2^1 & \dot{y}_2^2 \end{bmatrix} - \begin{bmatrix} 0 & 1 \\ -b_1 & 0 \end{bmatrix} \begin{bmatrix} y_1^1 & y_1^2 \\ y_2^1 & y_2^2 \end{bmatrix} = \begin{bmatrix} 0 & 0 \\ -z_1 & 0 \end{bmatrix}$$

$$\begin{bmatrix} y_1^1 & y_1^2 \\ y_2^1 & y_2^2 \end{bmatrix}_{(t^1)} = \begin{bmatrix} 0 & 0 \\ 0 & 1 \end{bmatrix} \tag{2.20}$$

Substituting the closed form solution of Eq. 2.16 into Eq. 2.20 for z_1, the initial-value problem of Eq. 2.20 may also be solved in closed form to obtain

$$y_1^1 \equiv (z_1)_{b_1} = \left(\frac{b_2 t}{2b_1}\right) \cos\sqrt{b_1}\ t - \left(\frac{b_2}{2(b_1)^{3/2}}\right) \sin\sqrt{b_1}\ t$$

$$y_2^1 \equiv (z_2)_{b_1} = \dot{y}_1^1$$

$$y_1^2 \equiv (z_1)_{b_2} = \left(\frac{1}{\sqrt{b_1}}\right) \sin\sqrt{b_1}\ t \tag{2.21}$$

$$y_2^2 \equiv (z_2)_{b_2} = \dot{y}_1^2$$

These results may be substituted, with Eq. 2.19, into Eq. 2.6 to obtain

$$\frac{d\psi}{d\mathbf{b}} = [1,0]\mathbf{z_b}(\pi/2) = \left[y_1^1(\pi/2)\ ,\quad y_1^2(\pi/2)\right]$$

$$= \left[\frac{\pi b_2}{4b_1} \cos\frac{\pi\sqrt{b_1}}{2} - \frac{b_2}{2(b_1)^{3/2}} \sin\frac{\pi\sqrt{b_1}}{2}\ ,\quad \frac{1}{\sqrt{b_1}} \sin\frac{\pi\sqrt{b_1}}{2}\right] \tag{2.22}$$

A direct calculation of the partial derivatives of ψ of Eq. 2.17, using the solution of Eq. 2.16, shows that the result obtained by the direct differentiation method is exact.

Adjoint Variable Method: Next, the adjoint variable method is applied. From Eq. 2.13, the first design derivatives of ψ are given by

$$\frac{d\psi}{d\mathbf{b}} = -\boldsymbol{\lambda}^T(0)\mathbf{h_b} - \int_0^{\frac{\pi}{2}} \boldsymbol{\lambda}^T \mathbf{f_b} dt \tag{2.23}$$

where

$$\mathbf{h_b} = \begin{bmatrix} 0 & 0 \\ 0 & 1 \end{bmatrix} \tag{2.24}$$

$$\mathbf{f_b} = \begin{bmatrix} 0 & 0 \\ -z_1 & 0 \end{bmatrix}$$

and $\boldsymbol{\lambda}$ is the solution of Eqs. 2.11 and 2.12. In this case,

$$\begin{bmatrix} \dot{\lambda}_1 \\ \dot{\lambda}_2 \end{bmatrix} + \begin{bmatrix} 0 & -b_1 \\ 1 & 0 \end{bmatrix} \begin{bmatrix} \lambda_1 \\ \lambda_2 \end{bmatrix} = 0 \tag{2.25}$$

$$\boldsymbol{\lambda}\left(\frac{\pi}{2}\right) = - \begin{bmatrix} 1 \\ 0 \end{bmatrix}$$

The solution of Eq. 2.25 for $\boldsymbol{\lambda}$ is obtained in closed form as

$$\begin{bmatrix} \lambda_1 \\ \lambda_2 \end{bmatrix} = \begin{bmatrix} -\cos\sqrt{b_1}\left(\frac{\pi}{2} - t\right) \\ -\frac{1}{\sqrt{b_1}} \sin\sqrt{b_1}\left(\frac{\pi}{2} - t\right) \end{bmatrix} \tag{2.26}$$

Substituting Eqs. 2.24 and 2.26 into Eq. 2.23,

$$\frac{d\psi}{d\mathbf{b}} = \begin{bmatrix} \cos\frac{\pi\sqrt{b_1}}{2} , & \frac{1}{\sqrt{b_1}} \sin\frac{\pi\sqrt{b_1}}{2} \end{bmatrix} \begin{bmatrix} 0 & 0 \\ 0 & 1 \end{bmatrix}$$

$$+ \int_0^{\frac{\pi}{2}} \begin{bmatrix} \cos\sqrt{b_1}\left(\frac{\pi}{2} - t\right), & \frac{1}{\sqrt{b_1}} \sin\sqrt{b_1}\left(\frac{\pi}{2} - t\right) \end{bmatrix} \begin{bmatrix} 0 & 0 \\ -z_1 & 0 \end{bmatrix} dt$$

Using the solution for z_1 in Eq. 2.16 and integrating,

$$\frac{d\psi}{d\mathbf{b}} = \begin{bmatrix} -\frac{b_2}{2b_1^{\frac{3}{2}}}\sin\frac{\pi\sqrt{b_1}}{2} + \frac{\pi b_2}{4b_1}\cos\frac{\pi\sqrt{b_1}}{2} , & \frac{1}{\sqrt{b_1}}\sin\frac{\pi\sqrt{b_1}}{2} \end{bmatrix} \tag{2.27}$$

which is indeed the result obtained by direct differentiation of Eq. 2.17.

3. SECOND ORDER DESIGN SENSITIVITY ANALYSIS FOR SYSTEMS DESCRIBED BY FIRST ORDER DIFFERENTIAL EQUATIONS

In some situations, first derivatives with respect to design are inadequate. For example, optimization algorithms

that use second order derivatives are generally superior to gradient based methods. Of more direct importance in design, bounds may be placed on the sensitivity of system performance to variation in some parameter. In this case, the derivative with respect to the parameter must be bounded. The designer thus needs second derivatives, in order to adjust the design to stay within an acceptable range of first order parameter sensitivity. These and other design requirements motivate the desire to calculate second design derivatives.

Three approaches to second order design sensitivity analysis are presented in Ref. 12. Only the hybrid second order method based on an idea introduced by Haftka for structural design sensitivity analysis [8], is presented here.

To avoid notational difficulties that are associated with defining the derivative of a matrix with respect to a vector, consider one component of Eq. 2.6; i.e., the derivative of ψ with respect to the i-th component of $\mathbf{b}$,

$$\psi_{b_i} = \mathbf{G}^T(t^2,\mathbf{z}(t^2),\mathbf{b})\mathbf{z}_{b_i}(t^2) + g_{b_i} + \int_{t^1}^{t^2} [F_{\mathbf{z}}\mathbf{z}_{b_i} + F_{b_i}]dt \tag{3.1}$$

To further simplify notation, attention will be limited here to the case in which the condition $\Omega(t^2,\mathbf{z}(t_2)) = 0$ does not depend explicitly on design.

Since ψ_{b_i} is a scalar quantity, it may be differentiated with respect to design, using the chain rule of differentiation and Eqs. 2.1 and 2.5, to obtain

$$\begin{aligned}\psi_{b_i\mathbf{b}} = {} & \mathbf{z}_{b_i}^T(t^2)\left\{-\mathbf{G}_t^2\left[\frac{\Omega_{\mathbf{z}}}{\dot{\Omega}(t^2)}\right] + \mathbf{G}_{\mathbf{z}}\right\}\mathbf{z}_{\mathbf{b}}(t^2) + \mathbf{z}_{b_i}^T(t^2)\mathbf{G}_{\mathbf{b}} \\ & - \mathbf{G}^T\mathbf{f}_{b_i}\left[\frac{\Omega_{\mathbf{z}}}{\dot{\Omega}(t^2)}\right]\mathbf{z}_{\mathbf{b}}(t^2) + \left\{g_{b_i\mathbf{z}} - g_{b_i t^2}\left[\frac{\Omega_{\mathbf{z}}}{\dot{\Omega}(t^2)}\right]\right\}\mathbf{z}_{\mathbf{b}}(t^2) \\ & + g_{b_i\mathbf{b}} - [F_{\mathbf{z}}\mathbf{z}_{b_i} + F_{b_i}]\left[\frac{\Omega_{\mathbf{z}}}{\dot{\Omega}(t^2)}\right]\mathbf{z}_{\mathbf{b}}(t^2)\end{aligned}$$

$$+ \int_{t^1}^{t^2} \left[(F_{\mathbf{z}} \tilde{\mathbf{z}}_{b_i})_{\mathbf{z}} \mathbf{z}_{\mathbf{b}} + F_{b_i \mathbf{z}} \mathbf{z}_{\mathbf{b}} + F_{b_i \mathbf{b}} \right] dt$$

$$+ \left\{ \mathbf{G}^T(t^2) \mathbf{z}_{b_i \mathbf{b}}(t^2) + \int_{t^1}^{t^2} F_{\mathbf{z}} \mathbf{z}_{b_i \mathbf{b}} dt \right\} \tag{3.2}$$

where ˜ denotes a variable that is to be held fixed for the partial differentiation indicated.

In order to evaluate terms on the right side of Eq. 3.2, the first partial derivatives of $\mathbf{z}$ with respect to $\mathbf{b}$ may be evaluated, using the direct design sensitivity analysis method of Section 2. Second derivatives of $\mathbf{z}$ with respect to design, however, arise and must be evaluated. To extend the direct design sensitivity analysis method, one component of Eq. 2.8 can be differentiated to obtain

$$\begin{aligned} \dot{\mathbf{z}}_{b_i \mathbf{b}} - \mathbf{f}_{\mathbf{z}} \mathbf{z}_{b_i \mathbf{b}} &= (\mathbf{f}_{\mathbf{z}} \tilde{\mathbf{z}}_{b_i})_{\mathbf{z}} \mathbf{z}_{\mathbf{b}} + (\mathbf{f}_{\mathbf{z}} \tilde{\mathbf{z}}_{b_i})_{\mathbf{b}} + \mathbf{f}_{b_i \mathbf{z}} \mathbf{z}_{\mathbf{b}} + \mathbf{f}_{b_i \mathbf{b}} \\ \mathbf{z}_{b_i \mathbf{b}}(t^1) &= \mathbf{h}_{b_i \mathbf{b}} \end{aligned} \tag{3.3}$$

Equation 3.3 for i = 1,...,k could be solved to obtain all second derivatives of state with respect to design and the result could be substituted into Eq. 3.2 to complete calculation of the matrix of second derivatives of ψ with respect to design. While this is mathematically feasible, an exceptionally large number of computations would be required. First, the system of k first order equations of Eq. 2.8 must be solved for $\mathbf{z}_{\mathbf{b}}$, which is then substituted in the right side of Eq. 3.3. Then, the system of k^2 equations of Eq. 3.3 must be solved for the second derivatives of $\mathbf{z}$ with respect to design. All of these results would have to be stored and the results substituted into Eq. 3.2, for evaluation of second derivatives of ψ with respect to design. Taken with the original state equations of Eq. 2.1, this constitutes a total of $1 + k + k^2$ systems of differential equations, each being n first order differential equations in

n unknowns. This approach is not pursued further, since it is clearly intractable.

An observation made by Haftka [8] allows for coupling the direct first order design sensitivity analysis method and the adjoint variable technique to efficiently solve the second order design sensitivity analysis problem. Consider that the direct first order design sensitivity analysis method of Section 2 has been used to obtain all first derivatives of state with respect to design. In this situation, the only terms in the second order design sensitivity formula of Eq. 3.2 that are not known are those involving second derivatives of state with respect to design; i.e.,

$$SOT_i \equiv G^T \mathbf{z}_{b_i \mathbf{b}} + \int_{t^1}^{t^2} F_{\mathbf{z}} \mathbf{z}_{b_i \mathbf{b}} dt \tag{3.4}$$

where SOT denotes second order terms that are to be computed.

As noted earlier, direct solution of the second order state sensitivity equations of Eq. 3.3 is impractical. As an alternative, multiply the differential equation of Eq. 3.3 by $-\boldsymbol{\lambda}^T$ and integrate from t^1 to t^2, using integration by parts, to obtain the following identity in $\boldsymbol{\lambda}$:

$$\begin{aligned} - \int_{t^1}^{t^2} [\boldsymbol{\lambda}^T(\dot{\mathbf{z}}_{b_i \mathbf{b}}) - \boldsymbol{\lambda}^T \mathbf{f}_{\mathbf{z}} \mathbf{z}_{b_i \mathbf{b}}] dt &= - \boldsymbol{\lambda}^T(t^2) \mathbf{z}_{b_i \mathbf{b}}(t^2) + \boldsymbol{\lambda}^T(t^1) \mathbf{h}_{b_i \mathbf{b}} + \int_{t^1}^{t^2} [\dot{\boldsymbol{\lambda}}^T + \boldsymbol{\lambda}^T \mathbf{f}_{\mathbf{z}}] \mathbf{z}_{b_i \mathbf{b}} dt \\ &\quad - \int_{t^1}^{t^2} \boldsymbol{\lambda}^T [(\mathbf{f}_{\mathbf{z}} \tilde{\mathbf{z}}_{b_i})_{\mathbf{z}} \mathbf{z}_{\mathbf{b}} + (\mathbf{f}_{\mathbf{z}} \tilde{\mathbf{z}}_{b_i})_{\mathbf{b}} + \mathbf{f}_{b_i \mathbf{z}} \mathbf{z}_{\mathbf{b}} + \mathbf{f}_{b_i \mathbf{b}}] dt \end{aligned} \tag{3.5}$$

Following exactly the same argument as in the adjoint variable method for first order design sensitivity analysis, select $\boldsymbol{\lambda}$ so that the coefficients of $\mathbf{z}_{b_i \mathbf{b}}$ in the second line of Eq. 3.5 and in Eq. 3.4 are identical; i.e.,

$$\begin{aligned} \dot{\boldsymbol{\lambda}} + \mathbf{f}_{\mathbf{z}}^T \boldsymbol{\lambda} &= F_{\mathbf{z}}^T \\ \boldsymbol{\lambda}(t^2) &= -\mathbf{G}(t^2) \end{aligned} \tag{3.6}$$

Note that the terminal-value problem of Eq. 3.6 is identical to the terminal-value problem of Eqs. 2.11 and 2.12 for first order adjoint design sensitivity analysis. Using this result, the second order terms of Eq. 3.4 may be evaluated, using Eq. 3.5 evaluated at the solution of Eq. 3.6, to obtain

$$\mathrm{SOT}_i = -\boldsymbol{\lambda}(t^1)\mathbf{h}_{b_i\mathbf{b}} - \int_{t^1}^{t^2} \boldsymbol{\lambda}^T \left[(\mathbf{f}_{\mathbf{z}}\tilde{\mathbf{z}}_{b_i})_{\mathbf{z}}\mathbf{z}_{\mathbf{b}} + (\mathbf{f}_{\mathbf{z}}\tilde{\mathbf{z}}_{b_i})_{\mathbf{b}} + \mathbf{f}_{b_i\mathbf{z}}\mathbf{z}_{\mathbf{b}} + \mathbf{f}_{b_i\mathbf{b}} \right] dt \quad (3.7)$$

This result may be substituted into Eq. 3.2 to evaluate all second derivatives of ψ with respect to design.

The remarkable aspect of this approach is that the adjoint equation of Eq. 3.6 does not depend on the index i. Therefore, only a single backward adjoint equation must be solved. Thus, to evaluate the full matrix of second derivatives in Eq. 3.2, only the single state equation of Eq. 2.1, the system of k first order state sensitivity equations of Eq. 2.8, and the single adjoint equation of Eq. 3.6 need to be solved. This is a system of only 2 + k systems of first order equations, each in n variables.

Second Order Design Sensitivity of a Simple Oscillator

It is fascinating that precisely the information needed for second order design sensitivity analysis of the simple oscillator is available from the first order sensitivity analysis presented in Section 2. Specifically, $\mathbf{z}_{\mathbf{b}}$ is given in Eq. 2.21 and $\boldsymbol{\lambda}$ is given in Eq. 2.26. Thus, no new integrations are required. Substituting these results into Eqs. 3.7 and 3.2 yields the desired results,

$$\frac{d^2\psi}{d\mathbf{b}^2} = \begin{bmatrix} \left\{\frac{3b_2}{4b_1^{5/2}} - \frac{\pi^2 b_2}{16b_1^{3/2}} \sin\frac{\pi\sqrt{b_1}}{2} - \frac{3\pi b_2}{8b_1^2}\cos\frac{\pi\sqrt{b_1}}{2}\right\} , & \left\{-\frac{1}{2b_1^{3/2}}\sin\frac{\pi\sqrt{b_1}}{2} + \frac{\pi}{4b_1}\cos\frac{\pi\sqrt{b_1}}{2}\right\} \\ \left\{-\frac{1}{2b_1^{3/2}}\sin\frac{\pi\sqrt{b_1}}{2} + \frac{\pi}{4b_1}\cos\frac{\pi\sqrt{b_1}}{2}\right\} , & 0 \end{bmatrix} \tag{3.8}$$

This is the same result obtained by direct differentiation of Eq. 2.17, so the second order design sensitivity analysis method yields precise results.

As a numerical example, using the computational algorithm, the first and second design derivatives are calculated with $\mathbf{b} = [1.0,\ 1.0]^T$. Numerical results and exact values from Eqs. 2.27 and 3.8 are given in Table 3.1.

Table 3.1 First and Second Design Derivatives of Simple Oscillator, for $\mathbf{b} = [1.0,\ 1.0]^T$

Design Derivatives*	Numerical Results	Exact Values
ψ_1	-0.49999972724	-0.5
ψ_2	1.0000000000	1.0
ψ_{11}	0.13299722240	0.13314972501
ψ_{21}	-0.49963656519	-0.5
ψ_{22}	-0.49999972723	-0.5
ψ	0.0	0.0

$^*\psi_{ij} = \dfrac{\partial^2\psi}{\partial b_i\,\partial b_j}$

Using the first and second design derivatives obtained, first and second order approximations of the functional ψ may be calculated, using Taylors formula; i.e.,

$$\psi(\mathbf{b}^i) \simeq \psi(\mathbf{b}^0) + \frac{d\psi}{d\mathbf{b}}(\mathbf{b}^0)(\mathbf{b}^i - \mathbf{b}^0) \tag{3.9}$$

$$\psi(\mathbf{b}^i) \simeq \psi(\mathbf{b}^0) + \frac{d\psi}{d\mathbf{b}}(\mathbf{b}^0)(\mathbf{b}^i - \mathbf{b}^0) + \frac{1}{2}(\mathbf{b}^i - \mathbf{b}^0)^T \frac{d^2\psi}{d\mathbf{b}^2}(\mathbf{b}^0)(\mathbf{b}^i - \mathbf{b}^0) \tag{3.10}$$

where $\mathbf{b}^0 = [1.0, 1.0]^T$, and $\mathbf{b}^i = \mathbf{b}^0 + i\delta\mathbf{b}$, with $\delta\mathbf{b} = [0.05, 0.05]^T$. Results are compared with exact values of $\psi(\mathbf{b}^i)$ in Fig. 3.1. Note that second order approximation is much more accurate than first order approximation. This illustrates one potential value of second order design derivatives.

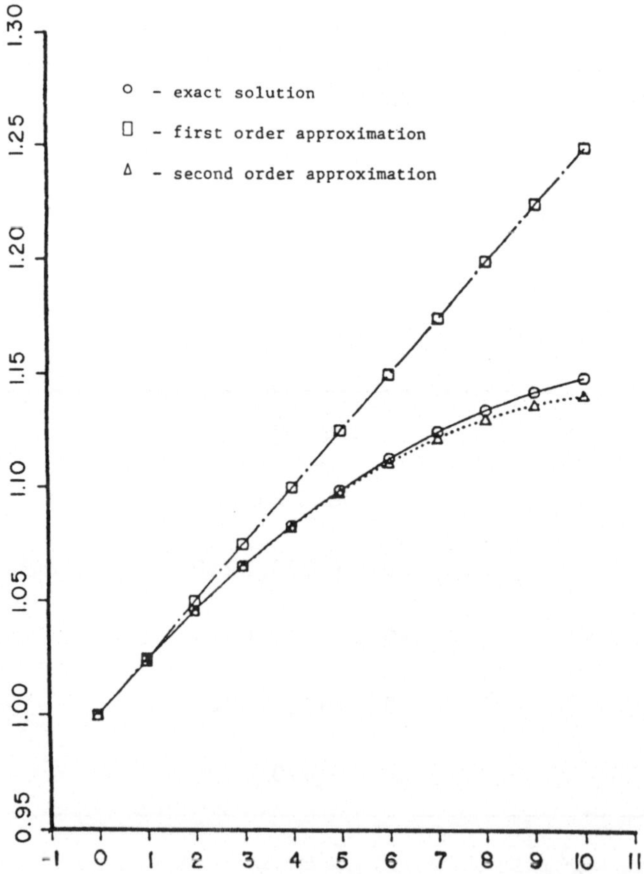

Figure 3.1 Linear and Quadratic Approximations of $\psi = z_1(t_2)$

4. DESIGN SENSITIVITY ANALYSIS OF A BURST FIRE AUTOMATIC CANNON

A burst fire automatic cannon can be modeled as shown in Fig. 4.1. The second order equation of motion of the recoiling mass is

$$m\ddot{x} - f_0 + cd(\dot{x}) + mg\sin\theta + F(t) = 0, \quad 0 \leq t \leq \tau \tag{4.1}$$

with initial conditions

$$\begin{aligned} x(0) &= 0 \\ \dot{x}(0) &= 0 \end{aligned} \tag{4.2}$$

where $cd(\dot{x})$ represents a general damping force.

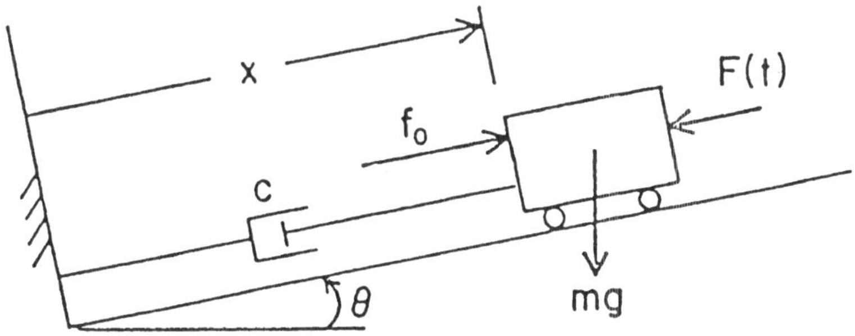

Figure 4.1 Model of Burst Fire Automatic Weapon

At t = 0, a latch is released, allowing the recoiling parts to move forward prior to firing, under the action of the actuator force f_0. Three shots are contained in a magazine that is attached to the recoiling parts and are to be fired in rapid succession, with a given time Δt between shots. The system is to be designed so that the impulse of the rounds fired brings the recoiling parts to rest at a fixed distance to the rear of the initial position, so that they can be latched again at the position x = 0.

The period of feeding and firing a three round burst is τ sec., with each round exiting the tube in a characteristic period ε sec., with an impulse I_0. The ballistic force F(t) acting on the recoiling parts is approximated by

$$F(t) = \begin{cases} \left(\frac{2I_0}{\varepsilon}\right) \sin^2(t - t_i)\frac{\pi}{\varepsilon}, & t_i \leq t \leq t_i + \varepsilon,\ i = 1,2,3 \\ 0 & , \text{otherwise} \end{cases} \tag{4.3}$$

as shown in Fig. 4.2

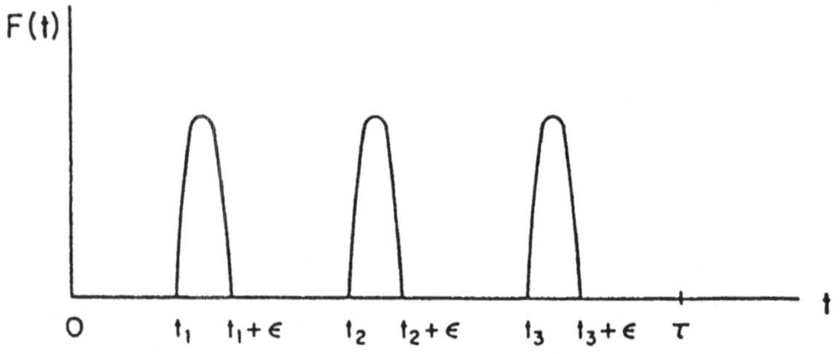

Figure 4.2 Ballistic Force

As a numerical example, let given data be as follows: mg = 1000 lb, m = 2.5892 lb/(in/sec^2), I_0 = 1000 lb sec, f_0 = 3,555 lb, and ε = 0.005 sec. Calculating the dynamic response with F(t) given in Eq. 4.3, it is found that for the given value of f_0 and with c = 0 and θ = 0, τ = 0.8461 sec and x(τ) = -1.103 in.

It is important to know how x(τ) varies with variations in the parameters f_0, c, and θ. If variations of these parameters that may occur in application cause x(τ) to be positive, then a failure of function will occur.

Linear Damping Case: In the case of linear damping, $cd(\dot{x}) = c\dot{x}$ in Eq. 4.1. Using a first order state variable formulation, Eq. 4.1 can be written as

$$\begin{bmatrix} \dot{z}_1 \\ \dot{z}_2 \end{bmatrix} = \begin{bmatrix} z_2 \\ \frac{1}{m}\{b_1 - b_2 z_2 - mg\sin b_3 - F(t)\} \end{bmatrix} \equiv \mathbf{f}(\mathbf{z},\mathbf{b}), \quad 0 < t < \tau \tag{4.4}$$

with initial conditions

$$\begin{bmatrix} z_1(0) \\ z_2(0) \end{bmatrix} = \begin{bmatrix} 0 \\ 0 \end{bmatrix} \equiv \mathbf{h}(\mathbf{b}) \tag{4.5}$$

where $\mathbf{b} = [f_0, c, \theta]^T$ and $F(t)$ is given in Eq. 4.3. The terminal time τ (playing the role to t^2 in earlier sections) is the time at which the extreme rearward position occurs, so it is determined by

$$\Omega\big(\tau, \mathbf{z}(\tau)\big) = z_2(\tau) = 0 \tag{4.6}$$

The functional for which design sensitivity is to be calculated is

$$\psi = z_1(\tau) \equiv g\big(\mathbf{z}(\tau), \mathbf{b}\big) \tag{4.7}$$

Non-zero terms in Eqs. 2.8 and 3.6 are

$$\begin{aligned} &\dot{\mathbf{z}}_{\mathbf{b}} - \mathbf{f}_{\mathbf{z}}\mathbf{z}_{\mathbf{b}} = \mathbf{f}_{\mathbf{b}} \\ &\mathbf{z}_{\mathbf{b}}(0) = \mathbf{0} \end{aligned} \tag{4.8}$$

$$\begin{aligned} &\dot{\boldsymbol{\lambda}} + \mathbf{f}_{\mathbf{z}}^T\boldsymbol{\lambda} = \mathbf{0} \\ &\boldsymbol{\lambda}(\tau) = -g_{\mathbf{z}}^T + \left[\frac{g_{\mathbf{z}}\mathbf{f}(\tau)}{\Omega_{\mathbf{z}}\mathbf{f}(\tau)}\right] \Omega_{\mathbf{z}}^T \end{aligned} \tag{4.9}$$

where

$$\mathbf{f}_{\mathbf{z}}^T = \begin{bmatrix} 0 & 0 \\ 1 & -\frac{b_2}{m} \end{bmatrix}, \quad \mathbf{f}(\tau) = \begin{bmatrix} 0 \\ \frac{b_1 - mg\sin b_3 - F(\tau)}{m} \end{bmatrix} \tag{4.10}$$

$$g_{\mathbf{z}}^T = \begin{bmatrix} 1 \\ 0 \end{bmatrix}, \quad \Omega_{\mathbf{z}}^T = \begin{bmatrix} 0 \\ 1 \end{bmatrix}$$

$$\mathbf{f}_{\mathbf{b}} = \begin{bmatrix} 0 & 0 & 0 \\ \frac{1}{m} & -\frac{z_2}{m} & -g\cos b_3 \end{bmatrix} \tag{4.11}$$

In terms of solutions of the state and adjoint equations, the first and second design derivatives are evaluated in Eq. 2.13 and Eqs. 3.2 and 3.7, respectively.

Using the nominal design $\mathbf{b}^0 = [f_0, 0, 0]^T$, first and second derivatives of ψ are calculated. From Eqs. 3.9 and 3.10, with the design variation $\delta\mathbf{b} = [0.05f_0, 0.05, 0.05]^T$ and $\mathbf{b}^i = \mathbf{b}^0 + i\delta\mathbf{b}$, $i = 1,2,\ldots, 10$, one and two term Taylor approximations of $\psi(\mathbf{b}^i)$ are calculated. Results are given in Fig. 4.3. As can be expected, the second order approximation is more accurate than the first order approximation.

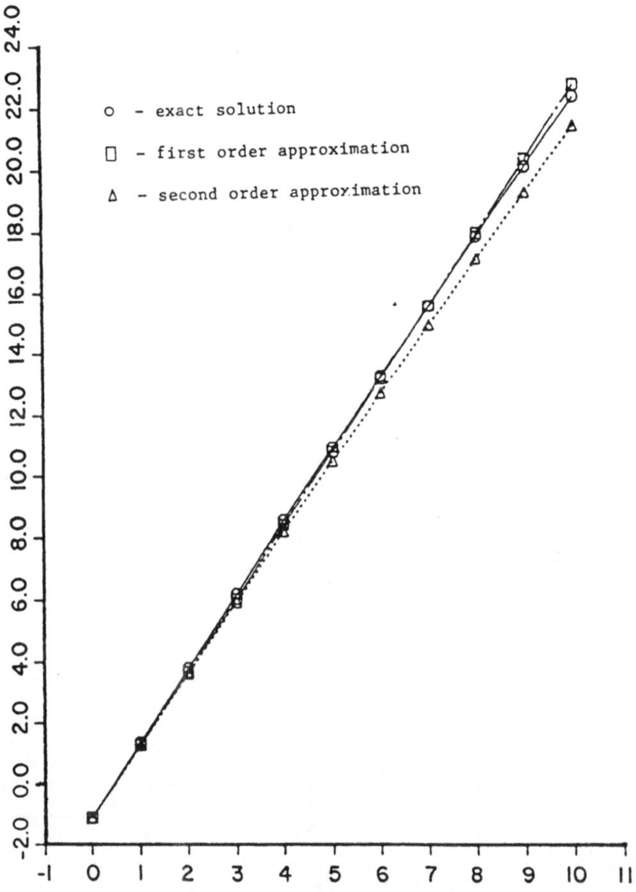

Figure 4.3 First and Second Order Approximation of ψ in Linear Damping Case

Quadratic Damping Case: In case of quadratic damping, $cd(\dot{x}) = c\dot{x}|\dot{x}|$ and the state equation becomes

$$\begin{bmatrix} \dot{z}_1 \\ \dot{z}_2 \end{bmatrix} = \begin{bmatrix} z_2 \\ \frac{1}{m}\{b_1 - b_2 z_2 |z_2| - mg\sin b_3 - F(t)\} \end{bmatrix} \equiv f(\mathbf{z},\mathbf{b}), \quad 0 < t < \tau$$

$$\begin{bmatrix} z_1(0) \\ z_2(0) \end{bmatrix} = \mathbf{0} \equiv \mathbf{h}(\mathbf{b}) \tag{4.12}$$

where the terminal time τ is determined by Eq. 4.6. The response functional is the same as in Eq. 4.7.

Terms in the state sensitivity and adjoint equations of Eqs. 4.8 and 4.9, in this case, are

$$\mathbf{f}_{\mathbf{z}}^T = \begin{bmatrix} 0 & 0 \\ 1 & -\frac{2b_2}{m}|z_2| \end{bmatrix}, \quad \mathbf{f}(\tau) = \begin{bmatrix} 0 \\ \frac{b_1 - mg\sin b_3 - F(\tau)}{m} \end{bmatrix} \tag{4.13}$$

$$g_{\mathbf{z}}^T = \begin{bmatrix} 1 \\ 0 \end{bmatrix}, \quad \Omega_{\mathbf{z}}^T = \begin{bmatrix} 0 \\ 1 \end{bmatrix} \tag{4.14}$$

$$\mathbf{f}_{\mathbf{b}} = \begin{bmatrix} 0 & 0 & 0 \\ \frac{1}{m} & -\frac{1}{m} z_2|z_2| & -g\cos b_3 \end{bmatrix} \tag{4.15}$$

Using the solutions of the state and adjoint equations, first and second design derivatives are calculated for $\mathbf{b}^0 = [f_0, 0, 0]^T$ and $\mathbf{b}^3 = b^0 + 3\delta\mathbf{b}$, where $\delta\mathbf{b} = [0.005f_0, 0.005, 0.05]^T$. The one and two term Taylor approximations of $\psi(\mathbf{b}^i)$ are calculated for these two nominal designs, with $\mathbf{b}^i = b^0 + i\delta\mathbf{b}$, $i = 1,2, \ldots, 10$, and $\mathbf{b}^i = b^0 + i\delta\mathbf{b}$, $i = -3, -2, \ldots, 7$, respectively. Results are given in Figs. 4.4 and 4.5. Note that in both cases the quadratic approximation is far more accurate than the linear approximation, for moderate design variations. For the first nominal design, Fig. 4.4

shows that for very large design perturbations, error in the quadratic approximation can be of the same order of magnitude as the first order approximation.

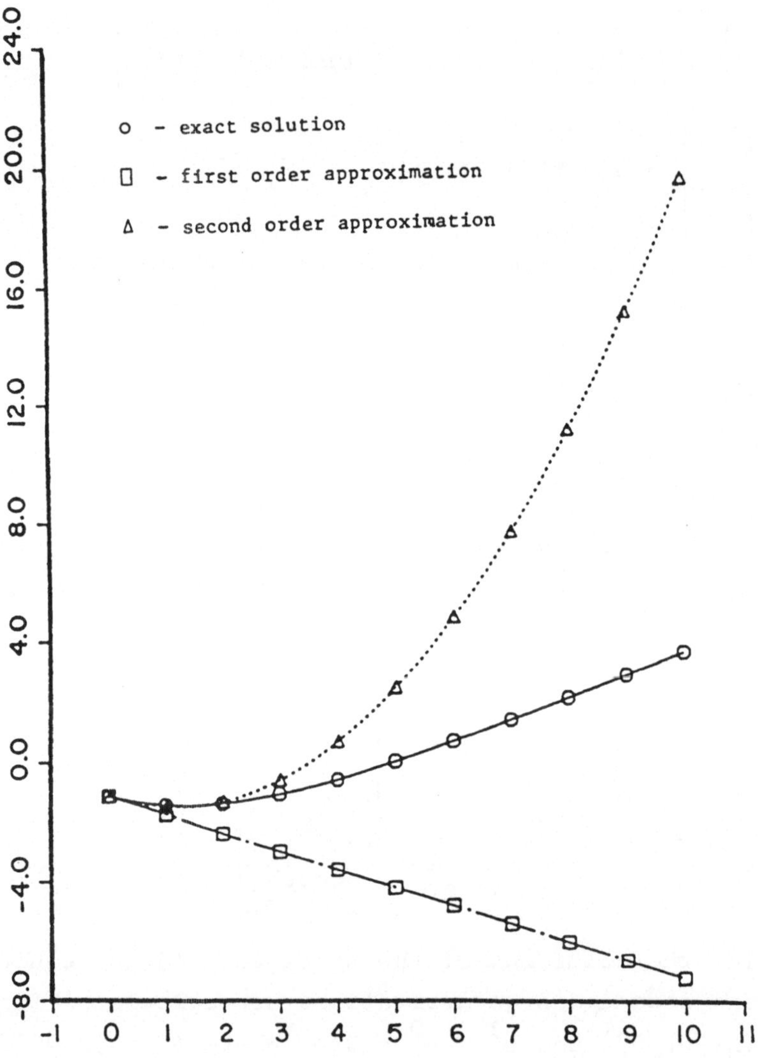

Figure 4.4 First and Second Order Approximations of ψ, with Quadratic Damping about $\mathbf{b}^0$

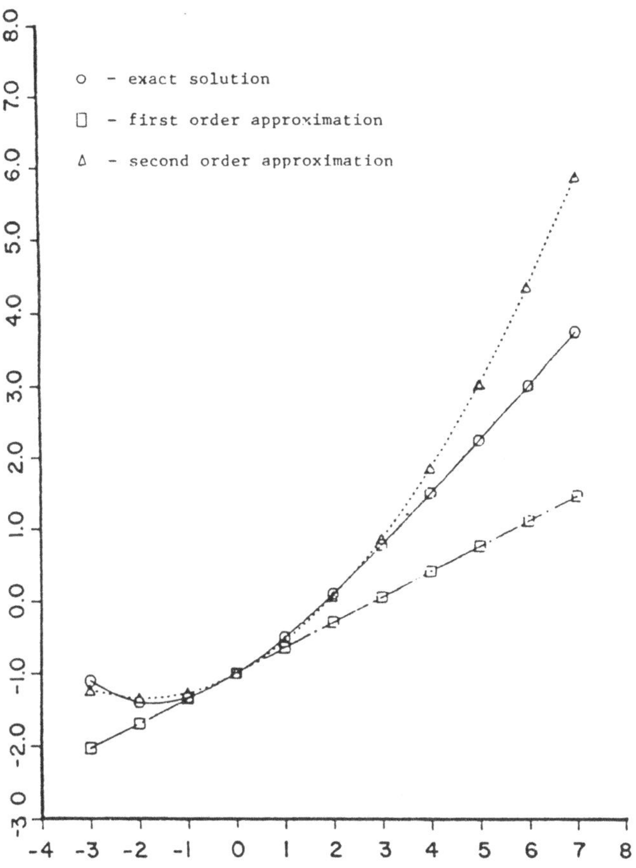

Figure 4.5 First and Second Order Approximations of ψ, with Quadratic Damping about $\mathbf{b}^3$

5. FIRST ORDER DESIGN SENSITIVITY ANALYSIS FOR SYSTEMS DESCRIBED BY SECOND ORDER DIFFERENTIAL-ALGEBRAIC EQUATIONS

The first order design sensitivity analysis methods presented in Section 2 are based on explicit differential equations of motion that are written in first order form. Many constrained mechanical systems involve numerous bodies that are connected by kinematic joints, which may be described by mixed systems of differential-algebraic equations. Automated formulation techniques are now available that provide computer generation of system equations of motion and numerical algorithms for their direct solution. The purpose of this section is to present a formulation that extends design sensitivity analysis methods to treat such classes of problems, with the ultimate objective of both computer generation and solution of equations of design sensitivity analysis.

A general formulation for constrained equations of mechanical system dynamics is presented in this section. Both the direct differentiation and adjoint variable methods of Section 2 are extended to treat these problems. Examples that have been solved with general purpose computer codes are presented.

Problem Formulation

Dynamic systems under consideration are presumed to involve constrained rigid body motion, under the influence of time varying forcing functions. Design of such systems is defined by a vector of k design parameters, denoted

$$\mathbf{b} = [b_1, \ldots, b_k]^T \tag{5.1}$$

These parameters are at the disposal of the designer and represent physical properties that define design of the system, such as dimensions, spring constants, damping coefficients, masses, or force magnitudes. In contrast to parameters that are specified by the designer, dynamic response of the system is described by a vector of n generalized coordinates $\mathbf{q}(t)$,

$$\mathbf{q}(t) = [q_1(t), \ldots, q_n(t)]^T \tag{5.2}$$

The generalized coordinates are determined by the governing equations of motion for the system, under the action of applied loads. Forces applied to the system are transformed, using the principle of virtual work [13], to obtain generalized forces that correspond to the generalized coordinates of Eq. 5.2. The generalized force vector treated here is of the form

$$\mathbf{Q} = \left[Q_1\left(t,\mathbf{q}(t),\dot{\mathbf{q}}(t),\mathbf{b}\right),\ldots,Q_n\left(t,\mathbf{q}(t),\dot{\mathbf{q}}(t),\mathbf{b}\right)\right]^T \tag{5.3}$$

The class of systems under consideration is subject to holonomic kinematic constraints of the form

$$\boldsymbol{\Phi}(\mathbf{q},\mathbf{b}) = \mathbf{0} \tag{5.4}$$

where $\boldsymbol{\Phi}(\mathbf{q},\mathbf{b})$ is a vector of constraint functions,

$$\boldsymbol{\Phi}(\mathbf{q},\mathbf{b}) = [\Phi_1(\mathbf{q},\mathbf{b}),\ldots,\Phi_m(\mathbf{q},\mathbf{b})]^T \tag{5.5}$$

For details concerning the form of such equations, see Refs. 5, 6, 7, or 13. These algebraic, time dependent constraints define dependencies among the state variables that must be accounted for in the equations of motion of the system. Note also that the constraints are design dependent.

Since dynamic systems under consideration are nonlinear, kinetic energy of the system is written in the form

$$T = \frac{1}{2}\,\dot{\mathbf{q}}^T\mathbf{M}(\mathbf{q},\mathbf{b})\dot{\mathbf{q}} \tag{5.6}$$

where $\mathbf{M}$ is a mass matrix that depends on the position of the system and design. Using the kinetic energy expression of Eq. 5.6 and the constraints of Eq. 5.4, the Lagrange multiplier formulation of the constrained dynamic equations of motion [13] may be written as

$$\mathbf{M}\ddot{\mathbf{q}} - \mathbf{S}(\mathbf{q},\dot{\mathbf{q}},\mathbf{b}) - \mathbf{Q} + \boldsymbol{\Phi}_{\mathbf{q}}^T\boldsymbol{\lambda} = \mathbf{0} \tag{5.7}$$

where $\boldsymbol{\lambda}$ is the Lagrange multiplier vector that is associated with the constraints of Eq. 5.4, $\mathbf{S}$ contains Coriolis and related terms, and a subscript denotes partial derivative. The reader is referred to the Appendix for definition of matrix calculus notation that is employed here. The differential equations of Eq. 5.7 and the algebraic constraint equations of Eq. 5.4 constitute the system equations of motion.

To simplify notation, the values of generalized coordinates at specific times t^i are denoted as

$$\mathbf{q}^i \equiv \mathbf{q}(t^i) \quad , \qquad i = 1,2 \tag{5.8}$$

It is presumed that the initial time t^1 is fixed and the final time t^2 of the dynamic event is determined by

$$\Omega(t^2, \mathbf{q}^2, \mathbf{b}) = 0 \tag{5.9}$$

which are defined by the engineer. The generalized coordinates must satisfy Eq. 5.4 at t^1 and t^2, so

$$\boldsymbol{\Phi}^i \equiv \boldsymbol{\Phi}(\mathbf{q}^i, \mathbf{b}) = 0 \quad , \qquad i = 1,2 \tag{5.10}$$

A complete characterization of the motion of the system requires definition of initial conditions on position and velocity. The initial conditions are specified by relations of the form

$$\mathbf{A}^1\mathbf{q}^1 = \mathbf{a}(\mathbf{b}) \tag{5.11}$$

$$\mathbf{A}^2\dot{\mathbf{q}}^1 = \mathbf{c}(\mathbf{b}) \tag{5.12}$$

where $\mathbf{A}^1$ and $\mathbf{A}^2$ are matrices that define initial conditions on position and velocity, respectively. Since Eqs. 5.4 and 5.11 are to determine the full set of initial generalized coordinates $\mathbf{q}^1$, the matrix

$$\begin{bmatrix} \mathbf{A}^1 \\ \boldsymbol{\Phi}_{\mathbf{q}}(t^1) \end{bmatrix} \tag{5.13}$$

must be nonsingular.

In addition to the initial conditions, the generalized velocities must satisfy the constraint velocity equation

$$\Phi_{\mathbf{q}}^{i}\,\dot{\mathbf{q}}^{i} = \mathbf{0} \quad , \qquad i = 1,2 \tag{5.14}$$

which is obtained by differentiating Eq. 5.4. Equation 5.12 and Equation 5.14 with i = 1 must determine the initial velocity, so the matrix

$$\begin{bmatrix} \mathbf{A}^2 \\ \Phi_{\mathbf{q}}(t^1) \end{bmatrix} \tag{5.15}$$

must also be nonsingular.

A necessary and sufficient condition [13] for Eqs. 5.4 and 5.7 to uniquely determine motion of the system is that the following matrix be nonsingular:

$$\begin{bmatrix} \mathbf{M} & \Phi_{\mathbf{q}}^{T} \\ \Phi_{\mathbf{q}} & \mathbf{0} \end{bmatrix} \tag{5.16}$$

Under the foregoing assumptions, once design **b** is specified, dynamic response of the system is uniquely predicted by the system constraints and equations of motion of Eqs. 5.4 and 5.7 and the initial conditions of Eqs. 5.11 and 5.12. A reliable numerical method of integrating these equations of motion is presented in Ref. 14, employing a generalized coordinate partitioning-constraint stabilization technique for automatic formulation and integration of the equations of motion.

Well-known results from the theory of initial-value problems [10] show that the dynamic response of this system is continuously differentiable with respect to the design variables, as long as the matrix of Eq. 5.16 retains full rank. It is therefore of interest to consider a typical functional that may arise in an optimal design problem of selecting the design $\mathbf{b}$ to minimize some cost function, subject to constraints on performance, each given in terms of functions of the general form

$$\psi = g(t^2, \mathbf{q}^2, \dot{\mathbf{q}}^2, \mathbf{b}) + \int_{t^1}^{t^2} F(t, \mathbf{q}, \dot{\mathbf{q}}, \boldsymbol{\lambda}, \mathbf{b})dt \qquad (5.17)$$

The function ψ is allowed to depend on all variables of the problem. Since the Lagrange multiplier $\boldsymbol{\lambda}$ uniquely determines reaction forces in the constrained system, such as reaction forces in bearings, bounds on force transmitted are included in the integral of the second term of Eq. 5.17.

Design Derivative of ψ

Derivatives of the functional of Eq. 5.17 with respect to design are to be calculated. Since t^2, $\mathbf{q}^2$, $\dot{\mathbf{q}}^2$, $\mathbf{q}(t)$, $\dot{\mathbf{q}}(t)$, and $\boldsymbol{\lambda}(t)$ depend on design, Leibniz rule [11] for the derivative of an integral and the chain rule of differentiation may be used to obtain

$$\psi_{\mathbf{b}} = g_{t^2} t^2_{\mathbf{b}} + g_{\mathbf{q}^2}(\mathbf{q}^2_{\mathbf{b}} + \dot{\mathbf{q}}^2 t^2_{\mathbf{b}}) + g_{\dot{\mathbf{q}}^2}(\dot{\mathbf{q}}^2_{\mathbf{b}} + \ddot{\mathbf{q}}^2 t^2_{\mathbf{b}}) + g_{\mathbf{b}} + F^2 t^2_{\mathbf{b}}$$

$$+ \int_{t^1}^{t^2} [F_{\mathbf{q}}\mathbf{q}_{\mathbf{b}} + F_{\dot{\mathbf{q}}}\dot{\mathbf{q}}_{\mathbf{b}} + F_{\boldsymbol{\lambda}}\boldsymbol{\lambda}_{\mathbf{b}} + F_{\mathbf{b}}]dt \qquad (5.18)$$

where $F^2 = F(t^2, \mathbf{q}^2, \dot{\mathbf{q}}^2, \mathbf{b})$ and the following relations are employed

$$\frac{d\mathbf{q}^2}{d\mathbf{b}} = \mathbf{q}^2_{\mathbf{b}} + \dot{\mathbf{q}}^2 t^2_{\mathbf{b}} \qquad (5.19)$$

$$\frac{d\dot{\mathbf{q}}^2}{d\mathbf{b}} = \dot{\mathbf{q}}^2_{\mathbf{b}} + \ddot{\mathbf{q}}^2 t^2_{\mathbf{b}} \qquad (5.20)$$

Integration by parts of the second term in the integral of Eq. 5.18 and rearranging terms yields

$$\psi_{\mathbf{b}} = (g_{t^2} + g_{\mathbf{q}^2}\dot{\mathbf{q}}^2 + g_{\dot{\mathbf{q}}^2}\ddot{\mathbf{q}}^2 + F^2)t_{\mathbf{b}}^2$$

$$+ (g_{\mathbf{q}^2} + F_{\dot{\mathbf{q}}}^2)\mathbf{q}_{\mathbf{b}}^2 + g_{\dot{\mathbf{q}}^2}\dot{\mathbf{q}}_{\mathbf{b}}^2$$

$$+ \int_{t^1}^{t^2} [(F_{\mathbf{q}} - \frac{d}{dt} F_{\dot{\mathbf{q}}})\mathbf{q}_{\mathbf{b}} + F_{\lambda}\lambda_{\mathbf{b}}]dt$$

$$+ g_{\mathbf{b}} + \int_{t^1}^{t^2} F_{\mathbf{b}}dt \qquad (5.21)$$

In order to make use of Eq. 5.21, partial derivatives of $\mathbf{q}$, $\dot{\mathbf{q}}$, λ, and t^2 with respect to $\mathbf{b}$ must be evaluated or rewritten in terms of computable quantities. This is done in the following two subsections, using direct differentiation and adjoint variable methods, respectively.

Direct Differentiation Method

For direct evaluation of Eq. 5.21, partial derivatives of all state related terms with respect to design must be calculated. This can be done by direct differentiation of the state equations. Beginning with the differential equations of motion of Eqs. 5.7 and 5.4, taking the derivative with respect to design,

$$\mathbf{M}\ddot{\mathbf{q}}_{\mathbf{b}} - \mathbf{S}_{\dot{\mathbf{q}}}\dot{\mathbf{q}}_{\mathbf{b}} - \mathbf{S}_{\mathbf{q}}\mathbf{q}_{\mathbf{b}} - \mathbf{Q}_{\dot{\mathbf{q}}}\dot{\mathbf{q}}_{\mathbf{b}} - \mathbf{Q}_{\mathbf{q}}\mathbf{q}_{\mathbf{b}}$$

$$+ (\Phi_{\mathbf{q}}^T\tilde{\lambda})_{\mathbf{q}}\mathbf{q}_{\mathbf{b}} + \Phi_{\mathbf{q}}^T\lambda_{\mathbf{b}}$$

$$= - (\mathbf{M}\tilde{\ddot{\mathbf{q}}})_{\mathbf{b}} + \mathbf{S}_{\mathbf{b}} + \mathbf{Q}_{\mathbf{b}} - (\Phi_{\mathbf{q}}^T\tilde{\lambda})_{\mathbf{b}} \qquad (5.22)$$

$$\Phi_{\mathbf{q}}\mathbf{q}_{\mathbf{b}} = - \Phi_{\mathbf{b}}$$

This is a system of linear second order differential-algebraic equations in the variable $\mathbf{q}_{\mathbf{b}}$. In order to solve these equations, a set of initial conditions for $q_b(t^1)$ and $\dot{q}_b(t^1)$ must be calculated.

Differentiating Eqs. 5.10 and 5.11 with respect to design yields

$$\begin{bmatrix} A^1 \\ \\ \Phi_q(t^1) \end{bmatrix} q_b^1 = \begin{bmatrix} a_b \\ \\ -\Phi_b(t^1) \end{bmatrix} \tag{5.23}$$

Similarly, differentiating Eqs. 5.12 and 5.14 with respect to design yields

$$\begin{bmatrix} A^2 \\ \\ \Phi_q(t^1) \end{bmatrix} \dot{q}_b^1 = \begin{bmatrix} c_b \\ \\ -(\Phi_q(t^1)\tilde{\dot{q}}^1)_b \end{bmatrix} \tag{5.24}$$

Since the coefficient matrices in Eqs. 5.23 and 5.24 are the same as in Eqs. 5.13 and 5.15, respectively, and are thus nonsingular, q_b^1 and $\dot{q}_b^1$ are uniquely determined.

Having calculated initial conditions, the mixed system of differential-algebraic equations of Eq. 5.22 can be solved for $q_b(t)$ and λ_b. Having calculated these quantities, Eq. 5.9 may be differentiated and solved for t_b^2 as

$$t_b^2 = -\,[\Omega_{q^2} q_b^2 + \Omega_b]/[\Omega_{t^2} + \Omega_{q^2}\dot{q}\,] \tag{5.25}$$

Since all terms on the right are known, Eq. 5.25 may be used to obtain t_b, yielding all terms that are required to evaluate the total derivitave of ψ with respect to design in Eq. 5.21.

Numerical Examples by Direct Differentiation

The direct differentiation algorithm presented here was coded in FORTRAN and implemented on a PRIME 750 supermini computer, as presented in Ref. 15. The program was tested on several problems, results of which are summarized here.

Verification Procedure: Results of design sensitivity calculations can be checked by methods based on perturbation theory. Two such methods are used to verify design sensitivity calculations by the present method, as follows:

(a) Check on Functional Design Sensitivity: The vector $\psi_{\mathbf{b}}$ of design sensitivity coefficients is checked by calculating state design sensitivity and constraint functional at a nominal design $\mathbf{b}$. The design is then given a small perturbation $\delta\mathbf{b}$, so that the new design becomes

$$\mathbf{b}^{*} = \mathbf{b} + \delta\mathbf{b} \tag{5.26}$$

The equations of motion are now solved at the new design $\mathbf{b}^{*}$ and the constraint functions are re-evaluated. Let the value of the constraint function at the original design be $\psi(\mathbf{b})$ and its value at the perturbed design be $\psi(\mathbf{b}^{*})$. The actual change in the value of the constraint function is given by

$$\Delta\psi = \psi(\mathbf{b}^{*}) - \psi(\mathbf{b}) \tag{5.27}$$

The design sensitivity prediction of the change in functional value for the design change $\delta\mathbf{b}$ is given by

$$\delta\psi = \psi_{\mathbf{b}}\delta\mathbf{b} \tag{5.28}$$

If design sensitivity analysis is correct and if significant digits are lost in the difference of Eq. 5.27, then the value of $\Delta\psi_i$ obtained from Eq. 5.27 should be approximately equal to the value of $\delta\psi_i$ obtained from Eq. 5.28.

(b) Check on State Design Sensitivity: As a check on state design sensitivity $\mathbf{q}_{\mathbf{b}}$, the system is solved at the original design $\mathbf{b}$ and the perturbed design $\mathbf{b}^{*}$. Consider the variation in acceleration of generalized coordinate at time t when the design is changed from $\mathbf{b}$ to $\mathbf{b}^{*}$. This variation may be written as

$$\Delta\ddot{\mathbf{q}}(t) = \ddot{\mathbf{q}}(t,\mathbf{b}^*) - \ddot{\mathbf{q}}(t,\mathbf{b}) \qquad (5.29)$$

For a design change $\delta\mathbf{b}$, the change in acceleration at time t is predicted by design sensitivity theory as

$$\delta\ddot{\mathbf{q}}(t) = \ddot{\mathbf{q}}_{\mathbf{b}}(t)\,\delta\mathbf{b} \qquad (5.30)$$

The perturbations of Eqs. 5.29 and 5.30 should be approximately equal. Acceleration design sensitivity accuracy is checked, since if it is accurate, then position and velocity sensitivities will be even more accurate.

For each example, the perturbation in design is chosen to be

$$\delta b_i = \begin{cases} 0.001 \text{ for } b_i > 0 \\ -\,0.001 \text{ for } b_i < 0 \end{cases} \qquad (5.31)$$

Example 5.1; Four-Bar Linkage under Self-Weight: The initial configuration of a four-bar linkage is shown in Fig. 5.1. The design variables are the coordinates of the revolute joints, as indicated in Fig. 5.1. The only loading is self-weight of the members. The simulation time is from 0 to 1 sec. Input data are as follows, in slug-inch units:

Masses;

$$m_1 = 0,\ m_2 = 8.0,\ m_3 = 8.0,\ m_4 = 8.0$$

Moments of Inertia;

$$J_1 = 0,\ J_2 = 8.0,\ J_3 = 8.0,\ J_4 = 8.0$$

The nominal design is

$$\mathbf{b} \equiv [-70.9107,\ 70.7107,\ -50.,\ -55.]^T$$

The performance functional is chosen to be

$$\psi \equiv \int_0^1 (\sin\phi_2 - 0.2389)^2 dt$$

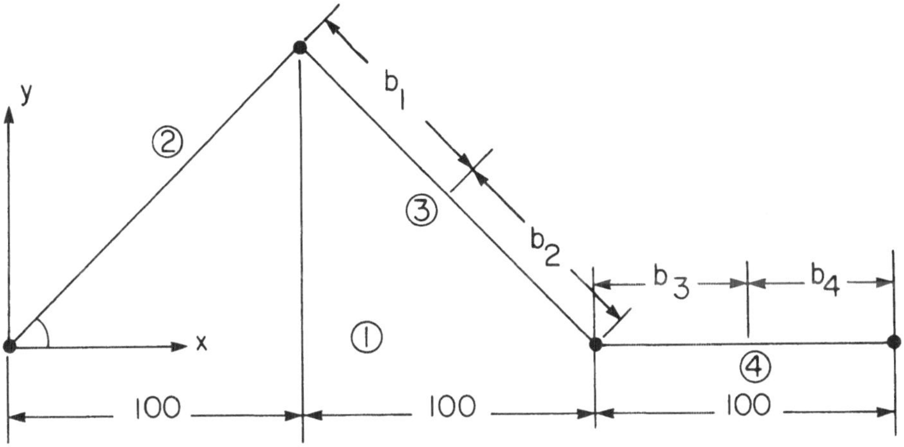

Figure 5.1 Four-bar Linkage under Self Weight

The design sensitivity vector for ψ is found to be

$$\psi_{\mathbf{b}} \equiv [-5.841,\ 1.616,\ -2.356,\ 2.601] \times 10^{-3}$$

Using this design sensitivity vector and the design perturbation of Eq. 5.31, the predicted change in cost is

$$\delta\psi = 1.24 \times 10^{-5}$$

Reanalysis with the perturbed design yields

$$\Delta\psi = 1.28 \times 10^{-5}$$

which is quite close.

The check on state sensitivity was carried out for $\ddot{C}_3 \equiv \frac{d^2}{dt^2}\cos\phi_3$, where ϕ_3 is the angle of link 3 in Fig. 5.1 with the x-axis. Plots of $\Delta\ddot{C}_3$ and $\delta\ddot{C}_3$ for this test are shown in Fig. 5.2. The curve obtained from state design sensitivity and the curve obtained by perturbation coincide to within numerical accuracy. The graph of $\ddot{C}_3$ is shown in Fig. 5.3., which indicates a substantial acceleration variation.

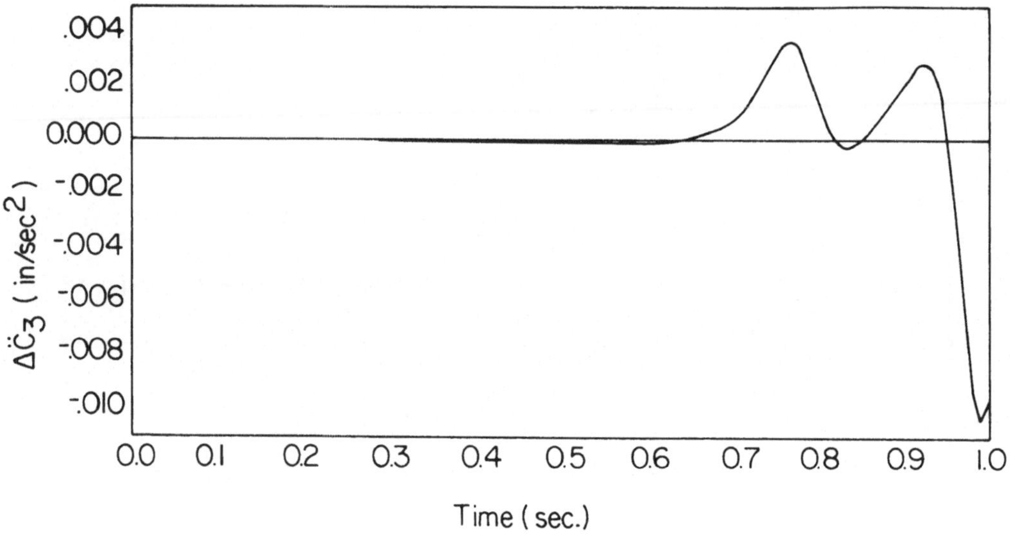

Figure 5.2 State Sensitivity Check for Example No. 1

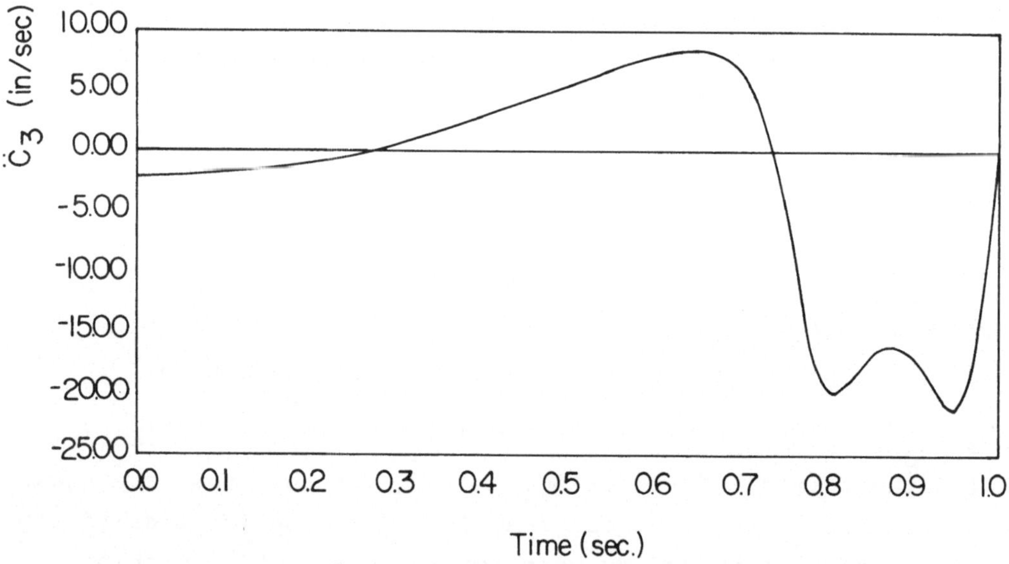

Figure 5.3 $\ddot{C}_3$ Versus Time Curve for Example 1

Example 5.2; Slider-Crank Mechanism: Figure 5.4 shows a slider-crank mechanism in its initial position. The loading in this case is the weight of the members and a constant force F = 125 lbs, which acts on the piston in the direction indicated in Fig. 5.4. The simulation time is from 0 to 1

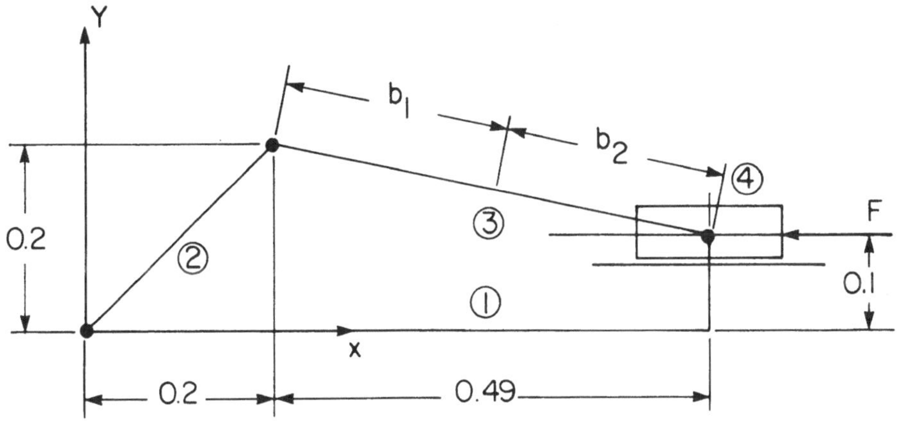

Figure 5.4 Slider-crank Mechanism

sec. The input data are given, in foot-pound units, as follows:

Masses;

$m_1 = 0,\ m_2 = 5.0,\ m_3 = 15.0,\ m_4 = 8.0$

Moments of Inertia;

$J_1 = 0,\ J_2 = 0.5,\ J_3 = 2.5,\ J_4 = 8.0$

The nominal design is

$$\mathbf{b} = [-0.25,\ 0.25]^T$$

and the functional selected for analysis is

$$\psi = \int_0^1 (x_4 - 20)^2 dt$$

For this problem, the functional design sensitivity vector was found to be

$$\psi_{\mathbf{b}} = [46.48,\ -34.73]$$

The predicted cost function variation is

$$\delta\psi = -0.0812$$

Perturbation and reanalysis yielded the comparable result

$\Delta\psi = -0.0814$

which again shows good agreement.

The state design sensitivity verification was done on $\ddot{x}_4$ and, again, coincident curves of Fig. 5.5 were obtained. Figure 5.6 is the plot of $\ddot{x}_4$ versus time.

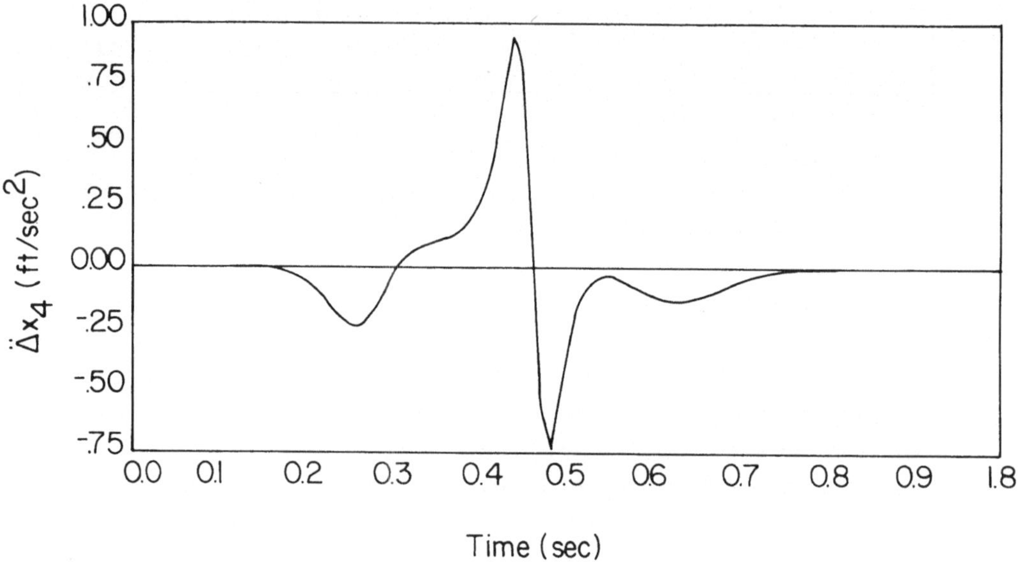

Figure 5.5 State Sensitivity Check of Example 3

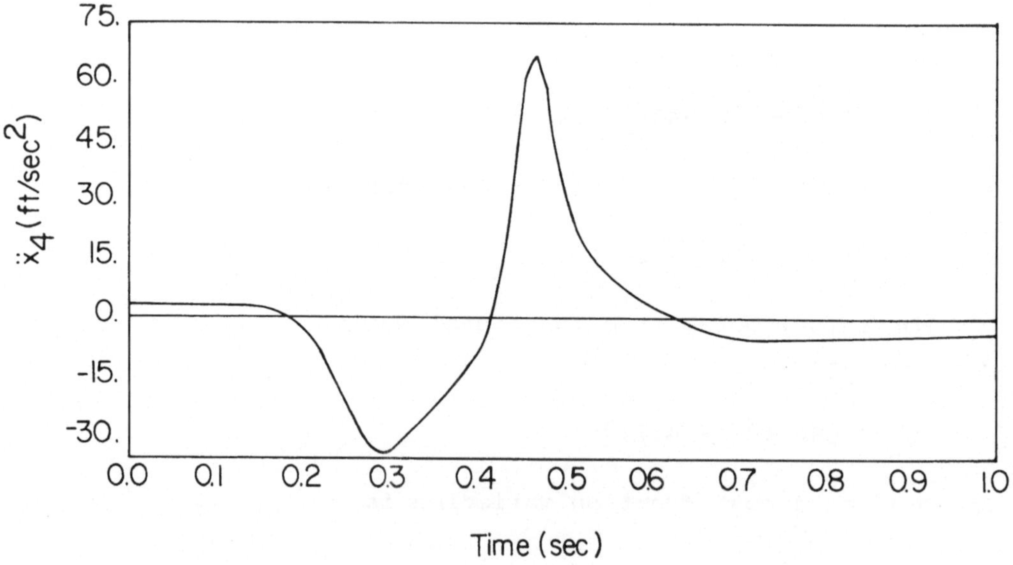

Figure 5.6 $\ddot{x}_4$ Versus Time Curve for Example 3

Adjoint Variable Method

Variations in state variables and initial and final times must be consistent with the system constraints and equations of motion. To implicitly account for this dependence and to avoid explicit computation of derivatives of state with respect to design, a sequence of adjoint relationships is now developed, using a method that has been widely applied in mechanical and structural design [3,4]. First, both sides of the equations of motion of Eq. 5.7 and the constraints of Eq. 5.4 are multiplied by arbitrary multiplier vector functions $\boldsymbol{\mu}(t)$ and $\boldsymbol{\nu}(t)$, to obtain the identities

$$\int_{t^1}^{t^2} \mu^T [\mathbf{M}\ddot{\mathbf{q}} - \mathbf{S} - \mathbf{Q} + \boldsymbol{\Phi}_{\mathbf{q}}^T \boldsymbol{\lambda}]dt = 0 \tag{5.32}$$

and

$$\int_{t^1}^{t^2} \nu^T \boldsymbol{\Phi} dt = 0 \tag{5.33}$$

These equations hold for all values of design, so the total design derivative of both sides of Eqs. 5.32 and 5.33 may be taken, yielding

$$\begin{aligned}\int_{t^1}^{t^2} \mu^T [\mathbf{M}\ddot{\mathbf{q}}_{\mathbf{b}} &+ (\mathbf{M}\dot{\mathbf{q}})_{\mathbf{q}}\mathbf{q}_{\mathbf{b}} - \mathbf{S}_{\mathbf{q}}\mathbf{q}_{\mathbf{b}} - \mathbf{S}_{\dot{\mathbf{q}}}\dot{\mathbf{q}}_{\mathbf{b}} - \mathbf{Q}_{\mathbf{q}}\mathbf{q}_{\mathbf{b}} - \mathbf{Q}_{\dot{\mathbf{q}}}\dot{\mathbf{q}}_{\mathbf{b}} \\ &+ (\boldsymbol{\Phi}_{\mathbf{q}}^T\boldsymbol{\lambda})_{\mathbf{q}}\mathbf{q}_{\mathbf{b}} + \boldsymbol{\Phi}_{\mathbf{q}}^T\boldsymbol{\lambda}_{\mathbf{b}} + (\mathbf{M}\tilde{\ddot{\mathbf{q}}})_{\mathbf{b}} \\ &- \mathbf{S}_{\mathbf{b}} - \mathbf{Q}_{\mathbf{b}} + (\boldsymbol{\Phi}_{\mathbf{q}}^T\tilde{\boldsymbol{\lambda}})_{\mathbf{b}}]dt = \mathbf{0}\end{aligned} \tag{5.34}$$

and

$$\int_{t^1}^{t^2} \nu^T(\boldsymbol{\Phi}_{\mathbf{q}}\mathbf{q}_{\mathbf{b}} + \boldsymbol{\Phi}_{\mathbf{b}})dt = \mathbf{0} \tag{5.35}$$

Integrating terms in the integrals of Eq. 5.34 involving $\ddot{\mathbf{q}}_{\mathbf{b}}$ and $\dot{\mathbf{q}}_{\mathbf{b}}$ by parts and using Eqs. 5.19 and 5.20 yields the identity

$$\begin{aligned}
\int_{t^1}^{t^2} \Big\{ [& \ddot{\mu}^T\mathbf{M} + 2\dot{\mu}^T\dot{\mathbf{M}} + \mu^T\ddot{\mathbf{M}} + \dot{\mu}^T\mathbf{S}_{\dot{\mathbf{q}}} + \mu^T\dot{\mathbf{S}}_{\dot{\mathbf{q}}} + \mu^T\mathbf{Q}_{\dot{\mathbf{q}}} + \mu^T\dot{\mathbf{Q}}_{\dot{\mathbf{q}}} + \mu^T(\mathbf{M}\ddot{\mathbf{q}})_{\mathbf{q}} \\
& - \mu^T\mathbf{S}_{\mathbf{q}} - \mu^T\mathbf{Q}_{\mathbf{q}} + \mu^T(\boldsymbol{\Phi}_{\mathbf{q}}^T\lambda)_{\mathbf{q}}]\mathbf{q}_{\mathbf{b}} + \mu^T\boldsymbol{\Phi}_{\mathbf{q}}^T\lambda_{\mathbf{b}} + \mu^T(\mathbf{M}\tilde{\ddot{\mathbf{q}}})_{\mathbf{b}} \\
& - \mu^T\mathbf{S}_{\mathbf{b}} - \mu^T\mathbf{Q}_{\mathbf{b}} + \mu^T(\boldsymbol{\Phi}_{\mathbf{q}}^T\tilde{\lambda})_{\mathbf{b}} \Big\} dt \\
& + \mu^T\mathbf{M}\dot{\mathbf{q}}_{\mathbf{b}} - \mu^T\mathbf{S}_{\dot{\mathbf{q}}}\mathbf{q}_{\mathbf{b}} - \mu^T\mathbf{Q}_{\dot{\mathbf{q}}}\mathbf{q}_{\mathbf{b}} - (\dot{\mu}^T\mathbf{M} + \mu^T\dot{\mathbf{M}})\mathbf{q}_{\mathbf{b}} \Big|_{t^1}^{t^2} = \mathbf{0}
\end{aligned} \tag{5.36}$$

Equation 5.9 may similarly be multiplied by an arbitrary multiplier ξ^2 to obtain the identity

$$\xi^2\Omega(t^2,\mathbf{q}^2,\mathbf{b}) = 0 \tag{5.37}$$

Since this equation must hold for all design, the total design derivative of both sides yields the identity

$$\xi^2\Omega_t t_{\mathbf{b}}^2 + \xi^2\Omega_{\mathbf{q}}(\mathbf{q}_{\mathbf{b}}^2 + \dot{\mathbf{q}}^2 t_{\mathbf{b}}^2) + \xi^2\Omega_{\mathbf{b}} = \mathbf{0} \tag{5.38}$$

Multiplying the kinematic constraint equation of Eq. 5.10 at t^2 by an arbitrary multiplier vector $\boldsymbol{\gamma}^2$ yields the identity

$$\boldsymbol{\gamma}^{2^T}\boldsymbol{\Phi}^2(\mathbf{q}^2,\mathbf{b}) = 0 \tag{5.39}$$

Taking the total design derivative of both sides of Eq. 5.39 yields the identity

$$\gamma^{2^T} \Phi_q^2 (q_b^2 + \dot{q}^2 t_b^2) + \gamma^{2^T} \Phi_b^2 = 0 \tag{5.40}$$

Velocity variations must satisfy Eq. 5.14 at t^2. Hence, Eq. 5.14 may be multiplied by an arbitrary multiplier vector η^2 to obtain the identity

$$\eta^{2^T} \Phi_q^2 \dot{q}^2 = 0 \tag{5.41}$$

Taking the total design derivative of both sides yields the identity

$$\eta^{2^T} (\Phi_q^2 \dot{q}^2)_q [q_b^2 + \dot{q}^2 t_b^2]$$
$$+ \eta^{2^T} \Phi_q^2 (\dot{q}_b^2 + \ddot{q}^2 t_b^2) + \eta^{2^T} (\Phi_q^2 \tilde{\dot{q}}^2)_b = 0 \tag{5.42}$$

The identities of Eqs. 5.35, 5.36, 5.38, 5.40, and 5.42 are relationships among design derivatives of state, Lagrange multiplier, and initial and final times. All of these identities are valid for arbitrary multiplier functions $\mu(t)$ and $\nu(t)$ and multiplier parameters ξ^2, γ^2, and η^2. The objective in use of these identities is to select the arbitrary multipliers in such a way that terms involving state derivatives with respect to design in Eq. 5.21 may be written explicitly in terms of computable quantities. The technique employed here is an extension of the adjoint variable method that has been used extensively in optimal control and optimal design literature [2-4]. The idea is simply to sum identities of Eqs. 5.35, 5.36, 5.38, 5.40, and 5.42, to obtain a linear expression in all the derivatives involved. The coefficients of each design derivative of state, Lagrange multiplier, and final time are equated to corresponding coefficients on the right of Eq. 5.21. This process yields the following set of adjoint relations:

Equating coefficients of q_b in the integrals yields

$$\mathbf{M}\ddot{\mu} + \left[2\dot{\mathbf{M}} + \mathbf{S}_{\dot{\mathbf{q}}}^{T} + \mathbf{Q}_{\dot{\mathbf{q}}}^{T}\right]\dot{\mu} + \left[\ddot{\mathbf{M}} + \dot{\mathbf{S}}_{\dot{\mathbf{q}}}^{T} + \dot{\mathbf{Q}}_{\dot{\mathbf{q}}}^{T}\right.$$

$$\left. + \left(\mathbf{M}\ddot{\mathbf{q}}\right)_{\mathbf{q}}^{T} - \mathbf{S}_{\mathbf{q}}^{T} - \mathbf{Q}_{\mathbf{q}}^{T} + \left(\Phi_{\mathbf{q}}^{T}\lambda\right)_{\mathbf{q}}^{T}\right]\mu = F_{\mathbf{q}}^{T} - \frac{d}{dt}\left(F_{\dot{\mathbf{q}}}\right)^{T} \qquad (5.43)$$

Equating coefficients of $\lambda_{\mathbf{b}}$ in the integrals yields

$$\Phi_{\mathbf{q}}\mu = F_{\lambda}^{T} \qquad (5.44)$$

Equating coefficients of $q_{\mathbf{b}}^{2}$ yields

$$- \mathbf{M}^{2}\dot{\mu}^{2} - \dot{\mathbf{M}}^{2}\mu^{2} - \mathbf{S}_{\dot{\mathbf{q}}}^{2^{T}}\mu^{2} - \mathbf{Q}_{\dot{\mathbf{q}}}^{2^{T}}\mu^{2} + \Omega_{\mathbf{q}^{2}}^{T}\xi^{2} + \Phi_{\mathbf{q}}^{2^{T}}\gamma^{2}$$

$$+ \left(\Phi_{\mathbf{q}}^{2}\dot{\mathbf{q}}^{2}\right)_{\mathbf{q}}^{T}\eta^{2} = g_{\mathbf{q}^{2}}^{T} + F_{\dot{\mathbf{q}}}^{2^{T}} \qquad (5.45)$$

Equating coefficients of $\dot{q}_{\mathbf{b}}^{2}$ yields

$$\mathbf{M}^{2}\mu^{2} + \Phi_{\mathbf{q}}^{2^{T}}\eta^{2} = g_{\dot{\mathbf{q}}^{2}}^{T} \qquad (5.46)$$

Equating coefficients of $t_{\mathbf{b}}^{2}$ yields

$$\left(\Omega_{t^{2}} + \Omega_{\mathbf{q}^{2}}\dot{\mathbf{q}}^{2}\right)\xi^{2} + \dot{\mathbf{q}}^{2^{T}}\Phi_{\mathbf{q}}^{2^{T}}\gamma^{2} + \dot{\mathbf{q}}^{2^{T}}\left(\Phi_{\dot{\mathbf{q}}}^{2}2\right)_{\mathbf{q}}^{T}\eta^{2}$$

$$+ \ddot{\mathbf{q}}^{2^{T}}\Phi_{\mathbf{q}}^{2^{T}}\eta^{2} = g_{t^{2}} + g_{\mathbf{q}^{2}}\dot{\mathbf{q}}^{2} + g_{\dot{\mathbf{q}}^{2}}\ddot{\mathbf{q}}^{2} + F^{2} \qquad (5.47)$$

Presuming that Eqs. 5.43 through 5.47 determine all multipliers, the resulting terms in the summation of Eqs. 5.35, 5.36, 5.38, 5.40, and 5.42 yield a formula for the sum of all terms in Eq. 5.21 involving design derivatives of state, Lagrange multiplier, and initial and final times. This identity is substituted into Eq. 5.21, yielding an explicit expression for the derivative of ψ with respect to $\mathbf{b}$ as

$$\psi_{\mathbf{b}} = g_{\mathbf{b}} + \mu^{1^T} \mathbf{M}^1 \dot{\mathbf{q}}^1_{\mathbf{b}} - (\mu^{1^T} \mathbf{S}^1_{\dot{\mathbf{q}}} + \mu^{1^T} \mathbf{Q}^1_{\dot{\mathbf{q}}} + \dot{\mu}^{1^T} \mathbf{M}^1 + \mu^{1^T} \dot{\mathbf{M}}^1) \mathbf{q}^1_{\mathbf{b}}$$
$$- \mathbf{y}^{2^T} \Phi^2_{\mathbf{b}} - \eta^{2^T} (\Phi_{\mathbf{q}} \tilde{\dot{\mathbf{q}}}^2)_{\mathbf{b}}$$
$$+ \int_{t^1}^{t^2} \left\{ F_{\mathbf{b}} - \mu^T [(\mathbf{M}\tilde{\ddot{\mathbf{q}}})_{\mathbf{b}} + (\Phi^T_{\mathbf{q}} \tilde{\lambda})_{\mathbf{b}} - \mathbf{S}_{\mathbf{b}} - \mathbf{Q}_{\mathbf{b}}] - \nu^T \Phi_{\mathbf{b}} \, dt \right\}$$

(5.48)

The design sensitivity analysis vector of Eq. 5.48 may be evaluated numerically, once the state and adjoint equations have been solved.

Adjoint Variable Design Sensitivity Analysis Algorithm

In order to evaluate the design sensitivity vector $\psi_{\mathbf{b}}$ in Eq. 5.48, it is important to define a practical computational algorithm that determines all variables that arise. For a nominal design **b**, the following sequence of computations uniquely determines all variables that are required for evaluating the sensitivity vector in Eq. 5.48:

Step 1: Integrate the equations of motion of Eqs. 5.4 and 5.7, with initial conditions of Eqs. 5.11, 5.12, and 5.15. Store $\mathbf{q}(t)$, $\dot{\mathbf{q}}(t)$, $\ddot{\mathbf{q}}(t)$, and $\lambda(t)$. The numerical method of Ref. 14 can be employed for this calculation.

Step 2: Equations 5.46 and 5.44 at t^2 are

$$\begin{bmatrix} \mathbf{M}^2 & \Phi^{2^T}_{\mathbf{q}} \\ \Phi^2_{\mathbf{q}} & 0 \end{bmatrix} \begin{bmatrix} \mu^2 \\ \eta^2 \end{bmatrix} = \text{known terms} \qquad (5.49)$$

Since the coefficient matrix is the nonsingular matrix of Eq. 5.16, $\boldsymbol{\mu}^2$ and $\boldsymbol{\eta}^2$ are uniquely determined.

Step 3: Equation 5.45 and $\boldsymbol{\Phi}_{\mathbf{q}}\dot{\mu} = -\left(\frac{d}{dt}\boldsymbol{\Phi}_{\mathbf{q}}\right)\boldsymbol{\mu} + \frac{d}{dt}F_{\boldsymbol{\lambda}}^T$ from Eq. 5.44 at t^2 are

$$\begin{bmatrix} \mathbf{M}^2 & \boldsymbol{\Phi}_{\mathbf{q}}^{2^T} \\ \boldsymbol{\Phi}_{\mathbf{q}}^2 & 0 \end{bmatrix} \begin{bmatrix} \dot{\mu}^2 \\ -\boldsymbol{\gamma}^2 \end{bmatrix} = \begin{bmatrix} \Omega_{\mathbf{q}^2}^T \\ 0 \end{bmatrix} \xi^2 + \text{known terms} \qquad (5.50)$$

Since the coefficient matrix is nonsingular, $\dot{\mu}^2$ and $\boldsymbol{\gamma}^2$ are uniquely determined as functions of ξ^2.

Step 4: Equation 5.47, with the solution of Eq. 5.50, determines ξ^2, hence $\dot{\boldsymbol{\mu}}^2$ and $\boldsymbol{\gamma}^2$.

Step 5: Backward integration of Eqs. 5.43 and 5.44, employing the numerical methods of Ref. 14, yields $\boldsymbol{\mu}(t)$, $\dot{\boldsymbol{\mu}}(t)$, and $\boldsymbol{\nu}(t)$, hence $\boldsymbol{\mu}^1$ and $\dot{\boldsymbol{\mu}}^1$. Uniqueness follows since these equations are linearizations of the constraints and equations of motion of Eqs. 5.4 and 5.7.

Step 6: Equations 5.23 and 5.24 yield $\mathbf{q}_{\mathbf{b}}^1$ and $\dot{\mathbf{q}}_{\mathbf{b}}^1$

Step 7: The design sensitivity vector $\psi_{\mathbf{b}}$ of Eq. 5.48 is evaluated, using results of Steps 1 through 6.

This algorithm has been implemented in the dynamic analysis program DADS. The program automatically assembles the constraint equations and the governing differential equations for the problem from user supplied data. The user identifies design variables and all derivatives that are needed in the adjoint equations are assembled. Integration forward in time is carried out for the state and backward integration in time is carried out for the adjoint variables. Backward integration requires that the state, Lagrange multipliers, mass matrix, and Jacobian matrix be stored on disk during forward integration, to be retrieved at the appropriate times during backward integration. When a successful time step is taken in forward integration, $\mathbf{q}$, $\dot{\mathbf{q}}$, $\ddot{\mathbf{q}}$, $\boldsymbol{\lambda}$, the Jacobian, and the mass matrix are written on disk. Polynomials that interpolate for $\ddot{\mathbf{q}}$ and $\boldsymbol{\lambda}$ are also generated and stored on disk. During backward integration, accelerations are obtained by interpolation and $\dot{\mathbf{q}}$ and $\mathbf{q}$ are obtained by integrating the polynomial for $\ddot{\mathbf{q}}$.

Numerical Examples by the Adjoint Variable Method

Several example problems have been analyzed with the software developed, two of which are discussed here.

Example 5.3; Two Degree Freedom Spring Mass System: As the first example, a simple two degree freedom spring mass system shown in Fig. 5.7 is considered. Two masses of 20 Kg and moment of inertia 125 Kgm^2 are connected by springs and dampers, as shown in Fig. 5.7. Body 1 is excited by a force F=1000 sin20t. The spring constants and damping coefficients are the design variables, as shown in Fig. 5.7. Analysis is carried out for a period of 1 sec and a nominal design $\mathbf{b}^0$ = $[3920, 10, 3920, 10]^T$.

The dynamic response functional for this problem is taken as

$$\psi = \int_0^1 (y_1 - y_1^1)^2 dt$$

where y_1^1 = 5 is the initial position of body 1. For this system, the vector of constraints is simply

$$\boldsymbol{\Phi} = [x_1, \theta_1, x_2, \theta_2]^T = \mathbf{0}$$

and the Jacobian matrix is

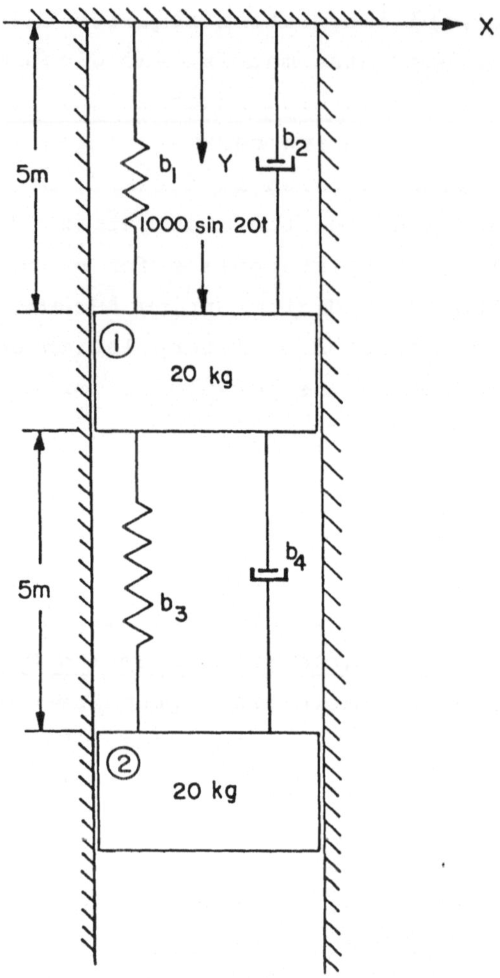

Figure 5.7 Elementary Two Mass Example

$$\Phi_q = \begin{bmatrix} 1 & 0 & 0 & 0 & 0 & 0 \\ 0 & 0 & 1 & 0 & 0 & 0 \\ 0 & 0 & 0 & 1 & 0 & 0 \\ 0 & 0 & 0 & 0 & 0 & 1 \end{bmatrix}$$

The mass matrix **M** is

$$\mathbf{M} = \text{diag}\ (20,\ 20,\ 125,\ 20,\ 20,\ 125)$$

Bodies 1 and 2 are initially at distances 5m and 10m from the global X-axis respectively. The terminal values of μ and η are obtained (Step 2) from

$$\begin{bmatrix} \mathbf{M} & \Phi_q^T \\ \Phi_q & 0 \end{bmatrix} \begin{bmatrix} \mu^2 \\ \eta^2 \end{bmatrix} = \begin{bmatrix} 0 \\ 0 \end{bmatrix} \tag{5.51}$$

which gives $[\boldsymbol{\mu}^2, \boldsymbol{\eta}^2]^T = \mathbf{0}$ and $\dot{\mu}^2$ and $\boldsymbol{\gamma}^2$ are obtained (Step 3) from

$$\begin{bmatrix} \mathbf{M} & \Phi_q^T \\ \Phi_q & 0 \end{bmatrix} \begin{bmatrix} \dot{\mu}^2 \\ \gamma^2 \end{bmatrix} = \begin{bmatrix} 0 \\ 0 \end{bmatrix} \tag{5.52}$$

which gives $[\dot{\boldsymbol{\mu}}^{2^T}, \boldsymbol{\gamma}^2]^T = \mathbf{0}$. Using these terminal conditions, the equations for $\boldsymbol{\mu}$ and $\boldsymbol{\nu}$ in Eqs. 5.43 and 5.44 are solved to provide $\dot{\mu}$ and $\ddot{\boldsymbol{\mu}}$ (Step 5). Substitution of these quantities in Eq. 5.48 gives the vector of design sensitivities. The design sensitivity vector was calculated from the program as

$$\psi_{\mathbf{b}} = [-0.1681, -5.707, -1.049, 6.08] \times 10^{-4}$$

The design sensitivity vector obtained approximately by perturbation (finite difference) analysis is

$$(\psi_{\mathbf{b}})_{pert} = [-0.1788, -6.259\ -1.19, 0.608] \times 10^{-4}$$

Good agreement is seen between the design sensitivities computed by the program and by differencing.

<u>Example 5.4; Four Bar Mechanism</u>: As a second example, a four bar mechanism (Fig. 5.8) that falls from rest under its own weight is considered. Links 1 and 2 are initially at rest at angles of 45° and -45°, with respect to the global X-axis. The design variables are locations of revolute joints, as shown in Fig. 5.8. The simulation was carried out for a period of one second.

In this problem, the dynamic response functional is taken as

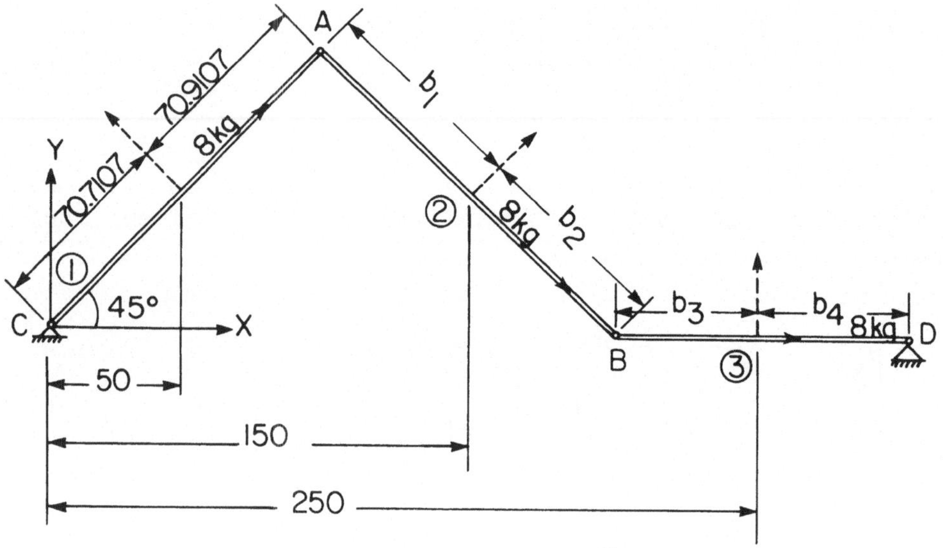

$b^0 \equiv [-70.9107, 70.7107, -50, 55]^T$

Figure 5.8 Four-Bar Linkage Example

$$\psi = \int_0^1 [\sin\theta_3 + 0.23893455]^2 dt$$

For this system, the constraint equations are the constraint equations for the four revolute joints,

$$\Phi = \begin{bmatrix} x_1 + 70.9107\cos\theta_1 - x_2 - b_1\cos\theta_2 \\ y_1 + 70.9107\sin\theta_1 - y_2 + b_1\sin\theta_2 \\ x_2 + b_2\cos\theta_2 - x_3 - b_3\cos\theta_3 \\ y_2 + b_2\sin\theta_2 - y_3 - b_3\sin\theta_3 \\ x_1 - 70.7107\cos\theta_1 \\ y_1 - 70.7107\sin\theta_1 \\ x_3 + b_4\cos\theta_3 \\ y_3 + b_4\sin\theta_3 \end{bmatrix} = 0$$

and the Jacobian matrix is

$$\Phi_q = \begin{bmatrix} 1 & 0 & -70.9107\sin\theta_1 & -1 & 0 & b_1\sin\theta_2 & 0 & 0 & 0 \\ 0 & 1 & 70.9107\cos\theta_1 & 0 & -1 & -b_1\cos\theta_2 & 0 & 0 & 0 \\ 0 & 0 & 0 & 1 & 0 & -b_2\sin\theta_2 & -1 & 0 & b_3\sin\theta_3 \\ 0 & 0 & 0 & 0 & 1 & b_2\cos\theta_2 & 0 & -1 & -b_3\cos\theta_3 \\ 1 & 0 & 70.71707\sin\theta_1 & 0 & 0 & 0 & 0 & 0 & 0 \\ 0 & 1 & -70.7107\cos\theta_1 & 0 & 0 & 0 & 0 & 0 & 0 \\ 0 & 0 & 0 & 0 & 0 & 0 & 1 & 0 & -b_4\sin\theta_3 \\ 0 & 0 & 0 & 0 & 0 & 0 & 0 & 1 & b_4\cos\theta_3 \end{bmatrix}$$

The mass matrix is a 9×9 diagonal matrix. The terminal values of the adjoint variables and their derivatives are computed as in Eqs. 5.51 and 5.52 of Example 5.3. The initial design was selected as

$$\mathbf{b}^0 = [-70.9107,\ 70.7109,\ -50,\ 55]^T$$

The design sensitivity vector from the program is

$$\psi_{\mathbf{b}} = [-0.9731,\ 0.272,\ -0.3897,\ 0.4125\times 10] \times 10^{-2}$$

The design sensitivity vector obtained by perturbation (finite difference) is

$$(\psi_{\mathbf{b}})_{pert} = [-0.96,\ 0.24,\ -0.35,\ 0.42] \times 10^{-2}$$

yielding good agreement between results obtained by the program and result obtained by differencing.

REFERENCES

1. Tomovic, R., and Vukobratonic, M., General Sensitivity Theory, American Elsevier, NY, 1972.

2. Bryson, A.E., and Ho, Y.C., Applied Optimal Control, Wiley, NY, 1975.

3. Haug, E.J., and Arora, J.S., Applied Optimal Design, Wiley, NY, 1979.

4. Haug, E.J., Choi, K.K., and Komkov, V., Design Sensitivity Analysis of Structural Systems, Academic Press, NY, 1986.

5. Haug, E.J., Wehage, R.A., and Barman, N.C., "Design Sensitivity Analysis of Planar Mechanism and Machine Dynamics", Journal of Mechanical Design, Vol. 103, No. 3, 1981, pp. 560-570.

6. Ehle, P.E., and Haug, E.J., "A Logical Function Method for Dynamic and Design Sensitivity Analysis of Mechanical Systems with Intermittent Motion", ASME Journal of Mechanical Design, Vol. 104, No. 1, 1982, pp. 247-255.

7. Haug, E.J., and Ehle, P.E., "Second Order Design Sensitivity Analysis of Mechanical System Dynamics", International Journal for Numerical Methods in Engineering, Vol. 18, 1982, pp. 1699-1717.

8. Haftka, R.T., "Second Order Sensitivity Derivatives in Structural Analysis", AIAA J., Vol. 20, 1982, pp. 1765-1766.

9. Mani, N.K. and Haug, E.J., "Singular Value Decomposition for Dynamic System Design Sensitivity Analysis", Engineering with Computers, Vol. 1, 1985, pp. 103-109.

10. Coddington, E.A., and Levinson, N., Theory of Ordinary Differential Equations, McGraw-Hill, NY, 1955.

11. Hildebrand, F.B., Advanced Calculus for Applications, Prentice-Hall, Englewood Cliffs, NJ, 1976.

12. Haug, E.J., Mani, N.K., and Krishnaswami, P., "Design Sensitivity Analysis and Optimization of Dynamically Driven Systems", Computer Aided Analysis and Optimization of Mechanical System Dynamics (ed. E.J. Haug), Springer-Verlag, Heidelberg, 1984, pp. 555-636.

13. Haug, E.J., Intermediate Dynamics, Allyn & Bacon, Boston, 1987.

14. Park, T.W., and Haug, E.J., "A Hybrid Numerical Integration Method for Machine Dynamic Simulation", J. of Mechanisms, Transmissions, and Automation in Design, to appear 1986.

15. Krishnaswami, P., Wehage, R.A., and Haug, E.J., Design Sensitivity Analysis of Constrained Dynamic Systems by Direct Differentiation, Technical Report No. 83-5, Center for Computer Aided Design, The University of Iowa, 1983.

APPENDIX

Matrix Calculus Notation

For $\mathbf{x} \in R^k$, $\mathbf{y} \in R^m$, $a(\mathbf{x},\mathbf{y}) \in R^1$, $\mathbf{A}$ an m×n constant matrix, $\mathbf{g}(\mathbf{x}) \in R^n$, and $\mathbf{h}(\mathbf{x}) \in R^n$, using i as row index and j as column index, define

$$a_{\mathbf{x}} \equiv \frac{\partial a}{\partial \mathbf{x}} \equiv \left[\frac{\partial a}{\partial x_j}\right]_{1\times k} \tag{A.1}$$

$$\mathbf{g}_{\mathbf{x}} \equiv \left[\frac{\partial g_i}{\partial x_j}\right]_{n\times k} \tag{A.2}$$

$$a_{\mathbf{xy}} \equiv \left[\frac{\partial^2 a}{\partial x_i \partial y_j}\right]_{k\times m} = \frac{\partial}{\partial \mathbf{y}}\left(\frac{\partial a}{\partial \mathbf{x}}^T\right) \equiv \left(a_{\mathbf{x}}^T\right)_{\mathbf{y}} \tag{A.3}$$

Using this notation, the following formulas are obtained:

$$\frac{\partial}{\partial \mathbf{x}}(\mathbf{Ag}) = \left[\frac{\partial}{\partial x_i}(A_{i\ell} g_\ell)\right] = \left[A_{i\ell}\frac{\partial g_\ell}{\partial x_j} = \mathbf{A}\mathbf{g}_{\mathbf{x}}\right] \tag{A.4}$$

$$\frac{\partial}{\partial \mathbf{x}}(\mathbf{g}^T\mathbf{h}) = \left[\frac{\partial}{\partial x_j}(g_\ell h_\ell)\right] = \left[\frac{\partial g_\ell}{\partial x_j} h_\ell + g_\ell \frac{\partial h_\ell}{\partial x_j}\right]$$

$$= \mathbf{h}^T\mathbf{g}_{\mathbf{x}} + \mathbf{g}^T\mathbf{h}_{\mathbf{x}} \tag{A.5}$$

where summation notation is used with repeated indices in the same term. A ~ over a term indicates that it is held fixed for purposes of partial differentiation; e.g.,

$$(\tilde{\mathbf{g}}^T\mathbf{h})_{\mathbf{x}} \equiv \mathbf{g}^T\mathbf{h}_{\mathbf{x}}$$

PART VI

KNOWLEDGE BASED SYSTEMS IN OPTIMAL DESIGN

KNOWLEDGE-BASED SYSTEMS IN OPTIMAL DESIGN

Panos Y. Papalambros
The University of Michigan
Ann Arbor, MI. 48109, USA

INTRODUCTION

In the early stages of development of new fields, it happens time and again that the imminent potential impact of the new field is exaggerated to the point of a new panacea, usually by workers at the fringe or outside the research area of the field. An attendant consequence often is increased scepticism, if not outright hostility, among the broader community of researchers who cannot separate fact from fiction. This appears to be somewhat the case currently with the emerging field of artificial intelligence (AI) and, in particular, with its applied branch of expert or knowledge-based systems. It is therefore with some caution that the present article, not addressing the mainstream research community of AI, examines the issues involved in developing knowledge-based systems for optimal design. By the nature of the present state of the art in the field, the exposition is to a large degree speculatory. There is yet very little done that would qualify as concrete results in the field. Still, the motivation for developing such systems is eminently strong and the apparent potential results are overwhelmingly desirable, so that at least some speculation is justified.

In the present article, the utilization of "knowledge" in AI is outlined first, followed by a description of the nature of knowledge-based systems. The potential application of these ideas to design optimization is examined next, followed by a description of current early work - to the best of the author's knowledge.

NATO ASI Series, Vol. F27
Computer Aided Optimal Design: Structural and Mechanical Systems. Edited by C. A. Mota Soares

KNOWLEDGE IN ARTIFICIAL INTELLIGENCE

A formal strict definition of knowledge in the usual sense seems impossible, much the same as a definition of intelligence. This is due not only to the complexity of representation and processing abilities involved but also to the philosophical attitude that one is compelled to assume when attempting such a definition. The latter points in fact to the possible necessity for including "beliefs" when describing knowledge, something that AI researchers do attempt to account for.

Operationally, it is more useful to view knowledge, within a specified domain, as consisting of descriptions, relations and operations in this domain. Descriptions pertain to ways of identifying and distinguishing objects in the domain. Relations are special descriptions giving associations among objects, such as defining classes and taxonomies. Operations are procedures which combine objects for the purpose of meeting a goal, such as solving a problem or generating new objects. In a given domain of knowledge, these objects constitute a knowledge base and they represent facts. Known relations among objects are also facts, while domain operations are procedures for manipulating facts.

Things should become clearer it we appreciate the fact that these ideas are expressed in the context of computers and computer programming. A knowledge base is indeed a database, the facts being data items. Procedures are just computer programs that manipulate the database. So "knowledge" can be simply viewed as the representation and processing of domain information in the computer. This appears, at first sight, no different than traditional computing. But this view is rather deceptive. In traditional computing detailed instructions are described by calculus and executed sequentially. Deviations from that are not significant. Now, however, the desire is for programs that can handle ideas, judgement and experience, and that can manipulate forms explicitly and symbolically.

We will proceed to examine some issues about how knowledge may be represented and manipulated.

Knowledge, or information, is not operationally useful unless it can be represented in an operationally useful way. The ability to use traditional computing for problem-solving exists to the extent that we can use mathematical modeling to represent the problem. For example, in order to apply optimization techniques to a design problem we must first develop a mathematical model which contains several things: the design variables which describe each design variation uniquely for each set of (arithmetic) values they assume; the design constraints which are precise functions of the variables judging unambiguously which designs are acceptable and which are not; the design objective which measures exactly how good each design is and expresses preferences stated in advance. In spite of the conceptual simplicity of such models, it is well recognized that good modeling is still a difficult task, often accounting for at least half of the expended effort in an optimization study. Apart from the difficulty of analyzing adequately the real world, there is the added complexity of modeling affecting the solution in terms of speed and/or quality. This is not true only for optimal design, but also for general knowledge representation in AI. Therefore the choice of representation of knowledge is related to the intended use of the knowledge base. This is a recurring issue in artificial intelligence.

Any method of knowledge representation contains also procedures for manipulating knowledge in order to derive new facts from known facts. This is referred to as an <u>inference mechanism</u> and it is an important consideration in judging representations. Keeping in mind that any representation must be coded and stored in the computer, key operational issues in representation are: amount of storage needed to represent the facts in the knowledge base and the relations among them, complexity of inferencing, and amount of information needed for special bookkeeping storage to facilitate inferencing. It is useful to keep in mind also the AI idea that the methods of inferencing are generally considered part of the knowledge in the system.

Let us now outline briefly some of the ways that may be used to represent knowledge.

State-Space. Here each object in the knowledge base is a data structure called state. Each state is described by a set of values given to the state variables, just like design points in a design optimization problem. Since for many problems it would be impossible to include every state in the database, the collection of all states, i.e. the state-space, can be described by a set of permissible operations with which all other states can be generated from some initial one(s).

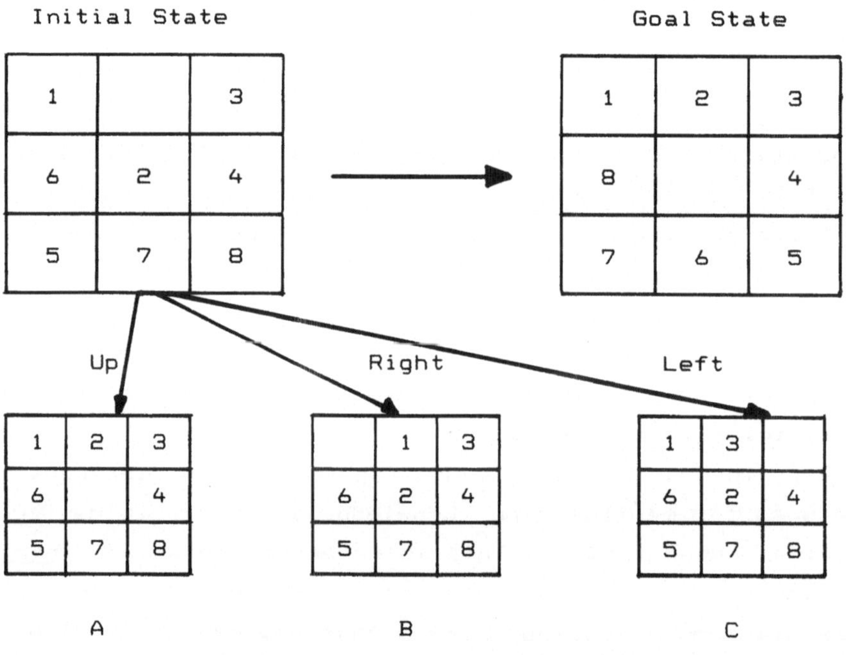

Figure 1. The Eight-Puzzle

As an example, consider the eight-puzzle, a popular example in AI textbooks. The objective is to rearrange a given initial configuration of eight numbered tiles on a 3x3 board so that a final desirable configuration is achieved. Rearrangement is allowed only by sliding a tile onto an empty square from an orthogonally adjacent position, Figure 1. In the state-space representation, the final configuration of the puzzle is called the goal state. The initial state can be used to generate three new states by applying the operators "up", "right", or "left". If the purpose of this state representation is to reach the goal state, then a search procedure must be developed. This state-space search is one of the simplest search formulations for problem solving. The presence of an objective raises the question of how it will be achieved. Clearly, the positioning rules above will allow the generation of the goal state eventually. But there is no information in the state description of the new states A, B, C about which one is closer to the goal state. Thus the state representation requires additional information (or knowledge) that defines the remaining subproblem. This can be a heuristic function that estimates the "distance" from the goal. In the eight-puzzle example, this function could be the number of tiles that are not in their proper position. This heuristic may not be a particularly good one and others can be proposed.

AND/OR Graphs. In the eight-puzzle example each state can be considered as a node in a graph and the task is represented as a procedure for finding a sequence or a path through the state-space graph. This graph is a disjunctive OR graph, since the generated states cannot occur together, Figure 2(a). There are classes of problems where a node is a conjunction of several others, Figure 2(b). In this figure node B has a single parent node (S: starting node) and two successors E and F. Node E has an "OR" successor D and two "AND" successors I and J. If this was a truth-seeking graph, B would be true if E and F were true together. If this was a problem-solving graph, problem B would generate two subproblems E and F, and both must be solved. Note however that each subproblem could be solved in isolation of its siblings, which is why we treat them as individual nodes. Presumably these subproblems are each easier to solve than the parent one, hence this graph is called a problem-reduction representation. Such graphs, containing OR and AND links, are called AND/OR graphs, or hypergraphs.

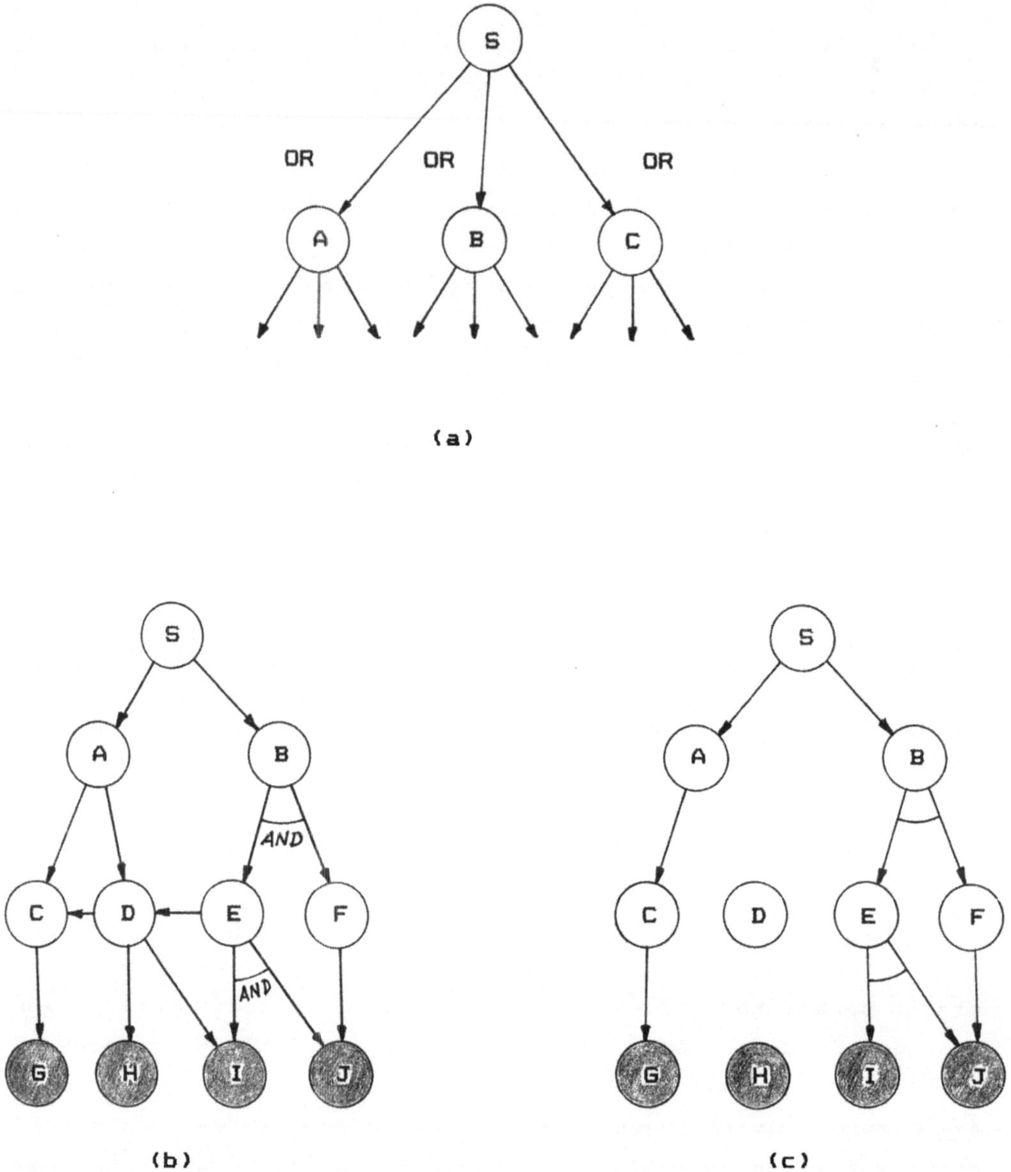

<u>Figure 2</u>. (a) OR Graph; (b) AND/OR Graph; (c) Two solution graphs for graph (b), the dark nodes representing goal states.

A solution for an AND/OR representation is not a path, as in state-space, but a subgraph of the entire graph. Such a solution graph for the problem in Figure 2(b) is given in Figure 2(c), where the dark nodes represent goal states.

Problems whose solution is a strategy rather than a single prescribed goal state, would be usually served best by knowledge representation in the form of AND/OR graphs. Examples are symbolic integration and theorem proving.

Knowledge representations that utilize OR graphs (state-space) or AND/OR graphs have received a lot of attention because they provide the simplest basis for problem-solving methods, i.e. in the form of graph searches. This is a point where AI and traditional operations research meet. It is also a point where the difference in viewpoint is quite evident. Operations research views a graph as a preexisting set of objects; a procedure such as branch-and-bound is considered a method for eliminating objects from the set until the solution object is isolated. Artificial intelligence views a graph as an object creation process, where a new node represents a newly generated object. Thus, the branch-and-bound method used for problem solving in AI is called a generate-and-test process. This difference in emphasis is due to the AI effort to treat problem solving in a human-like way, versus the operations research desire to prove optimality conditions (Pearl, 1984).

Descriptions for graph-searching procedures are given in most introductory AI texts (for a thorough treatment, see Pearl, 1984). Search methods for general problem solving are called weak methods, for example generate-and-test, hill climbing, breadth-first, depth-first. The methods can be data-directed, starting from the initial condition(s) and moving towards the goal(s) (forward searching), or goal-directed, starting from the goal state(s) and moving towards the initial condition(s) (backward searching). Bidirectional searching is used also sometimes. A depth-first method considers always a single successor to each state, "diving" into the graph. A breadth-first method considers all successors for all states at a given level, "spreading out" over each level. Uninformed searches can easily lead to combinatorial explosion. To avoid that, an _informed_ search is preferred, which uses a heuristic function estimating the chances of success among the candidate alternatives. The heuristic employs domain-specific information and is an example of domain knowledge representation and utilization. It is useful to recall here that by "heuristic" we define a strategy which aids, often substantially, the problem solving ability (of humans or machines) but with no guarantee that it will never lead in a wrong direction.

Predicate Calculus. Any attempt at automated logical reasoning would require a formal language that is able to state facts and make logical deductions. _Predicate calculus_ is such a system of logic able to express both mathematical and everyday natural language statements. Given a set of statements, predicate calculus contains general rules of inference which can be used to deduce new statements.

The syntax of predicate calculus involves an "alphabet" of _symbols_ and definitions of _expressions_ that can be constructed from symbols. Given a knowledge domain, a set of _semantics_ connect the language with the domain, so that language expressions represent assertions or facts about the domain. The basic symbols used are:

1. Punctuation marks: comma(,) and parentheses.
2. Logic symbols: not($\sim$), implies($\rightarrow$), and($\wedge$), or($\vee$), universal quantifier($\forall$), existential quantifier($\exists$).
3. Function symbols: f_i^n for $i \geq 1$, $n \geq 0$; if $n=0$, then f_i^0 are called constants.
4. Predicate symbols: P_i^n for $i \geq 1$, $n \geq 0$; if $n=0$, then P_i^0 are called propositions.

Given a domain of knowledge D that contains some elements (it is nonempty), a _constant_ is associated semantically with a _particular_ element of D; a _function_ is associated with a mapping of each n-tuple of elements of D into an element of D, for example the function "addition" in real numbers; a _predicate_ is associated with a particular _relation_ among elements of D, for example the relation "greater-than". The n-adic letters correspond to the n-adic functions or relations.

As an example consider the following statements about the nonlinear programming code VMCON (Crane et al., 1980):

1. VMCON is a Constrained Nonlinear Programming code (CNLP).
2. VMCON is a Numerical Algorithm (NA).
3. All CNLP's are NA's.
4. All common CNLP's contain some unconstrained NLP (UNLP) or some Step Size Algorithm (SSA).

We may write these as follows:

1. CNLP(VMCON)
2. NA(VMCON)
3. $\forall x$ CNLP(x) $\rightarrow$ NA(x)
4. $\forall x \exists y \exists z$ CNLP(x) $\wedge$ common(x), UNLP(y), SSA(z) $\rightarrow$ contain(x,y) $\vee$ contain(x,z)

Now we can use these to prove or disprove statements such as

common(VMCON) $\wedge \sim$ contain(VMCON,UNLP) $\rightarrow$ contain(VMCON,SSA).

This is a trivial example and backward or forward reasoning can be used to prove it. Note, however, that in order to do any reasoning we must scan through the <u>list</u> of statements to check for matching. The ability to provide primitives for manipulating lists and management for list storage is the idea behind the development of <u>list-processing</u> languages such as LISP.

A symbol processing language such as LISP uses a prefix syntax type of predicate calculus. For example, statements 1 and 3 above become

(ISA VMCON CNLP)

(All x) (IF (ISA x CNLP)(NA x)).

This tends to create more convenient symbol structures.

Symbols are used to create classes of expressions, defined recursively:

<u>Terms</u> describe names of things, so they are either constants or made up from functions of terms previously created, i.e. if t_i (i=1,...,n) are terms, then $f_j^n(t_1,...,t_n)$ are also terms.

Atomic Formulas describe propositions or are predicates of terms, i.e. if t_i (i=1,...,n) are terms, then $P_j^n(t_1,...,t_n)$ are atomic formulas.

Well-Formed Formulas (wff's) describe expressions built out of terms and atomic formulas by using connectives and quantifiers, i.e. the legitimate logic symbols of the calculus.

Statements 1,2 above are atomic formulas, while 3 and 4 are wff's. Of course, atomic formulas are wff's.

In statements 3 and 4 above we also introduced the symbol variable, e.g. x, y or z, which is used together with a quantifier. We say, then, that a formula is quantified over a variable. If quantification is allowed over terms but not over predicates the calculus is called first order.

Predicate calculus allows well defined representation of many involved statements. It also contains rules of inference that are generally applicable. Two often useful ones are

A, A → B yields B [modus ponens]

A, (($\forall$x), P(x)) yields P(A) [universal specialization]

The modus ponens rule infers the wff B from the wff's A and A→B. The universal specialization infers the wff P(A) from the wff (($\forall$x),P(x)) where A is any constant.

Inference using predicate calculus statements as is can be very involved. It is possible to create an efficient procedure that operates on a standardized form of statements called the clause form. A clause is a disjunction of literals, a literal being either an atomic formula, or its negation. The transformation procedure is quite staightforward (see, e.g. Nilsson, 1971, or Rich, 1983). Clauses are used in a problem-solving procedure called resolution which proves facts by contradiction (or refutation in AI jargon), see op. cit. for complete description.

Predicate calculus can be used also together with a state-space approach. For this, we need a set of wff's to have a complete state description. Then, computations which change one set of wff's to another will serve as state-space operators. The goal state will be a set of goal wff's and operators will be selected at each state based on a set of applicability wff's.

Production Systems. An IF-THEN, or condition-action rule of the form

IF < antecedent >, then < consequent >

is called a production. The antecedent, or IF part, of the rule may consist of many conditions and similarly the consequent, or THEN part, of the rule may have several actions. A production system is a rule-based representation consisting of three elements:

(i) a database (or knowledge base) which contains the information to be used by the system. This information, or knowledge, can be of permanent nature related to the system or of transient nature related to the specific problem currently addressed by the system.

(ii) a set of productions which operate on the database.

(iii) a control strategy which specifies how the rules are examined, i.e. sequencing of comparisons with the database, conflict resolution and other issues which may arise in the use of the rules.

The terms "rule-based" or "production" system are used interchangeably and describe a general and powerful way to represent and infer knowledge. Inferencing done with forward or backward reasoning is now called forward-chaining or backward-chaining respectively. In a data-driven domain a forward-chaining process may collect all rules whose antecedents match the current situation in the database (current state in a state-space representation) and proceed to execute the consequents implied by the rules. If any conflict exists is the antecedents, a conflict-resolution strategy is invoked. The process continues until a problem is solved (goal state is reached) or no rules exist with antecedents satisfied by the current situation (dead end, lack of sufficient knowledge etc.).

A rule whose antecedent is satisfied by the current situation is said to be triggered. A rule whose consequent is executed is said to <u>be fired</u>. Clearly, in the presence of conflict a triggered rule may not be fired.

As an example, in a program that tries to fit an appropriate optimization technique to a given NLP model, we may have a rule such as

If the model is unconstrained <u>and</u>
the objective is nonlinear <u>and</u>
the Hessian is sparse
then use <u>either</u> a discrete Newton method
<u>or</u> a conjugate-gradient method.

Another rule may be used to define what "sparse" means, while yet another may be introduced to measure the density of the Hessian matrix.

Statements in the antecedent or consequent parts may be expressed in predicate calculus, as was implied above with the "either, or" form in the consequent. This allows a complex yet flexible way to represent knowledge.

The control strategy in a production system determines the sequence of rule firing. A general requirement for any control strategy is that it be <u>systematic</u>, meaning that it does not overlook any rule and it does not trigger the same rule more than once. The former aims at not missing the generation of a possible desired outcome, and the latter aims at avoiding the inefficiency of repetitive computation. The search methods we discussed earlier for state-space or graphs could form the basis of a control strategy. In many cases the search will lead to more than one nodes for possible expansion, which in the production system means more than one rule will be triggered. A <u>conflict-resolution strategy</u> must then determine the firing selection. Such a strategy is usually heuristic, i.e. ad hoc with no particular science base. However, it can be effective provided it is systematic. Some examples of conflict-resolution strategies are (Winston, 1984):

<u>Rule Ordering</u>: All rules are arranged in a single priority list and the first one triggered, say starting from the top of the list, is fired - the others being ignored.

Data Ordering: All data (e.g., states) are arranged in a single priority list. Assuming forward chaining, the triggering rule with the highest priority antecedent is fired.

Size Ordering: Rules are assigned priorities according to the difficulty of satisfying the antecedent. Highest priority goes to the rule with the longer list of conditions in the antecedent.

Recency Ordering: The most recently, or least recently, used rule is assigned highest priority.

Specificity Ordering: If the antecedent of a triggering rule is a superset of the antecedent of another triggering rule, then the rule with the superset is fired. The idea behind this is that it is more specialized to the current state. This assumes a forward-chaining procedure; similarly for the consequents of backward-chaining procedures.

Context Limiting: The rules are separated in groups. Only some groups are allowed to be applicable at any time, said to be activated. A procedure for activating and deactivating groups is required.

Production systems are flexible because rules can be added or deleted without affecting the other rules and without changing the control strategy. They also seem to describe naturally human knowledge about what to do in a given situation. The basic rule-based systems do have many limitations also. For example, there is no ability of self-learning and no access to the reasoning behind the rules. As the knowledge base increases and the number of rules becomes large, problems of inefficiency may arise. Also, it may be difficult to detect truly conflicting productions, where, for example, the same antecedent calls for opposite consequents. This is related to the problem of monotonicity in reasoning. A reasoning system is called monotonic if the number of statements known to be true strictly increases over time of system usage. This means that a deduction, or new statement, does not cause an existing statement to become invalid. Moreover, in the process of arriving at a solution statement there is no need to keep track of all statements used to arrive at the solution. Most real systems require nonmonotonic reasoning and the basic production system must be modified to satisfy this need. Nonmonotonic logic is one of a class of techniques proposed for handling problems with uncertainty. Other techniques in that class include probabilistic reasoning, fuzzy logic and belief spaces. All these are sophisticated methods of knowledge representation as well as of inferencing, and they will not be expanded here. They are important in the development of expert systems designed to function in the presence of incomplete information, or dynamic (time-varying) databases (see e.g., Rich, 1983).

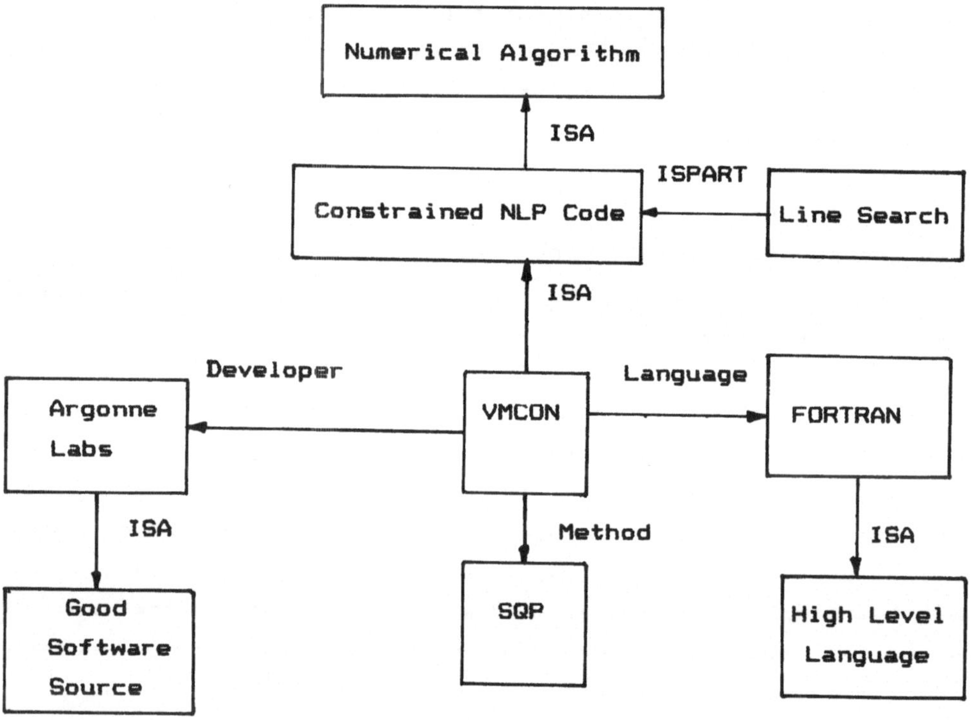

Figure 3. Part of a Semantic Network

Semantic Nets. A usual node-and-link graph may become more useful as a representation tool by introducing and labeling on the graph the semantics of the intended representation, i.e. the meaning embodied in the symbols and the symbol arrangements allowed by the syntax of the representation. A semantic net is a node-and-link representation with a simple syntax: nodes are objects and directed links are relations between pairs of objects. Meanings must be included in the net so that objects and relations are readily explained.

A fragment of a semantic net is given in Figure 3, where some information about the code VMCON is represented. Note how various attributes of the object "VMCON" are shown: VMCON is a constrained nonlinear programming (numerical) algorithm, it uses the optimization method of sequential quadratic programming, it is written in FORTRAN language and it has been developed at the Argonne National Laboratory. The node-and-link representation is given graphically only for illustrative purposes. In actuality a list processing representation would be used in the computer. Thus in LISP each node (object) would be an atom and each link (relation) a property, giving the following LISP description:

<u>Atom</u>	<u>Property</u>
CNLP Code	((ISA Numerical algorithm))
VMCON	((ISA Code CNLP) (Language FORTRAN) (Method SQP) (Developer Argonne Labs))
Argonne Labs	((ISA Good Software Source))
FORTRAN	((ISA High Level Language))
Line Search	((ISPART CNLP Code))

Semantic nets were originally developed for natural language understanding. In order to draw inferences, nodes were identified as known facts and an activation procedure spread out from each node through the links. Where activation procedures met, a matching was confirmed, hence the method was known as <u>intersection search</u>. More elaborate search procedures can be used which maintain the matching idea.

Developing good semantic nets requires skill in order to avoid inconsistencies. Special measures must be taken to distinguish between nodes representing a <u>class</u> of objects and an <u>instance</u> (i.e. a single object). Large nets which are intended for quantifiable relations may be partitioned in hierarchical levels each of which is associated with a variable or a small group of variables. This allows explicit representation of quantifiers such as $\forall$ and $\exists$. This greatly expands the utility of semantic nets (Rich, 1983; Findler, 1979; Winston, 1984).

Semantic nets are an example of representing complex structured knowledge in a declarative manner. Declarative methods of representation rely more on describing the facts and less on describing manipulations among them. Typical relationships in declarative structured representations are the ISA and ISPART relationships which give a partial ordering on the knowledge domain. Object attributes and associated values are given in a collective manner in a so-called slot-and-filler structure. Properties of a class may be also inherited to instances thus making transmission of descriptions easy. In nonmonotonic reasoning, when values for objects are not available, default values may be included. This is important to know, since default values may be inherited and propagated, just as known values.

Frames. When humans face a new situation, instead of trying to analyze it from scratch, they often attempt to fit it to a stereotype of previous experiences. In a similar way, special structures of knowledge may be stored in the memory of a computer. A new situation may be analyzed by selecting an appropriate stored structure, fill the details from the current situation and proceed with the analysis having the knowledge coded in this particular structure. This type of slot-and-filler representation is called a frame. So frames are semantic nets (usually complicated) with internal predetermined structure which describe a sterotyped object (or event).

Consider the example in Figure 4, where we want to report the results of testing constrained nonlinear programming codes (CNLP). The figure shows a semantic net where several frames are connected together. Each frame has a number of slots. How the slots are filled may be described with attached procedures.

Class inheritance is maintained through the net. For example, reporting the results of testing a CNLP problem with a

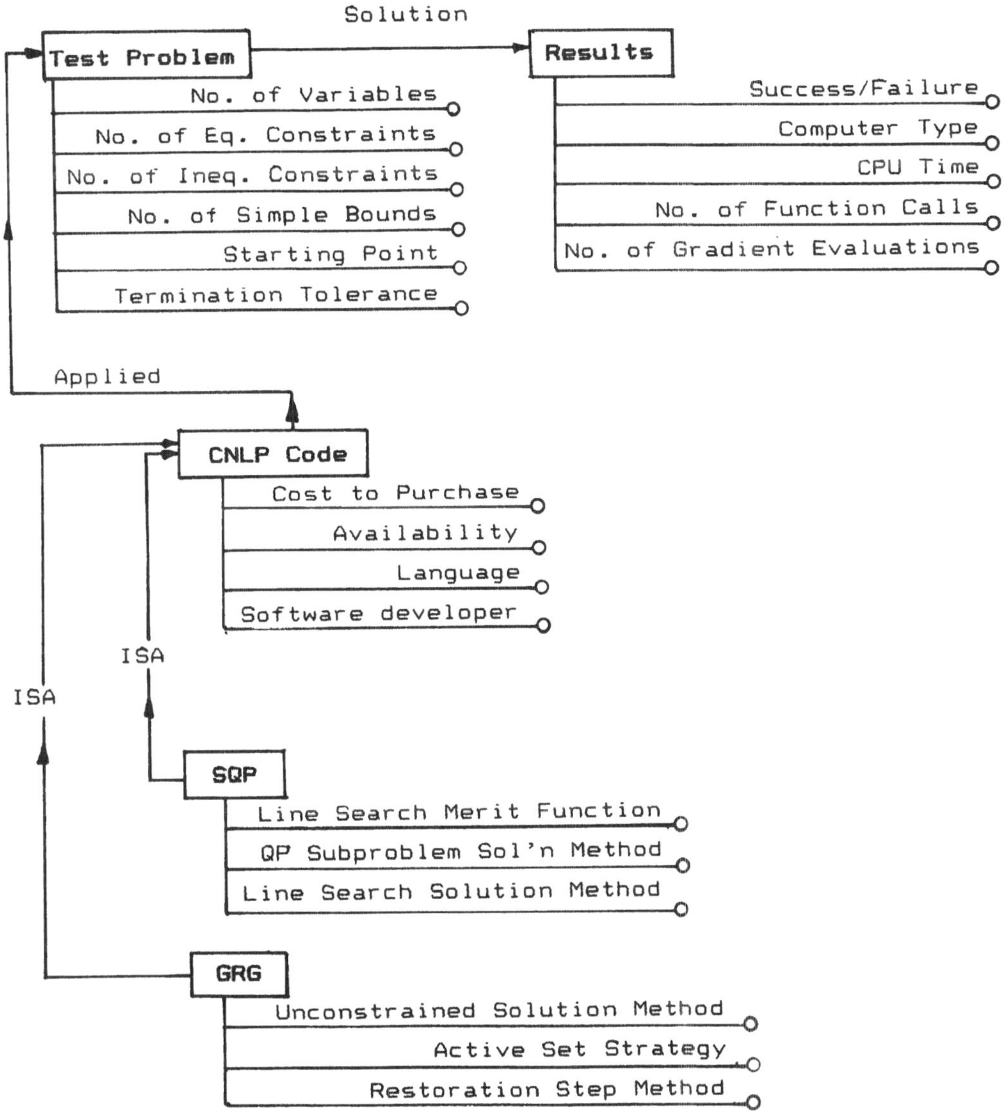

Figure 4. Frames connected in a net reporting results of a numerical CNLP code testing.

generalized reduced gradient method (GRG) requires 18 pieces of information, i.e. 18 slots to be filled, by following inheritance two levels up.

Frames are useful representations when the data can be structured in an organized albeit complicated manner, so that data items can be expected in a natural way. They can be used for inference in several ways. First, one may note that when a program invokes a particular frame there will be an automatic inference about the existence of objects within the frame. For example, in Figure 4 above, if the frame SQP is invoked, one is told immediately that a one-dimensional minimization of a merit function must be performed. Next, one may note that a frame representation gives a typical instance of the object (or event) that it represents. In the process of invoking a frame, the program tries to match the particular situation to the frame's slots. When a discrepancy is identified, an interesting aspect of the situation may be revealed. For example, in Figure 4 above, if the current situation describes the solution of a problem with a GRG method and a restoration step is not present, this may mean that the constraints are linear or that the code is incomplete.

A frame must be invoked in order to be used. Matching of a particular stored frame to a particular current situation can be a trial-and-error process. When an attempt for matching is made based on partial evidence on applicability, we say that a frame is <u>instantiated</u>. When a frame is instantiated, the program attempts to fill the slots of the frame with values from the current situation. If not enough values can be found, the instantiated frame may be abandoned and a new one selected for instantiation. Information about previous attempts on matching can be used to expedite a new selection, for example exploiting links between frames. For discussion on this as well as other aspects of frame representation, see Minsky (1975).

This concludes the discussion on knowledge representation. The next section describes the basic nature of knowledge-based systems.

KNOWLEDGE-BASED SYSTEMS

Traditional computing can be viewed as a data processing system employing a precise mapping of input data onto output data. Computing in artificial intelligence attempts to include some new features such as symbolic representation, symbolic inference and informed (heuristic) search. For a knowledge-based system one utilizes these AI features but places substantial emphasis on some very specific performance requirements. A knowledge-based system should act characteristically as an "expert" in a specific knowledge domain. Expertise here is thought primarily as the ability to solve a complicated but specialized problem efficiently and correctly most of the time, employing "personal" knowledge that is the result of experience. The emphasis on experience is not meant to diminish the importance of "text-book" knowledge, but rather to point out that human experts tend to employ shortcuts and heuristics that they have learned through long exposure to the domain. Therefore, a second characteristic of a knowledge-based system is that the problem-solving strategies tend to be domain-specific rather than general weak search methods. A third characteristic of such a system is the existence of metaknowledge, i.e. the ability to understand its own knowledge, reason about its own inferencing and explain how a conclusion is reached.

As of yet, there is no theory of expert systems or much more precise definition than what was presented above. The research approach has been experiential, through paradigms of specific programs that have been developed. Descriptions of these early programs with an attempt at generalization can be found in Hayes-Roth, Waterman and Lenat (1983), Davis and Lenat (1982), and Buchanan and Shortliffe (1984). In summary, knowledge-based or expert systems are computer programs which attempt to behave as human experts would, and which draw their power from exploiting specific knowledge of the problem domain rather than from general inferencing mechanisms.

Ideal Expert System. Any expert system should have certain components in order to perform its expected tasks. Since no present implementation appears to have all the desired components, we will refer to an ideal expert system as shown in Figure 5 (adapted from Hayes-Roth et al., 1983). These components are described below.

User Interface. This is a language processor providing the communication between the user and the system. The user supplies commands, questions or answers (information) in a more or less natural language, and the processor parses and interprets them to the system. The parsing is relatively

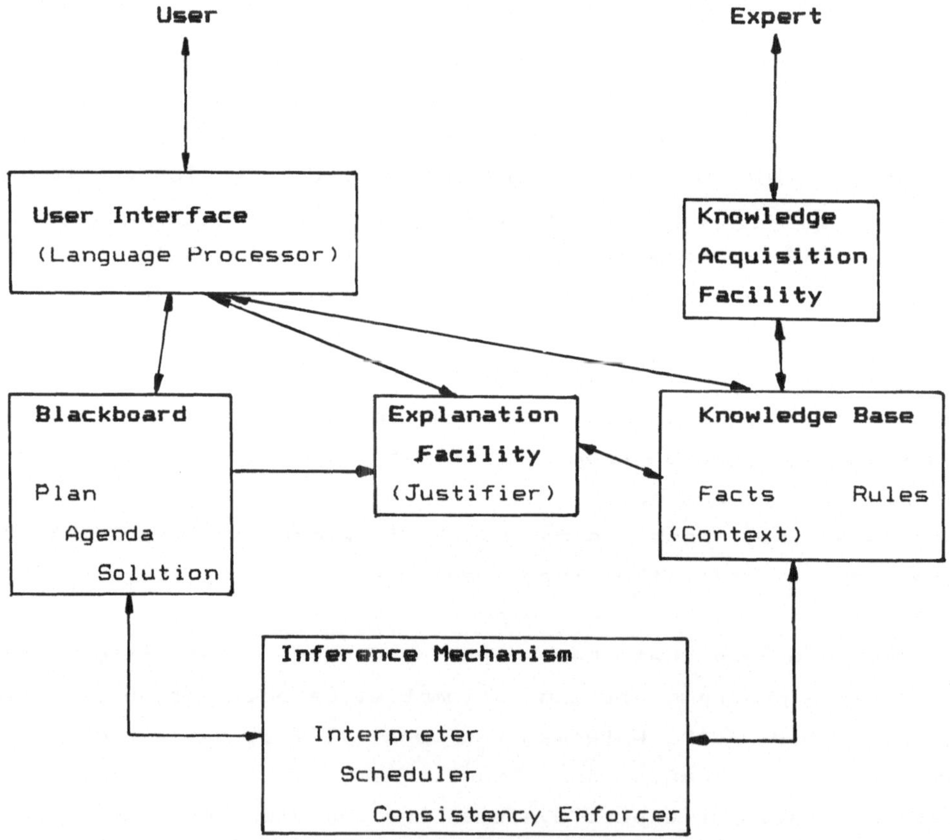

Figure 5. Ideal knowledge-based system schematic (adapted from Hayes-Roth et al., 1983)

sophisticated and is done with an AI language. The system response to the user is usually presented by generating messages in canned text, much like in traditional interactive programming. The information input may also use special editors utilizing graphics or structures.

Knowledge Base. This is a large database which contains two types of data. The first is a passive type of data, sometimes called context, which records all facts known to be applicable for the current problem facing the system. This includes not only facts existing already or introduced by the user, but also facts inferred during the solution process of the current problem. The second type of data is an active one; it comprises of a rule representation of knowledge. This representation is the most favored one for expert systems.

Inference Mechanism. This is the control strategy which uses the rules to manipulate the facts in the context and infer new ones. It is essentially a problem-solving strategy. There are typically three subcomponents, interpreter, scheduler and consistency enforcer, which will be discussed below in conjunction with the blackboard components.

Blackboard. During a problem-solving session the system makes a sequence of decisions and generates various hypotheses which are modified as progress towards the solution is realized. The blackboard is a recording component, in the form of a shared data structure, that keeps track of intermediate decisions and hypotheses. The plan elements on the blackboard describe the general strategy adopted for solving the current problem. In large complicated problems an expert system may be able to work only through a problem decomposition. Subproblems are solved separately and the union of partial solutions is the complete solution. A plan may be a decomposition strategy. When a problem is not decomposable but only weak links exist among the subproblems (it is so-called nearly decomposable), a plan may record links, or potential links, among these subproblems and offer ways to reconcile them at the solution state. Another way to view decomposition is by examining the amount of information that changes as the system proceeds through the state-space. It may not be necessary to recompute the entire new state, if only part of the state description has been altered from the old state to the new one. The question is how to decompose a state description in order to determine what will alter and what not during a subsequent state change. This is sometimes referred to as the _frame problem_. Plan elements may address also a decomposition of this sort. An _agenda_ is a list of tasks, or hypotheses, that the system considers as candidates for execution. Attached is also a list of justifications for each agenda item which explain why the item is on the agenda. There may also be an overall rating to prioritize the tasks. For example, an agenda may contain triggering rules in the knowledge base. The _solution_ elements record the candidate hypotheses and decisions towards solution that have been generated and the links among them. In some systems this may be considered as _the_ blackboard of the system.

In a typical cycle of the system, the most promising task or hypothesis from the agenda (a rule in a rule-based system) is chosen by the _scheduler_ of the inference mechanism. This agenda task is then executed by the _interpreter_ of the inference mechanism. The interpreter validates the task by proper matching of the task's requirements (e.g. rule conditions, variable values) with the knowledge base and the other elements on the blackboard. Typically the execution of a task will generate new tasks and/or indicate changes in the blackboard elements. For each new task the interpreter will examine if it is already in the agenda and insert it if it is not present. For a new task which was in the agenda, the interpreter will check its list of justifications and add the new evidence, if necessary. The new task will also receive a (possibly updated) rating. The _consistency enforcer_ is required when nonmonotonic reasoning is present, which is usually the case. Solution elements may be ascribed likelihood numbers, usually appropriate for diagnostic tasks. When logical deductions are solution elements, then a truth-value relationship checking is required. Enforcement of consistency is required continuously also for reasons of efficiency, i.e. to avoid inconsistent solutions as early as possible.

Explanation Facility. This important part of the system attempts to explain to the user the reasons for the actions taken by the system. It may answer questions about how a particular conclusion was reached, how another alternative was rejected, or why some information is requested from the user. This represents an attempt at emulating the justification that a human expert may offer about his actions. For an a priori justification (why is a fact necessary), the justifier can do a forward-chaining to identify a rule which failed because of inability to verify a premise, while no other rules were applicable. Thus it justifies to the user the need for the requested information. For an a posteriori justification (why a conclusion is reached), the justifier can do a similar backward-chaining along previous blackboard elements to trace the conclusion to the hypotheses or data that led to it. Thus the justifier is a bookkeeper of the inferences used by the system.

Knowledge Acquisition. This part of a knowledge-based system is a difficult and often controversial one. The difficulties stem from the very nature of expert systems and from their present state of the art. The knowledge domain to be "inserted" in the system is very specialized with its own terminology, idiomatic procedures and rules. Specialized knowledge domains are typically unintelligible to the non-experts. At the same time, current expert system programs require a good deal of knowledge about themselves in order to introduce the specialized domains in the databases and other components in an appropriate way. Developers of expert systems address this issue by introducing the job classification of a "knowledge engineer". This is a person familiar with the AI system who acts as an intermediary between the system and the human expert. It is immediately evident that this manner of knowledge transfer is at best very lengthy. Automated ways for satisfactory direct interaction between expert and program are still largely speculatory.

There are three main tasks involved in knowledge acquisition: transfer of data into the system, avoidance of initial data errors and inconsistencies, and finetuning of entered data to achieve desired performance. Special knowledge-base editors - the equivalent to text editors - are being developed to assist in the data transfer by automating some bookkeeping, correcting typographic and syntactic errors and attempting to enforce semantic consistency. The explanation facility described earlier can be used to test the system extensively and detect weaknesses and inconsistencies in the knowledge base. In any case, the knowledge acquisition facility is a critical area where more developments are needed before widespread use of such systems becomes feasible. For further discussion on this see Hayes-Roth et al., 1983.

This concludes the discussion on the components of an ideal expert system. It should be emphasized again that existing systems have parts of these components but no system appears to exist which has all the above components fully developed. A computer program can be seen as a legitimate expert system if it contains at least the knowledge base and inference mechanism components, and some reasonable user interface.

Theoretically, a properly constructed expert system can be used for several different applications (within a problem class) provided the knowledge base is changed to reflect the facts and rules of the specific domain. Often it is necessary to reprogram at least partially the inference mechanism, particularly the scheduler, to define domain-specific triggering and firing sequences for the rules. A system with empty knowledge base (and partially empty inference mechanism) is often referred to as a system shell. Description and evaluation of existing system shells can be found in Hayes-Roth et al. (1983) and in Buchanan and Shortliffe (1984). Design applications which contain also shell descriptions are given in Gero (1985a). Much of current development effort goes to improving commercial-grade system shells that reduce the need for mediation from a "knowledge engineer" and can be applied without excessive effort to similar knowledge domains.

Domain Classes. Experience from systems that have been developed to date shows that there can be wide differences in the way they operate. It appears that many of these differences are related to the different problem types that the system is built to address. We can classify the problem domains to the following few types. Within each class the relevant tasks that the system should perform are also briefly mentioned.

Diagnostic Systems. The problem is to find the fault(s) in a system based on given sampled data (information). For example, the system can be a malfunctioning machine or an ill living organism. A diagnostic expert system typically will reason backwards starting with data about the symptoms and working towards the potential faults that cause the symptoms. Often this leads to many triggering rules whose right sides match the current goal, e.g. several faults may have overlapping symptoms masking one fault by another. A system for diagnosis of, say, human disease must employ search procedures that will examine all possible faults. Thus, such systems will tend to spend a lot of their effort on search. In cases with less severe consequences a single fault assumption might be appropriate.

An additional characteristic of diagnostic problems is the uncertainty in the data that must be interpreted. The data may be themselves noisy and the data-sampling device (e.g. sensors) may itself be subject to error. Thus, reasoning must be usually tied to probabilities. Furthermore, apart from noisy data there may also exist data which are erroneous, difficult or impossible to sample, or expensive to acquire. This means that there may be contradictions which must be resolved, while decisions must be made about believable data with partial information. The reasoning chain behind a conclusion may be long. So diagnostic systems, to be credible, must be able to explain exactly how conclusions were reached, what evidence was used and what hypotheses were assumed true. Thus, the uncertainty in the data is properly traced.

Designing and Planning Systems. Solving a designing or a planning problem may be viewed in a similar manner. A set of planning goals or designing specifications is considered as given - let us call them design goals. Achieving these goals requires making decisions, and resultant actions, without violating problem constraints and using the smallest amount of resources possible. The result of these decisions is a plan or a design. A designing expert system will tend to reason forward since the condition-side of applicable rules will be specified and the main difficulty will be in picking the correct rule to apply. Moreover, in a typical design situation it is not necessary to find all possible solutions, or the best one, but only a sufficiently good one. So a search may quickly become a depth-first strategy which builds up an increasingly complete configuration, or plan. Thus most of the effort in such a system is expended in rule matching rather than search.

Uncertainty exist here also, in the sense that a plan or design will operate in the future which is never certain. Yet in the present context designing problems are substantially more deterministic than diagnostic ones mostly because the goals are relatively well-defined. The major difficulty comes from the size and complexity of the problem to be solved. The consequences of design decisions may become evident only at a later stage of the process, so the system must be able to make tentative decisions and have contingency plans available. This means again the need for an explanation facility which will trace a result back to the decision that originated it. Another ability required for solving large, complex problems is the isolation of important considerations from details, so that the system does not get overwhelmed when following a path. Sometimes the reduction of the problem to several (interdependent) subproblems to be solved separately is effective in dealing with size and complexity. Again, however, accounting for the interactions among subproblems may be a challenging task to perform.

There are two other problem classes that expert systems address, which can be viewed as special cases of the previous ones. <u>Monitoring</u> systems are diagnostic systems with a continuous interpretation of sampled signals. They should identify an anomaly and activate an alarm. Here, diagnosis of alarm conditions must be performed in real time. <u>Prediction</u> systems are required to forecast future events from past and present knowledge. This of course assumes incomplete information and absense of deterministic models. These systems are similar to planning ones (they can be components of a planning system) but they emphasize the dynamic, or time-dependent, nature of the state-space.

The classification presented above is not well-defined and it is based on observed trends in existing systems. It is very important for the development of commercial-grade shell systems to be able to understand better unifying characteristics, principles and methods for at least some classes of problem domains.

KNOWLEDGE-BASED OPTIMAL DESIGN

Optimization techniques have been recognized as useful design tools since about the early 1950's. Although it may be true that some significant advances of the last decade are just now reaching the wider public, one could easily argue that design optimization remains largely underused in the design practice. Most design practitioners recognize that what they do is in fact optimization, yet they do not adopt the use of formal tools - with very few exceptions. Though several explanations may be offered, it is undeniable that applying optimization techniques to realistic design problems is a complicated and difficult task requiring expertise in both optimization and the design problem domain. In what follows we are primarily concerned with optimal design problems formulated as nonlinear programming ones. Most of the issues discussed apply equally well to other models, such as variational ones. So the term "optimization techniques" will imply mathematical programming techniques without real loss in generality.

The difficulties in solving an optimization problem may be broadly separated to those related to modeling and those related to the optimizing procedure itself. By <u>model</u> here we mean the formalism: minimize $f(x)$, subject to $h(x)=0$ and $g(x)\leq 0$, where f, h and g are the vectors of the objective function, equality and inequality constraints respectively. The vector of design variables x takes values in a finite-dimensional Euclidean subspace. The most important phase in an optimization study is development of the model. The influence of the model is so great, that in all but the simplest problems, one or more model revisions are required before satisfactory results are obtained. Development of a good model requires engineering knowledge about the specific problem and mathematical knowledge about its properties and possible methods of solution. Typical questions in the modeling phase are:

- o Is the model consistent, bounded, feasible?
- o Are there any useful transformations? Which?
- o What should be done about discrete variables?
- o Are variables and constraints scaled?
- o Are there redundant constraints? Possible degeneracies?
- o What is the mathematical form (linear, quadratic, convex, monotonic, etc,)?
- o How many times will the problem be solved?
- o Is parametric or sensitivity analysis necessary?

Note that questions such as the aforementioned are generally handled by the analyst manually, rather than automatically (with partial exceptions,e.g. scaling) because of the sophistication required in their study. The knowledge to handle these questions generally exists although it is rather scattered and not well codified. Answering these questions will often require analysis of the model and result in reductions and simplifications.

The next important phase is the development of a solution strategy. Now typical questions that arise are:

- o What numerical method should be selected?
- o What submethod (e.g., line search, unconstrained), within the overall method, should be selected?
- o How should the user-defined program parameters (e.g., step size accuracy, penalties, termination) be selected?
- o What criteria for accepting a solution should be used?
- o Is it global or local solution?
- o Which are the active constraints?
- o What should be done, if solution errors are encountered?

Note again that many of these questions are handled manually, although the degree of automation is higher than in the modeling phase. The necessary knowledge is also better codified in this phase.

All these simply point out the well-known problem in engineering optimization, namely that the analyst must be an expert both in the technology of the model and in the mathematics of the optimization techniques. One of the hopes in applying AI here, is to reduce this burden to more attractive proportions, thereby encouraging wider and more correct use of optimization. We will now proceed to examine in more detail the types of difficulties which may be addressed effectively with knowledge-based systems.

Modeling Phase. The design optimization model is based on one or more analysis models which allow the mathematical description of the problem functions (objective and constraints). So all the inherent limitations of analysis are present in the classical optimization model. These limitations include a way of thinking about what constitutes a mathematical model. Thus, relations among design quantities which do not have a customary functional form tend to be ignored or handled separately.

Consider, for example, the situation where a constraint may be soft (i.e. its mild violations are tolerated) only in certain regions of the design space or under a particular combination of design circumstances (e.g, a specific set of other constraints being active). A soft simple bound constraint may be deleted from the constraint set and added as a penalty term in the objective (see e.g., Gill, et al., 1985). Now, however, this transformation should be done only if the algorithm generates points in the "soft" region. The model can be modified to account for this with a bit of programming effort. Some discontinuities may be introduced also. Another similar issue comes from the observation that constraints with small sensitivity coefficients should preferably be kept out of the active set, to avoid dealing with small Lagrange multipliers. If they can be classified as "soft", they could be removed from the active set with or without using the penalty transformation. This, of course, would require additional programming logic.

Further, suppose that some parameters (i.e. quantities considered fixed for the particular optimization statement - not variables) change values in a way similar to the one above for the soft constraint. If a parameter changes with the design variables, we may attempt to map this function and introduce it in the model. But the function may be discrete with large differences in values, making curve-fitting inadequate. Again, with some reprogramming (perhaps table interpolation) the situation can be addressed, although discontinuities may be also introduced.

In both of the above cases our knowledge is generally precise, but incorporating it in a traditional model brings complexities and possible errors. So another way of modeling may be more useful. In AI jargon, we need a different knowledge representation. One possibility is to put this knowledge in an IF-THEN production format and collect such rules in a production system. As iterations proceed, the context of the knowledge base is updated and rules are fired to cause the required model changes. Even discontinuities may be addressed with proper rule matching. We will describe later a production system that was developed to generalize active set strategies. The important point here is to recognize that a production system, even a very simple one, is a legitimate mathematical model, being another form of knowledge representation.

Techniques for assisting the construction of optimization models have received attention in problem areas where there is uniformity of structure, particularly linear and integer programming (Williams, 1985a and 1985b) and some network problems. Special programs, often called <u>matrix generators</u> or <u>modeling languages</u>, are employed to automate data organization (e.g. index relations) and assist in debugging. Some limited model transformations can be also handled. This is possible basically because of the linearity in the model. In nonlinear problems essentially none of these techniques can be extended.

Nonlinear programming (NLP) codes have been notorious for their unfriendliness to the user. Only in the last few years interactive programming has been introduced in several codes and efforts have been made to assist the user in model preparation. An example is the system EMP (Schittkowski, 1985b) which, although it is not a "real" expert system, has many long-awaited features for the modeling stage. The objective function can be given in several optional formats, e.g., L_1, L_2 or maximum norm of individual functions, sums of functions, general function. Constraints can be separated to linear, nonlinear and simple bounds. This aims primarily at efficient solution strategies. Since EMP uses different solution procedures for different NLP problem classes, it is necessary to have different data formats. An interesting feature of EMP appears to be a frame-like structure of the model knowledge base so that the user is asked to supply the facts (data) applicable to the current NLP problem class. This type of interactive knowledge input appears in other new codes usually revisions of older ones. Such an example is the program GINO (Liebman et al., 1986). These program developments exhibit elementary intelligence and are very useful in that they focus attention on the representation of knowledge, organization of a knowledge base with easy updating capabilities, and realization of the limitations of traditional programming environments.

A quite different application of expert systems to modeling comes from the use of symbolic mathematical processors, such as REDUCE (Hearn, 1983) and MACSYMA (MACSYMA Group, 1977). Application of the expert system MACSYMA to linear programming was reported by Anderson and Wand (1980) and Rajaram (1980). Application of REDUCE to design optimization was reported by Li and Papalambros (1985b) and Li (1985), with emphasis on modeling. It is advocated there that a modeling preprocessor for nonlinear programming can become the equivalent of a word processor in text editing. The motivation for the particular program comes from application of global monotonicity analysis (see e.g., Papalambros and Wilde, 1979; Li and Papalambros, 1985a). REDUCE is a LISP-based language allowing the definition of LISP-type procedures. The preprocessor called MODLAN (MODel ANalyzer) is described briefly below. This is given only as an example of what can be immediately created using existing AI software.

The input to MODLAN is the model of an optimization problem, the expressions of the objetive and constraint functions and a few dimensioning numbers. The output can be extracted at several intermediate levels of model changes. Thus, MODLAN can assist in solving problems where the optimal active set can be identified globally and complexity is not serious (see "Knowledge-Based Algorithms" below).

MODLAN is a collection of procedures called interactively. Currently there are 16 procedures developed. They are grouped as follows:

(i) Input and display: INSTRC, INPUTT, SHOMDL, DERIV;

(ii) Model manipulation: REDUND, VAREX, ELIMI, BEACT, RENUMB;

(iii) Intermediate or final file output: INTERM, PRODU1, PRODU2, PRODU3;

(iv) Optimal solution: MINSUB, MINRUN, SECOND.

A brief description of each procedure is given below. INSTRC gives a brief explanation to the usage of commands in MODLAN. INPUTT starts a model analysis session with options to work on either the original file or one of the intermediate case files. SHOMDL shows the current model in explicit algebraic form so that the user can inspect it visually and consider further manipulations. DERIV gives explicit expressions for any partial derivatives required by the user. If equality constraints are present, this procedure will give the constrained derivatives (reduced gradient). REDUND declares a constraint as redundant and will cause its deletion at the next revision performed by RENUMB. VAREX declares that variable with index i will take the same value as variable with index j so it will be replaced by jth variable. ELIMI eliminates a variable from the model by means of an equality or active inequality constraint, if the variable in the equation is solvable in closed-form. In the next model revision both the variable and the constraint will be discarded. BEACT declares an inequality constraint as globally active, so it will join the active set. A message is sent to RENUNB for proper arrangement of the constraint sequencing. RENUMB rearranges the sequencing of variables and constraints in accordance to various reductions that were performed by the procedures mentioned above. INTERM produces an intermediate file in the same format as the input file for further processing by MODLAN. This provides the user a chance of interrupting the execution while investigating the problem. PRODUI (I=1,2,3) produces a user's program in FORTRAN describing the optimization problem in the format requested by three specific NLP codes, with the option that gradient expressions are also produced. MINSUB produces a FORTRAN program running with Powell's hybrid method (Powell 1970) in MINPACK (Morè, Garbow, and Hillstrom, 1980) and the interfacing program USEMIN.O to solve the stationary point of an unconstrained or equality constrained problem from the first order optimality conditions. MINRUN applies several commands of the host operating system while staying in REDUCE. SECOND checks the second order optimality conditions at the solved stationary point.

Symbolic manipulations can be used in the development of analysis models which can subsequently form the basis for an optimization model. Such applications exist in dynamic system modeling (e.g., Krishnaswami and Bhatti, 1985) and finite element analysis. Special modeling procedures can be developed for generating NLP models and solution elements with a well-defined structure. An example of such application is in re-analysis problems in structural optimization (Hartmann, 1985a). For linear discrete systems, a series approximation procedure can be expanded to finite element formulations. A rule-based symbolic differentiation of the finite element matrices with respect to the design variables aids the optimization process. The system is written in a version of PROLOG.

So far in our discussion on modeling we concentrated on model development and some model transformations. There is a host of other issues related to modeling where a rule-based system would be a very desirable assistant. It is particularly important to have the ability to start with a tentative model, proceed to selecting a tentative solution method, examine (through a rule-based system) what modeling features affect the method, and iterate on the model formulation trying to meet the goals that will improve solution efficiency and quality. Issues that are suitable to be addressed by rules are: scaling, elimination or not of equality constraints, smoothing of problem functions, and advice on how to handle discrete variables. One can easily start formulating rules based on existing textbook information (e.g., Gill, Murray and Wright, 1981).

Solution Phase. Assuming that the modeling questions have been tentatively settled, solution of the model is undertaken. In this phase, knowledge-based systems - even basic production ones - can assist in three main areas:

(i) Selection of the appropriate optimization technique, including component options within the technique;

(ii) Consultation on the proper use of the selected method in the context of the specific model;

(iii) Diagnostics, interpretation of results and suggested remedies in the case of failure.

We will now examine briefly these three areas.

Selection of the appropriate optimization technique is probably the area that has received most attention, even in the simplest interactive optimization systems. This is due not only to the large number of techniques and variations that exist but also to the fact that some methods are in fact much better than others, at least within model classes. From the AI viewpoint this is a particularly interesting area because experts in the field do not generally appear to agree on method evaluation. This is not so much because of arguments about the theoretical algorithmic performance (when it can be proved analytically), but because of differences in interpreting observed performance and in prioritizing evaluation criteria. The scope of the issue is well-illustrated by Lootsma (1984, 1985). An expert system for selecting a technique should coordinate knowledge about the model at hand and about the available techniques. The model knowledge should go beyond simple classification of the problem class (e.g. quadratic, "mildly" nonlinear, etc., or availability of analytical derivatives). The system should also perform the following:

- o Assess the computational cost of function evaluations (possibly by pre-testing) and use the information not only for primary method selection but also in decisions such as type and accuracy of line searches.
- o Suggest and implement various scaling tasks associated with the specific method.
- o Check for nearly dependent constraints and suggest and implement countermeasures.
- o Interview the user about the intended use of the problem solution, assess the <u>user's</u> evaluation criteria and <u>then</u> select a method through matching with the knowledge base rules.
- o Assess the appropriateness of evaluating derivatives symbolically (e.g. with REDUCE) or numerically. If a numerical choice is made, several decisions must be made about proper finite differencing formulas, step size etc.
- o Assess the density of model matrices versus problem size and decide if special techniques should be employed.

This list is only indicative. The point here is that there is a fairly large amount of data that can be programmed and utilized automatically, if the ability to exercise judgement is available in the program. This is where intelligent programming becomes different from traditional interactive programming.

As mentioned earlier, attempts at automating method selection have been made. For example, the cited system EMP will use an SQP method for a general NLP model under normal situations. If failure due to several reasons (e.g., poor starting point or multiplier estimates) is detected, the model is treated first by a variant of the ellipsoid method until local convergence becomes apparent or movement away from an undesirable point is realized. Another example of a simple system which however employs a real expert system shell is given by Hartmann (1985b). The system is implemented on a commercial program (INSIGHT, 1985) and uses production rules to achieve two goals: rapid problem classification and rapid preliminary method selection.

The second area, consultation on the proper use of a selected method, is relatively easy to implement but, to this author's knowledge, hardly any effort has been expended to this end. The context (facts) of the specific model has immediate influence on the implementation of the method. The typical data that need consultation to be supplied are the often notorious "program parameters". These take, almost always, values that are given heuristically from empirical evidence. Such parameters include tolerances for a variety of termination criteria, tolerances for accepting constraints as active or violated, values for what is taken as "zero" and so on. Some program parameters are often transparent to the non-expert user - and rightly so, under normal programming circumstances. Yet they may be important to the success of the algorithm, and implementation of expert rules about their finetuning could be very beneficial. Once again, the emphasis here is at a uniform way of using context information (current model) in a domain of expertise (application of selected techniques).

The third area, diagnostics and interpretation of results, is an obvious one and most code developers will offer some diagnostic data at the end of a run together with run statistics. These data are generally completely inadequate for a user who is not already an expert. Reference to code manuals usually yield little more understanding. There is much development that could be done without using intelligent programming techniques. However, we may give some example issues where a knowledge based system would be appropriate.

First, recall that the nature of nonlinear programming expertise to date is such that a good code may terminate at a point indicating failure although the point is optimal, or indicating success at a non-optimal point. There is a variety of reasons that cause this situation. Deciding if an answer by a code is correct, is usually a result of judging different types of evidence, some of it computable and some of it not. A "smart" program must generate the evidence, assess it, reach a conclusion and explain to the user how this conclusion was reached. Apart from the usual termination criteria, other semi-heuristic information may be used: (i) sufficient decrease in the value of the gradient (or reduced gradient) of the objective function from the starting to the terminating point; (ii) verification of expected local convergence behavior (if better than linear) in the last few iterations; (iii) estimation of condition numbers for second order matrices; (iv) similar (or same) results with different program parameters; (v) comparable results with different model parameters; (vi) same results with different starting points. Some of these data may become costly to acquire so the system should consult with the user before proceeding to collect more evidence.

Second, consider a situation where failure has occurred. Causes of failure and possible remedies should be explored and tested. For example, in Gill et al. (1981), the following list (and suggestions for action) of failure causes is given: overflow in user-defined problem functions, insufficient decrease in the descent function (failure in step-length procedure, programming errors, poor scaling, too strict termination criteria, inaccuracy in finite difference approximation), consistent lack of progress, excessive number of iterations, lack of expected convergence rate, inability to maintain descent. Remedies for such failures are usually increasing the complexity of computation and they should be used only when needed. Thus, part of the diagnostic intelligence can be a strategy where relatively simple solution procedures are tried first and more powerful (and therefore usually more complicated and expensive) solution procedures are called if needed. This may be more cost-effective in the long run. Also, the diagnostic effort is smaller for simpler solution techniques.

The above arguments on the utility of expert systems both in the modeling and in the solution phase of an optimization procedure are certainly not profound and they have been elaborated before (e.g., Papalambros, 1984). The knowledge to implement such systems is available. The main difficulty is the coordination of experts. One may speculate that relatively small simple systems will be developed in the near future. An expert system in the above sense will require joint effort of a major software group with good grounding in numerical analysis, with an AI group that is willing to work in this possibly lackluster knowledge domain. However, it is the belief of this author that eventually all numerical analysis software of commercial quality will employ some form of expert system to assist in their use.

Knowledge-Based Algorithms. We now proceed to examine another approach of using expert system ideas in optimal design. As we will see, this implies a somewhat radical view of what may constitute an "optimization algorithm".

Active set strategies can provide us with an illustration of an interesting aspect of NLP solution methods. The decisions implied by a typical iteration $x_{k+1}=x_k+\alpha_k s_k$ are based primarily on information about the current point and they utilize limited memory, as for example the successively better approximation to the true Hessian in quasi-Newton type methods. Most active set strategies use current iteration information to make decisions and may also employ limited memory to prevent the same constraint entering the active set too many times. Global active set strategies are partially possible by using monotonicity principles (e.g., Azarm and Papalambros, 1984a). Here "global" means information that applies to all points in the design space, therefore usable in all iterations, not just the current one. This is an example observation that during the solution of a design optimization problem, different forms of knowledge are used: knowledge applicable to the current iteration only, to all iterations or to the sequence of iterations.

In real design problems, particularly when recognizing the inability or undesirability of very meticulous models, there is a subtler interaction between model and iterations. In a little-noticed article, Thompson and Hunt (1974) point out that many structural optimization formulations "lead inevitably to designs which exhibit the notorious failure characteristics often associated with the buckling of thin elastic shells". The rationale behind this assertion is easily recognized by the thoughtful analyst. Optimization models with simple objective functions, often monotonic with respect to several design variables, tend to demand simultaneous activity of several constraints at the optimum. This means designs deliberately made to fail simultaneously in several modes. Failures apparently unrelated may have nonlinear coupling with unexpected results. This coupling does not appear in traditional, less efficient designs. Such situations should be accounted for without unduly complicating the model. The bounds on constraints can be made tighter, say with higher safety factors, but only if certain combinations of active constraints occur. Otherwise the design would be overly conservative. This then is an example of global information which will influence local decisions.

We can recognize three forms of knowledge applicable during the design optimization process.

(i) Global Knowledge pertains to facts independent of a particular design point and equally true for all points in the design space. Most such knowledge is a result of analysis or experience.

(ii) Local Knowledge pertains to facts applicable to a particular design point. Such knowledge is mostly a result of computation.

(iii) Evolutionary Knowledge pertains to facts observed from a sequence of design points generated by a particular algorithm. Again most such knowledge is acquired computationally.

To cite some more examples, we may say that activity rules resulting from global monotonicity analysis, would represent global knowledge. Specific values of functions, gradients or multipliers at a point, would be local knowledge. A descent property of an algorithm, or constraint activity history such as identification of zig-zagging, would be evolutionary knowledge.

The idea of an "expert" active set strategy in the above sense was first proposed by Azarm and Papalambros (1984) and was implemented in a production system using only global knowledge by Li and Papalambros (1985a). A combined local-global active set strategy was recently developed (Li, 1985) and implemented for linear programming and also for a reduced gradient-based nonlinear programming algorithm. This latter algorithm, called KBRG (Knowledge-Based Reduced Gradient), is briefly outlined below. Note that many of the rules are motivated from experiences in monotonicity analysis (e.g., Papalambros and Wilde, 1979; Wilde, 1985; Papalambros and Li 1983).

A local strategy has the advantage of easy generation of the required data for decision making, but there are several disadvantages. In the general NLP case, the addition and deletion rules are completely heuristic and this may be cause for severe inefficiencies. Also, global convergence is theoretically possible only if the global optimum is found on each active set surface (i.e. a global solution is required for each equality constrained subproblem). A global strategy has the advantage of utilizing only knowledge known to be true at the optimum so no heuristics are required. The major disadvantage is that such global knowledge is both difficult to get and rarely complete enough for reaching a single decision. A compromise between the two should appear desirable. This is possible both in linear and in nonlinear programming (Li, 1985).

The basic idea behind a combined local-global active set strategy is as follow: any available global knowledge about constraint activity is processed first; when that knowledge is not sufficient for reaching conclusive decisions, then local knowledge is introduced as required, so that decision-making in each iteration is complete. All forms of activity knowledge are implemented in the "if-then" rule form. This is done to maintain algorithmic flexibility, as well as to enhance efficiency of data processing. The current system implementation is not strictly an expert one, since no elaborate inference mechanism is required at this stage and no

AI language is used (the program is all written in FORTRAN). However, it does have elements of a knowledge-based system and can be viewed as such.

In practice, global knowledge is organized in several rule forms which are applied to the constraints of a specific problem in order to generate working sets. The rule types are applicable also to the combined strategy and are summarized below. From an operational restatement of monotonicity principles in Li (1985) several rule forms have been found of practical use. These rules pertain to combinations of active and inactive constraints that are permissible and allow essentially a problem-reduction approach to the true active set identification. They are not included here for lack of space.

The algorithm KBRG uses an equality constrained minimization procedure (of the reduced gradient family) to search on the working set and an active set strategy to update the working set.

The major steps are as follows:

1. Initialize and input global knowledge, k=0;
2. At x_k, update the working set by local and/or global strategy (utilizing the rule format);
3. Solve the equality constrained subproblem with a descent reduced gradient procedure (here BFGS is used);
   if constraint violation occurs, set k=k+1 and go to 2;
   if a solution to the subproblem is feasible to the original problem, go to 4;
4. Check convergence criteria and multiplier estimates, if satisfied, terminate;
   otherwise, set k=k+1 and go to 2.

Step 2 is where the rules for global and local activity are implemented. Typical global information tells the user (and the program) if there is current available data in each rule type, the variable with respect to which they were introduced (bounding arguments) and which constraints are involved. The knowledge representation, given in FORTRAN, is very awkward for the user by AI standards. A rule search procedure retrieves the global knowledge from the database and builds up the working active set. If the global rules have ties, these are resolved by local information (local dominance strategy). Thus the addition part of the strategy uses global and local information. The deletion part uses multiplier estimates at check points for optimality (Karush-Kuhn-Tucker conditions to be satisfied). The constraint with the "most negative" multiplier is deleted provided the global rules are not violated. Global rule violation is also accounted for during the line search.

Further details, test results and discussion on this algorithm can be found in the cited reference (Li, 1985).

The combination of knowledge sources as was done in KBRG creates a departure from traditional NLP algorithms which utilize only local knowledge. A theoretical problem of what happens with convergence properties of the pure local strategy remains unsolved. However, all active set strategies for NLP are heuristic, so KBRG is not theoretically inferior. A practical consequence of introducing rule information based on expectation (as the one provided from experience) rather than rigorous knowledge (as the one provided by global monotonicity analysis), may bias the algorithm, more that what would be desirable, towards known and uninteresting solutions.

A NOTE ON AI LANGUAGES

Developing a knowledge-based system of any realistic size would require a special language, such as PROLOG or LISP. It is beyond the present scope to discuss these languages, or other variants in the AI practice. For an introduction to these tools, see e.g., O'Shea and Eisenstadt (1984), Hayes-Roth et al. (1983), Winston and Horn (1981). Connecting AI object-oriented programs with usual high level number processing languages is a challenge that must be addressed before systems of some practical complexity can be built. This is necessary in order to utilize the mass of existing engineering and mathematical software.

CONCLUSION

As promised in the introduction, the use of knowledge-based systems in optimal design was found to be still more speculation than reality. However, some first concrete steps are being taken and such systems will become increasingly available. The discussion on knowledge representation should stimulate ideas about how to organize optimization knowledge in a more rational way. In fact the most immediate benefit of trying to build a rule-based system either in the modeling or in the solution phase is the explicit and systematic codification of available knowledge in the field. Finally, symbolic processing and knowledge-based algorithms are examples of systems that can extend modeling capabilities beyond the traditional mathematical programming formulation.

ACKNOWLEDGEMENT

Work on this subject, part of which was briefly reported here, was supported by the U.S. National Science Foundation under grant DMC85-14721 at the University of Michigan. This support is gratefully acknowledged. The support of the Design Division at the University of Patras is also acknowledged.

REFERENCES

Anderson, N., and Wang, P.S., 1980. "MACSYMA Experiments with a modified Khachian-Type Algorithm in Linear Programming", SIGSAM Bulletin, Vol. 14, No. 1, pp. 8-13.

Azarm, S., and Papalambros, P., 1984. "A Case for a Knowledge-Based Active Set Strategy", Trans. ASME, J. of Mech., Transm. and Auto. in Des., Vol. 106, No. 1, pp. 77-81.

Barr, A., and Feigenbaum, E.A., (eds.), 1981. The Handbook of Artificial Intelligence, Kaufmann, Los Altos, California.

Buchanan, B.G., and Shortliffe, E.H., (eds.), 1984. Rule-based Expert Systems, Addison-Wesley, Reading, Massachusetts.

Crane, R.L., Hillstrom, K.E., and Minkoff, M., 1980. Solution of the General Nonlinear Programming Problem with Subroutine VMCON, Report ANL-80-64, Argonne National Laboratory, Argonne, Illinois.

Davis, R., and Lenat, D.B., 1982. Knowledge-Based Systems in Artificial Intelligence, McGraw-Hill, New York.

Findler, N.V., (ed.), 1979. Associative Networks - Representation and Use of Knowledge by Computer, Academic Press, New York.

Gero, J.S., (ed.), 1985a. Knowledge Engineering in Computer Aided Design, North-Holland, Amsterdam.

Gero, J.S., (guest ed.), 1985b. Special Issue: Expert Systems, Computer-Aided Design, Vol. 17, No. 9.

Gill, P.E., Murray, W. and Wright, M.H., 1981. Practical Optimization, Academic Press, London.

Hartmann, D., 1985a. Application of AI Tools for Re-analysis Within Structural Optimization, Technical Report, University of Dortmund, Dortmund.

Hartmann, D., 1985b. Selection and Evaluation of Structural Optimization Strategies by Means of Expert Systems, Technical Report, University of Dortmund, Dortmund.

Hayes-Roth, F., Waterman, D.A,. and Lenat, D.B., (eds.), 1983. Building Expert Systems, Addison-Wesley, Reading, Massachusetts.

Hearn, A.C., (ed.), 1983. REDUCE User's Manual, Version 3.0, The Rand Corporation, Santa Monica, California.

INSIGHT, 1985. Insight Knowledge System Manual, Level Five Research Inc., Melbourne Beach, Florida.

Krishnaswami, P., and Bhatti, M.A., 1985. "Symbolic Computing in Optimal Design of Dynamic Systems", ASME Paper No. 85-DET-76, New York.

Li, H.L., 1985. Design Optimization Strategies with Global and Local Knowledge, Ph.D Dissertation, Dept. of Mech. Engr. and Appl. Mech., The University of Michigan, Ann Arbor.

Li, H.L., and Papalambros, P., 1985a. "A Production System for Use of Global Optimization Knowledge", Trans. ASME, J. of Mech., Transm., and Auto. in Des., Vol. 107, No. 2, pp. 277-284.

Li, H.L., and Papalambros, P., 1985b. "REDUCE Applications in Design Optimization", Proc. CAD/CAM, Robotics and Auto. Inter. Conf., Tucson, Arizona, Feb 13-15, pp. 187-193.

Liebman, J.S., Schrage, L., Lasdon, L., and Waren, A., 1986. Applications of Modeling and Optimization with GINO, Scientific Press, Palo Alto, California.

Lootsma, F.A., 1984. Performance Evaluation of Nonlinear Optimization Methods Via Pairwise Comparison and Fuzzy

Numbers, Report No. 84-40 (revised), Dept. of Mathematics and Informatics, Delft Univ. of Technology, Delft.

Lootsma, F.A., 1985. "Comparative Performance Evaluation, Experimental Design, and Generation of Test Problems in Nonlinear Optimization", in K. Schittkowski, (ed.), 1985a, pp. 249-260.

MACSYMA Group, 1977. MACSYMA Reference Manual, Technical Report, Mass. Institute of Technology, Cambridge, Massachusetts.

Minsky, M., 1975. "A Framework for Representing Knowledge" in The Psychology of Computer Vision by P.H. Winston, (ed.), McGraw-Hill, New York.

Morè, J.J., Garbow, B.S., and Hillstrom, K.E, 1980. User's Guide for MINPACK-1, Applied Math. Division, Argonne National Laboratory, Argonne, Illinois.

Nilsson, N.J., 1971. Problem-Solving Methods in Artificial Intelligence, McGraw-Hill, New York.

O'Shea, T., and Eisenstadt, M., (eds.), 1984. Artificial Intelligence - Tools, Techniques and Applications, Harper and Row, New York.

Papalambros, P., 1984. "Utilizing Knowledge-Based Systems in Design Optimization", Presentation at the NATO ASI on Computational Mathematical Programming, Bad Windsheim, F.R. Germany, (July 23-Aug. 2).

Papalambros, P., and Li, H.L., 1983. "Notes on the Operational Utility of Monotonicity in Optimization", Trans. ASME, J. of Mech., Trans., and Auto. in Des., Vol. 105, No. 2, pp. 174-181.

Papalambros, P., and Wilde, D.J., 1979. "Global Non-Iterative Design Optimization Using Monotonicity Analysis", Trans. ASME, J. of Mech. Design, Vol. 101, No. 4, pp. 643-649.

Pearl, J., 1984. Heuristics - Intelligent Search Strategies for Computer Problem Solving, Addison-Wesley, Reading, Massachusetts.

Powell, M.J.D., 1970. "A Hybrid Method for Nonlinear Equations", in Numerical Methods for Nonlinear Algebraic Equations, by P. Rabinowitz, (ed.), Gordon and Breach, London.

Rajaram, N.S., 1980. "Random Oscillations Arising in the MACSYMA Implementation of Khachian-Type Algorithm in Linear Programming", SIGSAM Bulletin, Vol.14, No.2, pp. 32-35.

Rich, E., 1983. Artificial Intelligence, MaGraw-Hill, New York.

Schittkowski, K., (ed.), 1985a. Computational Mathematical Programming, NATO ASI Series F, Vol. 15, Springer-Verlag, Berlin.

Schittkowski, K., 1985b. EMP: An Expert System for Mathematical Programming, Mathematical Institute, University of Bayreuth, Bayreuth.

Thompson, J.M.T., and Hunt, G.W., 1974. "Dangers of Structural Optimization", Engineering Optimization, Vol. 1, No. 2, pp. 99-110.

Wilde, D.J., 1985. "A Maximal Activity Principle for Eliminating Over-Constrained Optimization Cases", ASME Paper, No. 85-DET-70, New York.

Williams, H.P., 1985a. Model Building in Mathematical Programming, (2d ed.), Wiley, New York.

Williams, H.P., 1985b. "Model Building in Linear and Integer Programming", in K. Schittkowski, (ed.), 1985a, pp. 25-53.

Winston, P.H., 1984. Artificial Intelligence, (2d ed.), Addison-Wesley, Reading, Massachusetts.

Winston, P.H., and Horn, B.K.P., 1981. LISP, Addison-Wesley, Reading, Massachusetts.

PART VII

INTEGRATED CAD/FEM/OPTIMIZATION TECHNIQUES AND APPLICATIONS

CAD TECHNOLOGY IN OPTIMAL DESIGN

Pierre BECKERS
Associate Professor

University of Liege, Belgium

1. INTRODUCTION

The insertion of CAD technology in optimal design is strongly related to the more general problem of integration.

The evolution of computer hardware and software is now giving to the analyst an opportunity to integrate the design and analysis tasks which are necessary to solve engineering problems. With such an integrated tool a company should be able to produce very quickly and at a lower cost, the best manufactured goods required by the community.

In this paper we will consider optimization as a part of the analysis task, but with the peculiarity that it is performed to improve the design of new products.

Initially, the main purpose of the analysis task was only to certificate products. Later, it was also to get a better understanding of their behavior and indirectly, to produce subsequent better designs.Nowadays, the analysis task in its optimization phase is able to give a direct feedback on the design. This fact give rise to a need for more integrated systems and more reliable tools for numerical simulations [1].

2. COMPARISON BETWEEN CAD SYSTEMS AND ANALYSIS PROGRAMS

Generally CAD technology mainly deals with the aspects of geometric definition of objects, the first objective being the manufacturing of these objects. Consequently, the most visible and spectacular characteristic of CAD systems is concerned with the graphical capabilities of such systems.

On the other hand, analysis programs generally contain a very poor graphical interface, because they are more concerned with numerical results of specific simulations, which often do not require a geometrical definition of the object, but a description of its behavior.

NATO ASI Series, Vol. F27
Computer Aided Optimal Design: Structural and Mechanical Systems. Edited by C. A. Mota Soares

Another characteristic of CAD systems is that they include very powerfull and very well designed user interfaces. The user interface is the part of the program which deals with all the communication aspects between user and program : in general it covers all the aspects of handling input and output data. This is the second difference between analysis programs and CAD systems : the latter include user friendly interface and are able to run interactively. On the other hand, analysis programs are generally difficult to use, input data are not easy to enter and output are mainly generated on listings.

The third difference between the two systems is concerned with data organization or data base. In CAD systems very sophisticated and efficient data management systems does exist while in the analysis programs more effort was devoted to the development of the algorithms. In the first CAD systems, the algorithmic part was quite reduced. Also it must be noticed that CAD systems generally come from the industrial environment while analysis programs come from the university community. For the first, the major constraints are efficiency, ease of use and good presentation, while for the latter the problems of choice of method and of algorithmic solution were the main concern of the developers.

But now the situation is evolving very quickly, because surface representation and solid modeling both require tremendous mathematical and algorithmic capabilities to be included in CAD systems. On the other hand analysis programs, and specially finite elements programs, are now completly mature and stable, the size of the problems increased very strongly and the profile of the users changed in such a way that they want to use black box systems for a large quantity of applications. Efficiency of data base management system and high quality of user interface are new requirements for analysis programs.

In conclusion the two aspects of design need the same type of tools. The demand for integration and for uniform computer handling, with as uniform and homogeneous command language as possible, must be satisfied.The increasing size of the applications and the complexity of the processed models also require an extensive use of graphical capabilities in the analysis step.

For these reasons, the following presentation will be divided in two parts :

- graphical aspects, problem of viewing,
- modeling aspects, some mathematical models of an object must travel from CAD world to Analysis world and must be adequatly transformed to meet the specific requirements of each step of the design process.

3. VIEWING: GRAPHICAL ASPECTS

3.1 Information display

The main feature arguing for the use of graphical displays is their effectiveness in communicating information. The power of the graphical displays lies not only in the speed with which they can generate images, but also in their flexibility and their capability to present the same data in a variety of ways [2,3].

This section discusses the techniques that can lead to a more effective display of informations; it is divided in two parts: first a short analysis of the properties of images in general, and then a presentation of the principal tools which are necessary to obtain an appropriate representation of three-dimensionnal objects on two-dimensionnal devices.

3.1.1. The properties of the image

Many works have been done on the aspect of psychology of perception. Some analyses give interesting results but they didn't got strong impact on the current developments in computer graphics because it is not possible at this time to deduce precise rules in order to effectively help the design of user interfaces.

On the other hand a pioneer and very usefull work has been done by Bertin on graphic design [4]. Bertin analyzes graphics as a medium of communication. He tries to define the characteristics of the image and to use them in order to obtain better graphics. His field of activities is mainly concerned with geographic and business environment, but the rules can be easily extrapolated to similar types of graphics in the technical world.

Basic elements of the image:

There are basically two different styles of drawings. The first is obtained by the technique of draftsmen. A picture consists only of lines drawn from point to point. The line is the simplest element of the drawing and any other thing that appears on the sketch is considered as an assembly of lines. Hardwares associated to this type of drawing include the vector displays and the pen plotters.

The second style is obtained with the technique of the painters which use a brush and fill areas in order to produce a picture. So the basic element of the picture is the area. The smallest area that can be displayed on the screen is called a PIXEL, short for picture element. The pixel surface depends of the resolution of the screen. Hardwares associated to this style include raster scan displays, similar to ordinary television sets, and electrostatic plotters.

It should be noticed that, with line drawing based systems, it is also possible to produce dots (very short lines) and to fill areas (with dots or by hatching techniques). On the other hand, with areas filling based systems, it is possible to draw lines, developing some linear interpolation schemes.

As a consequence, both systems can be used in any situation , but for a particular application, one system can be better suited than the other or can be more grateful for the user. The limiting case of both line and area systems is the point. For convenience we will say that any picture is composed of points, lines and areas. Below, we examine what are the attributes of these basic elements.

Bertin's eight variables of the image

It is possible to assign eight attributes or variables to the three entities defined above [4]. The first two characteristics of points, lines or areas are related to their location on the sheet. Two variables are necessary to define the position: X and Y coordinates.With some low cost devices, the only possibility to build an image is to define a collection of points (black or white pixels). No other attribute is available to define the image.

Two new variables can be used to represent the third dimension: size or value. Size is obtained by increasing the thickness of a line or the dimension of a point. Value refers to the possibilities of halftoning if they are available: for areas, it consists in varying the number of white dots on a black background. This is generally achieved by defining patterns with different numbers of black and white dots on monochromatic devices. These two attributes (size and value) are ordered, it means that the visibility increases with the size or with the value and consequently that the human brain intuitively associates a higher value to a higher visibility. A good example of this behavior can be found in the representation of oceans in cartography. The increasing depth of the sea is associated with darker blue, the small depth with light blue. Another property of these two variables is that size is quantitative while value is only qualitative. So, in order to evaluate the magnitude with the value variable, an additionnal scale of magnitudes must be associated to the picture. The use of the size variable is very common both in geographic and technical environment. Indeed, a set of variables of different possible values located in precise points of a map will generally be represented by points (circle or any symbol) of variable size (ex: country map with representation of cities).

The last four variables are: texture, form, orientation and color. They are defined as "separation" variables because their unique goal is to separate different constitutive elements of the picture. They can be implemented on points, lines and areas, but their ability to meet the criterion of separability will vary considerably according to the support.For areas and lines, the color variable induces the best separability. A well known and spectacular example is the artificial coloring of satellite photographs. The only problem occuring with color is the difficulty and the cost of documents reproduction. For points, the orientation variable is more suited (field velocities, etc...). The form variable is mainly used in business graphics to represent different types of points lying on curves.

3.1.2. The production of pictures

The most important characteristic of a computer graphic display environment is its two dimensionnality which requires the transformation from the three-dimensionnal scene to a two-dimensionnal image.

As a consequence two problems are to be solved:
-first : to correctly position the object in space and to project it on a plane. This is the general problem of geometrical transformations [2].
-second : to get more realism or to simply obtain a good visualization of the object. It can be achieved by detecting the visible parts of the object that must be displayed on the screen or reproduced on the paper sheet. Other techniques such as shading can be used to improve the visual rendering.

Geometrical transformations:

The homogeneous coordinates introduced by Maxwell in geometry and by Roberts in graphics allow to define a general transformation matrix able to handle and combine all the geometrical transformations. The principle of homogeneous coordinates is very simple, it consists in adding to the cartesian coordinates a supplementary coordinate that can scale the preceding ones. In a two dimensional space the x and y coordinates will be replaced by the three homogeneous coordinates hx, hy, and h, in such a way that :

$$x=hx/h \quad y=hy/h$$

With such a definition, cartesian coordinates (.25,.5) of a point can be defined in an homogeneous form either by (.25,.5,1) , or by (1,2,4) , etc.. In homogeneous form, the definition of a point is not univoque.

The principal advantage of this formulation is that all the rigid body motions can be defined as a matrix product. In fact, the transformation matrix can include rotations, translations and differencial scaling, this last one being able to perform symmetry operations. As all the operators are defined as matrix products, they can be combined to form more complex operators.

Note that with such a technique the spatial position of a part of an object can be updated without changing any coordinate of the object, by simply updating the transformation matrix attached to this object. If objects have to be displayed and if the terminal has some processing and storage capacity, moreover if we assume that the terminal has already stored the coordinates of the points, then it is sufficient to send the transformation matrix in order to modify the position of the object.The consequence is a very economic way to define and execute transformations.

Moreover this matrix can contain all the planar projections. Let us assume that coordinates are defined as line matrices, initially [P], and [P'] after transformation, and that the matrix product is defined as follows:

$$[P'] = [P] [T],$$

where [T] is the transformation matrix.

With this definition, removing one column of the matrix suppresses one dimension from the space and produce in a quite natural and very simple way, the orthographic projections in Y-Z, X-Z and X-Y planes, according to the position of the removed column.

Combinations of rotations and projections leads to axonometric projections (isometry, dimetry and trimetry). Oblique projections (cabinet or cavalier) are defined as a shear operator, while central projections are defined by introducing the inverse of the distance between observer and image plane in the fourth column of the matrix. This last projection give more realistic aspect and is quite good to simulate a camera image of the object.

<u>Visible surfaces detection</u>

According to Lucas [5], there are two types of algorithms to solve the problem of hidden parts removal.

With vector display hardwares, the purpose of the algorithms is to eliminate the hidden lines of the objects, while with raster scan based hardwares, it is more convenient to select a method of visible surface detection.

Many papers were published on this subject, and it is quite obvious that for each particular class of objects, it is possible to find a new algorithm that proves to be more efficient than the others. In fact it can take into account the peculiarities of the object in order to better use the internal coherence of the image [5,6].

Any comparison between algorithms is difficult because their performances are very sensitive with respect to the programming technique. Anyway, today, the subject is mature enough and we will very briefly present four families of algorithms as in [5].

The outline drawing algorithms are very close to the classical representations used by engineers, they give high quality images but remain quite abstract for non professional people. They include the classical Galimberti - Montanari algorithm which performs the comparison of all the edges of the object with all the faces. In its original form it assumes that the object is a solid defined as a polyhedron. This algorithm leads to an exact solution of the problem. It is quite expensive, well adapted to pen plotters and give high quality results.

The second family of algorithms or Warnock algorithm is based on a " divide and conquer " strategy. The whole screen is analyzed in order to know which types of polygon are present on the surface. If the situation is too complex the screen is subdivided in four parts and the process is repeated. The goal of the algorithm is to find a simple situation within the analyzed window : no polygon in the window, or one polygon, or one surrounding polygon eliminating those which are behind. When the resolution of the screen is reached some decision must be taken. As a consequence the quality of the image will depend on the resolution of the device and it is not so accurate as with the first method.

The next algorithms are based on filling area technique. Watkins algorithm is quite similar to the preceding one but it works on a single scan line. A cut of the 3-D scene is performed and the algorithm must detect the segments that can be displayed. If no segment or only one is present on a portion of the scan line, the situation is simple enough and the segment can be displayed with its corresponding color. Otherwise, the portion of scan line is divided in two parts and the process goes further. It is possible to save computation time using coherence property between two adjacent scan lines. This algorithm was the first implemented on hardware, it gives nice results but it is quite difficult to program.

At last, the Newell and Sancha algorithm is working as a painter does. Ordering all the polygons from background to foreground and painting them in this order can give the correct result. The only difficulty is to choose a criterion to sort the polygons. If the sizes of the different polygons are important and if the polygons are not convex, the so defined algorithm will fail. Refinements are to be included to solve conflicting situations. Easy to program and very fast, the simple sort of polygons is often used with finite element models and give generally good results because finite elements are small convex polygons.

3.1.3 To improve visual rendering

Two approaches are followed for improvement of visual rendering, they correspond to different disciplines.

In the first case, emphasis is laid on a better quality of the image, the challenge is to obtain a "true photograph". Improvement of quality is obtained by adding more texture, using shading techniques, and taking into account transparency and light reflection.

The other approach, mainly represented by flight simulators design, consists in trying to get real time visualization. In the simplest case, it is sufficient to move objects in rigid body motions. A more difficult problem covers the field of animation of deformable objects. Current applications concern vibration modes or static modes simulation in mechanics. This type of animation requires a lot of resources, powerful processor and large data storage, because the description of each configuration needs a large amount of paremeters (currently a finite element model contains more than ten thousand degrees of freedom). Simplified forms of representation have to be found. Presently, with the hardware available today, wireframe models of complex structures are easily animated, especially when the relevant edges of the model are extracted to simplify the representation.

3.2. Other aspects of graphical techniques

3.2.1. Design of the user interface

The user interface is the part of the program which handles interactions between the program and the user. It is now common to present user friendly interfaces: that means that user can very easily drive programs with the help of easy to use devices: mouses, tablets, etc., and with the help of well designed command languages including menus, etc..

Ergonomy is the speciality dealing with the facilities to pilot an application. It includes many aspects of workstations design: physical aspects and software aspects.

A careful summary of the aspects of user interface design can be found in [3], from a software point of view. The author points out four principal components of the user interface:

- user's model,
- command language,
- information display and
- feedback.

Information display characteristics has been presented in section 3.1. Nevertheless, other components also have a strong impact on the quality of the program. Specially, in optimization problems, the user's model which is the model the user has in mind to understand what he can ask to the program, will hardly influence the results of computations. For example, if well designed, a sensitivity analysis is an excellent tool to understand the behavior of the mathematical model and hence to get a better design.

Command language design is also important, because the user must manage several programs: Cad program, analysis programs, optimizer, etc... It should be more efficient for him to work with homogeneous and consistent command languages. This remark justifies a posteriori the search for an integration of all the programs used in the same company or at least in the same office. In this way integration will also facilitate communication and, as we hope, avoid misundertandings and errors.

Finally, the feedback exhibits the reaction of the program to the commands. A carefull design of this part will improve the efficiency of interactions, reduce the cost of process and decrease the response time. A compromise between, on one hand, hardware and developement cost and on the other hand, CPU time and velocity of interaction, has to be found.

3.2.2. Data base management aspects

The data base management system has to ensure the integrity and the recoverability of data throughout the full process of design and analysis. Efficient data management is essential, due to the quantity and the complexity of data that must be created, modified, processed and updated.

Conventional systems are generally unsuitable because in CAD environment the number of items is too high and the size of data that must be stored is too enormous. Another difficulty is that heterogeneous systems are often linked together. As a consequence, some bridges are to be build to ensure communication between systems.

In the restricted context of graphics, some standardization is also necessary to be able to transmit pictures or informations between different types of devices and programs. Some effort has been done and yielded concrete results. The most important norm in graphics is GKS, for Graphics Kernel System. This norm describes specifications for programming, designing graphic software and building files called metafiles whose characteristics allow their transfer from one workstation to any other, or from one computer center to another.

This standard and other ones are coming from computer graphic community, but other standard exist in the context of CAD. In certain opportunities they can also play the role of picture carriers. As they are able to understand and reproduce full geometrical models, they are not necessarily the best suited and the most economic vectors for communicating simple pictures.

Graphics standardization is important in the context of office automation and electronic publication, because graphical documents should be inserted in reports or should be available to produce presentation documents such as slides, videotapes, etc..

4. MODELING

The purpose of modeling is to create some representation of an object in order to manufacture it or to execute some simulation of its behavior. In fact we should distingish between mathematical model and numerical model. Here we are more concerned with the numerical model, but the mathematical model is always present behind the numerical model. Sometimes, mathematical and numerical models are very close to each other, sometimes, they are very different due to the discretization, or due to the different approximations introduced in the numerical model.

The same object will have several representations. Indeed, for the same object, we need a pure geometrical model and, in the present case, we also need a special model for manufacturing the object and a quite different model for analyzing it. The relevant attributes are different depending on the utilization of the model, but it is important that the fundamental definition of geometry should be identical. So it will be necessary to distinguish between universal parameters defining the geometry and specific parameters which are only useful for a given application. Anybody must have the ability to access the general or common parameter. Specific parameters will only be accessed by people which are concerned with their use.

4.1. Design models

This section mainly deals with the following subject : what are the different types of geometric models used in the design environnement and how to build these models ?

4.1.1. Traditional system

The first modeling techniques come from descriptive geometry techniques. Those techniques were extensively developed in the past century and are now well known. The theoretical base can be found in Monge geometry. In the industrial period this system led to the technical drafting techniques.

The basic elements of the model are generally circular arcs and lines. The draftsman must specify all the dimensions of the objects in each of the orthographic projections.

The three classical projections are most often sufficient to completely describe the object, but in some cases it is usefull or necessary to add some informations on details or on sections.

In the first CAD systems this technique was simply automated with some interesting improvements: possibility to automatically build symmetrical parts or to reproduce some details. So the process was faster and more accurate, compared to the manual work.

It is important to notice that, even with its weaknesses, this system proved to be very reliable and most of the objects of our actual environment were designed in this context.

4.1.2. Principal families of models

Generally, engineering activities are concerned with three types of models. For these models the basic constitutive elements are lines or truss members, surfaces or shells and solids or polyhedra.

Each of these families exhibits special properties and needs specific techniques for design and analysis. In fact the first steps in CAD were achieved with models composed of one dimensional elements, as well in analysis with the so called "matrix structural analysis", as in design with the wireframe models. The first optimization models were also concerned with 1-D elements as in the famous 5 bars, 72 bars trusses,etc.. Later, the finite element method introduced the concept of polygons with flat membrane elements, followed by plate elements as Clough-Tocher elements.

At the same time people mainly involved in car bodies and aircraft design found out the technique of sculptured surfaces. This tendency was emphasized when Coons, Bezier, Gordon, Ferguson and other pioneers replaced the lofting and clay modeling techniques traditionally used in the design of multicurved objects with mathematical techniques, defining curves and surfaces representable in computers. Very complex structures can be defined with this type of representation.

Finally with the increase of computer power and the introduction of super computers, full 3-D solid models are now accessible and their analysis is particularly important in mechanical engineering because most of the mechanical components are intrinsically 3-D objects.

Examples of these types of models are easy to find in the engineering environnement: towers, lattice masts, bridges, bulding frameworks, etc. are examples of one-dimensional elements, car bodies, ship's hulls, aircrafts, turbine blades etc. are examples of surfaces, and any mechanical component such as crankshaft is an example of solid.

4.1.3. Solid model construction

There are six major categories of methods for constructing solid models : wireframe representations, boundary representations, constructive solid geometry, also called building - block approach, sweep representations, instances or parametrized shapes and cell decomposition (including spatial occupancy enumeration) [7,8].

I) Wireframe representations

This technique is a generalization of the traditional method for producing drawings. Extended in three dimensions the drafting techniques allow to define objects by showing all their edges. The most achieved systems called 2.5-D systems allow to define directly the object in 3-D. However the description of an object is limited to the same points, line segments and curve elements as in the 2D case, but the possibility to use three-dimensional translations and rotations supplies a greater illusion of solidity.

The weaknesses of these models are well known :
- possibility of creating non sense objects ;
- ambiguity of the models ;
- no guarantee of uniqueness ;
- lack of visual coherence (silhouettes are not included in the model) ;
- long and tedious construction of the model ;
- impossibility to perform volumetric analysis.

In general this type of representation can be very effective in 2-D problems but becomes generally impracticable in complex 3-D problems. Wireframe model are not well suited to represent other shapes than polyhedra, such as cylinders, etc..

II) Sweep representations

These models follow a very simple and natural scheme. The basic idea is the following : to produce a curve, it is sufficient to define a point and to move it along a trajectory. It consists in the parametric definition of a curve, where the parameter can be assimilated to the time variable.

Generalizing this technique it is possible to move some curve along a trajectory to produce a surface and to move a surface in order to produce a volume.

The trajectory can be a line, an arc or another curve. This technique is particularly interesting to define axisymmetric objects and it is used in CAD systems as well as in finite element analysis.

The main quality of this technique is its efficiency and its conciseness, but in some cases, it is not obvious to check the validity of the model. Many of the present CAD systems offer sweep representations capabilities.

III) Instances or parametrized shapes

It is a very common and useful way to define objects. Indeed mechanical parts can completely be described with a small number of parameters. These parameters correspond to all the dimensions of the object. So it is possible to select the dimensions that may vary and to choose them as parameters. The definition of the object may include all the necessary tests to check the validity of the definition. This method is very efficient to define families of objects. A related technique, called group technology, was developed in concert with CAM techniques in order to promote standardization in part design and production.

This type of model is unquestionably concise, it is easy to validate and use and it theoretically allows to produce any type of shape. In practice, modeling systems based on this only type of representation are highly specialized and require a very large repertoire of generic primitives to deal with more general application.

Finally, let us notice that the parameters of a primitive or some of them can be identical to the design variables used in optimization. In this model the definition of design variable should be straightforward.

IV) Cell decomposition

This technique is regularly used in structural analysis. It is the basis for finite element modeling.

Any solid can be divided into a set of cells or elements whose union or sum supplies a representation of the solid itself. The advantage of this method is that any solid can be represented or approximated by a sufficient number of cells of very simple shapes. There are many ways of decomposing a solid, none is unique but all are unambiguous. In principle, it is possible to represent any object, using a reduced library of simple primitives (tetrahedra and/or hexahedra).

Simple shapes will be chosen in order to introduce simple mathematical models in analysis programs or to get simple algorithms for representation in CAD. The decomposition of the surface of a body in simple polygons (i.e. triangles) allows to use simple and fast algorithms for the representation. Two drawbacks may appear with this kind of technique. First it is not always obvious to decompose the object (it is the function of mesh generators). Second, an important amount of data is necessary to describe the object and any change in the model implies a rerun of the mesh generation process.

Spatial occupancy enumeration is a special case of cell decomposition, where cells are cubical in shape and located in a fixed spatial grid. With decreasing size of the cells, the representation approaches a solid as a set of contiguous points in space. One way to represent a solid is simply by listing the coordinates of the centers of cells. As the grid is fixed, the coordinates can be integer numbers, and the cell size defines the maximum resolution of the model.

In this kind of model, it is easy to access a given point and the spatial uniqueness is assured. But there is no explicit relation between the different parts of an object, and a large amount of data storage is required. Quadtree (in 2 dimensions) and octree (in 3-D) techniques provide a way of using more efficiently the spatial occupancy enumeration.

Octree technique of representation of 3-D objects is based on the recursive subdivision of a cube into 8 equal cubes. Each node of the tree represent a cube of the space. If any subcube of the cube is empty or full, it does not need further subdivision, otherwise, it is subdivided into eight cubes which are examined. The process is stopped either when all the cubes are totally full or empty, or when the assigned resolution is reached. In this case the partially full or empty cubes will be declared full or empty according to a convention. The leaf nodes of the tree have a standard size and position related to powers of 2.

All computations on these models are based on integer arithmetic. Algorithms that translate, rotate and scale octree models, that combine them using Boolean operators and that compute geometric properties are now available. Octree representations are also used as intermediate steps to produce finite elements meshes from other types of representations.

V) Constructive solid geometry

Constructive Solid Geometry (CSG) is a term used for modeling methods that definite complex solids as compositions of simpler ones (primitives).

Most of the concepts of CSG, were first introduced by Voelcker, Requicha and others on the Production Automation Project at the University of Rochester.

CSG can be viewed as a generalization of cell decomposition. In the latter case individual cells are combined using a gluing operation which is a limited form of the union operator. Boolean operators (union, intersection and difference) are used to realize the composition.

CSG representations of objects are ordered binary trees whose leaves or terminal nodes are primitives or geometrical transformations (rigid body motions or scaling) and whose internal nodes are boolean operators.

VI) Boundary representation

In boundary representation (B-rep.), the object is described by the surrounding surfaces of the solid. It is always present both in the design part and in the analysis part, to achieve visualization. The polygonal scheme is a particular simplified form of B-rep. In the analysis step, the B-rep. model is also necessary for representatión purposes. Another reason to build a B-rep. model in the analysis step, is that the design variables used in the shape optimization step, are controlling the outside surface. There is no particular difficulty to transform any model into a B-rep. model exept for the wireframe model. This transformation is easy to perform in 2-D but it is quite difficult to perform in 3-D, because the meaning of the lines is not unique, as they may represent edges or planes, or silhouette lines. Some usefull work has be done by Markowky and Wesley with the purpose of transforming old drafting into complete volume models [9,10]. The interest of their work can also be to check the validity of wireframe models.

Some caution must be taken during the transformation into B-rep.. An interesting approach is to first define an object in CSG, and then to build the associated finite element model. Let us notice that this transformation requires us to discretize the geometric model in some way because some F.E. models may involve only linear or bilinear surfaces. So the B-rep. model extracted from the F.E. model does not match anymore with the B-rep. extracted from the original model. This example shows the necessity to always refer to the original geometrical model.

4.1.4. Sculptured surfaces technique

A very complete and useful survey of the principles involved in this technique can be found in [11,12]. The different characteristics of the methods can be summarized as follows with the simple example of a circular arc whose radius is equal to the unity and whose center is at the origin of axes.

The first characteristic of the methods is to avoid the following implicit form:

$$x^2 + y^2 = 1$$

because, to compute points, it needs to solve non linear equation.

The explicit form will also be discarded because it is axis dependent and doesn't allow multi-valued fonctions:

$$y = \sqrt{(1 - x^2)}$$

The best suited form is the parametrized form; for a circular arc, it may be written :

$$x = \cos t, \; y = \sin t$$

where t is the parameter.

It would be preferable to use polynomials form to get more flexibility and more general treatment. In the case of the circular arc we can obtain a polynomial form, but it is no longer an exact solution. An example of approximation is given in [3] with a third degree polynomial

$$x(t) = 0.43\, t^3 - 1.466\, t^2 + 0.036\, t + 1$$

$$y(t) = -\,0.43\, t^3 - 0.177\, t^2 + 1.607\, t$$

Let us notice that if we want to return to an exact form, it is possible to use the rational parametric definition.

Using the homogeneous coordinates (hx, hy, h) presented in section 3.1.2, it is possible to define a circular arc with polynomials:

$$hx = 1 - t^2$$

$$hy = 2t$$

$$h = 1 + t^2$$

It is easy to see that x and y lie on the circular arc.

Let us notice that third degree polynomials will be very suitable, because they correspond to the lowest degree curves defined in space that can include twist. Moreover it has been pointed out that most of the practical situations can be accurately represented with third degree polynomials.

Till now we used algebraic forms, but in the field of design, the coefficients of polynomials are very difficult to handle. Consequently, it is preferable to use geometrical representations of curves for design purposes. In the case of a linear segment, the difference is obvious:

algebraic form:

$$x = a + bt$$

$$y = c + dt$$

geometrical form:

$$x = x_1 (1 - t) + x_2$$

$$y = y_1 (1 - t) + y_2$$

where (x_1, y_1) and (x_2, y_2) define the end points of the line.

The advantage of this presentation is that the coefficients have a physical meaning and that they can be directly defined in the model. Secondly, the limited segments can be directly computed using the last definition, because all the lines are bounded in design problems. Generally, the domain of variation of the parameter t is comprised between 0 and 1. Knots control or nodes control also help to ensure versatility.

The preceding rule is also used in F.E in the so called isoparametric elements, where the same blending functions, also called shape functions, are used for both geometric definitions and displacement fields approximations.

Many types of controls by nodes can be imagined. The first one also used in finite element method is the Lagrange interpolation technique where each node lies on the curve. But it is well known that if the spacing of nodes is not uniform or if the number of nodes increases, the degree becomes too bigh and the representation has a tendancy to amplify, rather than to smooth any small irregularity in the shape outlined by a set of control points. So other types of controls are introduced to obtain a variation diminishing property. The first and most famous was introduced by Bezier. In this case the curve only passes through the first and the last from a given set of points. The main features of the Bezier curves are :

- the curve passes through the first and the last points;
- the end derivatives are controled by the end segments;
- the curve is warranted to lie within the convex hull of the control points that define it (convex hull property).

Thanks to the preceding property the Bezier curve never wildly oscillates away from its defining points and the behavior of the curve is very similar to that of the control polygon.

The only drawback from this definition is that the control is global. The more points we have to define the curve, the higher the degree of the polynomial.A definition by piecing curves together naturally follows from the preceding considerations. The most common spline techniques provide this convenience at the expense of local control. The B-Spline formulation avoids this problem by using a set of blending functions that only have local support.

B-Spline technique which is a generalization of Bezier technique, provides the advantages of both of them : local or global control, continuity of curves and variation diminishing property.

This type of curve definition is also essential in optimization because the design variables used in shape optimization, are the parameters which control the shape of the curve and so, those two types of variables are identical. The properties of blending functions are essential for design, but they are even more essential for optimization, because they have proved to be the best design variables.

The case of surfaces is fundamentally identical. The same parametric forms are used, but two sets of blending functions with 2 parameters are now to be defined. Some techniques similar to those used for curves definition will be applied. According to the rule chosen for mixing the generating curves, several families of surfaces may be defined. The most important methods are cartesian products and transfinite surfaces (analogous to the eight modes serendipity element of Irons).

4.1.5. Conversions between models

The problem of conversion between models is the major difficulty to be solved in order to achieve a full integration of the different steps of design, analysis and optimization [13]. The basic idea of this section is that more than one model are needed in the different steps of design, because some models are more adequate for the construction step, or for the manufacturing step, or for one or another type of analysis. So it is mandatory to be able to convert one model into another, saving the basic charateristics of the object in a reference, or absolute description.

From the remarks outlined above, it is now obvious that some models are primary models, i.e. models build only for the purpose of construct a representation: sweep method, CSG method, wireframe method. Other are terminal models, such as F.E decomposition and B-rep.

4.2. The analysis models

The most important analysis models are:
A/ F.E.A, we saw that this model is a special case of cell decomposition.

B/ BEM, boundary element methods or integral methods : they are particularly well suited for B-rep.. In those methods, we only have to subdivide the external surface, or the boundary of the model. The resulting model is similar to the polygonal scheme used for graphical purpose, but it must include special refinement in singular points, if it is to be used for analysis purposes.

C/ Finite difference schemes are build on a fixed spatial grid. Some analogy can be found with quadtree or octree subdivision method.

D/ Fourier decomposition does not match exactly the preceding definitions although it is generally classed among F.E methods. This type of method allows to analyse pure geometrically axisymmetric structures when submitted to non symmetric loads or boundary conditions. The geometrical model of this type of structure is particularly well suited to sweep representation. The finite element model should be extracted from CAD system before the sweeping process. But to visualize the (non axisymmetric) results, a complete 3-D model must be build and the results must be combined and "plugged" on the 3-D model. This example exhibits very well the need for an integration of the different CAD tools, because in this case some typical CAD primitives are necessary after analysis of the model.

4.3. Communication between CAD and Analysis

4.3.1. Concept of parametrized model.

How to connect design to optimization? As reported by many authors, the design variables must be defined at the design level and by designers, they must be present in the analysis program and in the optimization process exactly with the same definition as in the design phase.

Consequently, it is necessary to clearly identify these design variables and to examine if they must be present in the analysis model or not. This statement is valid as well for size optimization as for discrete or shape optimization.

In the two first cases, the concept of design variable is quite clear because the sections of bars and/or the thicknesses of membranes are typically entered in the finite analysis program as data independent of the modelization, which is essentialy related to the mesh definition. This assertion is true for most finite element programs.

In shape optimization however the situation is quite different because no explicit variable is present in the input data to define the shape of the model. Typically, in the finite element method, the geometry is only defined by the nodes positions. So, in the first shape optimization test, it was tried to optimize the position of nodes which led to dramatic consequences on the results.

Two reasons explain these unacceptable results : first it is not convenient to work on the F.E. geometry, because it is only an approximation of the exact geometry. Second, the nodes coordinates are too sensitive to small variations and there is no natural mean to smooth the solution.

Now it is obvious that the nodal positions of the finite element mesh must be entered as dependent variables with respect to the master variables controlling the shape in a more global form. The knots of Bezier or B-Spline will give a very suitable definition of the design variables and of the parameters controlling the behavior of the F.E. mesh. Consequently, the nodes of the F.E. model cannot be independent variables, but the geometry of the mesh must be parametrized in terms of knots, which is not usually the case in the classical F.E. programs.

Using the same technique as for isoparametric finite elements, it is quite easy to define the design elements governing the mesh in a certain region of the model. The boundaries of these design elements should be defined exactly in terms of boundaries of CAD models. These design models are assumed to describe exactly the shape of the object. However the finite elements of the mesh contained in one design element will always approximate the exact shape because the blending functions of finite elements are generally low degree polynomials.At this stage it is obvious that some mesh generator capability is indispensable to build the model [14].

4.3.2. Parametrized and procedural models

When written in a procedural form, the data of a program are formulated in a special language exactly as we write a program. The so called "procedure" does not only include data or variables, but also primitives and tests.Data are replaced by the statements of the procedure. It is assumed that the program actuating as an interpreter is able to handle these statements.

As an example CSG method acts as a procedure, because indicating the Boolean operators and the geometrical transformations, the model is not described explicitly but through different actions to be performed ; so, the program has to execute the statements and maybe to process a boundary evaluation.

In the same way a finite element model can be defined by indicating boundaries, master nodes and the rules to be applied to interpolate the nodes in order to build the F.E. mesh. As mentioned above, the program has to evaluate the model before to compute results.

This method will be more expensive but it ensures more flexibility to correctly manage the step of optimization, independently of the particular problems.

Using this system and applying the finite difference scheme, the optimization should be completely independent of the F.E.A. programs.

This means that parts of mesh generation capabilities should be also present in the optimizer.

4.3.3. The actual model of communication

The IGES (Initial Graphics Exchange Specification) standardized format is a norm proposed by sellers and Federal government agencies to help in the communications between CAD systems. This standard allows different systems to speak the same language. With this tool engineers from different disciplines can readily exchange data in a quick and accurate way. The technical communication between different departments (or even different companies) is improved [15,16].

The new version of IGES now covers the finite element modeling data, including definition of nodes and connecting elements, along with material properties. More applications including solid models, piping, and sculptured surfaces are also covered.

However, it is not the unique language for interfacing dissimilar systems. Like IGES, commercial programs are neutral file translators that function as an intermediate step between systems.

In fact IGES can be better viewed as a geometrical model specification. On the other side GKS metafile will be a pure graphical specification. But the difference between the two is not always very clear. We can also imagine GKS metafile as a low level norm, while IGES is a high level one.

5. CONCLUSIONS

The application of CAD technology in optimization exhibits three main aspects.

The first one deals with interactive computer graphics. It is present in CAD systems as well as in analysis programs. It is concerned by geometric transformations in 3-D space, by hidden-line and hidden-surface removal and by general graphic rendering of objects and results.

The second aspect is related to modeling. In this case the fundamental problem is the conversion of models. The geometric models from CAD systems have to be converted in order to be used by the analysis and optimization programs.

The third aspect is the problem of the identification and of the use of the design variables from the early step of design to the final step of optimization. The design variables are expected to be the parameters of each program which must deal with them. This aspect must be carefully taken into account by developers in the field of shape optimization.

REFERENCES

[1] ENCARNACAO J.L. and SCHLECHTENDAHL E.G.
"Computer aided design. Fundamentals and System Architectures"
Springer-Verlag, 1983.

[2] FOLEY J.D. and VAN DAM A.
"Fundamentals of Interactive Computer Graphics"
Addison-Wesley Publishing Co., Inc., Reading, Mass., 1982.

[3] NEWMAN W.M. and SPROULL R.F.
"Principles of Interactive Computer Graphics"
McGraw-Hill Book Company, New York, 1979.

[4] BERTIN J.
"La graphique et le traitement graphique de l'information"
Flammarion, 1977.

[5] LUCAS M. and GARDAN Y.
"Techniques graphiques interactives et C.A.O."
Hermes Publishing, 1983.

[6] SUTHERLAND I.E., SPROULL R.F. and SCHUMAKER R.A.
"A Characterization of ten Hidden-Surface Algorithms"
Comput. Surv., 6(1) : 1, March 1974.

[7] REQUICHA A.A. and VOELCKER H.B.
"Solid modeling : A Historical Summary and Contemporary Assessment"
IEEE Computer Graphics, March 1982.

[8] MORTENSON M.
"Geometric modeling"
John Wiley & Sons, Inc. 1985.

[9] MARKOWSKY G. and WESLEY M.A.
"Fleshing out Wire Frames"
IBM J. Research and Development, Vol. 24, No 5, Sept. 1980.

[10] WESLEY M.A. and MARKOWSKY G.
"Fleshing out Projections"
IBM J. Research and Development, Vol. 25, No 6, Nov. 1981.

[11] BOHM W., FARIN G. and KAHMANN J.
"A survey of curve and surface methods in CAGD"
Computer Aided Geometric Design, 1 (1984), 1-60.

[12] FAUX I.D. and PRATT M.J.
"Computational geometry for Design and Manufacture"
Ellis Horwood Ltd, 1979.

[13] MYERS W.
"An Industrial Perspective on Solid Modeling"
IEEE Computer graphics, March 1982.

[14] WORDENWEBER B.
"Finite element mesh generation"
Computer-aided Design, Vol. 16, No 5, Sept. 1984.

[15] KROUSE J.K.
"Closing the Gap in CAD/CAM Communications"
Machine Design, Sept. 20, 1984.

[16] ENDERLE G., KANSY K. and PFAFF G.
"Computer graphics Programming - GKS - The Graphics Standard"
Springer-Verlag, 1984.

COMPUTER AIDED OPTIMAL DESIGN OF ELASTIC STRUCTURES

Part I

Integration of Optimization Concepts within CAD and FEM Technologies

Abstract This Section concentrates on two major objectives that are currently pursued at the research level, and that should soon be ready for implementation in practical Computer Aided Engineering systems. The first objective is to develop a general approach to shape optimal design of elastic structures discretized by the Finite Element Method (FEM). The key idea is to employ geometric modeling concepts typical of the Computer Aided Design (CAD) technology, in order to produce sensitivity analysis results. These sensitivity data can then be used by an optimizer to generate an improved design.

The second goal is to implement an interactive redesign system that integrates optimization methods within a flexible and efficient computational tool, easy to use by design engineers. This interactive module is intended to create the missing link between FEM and CAD technologies and therefore it should constitute one of the key elements in the complex chain needed to computerize the design cycle.

The approach followed can be summarized as follows. First the behavior of the structure is analyzed by using the finite element method. Subsequently a sensitivity analysis is performed to evaluate the first derivatives of the structural response quantities. These derivatives are used by an efficient optimizer, which selects an improved design. A reanalysis of the modified design is next performed after updating the finite element mesh. This iterative process is repeated until convergence to an acceptable optimum design has been achieved, which usually requires less than 10 FEM analyses.

The long term objective is to create a coherent interactive system that makes the best possible use of the respective capabilities of the engineer and the computer.

COMPUTER AIDED OPTIMAL DESIGN OF ELASTIC STRUCTURES

C. Fleury
Mechanical, Aerospace and Nuclear
Engineering Department
University of California
Los Angeles, California 90024

I. Integration of Optimization Concepts within CAD and FEM Technologies

II. Convex Approximation Strategies in Structural Synthesis

III. Aplication of Structural Synthesis Techniques

Acknowledgement

Part of the reserch results presented in this work was jointly supported by the MacNeal-Schwendler Corporation and the University of California under the MICRO program (Grant No. 85-204).

NATO ASI Series, Vol. F27
Computer Aided Optimal Design: Structural and Mechanical Systems. Edited by C. A. Mota Soares

INTRODUCTION

The last decade has been a period of considerable extension in the role of computer-based structural design. This can be seen both from the development of Computer Aided Design (CAD) systems for mechanical applications, and from the wide acceptance of the Finite Element Method (FEM) as the prevailing engineering analysis tool. Unfortunately, because there still exists a considerable gap between the CAD and FEM technologies, the design cycle is not yet fully computerized.

Much of the development of CAD systems has emphasized man-machine communication through interactive computer graphics and the use of data-base technology for the storage of information relevant to design. The realizations of CAD in Mechanics are still essentially limited to the production of digitalized conventional drawings and their transmission to other users. The design aspect itself is not largely covered, except perhaps in the case of elementary or specialized components. The design of complex mechanical systems remains highly dependent upon structural analysis methods. Yet these methods, either conventional or computerized, are not much integrated in the CAD chain, and the redesign steps following these analyses are usually not automated. Although today CAD systems are being extended to interface various analysis programs, the reverse is not true, and the actions of the design engineer, based on the structural analysis results, are usually not reflected in the CAD data base.

The finite element method (FEM) is now the most widely accepted computational tool in engineering analysis. Many FEM systems, however, only permits analyzing a given structure, without providing any information about how to improve the design. Typically the analysis is performed by a large system of the NASTRAN type. After examining the analysis results the design engineer introduces what he considers are suitable changes to the structure in order to obtain a satisfactory design. This use of FEM analysis represents a rather limited application of the power of the computer.

Considerable research efforts are still needed to integrate CAD and FEM technologies. The integration scheme that we propose to develop is based on the introduction of an optimization loop that would fully computerize the design cycle. Indeed the ultimate purpose of any designer is to obtain an "optimum" design by gradually bringing suitable modifications to the structure. In the past, these redesign steps were mainly based upon insight and experience. With the recent advances in CAD and FEM technologies, more rational approaches have been introduced. Computer resources are now routinely utilized to process simultaneously the many data required to improve a design without violating constraints. However the design changes still rely on the designer skills, which are obviously limited to very few modifications at each redesign step. In fact, when more than two design variables are considered, it becomes difficult for the human mind to guess appropriate new values on the basis of the analysis results. This dimensionality barrier can be broken by resorting to optimization techniques, which permits the designer to deal with problems involving many design variables and constraints. To this end the design goals must be expressed into a mathematical optimization problem. Such an optimization problem consists in minimizing an objective function, which represents some cost associated with the structural system, subject to inequality constraints which insure the design feasibility. Provided that the FEM system is capable of generating sensitivity analysis results, furnishing the required gradient information, a numerical optimization method can be employed iteratively to select the best possible design.

A general sensitivity analysis capability built in a FEM system and its CAD type pre-processor is the key to fully computerize the design cycle. However the few commercially available FEM systems providing a sensitivity analysis module can only address sizing problems (e.g. determining optimal thicknesses of structural components). Shape optimal design is still in a state where fundamental research is needed and, therefore, it constitutes the driving topic in this Chapter. The integration of shape optimization concepts within FEM and CAD frameworks should help bridging the gap between these two technologies. To be successful the proposed integration of software should lead to a system easy to use. From a practical point of view, the computational tool should indeed be employed by design engineers possessing only a superficial knowledge of the theoretical bases of each technique.

In the sequel we shall try to define what should be the foundations of a powerful Computer Aided Design tool in the true sense of the wording:

this interactive tool must assist any design engineer in creating an "optimal" design, i.e. a minimum cost design that satisfies many various requirements. Two main directions of research will be discussed. On one hand a true integration of CAD and FEM technologies is proposed, that is based on the introduction of shape optimization concepts. This part involves fundamental investigation of new ideas: creation of a suitable design-oriented structural model, generation of the corresponding finite element mesh, and implementation of a general sensitivity analysis capability. On the other hand an interactive optimization module will be described. Such a module, similar to conventional pre-and post processing FEM systems, can help the user in preparing the design model and in interpreting the optimization results through extensive use of graphics displays. Innovative visualization features must be devised in order to express the characteristics of the optimum design problems in a meaningful way.

DESIGN ORIENTED STRUCTURAL MODEL

Professor Lucien Schmit, the father of modern structural optimization methods, has emphasized for many years the need to distinguish between the analysis model and the design model (see e.g. Ref. [1]). Quoting him, "it should be recognized that analysis modeling and design modeling involve two distinct but interrelated sets of judgement decisions" [2]. This sentence summarizes well the inherent differences between CAD (design model) and FEM (analysis model) systems, as well as the necessity to integrate these two technologies.

Because optimal sizing problems have previously been largely covered (see e.g. Ref. [3,4]), this paper is focused on shape optimal design problems. When dealing with such problems, the design variables must be selected very carefully. The coordinates of the boundary nodes of the finite element model is a straightforward choice (a common practice in early work on shape optimization, e.g. Ref. [5])). This choice exhibits however many severe drawbacks. The set of design variables is very large and the cost and difficulty of the minimization process increase. It has a tendency to generate unrealistic designs due to the independent node movement and additional constraints avoiding such designs are difficult to cope with. Moreover an automatic mesh generator is necessary to maintain the mesh integrity throughout the optimization process. One obvious remedy is to avoid a one-to-one correspondence between the finite element model and the design variables.

One of the ways to achieve this goal is to use the "design element" concept. In this approach the structure is decomposed into a few subregions of simple geometry. These subregions are described in a compact way by using a limited number of control nodes (or master nodes). Each region consists of several finite elements. During the optimization process the geometry of conveniently selected subregions is allowed to change: these regions are called design elements. The movements of the corresponding control nodes are the design variables. The concept was initially introduced in Ref.[6] where two-dimensional isoparametric finite element interpolation functions were used to describe the design element boundary. Recently blending functions commonly used in computer graphics for interactive generation of curves and surfaces (Bezier, B-splines) have been proposed to describe the boundaries [7,8]. This is also the approach followed in the present paper. The shape variables are thus the positions of the master nodes which control two families of curves, whose cartesian product defines the design element.

This formulation lends itself well to shape optimal design problems. The blending functions provide a large flexibility for the geometric description. With the B-spline formulation, boundary regularity requirements are automatically taken into account. In addition a few design elements are generally sufficient to fully describe the regions that are modified during the optimization process. A direct benefit is that the optimization problem involves a reasonable number of design variables. A second important advantage lies in the ability to control the validity of the finite element mesh. Indeed it is relatively straightforward to generate a suitable finite element mesh and to maintain its integrity throughout the optimization process. The design element can be mathematically defined as the cartesian product of two families of curves. This provides an analytical interpolation scheme, which permits determining the coordinates of any point (finite element node) inside the design element or on its boundaries. A regular mesh is initially constructed in the curvilinear coordinate system of the design element. Next coordinate transformations are applied to obtain the mesh in the real design element.

To help fix ideas let us consider the design problem shown in Fig. 1. It corresponds to a quarter of a plate under a central force. The goal is to determine the shape of the side BC which minimizes the weight of the structure, with an upper bound on the displacement at node A. The geometric model of the quarter plate is made up of a subregion around the side BC with changing geometry (design element) and a subregion containing the rest of the structure with fixed geometry. The side BC, which represents the moving boundary of the design element, is described by a B-spline of order 4 with 6 control nodes. The design element boundary in the other direction is represented by a B-spline of order 2 with 2 control nodes (linear). The design element is defined as the cartesian product of these two families of curves and has thus 12 control nodes. As illustrated in Fig. 1. these control nodes belong to one of three different categories: fixed nodes, moving nodes and internal nodes.

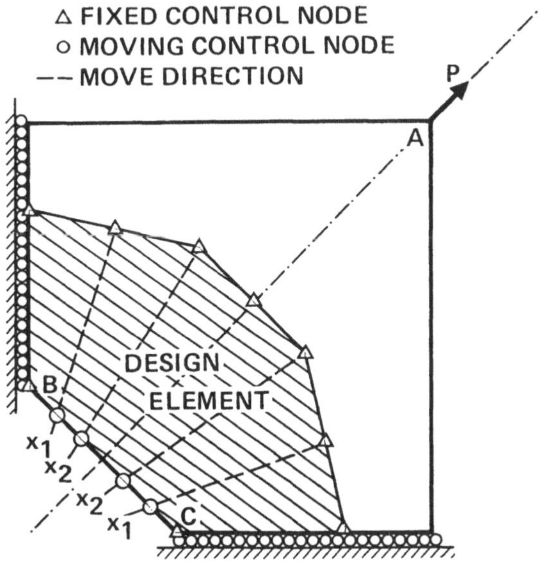

Fig. 1 Concept of Design Model

The unknowns of the problem are the positions of the 4 moving control nodes. These positions are determined by the distances from their respective fixed reference poles. Hence these 4 distances are the design variables of the optimization problem. Making use of the problem symmetry only two independent variables remain. The only design constraint is the displacement of the node A in the direction of the applied load. For practical reasons, this constraint can be replaced by an equivalent constraint on either the horizontal or the vertical displacement of node A, which are equal because of symmetry. In addition side constraints are specified for the design variables, preventing unreasonable large displacements of the moving control nodes. This leads to the following optimization problem statement:

$$\begin{aligned} & \min W(x_1,x_2) && \text{structural weight} \\ \text{s.t.} \quad & u_A(x_1,x_2) \leqslant \bar{u} && \text{displacement constraint} \\ & \underline{x}_1 \leqslant x_1 \leqslant \bar{x}_1 && \text{side constraints} \\ & \underline{x}_2 \leqslant x_2 \leqslant \bar{x}_2 && \end{aligned} \tag{1}$$

Before calling an optimization algorithm, a sensitivity analysis must be performed in order to evaluate the first derivatives of the weight W and the displacement u with respect to the design variables x_1 and x_2.

In the foregoing geometric representation the FEM mesh can be directly derived from the coordinates of the control nodes. This feature leads to the distinction between a design model and an analysis model. The design model is made up of the small number of design elements, whose geometry is determined by the control node positions, and of the fixed subregions. By entering a relatively small number of design elements, it is possible to create a compact design model that describes well the structure to be optimized. The analysis model is the finite element model, characterized by the node coordinates of the mesh, the types and material properties of the elements, the applied loads and boundary conditions, etc. The analysis model can directly be derived from the design model at any stage of the iterative optimization process, because of the adopted internal parametric representation. This feature considerably facilitates the task of implementing sensitivity analysis.

GENERAL SENSITIVITY ANALYSIS CAPABILITY

The purpose of sensitivity analysis is to evaluate the derivatives needed to solve structural optimization problems. The general strategy employed to get these derivatives is now well established (see e.g. [8]). Roughly speaking the equation solver in the FEM system must be modified to accommodate additional loading cases, often called pseudo-loads. The pseudo-loads depend upon the first derivatives of the stiffness matrix and consistent load vector with respect to the design variables. In the case of shape optimal design, the parametric modeling scheme described above provides an elegant way to perform the sensitivity analysis on the basis of finite difference techniques, following the so-called semi-analytical approach (see e.g. Ref. [6]). However it should be noted that an analytical formulation of the sensitivity derivatives can also be established [8].

The finite difference approach is introduced by perturbating each design variable by a small amount and regenerating the mesh for the modified structure. This first step is quite easy to accomplish, because the finite element mesh can be analytically derived from the geometric design model. Each element stiffness matrix must then be computed again as many times as the number of design variables. After assembling the regenerated element stiffness matrices, the global stiffness matrix of the perturbated structure is obtained. The pseudo-loads are then calculated as the difference between two vectors which result from the multiplication of the original displacement solution with respectively, the original assembled stiffness matrix, and the assembled stiffness matrix of the perturbated structure. The displacement derivatives can now be obtained by solving the equilibrium equations with these pseudo-loads as additional loading cases. The derivative of the structural weight can be approximated by the same finite difference technique. Finally the element stress derivatives are computed as follows. Using the original mesh and the displacement solution of the original structure, the stresses are evaluated at the Gauss points. This procedure can be repeated using the perturbated geometry and the approximated displacement solution for this modified structure. The first will give the exact values for the Gauss point stresses in the original structure, while the latter provides a good approximation for the values of the Gauss point stresses in the perturbated structure. The foregoing developments provides thus a means to compute the derivatives of the structural weight, of the displacements, and of the stresses by finite differencing.

The implementation of the foregoing technique is described in detail in Ref. [9]. It has been initially restricted to two-dimensional elastic structures in plane stress or plane strain, modeled with isoparametric eight-node elements. However it is important to emphasize that the same concepts can be employed to deal with more complicated design problems (e.g. solid models) or more sophisticated finite elements (e.g. three-dimensional structure involving plates and shells). In fact the finite difference approach presented in Ref. [9] is largely independent of the types of finite elements used in the analysis model, and therefore it could be readily implemented in any large scale general purpose FEM system. This is equally true for the fully analytical approach to shape sensitivity proposed in Ref. [8], also restricted to simple two-dimensional problems. However it would be an enormous task to generalize it to all the element types usually found in practical FEM systems.

The integrated CAD/FEM system of the future should be able to optimize virtually any structural design problem, provided that an appropriate data base is devised with optimization concepts in mind. This data base must contain all the information regarding the design oriented structural model, with a clear subdivision of data between design variables (to be modified in order to improve the design), and prescribed parameters (never affected by the optimization procedure). The FEM system should be capable of generating an analysis model from the analytic design model contained in the data base. A quite general sensitivity analysis capability is then easy to implement: a loop on the design variables permits creating a new analysis model for each small perturbation brought to the design model. The finite element matrices are computed for each perturbed model and their sensitivities to changes in the design variables can be approximated by a finite difference technique.

Although conceptually simple, the finite difference approach reveals very promising because of its generality and ease of implementation. It should be recognized, however, that coupling geometric modeling concepts and finite differencing suffers from some limitations. An obvious disadvantage is that the derivatives are not exact. However modern optimizers such as CONLIN (Section II) do not seem very sensitive to this potential lack of accuracy. A more serious drawback is that the method is computationally expensive. In the present stage of the study no special efforts were spent to increase the efficiency of the algorithm in this respect, as the primary concern was to check the validity of the approach. Further work is therefore planned to analyze how the CPU time is distributed over the different steps and to increase the computational efficiency. Major improvements could be gained by using advanced computational geometry techniques to speed up the derivation of the FEM mesh from the parametric design model.

As previously mentioned analytical derivatives for the shape optimal design problem can also be obtained, but in a very complex way, which is dependent on the type of finite element used. Introducing this analytical method in general purpose finite element packages, containing a vast library of element types, would require a huge development effort. In many cases it is obvious that the analytical method is much more efficient than the finite difference technique, simply because it does not require to generate again new element matrices. For sizing variables, the first derivatives of the stiffness matrices are easy to get explicitly. For shape design variables the analytical approach becomes much more difficult to implement, however it remains more efficient. Therefore a question that naturally arises is why would we employ an approximate and computationally expensive approach while we know that an exact and efficient method is available. Finally it is important to realize that the effort needed to develop an analytical sensitivity analysis could be spread over a few years. A reasonable approach would be to initially implement the finite difference scheme, and then to gradually replace it with analytical derivatives, starting with simple elements and design variables, and pursuing with more complex situations.

NUMERICAL OPTIMIZATION METHOD

The CONLIN optimizer described in Section II has been adopted to solve the numerical optimization problem. It is a specially well suited optimizer for structural optimization, based on a convex approximation scheme. The initial optimization problem is transformed in CONLIN into a sequence of explicit subproblems, which are solved in the dual space. The efficiency of this dual formulation is due to the fact that the dimensionality of the dual space is relatively low and depends on the number of active constraints at each design iteration. CONLIN advantages, which make it especially well suited to structural synthesis tasks, include:

- it does not demand a high level of accuracy for the sensitivity analysis results, because it is based on conservative approximation concepts allowing them to be obtained from finite difference techniques;

- it has an inherent tendency to produce a sequence of steadily improving feasible designs and usually generates the optimal design within less than 10 FEM analyses;

- it has a built-in constraint relaxation capability that allows the user to start from any infeasible initial design, and even to find a solution to infeasible problems (in the form of minimal relaxation);

- each CONLIN iteration is accomplished very rapidly, even for relatively large scale problems; this is an important feature within an interactive environment.

To illustrate the idea of optimizing a structural shape, let us return to problem (1), corresponding to the quarter plate example of Fig. 1. The design space corresponding to the first subproblem created by CONLIN is displayed in Fig. 2.a. The two axes represent the design variables x_1 and x_2. In this specific case the CONLIN optimizer generates a linear approximation of the structural weight, while the displacement constraint is linearized with respect to the reciprocals of the design variables. This procedure yields the following explicit approximate problem:

$$\begin{array}{lll} \min & W = 0.48\,x_1 + 0.52\,x_2 & \text{linear approximation} \\ \text{s.t.} & 0.65/x_1 + 0.16/x_2 \leqslant 1 & \text{reciprocal approximation} \\ & 0.05 \leqslant x_1 \leqslant 1.32 & \text{side constraints} \\ & 0.03 \leqslant x_2 \leqslant 1.28 & \end{array} \qquad (2)$$

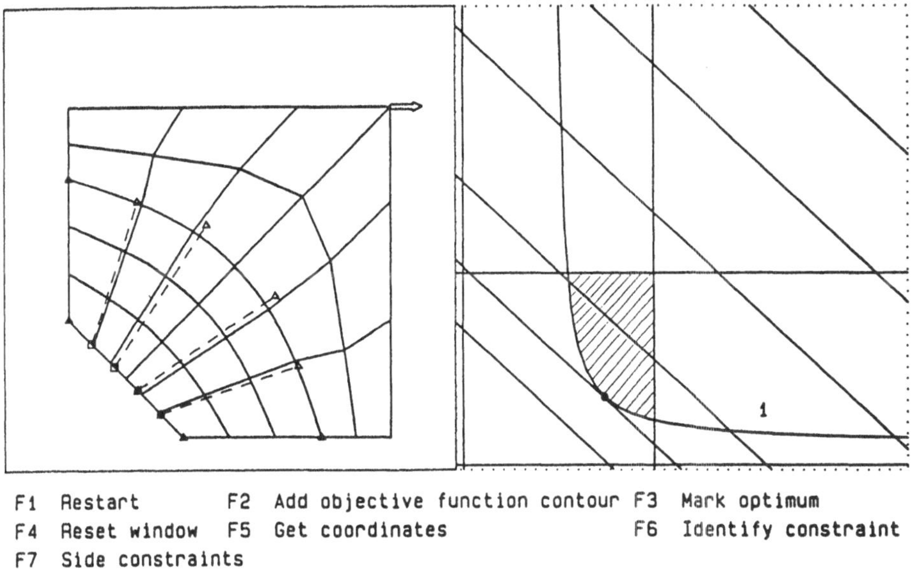

Figure 2a. Optimization Problem

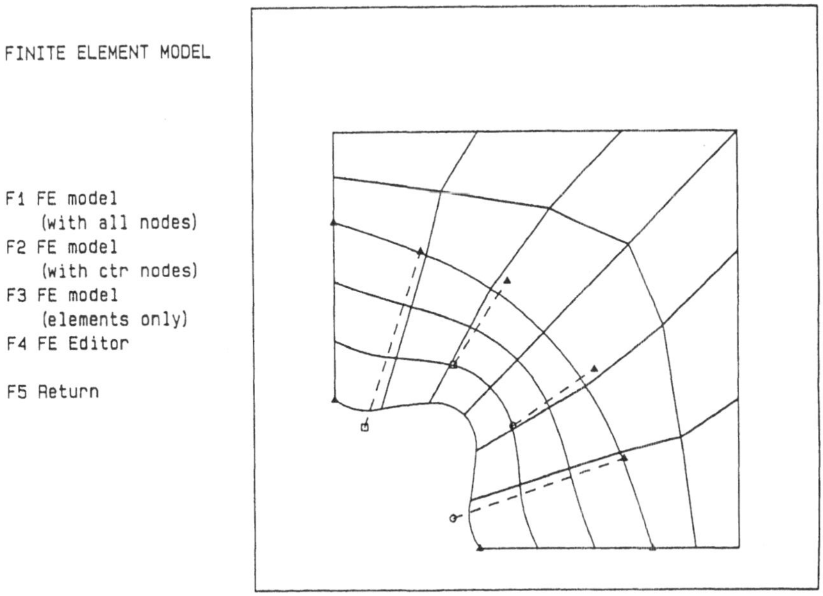

Figure 2b. Improved Shape

F1 OBJECTIVE FUNCTION

F2 DESIGNVARIABLE

F3 CONSTRAINTS

F4 RETURN

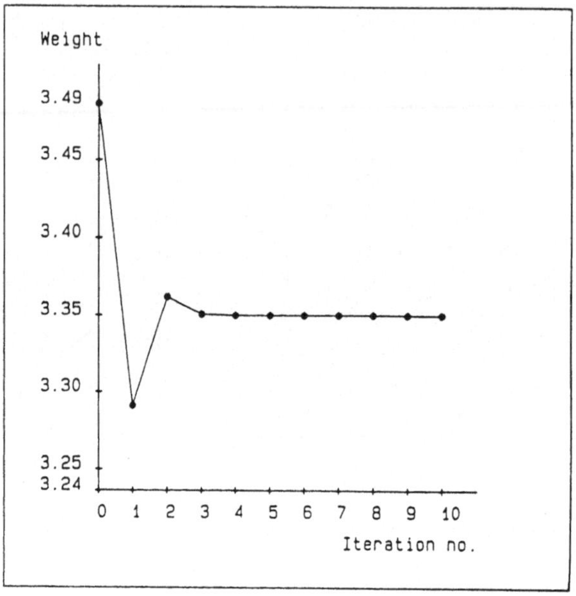

F1 OBJECTIVE FUNCTION

F2 DESIGNVARIABLE

F3 CONSTRAINTS

F4 RETURN

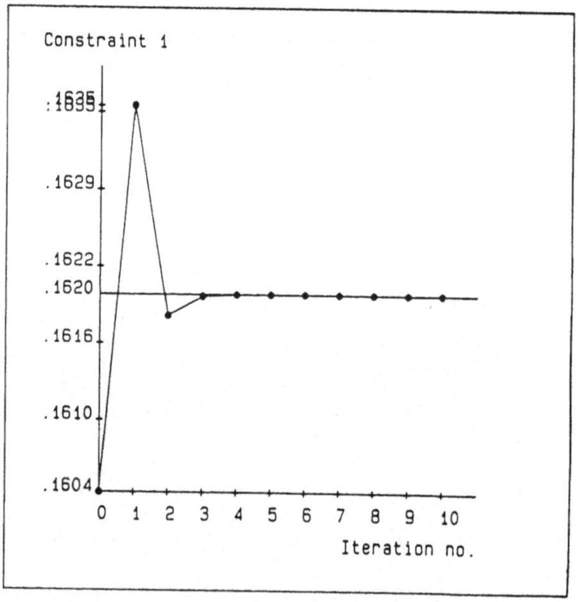

Figure 2c. Iteration History (CONLIN)

Hence the contours of the weight objective function are parallel straight lines while the constraint surface is drawn as a curve. Clearly the optimum lies at the point where this curve is tangent to a constant weight line ($x_1^* = 0.99$ and $x_2^* = 0.47$). This optimum point corresponds to a new design of the structure, i.e., the positions of the control points have been changed by the optimizer, leading to an improved shape of the side BC (see Fig. 2.b). From the modified design model, the finite element mesh is updated, the structure is reanalyzed, and the optimizer called again in order to still improve the design. This process is repeated until convergence is achieved to an optimal shape. The iteration history of the optimization process is summarized in Fig. 2.c. One can observe the quick convergence of the CONLIN optimizer.

INTERACTIVE DESIGN OPTIMIZATION MODULE

This Section deals with a general system for computer aided optimal design. Such a system should be distributed over various types of machines: main frame for FEM and design sensitivity analyses; engineering workstation for interactive optimization; PC for small problems arising in the preliminary design phase. The finite element analysis of a structure, as well as the associated sensitivity analysis, are time consuming tasks, which should be typically performed during the night in batch mode on a powerful computer. The task of redesigning on the basis of these results is typically an interactive job performed during the day on a graphics workstation. The designer has then the choice to accept the new design and send it back to the finite element optimization code for a new iteration, or to intervene in the optimization process and make some data adjustments. For example it is sometimes desirable to modify the design model, or to introduce additional geometric constraints such as tangency requirements. It is also quite interesting to examine intermediate analysis results and stop or correct the optimization procedure. Note however that performing several successive iterations should still be permitted, in order, for instance, to conclude an optimization process.

This scheme requires thus an interactive engineering design system, capable of displaying all the relevant information concerning the modified design. This interactive system can be viewed as a post-processing module, called when the design optimization process has been interrupted after a finite element analysis is completed (including the sensitivity analysis). The system should also contain pre-processing capabilities to allow the user to define the design and analysis data, and then to make adjustments as the optimization process continues. The optimizer should be contained both in the structural analysis module (for fully automatic redesign), and in its pre- and post-processing module (for interactive optimization). It is important to realize that a typical structural optimization job requires between 5 and 10 iterations. The 2 or 3 first iterations should be accomplished one by one, in a semi-interactive mode. In this way the designer can verify the results produced by the optimizer before starting a new finite element analysis. He has the opportunity to take some corrective actions (e.g. introduction of additional constraints) and to tune some control parameters (e.g. truncation factors related to constraint deletion). After these few interactive runs and restarts, the design problem is usually well posed and the automatic iterative mode can be switched on. This leads to a splitting of the tasks corresponding to the flow chart in Fig. 3.

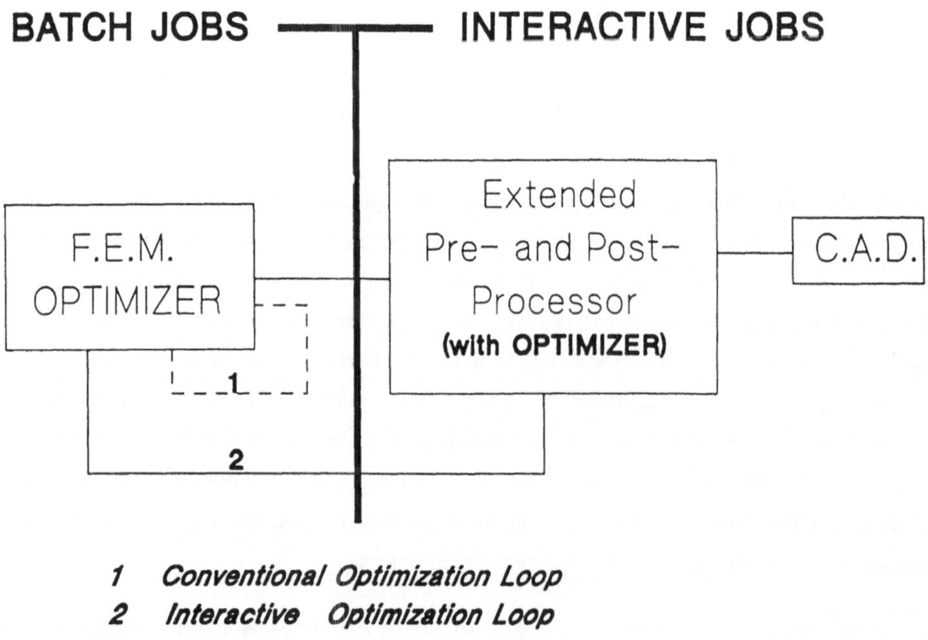

Fig. 3 Interactive Optimization System

As reliable optimization algorithms, such as CONLIN, are now available to solve shape optimal design problems, substantial development efforts must be devoted to implementing an appropriate user interface. The ultimate goal is to devise suitable tools for a designer to effectively incorporate optimization concepts into the real design cycle. Indeed an important aspect of finite element analysis and optimization capabilities should be their ability to really help the user in accelerating the design process. The interactive engineering design system of the future should be able to produce extensive graphical outputs, displaying in an expressive way meaningful results. These graphics display capabilities should be organized to be easy to use interactively. Industrial applications namely reveal that it is not generally possible to optimize a structure in one single computer job. Frequently it is necessary to improve the structural model, to modify the set of retained behavior constraints (because some constraints, initially not critical, become critical), or to relax some constraints if their maximum values are too severe and no feasible solution can be found.

At the present stage of the development efforts, the computer program largely functions as a post-processor. Is is envisaged that the pre-processing functions will be implemented in the near future. The system contains most of the conventional graphics displays usually found in pre- and post-processing programs: representation of finite element mesh, applied loads and boundary conditions; undeformed and deformed geometry plots; color-coding of elements based on specific stress components. In addition innovative graphics capabilities were implemented in connection with the new optimization concepts. The optimizer results can be immediately verified with the design model plot or analysis model plot of the modified geometry. The module can also produce evolution plots, representing the values of the objective function, design variables and constraints in terms of the number of iterations. Visualization of slices in the design space is possible: the user selects two significant design variables and the program plots the corresponding 2-D design space (contours of objective function, constraint surfaces defining the feasible domain, location of optimum design, etc...). These graphics displays are made on the basis of the convex linearization scheme, which constitutes an excellent explicit approximation.

The CONLIN optimizer can be supplied with the sensitivity data, and then called from the interactive optimization module. Note that this feature could be implemented because the CONLIN method is very fast, and it is simple enough to lead to a relatively small, well organized computer code. The foregoing functions allow the user to see the new shape generated after each iteration and they represent therefore a very valuable tool to verify the validity of the results produced by the optimizer. In addition, the user can interact with the system, by modifying the positions of some control nodes and examining the effect on the design model, as well as on the analysis model. This is exactly the way the batch program calculates the sensitivities through a finite difference scheme. It is also possible to produce simultaneously a plot of the design model and a plot of the design subspace. The user selects on the design model the design variables to form the design subspace. He has the opportunity of highlighting a behavior constraint both in the design model plot and in the design space plot. He can have the lower and upper limits on the design variables displayed both in the two plots.

APPLICATION: Optimization of a Hole in a Biaxial Stress Field

The foregoing optimization concepts have been implemented in a finite element system made up of two parts. The first module is used in batch mode to perform the structural analysis and its associated sensitivity analysis. The second one is an interactive optimum design system that uses the sensitivity coefficients produced by the first. This interactive module contains innovative graphics display capabilities that should considerably facilitate the task of a designer willing to optimize a structural shape. The purpose of this section it to demonstrate the effectiveness of the concepts presented in this paper. As previously mentioned these concepts are quite general and could be readily introduced into any large scale FEM system for the analysis and sensitivity analysis, and its CAD-type preprocessor for the creation of the geometric design model. However the current implementation is restricted to two-dimensional structures in plane stress or plane strain modeled with isoparametric eight-node elements. The example given below is not meant to represent a practical application. Rather it has been devised to illustrate the various functions of our interactive shape optimization system.

The second illustrative problem is concerned with the minimization of stress concentrations in a structural component. This example has been used

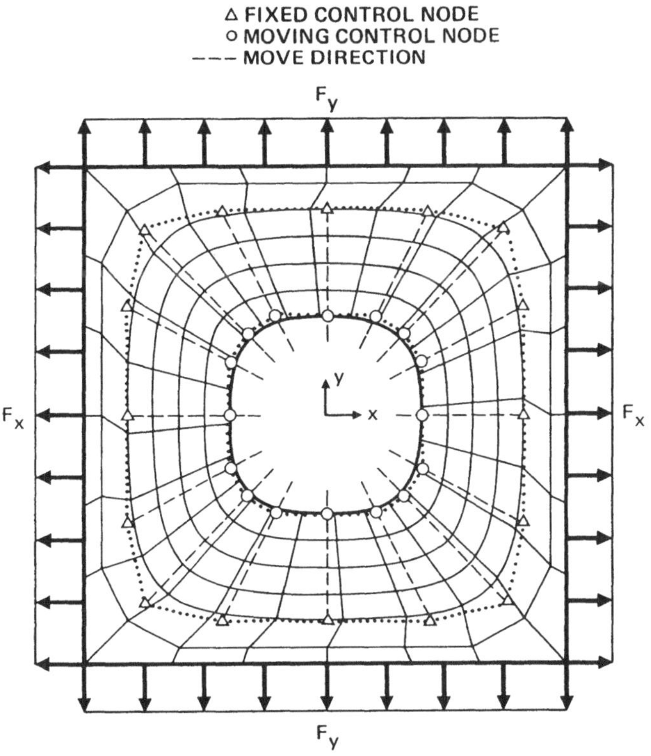

Figure 4a. Hole in Biaxial Stress Field

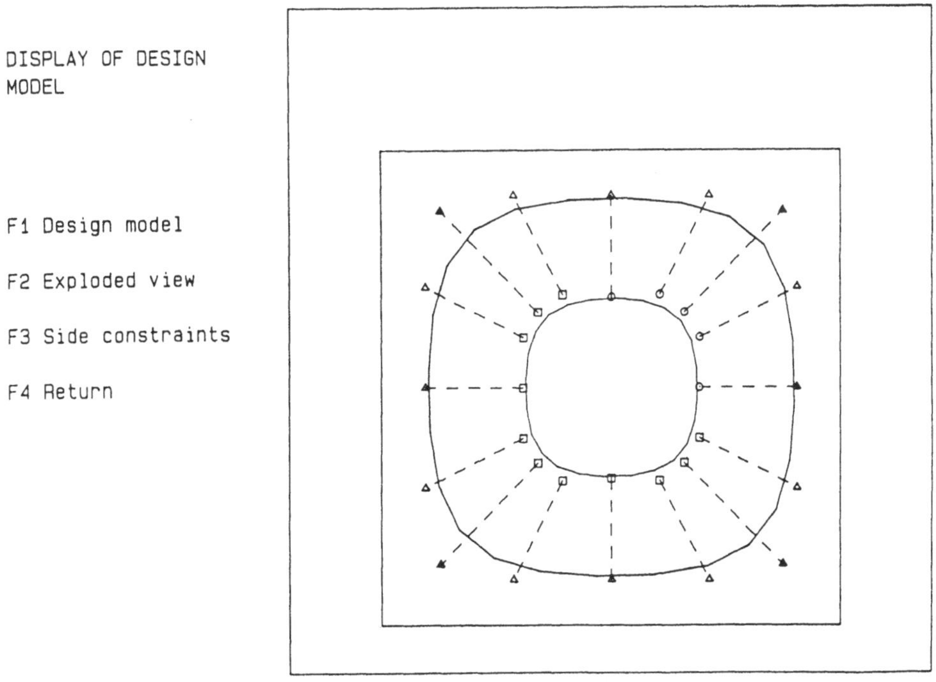

Figure 4b. Design Model

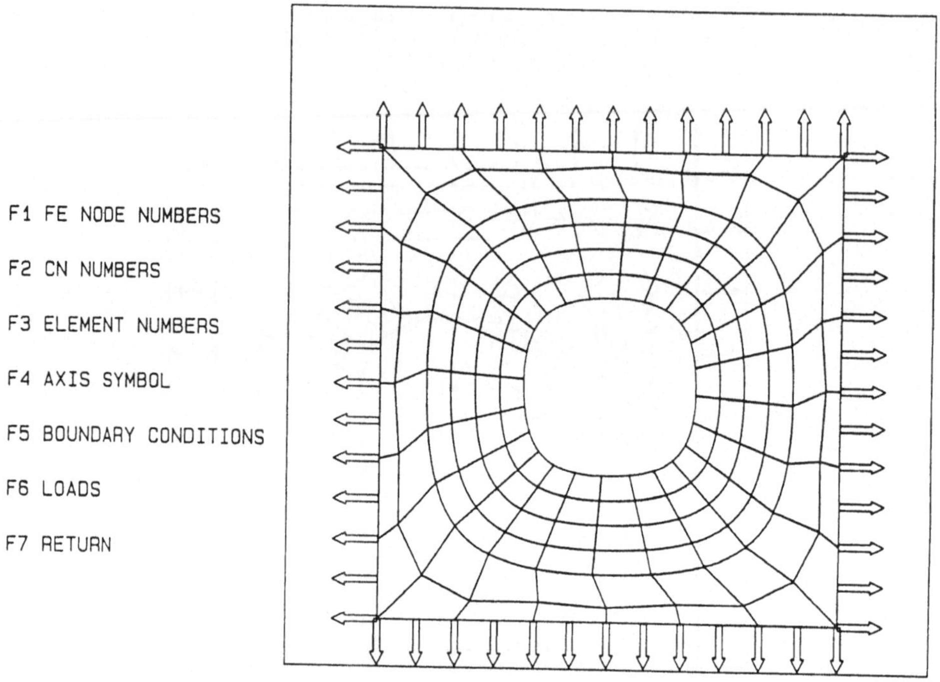

Figure 4c. Analysis Model

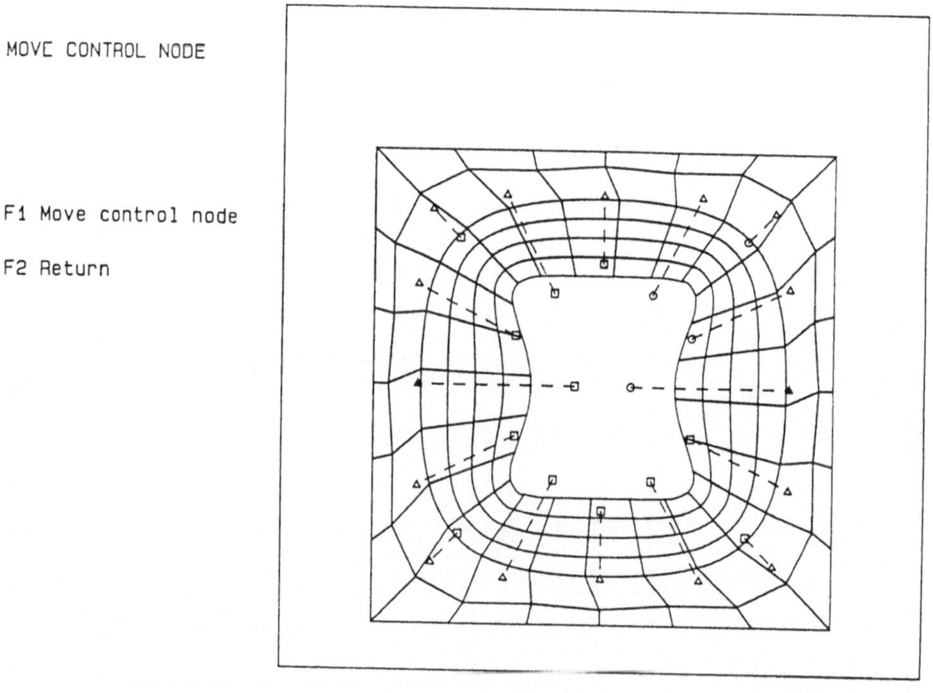

Figure 4d. Modified Shape

as key test-case to evaluate the CONLIN optimizer when supplied with finite difference gradients, because an analytical solution is available. The plate with hole displayed in Fig. 4.a is loaded with a combined tension in two perpendicular directions. For this initial configuration, stress concentrations will occur at the boundary of the hole. The problem is to determine the shape of the hole for which the tangential stresses around the hole are uniform. It can be proven that the shape of the hole with this property is elliptic, with the ratio of the axes being equal to the ratio of the applied tensile stresses [11]. The problem can be formulated under the following equivalent form: minimize the weight of the plate while constraining the maximum value of the tangential stress. Since the tangential stress is not available in the current version of the program, the Von Mises stress was used instead. It is believed that the Von Mises stress around the hole is close in value to the tangential stress since the radial stress is zero on a free boundary.

What follows is a description of how one would solve this problem using an interactive engineering design system, such as the one described in this paper. The first step which is required is the model description, using the design element concept. Since only the shape of the hole is to be changed, an adequate representation of the structure is to use one design element around the hole with one changing border, and one subregion containing the rest of the plate with fixed geometry. As the design elements are defined by their boundary curves, it is now up to the user to specify which type of curves he wants to employ and to locate the governing points. In this example the design element boundaries are 2 periodic B-splines of order 13 defined by 16 poles. So there are totally 32 control nodes, the 16 determining the inner contour are moving and the 16 poles shaping the outer contour are fixed. The design model as displayed by the program is shown in Fig. 4.b. This model description leads to a shape optimization problem, with the 16 distances between the moving control nodes and their respective fixed reference poles as design variables. Employing the double symmetry of the problem, this number can be reduced to only 5 design variables. The design model display of Fig. 4.b allows a clear visualization of these design variables as well as their associated side constraints.

The next step is to generate the analysis model from the design model description. As explained before, a mesh of isoparametric quadrilateral

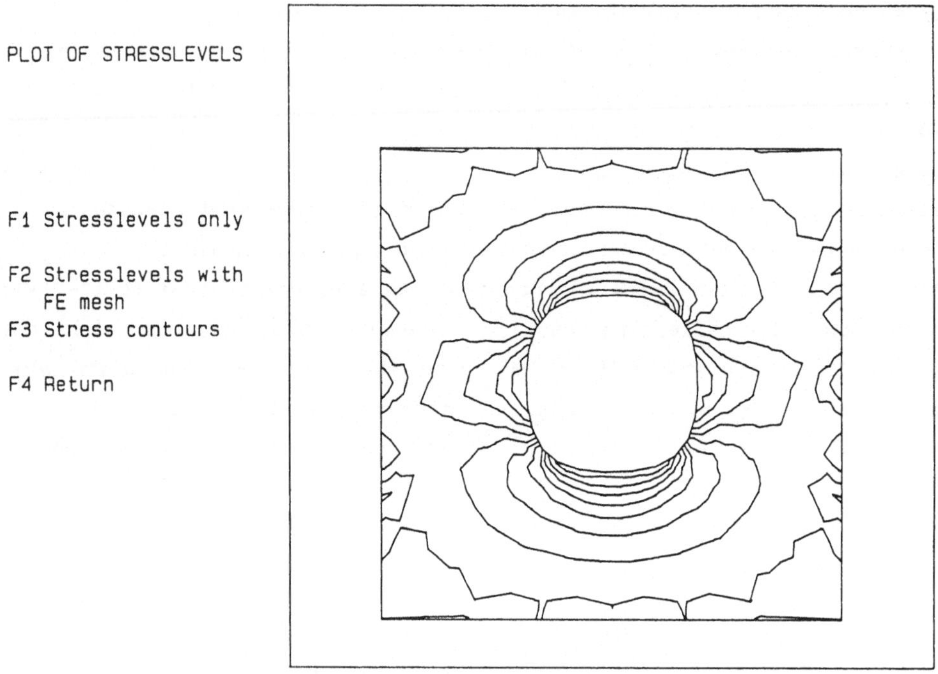

Figure 5a. Von Mises Stress

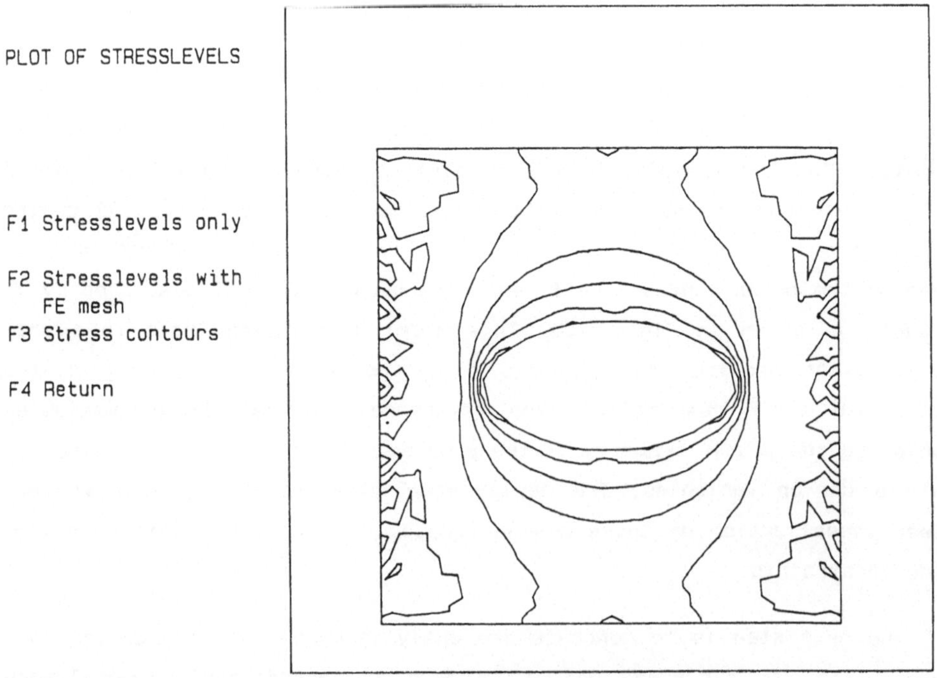

Figure 5b. Von Mises Stress

elements can be generated automatically inside each design element; the mesh inside the fixed subregion must be defined by the user. The interactive design system can display the mesh, as well as other basic information such as numbering, loads, support points, control nodes (Fig. 4.c). The user can then anticipate what will be done during the sensitivity analysis, by moving those control nodes which are design variables and examine the effect on the finite element model, that is the mesh deformation (see Fig. 4.d). The design constraints can be specified in terms of the analysis model. Using the mesh symmetry, the Von Mises stress in 12 of the 24 elements bordering the hole is constrained. Note that the mesh is not double symmetric, the reason being that in the case of periodic B-splines there is no simple way to impose the coordinates of the point corresponding to the initial value of the parameter describing the curve.

At this point the user is ready to submit the created data to the finite element optimization code for the finite element analysis and the sensitivity analysis of the initial design (one iteration). Upon completion of this job, the analysis results (displacements and stresses) can be visualized by the interactive system. From the Von Mises plot (Fig. 5.a) one can observe the important stress concentration associated with the initial design. Since the imposed stress limit (260 N/mm^2) is largely exceeded, the initial design is seriously infeasible.

Using the interactive shape optimization capability of the system, supplied with the calculated sensitivity results, the new shape can be computed and immediately verified with the design model plot and the analysis model plot.This is a critical point in the optimization procedure. If the modified shape is acceptable, the optimization process should be continued and the new analysis model can be submitted to the FEM-sensitivity program for another iteration. If the new shape is undesirable, the user can intervene through the system in the optimization by modifying constraints or parameters associated with the CONLIN optimizer. In particular, for this case, since it was seen that the initial design is seriously infeasible, no useful solution for the optimization was found unless the relaxation capability of CONLIN was activated. This requires the user to add acceptable increments to the constraint bounds, and so, to act on the CONLIN results.

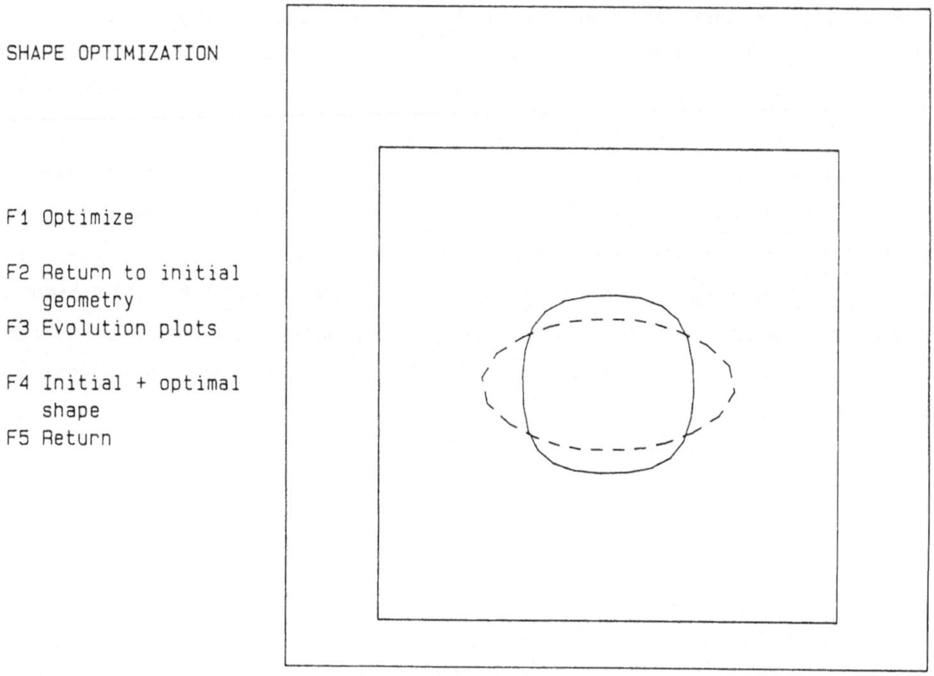

Figure 6a. Initial and Final Shapes

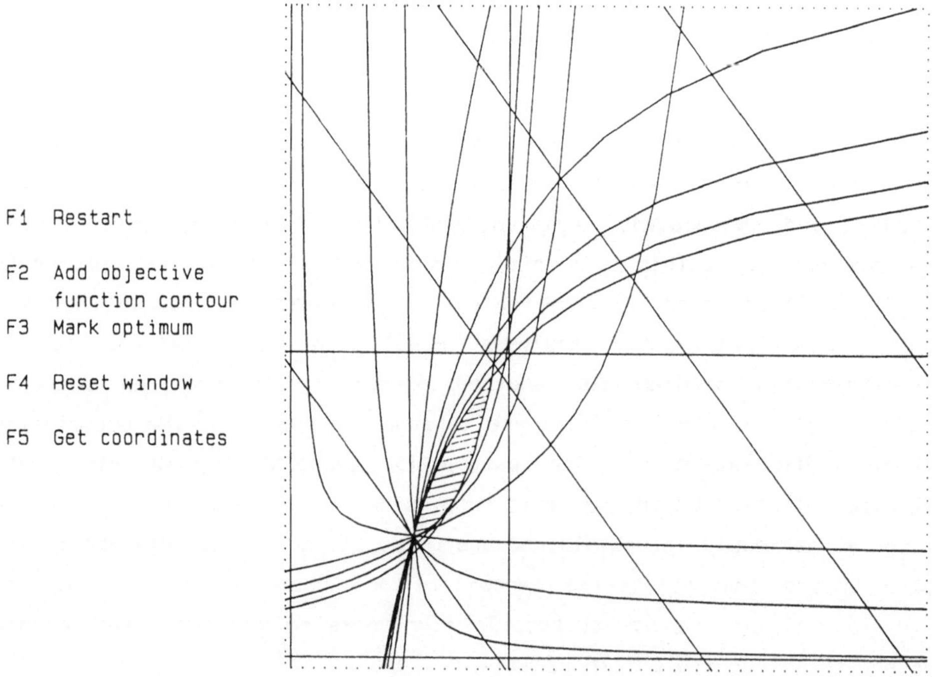

Figure 6c. Design Space

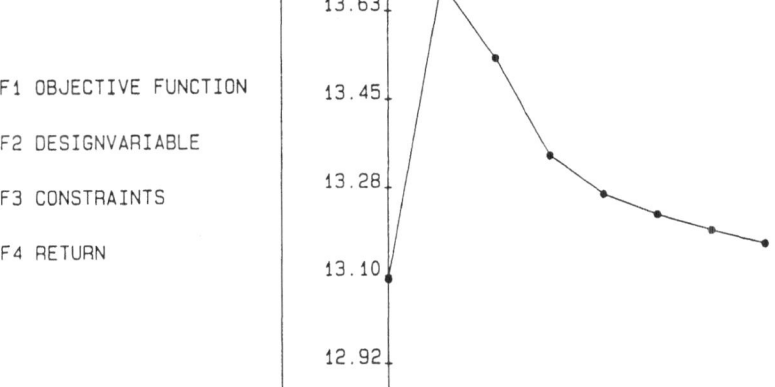

F1 OBJECTIVE FUNCTION

F2 DESIGNVARIABLE

F3 CONSTRAINTS

F4 RETURN

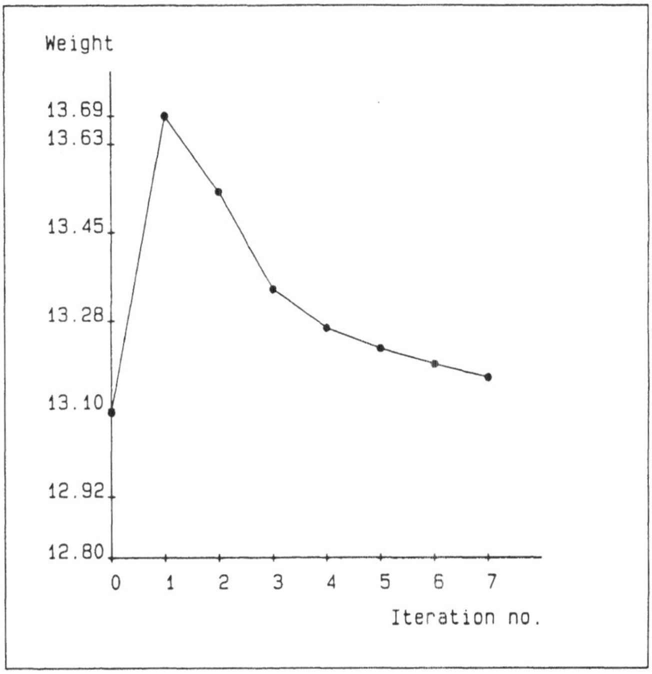

F1 OBJECTIVE FUNCTION

F2 DESIGNVARIABLE

F3 CONSTRAINTS

F4 RETURN

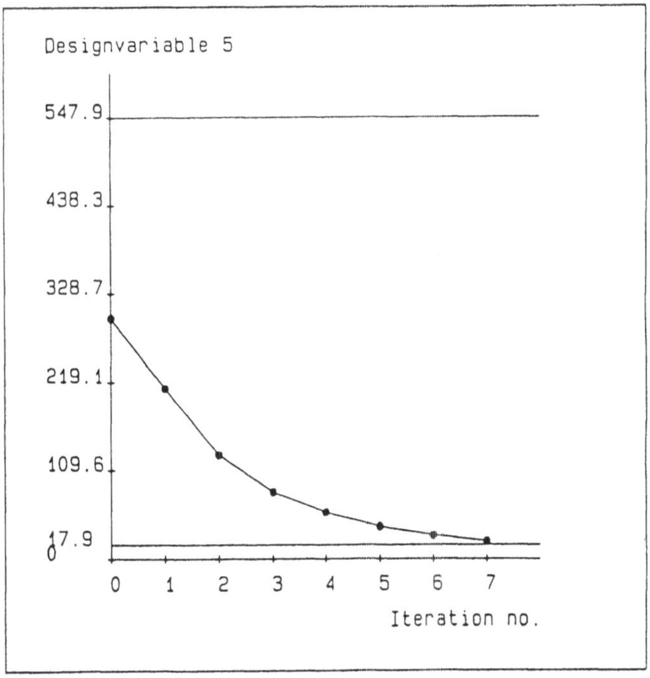

Figure 6b. Iteration History (CONLIN)

For this example with relaxation applied during the first optimization stage, the new shape is acceptable and additional iterations could be done. At each iteration, similar interactive handling can be done, though the user is mostly concerned with the display of the successive shapes. As the number of iterations increased, it became interesting to view the progress of the optimization process with the evolution plots of the objective function, design variables and constraints. After 7 iterations it was decided to stop the optimization process since satisfactory convergence was obtained to the optimal elliptic shape with axis ratio 0.5 . As shown in Fig. 6.a, the user can have both the initial and the optimal shape displayed simultaneously to assess the boundary changes. The stress plot (Fig. 5.b) reveals the constancy of the Von Mises stresses at the edge of this optimized hole: the constant stress lines run nearly parallel to the edge of the hole.

The user can ask for a graphical representation of the iteration history of all quantities of interest with the evolution plots. Fig. 6.b shows respectively the weight and design variable 5 in terms of the number of iterations; the user can assure himself of proper convergence with this graphical capability. From the weight evolution plot it can be seen that only at the first iteration relaxation had to be applied: it resulted in a feasible design, as are all the subsequent designs. This can be verified in the design space plot corresponding to the final design (Fig. 6.c), which shows a non-empty feasible region.

CONCLUDING REMARKS

The new concepts outlined above have proven to yield computationally efficient numerical methods, and therefore it is reasonable to expect their practical implementation into commercially available programs within the next decade. Nevertheless the research accomplished to date is encouraging. It demonstrates that the design cycle can be effectively computerized by implementing optimization capabilities into an interactive post-processor of the FEM analysis. As previously mentioned the proposed approach is quite general and it could be readily introduced into a large scale FEM system for the analysis and sensitivity analysis, and its pre-

and post-processing module for the creation of the geometric design model. However, in order to achieve this goal, several important topics related to Computer Aided Optimal Design need to be examined: accuracy of the sensitivity analysis; interactive optimization and innovative use of computer graphics; control of validity of analysis model; and generality of design model.

The two-dimensional pre-and post processor module described in Ref. [10], represents a first step to implement shape optimization concepts into the real design cycle. It is the interactive component in the logical division of the optimization task between a finite element optimization code, producing the analysis results in batch mode (including sensitivities), and an interactive design system, helping the designer to control the optimization process. At the present stage of the development effort the interactive module is mainly functioning as a post-processor. It is capable of interactive shape optimization by calling the CONLIN optimizer, and it exhibits some innovative visualization techniques that seem to constitute appropriate tools to facilitate the task of the designer. Future work should be directed toward the development of a pre-processing module, allowing for an efficient and user-friendly introduction of the analysis and design optimization data. Additional study should also be accomplished to devise a well organized data base for storing all the information involved in the shape optimization process. It is envisioned that a gradual extension of the module will be accomplished as new analysis and optimization capabilities become available. The ultimate goal is to provide the user with many visualization features for rapidly introducing input data and interpreting results related to complex three-dimensional structures.

Although conceptually simple, the finite difference approach presented in this paper reveals very promising because of its generality and ease of implementation. Analytical derivatives for the shape optimal design problem can also be obtained, but in a very complex way, which is dependent on the type of finite element used. Introducing this analytical method in general purpose finite element packages, containing a vast library of element types, would require a huge development effort. The finite difference approach is on the contrary highly general, in that the scheme is valid for any type of element, and can be implemented relatively easily.

In summary, then, although further research is still needed to fully automate the design process, Computer Aided Optimal Design is close to become a reality.

REFERENCES

[1] L.A. Schmit and H. Miura, "A New Structural Analysis/Synthesis Capability - ACCESS 1", AIAA Journal, Vol.14, No. 1, pp. 661-671, 1976.

[2] L.A. Schmit, "Structural Optimization - Some Key Ideas and Insights", Chapter 1 of New Directions in Optimum Structural Design (E. Atrek et al. eds.), John Wiley & Sons, New York, 1984.

[3] C. Fleury and L.A. Schmit, "Dual Methods and Approximation Concepts in Structural Synthesis", NASA Contractor Report, NASA-CR-3226, 1980, 196 pages.

[4] A.J. Morris (ed.), Foundations of Structural Optimization: A Unified Approach, John Wiley and Sons, New York, 1982.

[5] O.C. Zienkiewicz and J.S. Campbell, "Shape optimalization and Sequential Linear Programming", in Optimum Structural Design (Ed. R.H. Gallagher and O.C. Zienkiewicz), Wiley, New York, 1973, pp.109-126.

[6] M.E. Botkin, "Shape Optimization of Plate and Shell Structures", AIAA Journal, Vol.20, No. 2, pp. 268-273, 1982.

[7] V. Braibant and C. Fleury, "Shape Optimal Design using B-Splines", Computer Methods in Applied Mechanics and Engineering, Vol. 44, pp. 247-267, 1984.

[8] V. Braibant and C. Fleury, "Shape Optimal Design - A Performing C.A.D. Oriented Formulation", Proc. AIAA/ASME/ASCE/AHS 25th Structures, Structural Dynamics and Materials Conference, Palm Springs, CA., 1984.

[9] C. Fleury and D. Liefooghe, "Shape Optimal Design on an Engineering Workstation", Paper presented at the MSC/NASTRAN Users Conference, Universal City, California, March 20-21, 1986.

[10] D. Liefooghe, "An Interactive Graphics Capability for Shape Optimal Design", Master Thesis, UCLA, 1986.

[11] Muskhelishvili, N.I., "Some basic Problems of the Mathematical Theory of Elasticity", Noordhoff, Groningen -Holland, pp.320-372, 1953.

COMPUTER AIDED OPTIMAL DESIGN OF ELASTIC STRUCTURES

Part II

Convex Approximation Strategies in Structural Synthesis

Abstract In this Section innovative numerical methods relevant to Computer Aided Optimal Design are reviewed, with special emphasis on shape optimal design of elastic structures discretized in finite elements. After a short description of the approach followed to create an appropriate geometric model, involving a relatively small number of design variables, attention is mainly directed toward the selection of an adequate optimization algorithm. To this aim the paper will briefly present the various attempts that we have successively undertaken before adopting the convex linearization method as the basic optimizer, not only for shape optimal design problems, but also for all our other structural synthesis capabilities.

INTRODUCTION

The main goal of this Section is to discuss various optimization algorithms that have proven to be general and efficient to deal with structural synthesis problems. Attention will mainly be focused on shape optimal design, because an abundant literature is available for the case of structural sizing problems (including a Book collecting the Lectures presented at a previous NATO Advanced Study Institute) [1]. It should however be clearly recognized that the optimization strategies proposed in this paper are quite general, and they are thus applicable as well to optimal sizing problems.

The approach followed to describe the structural geometry is now well established and it can be summarized as follows (see Section I for more details). The method is based upon an internal parametric representation typical of modern techniques employed in computer aided geometric design. The structure is decomposed into a few subregions of simple geometry. These subregions are described in a compact way by using a relatively small number of control nodes (also named master nodes). During the optimization process the geometry of conveniently selected subregions is allowed to change: these regions are called design elements. The movements of the corresponding control nodes are the design variables. The design element boundaries are described by blending functions commonly used in Computer Graphics technology for interactive generation of curves (Bezier, B-splines) and surfaces (Coons patches). Such an internal parametric representation permits determining the coordinates of any point inside the design element or on its boundaries. This formulation lends itself well to shape optimal design problems. Only a few design elements are generally sufficient to fully describe the regions that are modified during the optimization process. In addition it is relatively straightforward to generate a suitable finite element mesh and to maintain its integrity throughout the optimization process. Finally, as explained in Section I, the sensitivity analysis may be formulated either in a fully analytical way or through finite differences at the element level.

Being capable of properly defining a design model, we can now envision efficient techniques to improve the characteristics of the structure by modifying its shape in an optimal way. Most often such a shape optimal design problem consists of minimizing some objective function subject to constraints insuring the feasibility of the structural design. Mathematically the numerical optimization problem considered in this paper can be written in the following general form:

$$\text{minimize } f(x) \tag{1}$$

subject to the constraints:

$$c_j(x) \leqslant 0 \qquad j=1,m \tag{2}$$

$$\underline{x}_i \leqslant x_i \leqslant \bar{x}_i \qquad i=1,n \tag{3}$$

The objective function (1) is a nonlinear function of the design variables x_i. It usually represents a structural characteristic to be minimized (e.g. the weight). The nonlinear inequalities (2) are the behavior constraints that impose limitations on structural response quantities (e.g. upper bounds on stresses and displacements under static loading cases). The design variables must also be bounded by the side constraints (3), where $\underline{x}_i$ and $\bar{x}_i$ are lower and upper limits that reflect manufacturing or analysis validity considerations. It should be noted that the side constraints (3) constitute a particular case of the more general constraints (2). However they are written separately in our optimization problem statement, because the dual method approach described later can handle them more efficiently when considered apart from the behavior constraints.

The nonlinear programming problem (1-3) can be solved iteratively by using numerical optimization techniques. Each iteration begins with a complete analysis of the system behavior in order to evaluate the objective function and constraint values along with their sensitivities to changes in the design variables (i.e. first derivatives). Most often the analysis capability is based on finite element discretization. A design iteration is concluded by employing the results of these behavioral and sensitivity analyses in a minimization algorithm which searches the n-dimensional design space for a new primal point that decreases the objective function value while remaining feasible (i.e. satisfying the constraints). Many such iterations are usually required before achieving the optimum design. This observation brings us to the essential difficulty in solving the nonlinear programming problem (1-3), which lies in the implicit character of the constraint functions $c_j(\mathbf{x})$. In other words these functions are not explicitly known in terms of the design variables. For each new design, they can only be evaluated numerically through a finite element analysis. The iterative nature of the optimization process implies that many structural reanalyses must usually be accomplished before finding an acceptable solution. Those repeated finite element analyses can lead to a prohibitive computational cost when dealing with large scale problems.

One widely used approach to shape optimal design problems is to join together by brute force, a general purpose optimizer and a finite element package having the required sensitivity analysis capabilities. During the initial development phase of our research efforts, this straightforward, though not highly efficient approach, was in fact the only possible choice. So, in a first step, direct nonlinear programming methods were compared to conventional linearization techniques using the Simplex algorithm. The numerical experiments conducted for that purpose have demonstrated, as it could be expected, that direct approaches like gradient projection or feasible direction methods are inadequate for shape optimization problems in view of the large number of iterations required for convergence. On the other hand recursive linear programming techniques, even though they necessitate difficult adjustments of move limits, prove to be computationally efficient. It was therefore decided to pursue the idea of sequential linearization, and to improve it by implementing more suitable approximation concepts similar to those used for sizing problems. The basic approach consists in replacing the primary optimization problem with a sequence of explicit approximate subproblems having a simple algebraic structure. Each subproblem is generated through Taylor series expansion of the objective function and constraints in terms of intermediate linearization variables. It was found that the efficiency of this approach is very sensitive to the mode of expansion employed for building the approximate subproblems: (1) first or second order expansion of the objective function in terms of the design variables; (2) first order expansion (i.e. linearization) of the constraints in terms of the design variables or their reciprocals.

Linearization of the constraints with respect to reciprocal variables is a well recognized technique to solve optimal sizing problems. Although this linearization scheme cannot be physically justified for optimum shape problems, surprisingly, it leads to remarkably good results. It seems, therefore, that reciprocal variables provide a miraculous tool to solve structural synthesis problems. To some extent, a mathematical justification can be found in a new and rather general optimization method, called the convex linearization method (CONLIN). The key idea in the CONLIN algorithm is to perform the linearization process with respect to mixed variables, either direct or reciprocal, independently for each function involved in the optimization problem. A convex, separable subproblem is therefore generated, that can be efficiently solved by a dual method formulation. Because it uses conservative approximations CONLIN has an inherent tendency to generate a sequence of steadily improving feasible designs.

Various examples of applications to optimum shape problems are offered in Section III to demonstrate the efficiency of this new algorithm. However, in order to give some flavor of its power, let us consider the widely used "plate with hole" example shown in Fig. 1. A complete description of the problem was given in Section I. The iteration history data provided by a general purpose optimizer (based on feasible directions) are compared to the results generated by the new CONLIN optimizer. It can be seen that, when using the "black box" approach, convergence is not yet achieved after 20 iterations. Moreover it should be noted that each iteration involves several structural analyses in order to determine a suitable step length along the selected feasible direction (line search process). Although this line search is implemented in a very efficient way, 53 structural reanalyses are needed to accomplish the 20 iterations shown in Fig. 1. The corresponding computational time is larger than 7 hours CPU on a VAX 11/780. On the other hand, when the convex linearization method is employed, only four structural analyses are sufficient to get an optimal design. Note that the theoretical ellipse is recovered by the numerical optimization algorithm.

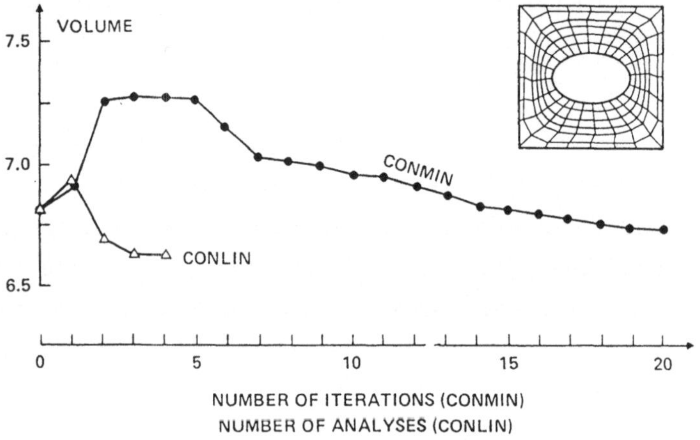

Fig. 1 Iteration History

APPROXIMATION CONCEPTS IN SHAPE OPTIMAL DESIGN

The approximation concepts approach to structural synthesis is now widely employed to solve optimal sizing problems [2]. Such problems consist in minimizing the weight of thin-walled structures modeled by bar and membrane elements. Because the geometry is fixed, the design variables reduce to the transverse sizes of the structural members (i.e. bar cross-sections and membrane thicknesses). This approach consists basically of the following steps:

- a finite element analysis is performed for the initial trial design;

- from the results of the current structural analysis, an approximate optimization problem is generated; this step implies that sensitivity analysis capabilities be available in the finite element code; the approximate subproblem is created by linearizing the behavior constraints in terms of the reciprocals of the design variables while the objective function keeps its exact linear form in terms of the direct sizing variables;

- because the subproblem is fully explicit, convex and separable, it can be efficiently solved by resorting to its dual formulation (when expressed in the reciprocal design variables, the objective function is strictly convex and the linearized constraints are linear);

- the solution of the approximate subproblem is adopted as a new starting point in the design space and the optimization process is continued until convergence is achieved.

From extensive numerical experiments it can be argued that the approximation concepts approach converges to an optimum design in usually less than ten iterations (i.e. finite element analyses). These remarkable convergence properties are generally attributed to the fact that the behavior constraints have a tendency to be much less nonlinear in the space of the reciprocal variables. The convex linearization method described later in this paper permits providing a more rigorous explanation.

When dealing with shape optimal design problems, the foregoing considerations on the intuitive choice of reciprocal variables or other intermediate linearization variables are no longer valid. There is indeed no reason why the constraints should be more shallow with respect to the reciprocals of nodal coordinates. Furthermore, although the structural weight keeps a simple explicit form, it is no longer a linear function of the shape design variables. For those reasons various approximation schemes have been successively experimented:

- first order expansion (i.e. linearization) of the objective function and constraint functions with respect to the direct variables;

- second order expansion of the objective function and linearization of the constraint functions with respect to the direct variables;

- second order expansion of the objective function with respect to the direct variables and linearization of the constraint functions with respect to the reciprocal variables;

- linearization of the objective function with respect to the direct variables and linearization of the constraint functions with respect to the reciprocal variables.

The numerical experiments reported in Ref. [3] show that the convergence properties of the optimization process mainly depend upon the quality of the constraint approximation. When the constraint approximation is too rough, divergence or oscillation frequently occur. It can also happen that the solution of an approximate problem becomes non feasible. In such a case many subsequent iterations are usually required in order to recover a feasible design. Because it has been found that no major benefit can be gained from using a second order expansion of the objective function, this paper will only consider the first and the last modes of approximation listed above.

Sequential Linear Programming

The sequential linear programming approach proceeds by replacing the primary nonlinear problem (1-3) with a sequence of linear subproblems. Each subproblem is generated by linearizing the objective function and the constraint functions with respect to the shape design variables x_i:

$$\tilde{f}(x) = f(x^0) + (x-x^0)\, \nabla f(x^0) \tag{4}$$

$$\tilde{c}_j(x) = c_j(x^0) + (x-x^0)^T \nabla c_j(x^0) \tag{5}$$

where x^0 denotes the current design point, that is, the point in the design space where the objective function and the constraints are linearized. The resulting approximate problem can be written in the following compact way:

$$\begin{aligned} &\text{minimize} && \sum_i f_i x_i \\ &\text{subject to} && \sum_i d_{ij} x_i \leqslant \bar{d}_j \\ & && \underline{x}_i \leqslant x_i \leqslant \bar{x}_i \end{aligned} \tag{6}$$

where f_i's represent the first derivatives of the objective function (i.e. the components of $\nabla f(x^0)$), and the d_{ij}'s denote the first derivatives of the constraint functions [i.e. the components of $\nabla c_j(x^0)$]. The upper bounds $\bar{d}_j$ collect the constant terms in the linearized constraints.

Because this explicit problem is a linear programming problem, it can be efficiently solved by using the Simplex algorithm. It is well known that the optimal solution point corresponding to each linear subproblem necessarily lies at a vertex of the design space. As a result the overall optimization process may either converge to a non optimal solution of the primary problem, or it may oscillate indefinitely between two or more vertices. In order to avoid this undesirable behavior, the so-called "move-limits" strategy can be implemented. It consists of adding temporarily some artificial side constraints to the linearized problem, so that the design point will not move too far away from the current linearization point x^0. It is however very difficult, even for an experienced user, to define adequate rules for setting the move limits, and the way they should be updated after each iteration. Therefore, even though some meaningful results have been obtained in the case of simple problems through the use of sequential linear programming [4], this approach has not been retained in the subsequent developments of our research efforts.

Constraint Linearization in the Reciprocal Space

It has now become almost certain that the best approaches to optimal sizing problems are those which make use of constraint linearization with respect to the reciprocal design variables [2]. There is a intuitive explanation for the success of these methods, in that stresses and displacements are exact linear functions of the reciprocal sizing variables in the case of a statically determinate structure. Unfortunately, for shape optimal design problems, there is no such physical guideline for the selection of intermediate linearization variables. Nevertheless, as explained in this Section, this change of variables continues to have a very beneficial effect on the convergence properties of the shape optimization process.

The approach now considered consists of keeping the linear approximation (4) of the weight objective function in terms of the original design variables, while linearizing the behavior constraints with respect to the reciprocals of the design variables:

$$z_i = 1/x_i$$

The approximate behavior constraints take on the following form:

$$\tilde{c}_j(z) = c_j(z^0) + (z-z^0)^T \nabla c_j(z^0) \qquad (7)$$

where z^0 denotes the current design point in the space of the reciprocal variables. This technique generates at each iteration an approximate problem of the following form:

$$\begin{aligned} &\text{minimize} && \sum_i f_i x_i \\ &\text{subject to} && \sum_i c_{ij}/x_i \leq \bar{c}_j \qquad (8) \\ &&& \underline{x}_i \leq x_i \leq \bar{x}_i \end{aligned}$$

where the c_{ij} coefficients denote the first derivatives of the constraint functions with respect to the reciprocal variables [i.e., the components of $\nabla c_j(x^0)$]. Note that:

$$c_{ij} = -d_{ij}(x_i^0)^2 \qquad (9)$$

This explicit subproblem is therefore generated from the primary problem (1-3) by linearizing the objective function with respect to the direct variables x_i, and by linearizing the behavior constraints with respect to the reciprocal variables z_i.

When restated in terms of the reciprocal variables z_i, the problem involves only linear constraints, and therefore it can be solved by using a gradient projection algorithm. All the numerical experiments conducted with this strategy tend to exhibit convergence properties similar to those previously obtained for optimal sizing problems: less than ten finite element analyses are usually sufficient to achieve stable convergence of the optimization process without having to resort to artificial move limits.

THE DUAL METHOD APPROACH

It can be recognized that the explicit problem (8) is convex and separable, and therefore it lends itself well to solution by a dual method approach. In fact this problem has exactly the same explicit form as the subproblem used in the approximation concepts approach to optimal sizing problems. It becomes thus possible to resort to the previously developed DUAL1 or DUAL2 optimizers, that have proven to be remarkably efficient [2].

The dual method approach, initially introduced in Ref. [5], has lead to a reconciliation of optimality criteria techniques and mathematical programming methods [6]. Let us here summarize the principles of the dual formulation. The solution of the primal problem (8) can be obtained by the following "Min-Max" two-phase procedure:

$$\text{maximize} \quad l(\lambda)$$

$$\text{subject to} \quad \lambda_j \geqslant 0$$

where the dual function $l(\lambda)$ results from minimizing the Lagrangian over the acceptable primal variables:

$$l(\lambda) = \min_{\underline{x}_i \leqslant x_i \leqslant \bar{x}_i} \left(\sum_i f_i x_i + \sum_j \lambda_j \left(\sum_i c_{ij}/x_i - \bar{c}_j\right)\right) \tag{10}$$

The separability of the primal problem implies that the minimum of the sum on the n individual functions can be expressed as the sum of the minimum of the n functions. On the other hand the convexity of the problem makes sure that each single-variable minimization problem

$$\min \quad L_i(x_i) = f_i x_i + \left(\sum_j c_{ij}\lambda_j\right)/x_i \tag{11}$$

$$\text{s.t.} \quad \underline{x}_i \leqslant x_i \leqslant \bar{x}_i$$

has a unique solution:

$$x_i = \underline{x}_i \qquad \text{if } \sum_j c_{ij}\lambda_j \leq f_i \underline{x}_i^2$$

$$x_i = \bar{x}_i \qquad \text{if } \sum_j c_{ij}\lambda_j \geq f_i \bar{x}_i^2$$

$$x_i = [(\sum_j c_{ij}\lambda_j)/f_i]^{1/2} \qquad \text{otherwise} \tag{12}$$

The last relationship is identical to the energy expression used in "optimality criteria" approaches to structural synthesis (see e.g. Ref. [11,12]). We shall come back with more details on the dual problem later in this paper.

Additional Linear Constraints

For various reasons the design variables may also be subjected to strictly linear constraints of the form (6). Such requirements often reflect manufacturing or "rule of thumb" considerations, such as a linear progressivity rule for the thicknesses of contiguous structural members in a sizing problem (e.g. increase in the number of layers in a laminated composite plate). In shape optimal design problems they may represent geometric considerations such as the imposition of tangential continuity along a moving boundary. It can also happen that the user wishes to linearize some of the behavior constraints with respect to the direct variables x_i, rather than with respect to the reciprocal variables z_i. When introducing such linear constraints, the explicit subproblem takes on a more general form, which can be viewed as a mix of the approximate problems (6) and (8):

$$\text{minimize} \quad \sum_i f_i x_i$$

$$\text{subject to} \quad \sum_i c_{ij}/x_i \leq \bar{c}_j$$

$$\sum_i d_{ij} x_i \leq \bar{d}_j \tag{13}$$

$$\underline{x}_i \leq x_i \leq \bar{x}_i$$

This problem can no longer be directly solved by using a primal projection method. Formulated either in terms of the direct variables x_i or in terms of the reciprocal variables z_i, it always involves nonlinear constraints. A possible strategy would be to restate the explicit constraints (13) in terms of the reciprocal variables, and to linearize them just as the behavior constraints (2). The resulting subproblem then keeps the same form as in the original approximation concepts approach (see Eq. 8), and it is thus amenable to solution by gradient projection or by dual algorithms. However it would be unfortunate, if not ridiculous, to treat approximately the only constraints that have a simple explicit form in the problem statement.

Simultaneously with our research efforts in shape optimal design, we were faced with another project aimed at improving the dual method approach to solve difficult optimal sizing problems that precisely involved many linear constraints of the form (13). The purpose was to handle these linear con-straints in an exact manner. The initial research effort has been focused on exploiting again the separable nature of the explicit subproblem to be solved, and to resort to a specially devised dual method approach. Unfortunately, as shown in Ref. [9], this research was not successful. Because the required convexity properties are not satisfied, the dual problem happens to be non differentiable, and therefore difficult to solve by conventional maximization algorithms.

With the addition of the linear constraints (13) the Lagrangian problem appearing in Eq. (10) becomes:

$$\underset{\underline{x}_i \leq x_i \leq \bar{x}_i}{\text{minimize}} \quad \Big(\sum_i f_i x_i + \sum_j \lambda_j \big(\sum_i c_{ij}/x_i - \bar{c}_j\big) + \sum_j \lambda_j \big(\sum_i d_{ij} x_i - \bar{d}_j\big)\Big)$$

Because this problem is still separable, it can again be replaced with n one-dimensional minimization problems of the form:

$$\begin{aligned} \min \quad & L_i(x_i) = a_i x_i + b_i/x_i \\ \text{s.t.} \quad & \underline{x}_i \leq x_i \leq \bar{x}_i \end{aligned} \tag{14}$$

where the coefficients

$$a_i = \sum_j d_{ij} \lambda_j + f_i$$

and

$$b_i = \sum_j c_{ij} \lambda_j$$

depend only upon the dual variables λ_j.

It is important to point out that nothing is known regarding the sign of the coefficients a_i and b_i (they may be positive or negative, because the constraint derivatives c_{ij} and d_{ij} are arbitrary). As a result various situations emerge when solving the Lagrangian problem.

Case 1: $a_i > 0$ and $b_i > 0$

The problem solution is obtained by stating that the first derivative of $L_i(x_i)$ must vanish (the second derivative is positive since $b_i > 0$).

$$L_i'(x_i) \equiv a_i - b_i/x_i^2 = 0$$

Because the side constraints must be satisfied, it comes:

$$x_i = (b_i/a_i)^{1/2} \qquad \text{if} \quad \underline{x}_i^2 \leqslant b_i/a_i \leqslant \bar{x}_i^2 \tag{15}$$

$$x_i = \underline{x}_i \qquad \text{if} \quad b_i/a_i \leqslant \underline{x}_i^2 \tag{16}$$

$$x_i = \bar{x}_i \qquad \text{if} \quad b_i/a_i \geqslant \bar{x}_i^2 \tag{17}$$

Case 2: $a_i > 0$ and $b_i \leqslant 0$

Since $L_i'(x_i) > 0$ for all values of x_i, $L_i(x_i)$ is monotonically increasing, so that the solution of the minimization problem lies at:

$$x_i = \underline{x}_i$$

Case 3: $a_i < 0$ and $b_i \geqslant 0$

Now $L_i'(x_i) < 0$ for all values of x_i. Hence $L_i(x_i)$ is monotonically decreasing, and the minimum occurs at:

$$x_i = \bar{x}_i$$

Case 4: $a_i < 0$ and $b_i < 0$

The function $L_i(x_i)$ is concave and it has a maximum at

$$x_i = (b_i/a_i)^{1/2}$$

Therefore the minimum is obtained by comparing the values $L_i(\underline{x}_i)$ and $L_i(\bar{x}_i)$:

$$x_i = \underline{x}_i \qquad \text{if} \qquad L_i(\underline{x}_i) < L_i(\bar{x}_i) \tag{18}$$

$$x_i = \bar{x}_i \qquad \text{if} \qquad L_i(\underline{x}_i) > L_i(\bar{x}_i) \tag{19}$$

$$x_i = \underline{x}_i \text{ or } \bar{x}_i \qquad \text{if} \qquad L_i(\underline{x}_i) = L_i(\bar{x}_i) \tag{20}$$

This latter case introduces a major difficulty: equation (20) demonstrates that the dual function is likely to be ill-conditioned. The dual problem, in fact, happens to be non-differentiable. Indeed, let us assume that at some particular dual point, $x_i = \underline{x}_i$ because condition (18) is just satisfied:

$$L_i(\bar{x}_i) = L_i(\underline{x}_i) + \epsilon$$

where ϵ is a small positive number. By slightly perturbating the dual variables the coefficients a_i and b_i will be modified, and so, the solution of the minimum problem (14) might jump suddenly to $x_i = \bar{x}_i$, because condition (19) is now satisfied. Such a jump causes discontinuities in the first derivatives of the dual function, which are given by the primal constraint values. As shown in Ref. [9], these discontinuities occur along hyperplanes in the dual space given by

$$\sum_j c_{ij}\lambda_j = \underline{x}_i\bar{x}_i\left(f_i + \sum_j d_{ij}\lambda_j\right)$$

For a given primal variable x_i, this hyperplane subdivides the dual space into a region where $x_i = \underline{x}_i$ and another region where $x_i = \bar{x}_i$. When crossing this plane, the first derivatives of the dual function are obviously discontinuous.

THE CONVEX LINEARIZATION METHOD

So the research reported in Ref. [9] was not capable of dealing with the linear constraints (13) within the framework of the dual formulation.

Nevertheless this study was essential for the continuation of our development efforts in structural optimization, because it finally lead to the convex linearization method that is now employed as well for sizing as for shape optimal design problems. Indeed one important conclusion of Ref. [9] was that further research was needed to convexify the explicit constraints (13), for example by linearizing them partially with respect to the reciprocal variables. When carefully examining the dual problem, it can be seen that the terms responsible for non differentiability are those which correspond to negative coefficients in the linear constraints (13). Hence, to obtain a sound dual formulation, it is sufficient to replace those negative terms with positive approximate ones.

The convex linearization method (CONLIN) is based on generalizing this idea to all the functions describing the mathematical programming problem to be solved. In problem (8) the objective function is linearized with respect to the direct variables x_i while the constraints are linearized with respect to the reciprocal variables z_i. In CONLIN each function defining the optimum design problem is linearized with respect to a properly selected mix of variables so that a convex and separable subproblem is generated. The selection of direct and reciprocal variables is made on the basis of the signs of the first partial derivatives. It is easily proven that, considering any differentiable function g(x), the following linearization scheme yields a convex approximation (hence the appellation "convex linearization") [10]:

$$g(x) \simeq g(x^0) + \sum_{+} g_i^0(x_i - x_i^0) - \sum_{-} (x_i^0)^2\, g_i^0(z_i - z_i^0)$$

where g_i denote the first derivatives of g(x) with respect to the original variables x_i. The symbol $\sum_{+}$ ($\sum_{-}$) means "summation over the terms for which g_i^0 is positive (negative)". One of the most interesting features of the convex linearization scheme is that it also leads to the most conservative approximation amongst all the possible combinations of mixed direct/reciprocal variables. This property was initially demonstrated in Ref. [11], where conservative approximation was employed to handle difficult buckling constraints.

The CONLIN algorithm proceeds by applying the foregoing convex linearization scheme to the objective function and to all the constraint functions, including the linear constraints (13). The resulting explicit approximations take on a simpler form if the design variables are normalized so that they become equal to unity at the current point x^0 where the problem is linearized [the factor $(x_i^0)^2$ disappears in Eq. 9). The following subproblem is then generated:

$$\text{minimize} \quad \sum_{+} f_i x_i - \sum_{-} f_i / x_i \tag{21}$$

subject to

$$\sum_{+} u_{ij} x_i - \sum_{-} u_{ij} / x_i \leqslant \bar{u}_j \tag{22}$$

$$\underline{x}_i \leqslant x_i \leqslant \bar{x}_i \tag{23}$$

where f_i and u_{ij} denote the first derivatives of the objective and constraint functions, respectively, evaluated at the current point x^0. These derivatives must be taken with respect to the same set of variables for each function. In other words the user must input the function derivatives with respect to either x_i or z_i, or any suitable mix of direct/reciprocal variables, but the choice of variables has to be unique (CONLIN will select by itself the mixed variables depending upon the sign of the function derivatives). The behavior constraints c_j and the linear constraints d_j [see Eqs. (2) and (13)] have been collected in a single set of constraints u_j. Note that the upper bounds $\bar{u}_j$ have been modified to contain the zero order contributions in the Taylor series expansion.

Therefore, in CONLIN, the initial problem is transformed into a sequence of explicit subproblems having a simple algebraic structure. Furthermore this subproblem is convex and separable. These remarkable properties make it attractive to solve the subproblem by using dual algorithms. In the dual approach, the constrained primal minimization problem is replaced by maximizing a quasi-unconstrained dual function depending only on the Lagrangian multipliers associated with the linearized constraints. These multipliers are the dual variables subject to simple non-negativity constraints. The efficiency of this dual formulation is due to the fact that maximization is performed in the dual space, whose dimensionality is relatively low and depends on the number of active constraint at each design iteration. The CONLIN approach can be viewed as a generalization of well established approaches to pure sizing structural optimization problems, namely "approximation concepts" and "optimality criteria" techniques [6], and as such it is capable of addressing a broader class of problems with considerable facility of use. collected in a single set of constraints u_j. Note that the upper bounds $\bar{u}_j$ have been modified to contain the zero order contributions in the Taylor series expansion.

The Lagrangian problem has the same form as in Eq. (14), with the coefficients a_i and b_i given in terms of the dual variables by:

$$a_i = \sum_{+} u_{ij}\lambda_j + f_i \geqslant 0$$

$$b_i = -\sum_{-} u_{ij}\lambda_j \qquad \geqslant 0$$

However there exists a fundamental difference with respect to the previously discussed case: the coefficients a_i and b_i are now always non-negative. Therefore the Lagrangian problem (14) has necessarily a unique solution given by the primal-dual relationships (15-17). Cases 2 through 4 will never have to be taken into account. As a result the dual problem

$$\max \quad \ell(\lambda) = \sum_{+} f_i x_i(\lambda) - \sum_{-} f_i/x_i(\lambda) + \sum_j \lambda_j [\sum_{+} u_{ij} x_i(\lambda) - \sum_{-} u_{ij}/x_i(\lambda) - \bar{u}_j]$$

$$\text{s.t.} \quad \lambda_j \geqslant 0$$

is well conditioned. The dual function does no longer suffer with a lack of C^1 continuity.

As previously mentioned the convex linearization scheme yields explicit approximations that are locally conservative (i.e. they tend to overestimate the values of the true functions). In other words, the approximate feasible domain corresponding to the explicit subproblem (21-23) is generally located inside the true feasible domain corresponding to the primary problem (1-3). As a result the CONLIN optimizer has tendency to generate a sequence of design points that "funnel down the middle" of the feasible region. Despite the use of a dual solution scheme, a primal philosophy is maintained, that is, the method often produces a sequence of steadily improving feasible designs. This represents an attractive property from an engineering point of view, since the designer may stop the optimization process at any stage, and still get an acceptable non critical design, better than its initial estimate.

CONSTRAINT RELAXATION TECHNIQUES

Although conservativeness is most of the time a desirable property, it is not when the initial starting point is seriously infeasible. In such a case is can happen that the approximate feasible domain be empty, so that the CONLIN method can no longer be applied. To cope with this difficulty, it is convenient to work in an expanded design space, by increasing the upper bounds assigned to the constraints. Denoting, v_j the maximal increments that the user accepts to add to the bound $\bar{u}_j$, and introducing an additional "relaxation" variable r, the following explicit subproblem must now be solved by the CONLIN optimizer:

$$\text{minimize} \qquad \tilde{f}(x) + r\, f(x^0)$$

$$\text{subject to} \qquad \tilde{u}_j(x) \leqslant \bar{u}_j + v_j(1-1/r)$$

$$\underline{x}_i \leqslant x_i \leqslant \bar{x}_i$$

$$r \geqslant 1$$

where $\tilde{f}(x)$ and $\tilde{u}_j(x)$ are the convex approximations of f(x) and $u_j(x)$ appearing in the subproblem statement. Note that convexity and separability are maintained with regard to the added variable r.

Clearly if the relaxation variable r hits its lower bound (r=1), nothing is changed in the problem statement, which will usually happen when the starting point x^0 is feasible of nearly feasible. On the other hand, if the starting point x^0 is serious infeasible, the algorithm will find a value of r greater than unity, which means that the approximate feasible domain is artificially enlarged. If the shape optimal design problem to be treated has a solution, this relaxation technique will usually be activated only once or twice. The subsequent iterations will then yield a unit relaxation variable. However if the feasible domain corresponding to the primary problem (1-3) is empty (for example because some of the requirements are incompatible), the optimization process will converge to the best possible design point. An infeasible design will probably be generated, however, it should correspond to a good compromise between the conflicting requirements. Note that the user can act on the CONLIN results by entering appropriate values for each increment v_j. For example if the kth constraint must be exactly satisfied because it corresponds to a particularly severe requirement, it is sufficient to input $v_k = 0$.

Other types of constraint relaxation techniques are worth being studied. For example a distinct relaxation variable can be added to the system for each constraint, as suggested in Ref. [12]. The one-variable relaxation technique proposed above is however numerically more attractive, because it can be introduced, external to the optimizer, through the addition of only one variable. To solve the modified explicit problem it is sufficient to increase the number of variables by one, and to input the following derivative information:

$$\left.\begin{aligned} f_i &= \tilde{f}(x^0) \\ u_{ij} &= -v_j \end{aligned}\right\} \quad \text{for } i = n + 1$$

Another possibility is to go inside the dual optimizer, and to explicitly introduce the effect of the relaxation variable in the convex linearization method. From the primal problem (21-23), it is easily seen that the Lagrangian problem has the form:

$$\min_{r \geq 1} \quad r\tilde{f}(x^0) + (\sum_j v_j\lambda_j)/r$$

From this minimum condition, r is given in terms of the dual variables by the relation:

$$r = [(\sum_j v_j\lambda_j)/\tilde{f}(x^0)]^{1/2} \qquad \text{if} \quad \sum_j v_j\lambda_j \geq f(x^0)$$

$$r = 1 \qquad \text{if} \quad \sum_j v_j\lambda_j \leq f(x^0)$$

As implemented herein, relaxation is uniformly applied to all the constraints (with the weighing factors v_j). Its purpose is merely to balance the effect of conservativeness in the convex approximation scheme. For this uniform relaxation to be effective, it is implicitly assumed that the feasible domain corresponding to the primary problem is non empty. If this is not the case, for example because two or more constraints are really in conflict, uniform relaxation is not a satisfactory technique. More sophisticated relaxation techniques are then needed, aimed at finding a minimal relaxation for an infeasible domain (see e.g. [13]).

CONCLUSIONS

The convex linearization method applied in this paper to shape optimal design problems has proven to be a highly efficient and reliable optimization tool. The CONLIN algorithm offers many attractive features that make it ideal for most of our research projects. CONLIN is especially adapted to structural synthesis problems and it has the ability of solving fairly large scale optimization problems (hundreds of design variables and constraints). The computational time needed is moderate, which is useful for an interactive use. It has a built-in constraint relaxation capability that allows the user to start from any infeasible initial design, and even to find a solution to infeasible problems (in the form of minimal relaxation).

Because of its many advantages, the CONLIN algorithm now forms the basis of all the optimization capabilities to be used in our research projects. At each successive iteration point, the CONLIN method only requires evaluation of the objective and constraint functions and their first derivatives with respect to the design variables. This information is provided by the finite element analysis and sensitivity analysis results. The CONLIN optimizer then selects by itself an appropriate approximation scheme on the basis of the sign of the derivatives. CONLIN benefits from many interesting features that make it specially well suited to structural synthesis problems:

- the CONLIN approach is very general, requiring only values and derivatives of the functions describing the optimization problem to be solved; it permits therefore straight interfacing to the finite element software;

- because it is based on conservative approximation concepts, CONLIN does not demand a high level of accuracy for the sensitivity analysis results, which can thus be obtained through a simple finite difference technique;

- CONLIN usually generates a nearly optimal design within less than ten finite element analyses;

- CONLIN has an inherent tendency to produce a sequence of steadily improving feasible designs;

- the CONLIN method is simple enough to lead to a relatively small computer code, well organized to avoid high core requirement ; this feature is interesting for interactive optimization.

CONLIN is not, however, the only possible choice, and many other good optimizers have now become commercially available. Therefore the critical issue is no longer which optimizer to select, but rather how to state properly the shape optimal design problems and how to implement optimization concepts in the context of an industrial environment. This important aspect of Computer Aided Optimal Design has been discussed in Section I.

In summary, then, reliable optimization algorithm such as CONLIN have now become available to solve shape optimal design problems. Of course further research is still needed to improve the optimizers and to expand the class of problems they can now address (e.g. increase in number of design variables and constraints, efficient treatment of nonlinear equality constraints, discrete variables, other convex approximation schemes, etc...). However, in order to facilitate the introduction of optimization concepts within the design cycle, the main development efforts in the future will likely be devoted to devising appropriate user's interfaces allowing for an easy handling of the optimizer.

REFERENCES

[1] A.J Morris (ed.), Foundations of Structural Optimization: a Unified Approach, John Wiley and Sons, 1982.

[2] C. Fleury and L.A. Schmit, "Dual Methods and Approximation Concepts in Structural Synthesis", NASA Contractor Report, NASA-CR 3226, 1980.

[3] V. Braibant and C. Fleury, "An Approximation Concepts Approach to Shape Optimal Design", Computer Methods in Applied Mechanics and Engineering, Vol. 53, pp. 119-148, 1985.

[4] V. Braibant and C. Fleury, "Aspect theoriques de l'optimisation de forme par variation de noeuds de controle", Conception Optimale de Forme, INRIA, Rocquencourt (France), pp. 303-395, 1983.

[5] C. Fleury, "Structural Weight Optimization by Dual Methods of Convex Programming", International Journal for Numerical Methods in Engineering, Vol. 14, No. 12, pp. 1761-1783, 1979.

[6] C. Fleury, "Reconciliation of Mathematical Programming and Optimality Criteria Approaches to Structural Optimization", chapter 10 of Foundations of Structural Optimization: a Unified Approach (A.J. Morris, ed.), John Wiley and Sons, 1982, pp. 363-404.

[7] L. Berke and N.S. Khot, "Use of Optimality Criteria Methods for large Scale Systems", AGARD Lecture Series LS-70, pp. 1-29, 1974.

[8] C. Fleury, "A Unified Approach to Structural Weight Minimization", Computer Methods in Applied Mechanics and Engineering, Vol. 20, No. 1, pp. 17-38, 1979.

[9] C. Fleury and V. Braibant, "Prise en compte de contraintes lineaires dans les methodes duales d'optimisation structurale", Rapport LTAS SF-107, University of Liege, 1982.

[10] C. Fleury and V. Braibant, "Structural Optimization - A New Dual Method Using Mixed Variables", International Journal for Numerical Methods in Engineering, Vol. 23, pp. 409-429, 1986.

[11] J.H. Starnes and R.T. Haftka, "Preliminary Design of Composite Wings for Buckling, Stress and Displacement Constraints", Journal of Aircraft, Vol. 16, pp. 564-570, 1979.

[12] K. Svanberg, "Method of moving Asymptotes - A new Method for Structural Optimization", International Journal for Numerical Methods in Engineering, to appear, 1987.

[13] R. Schnabel, "Determining Feasibility of a Set of nonlinear Inequality Constraints", Mathematical Programming Study", Vol. 16, No. 192, pp. 137-148, 1982.

COMPUTER AIDED OPTIMAL DESIGN OF ELASTIC STRUCTURES

Part III

Application of Structural Synthesis Techniques

Abstract The structural synthesis techniques described in Sections I and II are applied to various optimal design problems. Three structural sizing problems are first presented. Minimum weight design is sought by assuming that the design variables are restricted to the transverse sizes of the structural components (i.e. cross-sectional area of rods and thicknesses of panels). Next shape optimal design problems are considered. The examples offered are concerned with two-dimensional structures in plane stress. They have been selected to illustrate the geometric modeling concepts introduced in Section I. Finally a problem involving a mix of sizing and shape variables is briefly presented to demonstrate the generality and efficiency of convex linearization techniques.

APPLICATION TO STRUCTURAL SIZING

The convex linearization method was initially experimented on some simple problems, such as the 2-bar and 10-bar trusses classical in the structural optimization literature, by adding linear inequality constraints on the bar cross-sections. The results are not reported herein, because they are not very significant, no comparison with other methods being available. In this paper three examples are offered. The first one has been elaborated on the famous 10-bar truss problem. The two others are concerned with real-life aerospace structures.

10-Bar Truss

The first example has been specially devised to make the classical 10-bar truss problem difficult to solve by conventional methods (see Fig. 1). The displacements at nodes 4 and 5 are limited to 2 in. and 1 in. respectively. Instead of assigning a maximum allowable stress limit in the critical member 6, the stress flow (i.e. the force) in member 6 is limited to 2500 lbs.

Table 1. Iteration history for 10-bar truss example

Iteration	Weight	u_4	u_5	f_6
0	8393	1.898	0.8372	-40125.0
1	7290	1.635	0.7389	- 478.0
2	4856	1.980	0.8458	- 675.0
3	4221	1.968	0.972	- 2366.0
4	4095	1.994	0.9987	- 2458.0
5	4057	1.998	0.9992	- 2491.0
6	4058	1.999	0.9998	- 2498.0
7	4053	1.999	0.9999	- 2499.0
8	4050	2.000	1.000	- 2500.0

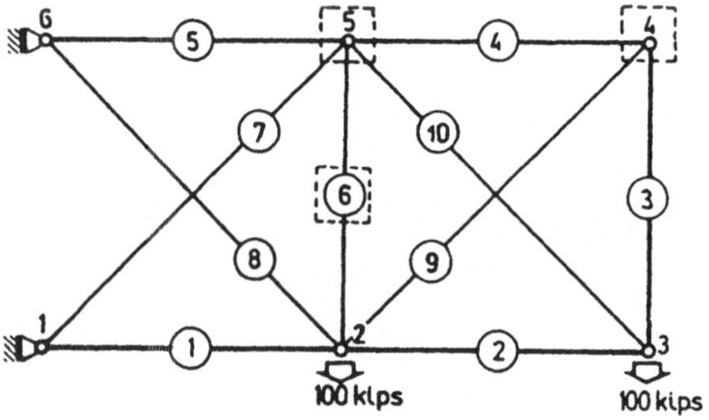

Figure 1. 10 Bar-Truss

Stress flow constraints are difficult to deal with. For a statically determinate truss, the bar forces are constant, and no change in the design can modify them. In the statically indeterminate case under consideration, stress flows are not affected by a scaling of the design variables, so that, if they were the only imposed constraints, the minimum weight design should be zero. In the design space, this means that a stress flow constraint is represented by a restraint surface that passes through the axes origin.

The initial design (x^o = 20 in.^{2} for each bar) is seriously infeasible so that the first explicit subproblem does not admit any solution. Therefore, the relaxation scheme discussed in Section II is employed in the convex linearization method. After this first difficult iteration the optimization process becomes normal, each subsequent feasible subdomain being non-empty. The iteration history data are given in Table 1. Note that except the initial design, all other designs are feasible.

Composite plate

The possibility exists in membrane elements to represent composite materials like fiber reinforced resins as the superposition of a number of layers with independent orthotropic properties. The thickness of each layer is then a separate design variable so that the superposition of results allows for the definition of the composite. This second example consists in weight minimization of such a composite plate. The plate represented in Fig. 2 is part of a floor for an Airbus plane.

The tensile sollicitation is applied by imposing prescribed displacements on one side of the plate. The other boundary conditions depicted in Fig. 2 result from symmetry considerations. The structure is made up of 0°, +45°, -45° and 90° high strength graphite-epoxy laminates. The laminates are represented by stacking four orthotropic membrane elements in each quadrangular region shown in Fig. 2. The layer thicknesses in each basis direction (0°, +45°, -45°, 90°) are the design variables. The finite element model involves 4 x 288 linear isoparameteric elements, and 946 degrees-of-freedom. After design variable linking according to the subdivision in regions of Fig. 2, it remains 39 independent design variables.

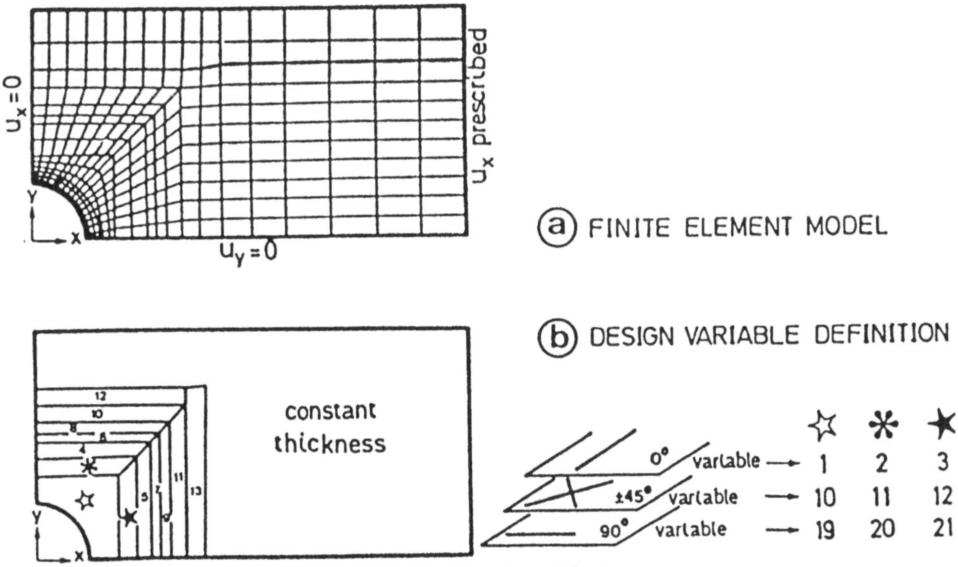

Figure 2. Composite Plate

The behavior constraints correspond to strength requirements based on the Tsai-Azzi failure criterion (with different tensile and compressive allowable stresses). This criterion is applied in the most critical element in each of the 39 linking regions. For manufacturing reasons 27 linear inequality constraints are assigned to the design variable. Typical linear constraints are as follows (see Fig. 2):

$$0 \leq -t_1 - 2t_{10} + 9t_{19} \leq 300$$

$$0 \leq -t_2 - 2t_{11} + 9t_{20} \leq 300$$

$$0 \leq -t_3 - 2t_{12} + 9t_{21} \leq 300$$

$$0 \leq t_1 + 2t_{10} + t_{19}$$

$$-t_2 - 2t_{11} - t_{20} \leq 0.325$$

$$0 \leq t_1 + 2t_{10} + t_{19}$$

$$-t_3 - 2t_{12} - t_{21} \leq 0.325$$

These constraints are linearly dependent and very sparse, which complicate the solution of the optimization problem. Finally, lower and upper bounds are assigned to the design variables.

It was previously shown in Section II that linear constraints make it cumbersome to use a dual method approach (see Eq. II.20). The essential difficulty is that the explicit subproblem leads to first-order discontinuities in the dual space (non-convexity). In fact, this second example is at the origin of the development of the convex linearization method presented in Section II.

When resorting to the convex linearization method, the constraint relaxation technique described in Section II must be activated, because the first explicit subproblem is infeasible. This explains the increase in weight after the first iteration, as indicated in Fig. 3 which plots the iteration-history data.

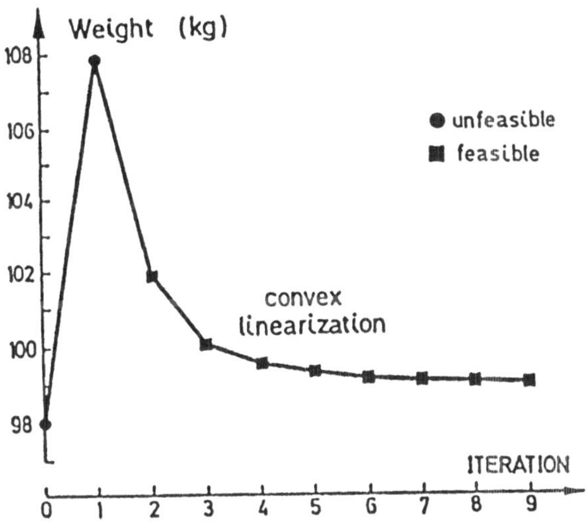

Figure 3. Composite Plate - Optimization History Data

Most of the application problems given in this Section have been solved by using the SAMCEF finite element system, which contains built-in capabilities for both sizing and shape design variables [1]. However we would like to provide additional results for the composite plate, recently obtained with MSC/NASTRAN [2]. Design sensitivity analysis for composite structures will soon be available in MSC/NASTRAN. In addition, as part of a research effort, the sensitivity analysis has been coupled with the CONLIN optimizer [3]. The composite plate (Fig. 2) was selected as a test case to validate the results of the new design sensitivity capability. The design constraints are the failure indices using the Tsai-Azzi criterion, selected for different lamina in specified elements. No linear constraints were considered. The results are given in Table 2.

It is worth mentioning that, after iteration 5, the failure indices of those elements not initially specified as critical were examined. It was found that some of them had exceeded the limit. The failure indices corresponding to the violated elements were input as new design constraints and the optimization loop was restarted at this point. This capability for the user to intervene in the optimization process, and to monitor the progress, is particularly important and convenient for realistic design of structures.

Table 2: **Iteration history for composite plate example (MSC/NASTRAN results)**

Analysis Number	Weight	σ_1 1(00)	σ_2 1(45)	σ_3 2(00)	σ_4 2(45)	σ_5 3(00)
1	.3575	1.1632	1.1421	-	-	-
2	.3562	.9446	.9076	-	-	-
3	.3545	.9886	.9238	-	-	-
4	.3541	.9948	.9160	-	-	-
5	.3540	.9982	.9164	-	-	-
6*	.3539	.9990	.9983	1.0999	1.1053	1.1634
8	.3552	.9552	.9651	.9855	.9994	.9805
9	.3552	.9585	.9690	.9854	.9999	.9796
10	.3552	.9594	.9700	.9853	1.0000	.9757

*User Intervention

Engine mount structure

The last example is concerned with a real-life application of optimization techniques to the European launcher Ariane 4. Four strap-on liquid boosters will be attached to a future version of the launcher in order to double the thrust. The Belgian company SABCA is working to design and build the three main structures of the booster: the forward skirt, the intertank skirt and the engine mount structure (see Fig. 4). From the beginning the interest of resorting to the optimization capabilities of the SAMCEF finite element system was recognized at various levels of the company. The main reason was the fundamental importance of obtaining a light weight structure: 1 kg gained on the booster permits increasing the payload by 0.14 kg. It was not possible to achieve this goal by conventional design techniques because of unusual specifications (stiffness requirements, stress flow limitations).

This application has led to many ups and downs, especially because the specifications were initially in conflict, so that no feasible design could be obtained. Due to the lack of space, attention is focused in this paper only on the engine mount structure.

The finite element model shown in Fig. 5 involves 4883 degrees-of-freedom and 1008 finite elements (second-degree displacement field). The

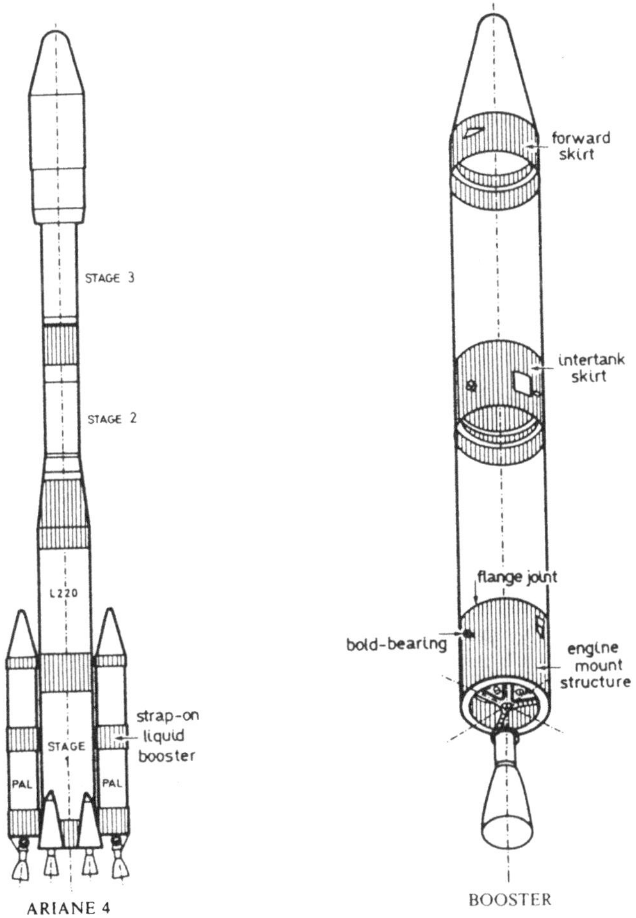

Figure 4. Real-Life Example

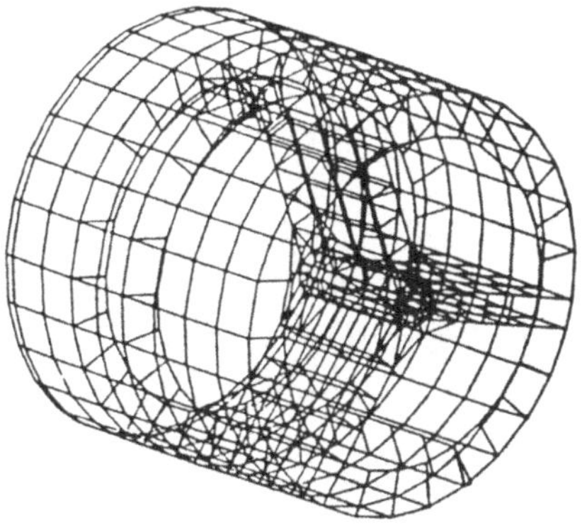

Figure 5. Engine Mount Structure (finite element model)

objective is to minimize the structural weight subject to the following behavior constraints.

- stiffness requirements at the bold-bearing joint, as well as at the point where the engine load is introduced, in order to take into account dynamics aspects;

- limitation of the normal stress flow in the upper ring, in order to diffuse the load transmitted to the upper flange joint;

- maximum allowable von Mises stresses under four loading conditions;

- in addition, local stiffness requirements must be taken into consideration at various critical points (e.g. where equipments are supported).

Note that all the stiffness constraints consist in fact in assigning upper limits to influence coefficients.

By using a simplified finite element model (half-cylinder), the optimization problem has shown that the normal stress flow limitation and one of the stiffness requirements were incompatible. After a while the responsible companies decided to reconsider these specifications and new stiffness constraints were imposed, which made it possible to obtain a feasible design. Several runs were then performed, with more and more accurate definitions of the manufacturing constraints (design variable linking, lower and upper bounds on the thicknesses). For the last optimization process, which involve 35 design variables and 20 behaviour constraints, convergence is achieved within 7 finite element analyses.

Finally, it is worth mentioning that, when the manufacturing process was started, it was found that the optimization results could not be used as such because of technological requirements that did not appear at first. The thickness distributions had to be modified. After analyzing the new design, the stress flow constraints in the upper ring were seen to be seriously violated at the bold bearing joint level. Therefore, it was decided to perform an ultimate optimization run with an appropriate design variable linking. The final problem involves 62 design variables and seven active behavior constraints out of 40; six variables reach their upper bounds, and one its lower bound. As shown in Fig. 6, the stress flow along the upper fing was properly cut to its limiting value. This additional run required four more structural analyses, each analysis demanding four hr CPU on a VAX 11/780 computer.

APPLICATION TO SHAPE OPTIMAL DESIGN

In this Section two examples of application of the convex linearization method to shape optimal design are offered. Although the two structures under consideration are simple from an analysis point of view, their geometric description is fairly complex when considering the design aspects. Therefore the present paper will try to emphasize with these examples, the fundamental concept of creating a design model well suited to shape optimization [4].

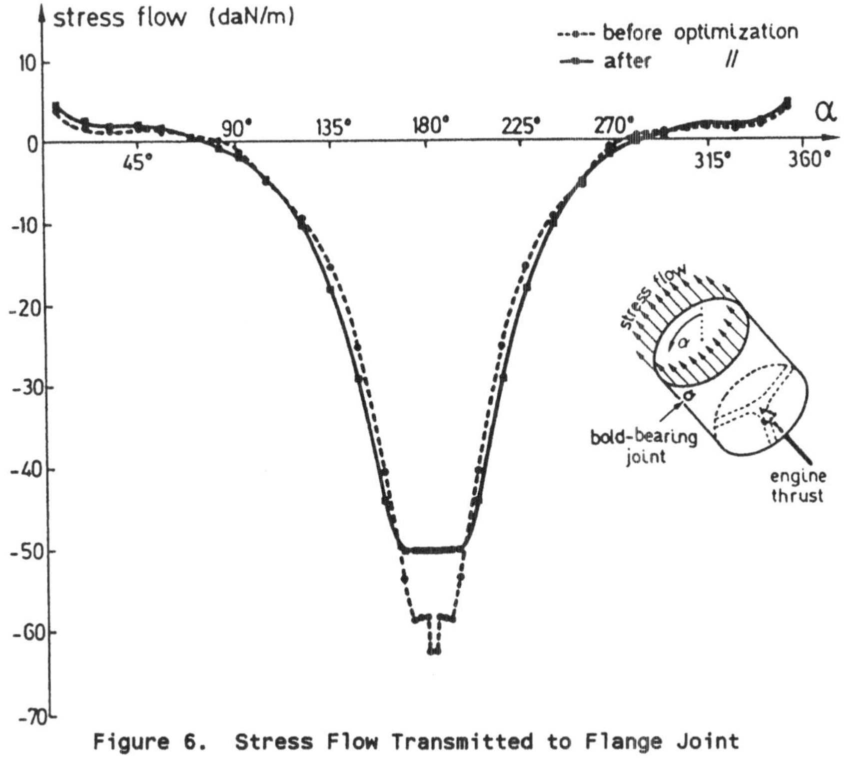

Figure 6. Stress Flow Transmitted to Flange Joint

Axisymmetric Disk

The first example consists of designing a axisymmetric turbine disk. The meridian cross-section of the disk is made up of four parts, as indicated in Fig. 7a: the hub (uniform thickness), the disk itself, the rim and the blades. The shape optimal design problem is posed as finding the shape of the disk and the thickness of the hub that yield minimum weight while satisfying upper limits on radial and circumferential stresses under thermal and centrifugal loads. The thicknesses of the rim and attached blades are considered to be fixed to predetermined values. Note that by symmetry only one half of the structure needs to be studied.

The geometric description of the design model begins with defining all its boundaries and selecting appropriate parametric curves. In the present case many parts of the boundary are required to be straight segments. Each segment can be considered, for example, as a particular case of a Bezier curve. The only curved part of the boundary will be represented by a B-Spline with 6th order continuity (i.e. degree five). To control it, nine master nodes are employed, as indicated in Fig. 7.b. The second step is to subdivide the structure into design elements, which is quite obvious for this problem. As shown in Fig. 7.b, two design elements are sufficient to fully describe the moving boundaries, corresponding to the disk and its hub. The two other subregions, representing the rim and the blades, have a fixed geometry. Next the design variables must be selected in order to monitor acceptable changes in shape of the two design elements. As a reminder, the design variables provide the positions of the master nodes describing the boundary curves. Here the hub thickness and the disk shape have to be determined while keeping the rim and blades thickness constant. Therefore only nodes numbered 2 through 10 in Fig. 7.b will be permitted to move. Finally the fourth step consists of expressing constraints restricting the control node displacements. For example the structure may not move into the negative side of the Z-axis. Also all the design elements must keep reasonable geometries.

To facilitate the introduction of these requirements, the design variables are defined as the distances separating each moving node from its corresponding reference node. In addition the move direction of each

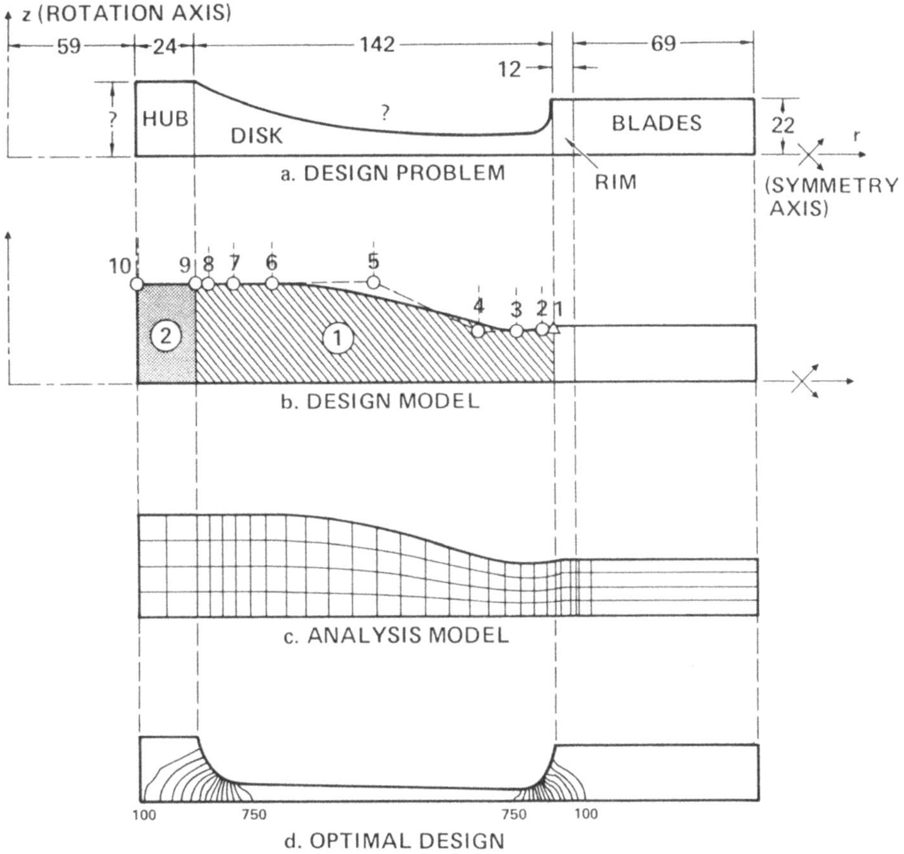

Figure 7. Optimization of a Turbine Disk

control node is kept constant. In the present case the control nodes are required to move in the Z direction. With this definition of the design variables, the geometric requirements can be easily stated and treated by the optimization algorithm:

- the hub must have a uniform thickness: it is sufficient to impose that the displacements of nodes 9 and 10 be the same; this is a simple equality constraint between two variables, which can be eliminated before entering the optimizer (design variable linking);

- in order to prevent the moving nodes to penetrate the negative Z-space, the design variables are imposed to remain positive (side constraints).

Having constructed a proper design model, involving only 8 independent variables, an analysis model can now be created. The finite element mesh is shown in Fig. 7.c. An optimal design is generated in five structural analyses, each followed by an optimization step using the CONLIN algorithm. The final design is given in Fig. 7.d, together with representation of typical stress contours. Figure 8 represents a three-dimensional view of the disk at iterations one (initial design), three, and five (final design).

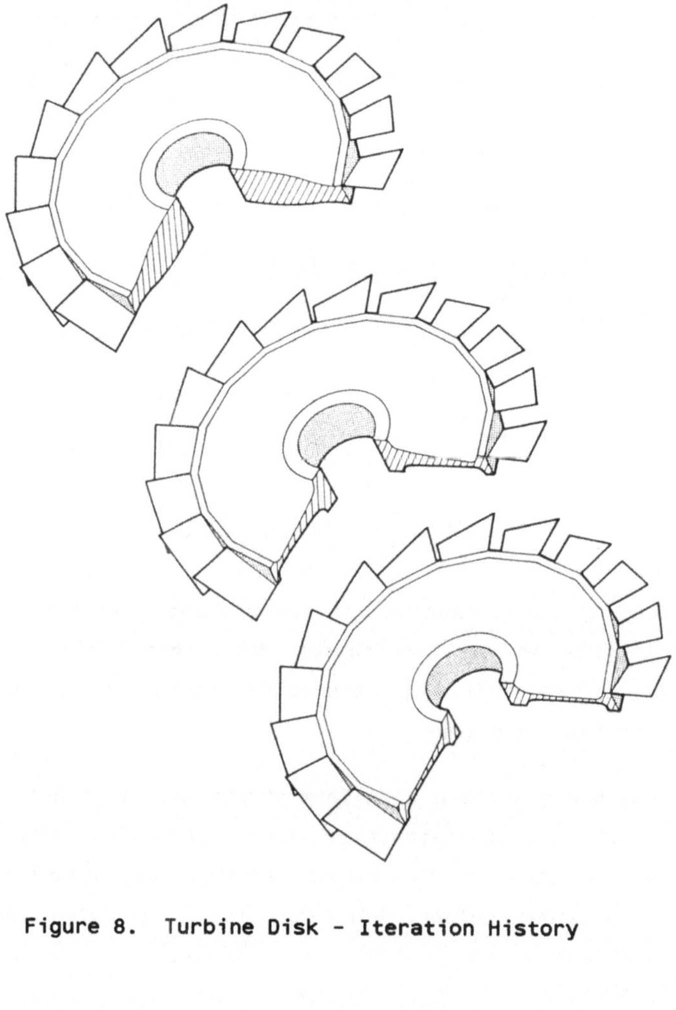

Figure 8. Turbine Disk - Iteration History

Bracket

Many shape optimal design capabilities are restricted to small changes in the geometry. The reason is that, very often, the design model definition is such that the possible shape modifications depend largely upon the initial geometry. In other words the subdivision into design elements requires to have an a priori estimate of the optimum shape, because many of the design element boundaries are fixed in advance. In order to avoid this important limitation, a more flexible design model must be envisioned. The purpose of the following example is to illustrate how Coons patches can be employed to permit larger changes in the geometry. Design elements defined by Coons patches become essentially characterized by their boundaries. As a result there is no longer a need to update the positions of the internal master nodes. Another important consequence is that all the boundaries of each design element are allowed to move.

The bracket shown in Fig. 9.a is clamped at its two lower ends to a rigid foundation, and acted upon by a concentrated load transmitted through a rigid axle. The objective is to minimize the structural weight while assigning maximum allowable values to the Von Mises stress. All the boundaries can be modified, except the internal circular hole (the radius of the axle is fixed). The design model is made up of the six subregions shown in Fig. 9.b. The corresponding analysis model (i.e. finite element mesh) is represented in Fig. 9.c for the initial design, and in Fig. 9.d for the optimal design generated by CONLIN.

For this particular example, it is helpful to provide a somewhat detailed explanation of the procedure followed to create a meaningful optimization model. The six design elements (referred to as A-F) are depicted in Fig. 10.a and the associated design variables are schematized in Fig. 10.b. Design element A has to be defined through a cartesian product of two families of curves, so that connection to elements B and D is possible. The first family of curves are described by cubic B-Splines with seven nodes. They represent, in particular, the external and internal curved boundaries (A_1-A_2 and A_3-A_4). The second family of curves are also cubic, and they define, in particular, the straight boundaries A_1-A_4 and A_2-A_3. As usual, the design variables measure the distances between each

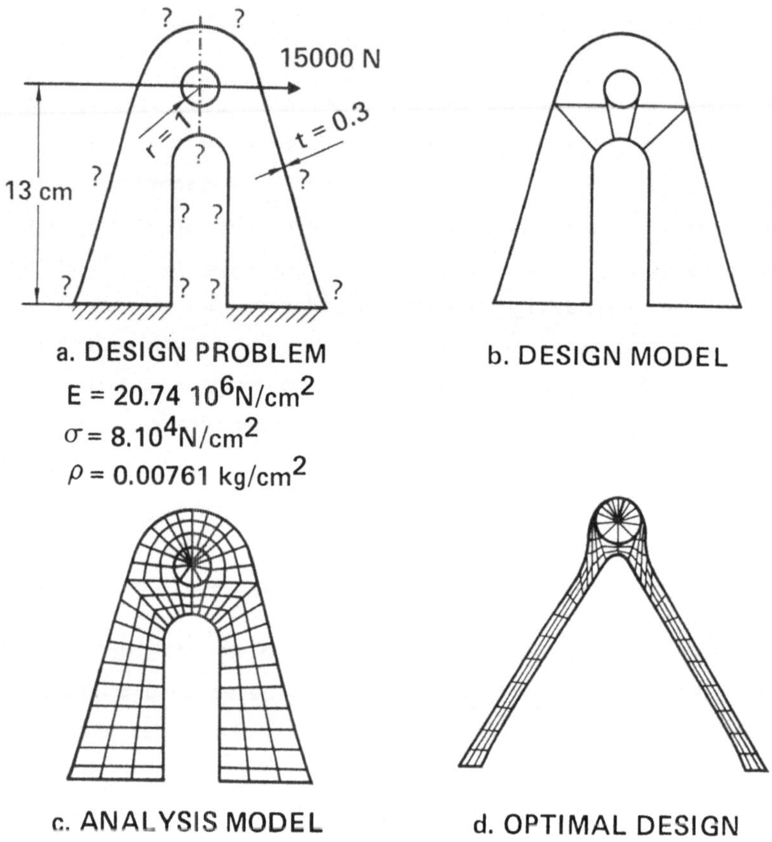

Figure 9. Optimal Design of a Bracket

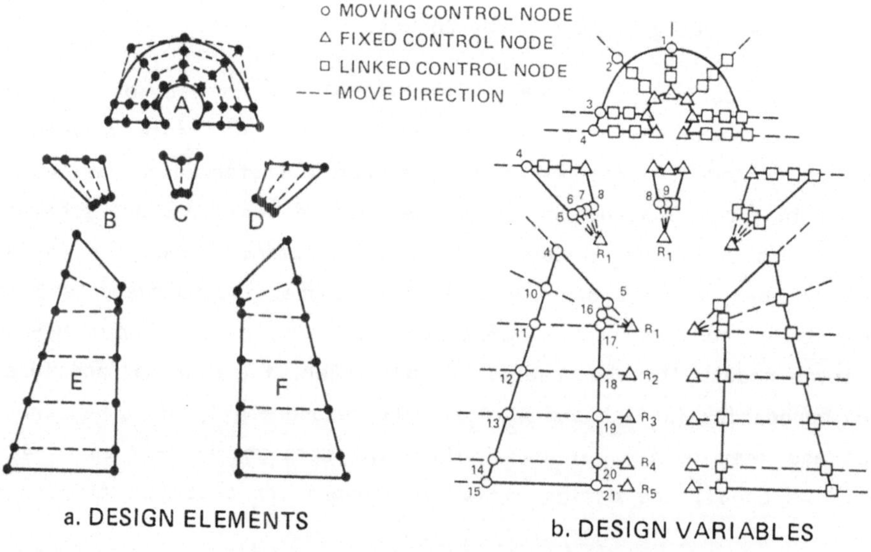

Figure 10. Detailed Design Model

fixed control node (i.e. attached to the internal hole) and the corresponding moving control node (i.e. attached to the external unknown boundary). Because of its definition by cartesian product, design element A possesses internal master nodes, which will be moved homothetically with respect to the external control nodes. Note that this will have to be reflected in the definition of the corresponding design variables associated with elements B and D. So seven design variables fully describe the element. However, because a symmetrical design is sought, only four independent variables are retained (see Fig. 10.b).

On the other hand, the remaining design elements B through F are defined by Coons patches, which means that all their control nodes can be considered as design variables. For these elements, two opposed edges of each patch are characterized by Bezier curves, the two other opposed edges being straight segments. Design variable linking is again employed to impose symmetry. Elements B and D are described each by eight control nodes (the curved boundaries are governed by cubic functions). However only four additional design variables are introduced, because the upper parts also belong to the previously considered element A. For the same reason, element C involves quadratic curves, but only one new design variable. Note that one single master node (denoted R_1 in Fig. 10.b) serves as reference node to define the design variables associated with elements B, C and D. Finally element E and its symmetrical element F are described by 14 control nodes (the boundaries are Bezier curves with 7 control nodes). All the control nodes are permitted to move along the directions shown in Fig. 10.b. These move directions are defined by introducing fixed reference nodes (denoted R_1 through R_5). Elements E and F involve 12 additional design variables, since variables 4 and 5 have been previously considered.

Therefore, by taking into account the foregoing elaborated linking scheme, only 21 independent design variables are sufficient to properly characterize a design model suitable to optimization. To terminate the description of the model, it should be mentioned that linear constraints must be added to the problem statement in order to impose tangential continuity of the curved boundaries at the junctions between design elements (points denoted 4, 5 and 8 in Fig. 10 b.) For example, the control nodes corresponding to the design variables 3, 4 and 10 are assigned to reside on a straight line.

The optimal design produced by CONLIN in a few iterations is represented in Fig. 9d. It can be seen that the finite element mesh remains quite reasonable, despite the large modifications in shape produced by the optimizer. Note that the final design looks very much like a 2-bar truss, which brings us close to topological design.

MIX OF SIZING AND SHAPE VARIABLES

The CONLIN optimizer has successfully been applied to problems involving both sizing and configuration optimization of planar trusses [5]. Using design variable linking to preserve symmetry, the 13-bar truss shown in Fig. 11 is described by 7 sizing variables (cross-sectional areas) and 3 configuration variables (node coordinates). The structural weight has to be minimized subject to stress limitations. The final design is given in Table 3 and Fig. 11. Figure 12 shows the iteration history. The convergence is extremely rapid with a majority of the weight change coming from the initial iterations.

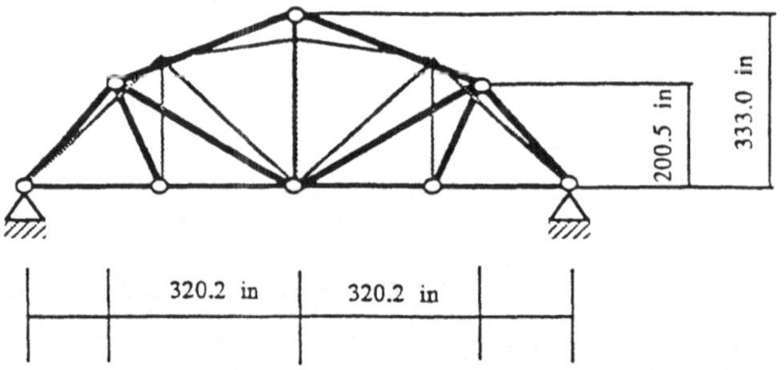

Figure 11. Configuration Optimization of a 13-bar Truss

One important goal of this investigation was to establish comparison with the Method of Moving Asymptotes (MMA) [6], which is a generalization of the CONLIN approach. The MMA method also relies on convex approximation techniques, however, it allows for the adjustment of the subproblem convexity through the use of "asymptotes" on the design variables. The addition of these asymptotes provides some control on the convergence properties. As can be seen in Fig. 11, both approaches furnish similar results. They provide significant improvement over more conventional nonlinear programming techniques such as employed in Ref. [7]. In the particular example discussed above, the total number of structural analyses performed before attaining an optimal design can be reduced by a factor of 20 (see Table 3).

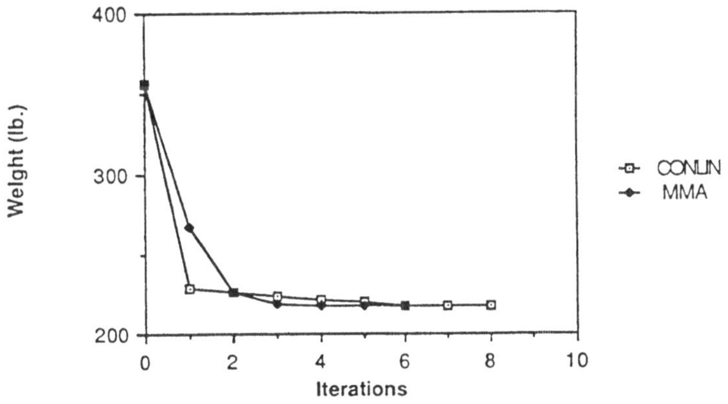

Figure 12. Iteration History for 13-bar Truss

Table 3. Thirteen Bar Truss Configuration and Sizing Problem

	Initial	CONLIN Final	MMA Final	Imai Final
A_1	1.0	1.28	1.28	1.01
A_2	1.0	0.10	0.10	1.00
A_3	1.0	0.54	0.54	0.72
A_4	1.0	1.17	1.18	0.85
A_5	1.0	0.10	0.10	0.10
A_6	1.0	0.10	0.10	0.62
A_7	1.0	0.89	0.89	0.92
X_2	240.0	320.2	319.7	458.9
Y_2	240.0	200.5	199.2	210.5
Y_4	280.0	333.0	330.4	506.0
Weight	356.4	216.8	216.9	269.8
Iterations		8	6	7
			C=1.00	(213 analyses)
			t=0.75	

REFERENCES

[1] SAMCEF, Systeme d'Analyse des Milieux Continus par Elements Finis, University of Liege, Belgium.

[2] MSC/NASTRAN User's and Application Manuals, The MacNeal-Schwendler Corporation, 1984.

[3] G.K. Nagendra and C. Fleury, "Sensitivity and Optimization of Composite Structures Using MSC/NASTRAN," paper presented at the NASA/VPI Symposium on "Sensitivity Analysis in Engineering," NASA Langley Research Center, Hampton, VA, 1986.

[4] V. Braibant, "Optimisation de forme des structures en vue de la conception assistee par ordinateur," Doctoral Dissertation, University of Liege (Belgium), 1985.

[5] S.P. Kuritz, "Configuration Optimization of Trusses using Convex Linearization Techniques," Master Thesis, UCLA, 1986.

Take-Off in Optimum Structural Design

H.R.E.M Hörnlein

MESSERSCHMITT-BÖLKOW-BLOHM GMBH
Helicopter and Aircraft Group
P.O. Box 801160, 8000 Munich 80
W. Germany

Abstract

This paper is intended for the potential developer and user of structural design software. A great number of recent publications have had a strong influence on the development and improvement of structural design software. New ideas have to be studied, realized and tested to develop or to use a state-of-the-art program. In this report I have attempted to make comments on some of the basic ideas. The practical knowledge and experience gained during the development of our in-house programming system, which is designated LAGRANGE, were taken into consideration. The features of about 30 internationally used program systems for structural design are listed in the appendix.

1. Introduction

How to find an optimum design? Take a few experienced, interdisciplinary scientists from the problem areas concerned and provide them with relevant analysis software, a large computer, an established CAE/CAD system, a data base and a few computer scientists. Create some suitable optimality criteria and add a mathematical programming library. Now put all these items together, break the project down into clear single steps by milestones and produce the necessary pressure by a close time schedule. Eventually, after a couple of years you will be surprised to see that your competitors have already finished their job.

Admittedly, this sounds silly and provocative although the 'ingredients' are right and necessary even when being more serious. Over the past twenty years, the problem of structural optimization has become a challenge for interdisciplinary cooperation among engineers, physicists and mathematicians.

The early days of variant comparisons and intuitive optimality criteria are over. Anybody who wants to deal with up-to-date optimum design tools has to become familiar with recent synthesis approaches as well as with established analysis procedures.

Not only the developer has to be well acquainted with the details but also the user should be able to tune the software to the problem specific properties. The existing program systems for designing special structural models, and more importantly, for designing the general FE structural models, have a long way to go before they can be used as 'black boxes'.

NATO ASI Series, Vol. F27
Computer Aided Optimal Design: Structural and
Mechanical Systems. Edited by C. A. Mota Soares

2. The Optimum Design Problem

In terms of physics, the problem of structural design is the optimization of the selected objective function, which is subject to certain restrictions, known as constraints. The essential structural responses, which are used to state the objective function and the constraints, are given in Table 1.

Structural responses	system equations and corresponding		requirements
• Structural weight	$W = c^T x$		$\underset{\sim}{W} \leq W(x) \leq \bar{W}$
• Displacements	$Ku = P$		$\underset{\sim}{u} \leq u(x) \leq \bar{u}$
• Stresses	$\sigma = Su$		$\underset{\sim}{\sigma} \leq \sigma_e(x) \leq \bar{\sigma}$
• Strains	$\varepsilon = Bu$		$\underset{\sim}{\varepsilon} \leq \varepsilon_e(x) \leq \bar{\varepsilon}$
• Aeroelastic efficiencies	$Ku = P + Qu$		$\underset{\sim}{\varrho} \leq \varrho(x) := \alpha + a^T u(x) \leq \bar{\varrho}$
• Aeroelastic divergence	$(K - \mu Q)u = O$		$\mu(v_d) \Rightarrow \underset{\sim}{v_d} \leq v_d(x)$
• Buckling	$(K - \mu K_g)u = O$		$\mu(P_b) \Rightarrow \underset{\sim}{P_b} \leq P_b(x)$
• Natural frequencies	$(K - \omega^2 M)u = O$		$\underset{\sim}{\omega} \leq \omega(x) \leq \bar{\omega}$
• Flutter speed	$(K - \omega^2 M - \mu Q)u = O$		$\mu(v_f) \Rightarrow \underset{\sim}{v_f} \leq v_f(x)$
• Transient response	$Ku + C\dot{u} + M\ddot{u} = F(t)$	*linear case*	
	$A\dot{u} = F(u, x, t)$	*nonlinear case*	

Table 1

Structural responses - i.e. state variables - which are used for modelling the problem, are given by the system equations, which are implicitly dependent on the design variables. System equations are set up as differential equations, difference equations or algebraic equations depending on the type of structure - continuous or discrete - and on the imposed constraints. FE modelling to deal with more general structures is now gaining ground.

One possible classification of the problems is to divide them according to the mathematical characterizations of the system equations, see Table 1. A second, more common classification would be by the design space. Figure 1 illustrates the potential of the design space using a bridge as an example.

The most difficult decision certainly is the choice of **building systems** in this hierarchy. To date, I know of no decision criteria that can be automated. The layout of elements - i.e. the **topology** - has been a subject of research for more than 100 years, with considerable success. Among the most important studies are those of Maxwell 1869 [40] and Michell 1904 [41], whose ideas were first transformed in 1964 with the aid of large computers into usable algorithms [17,33]. Originally, all the work done in answer to the question, 'which nodes are connected by which elements?' was limited to truss structures. Recent investigations in layout theory have been concerned with other building systems [39,53]. The design space of topological variables definitely offers the largest potential in structural improvements and must therefore be seen as a real challenge in the future. A simple example of optimum bridge topology is confirmed by real life structures, see Fig. 2.

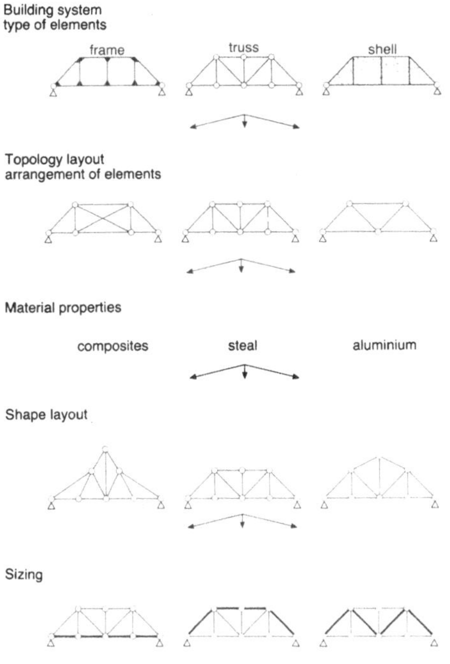

Fig. 1 **Hierarchy of Design Variables**

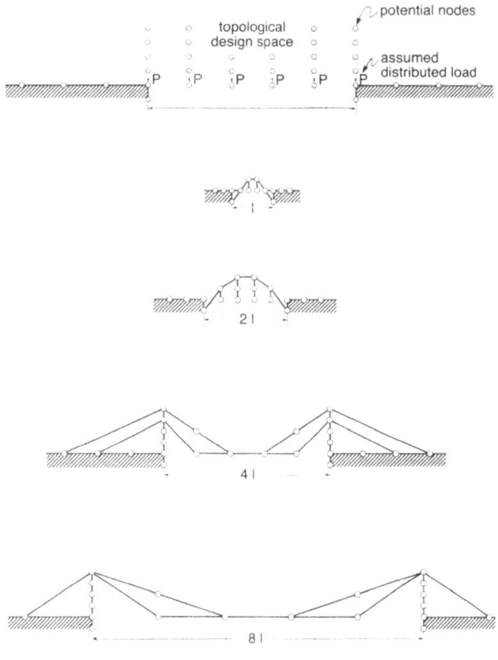

Fig. 2 **Simplified Optimum Bridge Layout with Respect to the Span**

The layout theory is also concerned with the **optimum shape** of structures. Shape variables or coordinates of nodes have been applied extensively to FE structures for some years [9,10,19,22,50,67,75,79]. In this context, mathematical programming is applied mainly to FE models. Problems arise in the shape optimization due to **adaptive mesh refinement** and **grid design.**

The lowest level in this hierarchy is given by the **sizing of elements.** The thickness and cross-sectional variables of the elements appear quite simply in most constraints, so the calculation of derivatives can be done analytically.

In many programming systems the treatment of **composite material** is already possible. However, there are varying opinions as to which parameters should be treated as variables. On the one hand, the discrete fibre orientation angles can be preset, and only the thicknesses of the individual layers need be varied to determine the suitable fibre direction. On the other hand, the fibre orientation angles can also be defined as continuous design variables. But the latter leads to serious nonlinearities and therefore to multimodal constraints, resulting in feasible domains that are neither convex nor simply connected.

3. The Design Model

The mathematical formulation of the structural optimization task is the general nonlinear programming problem (NLP):

$$\begin{array}{llll} \text{minimize} & f(x) & & \text{(objective function)} \\ \text{subject to} & g_j(x) \geq 0 & j \in J & \text{(behavioural constraint)} \\ & \underset{\sim}{x}_i \leq x_i \leq \tilde{x}_i & i = 1(1)m & \text{(side constraints)} \end{array} \tag{1}$$

New program systems offer the user a multitude of possibilities of modifying the variables, constraints, and objective functions in a suitable way.

Variable linking reduces the m structural variables x_i by affine linear transformation

$$x(t) = a + At \qquad \text{where } x,a \in \mathbb{R}^m;\ t \in \mathbb{R}^{\tilde{m}};\ A \in \mathbb{R}^{m \times \tilde{m}} \tag{2}$$

to the $\tilde{m}$ design variables t_i . This allows the utilization of serial parts, structural symmetry, fabricational requirements, and the fixing of structural variables.

The **transformation of variables**, e.g. the frequently used reciprocal variable, changes the characteristics of the constraint functions and thus increases the efficiency of particular synthesis procedures. The stiffness and strength requirements become 'more linear' by the reciprocal variables and thus can be dealt with by a sequence of linearized subproblems.

The **approximation of constraints** reflects the basic idea of approximating the implicit, complex, nonlinear constraints locally by suitable explicit functions. Linear, hyperbolic as well as hybrid formulations are used for the approximation [51,63]. The local approximations should be at least of first order. If additional characteristics such as convexity or even separability are fulfilled, these can be utilized by the synthesis formulation. **Global approximations** by what is known as reduced basis functions are also used to represent the characteristics of the state variables in the extended domain of the design space.

The **multi-level** concept or the **decomposition** of the structural optimization problems has been used since the early seventies and deserves particular attention. Large-scale problems are split into small, nested problems. The normally few global constraints, such as flutter speed or aeroelastic efficiencies, are formulated in terms of global variables - variable linking or smearing out the local design variables - and are dealt with in an outer loop. The many local costraints, such as stress and buckling failure of elements, are essentially dependent to the local variables and are dealt with in an inner loop.

This forcible, intuitive decoupling of variables does, of course, not yield a rigorous model of analysis. The error, however, can be minimized by interactive mechanisms such as **optimum sensitiviity analysis** with respect to the parameters and other coupling information. This heuristic decomposition does not ensure that the true optimum is determined, but makes large-scale problems solvable and eventually leads very close to optimum designs. The success of the multi-level approach greatly depends on the user's physical understanding. This iterative scheme has been succeessfully applied to some practical structures, such as wings, frame designs, ships, etc. [34,58,68].

Vector optimization allows the simultaneous consideration of several requirements, known as **multi objectives.** Of course, nobody expects solutions that are optimal for all these requirements. **Pareto optimal** solutions are used as an orientation instead. These are designs which do not allow simultaneous improvement of all the objectives. In this situation, the user must decide on which of the requirements is to be sacrificed in favour of the remaining objectives.

This situation, which is often referred to as a trade-off study, reminds one of the interpretation of Lagrangian multipliers. In fact, there is a relation between the Pareto optimal solutions and the Lagrangian multipliers [3]. Practically all the calculation methods are based on the scalarization of the multi-criteria problems and the use of the necessary or sufficient conditions of Pareto optimality. In terms of decision theory, a scalar preference function replaces the descision maker Refs. 3,7, 12,18,70.

4. The Design Process

The various methods available for the formulation of the physical model and the objectives can be supplemented by numerous user options of modern FE analysis tools. What is more important and interesting are the tools for the implementation of the actual design process.

The **active set strategies** divide the constraints J into active JA and passive JP. During the iterative design process, a permanent change of active $JA := \{j \in J : g_j = 0\}$ and passive constraints $JP := \{j \in J : g_j > 0\}$ takes place. The 'less active' constraints have less influence on the current design change and are therefore temporarily neglected. Suitable deletion of passive costraints accelerates the design process, but this is not easy to manage. Minor design changes may lead to a change from passive to active constraints and vice versa, which may cause an oscillation and thus a deterioration of convergence. Near the optimum, the estimated Lagrangian multipliers may be used as indicators. A simple interpretation of the first order necessary Kuhn-Tucker conditions reveals that the active constraints are characterized by positive Lagrangian multipliers.

The determination of active or potentially active constraints is one of the main problems of all design procedures. The information of the active constraints is more or less directly used for the actual modification of the current design.

The **school of optimality criteria** uses the following iterative approach to solve the problem:

OC	Optimality criteria
	$x_i^{\nu+1} := \psi(x_i^\nu, \lambda^{\nu+1}) \qquad i = 1(1)m$ Where $\lambda^{\nu+1}$ is an estimation of Lagrangian multipliers and ψ is a suitable recurrence relation.

The estimation of the Lagrangian multipliers is directly derived from the active constraints. On closer examination, both the derivation of recurrence relation ψ and the estimation of Lagrangian multiplies λ represent nothing but the more or less tricky attempt to fulfil the necessary nonlinear Kuhn- Tucker conditions [8]:

$$\nabla_x L(x^*,\lambda^*) = 0 \qquad \nabla_\lambda L(x^*,\lambda^*) \geq 0 \tag{3.1}$$
$$\lambda_j^* \, g_j(x^*) = 0 \qquad \lambda_j^* \geq 0 \quad \forall j \in J \tag{3.2}$$

Where $L = f - \lambda^T g$ is the Lagrangian function and x^* has to satisfy a certain regularity, which can also be guaranteed by what is known as constraint qualification.

The **school of mathematical programming** uses the following iterative formulation:

MP	**Mathematical programming**
	$x^{\nu+1} := x^\nu + \alpha^\nu s^\nu$ where $s^\nu \in \mathbb{R}^m$ is the search direction and $\alpha^\nu \in \mathbb{R}^+$ is the step size

Also here, the knowledge of the active constraints is used for the calculation of the search direction s^ν. The calculation of the step size α^ν is a relatively simple matter, which, however, requires many structural analyses depending on the problem and the applied optimizer and must therefore be carried out very efficiently. Without going into detail, a classification of mathematical programming methods for solving the constraint problems (1) is given below:

Transformation methods

- o Penalty functions
- o Barrier functions
- o Method of mutlipliers

Primal methods

- Indirect methods
  - o sequential linear programming (SLP)
  - o sequential quadratic programming (SQP)
- Direct methods
  - o gradient projection method (GPM)
  - o generalized reduced gradients (GRG)
  - o method of feasible directions (MFD)

Dual concepts

This comprises all methods for the solution of the **dual problem:**

$$\begin{aligned} &\textit{maximize } \ell(\lambda) \\ &\textit{subject to } \lambda \geq 0, \qquad \ell(\lambda) = \underset{\underline{x} \leq x \leq \bar{x}}{\textit{minimum}}\, L(x,\lambda) \end{aligned} \tag{4}$$

Problem (4) is dual with respect to the **primal problem** (1) in the sense of an indentical solution

$$f(x^*) = \ell(\lambda^*) = L(x^*,\lambda^*) \tag{5}$$

if the relatively strict conditions

NLP convex, differentiable with
non-singular Hessian matrix $\nabla\nabla_x L(x^*,\lambda^*)$ (6)

are fulfilled.

Compared with the numerous text books on mathematical programming there are only a few books on structural optimization which describe the complete optimization process Refs. 83,84,85,86,87,88,89,90,91,92,93,94.

5. Monitoring, Terminiation and Interaction

The significance of the duality theory can also be seen in the calculation and monitoring of the active constraints as well as in the application of an efficient termination criterion, the **'dual gap'** [4]. In addition, also the **optimum sensitivity analysis** is explained by duality.

For initial calculation or estimation of the Lagrangian multipliers frequent use is made of the linearized dual problem (4). During the design iteration these values are updated by repeatedly checking the Lagrangian function for stationary condition and the complementary conditions (3). The active constraints are identified by the positive Langrangian multipliers as has already been mentioned. The quantities of the Lagrangian multipliers have at least the same importance since this provides for an additional termination criterion. In conjunction with (5) the well-known saddle point condition provides a bounding of the optimum:

$$L(x^*,\lambda) \leq f(x^*) \leq L(x,\lambda^*) \tag{7}$$

Apart from the relative errors:

$$| f(x^\nu) - f(x^{\nu+1}) | \, / \, | f(x^{\nu+1}) | \leq \varepsilon_f \tag{8.1}$$
$$\| x^\nu - x^{\nu+1} \| / \| x^{\nu+1} \| \leq \varepsilon_x \tag{8.2}$$

also corresponding vectors $\lambda^\nu, x(\lambda^\nu) := arg\,\{\underset{\underline{x} \leq x \leq \bar{x}}{minimum}\, L(x,\lambda^\nu)\}$ and $x^\nu, \lambda(x^\nu) := arg\,\{\underset{\lambda \geq 0}{maximum}\, L(x^\nu,\lambda)\}$ may be used to calculate upper and lower bounds for $f(x^*)$ respectively:

$$L_L := L(x(\lambda^\nu),\lambda^\nu) \leq f(x^*) \leq L(x^\nu,\lambda(x^\nu)) =: L_U \tag{9}$$

The design optimality is defined with this dual gap as additional termination criterion

$$L_L / L_U \geq desired\ optimality \tag{10}$$

to realize the convergence towards non-optimum structures, e.g. FSD. Refs. 4,35,43,72.

If an optimum structure $x^* \in DS(p)$ has eventually been determined within the respective design space $DS(p) \subset \mathbb{R}^m$, potential improvements in the optimum $f(x^*)$ with respect to the specifications p are frequently desired.

The assessment of a change of the parameters p is carried out by means of the derivatives of the optimum $f(x^*(p))$ with respect to p. This process is called optimum sensitivity analysis and is again based on the calculation of the Lagrangian multipliers.

The Kuhn-Tucker conditions (3) of the problem:

$$\begin{aligned} &\textit{minimize } f(p,x(p)) \\ &\textit{subject to } g_j(p,x(p)) \geq 0, \quad j \in J \end{aligned}$$

yield with a differentiable dependence on

$$\frac{d}{dp}(D_xL) = 0 \Leftrightarrow \frac{d}{dp}\left\{ D_xf(p,x(p)) - \lambda^T(p)\, D_xg(p,x(p)) \right\} = 0 \Leftrightarrow$$

$$\frac{\partial}{\partial p}(D_xf) + D_x^2f \cdot D_px - D_p^T\lambda \cdot D_xg - \lambda^T \left\{ \frac{\partial}{\partial p}(D_xg) + D_x^2g \cdot D_px \right\} = 0 \tag{11}$$

and with the active constraints JA

$$\frac{d}{dp}(D_\lambda L) = 0 \Leftrightarrow \frac{d}{dp}\left\{ g_{JA}(p,x(p)) \right\} = 0 \Leftrightarrow$$

$$\frac{\partial}{\partial p}(g_{JA}) + D_xg_{JA} \cdot D_px = 0 \tag{12}$$

where the Cauchy symbol D was used for the derivatives. The desired derivatives can be calculated using the equation systems (11) and (12) [69].

The special case df/dp is derived directly from $df/dp = \partial f/\partial p + D_xf \cdot D_px$ with the Kuhn-Tucker condition $D_xf = \lambda^T D_xg_{JA}$ and (12):

$$\frac{df}{dp} = \frac{\partial f}{\partial p} + \lambda^T D_xg_{JA} \cdot D_px = \frac{\partial f}{\partial p} - \lambda^T \frac{\partial}{\partial p}(g_{JA}) \tag{13}$$

A simple interpretation of (13) shows why the Lagrangian multipliers are often referred to as the 'price' of constraints. When considering a single constraint $g(x) \geq p \Leftrightarrow g_j(x,p) := g(x) - p \geq 0$ whereas f is independend of p, (13) yields $df/dp = \lambda_j$. This information of the Lagrangian multipliers can be used for the interactive change of specifications p. Refs.59,69,76.

6. Pros and Cons: OC - MP

The initially parallel development of the two competing design philosophies - **optimality criteria** (OC) and **mathematical programming** (MP) - have long been reduced to a common basis thanks to the equivalence of the primal and dual optimization problem [21,25,29,73]. For this reason, both concepts are used in modern programming systems. The successful application of both schools of thought has given rise to mutual acceptance.

OC methods are oriented to the **global** criteria of the putative optimum - like a pilot flying above the clouds will consider the weather forecast for the potential landing sites. The OCs exploit the particular characteristics of the active constraints at the optimum for fast convergence, which is relatively independent of the initial design and the number of design variables. But they converge in general toward a **non-optimum design,** if the active constraints of the optimum have not been anticipated properly - like the pilot, who will miss the runway if fog has come down in the meantime.

MP-methods, in contrast, are oriented to the **local** information of the analysis and the sensitivity analysis in the vicinity of the current design - as a blind man would grope his way down to the valley using his cane. They converge relatively fast, even without prediction of the active constraints, but governed by the problem size, toward a **local optimum** still far from the actual global optimum, - in the same way the blind man will get stuck in the nearest hole, because the **unimodality,** which is the general assumption of any mathematical programming theory, is not fulfilled.

The realization of these different properties has obviously led to the emergence of mixed concepts. It is, therefore, advisable to carry out the global approach, the preliminary design, using OC methods and then continue the process locally with the MP methods. The crucial question of the most suitable solution method is controversial and certainly not to be answered in general. It would be better to ask for suitable combinations of synthesis ideas and to investigate the interaction criteria. Refs. 27,28,37.

7. New Trend: GOC

The OC algorithms are only valid for special types of constraints, but solve them faster than the more general MP algorithms. The success with OC procedures shows that in the future, MP procedures should also exploit the features of the structural mechanical system equations. An attempt to unify the advantages of both is a recent development. The earlier intuitive optimality criteria were directly derived from the physical characteristics of the optimum design. Today, they are interpreted as the more rigorous mathematical criteria, the Kuhn-Tucker conditions.

The more or less arbitrary choice of the active constraints at the optimum causes frequent failure of the OC methods. While studying the literature, it is possible to find at least as many arguments against as for the use of the OC procedures. A more careful assessment of the active constraints extends the OCs and makes them more reliable. As already mentioned, the **dual concepts** provide estimated values for the Lagrangian multipliers and facilitate an iterative update of the active constraints. Thus, the 'more consequent' dual concept use is also often referred to as a **generalized optimality criteria** (GOC) method.

In this context, the promising trend of **explicit constraint approximation** should be mentioned. A suitable approximation of the objective function and the constraint functions leads to a highly efficient formulation of the dual problem. For a sginificant class of standard problems, the primal problem can be approximated fairly well by convex, separable functions, which leads to an attractive dual problem structure. For the sequence of the approximated separable functions, the Lagrangian functions are also separable, so that for the subproblems - see (4) - only one-dimensional minimization problems need to be solved:

$$\underset{\underset{\sim}{x} \leq x \leq \tilde{x}}{minimize} \left\{ L(x,\lambda) = \sum^{i} \left(f_i(x_i) + \sum^{j} \lambda_j \cdot g_{ji}(x_i) \right) \right\} \tag{14}$$

This approach has also been used successfully with discrete variables. Refs. 11,23,24,25,26,28,29,61,71,72.

8. Design sensitivity analysis

Provision of derivatives for the objective function and constraints with respect to the design variables is knwon as **design sensitivitiy analysis** (DSA). Once it is realized how 'expensive' it is to calculate gradients using finite differences, then it is obvious that this is an important subject. By exploiting the analytical relations of the system equations, computing time can be reduced in proportion to the number of design variables: 'the more analytical the DSA, the cheaper the calculation'. Furthermore, an improvement results from exploiting the synthesis relations for a particular class of mathematical programming, which again leads to a reduction in computing time of the same order.

Of course, it is possible to employ synthesis procedures which do not need sensitivity derivatives. The 'price' of this waste of valuable information is a degraded convergence. Apart from the synthesis method, the first and second order information is used for problem approximations, process monitoring and termination. Even more, in the final stage of the design process, the derivatives with respect to the parameters provide the user with valuable information for a revision of the parameters. For these reasons the sensitivity analysis has become an indispensable tool of structural optimization. Refs. 13,14,15,16,36,49,81.

Only for displacement related constraints: $\delta(x) := s + S(x) \cdot u(x)$ such as displacements, camber, twist, stress, and also aerolastic efficiencies, is the fast FE-based derivative calculation mentioned here. The normalized constraints $g(\delta(x,u(x))) := 1-\delta(x,u(x))/\bar{\delta} \geq 0$ of the equivalent physical restrictions: $0 \leq \delta(x,u(x)) \leq \bar{\delta}$; where $\bar{\delta}$ are specifications, and $u(x)$ is the displacement vector, require the calculation of the Jacobian matrix

$$D_x g = D_\delta g \left\{ \frac{\partial \delta}{\partial x} + D_u \delta \cdot D_x u \right\} \tag{15}$$

where the matrices $D_\delta g$, $(\partial\delta/\partial x)$ and $D_u \delta$ are easy to calculate. The substantial numerical effort is in the calculation of matrix $D_x u$. By means of formal partial differentiation, the system equation for FE structures follows as:

$$K(x) \cdot u(x) = P(x) + Q \cdot u(x) \tag{16}$$

$$K \left\{ \frac{\partial u}{\partial x_i} \right\} = \left(\left\{ \frac{\partial P}{\partial x_i} \right\} - \left(\frac{\partial K}{\partial x_i} \right) u \right) + Q \left\{ \frac{\partial u}{\partial x_i} \right\} \quad \forall\, i = 1(1)m \tag{17}$$

where $Q \neq 0$ for aeroslastic loads.
The equation systems (16) and (17) have the same structure and can be solved using the iterative scheme

$$K y^{\mu+1} = b + Q y^{\mu} \qquad K = LL^T \qquad \mu = 1, 2, \ldots \tag{18}$$

with the stiffness matrix $K(x)$ which has been already factorized $K = LL^T$ for the analysis. With an eigenvalue transformation $\lambda_t = \omega\lambda + (1-\omega)$ of the matrix $K^{-1}Q$, good convergence acceleration can be attained through a suitable choice of relaxation parameters ω.

$$K y^{\mu+1} = \omega(b + Q y^{\mu}) + (1-\omega) K y^{\mu} \qquad \mu = 1, 2, \ldots \tag{19}$$

Substitution of Eq. (17) in (15) yields

$$D_x g = B + C(K-Q)^{-1}R \tag{20}$$

where $B = D_\delta g \left(\frac{\partial \delta}{\partial x}\right)$, $C = D_\delta g \cdot D_u \delta$, $R = \left(\frac{\partial P}{\partial x}\right) - \left[\left(\frac{\partial K}{\partial x_i}\right) u\right]_{i=1(1)m}$

The essential second terms on the right-hand side of (20) can now be calculated either with columns r_i of R (pseudo loads)

$$V := (K-Q)^{-1}R \Longleftrightarrow Kv_i = r_i + Qv_i \quad \underset{i=1(1)m}{\forall} \tag{21}$$

or with rows c_i of C (virtual loads)

$$W^T := C(K-Q)^{-1} \Longleftrightarrow Kw_i = c_i + Q^T w_i \quad \underset{i=1(1)|JA|}{\forall} \tag{22}$$

using the accelerated iteraton (19). This procedure is known as the 'design space' (21) or 'state space' (22) method [1]. Here, equation solutions m and $|JA|$ respectively are required. Refs. 1,20,31,32,55,56,78.

In many synthesis procedures, subproblems of the following type have to be solved:

$$\textit{minimize} \ \left\{\varphi(x) := f(x) + G(g(x))\right\} \tag{23}$$

Formal differentiation with the row vector $\beta^T := D_g G$ gives the gradient

$$\nabla^T \varphi = D_x \varphi = D_x f + \beta^T D_x g \tag{24}$$

By substituting Eq. (20) in (24) and using the 'state space' method (22) with the adjoint variable $\eta^T := \beta^T C$

$$D_x \varphi = D_x f + \beta^T B + \eta^T (K-Q)^{-1} R \tag{25}$$

give the gradient $\nabla\varphi$ by one single equation solution (22), (19). In the most common case, for the static loads $Q = 0$, the equation solution consists of one single forward-backward substitution. The subproblems (23) appear in the transformation methods and also in the dual concepts.

9. Test Studies

The circumstances in which a synthesis idea proves to be successful can only be discovered by thorough testing. There are a multitude of tests available which assess MP methods [57]. Here, computing time (efficiency), accuracy, reliability, ease of use (tuning) and the sensitivity to the initial designs are compared as evaluation criteria. But since the test examples have very general properties, the results from these tests are not immediately applicable to the problems of structural design. Special tests for structural design are rarely carried out [5, 6].

At the moment, a large-scale test for wing-like FE structures with stiffness and strength constraints is running at MBB. This test is not exhaustive, since only a few transformation methods compete with sequential linear programming. A further expansion of this test model is expected to result in a continuous improvement of the user guidelines, i.e. to a kind of **'expert system'**. Inclusion of OC procedures is important,because many program systems already use OC optimal structures as initial design.

10. A challenge to the MP experts

The use of mathematical programming experts' methods and ideas is becoming increasingly important in the field of structural optimization problems. Linear and quadratic problems are solved in a finite number of steps. However the procedures for solving general nonlinear problems are iterative, thus there is a healthy interest in the convergence analyses of these procedures. Unfortunately, these convergence properties have only been proved for some restrictive assumptions, such as for quadratic or, at the most, convex problems. For example, the m-step-quadratic convergence could be proved by Stoer (1977) for the widely used quasi-Newton method using asymptotically exact line search. This property shows that the convergence speed is dependent on the number of variables. It is percisely this dependency which is often confirmed by practical application to large-scale problems.

For a long time, attempts have been made to use the problem inherent properties in constructing the design model. Hence, the use of reciprocal variables, suitable constraint approximations, the linking of variables, even the application of 'move limits' are more or less desperate attempts to make the 'real life' problems more attractive to the established MP methods. Why not the other way round?

Thanks to many years of experience in using MP methods, the experts in structural optimization now have such an intimate knowledge of MP theory that they ought to meet the MP experts in order to develop problem-specific procedures. In the past the structural experts have tried out almost the entire range of MP models, such as 'Dynamic Programming, Geometric Programming, Fractional Programming, Integer Programming' etc. Refs. 42,43, 44,48,74.

The school of mathematical programming ought to accept this challenge, and signal its readiness for interdisciplinary collaboration. It is no surprise that Haftka and Kamat end their book 'Elements of Structural Optimization' (1985) with the statement:
"There are also several packages available from mathematical programming specialists. However, these programs do not enjoy as much populartiy in structural optimization application as those developed by engineers."
A good survey of structural optimization can be gained from the many excellent review papers Refs. 2,30,38,45,46,47,52,54,60,66,82.

Appendix

The following survey of approximately 30 internationally used program systems contains some information about their field of applications and solution concepts. The data was acquired by means of a questionnaire sent to the authors and users of these program systems in November 1983. This data was updated by means of another questionnaire in 1985/86. This list is by no means complete, neither are all the important features summarized nor all the currently used program systems listed. For instance, no investigation was carried out on the frequently used approximation techniques or the variable transformation and the variable linking possiblities. Quantitative characteristics of te analysis modules as well as the maximum number of design variables and constraints were deliberately omitted since these limitations can be extended by using larger computers.

Programming System	Name of program	ACCESS 1-3	DDDU	DESAP1 [DESAP2]	DOCS	ELFINI	FASTOP [FASTOP3]	JOPT/EAL	LAGRANGE
	Period of development	? 1974 →	3 years	3-4 years	1978 - 1982	1976 →	? 1976 →	83 →	84 →
	Developer	UCLA Schmit et al.	UNI DALIAN Qian	NASA Kiusalaas Reddy	UNI IOWA Arora et al.	DASSAULT Lecina Petiau	GRUMMAN	NASA Walsh Adelmann	MBB Hörnlein et al.
Structures	Type of structures	FE	FE	FE	FE	FE	FE	FE	FE
	Composite material	x	x	x [x]	x	x	[x]		x
Design-variable	Sizing	x	x		x	x	x [x]	x	x
	Fiber orientation angles					x			(x)
	Shape variable		(x) node			(x) node			(x)
	Remarks (x) means in development	discrete variables							
Obj. Funct.	Weight (Volume)	x	x	x [x]	x	x	x [x]	x	x
	Multiobjective					?		x	
Constraints	Stress	x	x	x [x]	x	x	x [x]	x	x
	Displacement	x	x	x	x	x	[x]	x	x
	Buckling of elements		x	x [x]	x	x	x [x]	x	(x)
	Natural frequencies	x			x	x		x	x
	Buckling of structure			[x]		x			
	Flutter speed					x	x [x]		(x)
	Divergence speed					x	[x]		(x)
	Aeroelastic efficiencies					x	[x]		x
	Remarks (x) means in development	approximation concepts			fail save constraints				
Synthesismodul	Simultaneous redesign	x	x		x	x		x	x
	Interactive redesign			x [x]			x [x]		
	MP: mathematical programm.	x	x		x	x		x	x
	OC: optimality criteria		x	x [x]		x	x [x]		x
	Optimizer strategy	primal/dual	SQP			SLP/SQP	SRM	SLP	SLP/SUMT/MOM
	Optimizer method	1st & 2nd order			PG	PCG/QN		MFD	SQP/GRG/QN
	Remarks	research program	dual concepts			dual concepts	optimality concepts	CONMIN	

SLP = sequential linear programming
SCP = sequential convex programming
SQP = sequential quadr. programming
SUMT = sequence of unconstraint minimization technique

VPM = variable penalty method
MOM = method of multipliers
GRG = generalized reduced gradients
PCG = projected conjugate gradients

MFD = method of feasible directions
CD = conjugate directions
CG = conjugate gradients
QN = Quasi Newton

QP = Quadratic programming
PG = projected gradients
SRM = stress rationing method

	Name of program	MAESTRO	OASIS	OTPCOMP	OPTFORCE II	OPTI/SAMCEF	OPTIMA	OPTNSR	OPTSYS
Programming System	Period of development	1980-83	1981 ⟶	?	1977-82	1973 ⟶	1976 ⟶	1978-80	?
	Developer	UNI NEW SOUTH WALES Hughes	UNI STOCKHOLM Esping	AFWAL/FIBRA Khot	BELL Gellatly	UNI LIEGE Fleury et al.	UNI STUTTGART Mlejnek et al.	UNI BERKELAY Bhatti Pister/Polak	SAAB-SCANIA FFA
Structures	Type of structures	FE	FE	FE	FE	FE	FE	FE	FE
	Composite material		x	x	x	x			x
Design-variable	Sizing	x	x	x	x	x	x	x	x
	Fiber orientation angles		x		x	(x)			x
	Shape variable		x		x	x	(x)		x
	Remarks (x) means in development								
Obj. Funct.	Weight (Volume)	x	x	x	x	x	x	x	x
	Multiobjective	x	x				(x)	x	x
Constraints	Stress	x	x	x	x	x	x	x	x
	Displacement	x	x		x	x	x	x	x
	Buckling of elements	x		x		x	x	x	(x)
	Natural frequencies		x		x		(x)		x
	Buckling of structure	x					(x)		
	Flutter speed								(x)
	Divergence speed								(x)
	Aeroelastic efficiencies								x
	Remarks (x) means in development			strain					
Synthesismodul	Simultaneous redesign	x	x		x	x	x		x
	Interactive redesign							? x	
	MP: mathematical programm.	x	x		x	x	x	x	x
	OC: optimality criteria			x	x	x	x		
	Optimizer strategy	SLP	SCP			SLP/SCP	SLP/SQP/VPM		SCP
	Optimizer method		QN			GRG/CG	MFD/CG	MFD/QP	QN
	Remarks		dual concepts		dual concepts	dual concepts	dual concepts		dual concepts

SLP = sequential linear programming
SCP = sequential convex programming
SQP = sequential quadr. programming
SUMT = sequence of unconstraint minimization technique

VPM = variable penalty method
MOM = method of multipliers
GRG = generalized reduced gradients
PCG = projected conjugate gradients

MFD = method of feasible directions
CD = conjugate directions
CG = conjugate gradients
QN = Quasi Newton

QP = Quadratic programming
PG = projected gradients
SRM = stress rationing method

	Name of program	PARS/EAL	PROSSS	SADDLE	STARS		ACTUATOR	AXYOPT	DEWAR
Programming System	Period of development	1980-84	1978-84	1980-83	1975 →		1981 →	1983-86	1985 →
	Developer	FORD MOTOR Haftka Prasad	NASA Rogers Sobieski	UNI IOWA Rajan Bhatti	RAE Bartholomew Morris		LOCKHEED Bushnell	INSA Trompette	LOCKHEED Bushnell
Structures	Type of structures	FE	FE	FE	FE		actuated mirrors	axisymmetric structures	truss
	Composite material	x	x		x		x		x
Design-variable	Sizing	x	x	x	x		x		x
	Fiber orientation angles		x						
	Shape variable	x	x	x	(x)		x	x	
	Remarks (x) means in development		integrated design				actuator locations		payload support truss
Obj. Funct.	Weight (Volume)	x	x	x	x		surface error	stress concentr	conductance
	Multiobjective	x	x				x	x	
Constraints	Stress	x	x	x	x				x
	Displacement	x	x	x	x		x		x
	Buckling of elements	x	x		x				x
	Natural frequencies	x	x	x	x				x
	Buckling of structure		x						x
	Flutter speed	x			(x)				
	Divergence speed				(x)				
	Aeroelastic efficiencies				(x)				
	Remarks (x) means in development	strain					actuator stroke/force	weight	clearance constraints
Synthesismodul	Simultaneous redesign	x	x		x		x	x	x
	Interactive redesign			? x				x	
	MP: mathematical programm.	x	x	x	x		x	x	x
	OC: optimality criteria		x		x				
	Optimizer strategy	VPM	SLP/SRM	SOP	SQP/SRM			MOM	
	Optimizer method	QN	MFD	MFD/PG	QP/PG		MFD	QN	MFD
	Remarks		CONMIN	interactive optimizer	dual concepts		CONMIN		CONMIN

SLP = sequential linear programming
SCP = sequential convex programming
SQP = sequential quadr. programming
SUMT = sequence of unconstraint minimization technique

VPM = variable penalty method
MOM = method of multipliers
GRG = generalized reduced gradients
PCG = projected conjugate gradients

MFD = method of feasible directions
CD = conjugate directions
CG = conjugate gradients
QN = Quasi Newton

QP = Quadratic programming
PG = projected gradients
SRM = stress rationing method

	Name of program	MAST 1-3	ODORCS	ODYSSEY	PANDA [PANDA2]	PASCO	TRUOPT	TSO	
Programming System	Period of development	1976-80	2 years	78 →	1981 →	1976-83	4 years	? 1975 →	
	Developer	SINTEF Hanssen	UNI CARDIFF Majid	GENERAL MOTORS Bennett Botkin	LOCKHEED Bushnell	NASA Stroud	UNI TUCSON Wang et al.	GENERAL DYNAMICS	
Structures	Type of structures	transmission towers	reinforced concrete	beams plates	stiffened plates/shells	stiffened plates	space trusses	plates	
	Composite material				x [x]	x	x	x	
Design-variable	Sizing	x	x	x	x [x]	x	x	x	
	Fiber orientation angles				x [x]	x		x	
	Shape variable	x		x		x			
	Remarks		also topology		stiffener spacing				
Obj. Funct.	Weight (Volume)	x	x	x	x [x]	x	x	x	
	Multiobjective						x	x	
Constraints	Stress	x	x	x	x [x]	x	x	x	
	Displacement		x	x			x	x	
	Buckling of elements	x		x beam	x [x]	x	x		
	Natural frequencies			x		x		x	
	Buckling of structure				x [x]	x			
	Flutter speed								
	Divergence speed							x	
	Aeroelastic efficiencies							x	
	Remarks				plastic [post] buckling	strain			
Synthesismodul	Simultaneous redesign	x	x	x	x [x]	x	x	x	
	Interactive redesign				x				
	MP: mathematical programm.	x	x	x	x [x]	x	x	x	
	OC: optimality criteria	x							
	Optimizer strategy		SLP					SUMT/MOM	
	Optimizer method	MFD/CD		MFD	MFD	MFD	GRG/CG	QN	
	Remarks			CONMIN	CONMIN	CONMIN			

SLP = sequential linear programming
SCP = sequential convex programming
SQP = sequential quadr. programming
SUMT = sequence of unconstraint minimization technique

VPM = variable penalty method
MOM = method of multipliers
GRG = generalized reduced gradients
PCG = projected conjugate gradients

MFD = method of feasible directions
CD = conjugate directions
CG = conjugate gradients
QN = Quasi Newton

QP = Quadratic programming
PG = projected gradients
SRM = stress rationing method

REFERENCES:

[1] **J.S.ARORA,E.J.HAUG**(1979) 'METHODS OF DESIGN SENSITIVITY ANALYSIS IN STUCTURAL OPTIMIZATION' AIAA J.,VOL.17,NO.9,PP.970-974.

[2] **H.ASHLEY**(1982) 'ON MAKING THINGS THE BEST-AERONAUTICAL USES OF OPTIMIZATION' J. OF AIRCRAFT,VOL.19,NO.1,PP.5-28.

[3] **H.BAIER**(1978) 'MATHEMATISCHE PROGRAMMIERUNG ZUR OPTIMIERUNG VON TRAGWERKEN INSBESONDERE BEI MEHRFACHEN ZIELEN' DOCTORAL DISSERTATION,TH DARMSTADT,GERMANY.

[4] **P.BARTHOLOMEW**(1979) 'A DUAL BOUND USED FOR MONITORING STRUCTURAL OPTIMIZATION PROBLEMS' ENGNG. OPTIMIZATION,VOL.4,PP.45-50.

[5] **A.D.BELEGUNDU,J.S.ARORA**(1984) 'A COMPUTATIONAL STUDY OF TRANSFORMATION METHODS FOR OPTIMAL DESIGN' AIAA J.,VOL.22,NO.4,PP.535-542.

[6] **A.D.BELEGUNDU,J.S.ARORA**(1985) 'A STUDY OF MATHEMATICAL PROGRAMMING METHODS FOR STRUCTURAL OPTIMIZATION. PART I: THEORY / PART II: NUMERICAL RESULTS' INT. J. NUM. METH. ENGNG.,VOL.21, PP.1583-1599/PP.1601-1623.

[7] **M.P.BENDSOE,N.OLHOFF,J.E.TAYLOR**(1984) 'A VARIATIONAL FORMULATION FOR MULTICRITERIA STRUCTURAL OPTIMIZATION' J. STRUCT. MECH., VOL.11, NO.4, PP.523-544.

[8] **L.BERKE,V.B.VENKAYYA**(1984) 'REVIEW OF OPTIMALITY CRITERIA APPROACHES TO STRUCTURAL OPTIMIZATION' ASME STRUCT. OPT. SYMP. , EDITOR L.A. SCHMIT, PP.23-34.

[9] **M.E.BOTKIN,J.A.BENNETT**(1984) 'SHAPE OPTIMIZATION OF THREE-DIMENSIONAL FOLDED PLATE STRUCTURES' AIAA J.,VOL.23,NO.11,PP.1804- 1810.

[10] **V.BRAIBANT,C.FLEURY**(1984) 'SHAPE OPTIMAL DESIGN - A PERFORMING CAD ORIENTED FORMULATION' REPORT SA-114,UNIVERSITY OF LIEGE,BELGIUM.

[11] **V.BRAIBANT,C.FLEURY**(1984) 'AN APPROXIMATION CONCEPTS APPROACH TO SHAPE OPTIMAL DESIGN' REPORT SA-112, UNIV. OF LIEGE, BELGIUM.

[12] **D.G.CARMICHAEL**(1981) 'MULTICRITERIA OPTIMIZATION' CHAPTER 10 IN REF.[84].

[13] **K.K.CHOI,E.J.HAUG**(1983) 'SHAPE DESIGN SENSITIVITY ANALYSIS' J. STRUCT. MECH.,VOL.11,NO.2,PP.231-269.

[14] **C.T.CHON**(1984) 'DESIGN SENSITIVITY ANALYSIS - STRAIN ENERGY VIA DISTRIBUTION' AIAA J.,VOL.22,NO.4.

[15] **K.DEMS,Z.MROZ**(1985) 'VARIATIONAL APPROACH TO FIRST- AND SECOND ORDER SENSITIVITY ANALYSIS OF ELASTIC STRUCTURES' INT. J. NUM. METH. ENGNG., VOL.21, PP.637-661.

[16] **K.DEMS,Z.MROZ**(1983/1984) 'VARIATIONAL APPROACH BY MEANS OF ADJOINT SYSTEMS TO STRUCTURAL OPTIMIZATION & SENSITIVITY ANALYSIS. PART I: VARIATION OF MATERIAL PARAMATER WITHIN FIXED DOMAIN / PART II: STRUCTURAL SHAPE VARIATION' INT. J. SOLIDS & STRUCT.,VOL.19,NO.8, PP.677-692 / VOL.20, NO.6, PP.527-552.

[17] **W.S.DORN,R.E.GOMORY,H.J.GREENBERG**(1964) 'AUTOMATIC DESIGN OF OPTIMAL STRUCTURES' JOURNAL DE MECANIQUE,VOL.3,NO.1,PP.25-52.

[18] **H.ESCHENAUER**(1985) 'RECHNERISCHE UND EXPERIMENTELLE UNTERSUCHUNGEN ZUR STRUKTUROPTIMIERUNG VON BAUWEISEN' FORSCHUNGSBERICHT DER DFG DES INST. FUER MECHANIK UND REGELUNGSTECHNIK DER UNIVERSITAET- GESAMTHOCHSCHULE SIEGEN,PP.9-22.

[19] **B.J.D.ESPING**(1983) 'A CAD APPROACH TO THE MINIMUM WEIGHT DESIGN PROBLEM' REPORT 83-14,ROYAL INST. OF TECHN. STOCKHOLM,SWEDEN.

[20] **B.J.D.ESPING**(1983) 'ANALYTICAL DERIVATIVES OF STRUCTURAL MATRICES FOR AN EIGHT NODE PARAMETRIC MEMBRAN ELEMENT WITH RESPECT TO A SET OF DESIGN VARIABLES' REPORT 83-7, ROYAL INST. OF TECHNOLOGY STOCKHOLM, SWEDEN.

[21] **J.E.FALK**(1967) 'LAGRANGE MULTIPLIERS AND NONLINEAR PROGRAMMING' J. MATH. ANAL. APPL.,VOL.19,PP.141-159.

[22] **C.FLEURY**(1985) 'SHAPE OPTIMAL DESIGN BY THE CONVEX LINEARIZATION METHOD' INT. SYMP.'THE OPTIMUM SHAPE: AUTOMATED STRUCTURAL DESIGN', WARREN, MICHIGAN.

[23] **C.FLEURY,V.BRAIBANT**(1984) 'STRUCTURAL OPTIMIZATION - A NEW DUAL METHOD USING MIXED VARIABLES' REPORT SA-115, UNIV. OF LIEGE, BELGIUM.

[24] **C.FLEURY,G.SANDER**(1983) 'DUAL METHODS FOR OPTIMIZING FINITE ELEMENT FLEXURAL SYSTEMS' COMP. METH. APPL. MECH. ENGNG.,VOL.37, PP.249-275.

[25] **C.FLEURY**(1982) 'RECONCILIATION OF MATHEMATICAL PROGRAMMING AND OPTIMALITY CRITERIA METHODS' CHAPTER 10 IN REF.[91].

[26] **C.FLEURY,L.A.SCHMIT**(1980) 'DUAL METHODS AND APPROXIMATION CONCEPTS IN STRUCTURAL SYNTHESIS' NASA REPORT 3226, UNIV. CALIF.,LA.

[27] **C.FLEURY**(1979) 'STRUCTURAL WEIGHT OPTIMIZATION BY DUAL METHODS OF CONVEX PROGRAMMING' INT. J. NUM. METH. ENGNG.,VOL.14, PP.1761-1783.

[28] **C.FLEURY,M.GERADIN**(1978) 'OPTIMALITY CRITERIA AND MATHEMATICAL PROGRAMMING IN STRUCTURAL WEIGHT OPTIMIZATION' COMP. & STRUCT.,VOL.8,PP.7-17.

[29] **C.FLEURY,G.SANDER**(1977) 'RELATIONSHIPS BETWEEN OPTIMALITY CRITERIA AND MATHEMATICAL PROGRAMMING IN STRUCTURAL OPTIMIZATION' PROC. SYMP. APPL. COMP. METH. ENGNG., UNIV. OF SOUTHERN CALIF.,PP.507-520.

[30] **R.A.GELLATLY**(1972) 'SURVEY OF THE STATE-OF-THE-ART OF OPTIMIZATION TECHNOLOGY WITHIN NATO COUNTRIES' AGARD WORKING GROUP ON OPTIMIZATION,PP.2.1-2.20.

[31] **R.T.HAFTKA**(1982) 'SECOND ORDER SENSITIVITY DERIVATIVES IN STRUCTURAL OPTIMIZATION' AIAA J.,VOL.20,PP.1765-1766.
[32] **E.J.HAUG**(1981) 'SECOND ORDER DESIGN SENSITIVITY ANALYSIS OF STRUCTURAL SYSTEMS' AIAA J.,VOL.19,NO.8,PP.1087-1088.
[33] **W.S.HEMP**(1964) 'STUDIES IN THE THEORY OF MICHEL STRUCTURES' PROC. INT. CONG., APPL. MECH., MUNICH.
[34] **O.F.HUGHES,F.MISTREE,V.ZANIC**(1980) 'A PRACTICAL METHOD FOR RATIONAL DESIGN OF SHIP STRUCTURES' J. SHIP RES.,VOL.24,NO.2, PP.101-113.
[35] **A.H.JAWED**(1985) 'SENSITIVITY APPLICATIONS IN STRUCTURAL DESIGN SYNTHESIS' PH.D. THESIS, CRANFIELD INST. TECHN., COLLEGE OF AERONAUTICS AIRCRAFT DESIGN DIVISION, U.K.
[36] **A.H.JAWED,A.J.MORRIS**(1985) 'HIGHER-ORDER UPDATES FOR DYNAMIC RESPONSE IN STRUCTURAL OPTIMIZATION' COMP. METH. APPL.MECH. ENGNG., VOL.49, NO.2, PP.175-201.
[37] **D.W.KELLY,A.J.MORRIS,P.BARTHOLOMEW**(1977) 'A REVIEW OF TECHNIQUES FOR AUTOMATED STRUCTURAL DESIGN' COMP. METH. APPL. MECH. ENGNG., VOL.12, PP.219-242.
[38] **W.LANSING,E.LERNER,R.F.TAYLOR**(1977) 'APPLICATIONS OF STRUCTURAL OPTIMIZATION FOR STRENGTH AND AEROELASTIC DESIGN REQUIREMENTS' AGARD REPORT NO.664, 45TH STRUCT. AND MATERIALS PANEL MEETING, VOSS, NORWAY.
[39] **K.I.MAJID,D.W.C.ELLIOTT**(1973) 'TOPOLOGICAL DESIGN OF PIN JOINTED STRUCTURES BY NON-LINEAR PROGRAMMING'PROC. INST. CIVIL. ENGRS.,VOL.55,PART 2, PP.129-149,LONDON.
[40] **J.C.MAXWELL**(1869) SCIENTIFIC PAPERS,VOL.2,P.175, CAMBRIDGE UNIV. PRESS.
[41] **A.G.M.MICHELL**(1904) 'THE LIMITS OF ECONOMY OF MATERIAL IN FRAMED STRUCTURES' PHIL.MAG.,SERIES 6,VOL.8,PP.589-597.
[42] **A.J.MORRIS**(1982) 'STRUCTURAL OPTIMIZATION BY GEOMETRIC PROGRAMMING' CHAPTER 17 IN REF.[91].
[43] **A.J.MORRIS**(1978) 'GENERALIZATION OF DUAL STRUCTURAL OPTIMIZATION PROBLEMS IN TERMS OF FRACTIONAL PROGRAMMING' QUARTERLY OF APPL. MATH., VOL.36, NO.2.
[44] **A.J.MORRIS**(1973) 'THE OPTIMIZATION OF STATICALLY INDETERMINATE STRUCTURES BY MEANS OF APPROXIMATE GEOMETRIC PROGRAMMING' AGARD 2ND SYMP. ON STRUCT. OPT.,CP-123, MILAN.
[45] **F.I.NIORDSON,P.PEDERSON**(1973) 'A REVIEW OF OPTIMAL STRUCTURAL DESIGN' PROC. 13TH INT. CONG. THEORY & APPL. MECH., PP.264-278, MOSCOW, SPRINGER-VERLAG.
[46] **N.OLHOFF,J.E.TAYLOR**(1983) 'ON STRUCTURAL OPTIMIZATION' J. APPL. MECH., VOL.50, PP.1139-1151.
[47] **O.E.LEV**(1981) 'STRUCTURAL OPTIMIZATION RECENT DEVELOPMENTS AND APPLICATIONS' ASCE, ISBN 0-87262-281-9
[48] **A.C.PALMER**(1968) 'OPTIMAL DESIGN BY DYNAMIC PROGRAMMING' J. STRUCT. DIV., ASCE, VOL.94,NO.8.
[49] **P.PEDERSEN,A.P.SEYRANIAN**(1983) 'SENSITIVITY ANALYSIS FOR PROBLEMS OF DYNAMIC STABILITY' INT. J. SOLIDS & STRUCT.,VOL.19,NO.4, PP.315-335.
[50] **P.PEDERSEN**(1973) 'OPTIMAL JOINT POSITIONS FOR SPACE TRUSSES' J. STRUCT. DIV. ASCE, VOL.99, NO.12, PP.2459-2476.
[51] **B.PRASAD**(1983) 'POTENTIAL FORMS OF EXPLICIT CONSTRAINT APPROXIMATIONS IN STRUCTURAL OPTIMIZATION - PART 1: ANALYSIS AND PROJECTIONS / PART 2: NUMERICAL EXPERIENCE' COMP. METH. APPL. MECH. ENGNG., VOL.40, PP.1-26 / VOL.46, PP.15-38.
[52] **L.X.QIAN**(1982) 'STRUCTURAL OPTIMIZATION RESEARCH IN CHINA PROC. INT. CONF. ON FINITE ELEMENT METH.,PP.16-24, SHANGHAI,CHINA.
[53] **G.I.N.ROZVANY**(1984) 'STRUCTURAL LAYOUT THEORY - THE PRESENT STATE OF KNOWLEDGE' CHAPTER 7 IN REF.[83].
[54] **G.I.N.ROZVANY,Z.MROZ**(1977) 'ANALYTICAL METHODS IN STRUCTURAL OPTIMIZATION' APPL. MECH. REVIEWS,VOL.30,NO.11.PP.1461-1470.
[55] **C.S.RUDISILL,K.G.BHATIA**(1972) 'SECOND DERIVATIVES OF FLUTTER VELOCITY AND THE OPTIMIZATION OF AIRCRAFT STRUCTURES' AIAA J., VOL.10, NO.12, PP1569-1572.
[56] **C.S.RUDISILL,K.G.BHATIA**(1971) 'OPTIMIZATION OF COMPLEX STRUCTURES TO SATISFY FLUTTER REQUIREMENTS' AIAA J.,VOL.9,NO.8.
[57] **K.SCHITTKOWSKI**(1980) 'NONLINEAR PROGRAMMING CODES. INFORMATION, TEST, PERFORMANCE' LECTURE NOTES IN ECONOMICS AND MATHEMATICAL SYSTEMS, SPRINGER-VERLAG.
[58] **L.A.SCHMIT,K.J.CHANG**(1984) 'A MULTI-LEVEL METHOD FOR STRUCTURAL SYNTHESIS' 25TH AIAA/ASME/ASCE/AHS STUCTURES, STRUCTURAL DYNAMICS AND MATERIAL CONF.,PALM SPRINGS.
[59] **L.A.SCHMIT,K.J.CHANG**(1984) 'OPTIMUM DESIGN SENSITIVITY BASED ON APPROXIMATION CONCEPTS AND DUAL METHODS' INT. J. NUM. METH. ENGNG., VOL.20.
[60] **L.A.SCHMIT**(1981) 'STRUCTURAL SYNTHESIS ITS GENESIS AND DEVELOPMENT' AIAA J., VOL.19, NO.10, PP.1249-1263.
[61] **L.A.SCHMIT,C.FLEURY**(1980) 'STUCTURAL SYNTHESIS BY COMBINING APPROXIMATION CONCEPTS AND DUAL METHODS' AIAA J.,VOL.18,NO.10, PP.1252-1260.
[62] **L.A.SCHMIT,C.FLEURY**(1980) 'DISCRETE -CONTINUUM VARIABLE STRUCTURAL SYNTHESIS USING DUAL METHODS' AIAA J.,VOL.18,NO.12, PP.1515-1524.

[63] **L.A.SCHMIT,B.FARSHI**(1974) 'SOME APPROXIMATION CONCEPTS FOR STUCTURAL SYNTHESIS' AIAA J.,VOL.12,NO.5,PP.692-699.
[64] **L.A.SCHMIT,R.H.MALLET**(1963) 'STRUCTURAL SYNTHESIS AND DESIGN PARAMETER HIERARCHY' J.STRUC.DIV. ASCE,VOL.89,NO.4,PP.269-299.
[65] **C.F.SHEFFEY**(1960) 'OPTIMIZATION OF STRUCTURES BY VARIATION OF CRITICAL PARAMETERS' PROC. 2ND NAT. CONF. STRUC. DIV. ASCE,PP.133-144.
[66] **C.Y.SHEU,W.PRAGER**(1968) 'RECENT DEVELOPMENT IN OPTIMAL STRUCTURAL DESIGN' APPL. MECH. REVIEWS,VOL.21,NO.10,PP.985-992.
[67] **C.A.M.SOARES,H.C.RODRIGUES,L.M.O.FARIA**(1983) 'BOUNDARY ELEMENTS IN THE SHAPE OPTIMAL DESIGN OF SHAFTS' CONF. OPTIMIZATION IN CAD, GROUP 5.2,LYON.
[68] **J.SOBIESKI,B.B.JAMES,M.F.RILEY**(1985) 'STUCTURAL OPTIMIZITION BY GENERALIZED, MULTILEVEL OPTIMIZATION' AIAA/ASME/ASCE/AHS 26TH STRUCTURES, STRUCTURAL DYNAMICS AND MATERIAL CONF.,ORLANDO,FLORIDA.
[69] **J.SOBIESKI,J-F.M.BARTHELEMY,K.M.RILEY**(1982) 'SENSITIVITY OF OPTIMUM SOLUTIONS TO PROBLEM PARAMETERS' AIAA J.,VOL.20,NO.9, PP.1291-1299.
[70] **W.STADLER**(1984) 'MULTICRITERIA OPTIMIZATION IN MECHANICS (A SURVEY)' APPL. MECH. REVIEW,VOL.37,NO.3,PP.277-286.
[71] **K.SVANBERG**(1982) 'AN ALGORITHM FOR OPTIMUM STRUCTURAL DESIGN USING DUALITY' MATH. PROGR. STUDY,VOL.20,PP.161-177.
[72] **A.B.TEMPLEMAN**(1979) 'OPTIMAL SIZING OF LARGE ENGINEERING SYSTEMS BY SEPARABLE DUALS' MATH. PROR. STUDY,VOL.11,PP.108-115.
[73] **A.B.TEMPLEMAN**(1976) 'A DUAL APPROACH TO OPTIMUM TRUSS DESIGN' J. STRUCT. MECH.,VOL.4,NO.3,PP.235-255.
[74] **A.B.TEMPLEMAN**(1970) 'STRUCTURAL DESIGN FOR MINIMUM COST USING THE METHOD OF GEOMETRIC PROGRAMMING' PROC. ICE,VOL.46,PP.459-472.
[75] **B.H.V.TOPPING**(1983) 'SHAPE OPTIMIZATION OF SKELETAL STRUCTURES - A REVIEW' ASCE J. STRUCT. ENGNG.,VOL.109,PP.1933-1951.
[76] **G.N.VANDERPLAATS,N.YOSHIDA**(1984) EFFICIENT CALCULATION OF OPTIMUM DESIGN SENSITIVITY' AIAA PAPER 84-0855.
[77] **G.N.VANDERPLAATS**(1981) 'STRUCTURAL OPTIMIZATION-PAST, PRESENT AND FUTURE' AIAA J.,VOL.20,NO.7,PP.992-1000.
[78] **G.N.VANDERPLAATS**(1980) 'COMMENTS ON: METHODS OF DESIGN SENSITIVITY ANALYSIS IN STRUCTURAL OPTIMIZATION' AIAA J.,VOL.18,NO11,PP.1406-1408.
[79] **G.N.VANDERPLAATS,F.MOSES**(1972) 'AUTOMATED DESIGN OF TRUSSES FOR OPTIMUM GEOMETRY' J.STRUCT. DIV. ASCE,VOL.89,NO.6,PP.671-690.
[80] **V.B.VENKAYYA**(1978) 'STRUCTURAL OPTIMIZATION:A REVIEW AND SOME RECOMMENDATIONS' INT. J. NUM. METH. ENGNG.,VOL.13,PP.203-228.
[81] **S.Y.WANG,Y.SUN,R.H.GALLAGHER**(1985) 'SENSITIVITY ANALYSIS IN SHAPE OPTIMIZATION OF CONTINUUM STRUCTURES' COMP. & STRUCT., VOL.20, NO.5, PP.855-867.
[82] **Z.WASIUTYNSKI,A.BRAND**(1963) 'THE PRESENT STATE OF KNOWLEDGE IN THE FIELD OF OPTIMUM DESIGN OF STRUCTURES' APPL. MECH. REVIEWS,VOL.16,NO.5,PP.341-350.

TEXTBOOKS:

[83] **E.ATREK,R.H.GALLAGHER,K.M.RAGSDELL,O.C.ZIENKIEWICZ: EDITORS** (1984) 'NEW DIRECTIONS IN OPTIMUM STUCTURAL DESIGN' JOHN WILEY & SONS LTD.
[84] **D.G.CARMICHAEL**(1981) 'STRUCTURAL MODELLING AND OPTIMIZATION A GENERAL METHODOLOGY FOR ENGINEERING AND CONTROL' ELLIS HORWOOD LTD., SERIES IN CIVIL ENGINEERING.
[85] **J.FARKAS**(1984) 'OPTIMUM DESIGN OF METAL STRUCTURES' ELLIS HORWOOD LTD, PUBLISHERS.
[86] **R.H.GALLAGHER,O.C.ZIENKIEWICZ: EDITORS**(1977) 'OPTIMUM STRUTURAL DESIGN - THEORY AND APPLICATIONS' JOHN WILEY & SONS.
[87] **R.T.HAFTKA,M.P.KAMAT**(1985) 'ELEMENTS OF STRUCTURAL OPTIMIZATION' MARTINUS NIJHOFF PUBLISHRES, NETHERLANDS.
[88] **E.J.HAUG,J.S.ARORA**(1979) 'APPLIED OPTIMAL DESIGN /MECHANICAL AND STRUCTURAL SYSTEMS' JOHN WILEY & SONS.
[89] **P.HUPFER**(1970) 'OPTIMIERUNG VON BAUKONSTRUKTIONEN' B.G.TEUBNER, STUTTGART.
[90] **U.KIRSCH**(1981) 'OPTIMUM STRUCTURAL DESIGN' MC GRAW-HILL, INC.
[91] **K.I.MAJID**(1974) 'OPTIMUM DESIGN OF STRUCTURES' NEWNES-BUTTERWORTH, LONDON.
[92] **A.J.MORRIS: EDITOR**(1982) 'FOUNDATIONS OF STRUCTURAL OPTIMIZATION: A UNIFIED APPROACH' JOHN WILEY & SONS.
[93] **G.V.REKLAITES,A.RAVINDRAN,K.M.RAGSDELL** (1983) 'ENGINEERING OPTIMIZATION: METHODS AND APPLICATIONS' JOHN WILEY & SONS.
[94] **J.N.SIDDALL** (1982) 'OPTIMAL ENGINEERING DESIGN PRINCIPLES AND APPLICATIONS' MARCEL DEKK, INC.

STRUCTURAL OPTIMIZATION OF SHIP STRUCTURES

by

P. Terndrup Pedersen and N-J. Rishøj Nielsen*

Department of Ocean Engineering
Technical University of Denmark
DK-2800 Lyngby

ABSTRACT

Based on: - a rapid design-oriented finite element method for analysis of large complex thin-walled steel structures such as ship hulls, - a comprehensive mathematical model for the evaluation of the capability of such structures, and - analytical expressions for the derivatives of the response and the capability with respect to the governing design variables, the paper presents a procedure for minimizing a linear combination of structural weight and production costs for a number of loading conditions. The large, highly constrained non-linear optimization problem is solved by sequential linear programming.

1. INTRODUCTION

It is the object of the paper to describe a procedure for structural optimization of marine structures in the form of large thin-walled steel constructions.

Marine structures such as ships and offshore platforms are of great complexity and size. Perhaps of greater complexity and size than any other type of man-made structure. They are built of elements which possess unknown residual stresses before they are ever joined, and in the building process they are subjected to processes that introduce deformations, stresses and strains which escape computation by present analysis procedures. On top of this, the system is launched and put into service in the hostile environment of the sea which can only be described by statistical methods.

Considering these facts, one is led to recognize that a fairly objective assessment of the strength of a marine structure is a very difficult task and to optimize such a structure even more difficult.

---

* Now: Maersk Oil and Gas A/S, Copenhagen

NATO ASI Series, Vol. F27
Computer Aided Optimal Design: Structural and Mechanical Systems. Edited by C. A. Mota Soares

This is probably one of the reasons why until now structural optimization procedures have had limited applications in practical design of marine structures in spite of the fact that considerable effort has been put into the development of such programs.

A review of the work carried out in this field up to 1981 is presented in /1/.

Among the more significant research work on optimization of marine structures reference can be made to work at the Technical University of Trondheim, Norway, on Indets (The Integrated Design of Tanker Structures) /2/. In this program system all the automated design programs employ the sequential unconstrained minimization technique (SUMT) for the non-linear optimization problems.

At the University of Newcastle, U.K., there has been an ongoing effort towards development of optimization procedures for ship hull structures /3/. The approach taken here is simplified analysis using the displacement method and an optimization procedure based on a sequential linear programming algorithm.

Probably the most successful structural optimization programs for marine structures until now are those developed by O. Hughes and coworkers /4/, /5/. These programs are based on a synthesis of finite element analysis, strength constraints, and optimization using sequential linear programming (SLP). The objective may be any continuous non-linear function of the design variables such as weight or cost.

The optimization procedure to be described in the present paper has adopted many of the features described in /4/ and /5/ concerning the segmentation of the structure and the element representation.

The main loading on marine structures is due to wave action. Normally, a wave response analysis is performed in two steps. In the first step an overall response analysis is done. Currently, most methods used for practical calculations of the global structural response of ships and offshore structures in waves are based on linear beam models. This overall analysis serves to give boundary conditions to the second step which is a more detailed analysis of submodules.

A submodule is a region of structure in which a sufficient number of the scantlings are linked such that the region forms a logical entity. An example of such a submodule may be one hold in a cargo ship.

Here we shall describe a procedure which can analyze and optimize a submodule in the form of a ship compartment, i.e. a cargo hold between bulkheads.

The response analysis is based on the finite element procedure, and the presentation will focus on simulation of the interaction between the submodule and the surrounding structure through boundary conditions.

In the evaluation phase some of the constraints for the optimization problem is established with the requirement that the demand must not exceed the capability of the structure. These requirements are to a large extent based on strength analysis of structural elements such as stiffened or unstiffened plates and beams. Other constraints to the problem are due to fabrication and functional requirements.

The optimization procedure combines the finite element method, the constraints, and sequential linear programming (SLP) with the purpose to minimize a function of cost and weight. The optimization problem is characterized by

- many design variables
- a large number of design criteria
- most of the design criteria are strongly non-linear in the design variables, and
- the objective function is non-linear.

The SLP procedure used in the present paper is based on variable move limits and analytical sensitivity analyses.

In the final section an application of the optimization procedure is presented.

2. RESPONSE ANALYSIS

In order to reduce the computational effort, a design oriented finite element program has been developed especially for structural analysis of prismatic thin-walled structures. The main features of the mathematical model are

- There is no variation in the longitudinal direction of the structural member sizes.
- A stiffened panel between major longitudinals and transverse beams is modeled with only one special ribbed rectangular membrane element.
- Major longitudinals and transverse beams are modeled with Timoshenko beam elements, and the effect of brackets is taken into account by introducing rigid end zones.
- To simplify the structural modeling and to speed up the redesign analysis, the structure is subdivided into strakes and girders. Strakes are defined as prismatic sections between two guidelines having frame segments at the transverse planes and stiffened plating between the frame segments. The length of each strake (number of bays) can be varied. Girders are defined as longitudinal beams positioned along the guidelines.

Figure 1 visualizes the structural modeling.

An important consequence of the prismatic modeling technique is that it reduces the number of design variables and thereby decreases the optimization effort. As shown in Figure 1, eight design variables have been chosen for each strake, and four design variables have been chosen for each girder. Furthermore, the total redesign problem can be divided into a series of subproblems corresponding to strake-redesigns and girder-redesigns, which are "easier" to handle for the program and at the same time reduce computational cost.

The analysis is performed for a number of different loading conditions where the loads are specified together with statistical information describing the uncertainty of the loads.

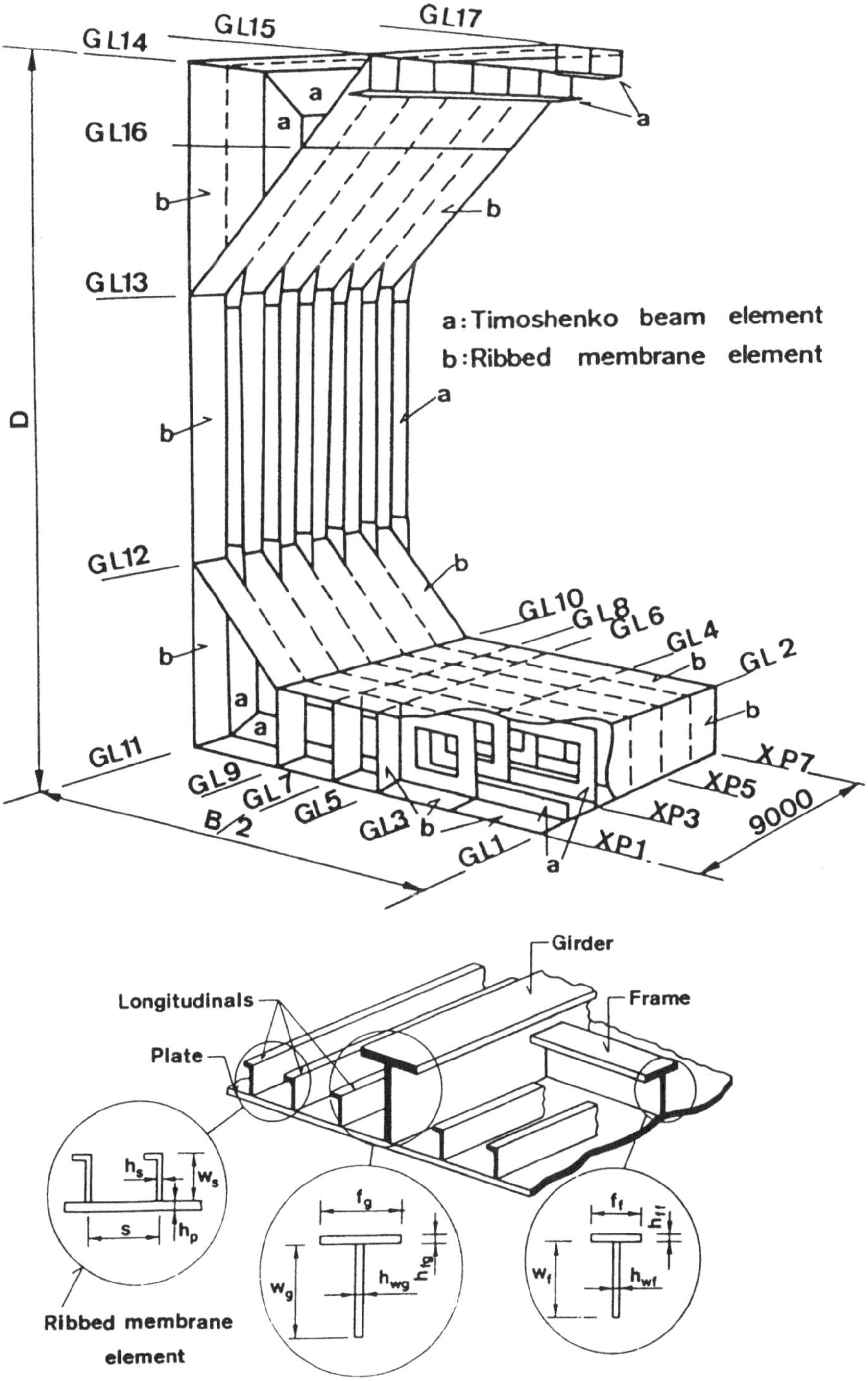

Figure 1. Structural modeling of one quarter midships cargo hold in a bulk carrier.

Each load case consists of local transverse loading due to cargo and hydrostatic pressure, see Figure 2, in combination with a set of sectional beam forces acting on the longitudinal boundaries. The sectional forces can be obtained from a calculation of the loading on the ship hull in irregular waves, see /6/, followed by a response calculation based on a Timoshenko-Vlasov beam theory /7/.

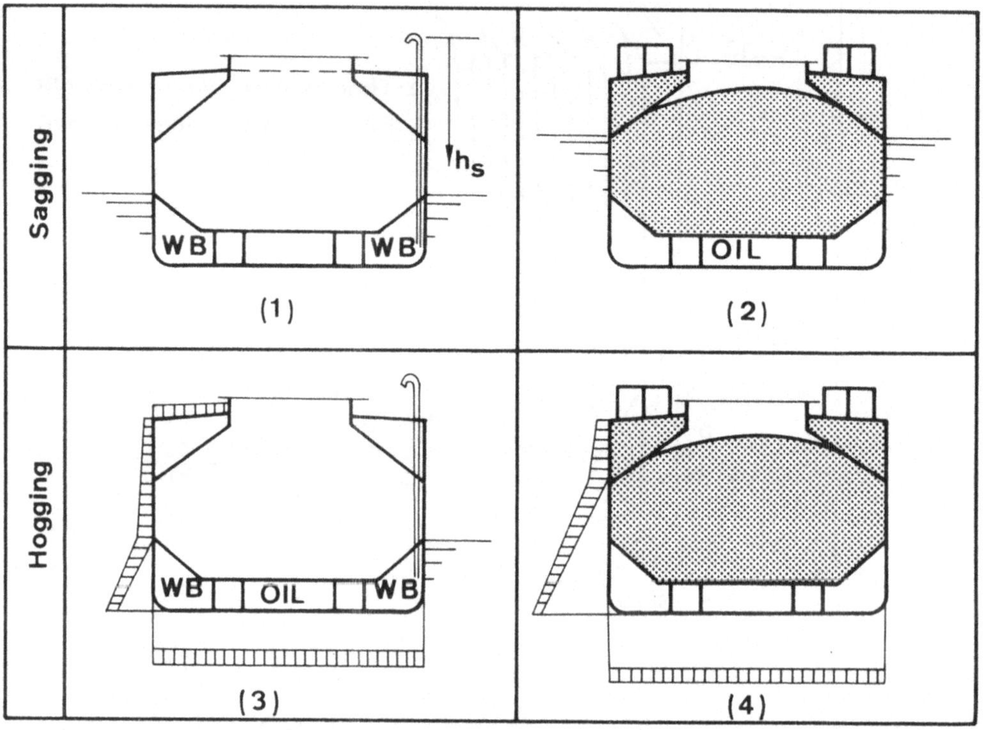

Figure 2. Load cases used in a cost optimization of the midship cargo hold in a bulk carrier.

Included in the present analysis and optimization procedure is a calculation of the distribution of normal stresses and shear stresses at the longitudinal boundaries based on specified sectional forces at the ends of the prismatic section. See Figure 3. This calculation of stress distributions is based on thin-walled beam theory using a procedure which automatically localizes closed cells.

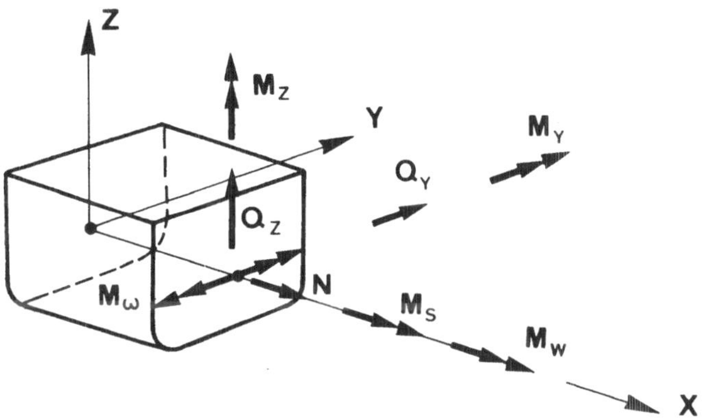

Figure 3. Definition of hull girder sectional forces and moments.

In order to model the interaction between the hull segment and the surrounding structure as well as possible, we have also introduced kinematic restraints for in-plane displacements at the longitudinal boundaries. Thereby the response analysis can be restricted to relatively short hull segments and still yield meaningful results.

The in-plane displacements at these boundaries are restrained such that they are compatible with thin-walled beam theory. That is, in-plane deformations are only allowed in linear combinations of the following four modes: plane axial displacements, vertical and horizontal plane bending displacements, and finally in a classical warping mode. This is achieved by introducing linear couplings between the longitudinal degrees of freedom of the nodal points at the boundaries in the form

$$u_j = \sum_{i=1}^{4} r_{ji} u_i \quad , \qquad j = 5,N \tag{1}$$

where N is the number of nodal points at the hull girder cross-section.

To determine the constants, r_{j1}-r_{j4}, $j = 5,N$, four nodal points are taken as reference points assuming that their longitudinal displacements can be described by

$$\begin{bmatrix} z_1 & y_1 & \Omega_1 & 1 \\ z_2 & y_2 & \Omega_2 & 1 \\ z_3 & y_3 & \Omega_3 & 1 \\ z_4 & y_4 & \Omega_4 & 1 \end{bmatrix} \begin{Bmatrix} a_1 \\ a_2 \\ a_3 \\ a_4 \end{Bmatrix} = \begin{Bmatrix} u_1 \\ u_2 \\ u_3 \\ u_4 \end{Bmatrix} \tag{2}$$

Here z_i, y_i are the distances from the horizontal and vertical bending neutral axes to the nodal points, Ω_i is the sector coordinate at the nodal points, and $[a_1, a_2, a_3, a_4]$ can be identified as generalized coordinates. By inverting the equations (2) the generalized coordinates can be determined from

$$\begin{Bmatrix} a_1 \\ a_2 \\ a_3 \\ a_4 \end{Bmatrix} = \begin{bmatrix} A_{11} & A_{12} & A_{13} & A_{14} \\ A_{21} & A_{22} & A_{23} & A_{24} \\ A_{31} & A_{32} & A_{33} & A_{34} \\ A_{41} & A_{42} & A_{43} & A_{44} \end{bmatrix} \begin{Bmatrix} u_1 \\ u_2 \\ u_3 \\ u_4 \end{Bmatrix} \qquad (3)$$

$$= [A]\{u\}$$

Assuming that also the remaining N-4 longitudinal displacements at the loaded cross section must satisfy the geometrical condition

$$u_j = a_1 z_j + a_2 y_j + a_3 \Omega_j + a_4 \quad , \qquad j = 5,N \qquad (4)$$

and introducing the generalized coordinates (3), the j'th displacement u_j can be put into the form

$$u_j = \sum_{i=1}^{4} (A_{1i} z_j + A_{2i} y_j + A_{3i} \Omega_j + A_{4i})\, u_i \quad , \qquad j = 5,N \qquad (5)$$

or by using matrix notation

$$u_j = \{u\}^T [A]^T \{x\}$$

where

$$\{x\}^T = \{z_j, y_j, \Omega_j, 1\}$$

Comparing Equations (5) and (1), it is seen that the constants r_{ji} in Equation (1) are given by

$$r_{ji} = A_{1i}\, z_j + A_{2i} y_j + A_{3i} \Omega_j + A_{4i} \quad , \quad i = 1,4, \quad j = 5,N \qquad (6)$$

3. STRUCTURAL ASSESSMENT

In the structural assessment procedure each strength member is checked against various possible modes of failure. The failure modes which determine structural capability include: excessive yielding; buckling of columns, plates, and panels; and excessive permanent lateral deflections of plates.

It was mentioned in the Introduction that in ship structures most of the principal factors which determine the structural demand and structural capability display statistical variations. Hence it has been found appropriate to use a probabilistic basis which can reflect uncertainties in estimating the structural demand and capability. In the present procedure the measure of structural safety is achieved by means of load factors and capability reduction factors in the form

$$\alpha G_D(\beta D_K) \leqq \frac{1}{\delta} G_C(C_K) \tag{7}$$

where

- G_D denotes the responses in the individual structural elements (forces, moments, deflections)
- G_C denotes the various limit states in the structural element (yielding capability, buckling strength, etc.)
- D_K represents a characteristic load on the structure
- C_K represents a characteristic strength of the structure (material properties)
- α symbolizes a load coefficient which reflects the seriousness in exceeding the limit states. In the program a distinction is made between a load factor α_c for collapse and a load factor α_s for serviceability
- β symbolizes a load coefficient which accounts for the subjective uncertainties in determining the loads on the structure
- δ symbolizes a material factor reflecting subjective uncertainties such as the accuracy of fabrication and the accuracy of calculating the limit states.

The characteristic values D_K, C_K in Equation (7) are defined by

$$D_K = \mu_D + k_{p_D} \cdot \sigma_D$$
$$C_K = \mu_C - k_{p_C} \cdot \sigma_C \qquad (8)$$

Here μ denotes the mean value, k_p is the p-fractile in the normalized distribution functions for demand and capability, and σ denotes the standard deviation. In Figure 4 is shown the principle of the above described method of using load factors and capability reduction factors in structural design to obtain an acceptable safety margin.

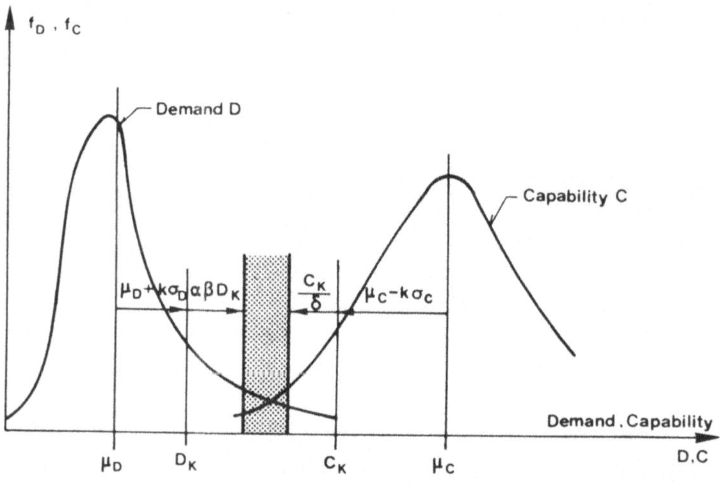

Figure 4. Illustration of safety margin using load factors and capability reduction factors.

4. SENSITIVITY ANALYSIS AND OPTIMIZATION

Mathematically the non-linear optimization problem can be expressed as

- Determine a set of design variables $\{x\}$ which
- minimizes $\Phi = f(\{x\})$ (9)
- subject to $g_j(\{x\}) \geq 0 \qquad j = 1,2,\ldots,s$ (10)

Here $\Phi = f(\{x\})$ is a scalar function which has to be minimized. In the present case it is a combination of cost and weight expressed as a general differentiable function of a set of n design variables

$$\{x\}^T = [x_1, x_2, \ldots, x_n]$$

The constraints are expressed through the functions $g_j(\{x\})$ of the design variables.

To solve this non-linear programming problem an iterative application of linear programming has been applied. Linear approximations of non-linear functions are accomplished by replacing the non-linear functions of the problem with their first-order Taylor series approximations expanded at the point of interest.

The derivatives needed to get the Taylor series approximation can be obtained as differences obtained by re-analyses or by an analytically based procedure. Using an approach suggested in /8/, we shall in the following show that it is possible to get analytically based expressions for the gradients and thereby reduce the number of re-analyses.

In the stiffness method the nodal point displacements $\{r\}$ are related to the nodal forces $\{R\}$ through the stiffness matrix $[K]$

$$[K]\{r\} = \{R\} \tag{11}$$

If x_i denotes the i'th design variable in the vector $\{x\}$, then partial differentiation yields

$$[K]\{r\},_{x_i} = \{R\},_{x_i} - [K],_{x_i} \{r\} \tag{12}$$

This expression shows that if relatively simple derivatives of the load vector and the stiffness matrix can be obtained, then the derivatives of the displacements with respect to the design variables can be obtained by adding a number of right-hand sides to (11).

The right-hand side of (12) can also be expressed as

$$\{R\}_{,x_i} - [K]_{,x_i}\{r\} = \sum_{\text{over all elements}} \{Q\}^{global}_{,x_i} - [k]^{global}_{,x_i}\{q\}^{global} \quad (13)$$

where $\{Q\}^{global}$ and $[k]^{global}$ denotes element loads and stiffness matrices, respectively.

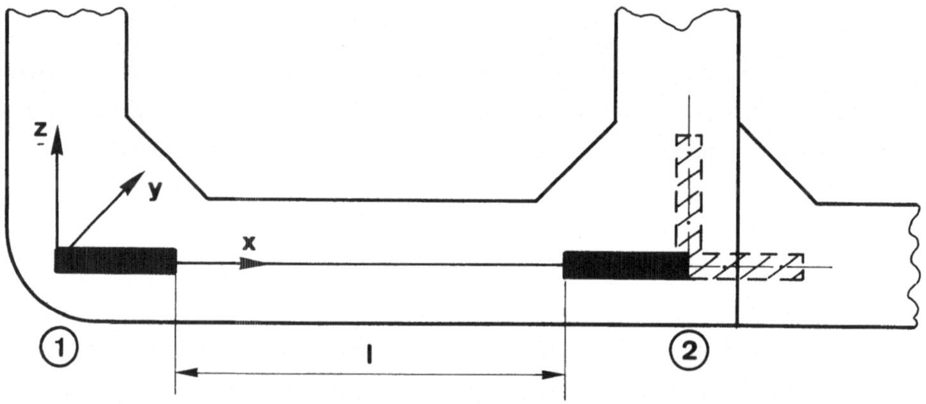

Figure 5. Timoshenko beam element with stiff end zones in a local x,y,z coordinate system.

For a Timoshenko beam element with brackets modeled as stiff end zones, see Figure 5, we have

$$[k]^{global}_{,x_i} = \frac{\partial}{\partial x_i}([S]^T [T]^T [E]^T [k] [E] [T] [S])$$

$$\{Q\}^{global}_{,x_i} = \frac{\partial}{\partial x_i}([S]^T [T]^T [E]^T \{Q\}_l + [T]^T \{Q\}_s)$$

Here the rotation matrix [T] is independent of the design variables. Also the transformation matrix [S] which transfers stiffnesses and loads from the elastic beam ends to the nodes is here assumed to be independent of the design variables (the scantlings of the brackets are not included in the optimization procedure). But the matrix [E] which transfers stiffnesses and loads from the torsional center of beam cross sections to the center of flexure is a function of the design variables indicated in Figure 1, and so is of course also the local stiffness matrix [k] for the flexible part of the beam. Finally, $\{Q\}_s$ denoting the nodal forces due to loading on the rigid beam ends is assumed independent of the design variables. Therefore, we have

$$[k]^{global}_{,x_i} = [S]^T\ [T]^T\ [\tilde{k}]_{,x_i}\ [T]\ [S]$$

$$\{Q\}^{global}_{,x_i} = [S]^T\ [T]^T\ \{\tilde{Q}\}_{,x_i}$$

where

$$[\tilde{k}]_{,x_i} = [E]^T_{,x_i}\ [k]\ [E] + [E]^T\ [k]_{,x_i}\ [E] + [E]^T\ [k]\ [E]_{,x_i}$$

$$\{\tilde{Q}\}_{,x_i} = [E]^T_{,x_i}\ \{Q\}_1 + [E]^T\ \{Q\}_{1,x_i}$$

Let the generalized local nodal displacements {q} be arranged such that the succession is

$$\{q\}^T = [u_{x1},\ u_{y1},\ u_{z1},\ \theta_{x1},\ \theta_{y1},\ \theta_{z1},\ u_{x2},\ u_{y2},\ u_{z2},\ \theta_{x2},\ \theta_{y2},\ \theta_{z2}]$$

where u_{x1}, θ_{x1} denote translation and rotation, respectively, in the local x direction at Node 1 and so on. Then the differentiated local stiffness matrix can be expressed simply as

$$[k]_{,x_i} = [a][k] + [b]$$

$$= \begin{bmatrix} [a_{11}] & [0] \\ [0] & [a_{22}] \end{bmatrix} \begin{bmatrix} [k_{11}] & [k_{12}] \\ [k_{21}] & [k_{22}] \end{bmatrix} + \begin{bmatrix} [b_{11}] & [b_{12}] \\ [b_{21}] & [b_{22}] \end{bmatrix} \qquad (14)$$

Here [a] is a diagonal matrix with the elements

$$[a_{11}] = [a_{22}] = \begin{bmatrix} a_1 & 0 & 0 & 0 & 0 & 0 \\ & a_2 & 0 & 0 & 0 & 0 \\ & & a_3 & 0 & 0 & 0 \\ & \text{symm.} & & a_4 & 0 & 0 \\ & & & & a_3 & 0 \\ & & & & & a_2 \end{bmatrix}$$

where

$$a_1 = \frac{\frac{\partial}{\partial x_i} A}{A} , \quad a_2 = \frac{\frac{\partial}{\partial x_i} I_z}{I_z} , \quad a_3 = \frac{\frac{\partial}{\partial x_i} I_y}{I_y} , \quad a_4 = \frac{\frac{\partial}{\partial x_i} K}{K}$$

with A denoting the cross-sectional area, I_z, I_y the cross-sectional moments of inertia, and K the Saint-Venant torsional stiffness factor.

The matrix [b] is given as

$$[b_{\alpha\beta}] = \begin{bmatrix} 0 & 0 & 0 & 0 & 0 & 0 \\ & b_1 & 0 & 0 & 0 & b_2 \\ & & b_3 & 0 & b_4 & 0 \\ & \text{symm.} & & 0 & 0 & 0 \\ & & & & b_5 & 0 \\ & & & & & b_6 \end{bmatrix} , \quad \begin{matrix} \alpha = 1,2 \\ \beta = 1,2 \end{matrix}$$

where with

$$\eta_y = \frac{E\, I_y}{k_z\, GA_{ez}\, l^2} , \quad k_z\, GA_{ez}$$ denoting the shear stiffness,

and with

$$\psi_y = \frac{\frac{\partial}{\partial x_i} I_y}{I_y} - \frac{\frac{\partial}{\partial x_i} A_{ey}}{A_{ey}} \tag{15}$$

the elements are given as

$$b_1 = k_{\alpha\beta}^{(2,2)} \; \psi_z \left(\frac{1}{1 + 12\eta_z} - 1 \right)$$

$$b_2 = k_{\alpha\beta}^{(2,6)} \; \psi_z \left(\frac{1}{1 + 12\eta_z} - 1 \right)$$

$$b_3 = k_{\alpha\beta}^{(3,3)}\ \psi_y \left(\frac{1}{1+12\eta_y} - 1\right)$$

$$b_4 = k_{\alpha\beta}^{(3,5)}\ \psi_y \left(\frac{1}{1+12\eta_y} - 1\right)$$

$$b_5 = \begin{cases} k_{\alpha\beta}^{(5,5)}\ \psi_y \left(\dfrac{1}{1+12\eta_y} - \dfrac{1}{1+3\eta_y}\right) & \text{for } \alpha = \beta \\ k_{\alpha\beta}^{(5,5)}\ \psi_y \left(\dfrac{1}{1+12\eta_y} - \dfrac{1}{1-6\eta_y}\right) & \text{for } \alpha \neq \beta \end{cases}$$

$$b_6 = \begin{cases} k_{\alpha\beta}^{(6,6)}\ \psi_z \left(\dfrac{1}{1+12\eta_z} - \dfrac{1}{1+3\eta_z}\right) & \text{for } \alpha = \beta \\ k_{\alpha\beta}^{(6,6)}\ \psi_z \left(\dfrac{1}{1+12\eta_z} - \dfrac{1}{1-6\eta_z}\right) & \text{for } \alpha \neq \beta \end{cases}$$

As an example, on the outcome of a differentiation of the load vector for a Timoshenko beam element we can consider the consistent node force Q_{y1} resulting from a linearly varying line load in the local y-direction with intensities q_{y1} and q_{y2} at the ends of the elastic zones of the beam element

$$Q_{y1} = -\ \rho_z \ell \left(\frac{7}{20} + 4\eta_z\right) q_{y1} - \rho_z \ell \left(\frac{3}{20} + 2\eta_z\right) q_{y2}$$

where

$$\rho_z = \frac{1}{1 + 12\ _z}$$

Partial differentiation of this expression with respect to the design variables $\{x\}^T = [u,\ h_w,\ f,\ h_f]$, see Figure 1, leads to

$$\frac{\partial}{\partial x_i} Q_{y1} = \left(Q_{y1} + \frac{7}{20}\,\ell\, q_{y1} + \frac{3}{20}\,\ell\, q_{y2}\right) \psi_z \frac{1}{1+12\eta_z}$$

where ψ_z is given by an expression analogous to (15).

The first-order derivatives of stresses with respect to the design variables can also be expressed. The stresses at the beam ends can be determined as

$$\{\sigma\} = [C]\ \{Q\}_t$$

where the sectional forces are given as

$$\{Q\}_t = \{Q\} + [k]\ \{q\}$$

By differentiation we get the stress derivatives

$$\{\sigma\},_{x_i} = [C],_{x_i}\ \{Q\}_t + [C]\ \{Q\}_{t,x_i}$$

where

$$\{Q\}_{t,x_i} = \{Q\},_{x_i} + [k],_{xi}\ \{q\} + [k]\ \{q\},_{x_i}$$

Also, for the stiffened membrane plate elements analytical expressions have been derived for partial derivatives of displacements and stresses with respect to the design variables indicated in Figure 1.

So, in each step the linearized optimization problem can be expressed as

- Determine a set of design variables
  $\{x\}^T = [x_1, x_2, x_3, \ldots, x_N]$

- which minimizes

$$\tilde{\Phi}(x) = \sum_{j=1}^{N} \left(\frac{\partial f}{\partial x_j}\right)_o x_j$$

- with the constraints

$$\sum_{j=1}^{N} \left(\frac{\partial g_i}{\partial x_j}\right)_o x_j \geq \sum_{j=1} \left(\frac{\partial g_i}{\partial x_j}\right)_o x_{jo} - g_i\ (\{X\}_o) \qquad i = 1,2,\ldots,S$$

where all the necessary derivatives can be found with just one triangularization of the overall stiffness matrix.

The SLP is solved using variable move limits.

As seen from Figure 6, the optimization problem contains a sub-optimization cycle where first all strakes and then all girders are redesigned. As already

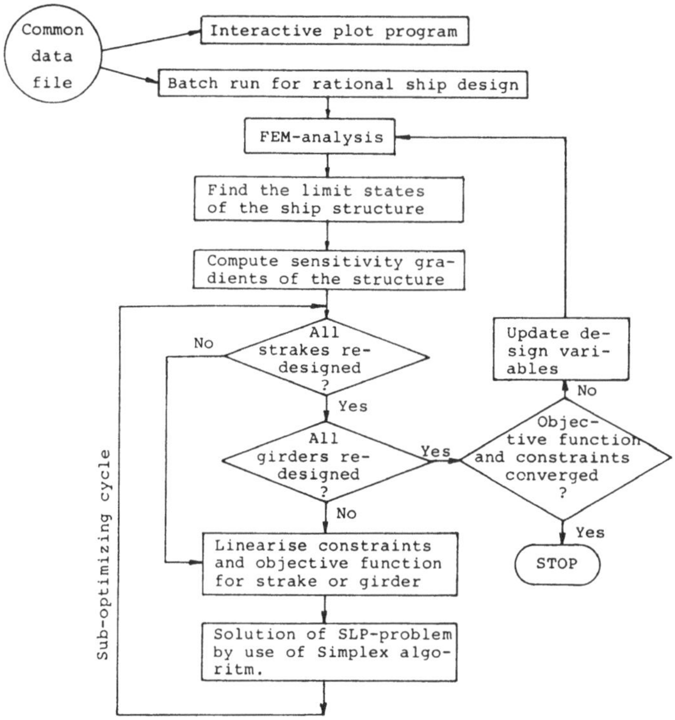

Figure 6. Summary of optimization procedure.

mentioned in Section 2, a strake is defined as a panel with secondary longitudinals and transverse beams between two guidelines as indicated in Figure 7. In such a strake all the stiffened membrane elements are of the same type, and all the transverse beam elements are of the same type.

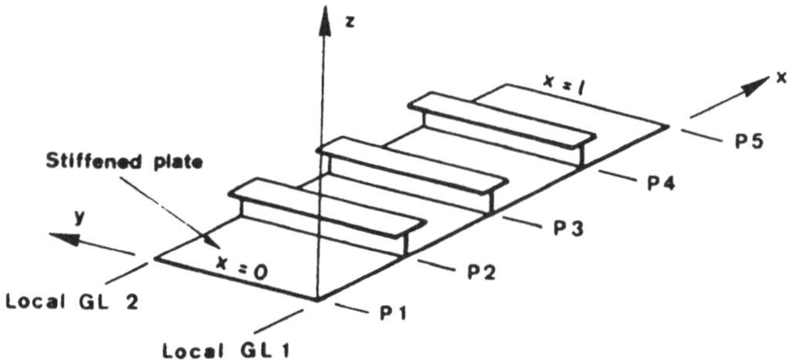

Figure 7. Sub-element in the form of a strake in a local coordinate system.

4. NUMERICAL EXAMPLE

The program has been used for optimization of ship structures of the type indicated in Figure 1 and for optimization of structural elements such as double bottom and bulkheads. Due to lack of space we shall here illustrate the procedure by a very simple example in the form of an 8 x 5 m steel bulkhead subjected only to the uniform loading pressure shown in Figure 8. The same panel has been optimized in /9/. The finite element model of this bulkhead consists of one beam and two rectangular ribbed membrane elements. One on each side of the beam element.

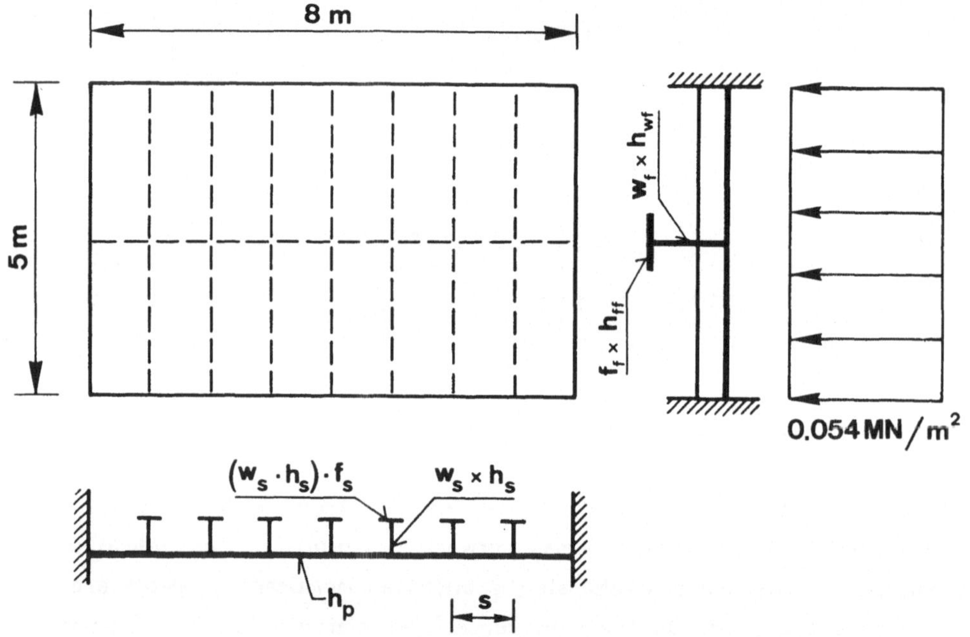

Figure 8. Pressure loaded bulkhead with indication of dimensions, design variables, and boundary conditions.

The design criteria express that the von Mises reference stress in the stiffeners is limited to 108 MN/m^2, that the plating shall not have noticeable permanent deflections after unloading, and that web thicknesses shall be larger than 1/60 of the corresponding web height.

The object function is given as

$$\Phi = \frac{W}{W_o}\,\theta + \frac{C}{C_o}\,(1 - \theta)$$

where $W = W(\{x\})$ is the panel weight, W_o a reference weight, $C = C(\{x\})$ is the cost based on steel cost 10,770 kr. per m^3 and welding cost 107.70 kr. per m weld, C_o is a reference cost, and finally θ is a weighting factor. The results of the optimization are given in the following table and in Figure 9. Figure 9 also includes the effect of omitting the center beam so that the structure becomes a prismatic panel.

Table 1. Optimization results for orthogonally stiffened bulkhead

	Object function $\Phi = \frac{W}{W_o}\theta + \frac{C}{C_o}(1-\theta)$		
	Minimum weight $\theta = 1.0$ $W_o = 1.0$	$\theta = 0.3$ $W_o = 2.89$ t $C_o = 12{,}200$ kr.	Minimum cost $\theta = 0.0$ $C_o = 1.0$
No. of stiffeners	47	16	11
s (mm)	165	462	635
h_p (mm)	1.8	4.6	6.0
w_s (mm)	156	229	379
h_s (mm)	2.6	3.8	6.3
A_f $(mm^2) = (w_s \cdot h_s) f_s$	199	428	1174
w_f (mm)	984	872	853
h_{wf} (mm)	16.4	14.5	14.2
$f_f \cdot h_{ff}$ (mm^2)	4515	5693	5854
Total cost C (1000 kr)	29.40	13.30	12.20
Total weight W (ton)	2.89	3.33	4.46

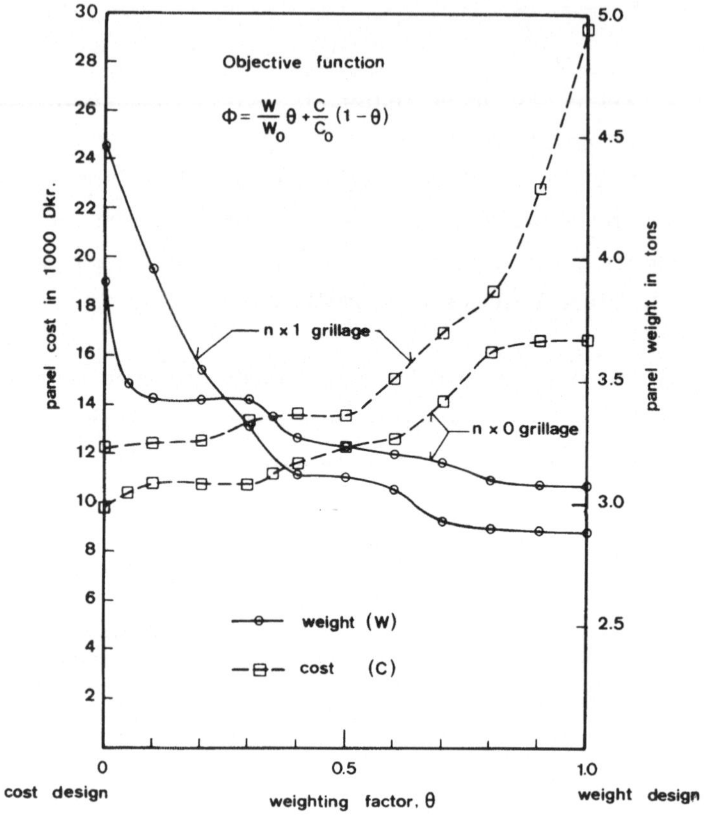

Figure 9. Weight and cost of prismatic bulkhead (n×0 grillage) and orthogonally stiffened bulkhead (n×1 grillage) as functions of the weighting factor θ. For the prismatic bulkhead the factor f_s in Figure 8 is taken as 0.6318 and for the orthogonally stiffened bulkhead $f_s = 0.4918$.

5. CONCLUSIONS

The work presented here should be regarded as the authors' first step towards a practical procedure for automated structural design of marine structures. The approach is promising. But, among the needed improvements are guidelines for substructuring procedures that guarantee a reasonably fast convergence. Based on the experience we have gained so far, a new program is being developed at the Technical University of Denmark which hopefully will be so flexible and robust that at some time in the future it can be used outside University walls.

6. REFERENCES

/1/ Mansour, A.G., and Thayamballi, A.: "Computer Aided Preliminary Ship Structural Design", Ship Structures Committee Report 302, 1981.

/2/ Moe, J.: "Integrated Design of Tanker Structures", Numerical and Computer Methods in Structural Mechanics, S.J. Fenves N. Perrone, A.R. Robinson, and W.C. Schnobrich, editors. Academic Press, New York, 1973, pp. 415-437.

/3/ Chowdhury, M., and Caldwell, J.B.: "An Automated Design Scheme: Weight Minimization of a Single Hulled Ship Compartment", Int. J. for Numerical Methods in Engineering, Vol. 20, 1984, pp. 1763-1790.

/4/ Hughes, O.: "Ship Structural Design", Wiley-Interscience, 1983.

/5/ Lin, S., Hughes, O.F., and Mahowald, J.G.: "Applications of a Computer-Aided Optimal Preliminary Ship Structural Design Method", Trans. SNAME, 1981.

/6/ Jensen, J. Juncher, and Pedersen, P. Terndrup: "Bending Moments and Shear Forces in Ships Sailing in Irregular Waves", Journ. of Ship Research, Vol. 25, 4 December 1981, pp. 250-287.

/7/ Pedersen, P. Terndrup: "Torsional Response of Containerships", Journ. of Ship Research, Vol. 29, No. 3, September 1985, pp. 194-205.

/8/ Pedersen, Pauli: "The Integrated Approach of FEM-SLP for Solving Problems of Optimal Design", in "Optimization of Distributed Parameter Studies" (Eds. G.J. Haug and J. Cea), Sijthoff and Nordhoff, 1981, pp. 757-780.

/9/ Hewitt, A.D.: "Production Oriented Design of Ship Structures", Thesis, University of Newcastle, U.K., September 1976.

ADVANCES IN OPTIMAL DESIGN WITH COMPOSITE MATERIALS

By

G. LECINA & C. PETIAU

AVIONS MARCEL DASSAULT - BREGUET AVIATION

78, Quai Carnot - 92214 SAINT-CLOUD FRANCE

ABSTRACT

It has been demonstrated that optimal design is indispensable for the sizing of complex structures using composite materials. On a significant application to the RAFALE wing optimisation saved 10 % on structural weight and obtained a 25 % more stiffened structure in comparaison with a classical sizing. Sensitivity analysis gives a great knowledge on the structure.

Multimodel optimisation, the most recent advance in our system, managing the synthesis of calculations on several static and dynamic models is described. A significant application to the lay-up of a composite wing panel is detailed (2 F.E. models with 3500 and 13000 DOF, 80 models for panel buckling, 480 design variables, 1000 static and aeroelastic constraints, 24 load cases).

Then it clearly appears that the passage between the rough results of optimisation to the final shape of each lay is impossible to solve with classical drawing means. An algorithm computing the optimal shape of each lay now integrated in our C.A.D. system CATIA is described. An application on the previous wing panel is detailed.

Close integration of C.A.D. systems and FEM systems, including geometry, mesh generation, analysis and optimisation is an obliged way for greater developments and applications of F.E.M.. It will enable FEM system to optimize the shape of structures. Integration of CATIA and ELFINI is described.

Multimodel optimisation on complex structures needs a nearly prohibitive CPU time on classical computers. The interest of vectorial multi-processor architectures with extended core memory is showed.

NATO ASI Series, Vol. F27
Computer Aided Optimal Design: Structural and Mechanical Systems. Edited by C. A. Mota Soares

1 - INTRODUCTION

The optimisation module of ELFINI code has been developped since 1975 by Dassault-Breguet. The algorithm has been detailed in earlier papers (Ref. 1).

Optimisation method minimise the weight by a finite element model. The optimisation design variables are multiplicative factors of the caracteristics of linked finite elements. Optimisation constraints can be of different types :

- Technological minimum thicknesses, simple tooling rules, lay-up rules for composite materials, etc....
- Limited displacements, stresses and strains,
- Miscellaneous failure and buckling criterias on metallic and composite materials.
- Limitations on static aeroelastic coefficients, surface control efficiencies, divergence speed.
- Limitations on frequencies, damping and flutter speed.

The optimisation process is iterative, each iteration including three steps :

- Analysis : static, dynamic, and aeroelastic.
- Derivation of constraints with regard to the design variables.
- Explicit optimisation using an inverse variable formulation.

Convergence is insure in about 5 iterations.

We detailed the multimodel optimisation, managing several analysis models, the most recent advance in our code.

The optimisation of the shape of lays on composite panels is described on a significant application on a wing box.

Future developments in optimisation are exhibited mainly close integration between CAD system and FEM code and influence of supercomputers on optimisation CPU time.

2 - MULTIMODEL OPTIMISATION

Optimisation has to provide the single physical caracteristics of a structure and must take in account all that could size it. So many F.E. models (or other type) are necessary, depending on the studied phenomenons. So optimisation must insure :

- identification between design variables defined on several models
- data transferts between models (caracteristics, boundary conditions, loads, ...)
- management of calculation and derivation of constraints defined on several models.
- the linking of all the design variables and constraints (values and derivatives) in the single explicit optimisation step giving the optimum.

The vertical organisation (Fig.1) of the models has needed some software investment and is able to manage several FE meshes with several boundary conditions, mass configurations (modal and flutter analysis), Mach number (aeroelasticity and flutter analysis). Other models are used for panel buckling analysis.

The horizontal organisation (Fig.1), i.e. the data transfert between models, can be difficult. We give the exemple of transfert between a F.E. model and a Raleigh-Ritz model for panel buckling analysis using composite materials.

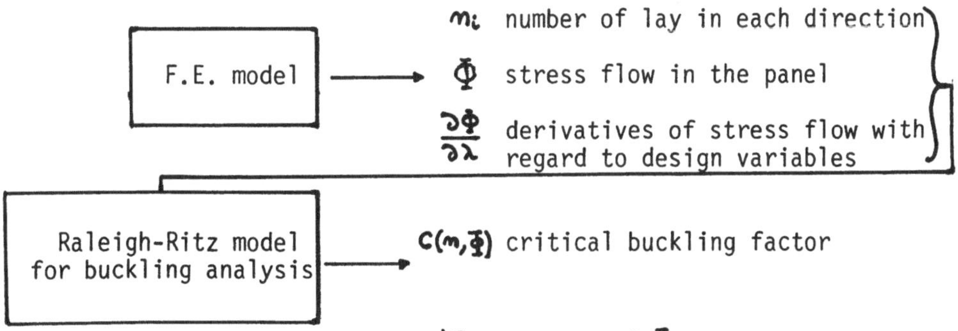

$$\frac{dC}{d\lambda} = \frac{\partial C}{\partial \lambda} + \frac{\partial C}{\partial \Phi} \cdot \frac{\partial \Phi}{\partial \lambda} \quad \text{derivatives}$$

C and $dC/d\lambda$ are recovered by the optimisation monitor in the explicit optimisation step.

3 - OPTIMAL DESIGN OF A VERTICAL FIN

We present the optimisation of the carbon-expoxy panels of the main box and the rudder of a vertical fin. Design variables represent the thickness of each ply (four design variables in the same mesh Fig. 2) and the main constraints are failure and buckling criterias on composite materials, rudder efficiencies and frequencies. The exact configuration of this optimisation follows :

	Model 1	Model 2
F.E. Models	Fin model (1800 DOF) with a super-element of the whole aircraft (Fig.2)	Fin model with a deflected rudder (Fig. 2)
Design variables	237 design variables on the number of lay, spars and ribs flanges (area) and web (thickness)	
Static Load cases	3	1
Failure criteria	190 failure criterias on composite materials with holes	190
Buckling criteria	98 buckling criterias computed from Raleigh-Ritz models for panel buckling analysis	82
Displacements	1 on the step between main box and rudder	0
Aeroelasticity	8 constraints on fin and rudder yaw efficiencies for two Mach number	0
Dynamic	Frequencies of 1^{rst} flexion mode and rudder mode	0
Technology	107 constraints on plies repartitions and on minimum distance between lay interruptions	

Explicit optimisation solves a problem with 237 design variables and 678 constraints with a conjugate projected gradient. The artificial "sparsity" of the partial derivatives matrix reach up 90 % of the 160.686 potential derivatives.

4 - LAY-UP OPTIMAL DESIGN ON COMPOSITE PANELS

Intensive use of optimisation on composite structures has showed that the passage between the rough results of optimisation, with a patchworh type presentation, to the final shape of each lay is a long and difficult step very often impossible to solve with classical drawing means.

Now this passage has been speed-up with two complementary tools :

1) An algorithm gives in interactive mode the optimal shape of each lay using the rough results of optimisation (number of lay in the basic directions for each mesh element of the patchwock) and lay-up rules as distances between cut-lines of the lays and mixture of directions interrupted on a cut-line.

   This algorithm is now integrated in our C.A.D. system CATIA and we present a significant application on the lay-up of a wing panel (Fig. 3 - 4).

2) An interactive fonction for lay-up drawing completely integrated to CATIA with all classical drawing means and composite specific features (staking sequence, thickness law, shape for sub-structure).

   It must be noticed that the small size of color raster video screens as IBM5080 and Tektronix 4125 unabled the drawing engineer to interactively design a lay-up because of a too small distance between cut-lines of lays (0,25 mm on the screen).

5 - IMMEDIATE DEVELOPMENTS

5.1 - Integration to C.A.D. system

Close integration of CAD and FEM systems is an obliged way for greater developments and applications of FEM nowadays too much reserved to specialists. These new systems enable the design engineers to :

- design the structure (geometry)

- generate FE mesh and solve F.E. analysis (interactive or batch mode)

- make optimisation with intensive use of sensitivity analysis (partial derivatives and Lagrange multipliers).

So the design variables will be extend from simple caracteristics (areas, thicknesses, number of lay) to the real shape of the structure - First test data showed that the great problems for shape optimisation are the definition of design variables on a shape and automatic and adaptative mesh generation. The CAD systems with a parametric geometry will be very usefull.

5.2 - Optimisation of un-linear structures

Post buckling analysis of composite structures is one of the most significant advance of the last years (Ref. 2). The next challenge is the optimisation of un-linear structure including post-buckling behavior.

The most problems can be solved with a similar algorithm to those of linear structures with a sequence of analysis and partial derivatives. Derivatives computation are relatively less expensive than in linear structures :

$$K_{tg}(\lambda).X - F_{ext} = 0 \quad ; \quad \sigma = L.X$$

$$\frac{dF_{int}}{d\lambda} = \frac{\partial K_{tg}}{\partial \lambda}.X + K_{tg}.\frac{\partial X}{\partial \lambda} = 0$$

$$\frac{\partial X}{\partial \lambda} = - K_{tg}^{-1}.\frac{\partial K}{\partial \lambda}.X \quad ; \quad \frac{\partial \sigma}{\partial \lambda} = L.\frac{\partial X}{\partial \lambda}$$

But this type of algorithm could lead to bad convergence on post-buckled structures with snap-through behavior.

So a simultaneous resolution of the analysis and optimisation problems can be considered.

$$\begin{cases} Min \; M(\lambda) \\ \sigma(X,\lambda) < \sigma_{max} \\ \nabla W_{tot} = 0 \end{cases}$$

λ = design variables
X = displacements
W_{tot} = un-linear potential

Recent advances in minimisation method based upon preconditionned matrices and explicit line-search (Ref.2) will be intensively used.

5.3 - Super-computers and optimisation

Multimodel optimisation on complex structures needs a nearly prohibitive CPU time on classical computers. Multimodel optimisation is very well adapted to multiprocessor computers. The cost of optimisation is made by the computation of partial derivatives. As detailed in Ref.1, partial derivatives consist in resolution of the FE problem with several virtual loads and the computational cost can be hardly reduced with vectorial computers (pipe-line).

6 - CONCLUSION

Multidisciplinary optimisation including aerodynamics, flight control, performance and structures is a very important developement where multilevel optimisation can be considered (Ref. 3).

Aside the computational cost of these applications, the choice of a multidisciplanary objective fonction is very delicate.

Ref.1 : FOUNDATIONS OF STRUCTURAL OPTIMISATION
A unified approach
NATO - ASI - Liège 1980 - Edited by A.J.Morris
John Wiley & Sons

Ref.2 : Tendances actuelles en calcul de structures
Efficient algorithms for post-buckling analysis
C.PETIAU and C.CORNUAULT Editions Pluralis

Ref.3 : STRUCTURAL OPTIMISATION BY MULTILEVEL DECOMPOSITION
J.SOBIESKY - AIAA.JOURNAL - VOL23 - NOVEMBER 1985

MULTIMODEL OPTIMISATION MONITOR

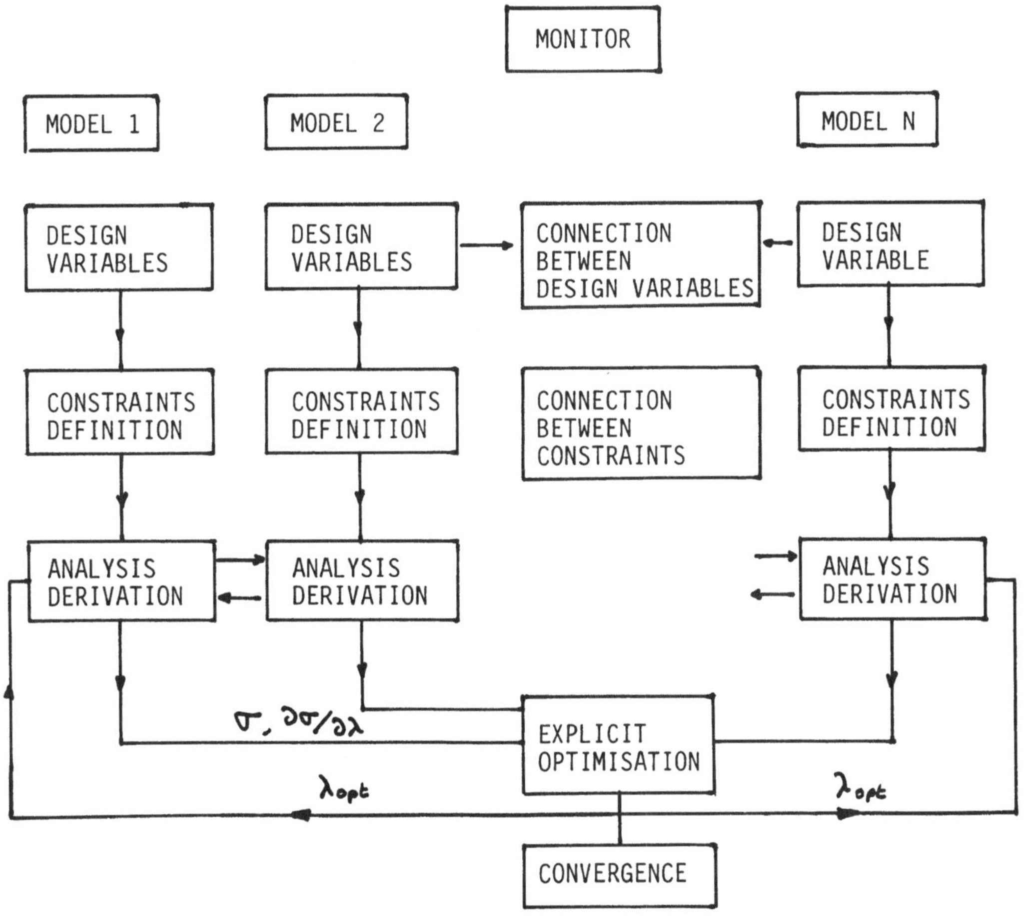

Fig. 1

OPTIMISATION OF A VERTICAL FIN
WITH COMPOSITE MATERIALS

MODEL 1 (1800 DOF)

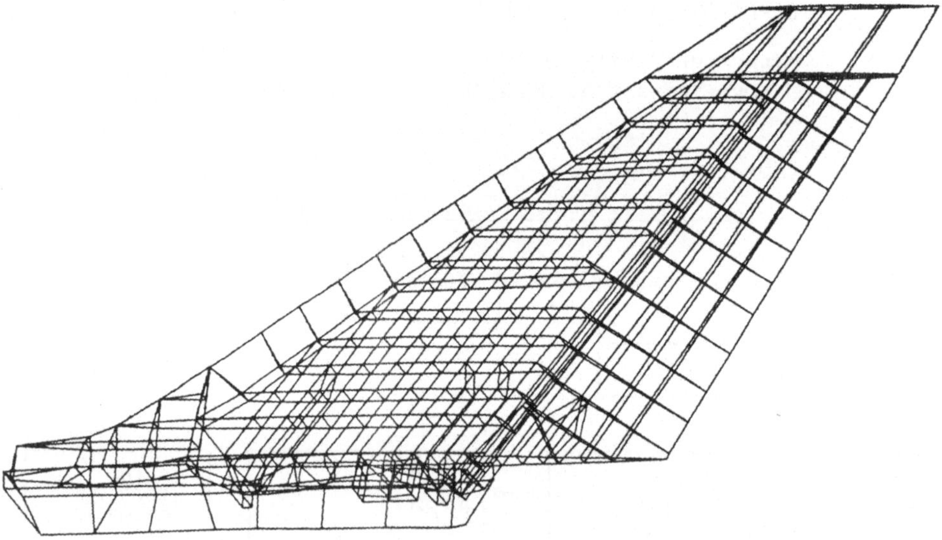

MODEL 2 (DEFLECTED RUDDER)

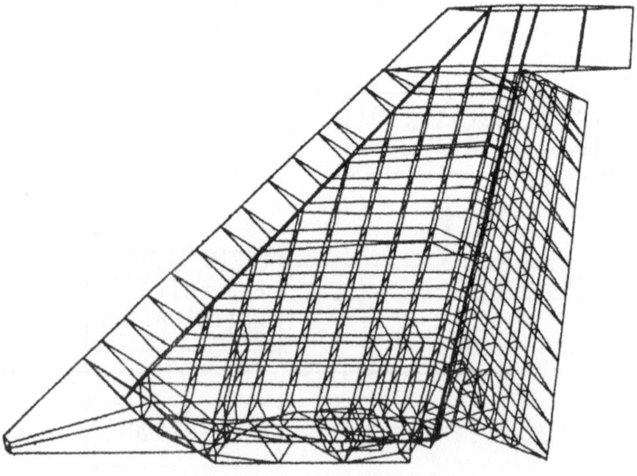

Fig. 2a

OPTIMISATION OF A VERTICAL FIN WITH COMPOSITE MATERIALS

OPTIMUM DESIGN VARIABLES

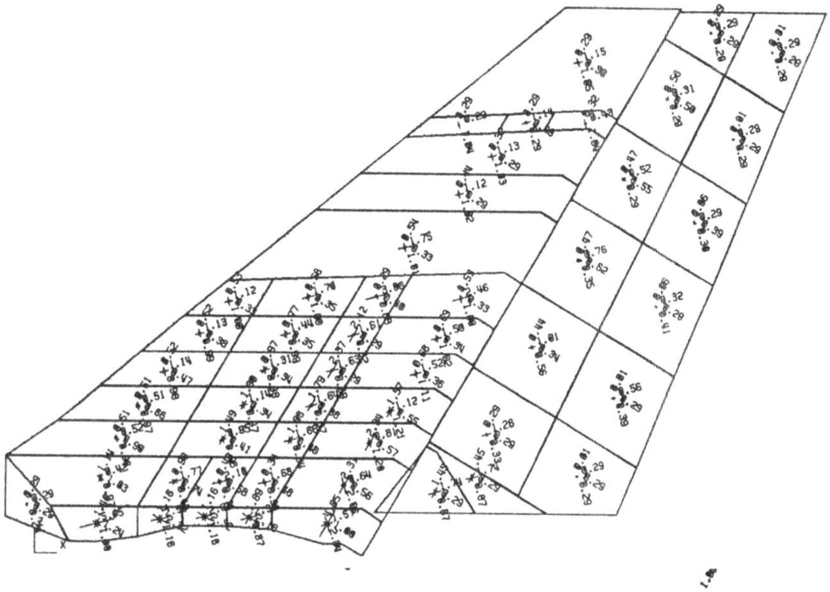

ACTIVE CONSTRAINTS

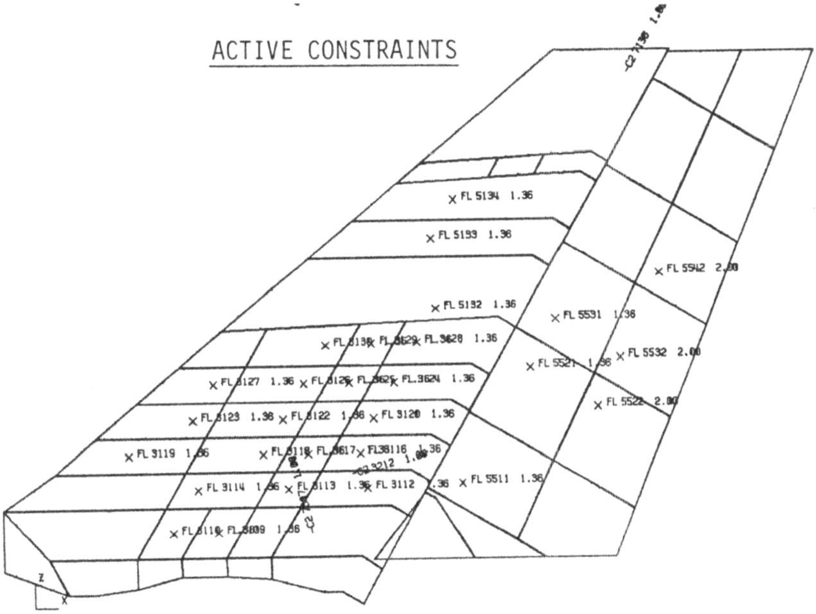

Fig. 2b

OPTIMISATION OF WING PANEL WITH COMPOSITE MATERIALS

- MODEL 1 : COMPLETE AIRCRAFT : 13000 DOF
- MODEL 2 : WING : 3500 DOF
- 476 DESIGN VARIABLES
- 1000 CONSTRAINTS (24 LOAD CASES, 476 FAILURE CRITERIAS, 144 BUCKLING CRITERIAS, AEROELASTICITY).

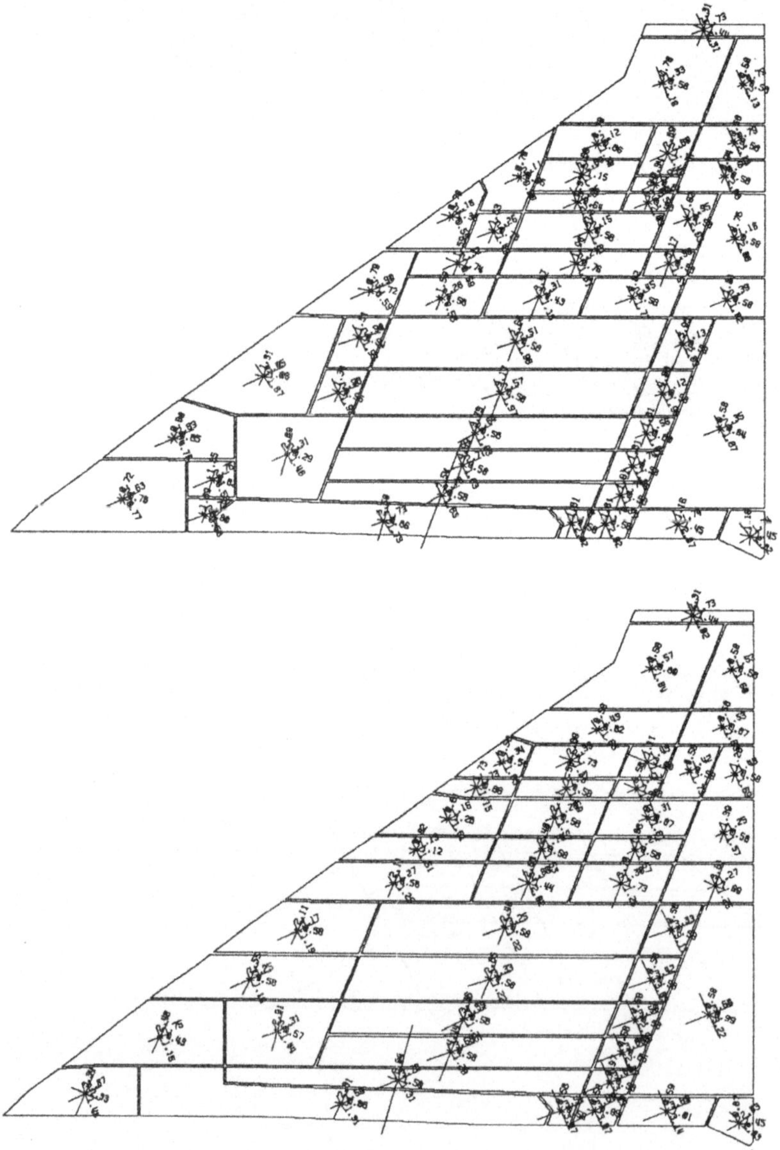

Fig. 3

OPTIMAL LAY-UP DESIGN

WING PANEL

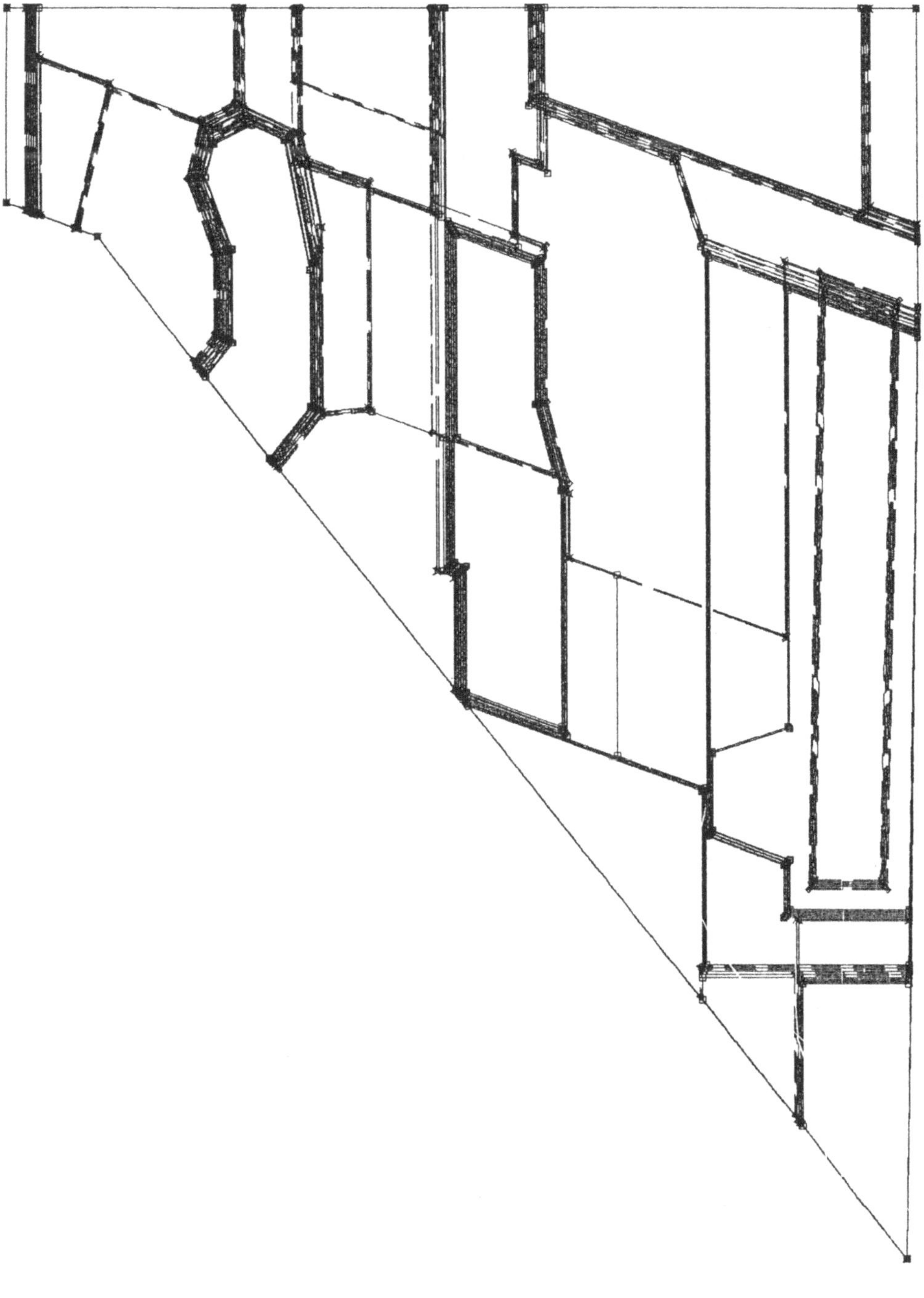

Fig. 4

Structural Optimisation in Aircraft Construction

Heinrich Wellen * and Peter Bartholomew **

* MBB-UT Bremen, TE 544
Hünefeldstr. 1–5
D-2800 Bremen 1

**Royal Aircraft Establishment
Materials & Structures Dept.
Farnborough, Hants.

Demands within the international aerospace industry for the design of primary components of ever higher quality, at minimum weight and low cost, are growing constantly while pressure to achieve the shortest time scales remains strong. As a result, the use of computers in design and for structural mechanics is of growing importance.

Structural optimisation is a suitable tool for the design of weight-optimised components, reducing both the time required and cost. Computer programs for structural optimisation have been in use for some years in the aircraft industry [1, 2, 3]. This paper shows how research originated at RAE has been carried through and applied in an industrial context at MBB-UT. Some of the principle methods are outlined and extensive applications are described.

1 COMPUTER–AIDED STRUCTURAL OPTIMISATION

The design of components is an iterative process in which the aim is to achieve a design that is

- light
- stiff
- favourable to manufacture
- inexpensive

That is, in some sense, an optimum design.

The design procedure, however, is very long if the conventional design process is applied. Lack of time often precludes a component being weight-optimised down to the last detail according to all important criteria.

The intensive use of existing data-processing possibilities during analysis, in connection with the Finite Element Method (FEM), with

- display workstations
- pre- and post-processors
- graphics and evaluation programs

constitutes an important step towards shortening the design process.

Structural optimisation 'by hand' is a lengthly process even in the relatively simple case of the stress optimisation of a statically indeterminate structure. The work becomes much more extensive and complicated if further requirements such as stiffness and frequency constraints apply. Structural optimisation is a valuable aid in this connection.

NATO ASI Series, Vol. F27
Computer Aided Optimal Design: Structural and Mechanical Systems. Edited by C. A. Mota Soares

Structural optimisation is generally understood to mean the fully automatic, weight-optimised dimensioning of components using programmed mathematical methods on an iterative basis (analysis and redesign). Different objective functions can be used for this purpose [1]. Here structural optimisation is understood as the weight-optimised dimensioning of components idealised as an FE-model of given geometry. This means that only the cross-sectional values – and not the geometry – of the structural elements are changed.

The aircraft industry has often been a pioneer in using new technologies, as was the case with the Finite Element Method (FEM) and Computer-Aided Design (CAD). Both these processes have fundamentally changed the manner and form of structural design, making both development steps more effective in themselves.

Today the situation with structural optimisation is much the same as it was with FEM and CAD some years ago. Structural optimisation is being increasingly used as a rational tool offering the possibility to design weight-optimised components with cost and time savings in the preliminary and main design phase.

2 STRUCTURAL OPTIMISATION PROGRAM STARS

The method and examples described in this paper relate to a particular software package called STARS – Structural Analysis and Redesign System. It is a highly modular program system for structural optimisation of thin-walled components (Fig. 1) [4, 5, 6].

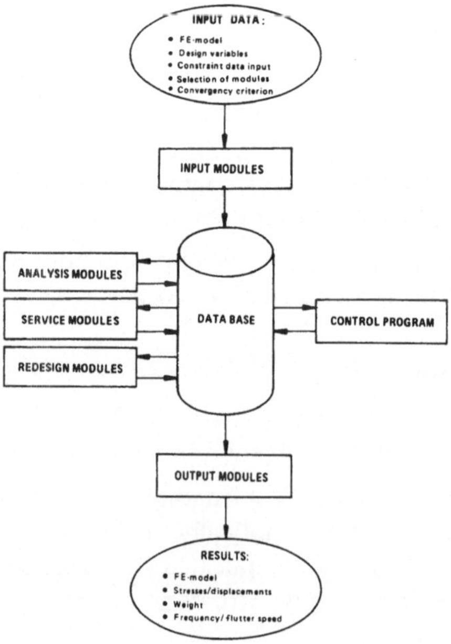

Fig. 1: Modular set-up of STARS

Program Overview

STARS accomplishes fully-automatic weight-optimised dimensioning of thin-walled components. For this purpose, the component must be available in form of an FE-model. Only the cross-sectional values (element thickness and cross-sectional areas) of the FE-model are varied in the course of the optimisation, the geometry remaining unchanged. The structural optimisation problem can be outlined as follows:

The dead weight W(x) is the objective function of optimisation. It depends on the individual design variables x, the cross-sectional areas or the thicknesses of the elements The objective function has to be minimised whilst at the same time fulfilling various constraints.

The following constraints can presently be taken into consideration simultaneously:

- Stresses
- displacements
- vibration amplitudes
- natural frequencies
- bounds on thickness and cross-sectional areas
- flutter speed.

The flutter optimisation is achieved by interfacing STARS to the MBB-UT program AEROOPT [7] for the calculation of the flutter speed and its derivatives.

Depending on the constraints to be considered, the following optimisation methods can be used to solve the problem [6]:

- Stress Ratio Method (SRM)
- Pseudo Newton Method (PNM)
- Optimality Criteria (strain energy density) (OPCR)
- Non-Linear Optimisation Process (NLOP)

Structural optimisation consists of a fully automatic interaction between analysis and redesign. The standard analysis program used by MBB is MSC-NASTRAN [8], an FE-program based on the displacement method, which is widely used in the aerospace industry. SRM and PNM have proved to be very successful for redesign purposes. Moreover, experience has shown that a combination of both procedures can be very advantageous for many practical problems [9].

Present Development Status

Fig. 2 briefly outlines the present development status of STARS.

Components made from metal as well as from fibre-reinforced materials can be optimised for several load cases whilst simultaneously allowing for stress, thickness and displacement constraints. Furthermore, components subjected to periodic loads can be optimised and frequency optimisation and flutter optimisation of components are possible.

The NASTRAN elements used for such computations are listed in the figure.

Linearly elastic FE-models of any size can be optimised, although the natural limits of the computer capacity and the costs of computing do set limits to the maximum size of components to be optimised. Fig. 3 shows the most important values for the present STARS dimensioning on the IBM-3081-K at MBB in Bremen.

The structural optimisation can be performed either interactively or in batch operation. Data pre-processing and post-processing can be accomplished automatically with the aid of various pre- and post-processors.

The largest optimisation model hitherto processed with STARS at MBB-UT was a composite test example with 1500 elements, 1500 design variables and 1500 constraints, approximately 3500 degrees of freedom and one load case.

• FE-Analysis	. MSC-NASTRAN . RAE Analysis	} (NASTRAN input)
• FE-Types	. Bar . Beam-box . Triangular . Quadrilateral . Shear element . Mass element . Spring element	CROD / CONROD CBAR / CBEAM CTRIA3 CQUAD4 CSHEAR CONM2 CELAS2
• FE-Models	. Linearly elastic . Fixed elements possible . Magnitude = f (computer, computing costs)	
• Load	. Static . Forced vibrations	
• Materials	. Isotropic . Orthotropic . Anisotropic	
• Constructions	. Metallic . Composite	
• Optimisation types	. Stress optimisation . Stiffness optimisation . Frequency optimisation . Flutter optimisation	
• Optimisation Modules.	. Stress Ratio Module . Pseudo-Newton Module . Optimality Criterion Module . Non-linear Optimisation Module	
• Particular features of optimisation	. DV-linking . Active set strategy . Fixed elements	
• Utilization	. Interactive/Batch . Input } automated, pre/post processor . Output } . Possibility of RESTART	
• Post-processor	. Profile selection program	
• Compatible with	. MSC-NASTRAN . Existing pre/post processor software (FEMGEN, FEMVIEW, SUPERTAB)	

Fig. 2: Present development status of STARS

Maximum number of nodes	**3000**	**Maximum number of load cases**	**20**
Maximum number of elements	**8000**	**Maximum number of materials**	**100**
Maximum number of design variables	**1500**	**Maximum number of active constraints**	**2000**
Maximum number of constraints	**16000**	**Maximum number of natural frequencies**	**60**

Fig. 3: Current dimensioning of STARS at MBB

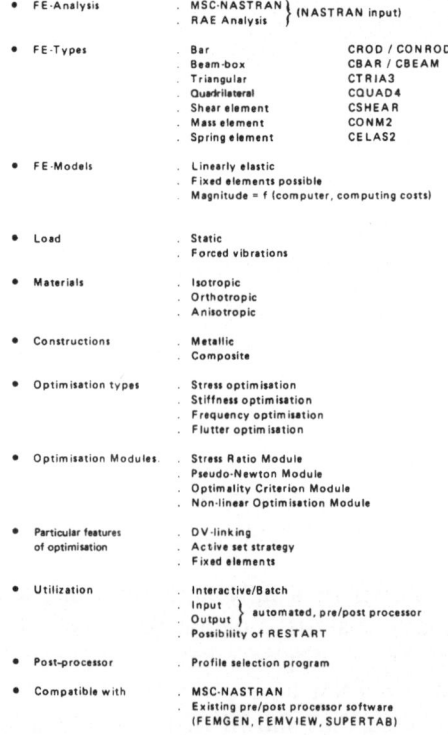

3 PRINCIPLE METHODS

Mathematical Foundations

Mathematically, the goal is simply expressed as:

$$\text{Minimise} \qquad W(z),$$

$$\text{subject to} \qquad g(z) \leqslant c_i, \quad i = 1 \ldots, m,$$

where z is the vector of n design variables and the constraints establish the behavioural response considered acceptable for the structure.

Some mathematical programming techniques based on hill-climbing approaches address this problem directly. However, for the present purposes, it is better to employ an equivalent formulation based on use of the Lagrangian function L (z, λ), where

$$L(z, \lambda) = W(z) + \sum_{i=1}^{m} \lambda_i (g_i - c_i).$$

The Lagrangian function depends on two sets of variables: the primal variables z, and the dual variables (otherwise known as Lagrange undetermined multipliers). A necessary condition for the minimisation of the original constrained optimisation problem is that the Lagrangian function should be stationary with respect to both primal and dual variables. Differentiating yields the well-known Kuhn-Tucker conditions.

$$\nabla W(z) + \sum_{i=1}^{m} \lambda_i \nabla g_i = 0$$

$$\left.\begin{aligned} \lambda_i (g_i - c_i) &= 0 \\ \lambda_i &\geqslant 0 \end{aligned}\right\} \quad i = 1 \ldots, m$$

The location of a stationary point thus requires the solution of a system of (n + m) simultaneous non-linear equations, as difficult a task as the original minimisation problem!

Structural Considerations

For the mathematical formulation above to applied in a structural context it is first necessary to define the design freedoms considered. In the following, elements are linked into groups, the stiffness of each being assumed to scale linearly with a single controlling design variable x . Thus the areas A of bars or the thickness of plates are given by.

$$A_k = \sum_{j=1}^{n} a_k B_{k_j} x_j, \qquad k = 1 \ldots, N,$$

where **B** is a Boolean matrix and a is a reference area for each of the N elements. The structural weight

$$W = \sum_{k=1}^{N} \varrho_k A_k l_k = \sum_{k=1}^{N} \sum_{j=1}^{n} \varrho_k a_k l_k \mathbf{B}_{kj} x_j$$

reduces to

$$W = \sum_{j=1}^{n} w_j x_j = \sum_{j=1}^{n} w_j / z_j$$

where the coefficients w are component masses given by

$$w_j = \sum_{k=1}^{N} \varrho_k a_k l_k \mathbf{B}_{kj}$$

and z are reciprocal variables, $z_j = 1 / x_j$, used above. The use of the reciprocal design space is adopted as a well-known device to linearise constraints arising from the design of statically loaded structures.

The Pseudo-Newton method is based on the direct solution of the Kuhn-Tucker equations and requires knowledge of design sensitivities with respect to these variables. STARS maintains a tight active-set strategy and therefore requires relatively few sensitivities to be calculated at any iteration. It uses fully analytic derivatives, to be contrasted with the semi-analytic approach employed within NASTRAN, and the calculation employs the adjoint, or dummy load, method [10].

Thus for a constraint on $g(z) = e^T u$ the combined effect of the use of reciprocal variables and design-variable linking is to give a form for the derivatives

$$\nabla g = v^T \left[\sum_{k=1}^{N} \frac{a_k \mathbf{B}_{kj} \mathbf{K}_k x^2}{A_k} \right] u$$

involving a summation over all elements linked to a given design variable, where K_k is the k-th element stiffness matrix and v is the adjoint displacement vector satisfying

$$\mathbf{K}^T v = e.$$

Newton Method

The approach taken is that employed previously [11] for the solution of strength critical problems for which the solution is known not to be fully-stressed. That is, it is assumed that the critical stresses have been identified, and so the Kuhn-Tucker equations are formulated for a set of active equality constraints.

As a first step towards solving this set of non-linear simultaneous equations, a linear approximation is formed about the current point z^o, λ^o, giving

$$\left[\nabla^2 W + \sum_{i=1}^{m} \lambda^o_j \ \nabla^2 \ g_i\right](z-z^o) + \sum_{i=1}^{m} \nabla g_i (\lambda-\lambda_i^o) = -\left\{\nabla W + \sum_{i=1}^{m} \lambda_i \nabla g_i\right\}$$

$$\nabla g_i \quad (z-z^o) \qquad\qquad = -(g_i - c_i)$$

Like any application of Newton's method, the repeated solution of this set of linear equations does not necessarily converge: but provided the start point lies within the domain of convergence, then that convergence will be quadratic. Unfortunately the requirement that second derivatives should be provided for all constraints in the active set requires excessive computation.

Thus, rather than emploing an exact Newton step, the equations are further approximated by neglecting second derivatives of the constraints. Such approximations are already implicit in both the stress-ratio and optimality criterion methods, and are known to be exact for statically determinate structures, optimised with respect to reciprocal variables.

$$\left[\frac{2 w_j}{z_j^3}\right] (z-z^o) + \sum_{i=1}^{m} \nabla g (\lambda_i - \lambda_i^o) = \left\{\frac{w_j}{z_j^2}\right\} - \sum_{i=1}^{m} \lambda_i^o \nabla g_i$$

$$\nabla g_i \quad (z-z^o) \qquad\qquad = -(g_i - c_i)$$

For a more general class of problems, this need to depart from the strict Newton form will lose the quadratic convergence properties, indeed it is quite possible that the iteration may deverge from any solution. In practice, however, many structural problems appear to be exceptionally well-behaved, giving good convergence to minimum weight designs.

Omitting the second derivatives of constraints, in fact, gives a very simple form for the linear equations. The weight as objective function is convex and separable, giving a diagonal Hessian matrix with positive coefficients. Thus, the first equation may be used to explicitly eliminate the primal variables from the constraint set, giving a reduced system of equations in which the dual variables are the unknowns.

At first the intention was to complement the optimality criterion method, and so it was assumed that such a method had been applied until convergence is reached. That is until the optimality conditions are satisfied; giving zero on the right-hand side of the first of these equations.

If this is done, the step actually taken in the primal space may be interpreted as a weighted least-squares restoration step, shown as SO in fig. 4. The step moves towards satisfaction of the constraints while minimising the length of the step, using the Hessian of the objective function as the metric.

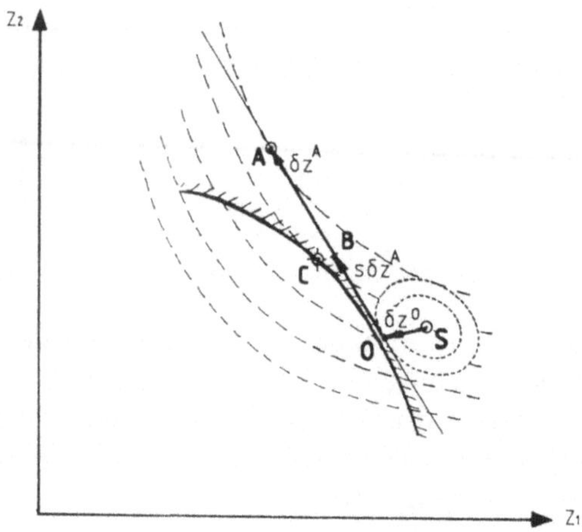

Fig. 4: Step length control

The use of this restoration step alternating with standard optimality criterion steps was shown to be effective; indeed the precursor program to STARS had such a strategy 'hard-wired' into it. While this procedure has some elegance, the assumption that the alternating steps of optimality criterion are necessary to satisfy the Kuhn-Tucker equations is not strictly true. Simply restoring the right-hand sides to the first set of equations removes the need for any such assumption. The additional increment OA to the step taken may be shown to lie in a surface parallel to the constraints. This combined step SA may then be alternated with the optimality criterion step or, more frequently, it is used by itself.

Step-Length Control

Even the simple problem of static strength design already described exhibits some non-linearity for indetermine structures and the algorithms based on the local linearisation of constraints could, in theory, diverge from the solution. To improve the convergence characteristics some form of step-length control is required.

To avoid a line search invoking repeated analyses, higher order derivatives are employed. In the Pseudo-Newton method the step-length control is based on second derivatives, which are calculated by differencing the analytic first derivatives. The restoration step is left unaltered because we still wish the combined step to satisfy the linearised constraints. Thus it is only the optimisation step, lying in the tangent space to the constraints, that is scaled.

By considering change

$$dz = dz^0 + s\, dz^A$$

and requiring the Lagrangian to be stationary in the direction dz^A, the first of the Kuhn-Tucker equations reduces to

$$s\, dz^A\ \nabla^2 W\, dz^A + dz^A \sum_{i=1}^{m} \lambda_i\ \nabla^2 g_i\, dz^0 + s\, dz^A \sum_{i=1}^{m} \lambda_i\ \nabla^2 g_i\ dz^A = -\nabla W\, dz^A$$

giving

$$s = \frac{dz^A \; \nabla^2 W \, dz^A \; - \; dz^A \, \Sigma \, \lambda_i \; \nabla^2 g_i \, dz^0}{dz^A \; \nabla^2 W \, dz^A \; + \; dz^A \sum_i \lambda_i \; \nabla^2 \, g_i \, dz^A}$$

It is a measure of the effectiveness with which the original scope of the problem was specified, that having implemented the step-length control, few examples could be found to illustrate its effectiveness.

One example which has been found recently with the incorporation of plate-bending structure is based on a simple cantilever problem, shown in fig. 5. The improvement achieved by scaling is considerable, though this must be in part due to the small number of design freedoms making the selection of the search direction trivial.

More stringent step-length controls are required if the goal is to achieve global convergence to a local minimum [12]. To date our experience is that the algorithms described here perform well on structural problems without such safe-guards. Nonetheless they should be included, provided they do not require excessive computation.

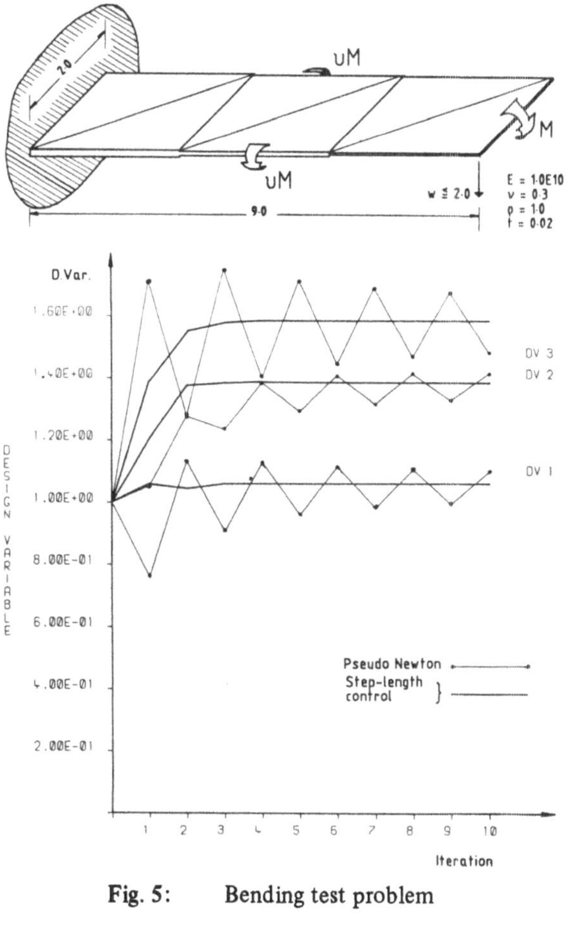

Fig. 5: Bending test problem

4 INDUSTRIAL APPLICATIONS

This chapter describes some characteristic examples of the practical application of STARS at MBB. These examples relate the stress, stiffness and frequency optimisation of metal and composite components as well as an optimisation of the track positions of a flap.

STARS is applied not only in the preliminary design phase but also in the main design phase. Consequently the FE-models used for optimisation will differ in the level of detail employed. The characteristic values of the FE-models and optimisation models are shown in the different illustrations of the application examples.

Fig. 6 shows a simplified FE-model (preliminary design) of the aluminium inner flap of the Airbus A310 with the constraints and a sketch of the load case considered [9]. In this optimisation run, not only the allowable stresses were taken into consideration as constraints but also the displacement constraint shown in fig. 6. 27 DV's were used in this model. The optimisation results in fig. 8 show the typical result for a SRM/PNM combination. On the basis of the DV history plots and fig. 9 and 10, it is seen that the SRM converges very quickly for element areas which have been dimensioned stress-wise (e.g. DV 8, 9, 10 an 11; element 643). The plots for the areas to be dimensioned with regard to the stiffness (e.g. DV 12, 13, 14, 15; mode 62) show, however, that the SRM is not suited for stiffness optimisations. In this case, the PNM reaches the optimum in just a few steps on the basis of the stress pre-optimised FE-model.

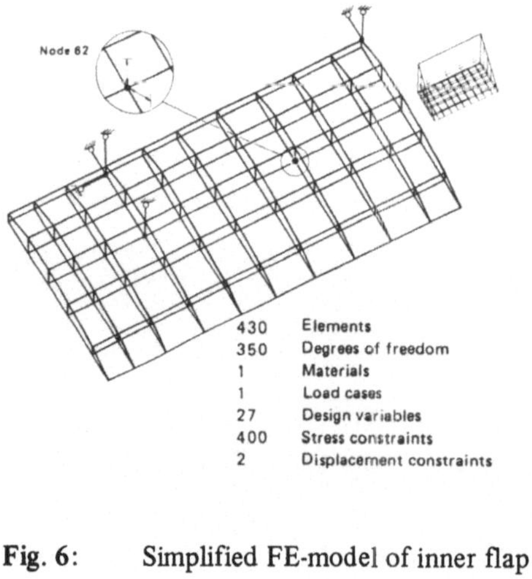

Fig. 6: Simplified FE-model of inner flap

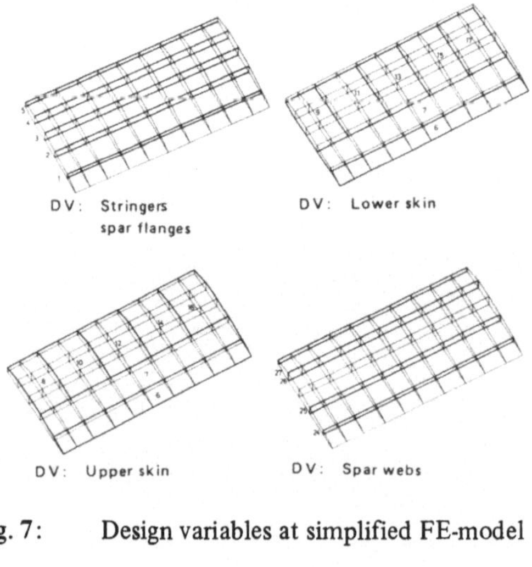

Fig. 7: Design variables at simplified FE-model

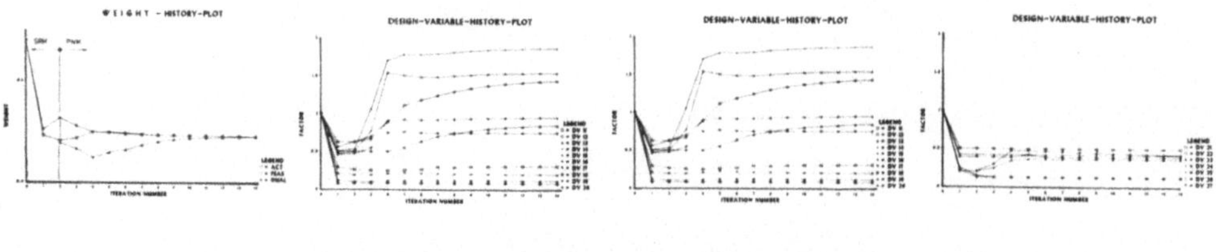

Fig. 8: Results for simplified FE-model

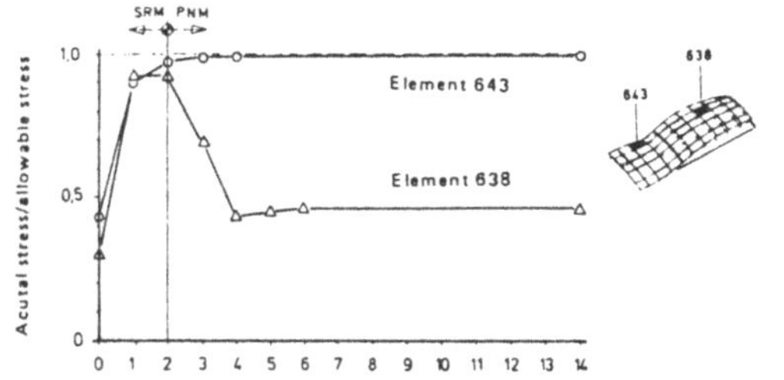

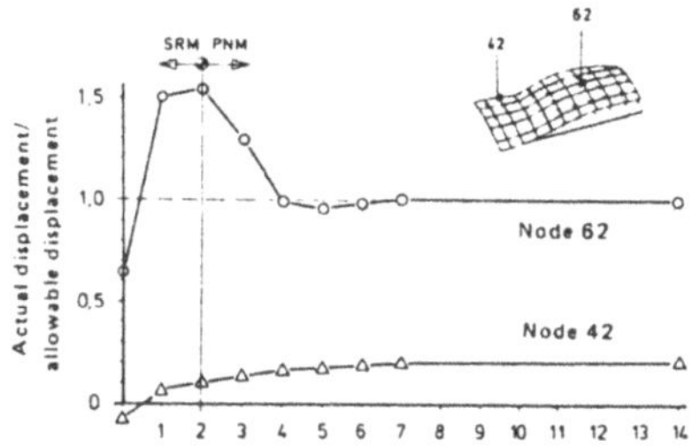

Fig. 9: Displacement of nodes versus iterations **Fig. 10:** Element stresses versus iterations

In contrast, fig. 11 illustrates the FE-model of the aluminium inner flap of the Airbus A310 during the main design phase [9]. With due consideration to the two load cases which determine design, this model has been optimised using 92 design variables.

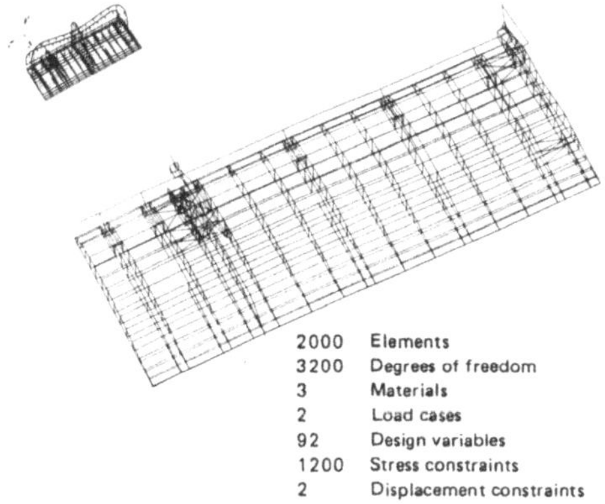

Fig. 11: FE-model of inner flap of Airbus A310

Fig. 12 shows the FE-model of the frame for a modern fighter aircraft in the preliminary design phase. This aluminium-titanium frame is subjected to the wing attachment forces. The allocation of DV's is given in fig. 13.

From the optimisation results obtained, fig. 14 plots the weight curve and some DV's versus the iterations. The weight curve shows clearly that the subsequent PNM leads to a lower optimum' weight than when SRM only is applied.

The physical reason for the difference in weight is as follows:

The SRM gives preference to the frame reinforcement (rod elements linked to DV 70) for load transfer at the wing attachment frame. By using the sensitivity analysis, the PNM determines that a load transfer through the frame wall (membrane elements linked to DV 90) is the best solution with regard to the weight and therefore changes the initial design determined by the SRM.

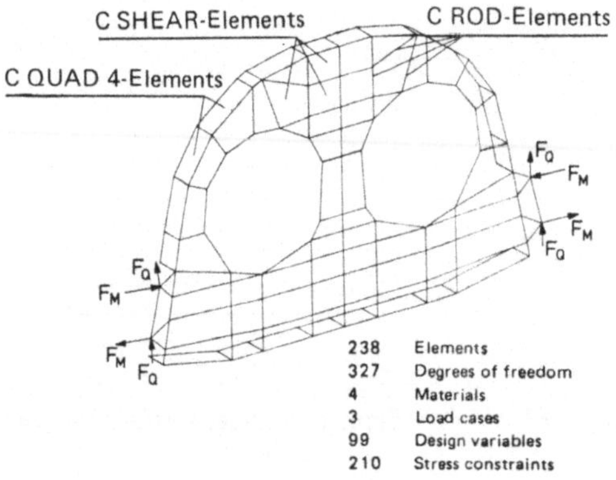

Fig. 12: FE-model of frame

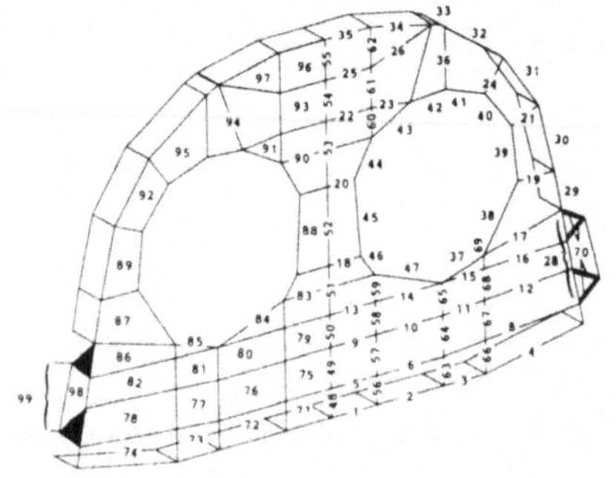

Fig. 13: Design variables at FE-model

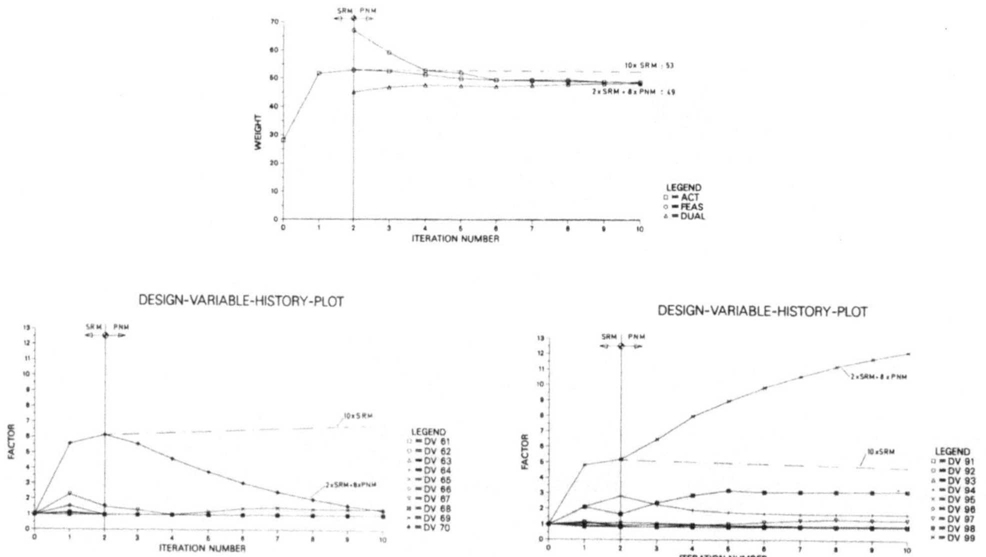

Fig. 14: Optimisation results for frame

Fig. 15 and 16 give two further practical applications. The FE-model of the wing box of a modern airliner is shown in fig. 15. This FE-model (metal inner wing/composite outer wing) served as the basis for stress optimisations at the inner and outer wing within the scope of a study. In these optimisation runs the number of DV's was systematically increased in successive runs.

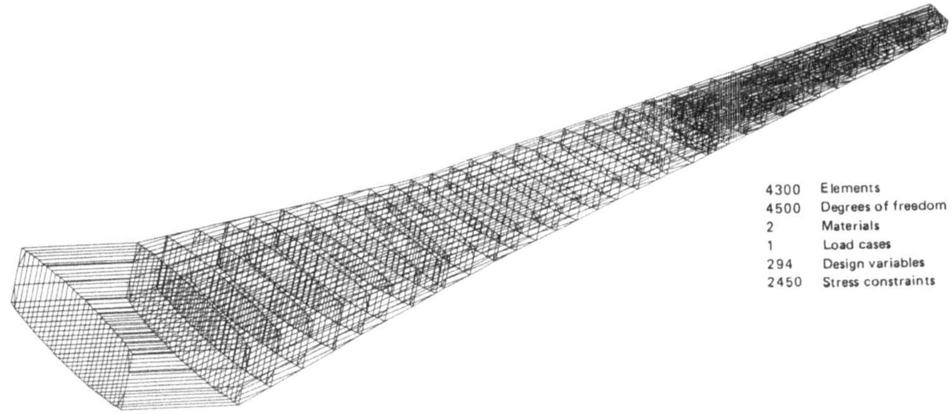

Fig. 15: FE-model of wing box

The FE-model of the fin of a modern fighter aircraft (composite) is shown in fig. 16 and the weight history plot of a stress optimisation with the two selected load cases is given in fig. 17.

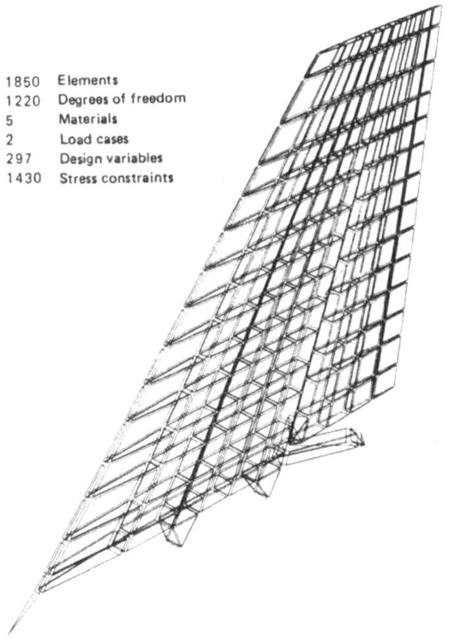

Fig. 16: FE-model of fin

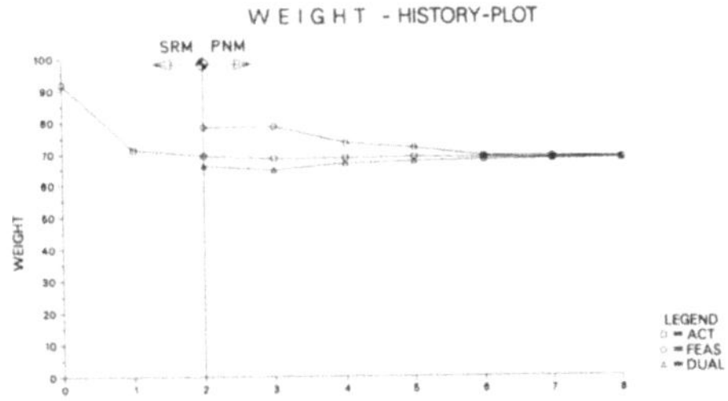

Fig. 17: Optimisation result for fin

Fig. 18 shows the FE-model of the Short Supply Mast (SSM) of the planned space station COLUMBUS. MBB-ERNO is examining this composite tube construction in a study [13]. After stress optimisation, this FE-model was additionally frequency optimised with PNM using 7 DV's (which correspond to seven tube types). The results show that the frequency constraint for the first eigenvalue is fulfilled in four iterations (fig. 19 and 20).

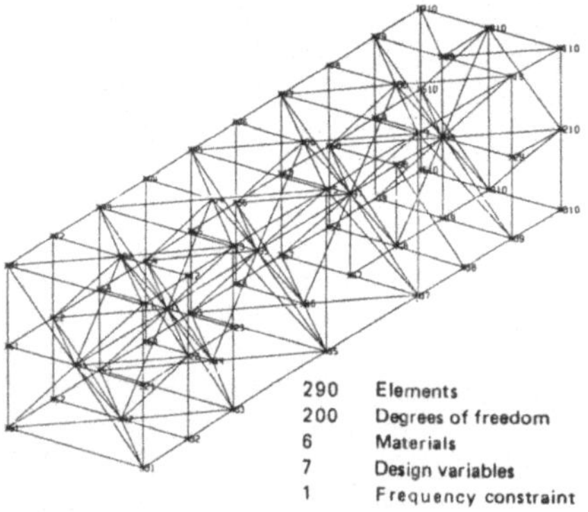

Fig. 18: FE-model of Short Supply Mast

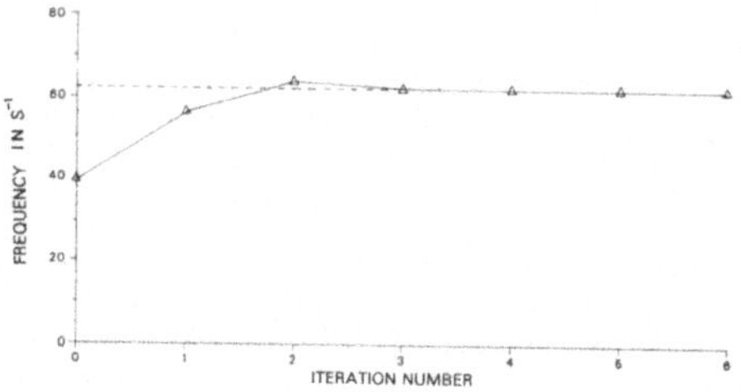

Fig. 19: Optimisation result: weight of SSM

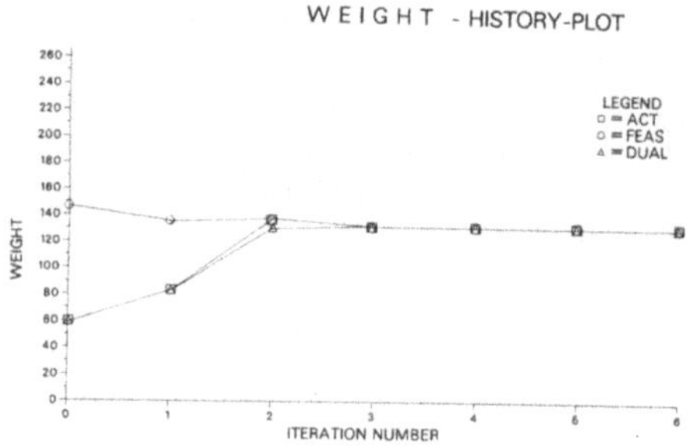

Fig. 20: Optimisation result:
1st eigenfrequency of structure

In all the applications so far mentioned, the geometry and the support of the FE-models were not considered variable. The simple beam model of a flap system sketched in fig. 21 shows that structural optimisation can even be used to optimise the position of the flap tracks.

The aim of the investigation [13] was to change the positions of two track stations in such a way as to give the flap system a minimum weight at a specified geometry and load as well as at predetermined stress and displacement constraints. Six different FE-models of systematically selected combinations of the track positions were prepared for this purpose (fig. 22). A combined stress and stiffness optimisation was achieved for each FE-model. The optimum positions of the tracks can be deduced from the graphic display of all established optimum weights (fig. 23).

The interaction between designer and structural optimisation as described here for optimum placement of supports can naturally also be applied to other types of components.

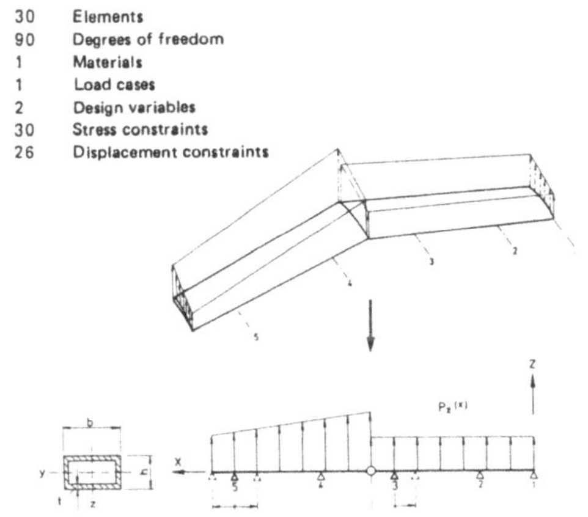

Fig. 21: Flap and equivalent computation model

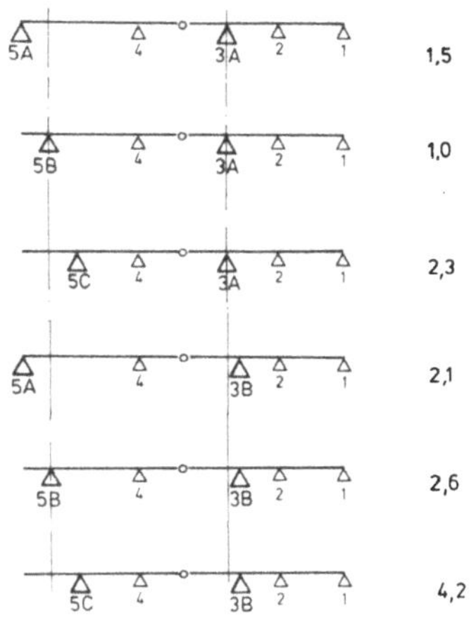

Fig. 22: Variants of computation model

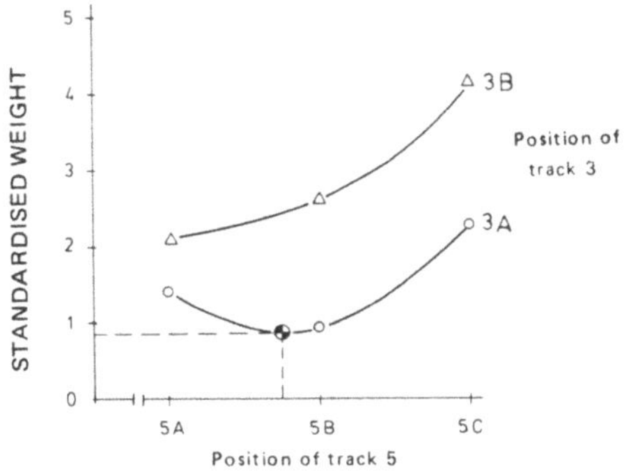

Fig. 23: Standardised weight versus track positions

5 RESULTS AND CONCLUSIONS

The industrial application of structural optimisation in aircraft construction has convincingly demonstrated its many advantages. Conclusively, the following can be said:

- Large and complicated components made of metal and composite materials can be weight-optimised with the optimisation methods used
- Weight savings over conventionally designed components are possible
- The development costs can be reduced and shorter process times achieved
- The computing costs for optimisation runs, per iteration step, amount to approx. one to three times of the costs of stress analysis alone (depending on the optimisation method employed).

On the whole, structural optimisation has become an efficient design tool for the preliminary design and main design phases. Further development, integrating the process more closely with standard design practice, will improve the possibility of developing components of even better quality in less time.

ACKNOWLEDGEMENTS

The authors wish to thank Messrs. Hertel, Kathmann, Kraul and Kröber for their assistance and constructive ideas in connection with the optimisation tasks described here.

REFERENCES

[1] Ed: Morris, A.J. Foundation of Structural Optimisation: A Unified Approach John Wiley, London, 1982

[2] Ed: Eschenauer, H.; Olhoff, N. Optimisation Methods in Structural Design Euromech-Colloquium 164. B.I., Mannheim/Wien/Zürich, 1982

[3] Ed: Atrek, E.; Gallager, R.H.; Ragsdell, K.M., Zienkiewicz, O.C. New Directions in Optimum Structural Design. John Wiley & Sons Ltd., New York, 1984

[4] Morris, A.J., Bartholomew, P.; Dennis, J. A Computer Based System for Structural Design, Analysis and Optimisation. AGARD Conf. Proc. No. 280. The Use of Computer as a Design Tool. Munich 1979

[5] Bartholomew, P., Morris, A.J. STARS: A software package for structural optimisation. Proceedings of the Intern. Symposium on Optimum Structural Design. Oct. 1981, University of Arizona, USA

[6] STARS USER'S GUIDE Scicon Consultancy International Ltd., London, July 1985

[7] Hinrichsen, J. Beschreibung des Programmes AEROOPT. MBB-UT-interner Bericht

[8] MSC – NASTRAN USER'S MANUAL Caleb W. McCormick (Editor) The MACNEAL – Schwendler Corp., May 1976 (Revised April 1981)

[9] Wellen, H. Anwendung der rechnergestützten Strukturoptimierung bei der Auslegung von Flugzeugbauteilen. Konstruktion, Springer-Verlag, Heft 37, 1985

[10] Hafka, R.T. Finite elements in optimum structural design. Paper in this publication.

[11] Bartholomew, P. and Morris, A.J. A unified approach to fully-stressed design. Engineering Optimisation. Vol. 2, pp3-15, 1976

[12] Han, S-P. A globally convergent method for non-linear programming. J. Optim. Theory Applic., Vol. 22,297-305 (1977)

[13] Warnaar, D.B. Mass Optimisation Study of the Short Supply Mast. MBB-ERNO-interner Bericht, 1985

WEIGHT OPTIMIZATION OF AIRCRAFT STRUCTURES

Torsten Brâmå
Saab-Scania AB
Saab Aircraft Division
S-581 88 Linköping, Sweden

ABSTRACT

The general structural optimization system OPTSYS has been developed. In this system software for mathematical programming, structural, aeroelastic and aerodynamic analysis are integrated. The system design and methods used in OPTSYS are presented and applications to aircraft structures are supplied to show the capabilities of the system.

BACKGROUND

Saab Aircraft Division and the Aeronautical Research Institute in Stockholm (FFA) are since 1982 developing a structural optimization system called OPTSYS. We decided to build our work on an existing system, developed by B. Esping who had been working with structural optimization for several years at the Royal Institute of Technology in Stockholm, ref (4). In late 1983 we got the CRAY 1 computer and a few months later the first version of OPTSYS was tested. At Saab Aircraft Division OPTSYS is linked to the ASKA FE-system, while FFA is using an inhouse FE-code named BASIS. We have however the major parts of the software in common. OPTSYS is today running on VAX and CRAY.

PROBLEM STATEMENT

Consider a structural optimization problem in the following general form:

NATO ASI Series, Vol. F27
Computer Aided Optimal Design: Structural and Mechanical Systems. Edited by C. A. Mota Soares

$$\text{Minimize} \quad W(X) \quad , \; X = (x_1, x_2, \ldots, x_n) \qquad \ldots (1)$$

$$\text{subject to} \quad g_j(X) \leq \bar{g}_j \quad , \; j=1,m \qquad \ldots (2)$$

$$\text{and} \quad \underline{x}_i \leq x_i \leq \bar{x}_i \quad , \; i=1,n \qquad \ldots (3)$$

$$\sum_i d_{ij} x_i \leq \bar{d}_j \; , \; j > m \qquad \ldots (4)$$

The objective function (1) (eg. weight) and the constraints (2) (eg. stresses, deflections) are implicit nonlinear functions of the design variables, X.

The variables are also bounded by upper and lower limits (3), corresponding to, for instance, fabrication considerations. Sometimes you also need an additional set of linear constraints (4),(eg. upper limit for total thickness of a composite stack when layers are associated to individual variables).

PROBLEM SOLUTION

The general approach used for solving this implicit problem is to generate a sequence of explicit subproblems according to the following iterative scheme:

i Choose a starting set of variables X_0.

ii Calculate the current values of objective and constraint functions, using for instance a FE-model of the structure. Select an active set of constraint functions and calculate their gradients.

iii Generate an explicit and convex subproblem by approximating (1) and (2) through a first order expansion based on the calculations in the previous step.

iv Solve this subproblem and let the solution be the next iteration point X_k. Go back to step ii.

This process is interrupted when some convergence criteria are fulfilled or when the user is satisfied with the current solution.

There are three basic tasks in each iteration, analysis, gradient calculation and solution of the subproblem. These task are expensive, therefore it is of great importance that the subproblem as good as possible approximates the original problem in order to get few iterations.

We can today choose between three different algorithms, the DUAL-2 by Fleury, ref (2), one by Esping, ref (4, 5) and the MMA by Svanberg, ref (3). All of these are using a dual formulation to solve the subproblem. The MMA algorithm which was developed and implemented during 1985 has been used since and so far behaved well in all applications.

OPTSYS SYSTEM

The system consists of a number of stand alone programs, each with a well defined task, and a central optimization database, ODB, which is an application of a general database system MEMCOM, ref (15).
Some programs are standard inhouse analysis programs only communicating with its own database, others transfer information between two databases through subroutine calls.
This approach requires that you have access to the internal data structure in the FE-database. In this sense the ASKA system is very suitable as it is a highly modular FE code and the user has access to a large library of subroutines.

The MONITOR function is used to control the optimization process, eg. get the iteration history, redefine or modify the problem statement, create the proper execution procedures as the calling sequence of programs depends on the application. In VAX you also have a complete interactive access to ODB through a standard monitor included in MEMCOM.

The system sketch below indicates the OPTSYS system functions, where each box consists of one or several stand alone programs.

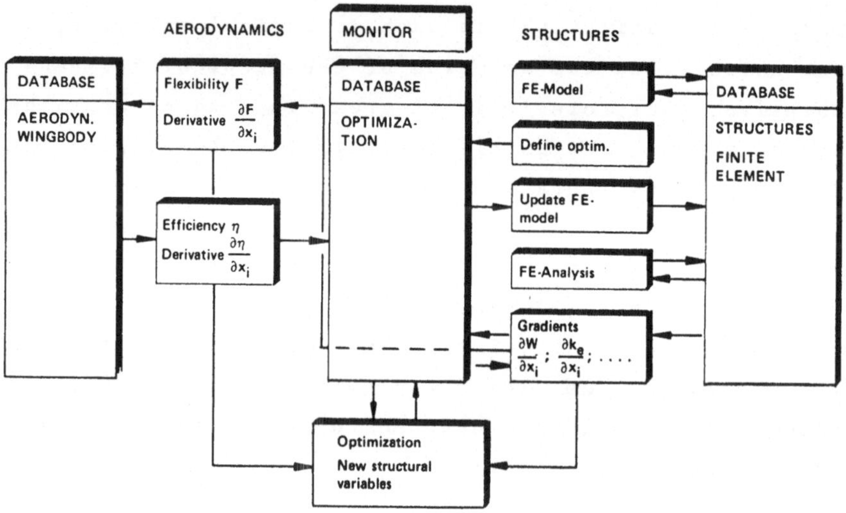

Fig 1 OPTSYS system

OPTSYS FEATURES

The features of OPTSYS are listed below. The options within parenthesis are currently under development and have so far not been tested on production like applications.

Objective function

Structural weight or the moment of inertia around an axis. (Linear combinations of the quantities above and constraint functions).

Constraints

General deflection constraint expressed as a linear combination of displacement components. Stress and strain constraints. Lower bound on eigenfrequency. Aileron efficiency constraint. Local buckling. (Flutter.)

Variables

Sizing variables as cross section areas in rods, thicknesses in membranes. In the case of a composite stack, the ply thicknesses and the fibre direction. Location of nodal points. (Shape variables using a CAD approach ref. (4)). Variables can be continous or discrete. Variables are linked to meet practical requirements. Variables are given upper and lower limits.

Elements included

All element types can be included in the FE model but variables and element dependent constraints like stresses can today only be associated to the following set of finite element types. Namely the 2 and 3 node bar, 6 and 8 node membrane, 3 and 4 node shell (only membrane stress) and 15 and 20 node solid elements. This set can however easily be extended.

Pre and post processing

There are tools for creating the input file defining the optimization problem and for documentation of the optimization process. The standard inhouse pre- and post processors can of course be used for the ASKA FE-model.

GRADIENT CALCULATION

Derivatives on element quantities like weight, stiffness and mass matrices are derived numerically using finite differences and stored in ODB.

$$dk_e/dx = [\ k_e(x+\Delta x) - k_e(x)\ ] / \Delta x \qquad \ldots (5)$$

These derivatives are later used when calculating gradients for objective and constraint functions. This approach makes it especially easy to implement new elements.

Weight gradient

The complete weight gradient is evaluated by assembling contributions from all elements affected by the design variable.

$$dW/dx = \sum_e dW_e/dx \qquad \ldots (6)$$

Deflection gradient

The static problem can be formulated,

$$K\ u = p.$$

The general deflection constraint can be expressed as a linear combination of components in the displacement vector.

$$d = u^t\ q < d_{max}$$

The gradient is now assembled, using (5), as

$$d(d)/dx = -\sum_e (u_e^t\ dk_e/dx\ v_e),\ \text{if}\ \ dp/dx = 0 \qquad \ldots (7)$$

where v is the solution to the new equation system,

$$K\ v = q\ .$$

Strain and stress gradients

These gradients are in the simple isotropic case evaluated in a similar way as

$$d\varepsilon/dx = -\sum_e (u_e^t\ \ dk_e/dx\ v_e) + \ dq_e/dx \qquad \ldots (8)$$

with $\varepsilon = u^t q$ and $K v = q$

ε being the element strain vector.

Eigenfrequency gradients

With the eigenvalue problem formulated as

$$(K - \omega^2 M)a = 0 \quad , \text{and} \quad a^t M a = 1$$

gradients for ω^2 ,using (5) are given by

$$d\omega^2/dx = \sum_e (a_e^t \, dk_e/dx \, a_e) - \omega^2 \sum_e (a_e^t \, dm_e/dx \, a_e) \qquad \ldots (9)$$

The constraint is then stated as $\omega^2 \geq \omega^2_{min}$ or $-\omega^2 \leq -\omega^2_{min}$

Aeroelastic gradients

For aerodynamic and aeroelastic analysis the FFA-WINGBODY program is used ref. (9). The program solves a linearized potential equation using a panel method. Among other quantities the FFA-WINGBODY can calculate aeroelastic efficiency parameters which are important for aircraft performance. This parameter η is defined as the ratio of an aerodynamic response quantity for an elastic structure over a rigid, ref (7).

Gradients of η with respect to design variables x can be calculated if the program is provided with a structural influence matrix (EM) and its derivatives (d(EM)/dx), computed at the panel points. (EM) is defined by the relation

$$S = - (EM) \, F \qquad \ldots (10)$$

where F is the vector of forces in panel points and S is corresponding slope changes. An application on aircraft structures is presented in ref (8).

Flutter gradients

Flutter is a serious vibration phenomenon which if it occurs might be disastrous. It is therefore vital to be able to avoid flutter during the optimization process. Analysis of this aerodynamic instability yields a nonlinear eigenvalue problem. The location of the eigenvalues in the complex plane indicate if the vibrations are stable or not. The method that is used at Saab Aircraft Division to solve this problem is described in ref.(10). This flutter analysis program has capabilities to calculate also the derivatives of the complex eigenvalues with respect to design variables, ref. (11). In order to reduce the number of unknowns, the flutter problem is expressed in a base ϕ of m selected structural eigenmodes. The information needed from the structural FE model is, besides the selected modes and corresponding frequencies, the stiffness and mass matrices and their derivatives transformed to the modal space. For example the reduced stiffness derivatives, dK_m/dx, can be evaluated like

$$dK_m/dx = \phi^t dK/dx\ \phi, \text{or expressed as an assembly operation}$$

$$= \sum_e \left(\phi_e^t\ dk_e/dx\ \phi_e \right),\ (m \times m) \qquad \ldots (11)$$

MMA, THE METHOD OF MOVING ASYMPTOTES

The demands for a good optimization algorithm could be summerized in the following qualities.

- Reliability. It should always converge to a good feasible solution.

- Efficiency in terms of few iterations.

- Generality. It should be able to treat all types of variables and constraints, eg. discrete variables, negative variable values, mixture of linear and highly nonlinear functions, and reasonably infeasible initial designs.

As we were not quite satisfied with the algorithms available at that time, we started in early 1985 a cooperation with K. Svanberg at the Royal Institute of Technology in Stockholm, who had some interesting ideas.

He came up with the MMA algorithm. It can be looked upon as a generalization of an algorithm proposed by Fleury using mixed direct/reciprocal design variables, ref (1). The expansion is now made in variables of type $1/(U_i-x_i)$ or $1/(x_i-L_i)$ depending on the sign of the gradient. L_i and U_i are the moving asymptotes. It can be shown that if you let $L_i = 0$ and U_i reach infinity you end up with the algorithm by Fleury.

The new idea here is firstly that each variable has its own upper and lower asymptote and secondly that these asymptotes can be moved between iterations. Rules for how to move the asymptotes are based on information about iteration history for the variables and on human experience and intuition. In this way the algorithm is automatically tuned to generate improved approximate functions during the optimization process.
The rules used today for how to move the asymptotes can briefly be described like this. If a variable tends to oscillate the process needs to be stabilized. This is accomplished by moving the asymptotes closer to the current variable value and in this way create a more conservative function approximation (ie. with more curvature). On the other hand when a variable is steadily going in the same direction the asymptotes are "moved away" to speed up the process.

There is a set of parameters for MMA controlling the rules for how to set the initial position of, and how to move, the asymptotes. These parameters can be set by the user in the input file or changed between iterations using a monitor function. However, the default parameters should be sufficient to solve any problem, although another set of parameters would have been more efficient. As the character of constraint functions is very difficult to estimate in advance, it is important to have an algorithm which can improve itself.

MMA has proved to solve large problems efficiently.

The treatment of discrete variables and equality constraints will be implemented in MMA during 1986.

APPLICATIONS

The fighter aircraft composite wing

Problem statement: Minimize the wing weight by varying ply thicknesses and the fibre directions in upper and lower skin including constraints on fibre strains and wing torsion stiffness. The FE-model contains 1046 elements, 1571 unknowns and 2 loading cases. A total of 92 independent variables, 90 ply thickness variables and 2 fibre directions corresponding to the zero degree ply in upper and lower skin. 432 strain constraints and wing torsion constraints at two wing sections.

With fixed angle variables the process converges with a weight reduction of 8.5 %. Some strain and torsion constraints has now reached their limit value. When letting the angles free the convergence is slow. To get a feasible solution the angles where fixed again at 68 degrees giving a total weight reduction of 11 %. The CPU time on the CRAY 1A is 100 seconds per iteration.

Recently this problem was restarted after iteration 10 with the new MMA algorithm giving the result indicated with the dotted lines. After 4 iterations we have a similar solution as previously.

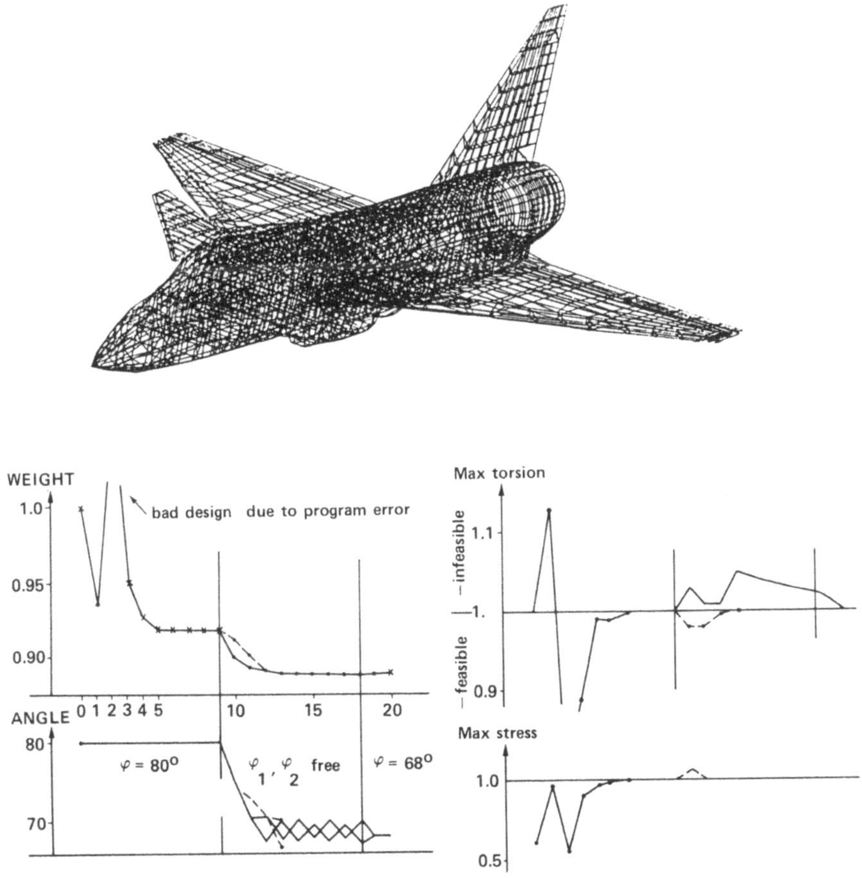

Fig 2 The fighter wing

The fin rudder

Problem statement: Minimize the moment of inertia around an axis a, by varying the ply thicknesses including a lower bound for the eigenfrequency. The FE model contains 346 elements and 1859 unknowns. A total of 393 independent design variables in 131 elements and only one constraint.

The initial design is infeasible as the eigenfrequency is to low. The wanted eigenfrequency is reached already after the first iteration and the objective function converges after 7

iterations. When the lower bound is raised another 7 Hz, the new optimum is reached after 4 iterations.
The CPU time is 44 seconds per iteration on the CRAY 1A.

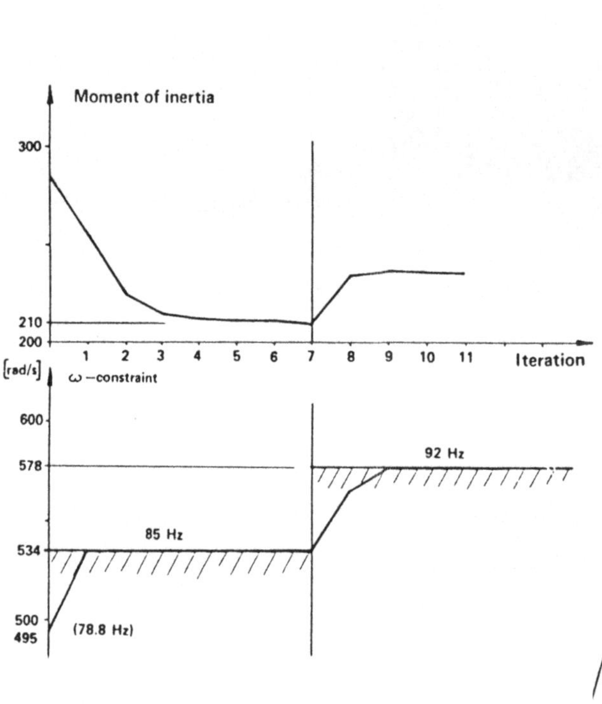

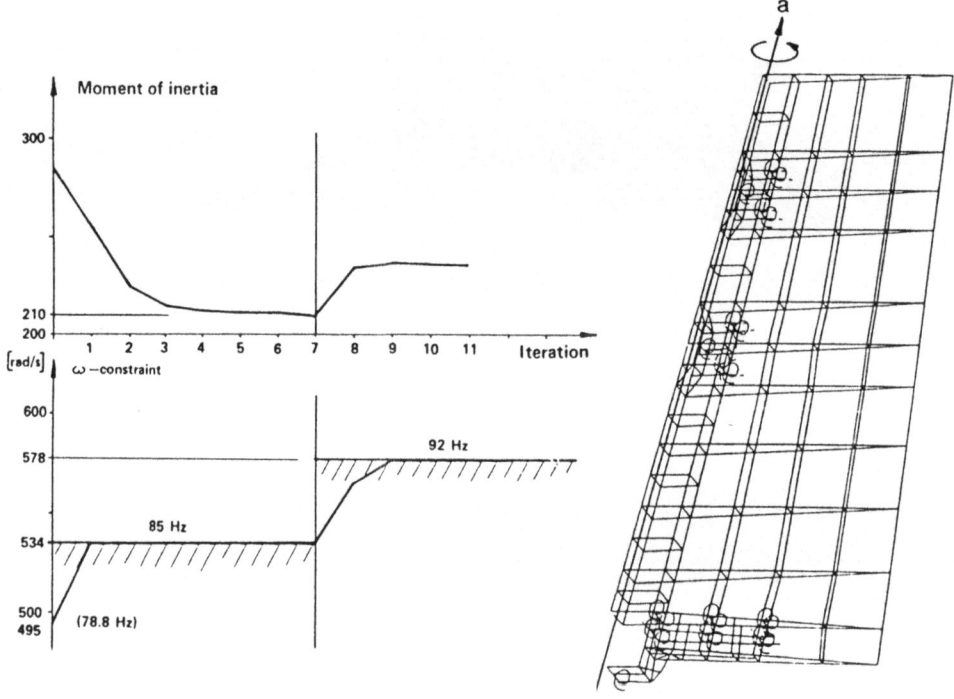

Fig 3 The fin rudder

Shape optimization of a cantiliver beam

Problem statement: Minimize the weight of a beam subjected to an eigenfrequency constraint of 1 rad/s by varying the cross section shape.

The beam has a constant length of 10, an initially quadratic cross section area of 1 and density 1. It is modelled by 10 solid 20 node elements. The node coordinate variables are linked to keep element edges straight and the cross section rectangular. This means two independent variables in each cross section, giving a total of 22 shape variables.

As the eigenmodes occur as pairs due to the symmetry in this problem the quadratic shape of the cross section is automatically kept during the process. The shape has converged after 10 iterations having reached the upper limit of the shape variables, corresponding to the minimum area, in the three outermost sections. The number of iterations needed to solve this type of problems using MMA is in the range of 7 to 15, ref (12).

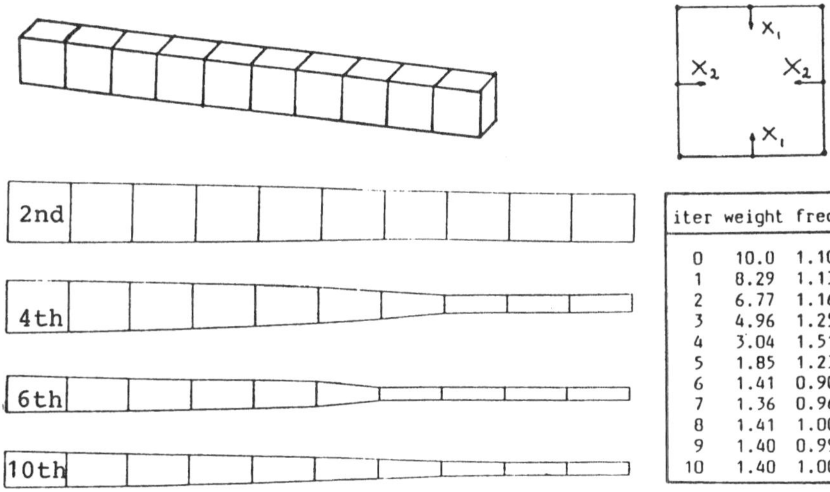

iter	weight	freq
0	10.0	1.10
1	8.29	1.13
2	6.77	1.16
3	4.96	1.25
4	3.04	1.51
5	1.85	1.23
6	1.41	0.908
7	1.36	0.961
8	1.41	1.00
9	1.40	0.999
10	1.40	1.00

Final area distribution: .307, .275, .242, .204, .168, .125
.089, .052, .040, .040, .040

Fig 4 Shape optimization

PRACTICAL EXPERIENCE

When you start to use a system like OPTSYS on real life problems you are immediately confronted with demands for further development and practical considerations that has to be taken care of. It is therefor very important to have a flexible system design which as far as possible is prepared for future extensions.

Systems like this tend to become very complex as software from several sources are linked together to an integrated unit. The need for careful planning and documentation is obvious.

The optimization system is not limited to solve the classic weight minimization problem. It can be a useful tool in a wide range of investigations. Already the basic gradient information is useful, ie. what is the influence of each variable on objective and constraint functions?
Is there a feasible design to my problem?
What is the cost in weight to change the limit of a constraint?
What is the potential weight reduction of a change of material?
Etc.

CONCLUSIONS

Using a good optimization algorithm together with efficient analysis tools and a general database based program system, it is possible today to create an efficient and general system for structural optimization. Providing you have a fast computer, it is possible to treat large and complex structural optimization problems within resonable time. The OPTSYS program is developed continously and has so far proved to have the flexibility needed.

REFERENCES

(1) Fleury C. and Brabiant V. "Structural optimization. A new dual method using mixed variables", Report SA-115, University of Liege, 1984.

(2) Fleury C. "Structural weight optimization by dual methods of convex programming", Int. Journal for Numerical Methods in Engineering, Vol. 14, pp 1761-1783, 1979.

(3) Svanberg K. "METHOD OF MOVING ASYMPTOTS - A new method for structural optimization", TRITA-MAT-1985-14, Royal Institute of Technology, Stockholm (to appear in Int Journal of Numerical Methods in Engineering).

(4) Esping B.J.D. "Structural optimization using numerical techniques", Report No. 85-8, Royal Institute of Technology, Stockholm.

(5) Esping B.J.D. "Minimum weight design of membrane structures", Computers and structures, Vol. 19, No 5/6, 1984.

(6) Fleury C. "Shape optimal design by the convex linearization method", Paper presented at the Int. Symp. "THE OPTIMUM SHAPE: AUTOMATED STRUCTURAL DESIGN ", Warren, Michigan, Sept. 1985.

(7) Hedman S.G. "Aeroelastic constraints for a wing weight minimization procedure", FFA TN 1985-4, The Aeronautical Research Institute.

(8) Johansson O. "Weight optimization under structural and aeroelastic constraints of a wing for a general aviation aircraft", FFA TN 1985-3, The Aeronautical Research Institute.

(9) Hedman S.G. and Tysell L.G. " The FFA Wingbody 83 Computer Program. A panel method for determination of aeroelastic characteristics at subsonic and supersonic speeds", FFA TN 1983-3, The Aeronautical Research Institute.

(10) Stark V.J.E. "Flutter calculations by a new program", The second International Symposium on aeroelastcity and structural dynamics, Aachen, W. Germany, April,1985.

(11) Stark V.J.E "Derivatives of damping and frequency with respect to design variables", Saab Aircraft Division, TKLF.
(to be completed).

(12) Rosengren R. "Shape optimization of solid structures in OPTSYS", Saab Aircraft Division, Report TKH R-3492.

(13) OPTSYS, User´s manual, Saab Aircraft Division.

(14) OPTSYS, Programmer´s manual, Saab Aircraft Division,

(15) MEM-COM User´s manual, SMR Corp., Switzerland.

(16) Dahlberg M. "Buckling Constraints in Truss Optimization. An Implementation in OPTSYS", Saab Aircraft Division, Report TKH R-3540.

A CAD approach to structural optimization

B J D Esping and D Holm
Dept. of Aeronautical Structures and Materials
The Royal Institute of Technology, Stockholm, Sweden

Abstract

A general approach to shape optimization based on CAD formulations is proposed and implemented in the OASIS-ALADDIN system for structural optimization. Two large scale examples demonstrates the ideas.

Introduction and design concept

CAD, in a very wide meaning, involves not only drafting, but also analysis and redesign (fig 1)

Fig. 1 CAD including optimization

NATO ASI Series, Vol. F27
Computer Aided Optimal Design: Structural and Mechanical Systems. Edited by C. A. Mota Soares

The design concept that we worked on looks like fig 2:

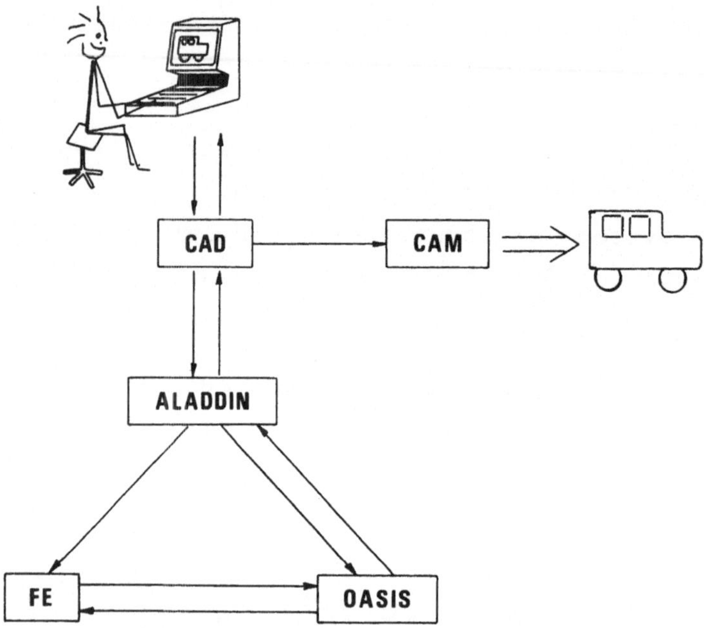

Fig. 2 Design concept

First we would like to briefly explain the different boxes in the concept

CAD - is the traditional CAD system used for drafting, etc. We assume that a wire frame model is used

CAM - is the manufacturing phase

ALADDIN - is our link to analysis and design. ALADDIN uses the wire frame data from the CAD system. The wire frame model can also be defined directly in ALADDIN.

FE - is a Finite Element program for analysis. Its input is produced by ALADDIN

OASIS - is the optimizer and adminstrator of the redesign process. OASIS includes some subroutines from the ALADDIN and FE programs and also uses their databases, see ref [7]. OASIS is using the data base system MEMCOM [9] for external storage.

Let us explain the process a little bit further:

1. The designer will create a CAD model of his structure using his ordinary CAD system. The model is temporarily since it will be changed during the optimization process
2. Information concerning the shape definition is transferred to the ALADDIN program. ALADDIN is using the wire frame concept from CAD and is basically using a set of controlpoints, lines, surfaces and solids to describe the shape. The designer will now add design variables to some of the control points. The shape can now be plotted, either as a CAD model or as a FE model, for different sets of design variables. Once the designer is pleased he will complete the structure with data such as loads, boundary conditions, material properties etc. Complete data for FE-analysis and optimization is transferred to the FE program and the OASIS program. ALADDIN is also creating a database that is used by OASIS.
3. OASIS takes over. OASIS computes function and gradient values of the objective and constraint functions using the FE and ALADDIN systems. A sequence of subproblems will be solved and result in a new set of design variables which are used to update the CAD and FE databases. The results can be inspected or interactively changed after each iteration using the optimization postprocessor GANDALF, ref [10]. The present design can also be visualized on a data screen using ALADDIN. A FE-postprocessor can also be used to present intermediate stress and displacement distributions. Corrections of the FE-mesh due to unexpected shapes can be made in ALADDIN before the optimization continues.
4. The optimized design can be checked using GANDALF, ALADDIN and FE postprocessers before the updated CAD model is transferred back to the CAD system.
5. The designer will now, on the CAD level, inspect his new design. Maybe he is not pleased with his result and

will thus do some modifications and another optimization. Once he accepts the design he will complete the data with measurements, surface finish etc before the CAM system takes over to produce the structure.

Up until today is the system not complete. Operative is now the ALADDIN-OASIS-FE system including GANDALF and FE-post-proccessing.

The following will present a more detailed description of the ALADDIN-OASIS-FE system.

Shape description and design variables

A wire frame model is basically defined by its control points, P_k, k = 1,K, and their coordinates. However, a general mechanical structure is a continuum. It can be described with a parametric formulation:

$$R = F(r,s,t,P)$$

where R is the position vector in the real space, r,s,t are the parameters (values 0 → 1) and P is the set of control point coordinates. F is the transformation fucntion, see fig 3.

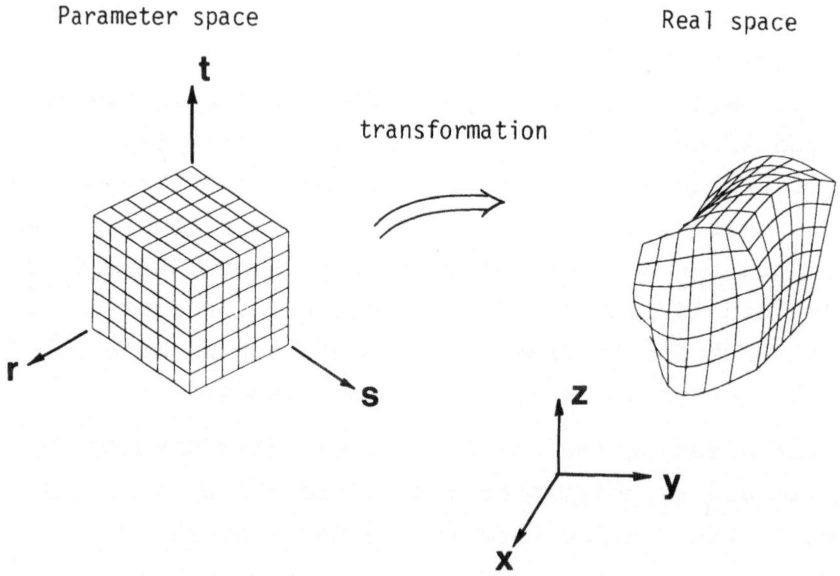

Fig. 3 Transformation from parameter space to real space

A FE mesh is easily defined in the parameter space. It is transformed to the real space using the mapping function F. Let C_l, l=1,L be the FE node points and $(r,s,t)_l$ their parameter values. If we want to go from a rough mesh to a more refined one, this is easily done. We simply introduce a new coordinate set $(r,s,t)_l$ in the parameter space. The mapping function is unchanged.

Another interesting feature is that the mathematical expression for the mapping function F is not affected by the values of the control points P_k. The value of F is of course changed but not its mathematical form. This implies that all the real space node points C_l are affected by a small change of a control point P_k. However the parameter values $(r,s,t)_l$ of the nodes will be unchanged. With a fixed set of parameters, the structure and the mesh is entirely defined by its control points. It is now logical to attach design variables X_j, j = 1, J to some of those control points. Parametric shape functions F can be found for most structures (see ref [2]). Several such structures or separately defined bodies can also be combined in order to describe more complex structures.

It shall be noticed that we can use the mapping function F not only for the geometrical shape but also for any kind of distribution, ie load, temperature, thickness, boundary condition etc.

The work described in this section is performed by the ALADDIN program, see ref [8].

Optimization algorithm and approximations

The design problem can be expressed as an optimization problem:

$$\begin{array}{lll} \min & w(x) & \\ \text{subject to} & & \\ & g_i(x) \leq 0 & i = 1,I \\ x_j \leq x_j \leq \bar{x}_j & & j = 1,J \end{array}$$

w is the objective function which will be minimized. g_i, i=1, I is the set of constraints and x_j, j=1,J the set of design variables. x is the vector of design variables x_j. w and g_i are, in general, nonlinear implicit functions of x. Some type of approximation has to be done.

Prasad and Haftka, see ref [3], have proposed a first order hybrid approximation where w and each g_i are linearized in 1/x or x_j, depending on the sign of the functions derivative. Let us consider one constraint g_i. It will be linearized in $1/x_j$ if $\partial g_i/\partial x_j \leq 0$ or linearized in x_j if $\partial g_i/\partial x_j > 0$. As a result the approximate problem will always be convex (for $x_j > 0$) and separable. It can be solved by many techniques, for instance dual methods, see Fleury and Braibant ref [4]. A sequence of hopefully converging approximate subproblems will thus be solved in order to solve the original problem.

We will in this paper use a further development of the hybrid method. It is called the Method of Moving Asymptotes (MMA) as proposed by Svanberg ref [5].

Let us introduce new variables $y_j = 1/(x_j - L_j)$ and $z_j = 1/(U_j - x_j)$. We shall notice that $L_j \to 0 \Rightarrow y_j \to 1/x_j$ and $U_j \to \infty \Rightarrow z_j \to x_j$, eg the hybrid method approximations. L_j and U_j are the lower and upper asymptotes to x_j, and x_j will always lie somewhere in between. We will now make convex approximations in these new variables, that is linear in y_j if $\partial g_i/\partial x_j \leq 0$ or in z_j if $\partial g_i/\partial x_j > 0$. L_j and U_j will successively be adjusted in order to improve the approximations. If the subproblem solutions are oscillating in a variable x_j, then L_j and U_j will be forced closer together. On the other hand if the variable x_j is monotonic, increasing or decreasing, they will be pushed further apart.

OASIS is using (k is here the iteration number):

$$L_j^{(k)} = x_j^{(k)} - (\bar{x}_j - \underline{x}_j) \qquad \text{for} \quad k = 1,2$$
$$U_j^{(k)} = x_j^{(k)} + (\bar{x}_j - \underline{x}_j)$$

and

$$L_j^{(k)} = x_j^{(k)} - s(x_j^{(k-1)} - L_j^{(k-1)}) \qquad \text{for} \quad k > 2$$
$$U_j^{(k)} = x_j^{(k)} + s(U_j^{(k)} - x_j^{(k-1)})$$

where $s = 0{,}7$ for oscillating variables, ie if $A = (x_j^{(k)} - x_j^{(k-1)})(x_j^{(k-1)} - x_j^{(k-2)}) < 0$ and $s = 1/\sqrt{0.7}$ for monotonic variables, ie if $A > 0$.

The approximated subproblem will always be convex, so dual methods can be applied. The resulting equations are uncoupled, due to the separability, and of second degree which implies that x_j can be expressed explicitly by the Lagrangian multipliers. The convexified problem can sometimes be too conservative, which results in a problem with no feasible domain. This situation can be avoided if we introduce relaxation factors h_i for each constraint, eg

$$\begin{array}{lll} \min & \tilde{w} + \sum_i a_i(h_i + h_i^2) & \\ \text{s.t.} & \tilde{g}_i \leq h_i & i = 1,I \\ & \underline{x} \leq x_j \leq \bar{x}_j & j = 1,J \end{array}$$

$\tilde{w}$ and $\tilde{g}_i$ are the approximations of w and g_i respectively, and a_i are constants.

Derivatives

The MMA needs information about the derivatives $\partial w/\partial x_j$ and $\partial g_i/\partial x_j$.

Let us consider a displacement constraint, ie:

$$g_i = d - \bar{d} \leq 0$$

$$d = u^T q$$

$$\frac{\partial d}{\partial x_j} = - \sum_e u_e^T \frac{\partial K_e}{\partial x_j} v_e$$

where $d, \bar{d}$ is the constrained displacement and its upper bound. q is the virtual load and u is the displacement vector in the FE problem Ku=p where K is the assembled stiffness matrix and p the load-vector. v is the virtual displacement vector defined by Kv=q.

u_e, K_e and v_e are the displacements, stiffness matrix and virtual displacement associated with element e, respectively.

We finally calculate the derivative of K_e numerically:

$$\frac{\partial K_e}{\partial x_j} = \frac{K_e(x_j + \Delta x_j) - K_e(x_j)}{\Delta x_j}$$

The data flow will be:

1. The optimizer OASIS gives x_j and an accurate step Δx_j to the shape definition system ALADDIN
2. ALADDIN computes the associated element node coordinates and their resulting steps, and brings this information over to the FE system
3. The FE system computes $K_e(x_j)$ and $K_e(x_j + \Delta x_j)$
4. OASIS finally calculates the derivative $\partial K_e/\partial x_j$

The FE system is also used to solve the equation Kv = q.

Examples

We will very briefly present three cases where the FE system BASIS, ref [6], has been used.

Truck front beam

This is a contract job for Saab-Scania truck division, Södertälje, Sweden. The forged steel beam is shown in fig 4.

We have in this example used Coons 3-D solids bodies to create the CAD model. Each solid body is described by its opposing cross section surfaces which are connected by alongside splines. Each cross section is defined by a combination of straight lines and splines. Design variables are then attached to some of the control points in those cross sections and alongside splines. Many of those variables are linked to each other. We also have requirements on "positive" slopes of all surfaces as the beam is going to be forged. This implies explicit, linear constraints on many variables. The structure is also subjected to stress constraints and constraints on the rotations of the steering pivot pin. In order to reduce the problem size we consider only stresses on the beam surface.

The process was interupted after 3 iterations for inspection and after iteration 5 for discussions with the client. Each interation required 1 cp-hour on a CRAY-1 computer. Fig 5-8 show the CAD-model, FE-model and iteration histories.

fig 4 truck front beam

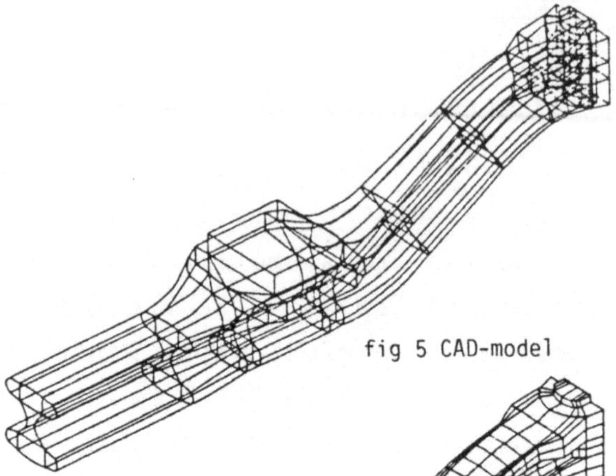

fig 5 CAD-model

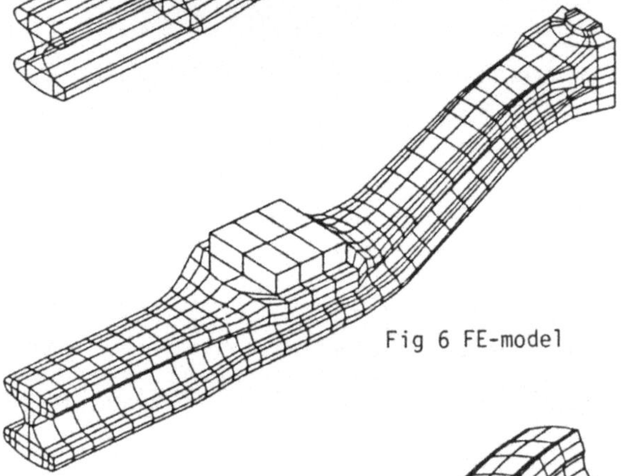

Fig 6 FE-model

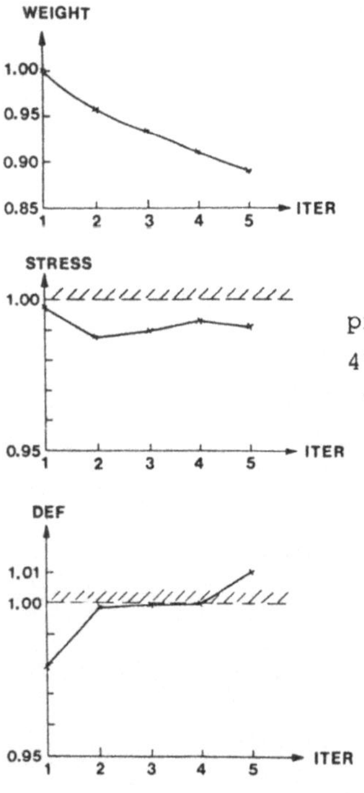

problem size:
4754 nodes
792 solid elements
(20-nodes)
3 loading cases
39 design variables

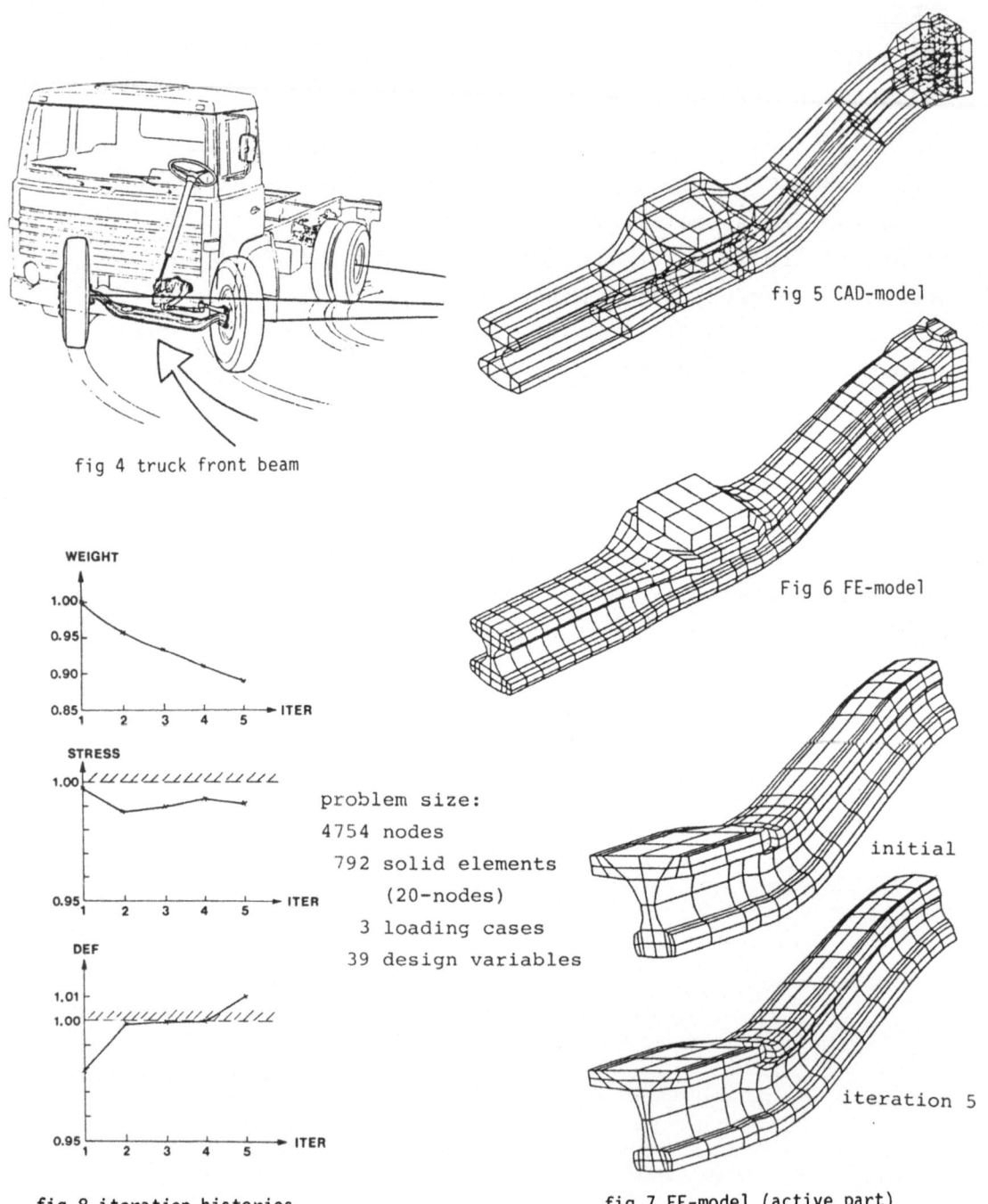

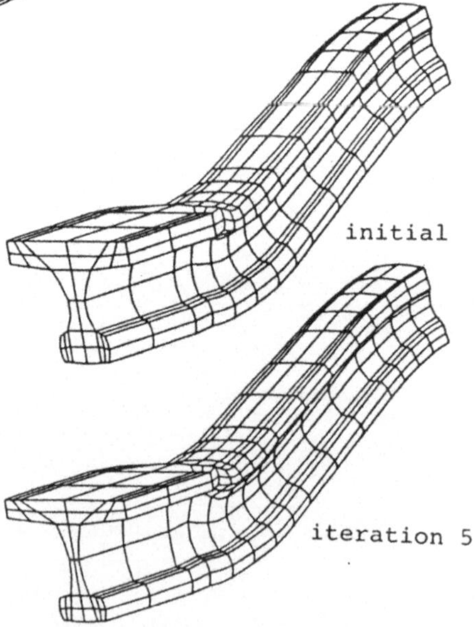

fig 8 iteration histories

fig 7 FE-model (active part)

Suspension arm

This is a contract job for Saab-Scania car division, Trollhättan, Sweden. The pressed steel suspension arm is shown in fig 9.

The CAD-model is defined by a combination of Coons 2-D and lofted spline surfaces, which in turn are composed of straight lines, circles and splines. We have in this case the uniform thickness as a design variable and also the height, width and overhang of the sides. The production requires "positive" slopes of the surface. Except for explicit, linear constraints we also have stress constraints. One iteration required 2 cp-hours on a VAX-750. Figs 10-12 show the CAD-model, FE-model and iteration histories.

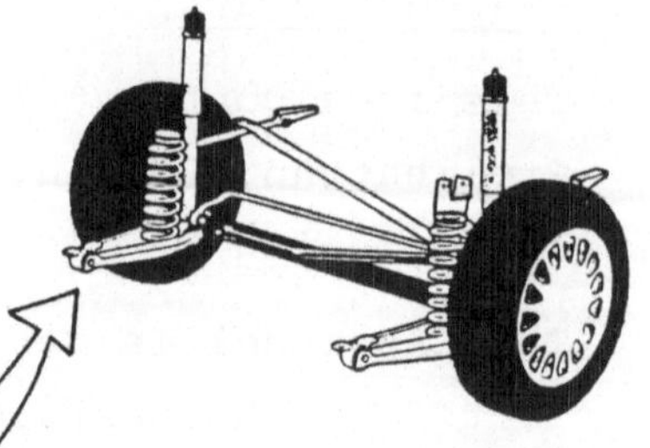

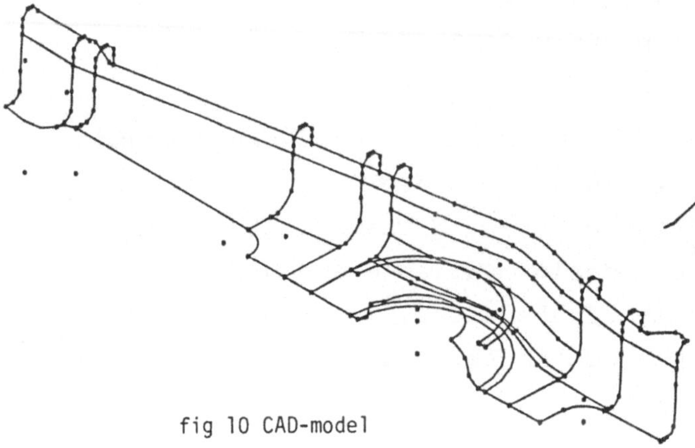

fig 9 suspension arm

fig 10 CAD-model

problem size:
819 nodes
223 shell elements
(8-nodes)
1 loading case
4 design variables

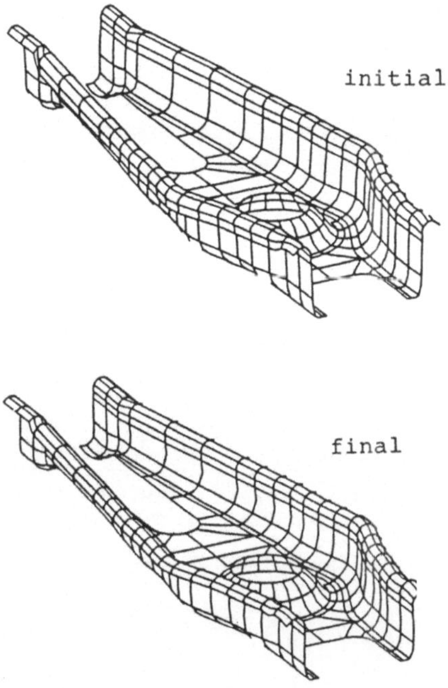

fig 11 FE-model

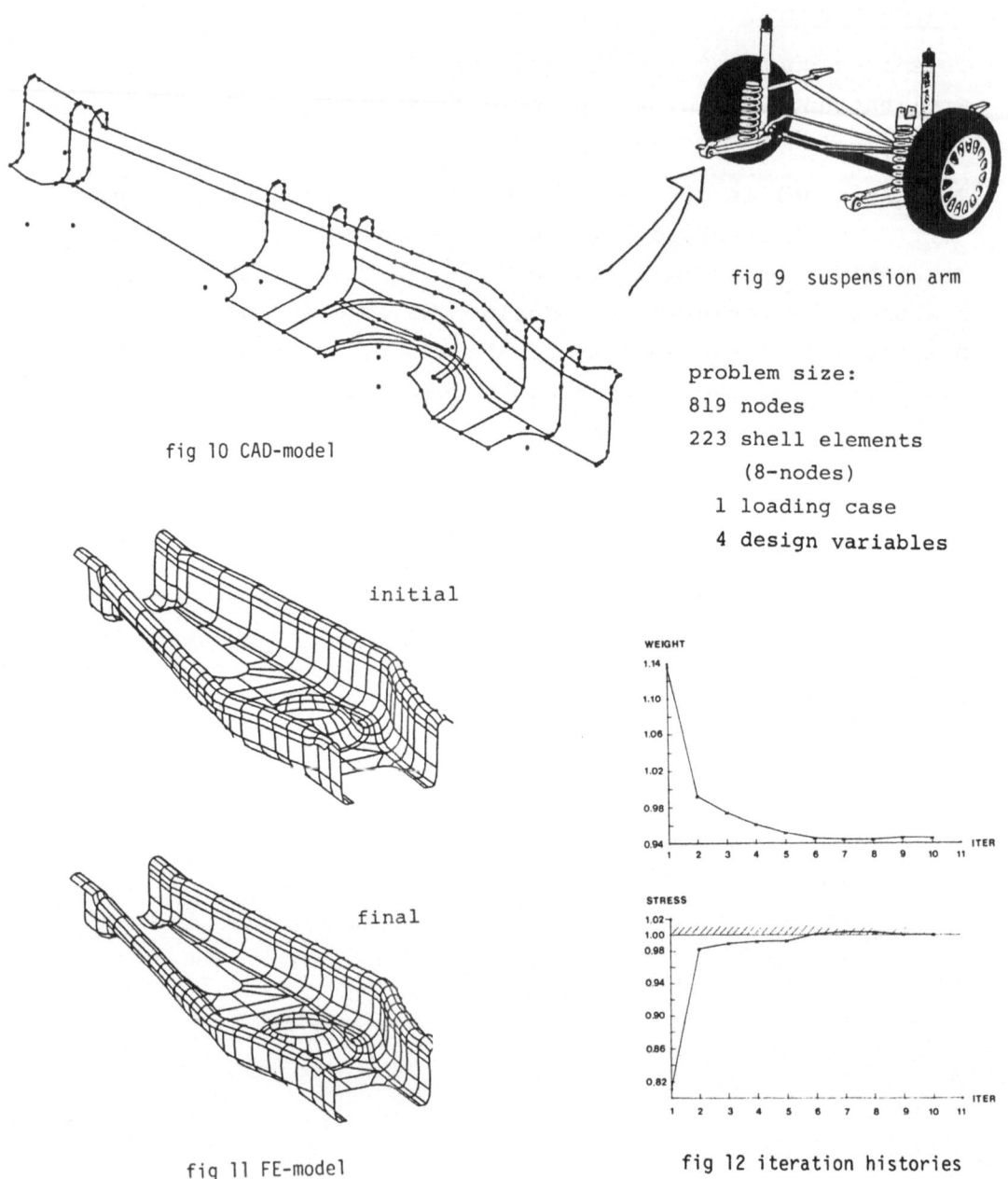

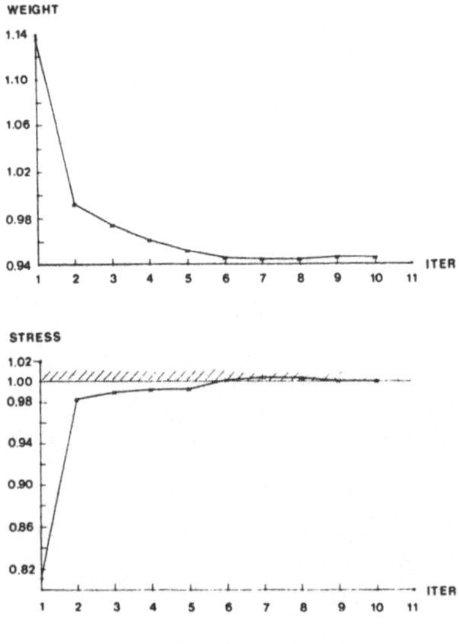

fig 12 iteration histories

Discussion

We have here been working with large scale problems on real structures which implies practical problems such as long execution times, very much I/O-operations, requirements on big disc size (300 Mbyte for the truck front beam) and need of big primary storage in the computer (1Mword on the CRAY-1).

We have also been confronted with practical requirements such as the positive slopes which implies explicit linear or nonlinear constraints. Other aspects are the forced linear dependency of many variables and mixtures of different kinds of variables such as thickness and shape. OASIS can also treat orientation of fibres in composite materials simultaneously [7].

In example 1 do we only consider surface stresses. We solved this problem by including surface stress elements with zero thickness. Stress constraints are only considered in those double curved membranes. OASIS can treat all kinds of elements simultaneously.

The steel plate used in the suspension arm is produced in certain discrete thicknesses only. We must therefore be able to solve mixed continous-discrete problems.

We can interactively inspect the results and do modifications after each iteraction. We do believe that this is very important for large scale problems where unexpected results may appear. We have, for certain examples, had problems with the element shapes after a number of iterations. So far we have solved this problem interactively using ALADDIN to modify the FE-mesh. We are right now working on adaptive mesh refinement and changes to overcome this problem.

On our work to link the CAD-system to ALADDIN we have noticed that sometimes we must use different CAD representations in the optimization then were originally defined in the CAD-system.

We have found that it is very important with good post-processing to evaluate the results. This includes plotting of iteration histories, shapes at different iterations, thickness distributions, stress distributions etc. The system must also be flexible to make it easy to include special requirements such as additional constraints for instance buckling, flutterspeed etc.

Finally we must say that the MMA method that we have used is very satisfactory. So far no problems.

References

1. B. J. D. Esping, A CAD approach to the minimum weight design problem, Int. Journal of Numerical Methods in Eng. Vol 21, 1985, pp 1049-1066.

2. I. D. Faux & M. J. Pratt, Computational Geometry for Design and Manufacture, Ellis Horwood, 1979.

3. B. Prased & R. T. Haftka, Optimum Structural Design with Plate Finite Elements, Journal of the Structural Division, Nov. 1979, pp 2367-2382.

4. C. Fleury & V. Braibant, Structural Optimization - A new dual method using mixed variables, report SA-115, Aerospace Laboratory of the University of Liège, Liège, Belgium, 1984, pp 1-11.

5. K. Svanberg, MMA - Method of Moving Asymptotes - A new method for structural optimization, TRITA-MAT-1984-14, Royal Institute of Technology, Stockholm, Sweden, 1985, pp 1-23.

6. BASIS - Finite Element Structural Analysis, Users Manual, The Aeronautical Research Institute of Sweden, Stockholm, Sweden, 1979.

7. B. J. D. Esping, The OASIS structural optimization system, Computers & Structures, Vol 23, No 3, pp 365-377, 1986.

8. D. Holm, ALADDIN - users manual, Dept. of Aeronautical Structures and Materials, The Royal Institute of Technology, Stockholm, Sweden.

9. MEMCOM, Users manual, SMR Corp., 2500 Bienne, Switzerland.

10. B. J. D. Esping, GANDALF - users manual, Dept. of Aeronautical Structures and Materials, The Royal Institute of Technology, Stockholm, Sweden.

APPLICATION OF STRUCTURAL OPTIMIZATION USING FINITE ELEMENTS

P. Ward, D. Patel, A. Wakeling, R. Weeks
Structural Dynamics Research Corporation
York House, Stevenage Road,
Hitchin, Herts, England.

Introduction

Until recently, structural optimization has not significantly impacted the wider analysis community. Several reasons may be advanced, the most prominent in the mechanical engineering field being the requirement for shell bending elements. In addition, the capability has generally been made available as ad hoc packages loosely connected to an analysis program. SDRC's OPTISEN offers fully integrated analysis and redesign with interactive graphics capabilities for both the specification of the design problem and the interpretation of results.

THE INTEGRATED CAE ENVIRONMENT

In order that structural optimization be usable for production applications, close integration is required between finite elment mesh definition, analysis, optimization and output interpretation software. In addition, the database environment must be sophisticated enough to recognize and support multiple design studies for a given component. These multiple design studies could arise form considering:

. Modified groupings of elements to form design variables
. Modified constraint limits
. Modified gauge limits

Figure 1 schematically illustrates the integration present within the SDRC I-DEAS[TM] software. A common application database allows free transfer of, for example, geometry, finite element mesh, loading definition and design variable groups between the various modules within the integrated re-design package.

NATO ASI Series, Vol. F27
Computer Aided Optimal Design: Structural and Mechanical Systems. Edited by C. A. Mota Soares

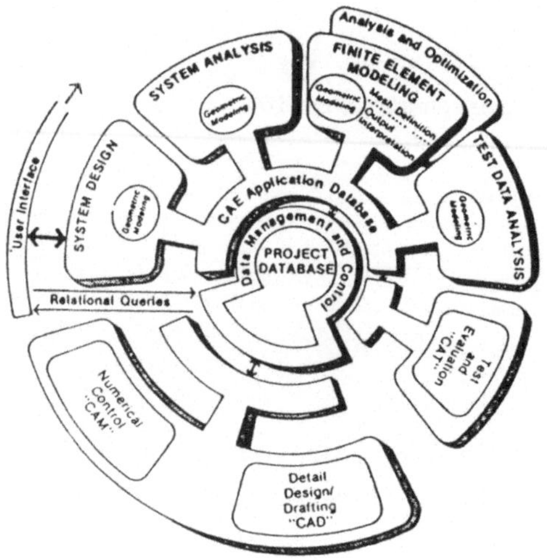

Figure 1. The I-DEAS Mechanical CAE System

Currently in OPTISEN optimization is available for linear static analysis for the following elements:

. 3, 4, 6 or 8 noded shell with membrane and bending effects
. Tube beam
. Box beam
. Bar
. Circular solid beam
. Channel beam

Other elements may be included in the model but they will not be resized.

Mathematical Foundation

OPTISEN seeks to minimise the mass of the structure $W(\underline{z})$ subject to performance constraints

$$G_j(\underline{z}) \leq 1$$

Where $\underline{z}$ is the vector of design variables, Z_i - typically thicknesses of shells. A group of finite elements is related to a design variable using the accepted technique of design variable linking [1].

In order to determine the conditions to be satisfied at the optimum we introduce a Lagrangean function

$$L(\underline{z}, \underline{\lambda}) = W(\underline{z}) + \sum_j \lambda_j (g_j(\underline{z}) - 1)$$

where λ_j are Lagrange multipliers which must be positive.

The Kuhn-Tucker conditions [2] may be derived as

$$\frac{\partial W(Z^*)}{\partial Z_i} + \sum_j \lambda_j \frac{\partial g_j}{\partial Z_i}(\underline{Z}^*) = 0$$

$$\lambda_j (g_j(Z^*) - 1) = 0$$

$$g_j(\underline{Z}^*) \leq 1$$

Solution Strategies

Two solution strategies are adopted in OPTISEN. The well known full stressing algorithm is available for stress constrained problems. In addition, a more general pseudo-Newton algorithm applicable to strength and/or stiffness controlled structures is included. The pseudo-Newton algorithm [3] seeks to satisfy the Kuhn-Tucker conditions in total by retaining linear expressions for the performance constraints and presuming that first and second derivatives of the objective function can be calculated. This approach may be interpreted as a second-order gradient projection algorithm.

i.e. Minimise

$$W(\underline{Z} + \underline{h}) = W(Z) + \underline{\nabla W}^t \underline{h} + \tfrac{1}{2} \underline{h}^t \underline{H}\, \underline{h}$$

Subject to

$$\underline{g}(Z) + \underline{Gh} \leq \underline{I}_c$$

where $\underline{I}_c$ is a column of unit values

This leads to the matrix equations

$$\lambda^{k+1} = \left[\underline{G} \quad \underline{H}^{-1} \quad \underline{G}^{t}\right]^{-1} \quad \{g^{k} - \underline{GH}^{-1} \quad \underline{\nabla W} - \underline{I}_{c}\}$$

$$\underline{h} = -\underline{H}^{-1} \left\{ \underline{G}_{t} \, \underline{\lambda}^{k+1} + \nabla \, \underline{W} \right\}$$

Where

$\underline{G}$ is a matrix of constraint gradients commonly referred to as design sensitivities.

$\underline{h}$ is a step in design space

The superscript refers to iteration level.

Active Set Strategy

A means of identifying active constraints is required to utilise this algorithm. OPTISEN uses a compound strategy based on constraint violation and information on Lagrange multipliers obtained by solving a linearised form of the dual problem [4] which requires maximisation of

$$L(\underline{z}, \underline{\lambda}) = W(\underline{z}) + \sum_{j} \lambda_{j} \, (g_{j}(\underline{z}) - 1)$$

Subject to

$$\frac{\partial L(\underline{z}, \lambda)}{\partial Z_{i}} = 0$$

and $\lambda_{j} \geq 0$

Constraints with positive Lagrange multipliers in the dual are added to the active set. An anti- zig-zag strategy is adopted by ensuring that a constraint is retained in the active set for at least two successive iterations.

It should also be noted that in the solution for the primal Lagrange multipliers if negative multipliers are obtained, the most negative is rejected and the solution repeated until only positive multipliers remain.

Sensitivity Analysis

The pseudo-Newton algorithm requires the calculation of the derivatives of the active constraints with respect to changes in the design variables - commonly referred to as design sensitivities.

OPTISEN computes sensitivities for static problems using either the dummy load method or the direct method [5,6]. Both methods use as a starting point differentiation of the equations of equilibrium.

Example 1 - The Fully Stressed Design of a Plastic Seat Tub

The plastic seat tub is fabricated in Azdel Thermoplastic laminate with the following properties.

Flexural strength:	152MPa
Tensile strength:	76MPa
Tensile Modulus:	5.52GPa
Density:	$1.19 \times 10^{-6} kg/m^3$

Three load cases were considerd (Figure 3)

1) Seat frame fatigue (678 N.m. about design H point)
2) Seat durability (2446 N through H point)
3) Racking load

Only the first two load cases were critical in designing material thicknesses.

The seat tub and the steel brackets to which it connects were idealized using 3 and 4 noded shell elements. The model geometry, topology, boundary conditions, material properties and load were defiend in SDRC Supertab. The finite element model was comprised of 763 nodes and 697 elements.

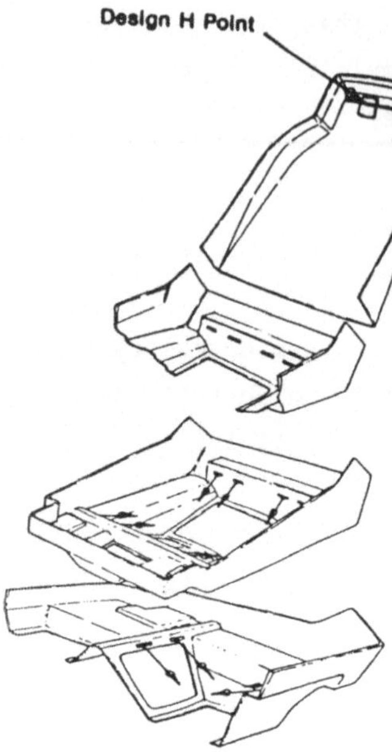

Figure 2 Three Load Cases were Critical to the Seat Tub Design

Additional data to define the optimization was specified in the OPTISEN module. Firstly, finite elements which were required to have a constant thickness were associated with an Optimization Group (Figure 3). This linked the thickness of the elements in an Optimization group to one design variable. Each group had the following parameters specified:

1) Allowable von-Mises stress in bending
2) Allowable von-Mises stress in membrane
3) Maximum and minimum allowable thickness (gauge limits)
4) Initial thickness

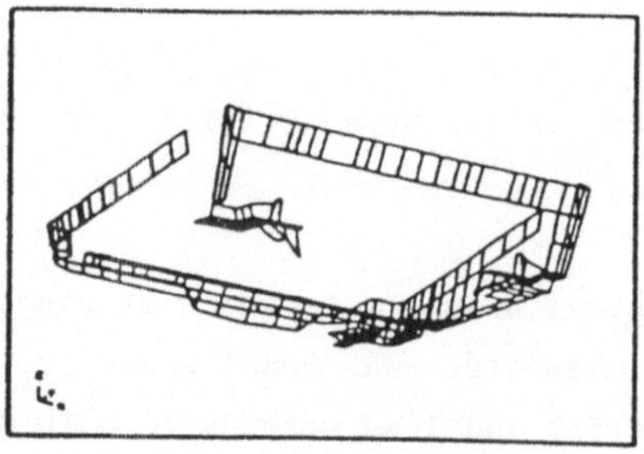

Figure 3 Optimization Group for Outer Edges Used to Maintain Constant Thickness

The optimization groups were defined based on a number of manufacturability and design criteria.

- Maintaining uniform thickness on outer edges for aesthetic purposes
- Maintaining constant thickness around structural connect points
- Promoting smooth flow of load and gradual thickness variation across the part

This led to the use of 10 optimization groups.

Groups of elements which were not to be redesigned, in this case the steel brackets, were specified as frozen elements.

Initially all plastic Optimization Groups were prescribed a minimum gauge limit of 2.54mm.

The Optimization strategy adopted was a classical fully stressing technique. In this approach the optimization groups are redesigned in an attempt to ensure that the most highly stressed finite element in the optimization group is either on its stress limit or its gauge limit.

For combined bending and membrane cases where different limits are specified the following failure law is adopted

$$\frac{\sigma_m}{\bar{\sigma}_m} + \frac{\sigma_b}{\bar{\sigma}_b} \leq 1$$

Where σ_m and $\bar{\sigma}_m$ are the actual and limit von-Mises Stresses for membrane and σ_b and $\bar{\sigma}_b$ are the actual and limit von-Mises Stresses for bending.

The described procedure is essentially iterative, requiring a finite element analysis at each sweep. The analysis forms the vast majority of the computational expense of each iteration.

In two iterations most of the optimization groups were on their lower gauge limits and only two groups were being designed. This suggests a possible route for efficiency improvements obtained by rapid re-evaluation of designs in succeeding iterations based on structural modification theory rather than a complex reanalysis.

The mass of the seat tub was further reduced by 16% from the manually optimized design. As the majority of the structure was held on its lower gauge limits a second design study using the minimum moldable thickness of

2mm was undertaken. The convergence history for this example is illustrated in Figure 4. This design produced a mass reduction of 24% from the starting figure.

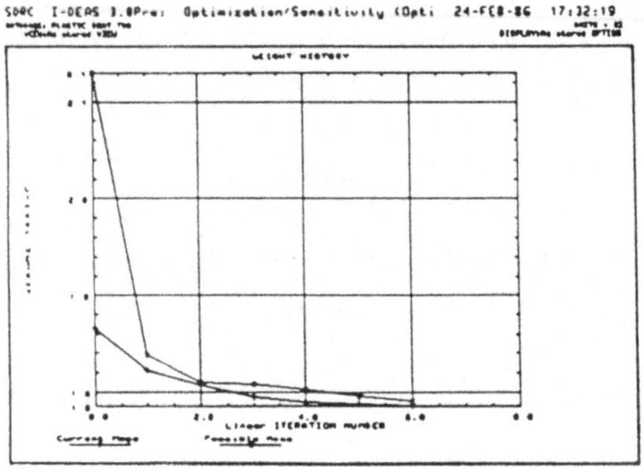

Figure 4 Weight versus Iteration using 2.0mm Minimum Gauge Limit

Example 2 - Fiat-Allis Excavator Arm

The objective of this example, undertaken jointly by SDRC and Fiat-Allis, was to improve the durability of a production arm while giving due consideration to product cost and manufacturability. Fatigue failures had been occurring in welds located in four critical areas.

A finite element model of the production arm was configured using 3 and 4 noded shell and linear beam elements. The mesh density of the model was consistent with an accurate representation of the overall stiffness distribution and the accurate prediction of stresses in the following critical areas.

A. Change in section between the foot casting pivots and the boom cylinder castings.
B. The area surrounding the attachment of the boom cylinder casting to the side panels.
C. The area surrounding the attachment of the stick cylinder to the toe panel.
D. The nose casting area.

The finite element model with these four critical regions highlighted is shown in Figure 5.

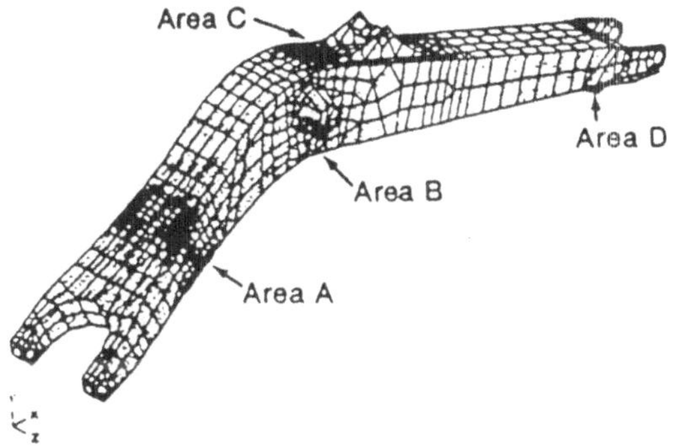

Figure 5 Finite Element Model of Excavator Arm

Local weld detail was modelled in critical Area A. In other critical areas the mesh was defined to give an accurate representation of nominal stresses in the vicinity of the welds.

The finite element model is comprised of:
1617 nodes, 1728 shell amd 176 beam elements
Three load cases were considered (Figures 6 - 8)
1) Fatigue
2) Yield - boom fully extended
3) Yield - boom vertical

Optimization groups were selected based on manufacturability considerations (specifically a reduced level of fabrication) and the production boom was analysed using OPTISEN for stress and displacement requirements. The displacement constraints were based on acceptable in operation behaviour and the stress constraints were based on achieving an acceptable fatigue life.

Various design studies were performed with, typically, 15 design variables.

The OPTISEN analyses showed a potential for weight reduction of 14% (437kg) when considering displacement constraints and 20% (611kg) when considering stress constraints.

OPTISEN provided a great deal of design insight regarding the most efficient locations for material redistribution. Many internal panels were reduced in thickness to the extent of becoming redundant. The sidewalls were also sized down significantly.

Based on the optimization results a new concept was developed. Optimization groups were chosen such that the top and bottom of the boom would be of uniform thickness. The sides of the booms were each divided such that there could be two different thicknesses. Each internal member was allocated a unique optimization group.

Stress constraints were applied to each of the optimization groups based on allowable stresses derived from fatigue calculations.

After five iterations OPTISEN had reduced the mass by 8% and produced a structure satisfying the constraints. The full potential for weight reduction was not achieved because of the restrictions imposed by ease of manufacture.

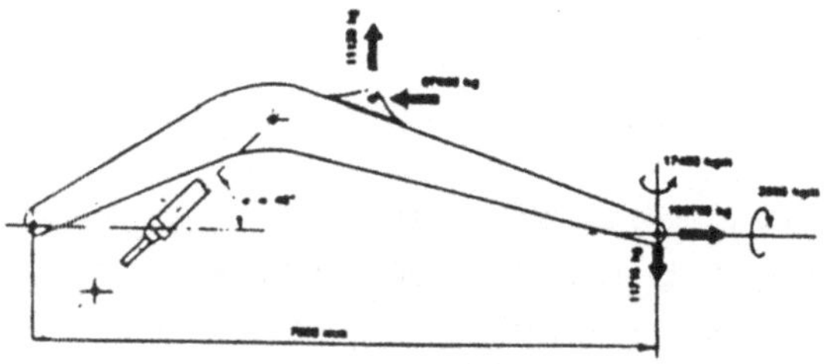

Figure 6 Load Case 1 Fatigue

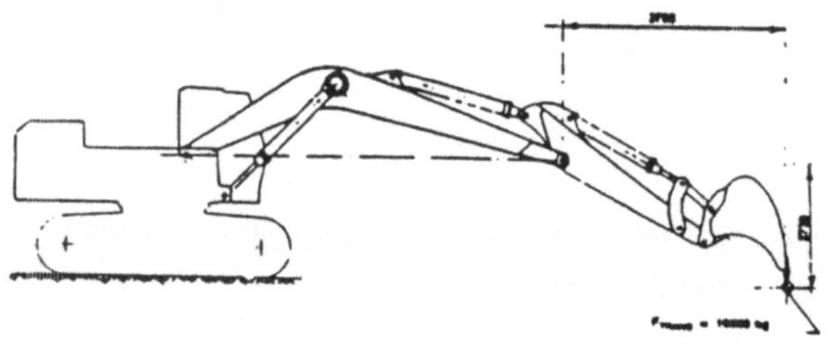

Figure 7 Load Case 2 Yield - Boom Fully Extended

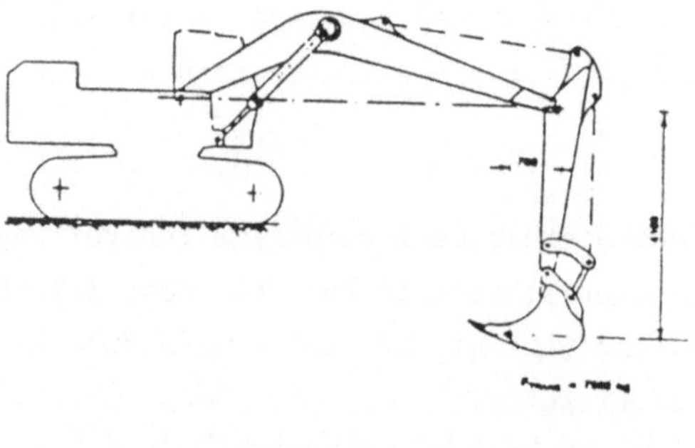

Figure 8 Load Case 3 Yield - Boom Vertical

The stress levels in the critical areas were reduced to acceptable levels. A comparison between the production and new concept level is shown in Table 1. Although the fatigue stress has increased for critical Area B the location of the maximum has been relocated away from the welds.

Fiat-Allis produced a prototype of the "new concept" arm in accordance with the analytical recommendations. Operating tests consistent with the analytical load cases were carried out while strain measurements were recorded.

The test measurements show a good level of correlation with the analytical results and thus demonstrates the effectiveness of such a technique.

Table 1: New Concept versus Production

Design	Area A Fatigue	Area A Yield	Area B Fatigue	Area B Yield	Area C Fatigue	Area C Yield	Area D Fatigue	Area D Yield
Production	273	1489	141	514	231	294	239	193
New Concept	91	447	199	448	186	435	211	133

Units in N/mm^2

Conclusions

Modern structural optimization techniques can significantly impact the design of complicated mechanical components. Application of these techniques has been demonstrated on practical problems. Current development includes extending the software to include dynamics constraints.

References

1. Morris, A.J. (Ed.), "Foundations of Structural Optimization, A Unified Approach", John Wiley & Sons (1982).
2. G.R. Walsh, "Methods of Optimization", J. Wiley & Sons (1975).
3. A.J. Morris, "An Introduction to the Techniques of Structural Optimization", Int. J. of Vehicle Design, 7, (1986), 231-241.
4. P. Bartholomew, "A Dual Bound Used For Monitoring Structural Optimization Programs", Engineering Optimization, 4, (1979), pp 45-50.
5. J.S. Arora, E.J. Haug, "Methods of Design Sensitivity Analysis in Structural Optimization", AIAA Journal, 17, (1979), pp 970-974.
6. H.M. Addman, R.T. Haftka, "Sensitivity Analysis of Discrete Structural Systems", AIAA, 24, (1986), pp 823-832.

PANEL DISCUSSION

TRENDS IN COMPUTER AIDED OPTIMAL DESIGN

PANEL DISCUSSION: TRENDS IN COMPUTER AIDED OPTIMAL DESIGN

Panel Discussion Moderator:
Dr. Jaroslaw Sobieski
NASA Langley Research Center

The ASI program culminated in a panel discussion. The panel was made up of the following eminent contributors to the theory and practice of optimization, introduced in alphabetical order:

Dr. L. Berke, NASA Lewis Research Center, USA
Prof. C. Fleury. UCLA, USA
Prof. E.J. Haug, University of Iowa, USA
Mr. H. Hornlein, MBB, FRG
Mr. G. Lecina, Avions Marcel Dassault, France
Prof. A. Morris, Cranfield Institute of Technology, USA
Prof. J.E. Taylor, University of Michigan, USA

The discussion format was defined by the Moderator at the outset of the session: the session was divided in two parts. The first part focused on four topics:

* Acceptance of optimization methods in engineering and the role of education in fostering that acceptance.
* Interactive use of information in optimization.
* Potential of the AI methods in optimization.
* Integration of optimization into design process.

Each topic was briefly introduced by one panel member who volunteered to do so. Following the introduction, the topic was declared open for discussion by the panel members and by the audience. The audience participation was not limited to the question adressed to the panel members; the audience's own comments and polemics were encouraged. The discussion's division into topics was treated flexibly, with the division lines acrossed freely often.

The second part of the discussion was entirely devoted to an open forum involving the panel members and the audience.

The Panel Session lasted 3 $\frac{1}{2}$ hours and there was 84 interventions. It was only possible to include a few of these in this book. The Editor apologizes for any inconvenience it may have caused. All discussion was recorded and anyone interested in the magnetic tape please contact the Editor.

NATO ASI Series, Vol. F27
Computer Aided Optimal Design: Structural and Mechanical Systems. Edited by C. A. Mota Soares

ACCEPTANCE OF OPTIMIZATION TECHNOLOGY

IS IT A PROBLEM? WHY? WHAT TO DO ABOUT IT?

ADVANTAGES

- o FEASIBLE DESIGN
- o OPTIMUM DESIGN
- o LABOR SAVING AUTOMATION

INHIBITORS

- o UNDERSTANDING METHOD (NIH)
- o TRAINING EXPENSE
- o CONFIDENCE IN AUTOMATED DESIGN

RANGES OF APPLICATION

- o CONCEPTUAL STUDIES
- o PRELIMINARY, o INTERMEDIATE, o FINAL DETAILED } DESIGNS
- o SPECIAL PROBLEMS

ACCEPTANCE BY WHOM?

- o MAJOR AEROSPACE COMPANIES
- o CONSULTING COMPANIES (CIVIL)
- o SMALL COMPANIES

MIND-SETS

MANAGERS
(WHAT DOES IT BUY ME?)

DESIGNERS
(CAN I TRUST IT?)

METHODS DEVELOPERS
(WHY NOT MY METHOD?)

CONCERNS

- o COMPETITIVE EDGE V.S. UNIFORM METHODOLOGY?
- o PROPRIETARY DESIGN SYSTEMS?
  (TRADITIONAL VS. NEW)
- o TRAINING/RETRAINING-JUNIOR/SENIOR
- o PRESSURES OF HIGH TECH (NO EXPERIENCE)

ROLE OF GOVERNMENT; UNIVERSITY; INDUSTRY RESEARCHERS

AI/EXPERT SYSTEMS: "USER SEDUCTIVE" CODES?

JUL 86 L. BERKE

The topic of the discussion is: Why do we have problems with optimization methods to be accepted by designers and used in daily practice? Well, we might or might not have a problem, and it is a matter of degree of acceptance, and for what kind of problems. There are major automated design capabilities, representing large coded systems, that are in use over a decade by the aerospace industry for detailed design studies. These are based mostly on classical optimality criteria approaches, such as in the case of the ASOP/FASTOP capabilities with FSD for stress constraints and "equal cost of improvement for all design variables" as the criteria for stiffness type constraints (displacements, flutter, vibration, divergence, etc.). Also, the TSO preliminary design capability, based on direct search techniques, is in industry-wide use for a decade. However, the newer capabilities might have some difficulty in being readily accepted and certain amount of aggressive marketing is needed, as occurring in the case of COPES/CONMIN and ADS.

Let us briefly look at the various aspects of the problem. The advantages are clear, feasible designs can be obtained where human designers may have problems finding isolated feasible regions that look like the cosmologist's worm-holes in the universe. Optimum design with labor saving automation is the most obvious benefit for a design office.

There are inhibitors, of course, such as confidence in automated design wisdom that is perhaps not clear to a non-specialist, and the expense in time and money to train designers in new methodology. The drivers towards acceptance are the broad application in the total range of design from conceptual studies to final detailed sizing once confidence is built. Then there are special problems perhaps not even solvable except by the infinite patience of an automated search procedure.

The question of acceptance is also different relative to what kind of company is to adopt new methods. Aerospace is perhaps more innovative, civil more conservative, small companies can go either way to survive and prosper.

Within a company one can think of three sets of people with entirely different attitudes. Managers weigh advantages-disadvantages, designers worry about the correctness of the design, method developers are the most flexible, they have nothing to loose, but prefer their own developments to be adapted.

There can be concerns by companies also relative to what is the competitive edge if government agencies distribute the same "best" method to everyone? What is the proprietory design system in that case giving real or imaginary competitive edge? Training is costly, but pressures of unusual High Tech. problems with no previous body of experience forces designers towards rational automated methods. One can also raise the question of the role of government, university and industry researchers in promoting acceptance of optimization methods. Aggressive demonstration of applicability through real world design problems is the most productive way. ASOP/FASTOP was demonstrated through developing designs parallel and concurrent with manual design activities. To ease the training needs and reduce aversion to learning to use a new code, modern "expert systems" approaches can make the first experience with a new code "user seductive". This topic is now open for discussion.

Computational effort in structural design

H.R.E.M. Hörnlein

The different algorithms which are used to solve the general nonlinear programming problems, are usually assessed by means of efficiency, accuracy, reliability and robustness. The efficiency is a measure of the processing time and is considered as the most important criterion in almost all test studies. Obviously the acceptance of a method is dominated by efficiency, because of excessive execution time all other criteria are superfluous.

The numerical effort for function evaluation and gradient calculation was never considered during the development of mathematical programming methods. Quite the opposite, the measures to accelerate the convergence speed are in most cases based on higher order information, i.e. the derivatives. In 1970 R.L. Fox argued in his book 'Optimization Methods for Engineering Design' for the choice of a method: "For very large, ill-conditioned problems (200 variables or more) with no gradients available, one is likely to need divine assistance, ..."

In structural design problems, the restricted state variables (stress, displacement, natural frequencies, aeroelastic efficiencies etc.) are given implicitly in the system equations. Therefore, the evaluation (analysis) and derivation (sensitivity analysis) of these constraint functions are very costly. The required computational cost for finite element structures increases polynomially with the number of design variables.

Subsequently, lower bound estimates of the processing time for displacement related constrained problems will be given. The main effort is the solution of linear equation systems. This will be measured by means of operation-counting (operation = multiplication + addition). The following three different methods for gradient calculation will be assessed and compared by the relative analysis effort.

DQ : Difference quotient via finite differences.
DSA* : Design sensitivity analysis by exploiting the analytical relationship.
SAXT*: Extended sensitivity analysis for special structured mathematical programming problems e.g. transformation methods.

The number of degrees of freedom for the finite element structure is chosen as reference number for counting the number of operations OP(n). Some further assumptions are based on factors of proportionality.

$n \sim m$:# structural elements $\Rightarrow m := n$
$m \sim \tilde{m}$:# design variable $\Rightarrow \tilde{m} := m/2$

These assumed factors might be discussed, but this does not change the final qualitative conclusions. The number of load cases is set to NLC= 4 and the relative bandwidth of the stiffness matrix K is assumed to be c = 10% of the number of degrees of freedom.

---

* see item 8. in 'Take-off in optimum structural design'. Lecture presented at the NATO Advanced Study Institute on Computer Aided Optimal Design, Troia, Portugal, July 1986

The static response requires one Cholesky factorization (CF) and NLC forward backward substitutions (FB) to obtain the displacement vectors u^l from the static equilibrium equations: $Ku^l = R^l$; $l = 1,...,NLC$

$$OP_A(n) = OP_{CF}(n) + 4 \cdot OP_{FB}(n) \tag{1}$$

This is assumed to be the main effort for the analysis. For the sensitivity analysis by finite differences DQ as many analyses are needed as design variables.

$$OP_{DQ}(n) = \frac{n}{2} \cdot OP_A(n) \tag{2}$$

The analytical calculation of the gradients DSA*, only as many forward backward substitutions are required as design variables (design space method) for each load case. The adjoined variable method (state space method) needs proportionally the same effort.

$$OP_{DSA}(n) = \frac{n}{2} 4 \cdot OP_{FB}(n) \tag{3}$$

Eventually the special sensitivity analysis SAXT* needs just one forward backward substitution for each load case.

$$OP_{SAXT}(n) = 4 \cdot OP_{FB}(n) \tag{4}$$

Because of the required operations for one standard Cholesky factorization and forward backward substitution respectively

$$OP_{CF}(n;c) = \frac{n^3}{6}(3c^2 - 2c^3) + \frac{n^2}{2}(3c-c^2) + \frac{n}{6}(6+5c)$$

$$OP_{FB}(n;c) = n^2(2c - c^2) + nc$$

the required computing time for equations (1) to (4) can be expressed by polynomials of second to fourth degree.
Fig.1 shows the processing times for a computer with 10^6 operations per second.

A comparison of the methods reveals that the gradient calculation via finite differences is prohibitive for large scale structures. The DSA and SAXT are actually not comparable. The DSA provides the primal mathematical programming methods with the entire Jacobian of the constraints (row-wise gradients), while the SAXT gives only a linear combination of the gradients as needed in the transformation methods. More obviously is the superiority of SAXT (if applicable) when considering the relative effort, Fig.2. It is evident that the computational effort becomes insignificant in comparison with the overhead calculations for large scale structures.

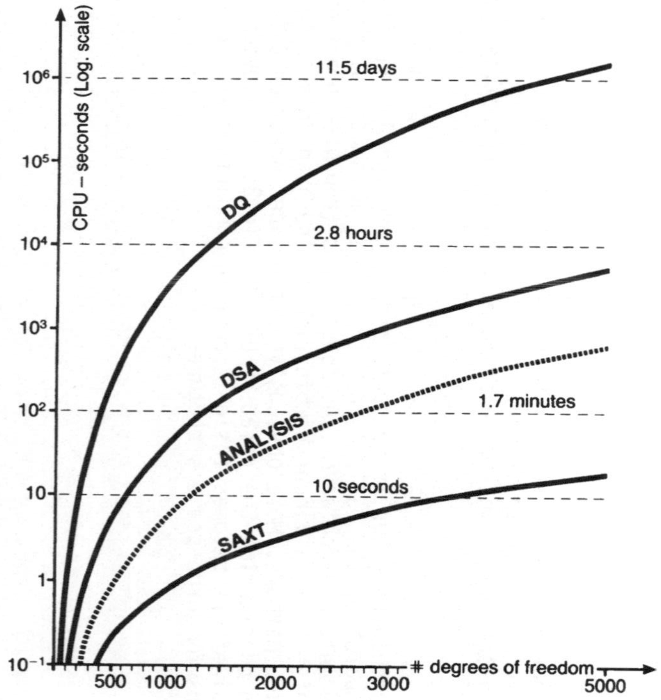

Fig. 1 Processing time (10^6 operations per second)

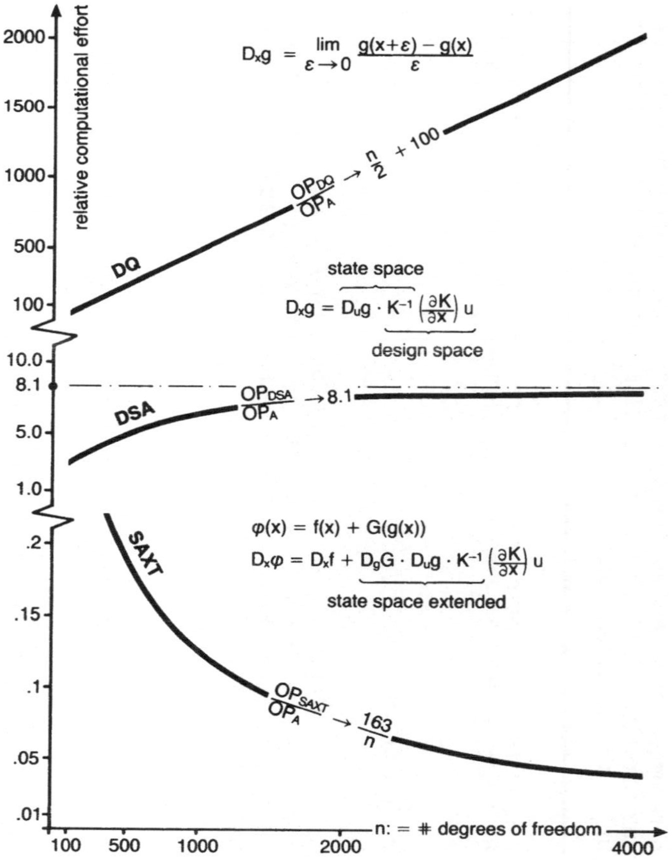

Fig. 2 Relative computational effort

ISSUES OF GENERALITY, RELIABILITY AND EFFICIENCY IN OPTIMUM DESIGN

Discussion by Jasbir S. Arora, University of Iowa

Optimal design of large systems in the interdisciplinary design environment needs reliable algorithms. Many algorithms have been developed and evaluated for structural optimization. In these algorithms, efficiency has been given priority over reliability and generality. To have efficiency, approximations are introduced into the algorithms. With such approximations, many algorithms loose their robustness and applicability to complex problems. In summary, approximate methods are efficient but unreliable, inaccurate and not general. Globally convergent (robust) methods are general, accurate and reliable but need more computational effort. My contention is that we should relax the efficiency "constraint" for optimum design of practical systems. I am not saying that we should use inefficient algorithms. What I am advocating is that reliability should be given more weightage over efficiency in practical design environment. We should concentrate on developing reliable algorithms that are generally applicable. For complex and interdisciplinary systems, reliability of algorithms is essential. Unreliable algorithms can be actually more expensive because they require more user interaction and time, resulting in overall inefficiency. Use of optimization in general design environment also needs reliable algorithms. This is essential to gain user confidence and promote use of design optimization. The efficiency aspect will be taken care of in the near future with spectacular advances in computer hardware. In addition, we can expect to have finite element methods, design sensitivity analysis methods and optimization algorithms on hardware making the design process very efficient. With parallel computers becoming available we should concentrate on developing robust algorithms that exploit this capability. Finally, implementation of optimization algorithms into a robust software is a tedious problem requiring considerable resources and time. Knowledge engineering concepts can help considerably in robust implementation of algorithms as well as the design process itself.

Some Factors Favoring Interactive Optimal Design

E.J. Haug

Optimal design, has, in the past, been viewed as a batch mode computational problem. Most iterative optimization methods make little or no provision for interactive control of the process by the experienced designer. A number of factors are emerging that suggest more emphasis should be placed on methods of optimial design that involve the designer, preferably an interactive mode.

Graphics based workstations have become commonly available and used in finite element modeling. Expansion of graphic definition of finite element models to include design variable definition would be extremely valuable in formulation of design optimization problems. It can also be extremely valuable in displaying information in a form that is interpretable by the experienced designer.

Most experienced designers can evaluate progress in design and optimization best if feasible, or near feasible design iterations are reported. Human experience in control of the optimization process can (1) aid in formulation refinement, (2) identify onset of mesh distortion in shape optimization, and (3) evaluate progress and/or difficulties and permit termination of the iterative design optimization process. Reporting trade-off information to the experienced designer can greatly assist in refinement of a formulation that may not have considered all factors that become important during the iterative optimization process. This is particularly important in complex systems, since modes of failure or classes of feasible design alternatives may not have been adequately considered prior to analysis of the structure.

Of critical importance in gaining acceptance of design optimization methods by experienced designers is providing them with design change control on an iteration by iteration basis. Through realization that optimization methods can resolve trade-offs and predict improved designs in complex situations that defy human intuition, over time the experienced designer may develop enough confidence in optimization algorithms to adopt them and turn control over to the algorithm for a few iterations or for convergence to an optimized design. Without interactive control, many experienced designers may never give optimization methods a fair evaluation.

Finally, artificial intelligence methods may be able to capture some forms of "design wisdom" and take advantage of experienced designers capabilities in formulating and solving design optimization problems.

Some Factors Favoring

INTERACTIVE OPTIMAL DESIGN

E.J. Haug

- Graphics to assist in design variable definition
- Report feasible (or near feasible) design iterations to user:
  - (1) aid in formulation refinement
  - (2) identify onset of mesh distortion
  - (3) evaluate progress/terminate
- Report trade-off information to assist in formulation refinement
- Permit experienced designers who do not trust or believe in optimization to control the process, to encourage adoption
- Enhance communication (AI)

Potential of AI Methods in Optimisation

A.J. Morris
Cranfield Institute of Technology, U.K.

AI has been "over-sold" in recent times and many exagerated claims are made on its behalf which seems unlikely to come to fruition in the near future. Nevertheless, some important developments are taking place and the Structural Optimisation community show view. therefore. view AI as representing a new "box of programing tools" some of which will be beneficial in making Structural Optimisation systems more user friendly.

Beginning with Natural Language developments it is possible to see a simple passer being incorporated into a structural optimisation program to allow a more flexible command language structure to be used. At the lowest level this would free the user from needing to employ exact sentences or word lists in a command statements in order to avoid syntax errors. The incorporation of declarative routines would permit the creation of effective explanation and tutoring facilities. On a higher level Expert System concepts could be used with a "first-level" structural optimisation routine to aid in the tasks of modelling the structure, order the algorithms and controlling the utility routines.

NATO-ASI-TROIA-1986
"COMPUTER AIDED OPTIMAL DESIGN"
PANEL DISCUSSION
VIEW PRESENTED BY :
G. LECINA
AVIONS MARCEL DASSAULT

DESIGN PROCESS WITH OPTIMISATION

PRESENT STATE

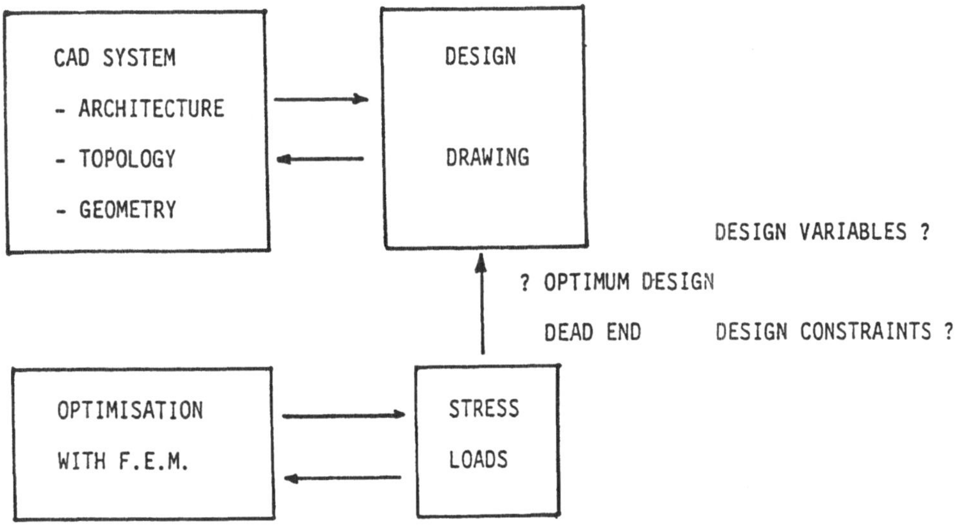

PREFERABLE

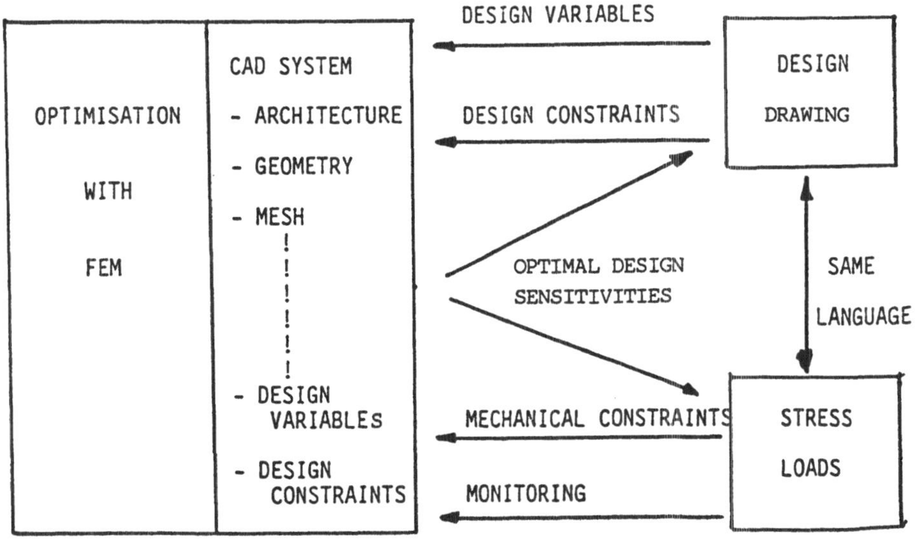

NATO-ASI-TROIA-1986
"COMPUTER AIDED OPTIMAL DESIGN"
PANEL DISCUSSION
VIEW PRESENTED BY :
G. LECINA
AVIONS MARCEL DASSAULT

DESIGN PROCESS WITH FINITE ELEMENT METHOD

PRESENT STATE

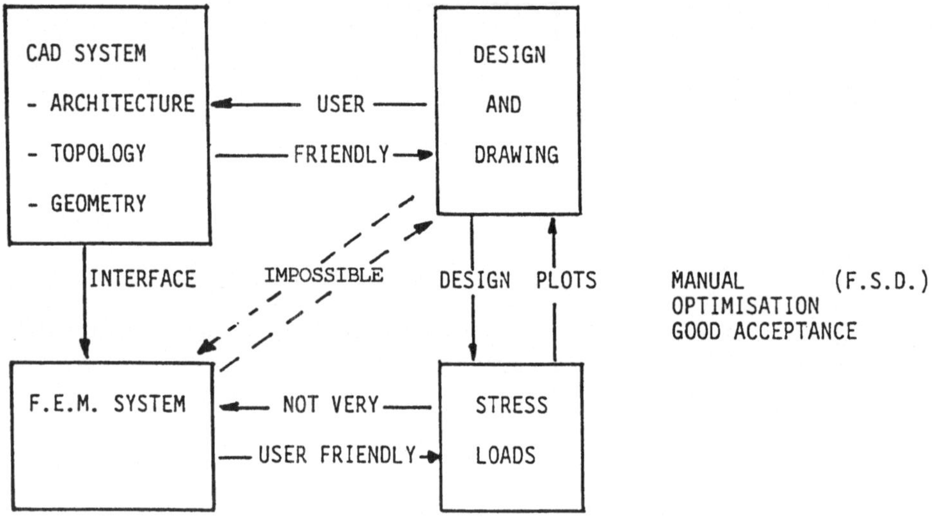

PREFERABLE

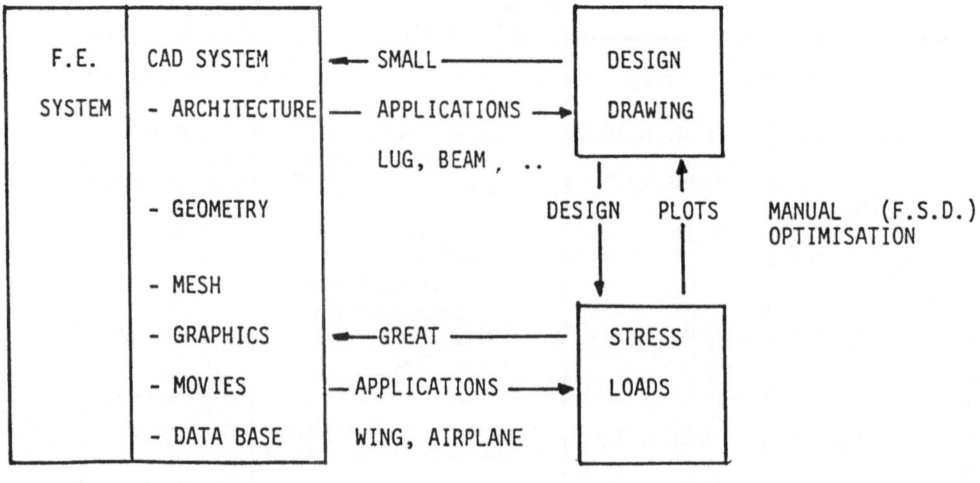

Future Trends in Computer Aided Optimal Design

Claude Fleury
Mechanical, Aerospace and Nuclear Engineering
University of California
Los Angeles

Structural optimization is now a mature discipline, and it begins to penetrate the industrial comunity. Several commercially available FEM systems will soon be released with design optimization capabilities. Therefore the most important area of development in the next few years will certainly reside in creating good user interfaces aimed at facilitating the task of the designers. In particular it will be extremely important to nicely integrated FEM and CAD technologies within an optimization loop, in order to fully computerize the design cycle. From the academic research point of view, considerable efforts should be redirected toward the following topics:

- development of appropriate "design oriented" structural models (for example, instead of optimimizing the thickness of membrane elements, it would be more suitable to optimize a design model from which the finite element mesh can be derived);
- fundamental study of geometric modeling concepts and methods is probably the key of successfully implement shape optmization capabilities (again the finite element mesh should be derived analytically from the parametric geometric model);
- an important and challenging research domain is concerned with error analysis and adaptive mesh refinement methods, which will be in the future key ingredients of FEM analysis and optimization systems (any user friendly FEM system should create and update the mesh by itself);
- the computer graphics aspects of CAD should now be exploited with design optimization concepts in mind (display of design model, representation of design space, iteration history, etc...);
- as a result interactive optimization capabilities should be developed, making an extensive use of innovative visualization features;
- finally, when we shall have enough experience and expertise in using user friendly Computer Aided Optimal Design Systems, it will become possible to gradually replace the interactive capabilities with more automated ones, by bringing artificial intelligence methods, as well as expert systems.

Of course more traditional approaches to structural optimization methods will continue to be studied, because there still exists many fundamental problems that have not been solved (optimal design of solid plates; dynamic response constraints,...). For the same reason, numerical optimization algorithms will still be the object of substancial development efforts (discrete variables, equality constraints, very large scale problems,...). However I believe that most of the research and development activities in the next decade will be devoted to the topics listed above.

NATO ASI Series F

Vol. 1: Issues in Acoustic Signal – Image Processing and Recognition. Edited by C. H. Chen. VIII, 333 pages. 1983.

Vol. 2: Image Sequence Processing and Dynamic Scene Analysis. Edited by T. S. Huang. IX, 749 pages. 1983.

Vol. 3: Electronic Systems Effectiveness and Life Cycle Costing. Edited by J. K. Skwirzynski. XVII, 732 pages. 1983.

Vol. 4: Pictorial Data Analysis. Edited by R. M. Haralick. VIII, 468 pages. 1983.

Vol. 5: International Calibration Study of Traffic Conflict Techniques. Edited by E. Asmussen. VII, 229 pages. 1984.

Vol. 6: Information Technology and the Computer Network. Edited by K. G. Beauchamp. VIII, 271 pages. 1984.

Vol. 7: High-Speed Computation. Edited by J. S. Kowalik. IX, 441 pages. 1984.

Vol. 8: Program Transformation and Programming Environments. Report on an Workshop directed by F. L. Bauer and H. Remus. Edited by P. Pepper. XIV, 378 pages. 1984.

Vol. 9: Computer Aided Analysis and Optimization of Mechanical System Dynamics. Edited by E. J. Haug. XXII, 700 pages. 1984.

Vol. 10: Simulation and Model-Based Methodologies: An Integrative View. Edited by T. I. Ören, B. P. Zeigler, M. S. Elzas. XIII, 651 pages. 1984.

Vol. 11: Robotics and Artificial Intelligence. Edited by M. Brady, L. A. Gerhardt, H. F. Davidson. XVII, 693 pages. 1984.

Vol. 12: Combinatorial Algorithms on Words. Edited by A. Apostolico, Z. Galil. VIII, 361 pages. 1985.

Vol. 13: Logics and Models of Concurrent Systems. Edited by K. R. Apt. VIII, 498 pages. 1985.

Vol. 14: Control Flow and Data Flow: Concepts of Distributed Programming. Edited by M. Broy. VIII, 525 pages. 1985.

Vol. 15: Computational Mathematical Programming. Edited by K. Schittkowski. VIII, 451 pages. 1985.

Vol. 16: New Systems and Architectures for Automatic Speech Recognition and Synthesis. Edited by R. De Mori, C.Y. Suen. XIII, 630 pages. 1985.

Vol. 17: Fundamental Algorithms for Computer Graphics. Edited by R.A. Earnshaw. XVI, 1042 pages. 1985.

Vol. 18: Computer Architectures for Spatially Distributed Data. Edited by H. Freeman and G.G. Pieroni. VIII, 391 pages. 1985.

Vol. 19: Pictorial Information Systems in Medicine. Edited by K. H. Höhne. XII, 525 pages. 1986.

Vol. 20: Disordered Systems and Biological Organization. Edited by E. Bienenstock, F. Fogelman Soulié, G. Weisbuch. XXI, 405 pages.1986.

Vol. 21: Intelligent Decision Support in Process Environments. Edited by E. Hollnagel, G. Mancini, D.D. Woods. XV, 524 pages. 1986.

Vol. 22: Software System Design Methods. The Challenge of Advanced Computing Technology. Edited by J.K. Skwirzynski. XIII, 747 pages. 1986.

NATO ASI Series F

Vol. 23: Designing Computer-Based Learning Materials. Edited by H. Weinstock and A. Bork. IX, 285 pages. 1986.

Vol. 24: Database Machines. Modern Trends and Applications. Edited by A.K. Sood and A.H. Qureshi. VIII, 570 pages. 1986.

Vol. 25: Pyramidal Systems for Computer Vision. Edited by V. Cantoni and S. Levialdi. VIII, 392 pages. 1986.

Vol. 26: Modelling and Analysis in Arms Control. Edited by R. Avenhaus, R.K. Huber, J.D. Kettelle. VIII, 488 pages. 1986.

Vol. 27: Computer Aided Optimal Design: Structural and Mechanical Systems. Edited by C.A. Mota Soares. XIII, 1029 pages. 1987.